PARASITIC BIRDS AND THEIR HOSTS

Oxford Ornithology Series
Edited by C. M. Perrins

1 *Bird Population Studies: Relevance to Conservation and Management* (1991)
Edited by C. M. Perrins, J.-D. Lebreton, and G. J. M. Hirons

2 *Bird-Parasite Interactions: Ecology, Evolution, and Behavior* (1991)
Edited by J. E. Loye and M. Zuk

3 *Bird Migration: A General Survey* (1993)
Peter Berthold

4 *The Snow Geese of La Pérouse Bay: Natural Selection in the Wild* (1995)
Fred Cooke, Robert F. Rockwell, and David B. Lank

5 *The Zebra Finch: A Synthesis of Field and Laboratory Studies* (1996)
Richard A. Zann

6 *Partnerships in Birds: The Study of Monogamy* (1996)
Edited by Jeffrey M. Black

7 *The Oystercatcher: From Individuals to Populations* (1996)
Edited by John D. Goss-Custard

8 *Avian Growth and Development: Evolution within the Altricial-Precocial Spectrum* (1997)
Edited by J. M. Starck and R. E. Ricklefs

9 *Parasitic Birds and Their Hosts: Studies in Coevolution* (1998)
Edited by Stephen I. Rothstein and Scott K. Robinson

PARASITIC BIRDS AND THEIR HOSTS

Studies in Coevolution

Edited by

STEPHEN I. ROTHSTEIN

SCOTT K. ROBINSON

New York • Oxford

Oxford University Press

1998

Oxford University Press

Oxford New York
Athens Auckland Bangkok Bogotá Buenos Aires Calcutta
Cape Town Chennai Dar es Salaam Delhi Florence Hong Kong Istanbul
Karachi Kuala Lumpur Madrid Melbourne Mexico City Mumbai
Nairobi Paris São Paulo Singapore Taipei Tokyo Toronto Warsaw

and associated companies in
Berlin Ibadan

Published by Oxford University Press, Inc.
198 Madison Avenue, New York, New York 10016

Library of Congress Cataloging-in-Publication Data
Parasitic birds and their hosts : studies in coevolution / edited by
Stephen I. Rothstein and Scott K. Robinson.
p. cm. — (Oxford ornithology series; 9)
Includes bibliographical references and index.
ISBN 0-19-509976-1
1. Parasitic birds Ecology. 2. Brood parasitism.
3. Host-parasite relationships. 4. Coevolution.
I. Rothstein, Stephen I. II. Robinson, Scott Kuehner.
III. Series.
QL698.3.P366 1998
598.16′5—DC21 97-18599

1 3 5 7 9 8 6 4 2

Printed in the United States of America
on acid-free paper

Dedication to Herbert Friedmann

The study of avian brood parasitism has been dominated by the contributions of Herbert Friedmann for over 60 years. Indeed, few areas of scientific inquiry can be said to have been so heavily influenced by the long-term work of a single person. It is a great honor for us to dedicate this book to Herbert. Besides being a remarkably productive and diverse scholar, he was also a kind and gentle person. In this dedication, we want to convey the respect we and the authors of this book's papers have for Herbert and also give a brief overview of his life and contributions.

Herbert published 17 books and some 315 articles, reviews, and monographs. He devoted one or more books to every group of parasitic birds (cowbirds, cuckoos, finches, honeyguides) except for parasitic ducks but published a theoretical review paper on the latter group in 1932. Although best known for his work on parasitic birds, Herbert published numerous systematic papers, especially on birds from North and South America and Africa. He also published many other papers, dealing with topics as diverse as molt in penguins and handedness in parrots. In addition to his accomplishments in biology, Herbert pursued a second career in art history, a field in which he received considerable acclaim and in which he published two books and 20 papers. His special contribution here was the use of his extensive natural history background to provide novel insights into the significance of animals and plants in well-known paintings.

The scholarly work Herbert did is notable for three reasons in addition to its breadth and quantity. First, he exhibited a great deal of prescience as to questions and phenomena that would be of long-term significance. For example, his first book, which was published in 1929 and dealt with the biology of all cowbird species, is still commonly cited. Second, Herbert was extremely open about his work and was always eager to see people apply new concepts and techniques to subjects he had worked on. Third, and finally, he was remarkably broad ranging and meticulous in his search for knowledge as shown by the numerous sources of information he consulted, especially for his frequent updates of known cowbird hosts.

Herbert had a wonderful wit and many of the humorous stories he told give some perspective on the long period of time over which he made his contributions. Those of us who have been a bit slow to bring computers into our work lives can take some solace in the fact that Herbert's experiences show that a similar reluctance existed earlier in this century with the typewriter, and with the notion of men and women working at close quarters. When Herbert was hired in the bird division at the Smithsonian Institution, he made his appointment contingent on two conditions, the purchase of a typewriter and agreement that he could hire as a secretary the most qualified person available, even if that person was a woman. The idea of talking to someone who could then type what you said was alien to the male-dominated hierarchy of the Smithsonian, as was the notion of working in the same room with a woman! Herbert was fond of pointing out that some of these men never did get used to using secretaries and working at close quarters with women. Those who complain about the scarcity of resources to support teaching and research might similarly find solace in Herbert's experience during his postdoctoral days at Harvard's Museum of Comparative Zoology. When his fountain pen ran dry, Herbert was told to see Henshaw, the museum director, for more ink. Henshaw told Herbert that he could not give him a bottle of ink but that if Herbert gave him his pen, he would fill it for him. Herbert had a large collection of such stories, which always made the time spent with him enlightening as well as fun.

Herbert was born in Brooklyn, New York on 22 April, 1900. He did his undergraduate work at City College in Manhattan and while there carried out the research for his first publication, which appeared in 1922. After obtaining his Ph.D. in 1923 under Arthur A. Allen at Cornell University, Herbert went on to a three-year postdoc at Harvard, and short teaching stints at Brown University and Amherst College. He joined the Smithsonian in 1929 as Curator of Birds, an institution he left in 1961 to become Director of the Los Angeles County Museum of Natural History. Although he retired from the latter museum in 1970 and became Director Emeritus, Herbert continued to publish on birds and on art history. His last biological publication, a major synthesis of known cowbird hosts done with Lloyd Kiff, appeared in 1985. Herbert's scientific contributions garnered a number of highly distinguished signs of recognition, notably the Brewster Medal of the American Ornithologists Union and membership in the National Academy of Sciences.

Herbert married Karen Juul Vejlo in 1937, an agricultural economist who had been visiting Washington, D.C. from her native Denmark. She and their daughter, Karen, maintain an interest in recent developments related to Herbert's ornithological work. Herbert's health began to falter in the early 1980s and he died on 14 May, 1987. More details on his biography can be found in a memorium in *The Auk* (**105**:365–368) and a complete bibliography of his publications is in the archives of the American Ornithologists Union stored at the Smithsonian Institution.

We close this dedication with two regrets. The first is the obvious one of our sadness that Herbert is not here to read the exciting advances made in the last 10 years and presented in this volume. This amazing increase in our knowledge would have delighted Herbert, even in cases where it departs from his own contributions on the same subjects for he was the kind of person who was much more interested in seeing good answers to important questions than in seeing his views adopted. Our second regret is that it has taken us so long to provide what we hope will be a fitting memorial to Herbert. The only consolation here is that the study of parasitic birds and their hosts is moving so rapidly that a book appearing now reports on many more advances than could a book appearing shortly after Herbert's death, and scientific advances were the things that Herbert most wanted to see. Despite these enormous advances in the past decade, it is clear that much of this progress could not have been made without the extensive foundation of knowledge and concepts that are the legacy of Herbert's remarkable career.

Preface

This book had its genesis shortly after Herbert Friedmann's death when I realized that the best memorial that researchers could provide for him was a book presenting recent developments in the study of avian brood parasitism. This was the field to which he devoted most of his scholarly attention and the subject he is most closely associated with among biologists. I also thought that the time was ripe for a book on brood parasitism because this subject had long attracted a great deal of interest for reasons discussed in the first part of chapter 1. Furthermore, it was evident from my communications with other researchers that a great deal of significant new research was about to appear.

The first step in the book's development was a symposium on parasitic birds held in honor of Herbert Friedmann at the 1990 meeting of the American Ornithologists Union in Los Angeles. Symposium participants were eager to contribute to a book honoring Herbert but as most of them were from North America, it was obvious that they would not cover the range of diverse research being done. It was therefore decided to broaden this project and invite additional contributors, especially from outside of North America.

As sometimes happens in projects such as this, the editorial responsibilities were somewhat overwhelming and this along with other commitments meant that progress on the book proceeded slowly. To aid in completion of the project, indeed to rescue it from what might have been oblivion, Scott Robinson graciously volunteered to become a co-editor in 1993. His efforts have been invaluable, not only for his superior organizational skills but for his keen insights. It has been a pleasure to work with him. In fact, I wanted to write this Preface myself not because this book is "my project" but because doing so gives me a mechanism for expressing my thanks to Scott. This book is fully a joint project and if anything, Scott's input has exceeded mine since he became a co-editor. Both of us are especially appreciative of the patience shown by the authors over the long time this project has taken. They have also been remarkably cooperative given the number of times they have been asked to revise and update their chapters as we all realized that this book had to be current if it was to make an impact.

The goal in inviting researchers to participate in this book was to include representative chapters on the largest possible variety of parasitic birds and of key issues in the study of these birds and their hosts. However, the six groupings of papers reflect the research that has been done worldwide in recent years, not necessarily all of the emphases that need to be pursued for a more complete understanding of brood parasitism. As a result, there is a heavy bias towards chapters on cuckoos and cowbirds. Although this book is oriented toward questions and issues and not toward dealing with each parasitic taxon individually, we have grouped the cuckoo and cowbird chapters in separate sections of the book because the features of these parasitic systems are so different. However, papers in part IV synthesize knowledge of both of these groups in attempts to answer what we consider to be the most important tractable question concerning the evolutionary aspects of brood parasitism, namely the reason many hosts accept parasitic eggs and young that are easily distinguishable from their own offspring. Later sections in the book focus on other important questions, namely the extent to which brood parasites affect host populations in ecological rather than evolutionary time frames and are therefore potential threats to the survival of host species (part V); the extent to which freedom from parental care has affected basic attributes of the biology of parasitic birds, such as their mating systems (part VI); and lastly the importance of conspecific brood parasitism (part VII), a phenomenon whose widespread prevalence has been recognized only in the past decade.

We conclude the book with a chapter that explores future research directions in the study of avian brood parasitism. The goal of the chapter is to identify crucial gaps in our knowledge of brood parasitism and the major questions that need to be answered. It seems fitting to end the book in this manner because doing so highlights the fact that brood parasitism is such an appealing phenomenon for research precisely because of the clearly defined questions it poses.

We begin this book with a long chapter of our own which has four purposes. First, it describes the features of brood parasitism that make it an especially interesting phenomenon for evolutionary biologists. Second, it sets the groundwork for the evolutionary interactions that occur between parasites and their hosts, namely the reproductive losses parasitism inflicts on hosts and the likelihood that host defenses will in turn inflict losses on parasites. Third, the opening chapter presents a brief overview of all of the clades of obligate parasites and of intraspecific parasitism. The former is especially needed because some of the major groups of parasitic birds such as the honeyguides are not covered in any of the papers contributed to this book, yet we felt that readers would want to have some easily accessible information on these groups.

Fourth, chapter 1 presents synthetic overviews of each of the six groups of papers that make up the major part of this book. We wrote these overviews because we wanted this book to be more than a series of essays and we felt that readers would prefer to see some discussion that synthesizes the material in related topics. These overviews briefly summarize and contrast the findings of the individual chapters of each section. They also discuss recent work that appeared in the literature after the authors submitted the final versions of their work. In a few cases in chapter 1, we present views that differ from those in the contributed work. In such cases, we have tried to pinpoint the sort of data needed to resolve these conflicting interpretations. Although few in number, these contrasting viewpoints demonstrate the heuristic, thought-provoking nature of research on brood parasitism and hence one of the reasons this subject is so interesting.

Stephen I. Rothstein
Santa Barbara, California
May 1997

Contents

Contributors

Luis Arias-de-Reyna—Departmento de Biologia Animal (Etologia), Facultad de Ciencias, Universidad de Cordoba, 14071 Cordoba, Spain
ba1armal@vco.es

Phoebe Barnard—Namibian Evolutionary Ecology Group, Box 5072, Windhoek, Namibia
pb@dea1.dea.met.gov.na

Michael de L. Brooke—Department of Zoology, University of Cambridge, Downing Street, Cambridge, CB2 3EJ, United Kingdom

Alexander Cruz—Environmental, Population and Organismal Biology Department, University of Colorado, Boulder, CO 80309, USA
alexander.cruz@colorado.edu

Nicholas B. Davies—Department of Zoology, University of Cambridge, Downing Street, Cambridge, CB2 3EJ, United Kingdom
nbd1000@hermes.cam.ac.uk

Rosendo M. Fraga—Asociacion Ornitologica del Plata, 25 de Mayo 749, 1002 Buenos Aires, Argentina
fraga@aorpla.org.ar

Brian J. Gill—Auckland Institute and Museum, Private Bag 92018, Auckland, New Zealand

Sharon A. Gill—Department of Biology, York University, North York, Ontario M3J 1P3, Canada
sgill@your.u.ca

Hiroyoshi Higuchi—Laboratory of Wildlife Biology, School of Agriculture and Life Sciences, University of Tokyo, Bunkyo-ku, Tokyo 113, Japan
higuchi@uf.a.u-tokyo.ac.jp

Keith A. Hobson—Canadian Wildlife Service, Prairie and Northern Region, 115 Perimeter Road, Saskatoon, Saskatchewan S7N 0X4, Canada

Wendy M. Jackson—Center for Evolutionary and Environmental Biology, The Field Museum, Roosevelt Road at Lake Shore Drive, Chicago, IL 60605, USA
jackson@fmnh785.fmnh.org

Gustavo H. Kattan—Wildlife Conservation Society—Colombia Program, P.O. Box 25527, Cali, Colombia
gukattan@cali.cetcol.net.co

Satoshi Kubota—Department of Biology, Faculty of Education, Shinsu University, Nishinagano, Nagano 380, Japan

Arnon Lotem—Department of Zoology, Tel Aviv University, Tel Aviv 69978, Israel
lotem@post.tau.ac.il

Richard F. Maloney—Department of Zoology, University of Canterbury, Christchurch, New Zealand

Juan G. Martinez—Departmento de Biología Animal y Ecología, Facultad de Ciencias, Universidad de Granada, 18071, Granada, Spain

Ian G. McLean—Department of Zoology, University of Western Australia, Nedlands, WA 6907, Australia
ianm@cyllene.uwa.edu.au

Arne Moksnes—Department of Zoology, University of Trondheim, N-7055 Dragvoll, Norway
arnem@alfa.itea.unit.no

Anders P. Møller—Laboratoire d'Ecologie, CNS URA 258, Université Pierre et Marie Curie, 7 quai St Bernard, Case 237, F-15252, Paris Cedex, France
amoller@hall.snv.jussieu.fr

Isla H. Myers-Smith—Department of Zoology, University of British Columbia, Vancouver, British Columbia V6T 1Z4, Canada

Hiroshi Nakamura—Department of Biology, Faculty of Education, Shinsu University, Nishinagano, Nagano 380, Japan

Tammie K. Nakamura—Environmental, Population and Organismal Biology Department, University of Colorado, Boulder, CO 80309, USA

Diane L. Neudorf—Department of Biology, Pomona College, 609 N. College Ave., Claremont, CA 91711, USA
dneudorf@pomona.edu

Catherine P. Ortega—Biology Department, Fort Lewis College, Durango, CO 81301, USA
ortega_c@grumpy.fortlewis.edu

Laura L. Payne—Museum of Zoology and Department of Biology, University of Michigan, Ann Arbor, Michigan 48109, USA

Robert B. Payne—Museum of Zoology and Department of Biology, University of Michigan, Ann Arbor, Michigan 48109, USA
rbpayne@umich.edu

William Post—Charleston Museum, 360 Meeting Street, Charleston, SC 29403, USA
grackler@aol.com

Harry W. Power—Department of Biological Sciences, Rutgers University, P.O. Box 1059, Piscataway, NJ 08855, USA
hpower@gandalf.rutgers.edu

John W. Prather—601 Science-Engineering Bldg, University of Arkansas, Fayetteville, AR 72701, USA

Scott K. Robinson—Illinois Natural History Survey, 607 East Peabody Drive, Champaign, IL 61820
scottr@denri.igis.uiuc.edu

W. Douglas Robinson—Smithsonian Tropical Research Institute, Unit 0948, APO, AA 34002-0948, USA

Eiven Røskaft—Department of Zoology, University of Trondheim, N-7055 Dragvoll, Norway
eiven.roskaft@ninatrd.ninaniku.no

Stephen I. Rothstein—Department of Ecology, Evolution and Marine Biology, University of California, Santa Barbara, CA 93106
rothstei@lifesci.ucsb.edu

Spencer G. Sealy—Department of Zoology, University of Manitoba, Winnipeg, Manitoba R3T 2N2, Canada
sgsealy@cc.umanitoba.ca

James N. M. Smith—Department of Zoology, University of British Columbia, Vancouver, British Columbia V6T 1Z4, Canada
smith@zoology.ubc.ca

Juan J. Soler—Departmento de Biología, Animal y Ecología, Facultad de Ciencias, Universidad de Granada, 18071, Granada, Spain

Manuel Soler—Departmento de Biología, Animal y Ecología, Facultad de Ciencias, Universidad de Granada, 18071, Granada, Spain
msoler@goliat.wgr.esp

Michael D. Sorenson—Department of Biology, Boston University, 5 Cummington St, Boston, MA 02215, USA
msoren@bio.bu.edu

Reiko Suzuki—Department of Biology, Faculty of Education, Shinsu University, Nishinagano, Nagano 380, Japan

Cheryl L. Trine—Department of Ecology, Evolution and Ethology, University of Illinois, Urbana, IL 61801, USA

James W. Wiley—Grambling Cooperative Project, Grambling State University, Grambling, LA 71245, USA
wiley_iw@vaxo.gram.edu

I

OVERVIEW AND COMMENTARY

1

The Evolution and Ecology of Avian Brood Parasitism

An Overview

STEPHEN I. ROTHSTEIN, SCOTT K. ROBINSON

The ludicrous sight of a fledgling cuckoo or cowbird being fed by an adult host less than half its size (cover of book) makes it easy to see why parasitic birds have attracted so much attention. A person would have to be devoid of all curiosity to not wonder how such a relationship arose between two very different species and why hosts allow themselves to be part of the whole process. Much of this book addresses such questions about brood parasites, which are individuals that lay eggs in the nests of other individuals. The latter individuals, the hosts, then raise the parasitic young, usually at the expense of part or all of their own clutch. This sort of breeding biology has generated a great deal of fascination since it was first described formally—in the western world by Aristotle, and even earlier, in about 2000 B.C., in India (Friedmann 1964*a*).

Brood or social parasitism is not limited to birds as it is widespread in some insect groups, especially ants (Davies et al. 1989) and also occurs in some fish (Sato 1986; Baba et al. 1990) but this book is limited to avian brood parasitism, which has so far received the most attention. Parasites can lay their eggs in nests of another species (interspecific brood parasitism) or in those of conspecifics (conspecific brood parasitism). The latter type of parasitism must, by definition, be facultative in that a single conspecific population must have both parasitic and nonparasitic individuals and/or individuals that follow a mixed strategy. Although it would be possible for a species to sometimes rear its own young and also to regularly parasitize other species but not conspecifics, no bird species shows this particular mixed strategy. In other words, all species that are strictly interspecific parasites are obligate parasites, nearly all of which are altricial. Precocial species appear to be less likely to evolve obligate parasitism because their fecundity is relatively high even in the absence of parasitism (Lyon and Eadie 1991).

Because humans tend to moralize about how individuals care for their own offspring, we often vilify brood parasites and draw parallels to human behavior. In 1995 alone, the *New York Times*, North America's most respected newspaper, ran two feature-length articles on avian brood parasites. One featured the brown-headed cowbird (*Molothrus ater*), which may be causing the decline of dozens of species of North American songbirds (Line 1993, 1995; but see Peterjohn and Sauer in press). The other *Times* article featured the great spotted cuckoo (*Clamator glandarius*), which may have a special strategy for defeating host efforts to resist parasitism, namely a "mafia"-like "protection" behavior of destroying eggs and nestlings in nests from which cuckoo eggs have been ejected. Three different film companies attempted to capture cowbird parasitism on film, one of which was successful (Kurtis Productions 1995) and whose production appeared widely on American TV.

Even some veteran scientists react judgmentally to sights such as a tiny host species trying to feed a parasitic nestling while perched on its back (see photo on pp. 78–79 in Line 1993). It takes a constant struggle to be objective and to try to keep the media from using such terms as "nefarious," "evil," "thuggish," and "heinous" when there is good evidence that some brood parasites could cause the extinction of certain endangered species if left unchecked. A current debate within scientific, management, and birdwatching communities pits advocates of cowbird "control" (even killing them by the millions in winter roosts) versus advocates of other possible solutions (Rothstein and Robinson 1994; Schram 1994; Smith 1994 and [*Birding Forum*]). It is sometimes easy to forget that the excessive cowbird parasitism observed in many parts of the Americas may be partly or wholly a human-generated problem even though the cowbird is a native songbird.

Within the scientific community, brood parasites have generated especially intense interest as models for coevolutionary processes (e.g., Rothstein 1990*a*; Redondo 1993) and coevolution is a major theme of this book. The parasite–host coevolution is thought to take the form of an evolutionary "arms race" in which hosts develop some sort of defense against parasitism which then selects for the evolution of a counterdefense in the parasite potentially leading to cycles of adaptation and counteradaptation (Dawkins and Krebs 1979; Davies and Brooke 1988, 1989*a,b*; Harvey and Partridge 1988; Davies et al. 1989). Because of likely recent evolutionary changes, the great spotted cuckoo in Europe (Lotem and Rothstein 1995; Soler et al. 1994) may come to parallel industrial melanism in the moth *Biston betularia* as a textbook case of rapid evolution. However, some authors in this book argue that these arms races may have ended in evolutionary equilibria in some to even most parasite–host systems (Lotem and Nakamura; McLean and Maloney; Røskaft and Moksnes). In contrast to the usual situation of harm to the host, a possible evolutionary scenario that is featured in virtually every ecology and evolution text suggests that some hosts of the giant cowbird (*Molothrus oryzivorus*) benefit from parasitism because the parasitic nestlings themselves remove harmful ectoparasites from their host nestmates (Smith 1968). But this is the only case known in which parasitism seems to be beneficial to hosts and the benefits may occur in only part of this parasite's range (Robinson 1988; Webster 1994).

There are yet other reasons for the widespread interest in avian parasites. Some of the strongest evidence for character displacement in birds comes from studies of cuckoos (Higuchi and Sato 1984; Higuchi, this vol.). Because parasites are raised by host young, their song development provides a classic case of genetic hardwired behavior that can nevertheless be modified by learning (West et al. 1981*a,b*; Rothstein and Fleischer 1987*a,b*; O'Loghlen and Rothstein 1995). The fact that brood parasites are freed from parental care makes them excellent subjects for studying sexual selection and the evolution of mating systems (Yokel and Rothstein 1991). In this context, it is interesting that most brood parasites are monogamous in spite of the absence of parental care (Barnard, this vol.).

The general purpose of this book is to elaborate on many of the ecological, behavioral, and evolutionary issues raised by the phenomenon of brood parasitism. Our goal was to offer as wide a perspective as possible on brood parasitism with papers from every major region of the world where brood parasitism has been studied. While the authors who have contributed to this book report on work from all six continents with parasitic birds, no book with only 21 contributed papers is likely to be comprehensive about such a vast and complex topic. For this reason, the remainder of this opening chapter reviews some of the basic concepts and facts related to brood parasitism. It then presents a brief description of all parasitic birds, with an emphasis on taxa that are not covered in detail in any of the papers (chapters 2–22) that make up the most important part of this book. Finally, the major part of this first chapter is devoted to overviews of each grouping of essays in the book.

NATURAL HISTORY OF BROOD PARASITISM

Adaptations of brood parasites

Many chapters of this book and the introductions to parts II and III describe adaptations of brood parasites and this topic has been reviewed frequently (e.g., Lack 1968; Payne 1977*a*, 1997*a*; Rothstein 1990*a*; Johnsgard 1997). Here we provide only a summary of these adaptations.

Brood parasites search for host nests, often by observing host behavior and following hosts to nests rather than by searching for nests directly (Hann 1941; Norman and Robertson 1975; Wyllie 1981; Smith et al. 1984). Once a nest is located, the parasite must time its egg laying with that of its host. Laying too early may cause the host to abandon the nest or eject the egg or bury it with nesting material. Laying too late would put the parasitic nestling at a competitive disadvantage with its nestmates (Lotem et al. 1995). Brood parasites often remove a host egg at or close to the time they lay one of their own and sometimes damage host eggs in the process, which further reduces host nestling success (Sealy 1992; Soler, this vol.; Arias-de-Reyna, this vol.). Most parasites approach nests stealthily and lay when hosts are absent, especially early in the morning (Scott 1991). In some parasites, however, females have access to unattended nests because males distract the hosts, which may pursue and mob cuckoos because of a resemblance to hawks (Duckworth 1990) or simply because they recognize cuckoos (McLean 1987; Arias-de-Reyna, this vol.) or honeyguides (Payne 1992; Short and Horne 1992) as special enemies.

Adaptative features of parasitic eggs include thick eggshells, small size relative to the size of the parasite, mimetic colors, and rapid developmental rates. These features benefit parasites by making their eggs difficult to puncture and eject (thick egg shells: Rahn et al. 1988; Rohwer and Spaw 1988), by incidentally damaging host eggs and therefore reducing nestling competition (thick egg shells: Blankespoor et al. 1982; Soler, this vol.), by increasing the difficulty of the hosts in identifying parasite eggs (mimetic colors and small egg size: Davies and Brooke, this vol.; Lotem and Nakamura, this vol.), and by giving parasitic nestlings a head start in competition with their nestmates (rapid development and early hatching: Briskie and Sealy 1990). In cuckoos, and possibly honeyguides, part of this headstart occurs because newly laid eggs already have small embryos due to development within the female's oviduct (Wyllie 1981; Payne 1992). The adaptations of parasitic eggs form the core of arguments that many hosts accept parasitic eggs because it is their best option (Rohwer and Spaw 1988; Lotem et al. 1995; see part IV). In general, specialist brood parasites that have only one or a few hosts lay mimetic eggs, whereas generalist brood parasites such as cowbirds lay nonmimetic eggs. But egg similarity in viduines and some honeyguides is probably due to common ancestry between parasites and hosts not mimicry (Payne 1977*a*, 1989). At least one brood parasite, the common cuckoo (*Cuculus canorus*), can be viewed as a generalist at the species level but a specialist at the individual level. Within a single region, each female belongs to one of several gentes with each gens mimicking the egg and specializing upon a different host species (Wyllie 1981).

Adaptations of parasitic nestlings include a variety of mechanisms that help them compete with host nestlings for food. The nestlings of most cuckoo species eliminate competition by pushing host nestlings out of the nest on their concave backs (Wyllie 1981). Honeyguide nestlings, and some cuckoo nestlings, have remarkable raptorial hooks on both mandibles that they use to kill nestmates (Friedmann 1955; Morton and Farabaugh 1979). Most viduine finches mimic the complex gape patterns of host nestlings (Nicolai 1964, 1974; Payne 1973*a*). Cowbird nestlings have relatively large mouths (Ortega and Cruz 1992), especially intense begging behavior (Gochfeld 1979*a*) and get a head start on host nestlings by hatching earlier; as a result, cowbird nestlings usually outcompete host nestlings for food delivered to the young. Nevertheless, some host young usually fledge from nests of many (but by no means all) species that are parasitized by cowbirds, a result that is vital to modeling coevolutionary and population dynamics in these systems (part IV).

Some brood parasites are remarkably fecund, whereas others may lay only a few more eggs in a season than many of their hosts. Estimates of cowbird fecundity range from 40 or more eggs per season in the brown-headed cowbird (Scott and Ankney 1983; Rothstein et al. 1986; Holford and Roby 1993) to over 100 eggs per season in the shiny cowbird (*Molothrus bonariensis*) (Kattan 1993). The high fecundity of these generalist brood parasites creates potential problems for many of their hosts in areas where cowbirds are abundant (part V). Estimates of cuckoo and honeyguide fecundity, on the other hand, are lower, usually less than 20 eggs per season (Payne 1974, 1992). Specialist brood parasites such as cuckoos may have fewer opportunities to lay eggs than generalists. One of the benefits of a parasitic lifestyle is that females can put all of their foraging efforts into producing eggs rather than raising young (Lyon and Eadie 1991).

Costs of brood parasitism for hosts

Host nests in which cuckoo and honeyguide eggs hatch usually fledge no host young. Hosts of viduines, on the other hand, suffer only slight losses from parasitism (Payne 1977*a*,*b*). Cowbird hosts vary enormously in the costs of parasitism; some raise mixed broods of cowbirds and host young, whereas others usually raise only the cowbirds (see part V). In general, smaller hosts and those with long incubation and nestling periods suffer more from parasitism.

Population-level consequences of brood parasitism are generally much less severe than the consequences for individual hosts that are parasitized. Frequencies of cuckoo parasitism, for example, are usually less than 10%, a value that may have few long-term consequences for host populations (Wyllie 1981). In some areas where cuckoos have only recently arrived, however, parasitism levels may be higher (Nakamura et al., this vol.), presumably because the host populations have little resistance (Soler et al. 1994). These cases of population level impacts, however, appear to be the exception and in at least one case, hosts have sometimes rapidly developed resistance (Wyllie 1981; Nakamura et al., this vol.). Some cuckoo hosts that are heavily parasitized in one brood may escape parasitism by laying a clutch when parasites are not breeding (Gill 1983). Specialist parasite–host interactions can be modeled using traditional host–parasite equations, but with overlapping generations (May and Robinson 1985). These models generally predict that the frequency of parasitism in specialist parasite–host systems will stabilize at a low level that will pose no direct threat to the host populations (see also Takasu et al. 1993).

Generalist brood parasites, on the other hand, can pose a direct threat to host populations (reviewed in Robinson et al. 1995*b*; Post et al. 1990; see also Smith et al., in press). Because cowbirds have multiple hosts in an area and are often more abundant than most of their hosts (Rich et al. 1994), they can drive some of their hosts nearly to extinction without necessarily suffering any significant negative feedback on their own populations (Mayfield 1977; May and Robinson 1985). As long as there are hosts in a community that can raise cowbirds as well as their own young, the supply of cowbirds will continue to be high. Paradoxically, parasitism levels of preferred hosts may increase as hosts become rare because there are just as many cowbirds seeking out fewer preferred hosts, although it is unclear that cowbirds exercise strong preferences amongst host species (Rothstein 1976a; Smith and Myers-Smith, this vol.). Cowbirds have been implicated in continent-wide declines of many North American birds (Böhning-Gaese et al. 1993; but see Wiedenfeld, in press; Peterjohn et al., in press). This topic is addressed in our overview to part V (p. 31) and in the essays in that part.

Defenses by hosts

As a result of the severe costs of brood parasitism for individual hosts, there is strong selection for host defenses (Rothstein 1975*a*). Counteradaptations of hosts to brood parasitism include a wide variety of tactics reviewed in parts II–IV of this book. In addition to the evolution of unusual or variable egg colors and nestling gape patterns (Nicolai 1964, 1974) usually associated with an evolutionary arms race, hosts have many behavioral defenses. Some hosts aggressively defend nests against brood parasites, and in the case of the colonial hosts, this mobbing may be successful (e.g., Clark and Robertson 1979; Wiley and Wiley 1980). Hosts with scattered nests, on the other hand, may be less successful at mobbing and the mobbing may alert brood parasites to the presence of a nest (Smith et al. 1984; McLean 1987). Because most brood parasites lay very early in the morning (but see Davies and Brooke 1988), hosts may defend against parasitism by arriving early at their own nests or spending the night there (Neudorf and Sealy 1994). Although small hosts may not be able to chase away parasites, they may be able to use the presence of the cowbird as a cue to abandon a nest (Burhans, in press), which may explain why nest desertion and burial of cowbird eggs is common in naturally parasitized nests (Graham 1988) but rare in response to eggs placed in nests by researchers (Rothstein 1975*a*,*b*; Sealy 1995). Abandonment and burial are especially likely if parasitic eggs are laid too early or if there are too many parasitic eggs. These are, however, only partial defenses as parasitic eggs properly synchronized with host laying are often accepted by hosts and renests can be parasitized (Clark and Robertson 1981; Sealy 1995).

A more direct defense is the ejection of the parasitic egg, either by grasping it or puncturing it. This avoids the major costs inflicted by parasitic nestlings but can not recover the usually lesser costs due to host eggs removed by adult parasites. Larger hosts that eject parasitic eggs can grasp the eggs whereas smaller hosts may have to puncture them (Rothstein 1975*a*; Rohwer and Spaw 1988). Because many hosts do not eject cowbird eggs, many authors (reviewed in Rothstein 1990*a*) have argued that this behavior has simply not had time to evolve (the "evolutionary lag" hypothesis). Others, however, have argued that egg ejection can be costly, especially for hosts that are too small to grasp eggs and might damage their own eggs while attempting to puncture-eject a thick-shelled cowbird egg. These arguments form the basis of the evolutionary equilibrium hypothesis, in which hosts are making the best of a bad situation by accepting parasitic eggs (Rohwer and Spaw 1988). Egg acceptance may also result from limitations in the host's ability to detect mimetic eggs (Lotem et al. 1995). These issues are discussed at length in part IV of this book.

Recent evidence that some brood parasites act as nest predators (Smith and Arcese 1994; Soler et al. 1995*a*) may bolster arguments based on an evolutionary equilibrium. Cowbirds, for example, may depredate unparasitized nests to increase the supply of new nests (renests) to parasitize (Arcese et al. 1996). Some cuckoos may penalize hosts that reject their eggs by depredating the contents of nests that do not have cuckoo eggs (Zahavi 1979; Soler et al. 1995*a*). Clearly, these behaviors would decrease the benefits of host egg rejection, but there are many questions about just how widespread these parasitic behaviors are as the examples still leave some questions unanswered and there is no overall strong trend for unparasitized nests to be more prone to nest predation (see part VIII).

Consequences of parasitism for the behavior and ecology of brood parasites

One of the primary advantages of brood parasitism is the freedom from the severe constraints of parental care, especially feeding dependent young (Lyon and Eadie 1991). The implications of brood parasitism for the ecology and behavior of parasites are reviewed in the overview to part VI p. 35) and by Barnard (this vol.).

A SURVEY OF THE KNOWN INTERSPECIFIC BROOD PARASITES

Obligate brood parasitism has evolved independently at least five to seven times as the 90–95 species of obligate parasites represent seven mostly unrelated taxa (Johnsgard 1997): the cowbirds (Icterinae), the cuckoos (Cuculinae), the ground cuckoos (Neomorphinae), the cuckoo-finch (Ploceidae: *Anomalospiza*), the viduine finches (Viduinae), the black-headed duck, *Heteronetta* (Anatidae), and the honeyguides (Indicatoridae). In contrast to current arrangements, Hughes (1996) has suggested that obligate parasitism evolved only once in the cuckoos (see below). Two of the parasitic taxa, the Cuculinae (parts II and IV) and cowbirds (parts III, IV, and V) are treated in detail in this volume. Part VI provides some details of other host–parasite systems, but mainly in the context of their mating systems. Because the Cuculinae and the cowbirds are treated in great depth later in this book, the following brief summary of the taxa of interspecific brood parasites gives relatively short shrift to them (especially in light of their extensive literatures) and instead focuses on the other taxa (for whom major sources of information are listed in table 1.1).

Cuculinae

The nearly 50 species in this subfamily are all obligate parasites that occur throughout the Old World. Although this group contains the most famous parasite, the common cuckoo, all of its members tend to be uncommon and many are poorly known. The hosts of several species remain unknown (Wyllie 1981). These cuckoos range from the sparrow-sized *Chrysococcyx* to the raven-sized *Scythrops* and also show a remarkable range in plumage and overall appearance. Some cuckoos parasitize mostly one or two host species and all are fairly specialized compared to some of the parasitic cowbirds.

Icterinae

In contrast to the parasitic cuckoos, DNA data indicate that the five species of parasitic cowbirds are a relatively young group that arose about 2.8 to 3.8 million years ago (Lanyon 1992; Rothstein et al., in prep.). The five species are all dark-colored birds ranging in size from a large sparrow to a small crow and are similar in basic form to many nonparasitic blackbirds in the Icterinae. Two genera are currently recognized although Lanyon's (1992) mitochondrial DNA-based phylogeny indicates that *Scaphidura* must be placed with the other parasitic cowbirds in the genus *Molothrus*. The latter contains one nonparasitic species, *M. badius*, which probably belongs elsewhere (Lack 1968; Lanyon 1992). The cowbirds are primarily Neotropical, although one species breeds throughout North America, another has recently colonized North America from the Caribbean (Cruz et al., this vol.) and a third straddles the separation between the Neotropics and the Nearctic. As discussed later, the five parasitic cowbirds exceed all other parasitic birds combined in terms of their range in numbers of host species. Cowbird–host systems are discussed further in the overview for part III (p. 16).

Table 1.1 Key references on brood parasites. See Johnsgard (1997) for a more extensive tabulation of references

A. References describing obligate parasites not covered by chapters in this volume.

Parasitic clade	References
Honeyguides	Friedmann (1955), Ranger (1955), Short and Horne (1985, 1992), Vernon (1987)
Viduines	Friedmann (1960), Morel (1973), Nicolai (1964, 1974), Payne (1967, 1969, 1973, 1976, 1977, 1982, 1985), Payne and Payne (1977), Shaw (1984); see also Barnard (this vol.)
Cuckoo-finch	Vernon (1964)
Black-headed duck	Weller (1968), see also Sorenson (this vol.)

B. Geographic distribution of descriptions of cuckoo–host systems.

Geographic region	References
Eurasia	Arias-de-Reyna (this vol.), Baker (1942), Braa et al. (1992), Chance (1922, 1940), Davies and Brooke (this vol.), Friedmann (1964, 1968), Herrick (1910), Higuchi (this vol.), Higuchi and Sato (1984), Jenner (1788), Jourdain (1925), Lack (1963), Lotem et al. (1995), Moksnes et al. (1993), Nakamura (1990), Nakamura et al. (this vol.), Neufeldt (1966), Owen (1933), Redondo and Arias-de-Reyna (1988), Seppä (1969), Soler et al. (1995), Soler and Martinez (this vol.), Swynnerton (1916, 1918), Ali (1931), Wyllie (1981), Yamagishi and Fujioka (1986)
India	Becking (1981), Phillips (1948), Wyllie (1981), Baker (1942), Lamba (1975), Gaston (1976)
Australia	Brooker and Brooker (1989*a*,*b*, 1992), Dow (1972), Friedmann (1968), Gill (1983, this vol.), Gosper (1964), Marchant (1972), McLean (this vol.), McLean and Waas (1987), Payne et al. (1985), Poiani and Elgar (1994)
Madagascar	Appert (1970)
Africa	Barnard (this vol.), Benson and Irwin (1972), Friedmann (1948, 1964, 1967), Fry et al. (1988), Gaston (1976), Jensen and Jensen (1969), Jensen and Clining (1974), Jensen and Vernon (1970), Liversidge (1955, 1971), Mundy (1973), Payne and Payne (1967, this vol.), Reed (1968), Rothstein (1992), Swynnerton (1916, 1918), Steyn (1973), Vernon (1984), Wyllie (1981)
New World	Hamilton and Hamilton (1965), Haverschmidt (1961), Morgensen (1927), Morton and Farabaugh (1979), Nolan and Thompson (1975), Potter (1980), Preble (1957), Ralph (1975), Spencer (1943), Sick (1993)

Neomorphinae

Only three of the 11 species in the Neomorphinae, or ground cuckoos, are known to be obligate brood parasites; this subfamily, however, is extremely poorly known. The parasitic species (*Tapera naevia* and *Dromococcyx pavoninus* and *D. phasianellus*) are rare inhabitants of the Neotropics and prefer disturbed habitat (Wetmore 1968). They tend mainly to parasitize just a few host species, nearly all which have domed nests. *Tapera* nestlings use special hooks on their bills to stab and kill off all of the host young, an example of convergence with honeyguides (Morton and Farabaugh 1979). Egg mimicry remains poorly known in this group as do aspects of their mating systems, coevolutionary adaptations, ecology, host use, and habitat selection. Although they contain no obligate parasites, two of the other four subfamilies in the Cuculidae, besides the Cuculinae and Neomorphinae, show features that may have predisposed this family to the evolution of obligate parasitism. *Coccyzus* spp. in the Phaenicophinae build their own rather flimsy nests but also lay in the nests of conspecifics and other congeners (Nolan and Thompson 1975), in the nests of unrelated species (Wyllie 1981), and even in artificial nests baited with quail eggs (S. Robinson, unpubl. data). Several genera in the Crotophaginae have cooperative breeding

(*Guira*, *Opisthocomus*) and even communal nesting (*Crotophaga*) (Macedo 1994). Thus nesting cuckoos have flexible breeding behavior, a possible precursor to the evolution of brood parasitism (Hamilton and Orians 1965). Most nesting species consume high proportions of hairy caterpillars (e.g., Wetmore 1968; Sick 1993), but feed their young on softer-bodied, less-protected caterpillars. Hughes (1996) suggests that parasitic members of the Neomorphinae and the genus *Coccyzus* are New World members of the subfamily Cuculinae, that obligate brood parasitism evolved only once among cuckoos and that *Coccyzus* reacquired nesting behavior after having evolved from a parasitic ancestor.

Cuckoo-finch (Anomalospiza)

This aberrant finch is known to parasitize about 11 species of warblers (*Cisticola* and *Prinia*) in grassland habitats of Africa (Friedmann 1960). Although all aspects of its biology, such as the possibility of egg mimicry, are poorly known, it should be relatively easy to study because it lives in open habitats (see Barnard, this vol.). Recent DNA data indicate that the parasitic viduines are the sister group to *Anomalospiza* and that parasitism evolved once in this clade (Payne 1997*b*).

Viduine finches

The 14 or more parasitic species of viduines have been studied extensively by Payne (1973*a*, 1982), Klein and Payne (in press) and Nicolai (1964, 1974) in the context of mating systems, vocal dialects, speciation, and coevolution of parasites and hosts. Viduines are remarkable in their close mimicry of host nestling gape patterns and in the relatively low cost of parasitism. Parasites often multiply parasitize nests and remove host eggs, but hosts have no problem raising both hosts and parasites from the same nest (Morel 1973). Many species incorporate the songs of their hosts into their own vocal repertoires and females choose males that have been raised by the same host (Payne 1973*b*; Payne et al., in press). Most *Vidua* spp. are host specific on different species of grassfinches (Estrildidae), which are their close relatives (Sibley and Ahlquist 1990). DNA evidence indicates that *Vidua* spp. did not speciate along with their hosts but have instead speciated more recently (Klein et al. 1993). All are seedeaters, an unusual diet among brood parasites, most of which tend to be insectivorous. At least one species, the pin-tailed whydah (*Vidua macroura*) is a host generalist. The intricate nestling mouth markings of the estrildid hosts may not have evolved in response to parasitism because they are present in all members of the family, even in Asia and Australia where estrildids are unaffected by brood parasites.

Honeyguides

Among African honeyguides, the nine species in the genus *Indicator* parasitize cavity-nesting species, most of which are nonpasserines, while the three *Protodicsus* species parasitize passerines with open cup nests. The hosts of three species in two other genera (*Melignomon* and *Melichneutes*) are unknown (Fry et al. 1988). The two Asian species of *Indicator* are assumed to be parasitic but their breeding biologies are totally unknown (Friedmann 1960). So far as is known, all species lay white eggs as do most of the cavity nesting hosts. Given the dark interiors of their nests, it is unlikely that cavity-nesting hosts exhibit egg recognition based upon coloration. But the hosts of *Protodiscus* have colored eggs that contrast with the white eggs of this parasite. Unlike most brood parasites, honeyguide eggs are usually smaller than those of the host (Payne 1989). Some species show male–female cooperation when parasitizing nests and may defend host pairs (Short and Horne 1985) against conspecifics. Most species are rare and inconspicuous. The most remarkable honeyguide adaptation for parasitism are the raptorial hooks on the bills of recently hatched nestlings, which they use to stab their nestmates to death. Some species also puncture host eggs when laying. Honeyguides also have remarkable diets (wax) and two species lead people and other mammalian bee predators to hives (Fry et al. 1988; Isack and Reyer 1989). Brood parasitism is probably older in the honeyguides than in any other group of parasites except the Cuculinae (Friedmann 1955).

Some honeyguides appear to have the most prolonged associations with their hosts among all brood parasites, as they monitor host pairs year-round and are attracted to host vocalizations (Short and Horne 1992). Hosts attack honeyguides, which may use the aggression to assess host breeding stages. Nestling development in honeyguides is slower than in any other brood parasite, but is typical of cavity nesters (Fry et al. 1988). In the only case in which fledging by a honeyguide was observed, the young parasite was immediately driven off by its barbet hosts, so young honeyguides may care for themselves after leaving host nests (Short and Horne 1992), as is the case for the only precocial species that is an obligate parasite (next subpart). This phenomeno of a negative response by host parents when the young parasite leaves the nest cavity and reveals its divergent appearance is reminiscent of a situation in cowbirds in which a host stops caring for nonmimetic parasitic nestlings once they have fledged (Fraga, this vol.). The two best-known species of honeyguides, *Indicator indicator* and *I. minor*, have parasitized 39 and 19 host species and overlap in host use (Fry et al. 1988). Several hosts are parasitized by at least three species of honeyguides. Most of the other brood parasites sympatric with honeyguides, cuckoos, and viduines, rarely or never parasitize cavity nesters.

Heteronetta

In contrast to all other obligate brood parasites, the black-headed duck is precocial and the only thing it "parasitizes" from its hosts is incubation warmth as young go off on their own shortly after hatching (Weller 1968). As such, the costs to the hosts are minimal and because incubation is essentially identical in all bird species, the black-headed duck can and does parasitize a diverse range of species such as other ducks and coots (its main hosts) but also ibises and gulls (Weller 1968). The obligate parasitism of this species is an extreme expression of the facultative interspecific and conspecific parasitism present in many species of waterfowl (Sorenson, this vol.).

CONSPECIFIC BROOD PARASITISM

The majority of research on brood parasitism has involved interspecific brood parasitism, but recent studies have shown that conspecific brood parasitism is also widespread (Yom-Tov 1980; Møller 1987; Rohwer and Freeman 1988; Petrie and Møller 1991). Conspecific brood parasitism is reviewed in the overview for part VII (p. 38).

GUIDE TO THE PARTS OF THIS BOOK

The remainder of this book is divided into six parts, each of which contains related chapters, and a final part that outlines major unanswered research questions. The chapters in each part are perhaps best read in conjunction with the corresponding overview and synthesis subpart in this chapter. These overviews make up the remainder of this chapter and review the basic literature relevant to each of the six parts with contributed chapters, with special emphasis on recent publications and on relevant subjects that are not covered by the chapters in that part. Our overviews include additional, original syntheses of certain subjects and in some cases present viewpoints alternative to those presented by authors of the individual chapters.

OVERVIEW OF PART II: COEVOLUTION BETWEEN CUCKOOS AND THEIR HOSTS

The diverse cuckoo–host systems are believed to be the oldest and among the most highly coevolved of avian parasite–host systems (Friedmann 1948, 1964*b*, 1968; Marchant 1972; Payne 1977*a*; Rothstein 1992). Most cuckoo species specialize on one or a few hosts (Friedmann 1948, 1964*a*,*b*, 1968; Lack 1968; Wyllie 1981; Brooker and Brooker 1992). Even within cuckoo species that parasitize a range of hosts, individual females specialize on a single host species (Brooker and Brooker 1992; Davies and Brooke, this vol.). Furthermore, although the common cuckoo, which is a host generalist relative to most cuckoos (but not most cowbirds), has about 100 known host species over its extensive Eurasian range, it regularly parasitizes only five to ten species in single regions such as Great Britain or Japan (Davies and Brooke, this vol.; Nakamura et al., this vol.). A congener, the pallid cuckoo (*C. pallidus*), also has about 100 known hosts but only 21 species regularly raise the cuckoos (Brooker and Brooker 1989*a*) and all belong to a single family, the Meliphagidae. Where cuckoo species coexist, they tend to show little overlap in host species use, providing evidence of resource partitioning resulting from competition (Lack 1968; Friedmann 1967; Payne and Payne 1967; Jensen and Clinning 1974; Higuchi and Sato 1984; Nakamura 1990; Higuchi, this vol.; Nakamura et al., this vol.). Most species parasitize passerine species smaller than themselves and have nestlings that eject all host eggs and young. But three genera (*Clamator*, *Eudynamis* and *Scythrops*) often parasitize hosts as large or larger than themselves and do not eject nestmates (Davies and Brooke 1988), although *Eudynamis* does eject in some areas (Brooker and Brooker 1989*a*).

The complexity of adaptations related to cuckoo parasitism, plus the large number of parasitic cuckoo species (about 50) and their diverse morphology, suggests that cuckoo parasitism is of very ancient vintage compared with other parasitic systems. Indeed, Sibley and Ahlquist's (1990) DNA data indicate that the cuckoos probably arose in the Cretaceous, or more than 60 million years ago. Furthermore, the maximum DNA sequence divergence of the mitochondrial cytochrome b gene is 12.6% among three *Cuculus* spp. (Rothstein et al., in prep.). By contrast, Lanyon's (1992) data show that the first species to branch off from the rest of the parasitic cowbird clade has an average divergence of 5.6% with the four other species. Because sequence divergence is proportional to time since divergence, these data show that cuckoo parasitism has occurred for at least twice as long as cowbird parasitism. This contrast will become even more extreme when other cuckoo genera, some of which are very different in morphology from *Cuculus*, are studied.

Given the antiquity of cuckoo–host systems, many may have achieved evolutionary equilibria (Kelly 1987; Takasu et al. 1993; May and Robinson 1985; Brooker and Brooker 1989*b*; Lotem et al. 1995, this vol.). Nevertheless, there is strong evidence of ongoing coevolution in cuckoos in Europe and Japan (Higuchi and Sato 1984; Davies and Brooke 1988, 1989*a,b*, this vol.; Soler and Møller 1990, this vol.; Higuchi, this vol.; Nakamura, this vol.). Indeed, recent changes in host defenses against the great spotted cuckoo may provide one of the clearest cases of microevolution in vertebrates (Lotem and Rothstein 1995; Soler et al., this vol.; but see Zuñiga and Redondo 1992).

Cuckoo–host systems are among the best-studied of all brood parasite systems, but the geographical coverage of these studies is spotty (table 1.1). The common cuckoo has been almost as thoroughly studied as the brown-headed cowbird with several books and monographs (e.g., Herrick 1910; Swynnerton 1916, 1918; Chance 1922, 1940; Baker 1942; Wyllie 1981) and papers from diverse parts of its huge range (Europe: reviewed in Davies and Brooke, this vol.; Japan: Lack 1963; Nakamura, 1990; Higuchi, this vol.; Nakamura et al., this vol.; Central Asia: Baker 1942; Wyllie 1981). The literature on this bird includes some of the earliest published studies of bird natural history (e.g., Jenner 1788), although little is known about this parasite's biology in large parts of its range, unlike the brown-headed cowbird, which has been the subject of studies in most of its range. The cuckoos of Asia, Australia and, especially, Africa on the other hand are much less well known, with the exception of Japan (table 1.1). Nevertheless, African hosts were the subjects of some of the earliest experimental field studies in behavioral ecology (Swynnerton 1916, 1918). Data for Australian cuckoos have increased greatly in recent years (Brooker and Brooker 1989*a*) and allowed Poiani and Elgar (1994) to do a comparative study of effects of host cooperative breeding on vulnerability to parasitism.

The seven cuckoo studies in this volume include data on host systems from diverse regions (Europe: Davies and Brooke; Soler et al.; Arias-de-Reyna), (Australia/New Zealand: Payne and Payne; Gill), and (Japan: Nakamura et al.; Higuchi). All of these essays deal with the often subtle coevolutionary interactions between host defenses and cuckoo counterdefenses. Because most cuckoo hosts rarely raise any of their own young from a parasitized nest (reviewed in Payne 1977*a*; Wyllie 1981), selection for host defenses can be very strong. Population-level threats of cuckoo parasitism for hosts, however, are slight compared with cowbirds (see part V; Lack 1963; Wyllie 1981; Brooker and Brooker 1989*a*, 1994), either because cuckoos are uncommon relative to their hosts or because of evolutionary equilibria reached over very long periods (May and Robinson 1985; Rothstein 1992; Kelly 1993; Takasu et al. 1993). In cases where there has been recent contact between cuckoos and hosts there is sometimes heavy parasitism and potential population-level threats (e.g., Molnar 1944; Glue and Morgan 1972; Yamagishi and Fujioka 1986; Brooke and Davies 1987; Soler and Møller 1990; Nakamura et al., this vol.; Soler et al., this vol.; Arias-de-Reyna, this vol.). Most potential cuckoo hosts in the Old World, including some that are rarely parasitized, reject cuckoo and other foreign eggs (reviewed in Rothstein 1992), which suggests that they faced cuckoo parasitism in the past. In contrast, most potential hosts of the New World cowbirds do not reject foreign eggs (Rothstein 1990*a*, 1992). Old World passerines that could serve as cuckoo hosts but that are geographically isolated from cuckoos, however, show relatively low levels of egg rejection (Brown et al. 1990; Soler and Møller 1990; Soler et al. 1994; Davies and Brooke, this vol.).

In the remaining part of this overview, we attempt to place the seven essays in this section into the larger context of adaptations and counteradaptations by summarizing relevant features of cuckoos and their hosts. We also highlight contrasts among these host–parasite systems. Details of the coevolutionary processes that are involved are modeled in part IV, which also includes the cowbird–host systems described in part III.

CUCKOO ADAPTATIONS

Behavioral adaptations

Most cuckoos approach nests of their hosts stealthily while the parents are away (Chance 1922, 1940; Baker 1942; Seppä 1969; Wyllie 1981; Gill, this vol.). Cuckoos lay their eggs quickly (less than 10 s: Chance 1922; Baker 1942; Arias-de-Reyna, this vol.) and may drop them from the nest rim onto host clutches to increase chances of breaking host eggs (e.g., Gaston 1976; Soler et al., this vol.). Cuckoo laying times vary from early morning (Liversidge 1971, 1955; Gaston 1976; Arias-de-Reyna, this vol.) to afternoon (Davies and Brooke 1988). Besides avoiding attacks by hosts, stealth may be adaptive because hosts that observe cuckoos at their nests have an increased likelihood of rejecting cuckoo eggs (Davies and Brooke 1988; Moksnes and Røskaft 1989). Instead of stealth, males in some cuckoo species attract host aggression and thereby draw the hosts away from nests, allowing females to enter nests unchallenged (Liversidge 1971; Steyn 1973; Gaston 1976; McLean 1987; Fry et al. 1988; Duckworth 1991; Arias-de-Reyna, this vol.). Individual females tend to specialize on particular hosts (Wyllie 1981; Brooker and Brooker 1992), but there is no known adult mimicry of the calls or songs of their hosts (Payne 1977*a*) as has been documented in viduine brood parasites (Payne 1967, 1973*a*, 1976; Payne and Payne 1977). However, the nestlings of some cuckoo species mimic their host's begging calls (Redondo and Arias-de-Reyna 1988; Payne and Payne, this vol.). The great spotted cuckoo imprints on its host species and expresses recognition as fledglings (Arias-de-Reyna, this vol.). It has long been thought that imprinting on hosts leads cuckoos to later parasitize the species that reared them. However, the only relevant experiment showed that adult common cuckoos exhibited no recognition of their foster species (Davies and Brooke, this vol.). Recent experimental evidence suggests that the great spotted cuckoo revisits nests it has parasitized and kills host eggs or young if the adult hosts have removed a parasitic egg (Soler et al. 1995*a*). This avian "mafia" effect (Zahavi 1979) reduces the adaptive value of host egg rejection and could even select for acceptance of cuckoo eggs. Similarly, Arias-de-Reyna (this vol.) suggests that this same cuckoo may help defend parasitized nests against predation. Davies and Brooke (this vol.) review evidence that cuckoos destroy unparasitized nests to increase host availability (see also Dow 1972).

Cuckoos presumably find most nests by observing host behavior from concealed perches (Chance 1922; Øien et al. 1996*a*; Arias-de-Reyna, this vol.). However, there is some evidence that cuckoos may also use the intensity of host aggression to indicate their distance to nests (Seppä 1969) and at least one cuckoo approaches nests while calling, a possible distraction display that enables easy determination of the host nest's stage (Arias-de-Reyna, this vol.). Laying is generally well synchronized with that of the host (Arias-de-Reyna, this vol.; Davies and Brooke, this vol.; Payne and Payne, this vol., Soler et al. this vol.; but see Gill, this vol.).

Most cuckoos lay only a single egg per nest (Wyllie 1981; Davies and Brooke, this vol.; Gill, this vol.), although multiple parasitism is common in the great spotted cuckoo, one of the few species in which nestlings do not actively kill off all of the other nestlings that are present (Arias-de-Reyna, this vol.; Soler et al., this vol.). Cuckoos defend territories and exclude other females, which minimizes costly multiple parasitism (reviewed in Payne 1977*a*; Liversidge 1971; Gaston 1976; Arias-de-Reyna, this vol.; Davies and Brooke, this vol.; Gill, this vol.). When hosts are parasitized by more than one female, the success of the first female is much higher than that of subsequent females (Soler et al., this vol.). Most cuckoos remove a host egg before laying one of their own (Brooker et al. 1990; Davies and Brooke, this vol.; Gill, this vol.; but see Soler et al., this vol.). Egg removal by cuckoos forms the basis for the hypothesis, endorsed by Brooker and Brooker (1989*b*) and Brooker et al. (1990), that egg mimicry results from competition among cuckoos, not from host egg rejection (see below). Removal of host eggs apparently has little or no effect on host responses to cuckoo eggs, and may instead be related to the efficiency of incubation and the nutrition cuckoos receive by eating the egg (Davies and Brooke, this vol.).

Plumage

Female cuckoos are generally less conspicuous than males, which may facilitate stealthy approaches to nests (reviewed in Payne 1977*a*). The bright colors and close resemblance to raptors by males of many cuckoos may help them distract hosts and enable females to gain access to nests (Duckworth 1991).

Fecundity

Cuckoos tend to lay more eggs per season than is typical of their hosts, presumably because they are freed from parental care (Payne 1974, 1977*a*). Typical estimates range from 8–25 eggs per season (Payne 1974; Arias-de-Reyna, this vol.; Davies and Brooke, this vol.). Eggs are laid every other day in "clutches" of 3–6 eggs, with clutches separated by several days (Arias-de-Reyna, this vol.).

Egg adaptations

The most important adaptations of cuckoo eggs are their relatively small size and mimetic patterns that closely match just a few of their potential hosts (Swynnerton 1916; Ali 1931; Baker 1942; Wyllie 1981; Brooke and Davies 1988; Brooker and Brooker 1989*a*,*b*). Egg mimicry has been the focus of experiments since the late 1800s (reviewed in Payne 1977*a*; Rothstein 1990*a*, 1992) and has provided some of the strongest evidence of coevolution in vertebrates (Lotem and Rothstein 1995). The selection on parasites to match host eggs can be very strong because of the rapidity with which egg rejection by hosts (see below) can increase (Rothstein 1975*a*,*b*; Kelly 1987; Takasu et al. 1993). Many of the chapters in part II describe the presence of egg mimicry (Arias-de-Reyna, this vol.; Higuchi, this vol.; Nakamura et al., this vol.), but mimicry is notably absent in the system described by Gill (this vol.). In debates over the coevolutionary process (see part IV), the primary focus has been on the adaptiveness of egg mimicry in cuckoos and on limitations in the host's ability to discriminate between host and parasitic eggs (Redondo 1993; Lotem et al. 1995; Lotem and Nakamura, this vol.; Røskaft and Moksnes, this vol.).

An exception to the general rule that cuckoo eggs mimic those of one to a few host species, occurs in cuckoos that parasitize enclosed nests where dark conditions make mimicry unnecessary (Gill, this vol.). In such cases, cuckoo eggs tend to be dark in color, perhaps to make them cryptic to the host or to other cuckoos (Brooker and Brooker 1989*b*). Cuckoo species that parasitize a range of host species have individually specialized females, with each such group of females or gens (plural gentes) mimicking the eggs of one to a few hosts with similar eggs (Davies and Brooke, this vol.). Recent contact between hosts and parasites usually means that cuckoo and host eggs are poorly matched (Davies and Brooke, this vol.: Nakamura et al., this vol.). As host rejection rates increase, egg matching may improve (Soler et al. 1994, this vol.; Nakamura et al., this vol.) under intense selection for mimicry (Davies and Brooke, this vol.; Nakamura et al., this vol.). In areas where multiple cuckoo species coexist, the different species have eggs adapted to different hosts (Lack 1968; Jensen and Clinning 1974; Brooker and Brooker 1989*b*; Gill, this vol.; Higuchi, this vol.). This form of resource partitioning is especially well documented by Higuchi (this vol.), who presents compelling evidence of character release.

As a further general rule, most cuckoos parasitize host species smaller than themselves and have small eggs in proportion to their size (Davies and Brooke, this vol.; Gill, this vol.; Higuchi, this vol.). The minority of cuckoos that parasitize large hosts lay eggs typical for birds of their size (Arias-de-Reyna, this vol.; Soler et al., this vol.). Cuckoo eggs are also thicker-shelled (Brooker and Brooker 1991) than host eggs, which may reduce breakage during rapid laying (Arias-de-Reyna, this vol.; Gill, this vol.). Cuckoo embryos have a more rapid developmental rate and hatch earlier than those of their hosts, which helps cuckoo nestlings compete, especially in cuckoos that do not evict host nestlings.

Nestling adaptations

The best-known adaptation of cuckoo nestlings is the eviction by nestlings of their nestmates (Jenner 1788; described in Davies and Brooke, and Payne and Payne, this vol.). Nestlings of some cuckoos have concave backs and push host nestlings or eggs up over the nest rim (see also Gill, this vol.; Payne and Payne, this vol.). Payne and Payne (this vol.) determined that nestling cuckoos use the temperature drop associated with the departure of an incubating parental host as a cue for initiating nestling eviction. Cuckoos, however, sometimes eject host nestlings or eggs even when the foster parent is incubating (reviewed in Payne and Payne, this vol.). Eviction behavior is absent in cuckoos of the genera *Clamator* and *Scythrops*, which parasitize relatively large hosts (Davies and Brooke 1988;

Arias-de-Reyna, this vol.; Soler et al., this vol.). The relation between host size and eviction behavior is shown most strongly by the koels in the genus *Eudynamis*. *E. scolopacea* and *E. taitensis* evict in Australia and New Zealand, respectively, where relatively small hosts are used, but the former species does not evict in Asia where it primarily parasitizes crows (Gosper 1964; Lamba 1975; Brooker and Brooker 1989*a*). Nestlings of cuckoos that do not evict compete successfully with those of their hosts by begging more loudly and incessantly, hatching earlier and developing more rapidly. As a result, most parasitized nests of magpies fledge only great spotted cuckoos (Arias-de-Reyna, this vol.; Soler et al., this vol.). Cuckoos that evict their nestmates do not develop more rapidly than host nestlings (e.g., Gill, this vol.).

Although mimicry of host begging calls occurs primarily in cuckoos that do not evict host nestlings (Courtenay 1967; Mundy 1973; Redondo and Arias-de-Reyna 1988; Redondo 1993), it also occurs in at least some evicting species (Payne and Payne, this vol.). Mimicry in some of the latter cases may occur because cuckoo nestlings often fail to evict hosts until 5–7 days after hatching because of poor laying synchrony by adult cuckoos (see also McLean and Waas 1987; McLean and Rhodes 1991; Gill, this vol.). Mimicry of gape patterns such as that reported in viduines (Nicolai 1964, 1974) has rarely been reported in cuckoos (but see Gill, this vol.). Cuckoo plumages may mimic those of the hosts in species that do not evict nest contents (Davies and Brooke 1988) or that delay eviction (Gill, this vol.). However, none of these cases of possible mimicry of host nestling plumages or calls is strong and host and cuckoo nestlings are always easily told apart. Mimicry is much stronger in the viduines and especially in the screaming cowbird (*Molothrus rufoaxillaris*) which was once considered to be indistinguishable from its host (Fraga 1979, this vol.). The adaptive value of cuckoo mimicry of the coloration and calls of host nestlings is unknown as host responses to mimetic and nonmimetic nestlings remain largely untested. Pilot experiments by Payne and Payne (this vol.) suggest that vocal mimicry may be important to the responses of two Australian host species whereas a larger set of experiments found that a New Zealand host was more responsive to conspecific than to cuckoo begging calls (McLean and Rhodes 1991) even though the latter appeared to mimic the former (McLean and Waas 1987; see also McLean and Maloney, this vol.).

Fledgling adaptations

Evidence that fledglings mimic the calls of hosts is summarized in Payne and Payne (this vol.; see also Courtenay 1967). Loud, persistent begging can attract both foster parents and other species (Hardcastle 1925). Great spotted cuckoo fledglings from different nests band together and, as a result, attract more attention and receive more food from their foster parents (Soler et al. 1995*b*).

HOST DEFENSES

Defense of the nest site against adult cuckoos

Hosts are more likely to respond aggressively to cuckoos than they are to "neutral" stimuli (Payne et al. 1985; Davies and Brooke 1988; Moksnes and Røskaft 1989; Duckworth 1991). Even though some cuckoo species resemble hawks, at least one host mobs cuckoos but avoids approaching hawk models (Duckworth 1991). Mobbing of cuckoos also ceases once the young fledge and the cuckoo is no longer a threat. Recognition of cuckoos clearly has a genetic basis as an island population isolated from cuckoos was responsive to cuckoo mounts. But there may also be learned aspects as well because responses were weaker than those of a mainland population exposed to cuckoos (McLean and Maloney, this vol.). It is not clear, however, that active defense of nest sites by hosts is adaptive because it may be ineffective in driving off cuckoos (Wyllie 1981) and may be easily exploited by cuckoos to locate nests (Seppä 1969; McLean 1987; McLean and Rhodes 1991). Communally nesting birds may be better able to protect their nests from cuckoos (Payne et al. 1985), but the increased activity near nests may attract brood parasites (Skutch 1961). A comparative study of Australian birds showed no net effect of social system (communal versus pairs) on susceptibility to parasitism (Poiani and Elgar 1994). Perhaps the primary advantage of aggressively defending a nest site is that hosts may use the presence of a cuckoo to trigger egg rejection (Davies and Brooke 1988; Moksnes and Røskaft 1989; Zuñiga and Redondo 1992).

Abandonment of parasitized nests

One effective response to cuckoo parasitism is nest abandonment followed by renesting (Lotem et al. 1995). Small cuckoo hosts are especially likely to abandon parasitized nests, presumably because they are too small to eject cuckoo eggs (Davies and Brooke 1989*a*; Higuchi 1989). Although renesting may incur much lower costs than rearing a cuckoo in Northern Hemisphere birds, Gill (this vol.) stresses that abandonment may be relatively costly for birds in New Zealand and Australia. Many of these species require especially large investments of time and effort to renest because they have elaborate nests and lay relatively large eggs at 48 h, rather than the 24 h intervals characteristic of most birds.

Ejection of cuckoo eggs

By far the best-documented defense by hosts is ejection of dissimilar eggs (Swynnerton 1918; Kelly 1987; Davies and Brooke 1989*a*,*b*, this vol.; Rothstein 1990*a*; Braa et al. 1992; Moksnes et al. 1993). Most hosts that have long coevolutionary histories of dealing with cuckoo parasitism show fine discriminatory abilities (Arias-de-Reyna, this vol.; Davies and Brooke, this vol.; Higuchi, this vol.; Nakamura et al., this vol.; Soler et al., this vol.). By contrast, egg recognition is absent in the small number of Australian (Brooker and Brooker 1989*b*) and New Zealand host species (Gill, this vol.) that have been tested. These species build domed nests whose dim interiors may retard the evolution of egg recognition. In most comparisons, populations that are not in contact with cuckoos but are conspecific with actual hosts or at least closely related to hosts, show less pronounced recognition behavior than parasitized populations (Brown et al. 1990; Soler and Møller 1990; Davies and Brooke, this vol.; Nakamura et al., this vol.; Soler et al., this vol.). But Nakamura et al. (this vol.) report two cases in which local populations appear to have retained high levels of rejection behavior in the absence of parasitism. The apparent rapid increase in egg rejection frequency in a newly parasitized host population (Soler et al. 1994, this vol.; but see Zuñiga and Redondo 1992; Arias-de-Reyna, this vol.) may provide the strongest evidence to date of microevolutionary processes related to brood parasitism (Lotem and Rothstein 1995). Nakamura et al. (this vol.) report even more rapid increases in rejection frequency but in this case the increase is too rapid to be due solely to increases in alleles coding for rejection. The host is a communal breeder and rejection behavior may be acquired through a mix of genetic factors combined with learned influences when individuals witness rejection by conspecifics.

Experimental evidence suggests that hosts learn to recognize their own eggs through a process resembling imprinting (Rothstein 1974, 1978*a*; Lotem et al. 1995). In some cases, eggs laid too early were more likely to be rejected (Arias-de-Reyna, this vol.; but see Lotem et al. 1995). As females grow older, they are more likely to reject eggs (Lotem et al. 1992, 1995). Host egg variability within clutches, however, may constrain the host's ability to recognize mimetic cuckoo eggs and increase the chances that birds will reject some of their own eggs if these are unusual in any way (Marchetti 1992; Lotem et al. 1995; Lotem and Nakamura, this vol.). If such recognition errors occur at unparasitized nests and if parasitism rates are low, they may make it likely that accepter and rejecter individuals will coexist in an evolutionary equilibrium or that acceptance will be the most adaptive option for all individuals (see part IV and Takasu et al. 1993). But documenting recognition errors is difficult. The occasional disappearance of some but not all host eggs from the unparasitized nests of rejecter species (Marchetti 1992) is not proof of recognition errors because such cases of partial clutch reduction also occur in accepter species (Rothstein 1982*a*, 1986).

Nestling recognition

The general scarcity of nestling mimicry in cuckoo–host systems is discussed at length by Davies and Brooke (this vol.), Gill (this vol.), McLean and Maloney (this vol.), and Lotem and Nakamura (this vol.). Lotem (1993) suggested that rejection of host nestlings is maladaptive because of the high likelihood of mistakes if hosts imprint on the first nestlings they experience. McLean and Maloney (this vol.) further argue that avian cognitive constraints might make the evolution of nestling recognition unlikely given the rapid changes in appearance as nestlings grow (see also Rothstein 1990*a*). Data on feeding rates of mimetic versus nonmimetic nestling cuckoos might reveal the benefits of mimicking host natal plumages and/or begging calls (Gill, this vol.). Nevertheless, it is puzzling that no known cuckoo host shows ejection of nonmimetic nestlings, even though simple innate decision rules based on invariant features of host and cuckoo nestlings could result in the acceptance of all of the former and none of the latter in many hosts (Davies and Brooke 1988; Rothstein 1990*a*).

OVERVIEW OF THE CHAPTERS ON CUCKOO–HOST SYSTEM

In this overview, we have taken elements from each of the chapters in part II and put them in the general framework of adaptations and counteradaptations in cuckoo–host systems. The different emphases of the chapters reflect both differences in host–parasite systems (e.g., enclosed versus open nesters, a single cuckoo versus several sympatric cuckoo species) and different interpretations of coevolutionary processes. Perhaps the most unusual system in this part of the book and the one that has received the least prior attention in the literature is the system described by Nakamura et al. Unlike other cuckoo–host systems, which generally appear to involve stable interactions, mimetic eggs and low rates of parasitism, the common cuckoo system described by Nakamura et al. in Japan is a dynamic one involving poor egg mimicry, the recent abandonment of one major host species, and the adoption of a new one with very different eggs. This system may also involve rates of parasitism high enough to limit the sizes of local host populations. In these respects, this system is more like that of the brown-headed cowbird of North America (treated in part III) than like that of conspecific populations of the common cuckoo in Europe described by Davies and Brooke (this vol.).

OVERVIEW OF PART III: COEVOLUTION BETWEEN COWBIRDS AND THEIR HOSTS

Although the New World parasitic cowbirds contain only five species, one of those species is as specialized as any brood parasite in the world and two others are the most generalized of all parasites as regards numbers of host species. The three essays in this section deal with both specialized and generalized cowbird species. Cowbirds have probably received more research attention than any other North American bird since Friedmann's (1929) pioneering monograph. A look at the year-end index of any North American ornithological journal will show that the brown-headed cowbird is nearly always one of the three or four most often cited birds. There are three reasons for this attention. First, cowbirds are intrinsically interesting and provide unique research opportunities. Besides the questions emphasized in this book dealing with parasite–host coevolution, the lack of parental care raises interesting questions and makes cowbirds attractive subjects of research on reproductive physiology (Höhn 1959; Dufty and Wingfield 1986) and behavioral development (Rothstein and Fleischer 1987a,*b*; West et al. 1981*a*; O'Loghlen 1995). Studies of these and other subjects are facilitated by the facts that cowbirds are abundant over much of North America and are easily trapped and maintained in captivity.

Second, cowbirds are so common and widespread and have such a large range of hosts (221 known species) that many breeding studies of North American passerines find significant rates of parasitism and so end up also being cowbird studies (regardless of a researcher's original intentions!). Third, cowbird parasitism has long been recognized as a potential threat to the survival of several endangered passerines and has more recently been identified as a possible factor in the declines of numerous other species (Mayfield 1977; Brittingham and Temple 1983; Robinson et al. 1995*a*,*b*). Accordingly, the management or limitation of cowbird numbers has become a widespread practice in North America (Rothstein and Robinson 1994). A volume of 28 papers on cowbird biology, management and host impacts will be appearing soon (Smith et al., in press). Conservation aspects of cowbird parasitism will be covered in part V, which also details the varying costs of cowbird parasitism for different host species.

The most specialized cowbird is the screaming cowbird, which was long thought to parasitize only a single species, the putatively closely related bay-winged cowbird. But the latter is now known to be no more closely related to the parasitic cowbirds than are a number of other nonparasitic members of the subfamily, Icterinae, to which cowbirds belong (Lack 1968; Lanyon 1992). Furthermore, two other hosts, both icterines, are now known. One of these, the brown-and-yellow marshbird (*Psuedoleistes virescens*), appears to be parasitized at a low (5%) rate (Mermoz and Reboreda 1996), whereas the other, the chopi blackbird (*Gnorimopsar chopi*), is heavily parasitized in the northern part of the screaming cowbird's range (Sick 1985; Fraga 1996). Recent reports of parasitism of the latter host are from areas lacking bay-winged cowbirds and it is unknown whether there are areas where both hosts are commonly parasitized. Among the four more generalized species, the largest, the giant cowbird (*M. oryzivorus*), has seven known host species, most of which are large, colonial icterines known as caciques (*Cacicus* spp.) and oropendolas

(*Psarocolius* spp.) (Smith 1968; Friedmann et al. 1977; Robinson 1988; Webster 1994). The numbers of species known to have been parasitized by the three smaller cowbirds are 81 (bronzed cowbird, *M. aeneus*), 201 (shiny cowbird, *M. bonariensis*, and 221 (*M. ater*) (Friedmann 1929, 1963; Friedmann et al. 1977; Friedmann and Kiff 1985; De Geus and Best 1991; Lowther 1995). The two extreme generalists were formerly allopatric, with shinies in South America and brown-headeds in North America. But both are undergoing range expansions and have become sympatric in Florida in the last 10 years (Cruz et al., in press, this vol.). Shiny cowbirds may soon occupy much of North America, which will expose new host populations to parasitism (Mayfield 1965; Robbins and Easterla 1981; Cruz et al. 1985, in press, this vol.; Rothstein 1994; Robinson et al. 1995*b*). Cowbird range expansions have provided ideal situations for testing hypotheses for the evolution of host defenses, especially the evolutionary lag model (Rothstein 1990*a*; see Kattan this vol., and chapters in part IV). Our summary below of the prominent features of cowbird–host systems is just a brief overview of a vast literature and focuses on the brown-headed cowbird, which has been by far the most heavily studied species.

ADAPTATIONS OF COWBIRDS FOR BROOD PARASITISM

Egg adaptations

Some aspects of cowbird life history differ dramatically from passerines that provide parental care. At least some cowbirds are extremely fecund, a reflection of the trade-off between parental care and production of eggs. Kattan (1993) estimated that shiny cowbirds lay over 100 eggs per year in the long tropical breeding season. Captive brown-headed cowbirds can lay as many as 77 eggs in a season (Jackson and Roby 1992; Holford and Roby 1993) and Scott and Ankney (1983) estimated that wild birds lay 40 or more eggs during the relatively short temperate breeding season. Contrary to expectations, the latter cowbird appears to lay eggs at the same rate of 0.7–0.8 eggs per day during the breeding season throughout its range, despite larage differences in breeding season length and host and food availability (Rothstein et al. 1986; Fleischer et al. 1987).

The eggs themselves tend to be larger, thicker-shelled, rounder, and faster to develop than those of the hosts (Blankespoor et al. 1982; Spaw and Rohwer 1987; Rahn et al. 1988; Rohwer and Spaw 1988; Picman 1989; Briskie and Sealy 1990; Kattan 1995). The lone exception appears to be the giant cowbird, which has proportionately normal-sized eggs and parasitizes hosts as large as or larger than itself (Smith 1968, 1979, 1980; Fleischer and Smith 1992). The three smaller generalist cowbirds mostly parasitize smaller hosts (Friedmann 1929) although many large hosts are also used. Thick shells make eggs more difficult to break and may therefore increase the difficulty of host ejection (see part IV). They may also increase the likelihood that cowbird eggs will cause the breakage of some host eggs, thereby deceasing the competition nestling cowbirds must face (Carey 1986; Rothstein 1990*a*; Kattan, this vol.). Rounded eggs may provide cowbirds with similar benefits because they are stronger than elongated eggs. However, at least one case is known in which rounded eggs have been selected for because of host discrimination (Mason and Rothstein 1986).

The only cowbird in which egg mimicry has been described is the giant cowbird (Smith 1968, 1979, 1980). However, Fleischer and Smith (1992) found significant differences in appearance among eggs of this cowbird and two host species and thereby weakened the case for mimicry. Although cowbird species lack the clear examples of egg mimicry that occur in many cuckoo species, egg coloration has undergone significant evolution during the cladogenesis of the cowbirds. The first cowbird to branch off (Lanyon 1992), the screaming cowbird, has a generalized spotted egg with a whitish ground color, as do the last two species to branch off (shiny and brown-headed cowbirds). The second species to branch off, the giant cowbird, has both spotted and immaculate eggs, while the third species, the bronzed cowbird, has only immaculate eggs. To further complicate matters, one of the last two species to appear, shiny cowbird, has both spotted and immaculate eggs in parts of its range. At least two hosts with spotted eggs accept spotted cowbird eggs and reject immaculate ones (Fraga 1985; Mason 1986*a*; Mermoz and Reboreda 1994) so intraspecific variation in egg coloration clearly has adaptive consequences for shiny cowbirds. Presumably, the egg color polymorphism in shiny cowbirds is maintained by one or more host species that reject spotted eggs but accept immaculate ones but no such species has yet been identified conclusively (see Friedmann et al. 1977). Although the polymorphism could be maintained by differential crypsis of the spotted and immaculate morphs in the nests of different host species,

Mason and Rothstein (1987) found no support for this alternative. A clearer case of mimicry occurs in egg shape and size. Where cowbirds have long parasitized a large host that has a domed nest with a dark interior and that rejects eggs on the basis of egg width, they have larger and more rounded eggs than in another area where this host has only recently become available (Mason and Rothstein 1986). Thus, although cowbirds lack the clear-cut egg mimicry present in many cuckoos, several lines of evidence indicate that cowbird egg features have undergone evolutionary changes.

Because egg mimicry is widespread among parasitic birds whereas nestling mimicry is rare, it seems likely that the former would precede the latter in a coevolved interaction. However, the only cowbird with mimetic nestlings, the screaming cowbird, does not lay mimetic eggs and its most important host does not reject even strongly nonmimetic eggs (Fraga 1983; Mason 1986*a*). This lack of host egg recognition and parasitic egg mimicry probably occurs because the host usually nests in enclosed areas that are too dark to allow egg coloration to be seen (Fraga 1983).

Cowbird behavioral adaptations

Cowbirds search for host nests primarily by observing the behavior of hosts (Hann 1941; Norman and Robertson 1975; Robertson and Norman 1976, 1977; Gochfeld 1979*b*; Smith 1981; Smith et al. 1984; Wiley 1988), although giant cowbirds probably do not have to search long for the extremely conspicuous nests of their colonial hosts. Cowbirds may also search actively for nests and may even attempt to flush hosts from nests (Norman and Robertson 1975; Wiley 1988). Most cowbirds approach nests very stealthily, usually early in the morning, and lay eggs very quickly (Hann 1947; Scott 1991; Neudorf and Sealy 1994; Sealy et al. 1995), often before the host has arrived. Giant cowbirds, which have to circumvent the defenses of colonial hosts, use a combination of distraction displays and stealthy visitation at any time of the day (Robinson 1988) but especially in late morning and early afternoon when hosts are least likely to be present (Webster 1994). Screaming cowbirds also visit host nests in groups, perhaps because their main host is a communal breeder (Mason 1987) but males do not help females with distraction displays designed to draw hosts from nests (Fraga, this vol.) unlike some cuckoos (Arias-de-Reyna, this vol.; Soler, this vol.). Even the generalist shiny cowbird may approach nests in groups if the host is large and vigorous in defending its nests (Fraga 1985). Male brown-headed cowbirds are rarely seen at nests but this is not always the case for male shiny cowbirds (Piper 1994). Proper timing is essential for successful parasitism as cowbird eggs laid before hosts begin to lay have increased chances of rejection (Clark and Robertson 1981; Rothstein 1986; Sealy 1992, 1995). Eggs laid too late may not hatch. Nevertheless, many cowbird eggs are not properly timed (Friedmann 1963; Clark and Robertson 1981; Kattan 1997; Fraga, this vol.).

The extent to which generalist brood parasites select the most favorable hosts within a community remains unclear. Cowbird hosts vary enormously in quality (e.g., Robertson and Norman 1976, 1977; Clark and Robertson 1979; Mason 1986*a*; Graham 1988; Weatherhead 1989). Nonpasserines that are completely unsuitable as hosts, but are within the size range of most passerines, such as doves and nonparasitic cuckoos, are clearly avoided (Friedmann 1963; Rothstein 1975*a*,*b*, 1976*a*; Wiley 1988). However, some passerines that are unsuitable because they do not feed their young on a high-protein diet of insects (Rothstein 1976*a*; Middleton 1977) or because they reject cowbird eggs (Friedmann et al. 1977; Scott 1977; Mason 1986*b*; Rohwer et al. 1989) are parasitized, sometimes quite heavily. In fact, the largest available data set on brown-headed cowbird parasitism rates—nearly 45,000 nests from Ontario (Peck and James 1987)—shows what seem to be some remarkably poor host choices. Two of the 10 species with the largest numbers of reported cases of parasitism are unsuitable hosts because of diet and/or rejection of cowbird eggs (these are *Carduelis tristus* and *Bombycilla cedrorum*). While these numbers may reflect host abundances more than cowbird preferences, they do suggest that a large proportion of the cowbird eggs that are laid have little chance of success. Furthermore, two of the five hosts with the highest percentages of nests parasitized in Ontario, which could reflect direct host preferences, are cardueline finches (*Carpodacus mexicanus* and *C. purpureus*) that are usually unsuitable hosts (but not always; Friedmann et al. 1977) because they feed their young on seeds. There has been debate on whether the use of unsuitable hosts is a maladaptive but necessary consequence of being a generalist (Rothstein et al. 1986) or whether it is adaptive because cowbird eggs are so cheap to produce that even parasitism of hosts with low prospects for success produces more benefits than costs (Ankney and Scott 1980). Because cowbirds sometimes revisit nests they have parasitized (Mayfield 1961; Friedmann 1963), they may adjust their preferences toward successful hosts and most of the parasitism of unsuitable hosts may be by young females (Rothstein 1976*a*).

Some data from local communities have shown no statistically significant preferences for particular hosts among species that are suitable but of varying host quality (Fleischer 1985). Nevertheless, cowbird breeding seasons may be adapted to the availability of high-quality host species (Payne 1973*b*) and different passerine species within the same community sometimes experience highly divergent parasitism levels, suggesting something akin to host preferences (Robinson et al. 1995*a*,*b*). It is unknown whether the proximate basis for such variable parasitism is due to preferences for particular host species or to other factors (certain nest types, particular microhabitats, host defenses, etc.). Clearly, though, observed levels of parasitism of hosts that are unsuitable because they regularly eject cowbird eggs are underestimates of actual parasitism rates because many cowbird eggs are removed before observers see them (Rothstein 1990*a*).

Summing data from thousands of nests parasitized by brown-headed cowbirds, Friedmann (1963) reported that about 40% contained two or more cowbird eggs. The per capita success rate of cowbird eggs is inversely proportional to the number of cowbird eggs in a nest (Friedmann 1963). Cowbirds should thus avoid laying in already parasitized nests but there is little evidence of this as the numbers of eggs per host nest conform to a Poisson distribution (Orians et al. 1989). Avoidance of multiple parasitism may depend upon whether or not cowbirds have exclusive use of an area. Nearly all studies have found that female cowbirds are aggressive towards one another (Dufty 1982*a*,*b*; Darley 1983; Rothstein et al. 1986, 1988) but they apparently find it difficult to exclude all other females from their very large home ranges (Rothstein et al. 1984; Thompson 1994). Thus, overlap among laying areas is common and is probably a function of the local density of female cowbirds (Rothstein et al. 1984, 1986). Multiple parasitism was rare on an island with only one or two female cowbirds (Smith and Arcese 1994) but was common on the nearby mainland of British Columbia (Smith and Smith, this vol.). Multiple parasitism was also very common (Elliott 1977) and was due largely to different females laying in the same host nest (Fleischer 1985) in Kansas, where cowbirds were especially abundant and may not even have defended home ranges (Elliott 1980).

Cowbirds often remove a host egg before or after laying one of their own eggs (Sealy 1992). Bronzed cowbirds puncture-eject host eggs (Clotfelter 1995) whereas brown-headeds either grasp- or puncture-eject eggs (Sealy 1992; S. Robinson, pers. obs.). Cowbirds at least occasionally eat the eggs they remove (Scott et al. 1992), which makes sense if they are calcium-limited (Holford and Roby 1993). But many eggs are not eaten and the function of host egg removal is unclear (Sealy 1992) as it has no effect on host responses toward eggs (Rothstein 1975*a*, 1986).

There is some recent evidence that cowbirds visit nests and depredate eggs and young. On Mandarte Island, British Columbia (Canada), late-arriving cowbirds appear to depredate unparasitized nests, apparently to increase the availability of renests that they can parasitize (Arcese et al. 1992, 1996; Smith and Arcese 1994). This predation may be unique to some situations, such as those on islands where there are few hosts and few cowbirds, as other studies have not shown consistent patterns as to the relative predation rates of parasitized and unparasitized nests (Trine et al., this vol.). Nevertheless, this situation is vaguely similar to that reported for great-spotted cuckoos in Spain (Soler et al. 1995*a*), which are said to penalize magpies for rejecting their eggs.

Nestling adaptations

Although nestling mimicry is absent or poorly developed in most brood parasite–host systems (Lotem 1993; Lotem et al. 1995), Fraga (1979, this vol.), shows that the screaming cowbird mimics almost perfectly the nestling and fledgling plumage of its main host. The lack of nestling and fledgling mimicry in shiny cowbirds, which also parasitize bay-wings, may explain why their fledglings are rarely fed by the host (Fraga, this vol.). Rothstein (1978*b*) reported that brown-headed cowbird nestlings have yellow rictal flanges in some parts of North America and white ones elsewhere (with both occurring in some regions [Fleischer and Rothstein 1988]). This nestling variation may relate to geographic variation in the use of host species because nonparasitic species rarely show such variation (Rothstein 1978*b*). Cowbirds, however, fledge from nests regardless of whether they match host nestlings' flange colors. Thus there is no evidence for rejection of nestlings based on flange color but the quality of parental care may be influenced by this trait (Rothstein 1978*b*; Stevens 1982). Shiny cowbirds show comparable variation in their flange colors

(Fraga 1978, 1985). Other cowbird species appear to have no such variation (Friedmann 1929; Rothstein 1978*b*) but more data are needed.

Nestling cowbirds beg very loudly (Gochfeld 1979*a*) and develop rapidly (Wiley 1986; Kattan 1996), presumably to help them compete with nestmates for food (Friedmann 1929; Payne 1977*a*; Rothstein 1990*a*). This begging probably increases the conspicuousness of nests, but there is no consistent evidence that predation rates of parasitized nests are higher than those of unparasitized nests (Gochfeld 1979*a* Robinson et al. 1995*b*). Loud begging by fledglings occasionally attracts non-foster parents that feed the fledglings (Friedmann 1963; Klein and Rosenberg 1986; S. Robinson, pers. obs.). Another possible adaptation for garnering a disproportionate amount of host feedings is the large mouth size reported for brown-headed cowbird nestlings relative to some host nestlings (Ortega and Cruz 1992), but data on relative mouth sizes are needed for many more host species. More data are also needed to determine whether Dearborn's (1996) report of a cowbird nestling ejecting a host nestling is a regular occurrence.

HOST DEFENSES AGAINST COWBIRDS

Sealy et al. (this vol.), Fraga (this vol.) and Kattan (this vol.) discuss host defenses against cowbird parasitism. Rothstein (1990*b*) distinguished behaviors that prevent cowbirds from gaining access to nests from those involving egg rejection once nests have been parasitized. Egg rejection includes abandonment of parasitized nests, egg burial, or outright ejection of eggs either by grasping or puncturing. Rejection of nestlings and fledglings is rare, although Fraga (this vol.) has strong evidence that bay-winged cowbirds feed only mimetic fledglings. Because some cowbirds are expanding their ranges, they are coming into contact with populations and even entire species that may not yet have evolved resistance to parasitism. This situation is therefore suitable for studies of rapid microevolutionary changes (Lotem and Rothstein 1995).

Defense of the nest site

Defense of host nests by the parents has been well documented (Slack 1976; Robertson and Norman 1977; Folkers and Lowther 1985; Graham 1988; Briskie and Sealy 1989; Neudorf and Sealy 1992; Mark and Stutchbury 1994; Sealy et al., this vol.). Coloniality sometimes provides protection (Wiley and Wiley 1980; Carello and Snyder, in press) and this protection can also apply to noncolonial species nesting within colonies of another species (Clark and Robertson 1979). The benefits of colonial nesting are best known in the red-winged blackbird (*Agelaius phoeniceus*), a species whose parasitism rates vary greatly according to the nesting density of local populations (Friedmann 1929, 1963; Facemire 1980; Wiley and Wiley 1980; Freeman et al. 1990; Carello and Snyder, in press). However, the major hosts of the giant cowbird are vulnerable to parasitism even though they are colonial (Smith 1968; Robinson 1988; Webster 1994). Among the latter hosts, oropendolas (*Psarocolius* spp.) sometimes leave colonies undefended when they are foraging, which provides giant cowbirds with windows of opportunity (Robinson 1988; Webster 1994). In contrast, cacique (*Cacicus cela*) colonies in southeastern Peru almost always have multiple males defending against parasites during the critical egg-laying period and rarely experience parasitism (Robinson 1988). Interestingly, *Cacicus cela* males often prevent parasitism of oropendola nests in mixed colonies (Robinson 1988).

Although it may reduce nest predation, aggressive defense of nests by solitary-nesting species may actually attract cowbirds to nest sites (McLean 1987). Smith (1981) and Smith et al. (1984) found that older, more experienced song sparrows defended nests more aggressively and were more likely to be parasitized (see also Arcese et al. 1992; Smith and Arcese 1994). Some birds that aggressively defend nests against cowbirds, however, are rarely parasitized even though they accept cowbird eggs (Briskie and Sealy 1987; S. Robinson, unpublished data). Smaller birds may be unable to chase cowbirds from their nests and their best defense may be to abandon when they detect a cowbird at their nest site (Robertson and Norman 1976; Burhans, in press). Alternatively, such birds may simply sit tightly on the nest, thereby denying cowbirds access (Hobson and Sealy 1989; Gill and Sealy 1996) although there are observations of cowbirds driving hosts from their nests (Hann 1941; Prescott 1947). Mark and Stutchbury (1994) hypothesized that recognition of cowbirds as a threat can evolve rapidly, which implies that it may have some benefit for the host. Briskie and Sealy (1989) found that defense against cowbirds varied within a season, being highest when cowbirds posed the greatest threat. Besides aggressive behavior, cowbirds may use vocalizations that accompany normal nesting activities to find nests. Uyehara and Narins (1995), for example, found that parasitized willow flycatchers (*Empidonax traillii extimus*) were noisier than unparasitized ones.

Nest site placement

An additional line of defense is nesting sites that are protected against cowbirds (Cruz et al. 1985). In North America, cavity nesters are less likely to be parasitized than open-cup nesters (reviewed in Robinson et al. 1995*b*), but there are exceptions (Friedmann 1963; Petit 1991). Other cowbirds show no such avoidance of cavity or enclosed nests (Friedmann and Kiff 1985; Cruz et al. 1985, in press, this vol.; Mason 1986*b*; Wiley 1986, 1988; Kattan, this vol.) although nests with very small cavity openings may be safe because cowbirds cannot enter them (Post and Wiley 1976; Woodward and Woodward 1979; Kattan, this vol.; J. P. Hoover, unpubl. data). Nesting in dense cover may keep cowbirds from finding nests (Uyehara and Whitfield, in press), but not always (Robinson et al. 1995*a*,*b*).

Defenses against cowbird eggs

Egg rejection is the best-documented form of defense against cowbird parasitism (reviewed in Rothstein 1990*a*). Nearly all potential cowbird hosts in both the Nearctic and Neotropics can be divided into "accepters," species that accept cowbird eggs nearly 100% of the time, and "rejecters," species that almost always reject them. Rejection occurs by abandoning parasitized nests, burying cowbird eggs or most commonly by ejecting them. Some birds eject eggs by puncturing them, whereas others grasp them in their bills (Rothstein 1975*a*, 1976*b*; Rohwer and Spaw 1988). Egg burial is a relatively uncommon defense, except in the yellow warbler, and occurs most frequently when cowbird eggs are laid before those of the host (Clark and Robertson 1981; Sealy 1995). Abandonment of parasitized nests by accepter species and subsequent renesting is most frequent when cowbird eggs are laid too early and hosts are too small to eject eggs (Rothstein 1975*a*; Nolan 1978; Graham 1988), when there are too many parasitic eggs (Friedmann 1963; Fraga, this vol.; Kattan, this vol.), or when too many host eggs are removed (Southern 1958; Hill and Sealy 1994).

Experimental simulations of the circumstances associated with egg burial and nest desertion indicate that birds are not responding to cowbird eggs per se but are instead responding to adult cowbirds at their nests or to overall clutch sizes highly divergent from the norm (Rothstein 1982*a*, 1986). Accepters appear to lack any egg recognition based on egg appearance as all tested so far will incubate clutches made up entirely of nonmimetic eggs, whereas rejecters always eject nonmimetic eggs (Rothstein 1982*a*; Ward et al. 1996). Indeed, because birds seem to perceive their clutch size by the combined volume or surface area of all eggs present, rather than by counting eggs, accepters with small eggs are more likely to desert a clutch reduced to two eggs of their own than one replaced with two cowbird eggs (Rothstein 1986)! This occurs because cowbird eggs, being larger, make up a larger proportion of the volume or surface area of a normal-sized host clutch. Although some accepter species desert naturally parasitized nests at high rates (>20%; Graham 1988) they almost never desert nests when experimentally parasitized with real or artificial cowbird eggs (Rothstein 1975*a*). This difference once again suggests that desertion is not due to egg recognition but to other behaviors that may or may not have evolved in the context of brood parasitism (Goguen and Mathews 1996), for example desertion after any animal has visited the nest because the intruder could be a predator that will return.

Regardless of its ultimate or proximate causation, desertion of parasitized nests may be adaptive because renesting gives a host a chance of rearing an unparasitized brood (Sedgewick and Knopf 1988; Pease and Grzybowski 1995). At least one host seems to abandon if it is chased off the nest by a cowbird and is subsequently parasitized (Burhans, in press). Thus, small hosts may benefit from sitting on their nests early in the morning, even if they are too small to chase away cowbirds, because they can determine that they have been parasitized (see also Burgham and Picman 1989; Neudorf and Sealy 1994). Therefore, it is possible that selection for host defenses has modified nest attendance behavior (even with no direct modifications to desertion behavior). In one cavity nester, the likelihood of desertion is related to the availability of alternative nest sites (Petit 1991).

With one exception, accepter and rejecter species show no geographic variation in responses towards cowbird eggs (Rothstein 1975*a*) even though recent cowbird range extensions (Mayfield 1965; Rothstein 1994) mean that many species have populations that have been exposed to parasitism for different lengths of time. In the exception, American robins sympatric with cowbird reject nearly all cowbird eggs (Rothstein 1975*a*) where Briskie et al. (1992) found that an allopatric population showed only a 66% rejection rate.

Egg rejection is based on birds recognizing their own eggs because rejecters will eject eggs that differ from both their own and from cowbird eggs (Rothstein 1982*a*,*b*). Although the tendency to reject seems to be innate, the type of egg a bird recognizes as its own appears to be learned (Rothstein 1974, 1978*a*). One cavity-nesting host rejects solely on the basis of tactile perception of egg size (Mason and Rothstein 1986), but usually recognition is by visual discrimination (Rothstein 1982*b*). Rejecter species vary in their tolerance of eggs that deviate from their own and eggs may have to differ in more than one way before they are rejected, a behavior that reduces the number of mistaken ejections of host eggs (Rothstein 1982*b*).

Egg rejection has been experimentally documented in many potential cowbird hosts (Rothstein 1975*a*; Finch 1982; Cruz et al. 1985; Carter 1986; Mason 1986*a*). Hosts that are too small to grasp cowbird eggs may be forced to puncture them to bring about ejection (Rothstein 1976*b*; Rohwer and Spaw 1988). Puncture-ejection sometimes results in damage to host eggs (Rothstein 1976*b*, 1977; Rohwer et al. 1989), which is a cost that could favor acceptance over rejection (see chapters in part IV). However, the smallest species known to puncture-eject incurs an ejection cost that is far below the cost of rearing a cowbird (Sealy 1996). In one puncture-ejecter, both males and females eject eggs, but males do it less often and cause more damage to host eggs (Sealy and Neudorf 1995).

Defenses against nestlings and fledglings

For a variety of reasons discussed in part IV, defenses against nestlings and fledglings appear to be lacking in cowbird–host systems except for the one described by Fraga (this vol.). Most hosts of generalist cowbirds feed cowbird nestlings and fledglings preferentially (Rothstein 1990*a*); only one species (*Vireo philadelphicus*: Rice 1974) shows any evidence of rejection of fledglings.

OVERVIEW OF THE PAPERS ON COWBIRD–HOST SYSTEMS

The three chapters in part III describe cowbird adaptations and host defenses in different cowbird–host systems. Fraga provides the first detailed account of the interaction between the highly specialized screaming cowbird, and its host, the bay-winged cowbird, since Friedmann's (1929) work. A striking feature of this system is that parasitism is extremely frequent (87%), intense (3.5 cowbird eggs/parasitized nest), and costly. Host defenses include nest guarding, use of cavity nests, and a highly variable period between the completion of nest building and the initiation of laying. The latter makes the optimal date of parasitization unpredictable, causing the parasite to lay almost half of its eggs before the host begins to lay. The host ejects all such early parasitic eggs when it begins to lay. The degree of host mimicry by nestlings and fledglings of screaming cowbirds is unsurpassed among parasitic birds. Only the viduines (Nicolai 1964, 1974; Payne 1982) approach this degree of mimicry and examples among cuckoos (Gill, this vol.; Payne and Payne, this vol.) involve general resemblances that still leave host and parasitic young easily distinguishable. The degree of specialization in this highly evolved parasite–host association makes it similar to those of cuckoos (see part II) and provides a remarkable contrast with more generalized cowbirds, the topic of the other two chapters in this section. However, the high rate of parasitism is a distinctly cowbird-like feature. In a remarkably convenient natural experiment, about 25% of bay-winged nests are also parasitized by the generalist shiny cowbird, enabling Fraga to compare the success of a generalist and a specialist in the same sample of nests. There is so much parasitism related to bay-winged nests and such a seemingly high level of chaos involving eggs going into and being removed from nests that one wonders how this host and its specialist survive. In his own experiences with this species, one of us (S.I.R.) recalls checking a nest with five eggs of two of the cowbird species. Each egg was dutifully marked. The next time the nest was visited there were again five eggs, but close examination showed that none of these was marked and that some of the marked eggs were strewn about on the ground near the nest. Fortunately, Fraga's study goes a long way towards making sense of this strange system.

The second chapter in this part reviews the myriad defenses used by hosts against brown-headed cowbirds, especially nest-guarding tactics that may reduce the likelihood that a bird will be parasitized at all. Unlike egg rejection, which has no function outside of parasitism, responses to nest intruders could be unrelated to parasitism (as discussed above) since most birds experience nest predation and will attack or mob animals that come too close to their nests. But Sealy et al.'s well-designed experiments show that many, but not all, species do indeed recognize cowbirds as posing a special threat. Sealy et al. argue that it is primarily rejecter species that do not recognize cowbirds as special enemies, perhaps because their egg recognition is an effective defense against cowbird parasitism. (This is different from cuckoo–host systems in which even species with well-developed egg rejection recognize and attack cuckoos [Davies and Brooke, this vol.].) In contrast, accepters clearly recognize cowbirds as a special threat and respond aggressively or in other ways, e.g. Sealy et al. argue that sitting on the nest may be the most effective defense for small hosts. Sealy et al. stress that defenses revealed by experimental presentation of mounted cowbirds will be useless if hosts are not present when real cowbirds visit nests. So they suggest that researchers must also assess the extent to which hosts monitor nests during periods when parasitism is likely to occur. Even large hosts that can easily drive off giant cowbirds leave their nests vulnerable at times (Webster 1994). Brown-headed cowbird counterdefenses to host nest guarding include a variety of sneaky and stealthy ways of approaching nests and the habit of laying shortly before dawn, when most host species are away from their nests.

The final chapter, by Kattan, shows that a heavily parasitized host rejects shiny cowbird parasitism, but only in fairly inexplicable circumstances. House wrens always accepted when parasitized with one or two cowbird eggs but nearly always deserted when parasitized with three or more eggs. These differential responses would make adaptive sense except for the fact that nests with two or more cowbird eggs suffered complete nest failure. Furthermore, even one cowbird egg results in the loss of most wren young (see also Kattan 1996, 1997) and Kattan's experiments show that this species is capable of ejecting cowbird eggs from its nests. This study provides an excellent example of poorly developed host defenses, due presumably to evolutionary lag (see part IV). Other notable features of this system are the cavity nesting of the host, which shows that although shiny and brown-headed cowbirds seem equally generalized in terms of the numbers of species they parasitize, the former much more readily uses species that nest in cavities or that build enclosed domed nests. Finally, most cowbird eggs were laid before wrens laid, which is significant for two reasons. First, while it would have been advantageous, wrens starting to lay their own clutch did not employ the effective eject-all-eggs strategy followed by bay-winged cowbirds. Second, parasitized wrens lost nearly two-thirds of their eggs when these broke due to being jostled against the stronger cowbird eggs during incubation, losses that could have been avoided had they ejected.

OVERVIEW OF PART IV: MODELS OF HOST–PARASITE COEVOLUTION: WHY DO SO MANY HOSTS ACCEPT THE EGGS AND YOUNG OF PARASITIC BIRDS?

It is clear that many of the complex adaptations in brood parasite–host systems result from coevolution between the parasites and their hosts (Rothstein 1975*a*, 1990*a*; Davies and Brooke 1989*a,b*) as described in parts II and III. Because parasitic counterdefenses such as egg mimicry have to match host characteristics, parasites cannot have effective defenses against more than a few hosts at any one time. As a result, the traditional view has been that parasites are likely to specialize on fewer and fewer hosts as a parasite–host system gets older and more hosts develop defenses. By contrast, Lanyon (1992) argued that the opposite trend has occurred in cowbirds because the more recently derived species are the most generalized. However, the phylogenetic trend noted by Lanyon actually supports the traditional view because the older cowbird species are likely to have been in contact with their host communities longer than have the more recently derived cowbirds (Davies and Brooke, this vol.; Rothstein et al., in prep.). A comparison between cowbirds and parasitic cuckoos also supports the traditional view because DNA evidence (cited above) shows that the latter group is much older and nearly all cuckoos are more specialized than most cowbirds. Although they may differ on the direction of long-term trends, the Lanyon and traditional views agree that parasites and hosts have undergone evolutionary responses to one another.

Brooker er al. (1990) have endorsed the contrary view that egg mimicry and host specialization by parasites are not responses to host adaptations but are due instead to competition among the parasites. This view assumes that a second cuckoo parasitizing a nest already parasitized by another cuckoo will preferentially remove the first cuckoo's egg because it is a threat to her own offspring. Under this scenario, egg mimicry or cryptic, darkly colored cuckoo eggs with hosts that have dark enclosed nests, is adaptive because they make it difficult for cuckoos to identify cuckoo eggs that are already present. This alternative to coevolution between parasites and hosts may be a viable hypothesis in some Australian systems where there is egg mimicry but no host rejection of nonmimetic eggs (Brooker and Brooker 1989*b*), but egg recognition has been tested in very few Australian birds. This cuckoo competition hypothesis is, however, unlikely to be of major importance in other systems in which cuckoos are uncommon and the likelihood of host rejection of nonmimetic eggs is demonstrably greater than the likelihood of more than one cuckoo using the same nest (e.g. Davies and Brooke, this vol.). Furthermore, available evidence indicates at best only a weak trend for cuckoos laying in already parasitized nests to selectively remove cuckoo eggs (Davies and Brooke 1988), even in parts of Japan where cuckoos are relatively common (H. Nakamura, personal communication).

EVIDENCE FOR COEVOLUTION BETWEEN PARASITES AND HOSTS

A summary of evidence for coevolutionary processes between parasites and hosts, as reviewed previously (Davies and Brooke 1988, 1989*a,b*; Rothstein 1990*a*; Lotem and Rothstein 1995), and in chapters in this volume is as follows: 1) The most recently evolved cowbirds (Lanyon 1992) are also the most generalized (Friedmann and Kiff 1985), whereas the oldest species, *M. rufoaxillaris*, has a highly coevolved host–parasite system characteristic of specialized brood parasites (Fraga, this vol.). 2) The costs of parasitism are usually very high (Rothstein 1975*b*; May and Robinson 1985; Davies and Brooke 1988; Kattan, this vol.; Trine et al., this vol.) except for hosts of *Vidua* (Morel 1973) and some hosts of cowbirds (Smith 1968; Freeman et al. 1990; Røskaft et al. 1993; Smith and Arcese 1994). 3) Many traits interpreted as host defenses are very effective against parasitism (Rothstein 1975*b*, 1990; reviewed in parts II and III), and have no detectable costs, although costs, generally small, have been demonstrated in some species (Rothstein 1976*a,b*, 1977; Davies and Brooke 1988; Rohwer et al. 1989; Lotem and Nakamura, this vol.; Røskaft and Moksnes, this vol.). Thus, the selective pressures on parasites are strong enough to continue an evolutionary arms race (Rothstein 1975*b*; Lotem and Rothstein 1995; Davies and Brooke, this vol.). 4) Several systems in which hosts and parasites have only recently come in contact have many hosts with few or no effective defenses. This is especially apparent in comparisons of entire avifaunas such as potential cowbird hosts in the Nearctic versus potential cuckoo hosts in Africa, Europe, and Asia (Rothstein 1992; Nakamura et al., this vol.). Nevertheless, some birds allopatric with parasites or only recently exposed to them do have high levels of rejection behavior (Cruz et al. 1985, in press, this vol.; Davies and Brooke 1989*a*; Post et al. 1990), retained perhaps from past episodes of parasitism. 5) Apparent microevolutionary responses to recent contact with parasites have been documented in a few hosts (Soler and Møller 1990; Soler et al. 1994; Lotem and Rothstein 1995; Arias-de-Reyna, this vol.; Nakamura et al., this vol.; Soler et al., this vol.). 6) Parasites have responded evolutionarily to host behaviors such as egg and nestling/fledgling recognition that would otherwise reduce their breeding success. Most notable here are parasitic egg mimicry and the small number of cases with nestling and fledgling mimicry (Nicolai 1964, 1974; Wyllie 1981; Davies and Brooke, this vol.; Fraga, this vol.; and many other studies). 7) Except for cowbirds, sympatric brood parasites rarely show overlapping host use, which suggests evolutionary divergence. Some of this divergence is likely due to parasite–host coevolution but interspecific competition among parasites is also a likely cause (Friedmann 1955, 1967; Payne and Payne 1967; Higuchi and Sato 1984; Brooker and Brooker 1989*b*; Higuchi, this vol.). 8) Some parasitized hosts with well-developed defenses have conspecific populations or closely related species that are allopatric with parasites and that show relatively low incidences of egg rejection (Cruz and Wiley 1989; Davies and Brooke 1989*a*, this vol.; Brown et al. 1990; Briskie et al. 1992). 9) Studies of cuckoo–host systems suggest (Davies and Brooke, 1989*a*, this vol.; Marchetti 1992; Moksnes and Røskaft 1992) or demonstrate (Nakamura et al., this vol.) that host shifts occur and that parasite–host coevolution is therefore dynamic.

Given all of this evidence for adaptation and counter-adaptation by parasites and hosts, why then do so many hosts accept easily distinguishable parasitic eggs? This question, to which the majority of this section overview is devoted, is especially apropos for cowbird hosts but applies also to some important cuckoo hosts (Davies and Brooke 1989*a,b*).

EVOLUTIONARY LAG

The evolutionary lag hypothesis attributes egg acceptance to a lag in the appearance of antiparasitic adaptations. In ecological time, we see only snapshots of the coevolutionary process in which the hosts may not have had time to evolve counterdefenses to parasitic adaptations (Lotem and Rothstein 1995). A host may lack defenses altogether because of a lack of genetic variation coding for defenses or it may have bona fide defenses that are so rare that they are unlikely to be detected by researchers. Acceptance of parasitic eggs by some hosts in heavily parasitized, unproductive populations may also result from gene flow from populations in areas where cowbirds are absent and reproduction is high (May and Robinson 1985; Trine et al., this vol.). Under these views, defenses would be adaptive and may eventually occur at high frequencies. Lag is especially applicable to situations in which parasites have only recently invaded an area. The lack of defenses in such cases can even threaten entire host populations or species (Yamagishi and Fujioka 1986; Post et al. 1990; Trine et al., this vol.; Cruz et al., this vol.; Nakamura, this vol.). But even some hosts with long histories of exposure to parasitism accept eggs (Rothstein 1975*a*; Davies and Brooke 1989*a*,*b*).

EVOLUTIONARY EQUILIBRIUM

An alternative to evolutionary lag are evolutionary equilibrium models, which view acceptance as a result of an equilibrium among various selective pressures (e.g., Takasu et al. 1993) or as an option that makes the best of a bad situation. In the latter case, acceptance is the most adaptive option because the high costs of rejection make the host's pay-off (in terms of its own young) from rearing a parasite greater than if the host attempts to reject a parasitic egg. All three of the essays in this part provide some support for such models, which emphasize both the possible costs and benefits of egg rejection and the cognitive limitations of hosts (Lotem and Nakamura, this vol.; McLean and Maloney, this vol.; Røskaft and Moksnes, this vol.). Among possible costs of rejection, Røskaft and Moksnes (this vol.) emphasize the cost of damaged host eggs that result from puncture-ejecting thick-shelled cowbird eggs (see also Rohwer and Spaw 1988; Rohwer et al. 1989). They also emphasize the costs of nest desertion and renesting for cowbird and cuckoo hosts that can not eject parasitic eggs or that would find ejection too costly. Lotem and Nakamura (this vol.) emphasize similar costs to rejecting cuckoo eggs.

A crucial additional cost touched upon in all three chapters is recognition errors, that is rejection of a bird's own eggs, especially when it is not parasitized (Rothstein 1982*b*; Marchetti 1992; Lotem et al. 1995). This last cost is especially significant for hosts that regularly lose their entire brood if parasitized (most cuckoo and honeyguide hosts and many cowbird hosts). For such hosts, only losses when not parasitized can shift selection in favor of acceptance. Other costs for hosts that suffer complete losses lower the adaptive value of rejection but can not make rejection less adaptive than acceptance (Rothstein 1990*a*; Lotem et al. 1995). For hosts able to raise all or at least some of their own young along with a parasite, the difficulty and costs of egg rejection, especially for small hosts, combined possibly with the chances that the parasitic egg will not hatch (because it is infertile or timed improperly), can make it adaptive for hosts to accept parasitic eggs regardless of parasitism rates. Another scenario for an equilibrium is that acceptance behavior is maladaptive but occurs primarily in first-time breeders that have not yet learned to recognize their eggs. In colonial hosts, parasitism could be preferentially directed towards such susceptible individuals if they nest in smaller aggregations than older birds and are thus less likely to benefit from communal defense of nests (Lawes and Kirkman 1996).

DISTINGUISHING BETWEEN LAG AND EQUILIBRIUM

Testing the models for acceptance of parasitism is a challenging and very frustrating process. It is difficult to test the lag hypothesis directly, yet lag is in accord with basic evolutionary theory because species can not have pre-existing adaptive responses to every new selection pressure that they might experience. Nevertheless, no behavioral ecologist has readily endorsed lag without first assessing possible adaptive explanations for acceptance (e.g. Rothstein 1975*a*, 1982, 1986; Davies and Brooke 1989*b*; Sealy 1996; Ward et al. 1996). Thus, lag is something of a default hypothesis to be adopted when no reasonable adaptive explanation is borne out. This means that lag is not easily falsified. Perhaps the best test of lag is a comparative one involving parasite–host systems of different ages. On this basis, lag is supported because molecular data show that cowbird parasitism evolved more recently than cuckoo parasitism (as discussed in our Introduction) and rejection is much less prevalent among potential cowbird hosts than among potential cuckoo hosts (Davies and Brooke 1989*a*,*b*; Rothstein 1990*a*, 1992).

Efforts to test the equilibrium model are just as frustrating. Measuring the cost of parasitism is easy in a species that accepts, but how is one to test the cost of rejection in such a species? Obviously the whole issue would be solved if we could train accepters to eject cowbird or cuckoo eggs, but no one has done that. Suitable conditioning regimes would probably require captive birds but just getting most bird species to breed in captivity is difficult. A much easier test is possible for renesting as one can readily make birds desert parasitized nests by removing all or most of their eggs (Rothstein 1982*a*, 1986). The success of birds that renest can then be compared with a control series of unmanipulated parasitized nests (Rohwer and Spaw 1988). This approach too has not been applied, but available data for some hosts are sufficient in the view of some workers to show that acceptance is likely to be due to lag. For example, parasitized eastern phoebe (*Sayornis phoebe*) nests that escape predation average only 0.32 host fledglings, as opposed to four or more for unparasitized nests (Rothstein 1975*b*; see also Klaas 1975). Although phoebe and cowbird eggs differ greatly in appearance, phoebes almost never desert parasitized nests. However, they usually desert when their typical four- to five-egg clutch is reduced to two eggs. In such cases they renest rapidly and replacement clutches are as large as original ones (Rothstein 1986). Furthermore, parasitism rates of phoebes go down as the season advances (Friedmann 1963; Klaas 1975). In such a system, it is hard to see how any reasonable range of outcomes for replacement nests could tilt the balance in favor of making acceptance more adaptive than renesting. In particular, if phoebes usually desert when left with two eggs of their own and therefore forsake a potential pay-off of two young, what else besides lag could explain staying with parasitized nests that have an average pay-off of 0.32 young? Nevertheless, the enforced desertion experiment described above should still be done.

In the absence of direct tests of potential rejection costs of accepters, researchers have assessed the likely costs by comparisons with individuals or species that reject. But even here there is plenty of room for disagreement. Rohwer and Spaw (1988), Rohwer et al. (1989) and Røskaft et al. (1993), for example, determined that Bullock's orioles (*Icterus bullockii*) lose 0.26 of their own eggs for each cowbird egg they puncture-eject. They then suggested that smaller hosts would lose even more eggs if they ejected. But Rothstein (1977) argued that these orioles are not indicative of the ejection capabilities of other species because ejection is difficult from the unusually deep, pendant nests of orioles.

Evidence for other sorts of rejection costs is likewise open to alternative interpretations. For example, Marchetti (1992) and Lotem et al. (1995) reported that one or two eggs, but not the entire clutch, disappeared from 4.4 and 14.4% of unparasitized nests of two species with well-developed rejection behavior. This is just the sort of evidence that recognition errors would leave. However, even species that exhibit no egg rejection behavior lose some but not all of their eggs at comparable rates, e.g. 8.7% of eastern phoebe nests (Rothstein 1986), and 10.6% of Abert's towhee (*Pipilo aberti*) nests (Finch 1983). Although predators usually take entire clutches, these instances of partial clutch reduction, which occur in virtually all nesting studies, are presumably due to unusual predation events or to removal of the eggs by the birds themselves after eggs become cracked (see Kemal and Rothstein 1988).

In the most in-depth analysis of recognition errors, Davies et al. (1996) developed a useful model comparing the pay-offs of reed warblers (*Acrocephalus scirpaceus*) that accept versus reject. Using data from a previous study (Davies and Brooke 1988), their model shows that this host should accept mimetic cuckoo eggs when the likelihood of being parasitized is less than 19–41%. Below this level of parasitism, recognition errors produce more costs than acceptance behavior, which produces costs only if parasitism actually occurs. However, the incidence of recognition errors used in the model was based on host eggs mistakenly rejected when the warblers were exposed to a mounted cuckoo or when a mimetic cuckoo egg was placed in the nest. Therefore, these data do not indicate the recognition cost in nests unaffected by parasites and it is that cost that is essential to determining whether a bird should exhibit rejection behavior.

Unfortunately, it will be very difficult to confirm that unparasitized rejecters commit recognition errors. A researcher would have to monitor nests and hope to see or video-tape birds rejecting one of their own eggs, a rare event that may take a matter of seconds. Even if such "self-rejection" were documented in the nests of a rejecter species, one would have to monitor nests of accepter species to confirm that such rejection is not some aberrant behavior unrelated to egg recognition.

Needless to say, the issues here are extremely complex. Røskaft and Moksnes (this vol.) list the variables that need to be measured to determine if the lag or the equilibrium models are likely to apply to any given system. They conclude that most data on cowbird hosts support the lag view but that most cases of acceptance by cuckoo hosts are probably due to an equilibrium.

Lotem and Nakamura (this vol.) and Røskaft and Moksnes (this vol.) further argue that conditional acceptance of parasitic eggs provides evidence of an evolutionary equilibrium. Petit (1991), for example, found that prothonotary warblers (*Protonotaria citrea*) were more likely to abandon parasitized nests if there were more than three alternative nest sites on a territory, although a study of this host in another area has found no evidence for such adaptive decision rules (J. Hoover, in prep.). Other studies have shown an increasing likelihood of parasitic egg acceptance as the season progresses and opportunities to renest become limited (Clark and Robertson 1981; Burgham and Picman 1989). Moksnes et al. (1993) found that cuckoo eggs laid late in a host's incubation are more likely to be accepted than properly timed eggs, presumably because they are less likely to hatch. Rothstein (1976*b*, 1977) found similar results for two cowbird rejecter hosts that incur ejection costs but found no conditional responses in the larger of number of rejecters that incur no detectable costs. Lotem et al. (1992, 1995) documented increasing host rejection of cuckoo eggs in older, more experienced breeders.

These sorts of conditional behavior suggest that egg acceptance may be adaptive under certain circumstances, which represent evolutionary equilibria. Lotem and Nakamura (this vol.) and Røskaft and Moksnes (this vol.), however, acknowledge that the crucial data to evaluate costs and benefits are lacking for all but a few species. This is an important point because rejection in only some circumstances does not necessarily mean that acceptance in others is adaptive. For example, Kattan (this vol.) reports that house wrens (*Troglodytes aedon*) always desert nests with three or more shiny cowbird eggs, but not ones with one or two cowbird eggs. This could represent an adaptively balanced conditional response except for the fact that wrens with two or more cowbird eggs always suffer complete loss of all of their young. An advocate of equilibria can use this evidence to argue that differentiating between two versus one or no cowbird eggs is so difficult that it would cause too many desertions of unparasitized nests. Because they nest in cavities, wrens presumably identify cases of multiple parasitism by an extreme increase in the size of their clutch as cowbird eggs are larger than wren eggs. Taking this information, an advocate of lag could argue that the wrens show their current behavior because it was relatively easy to evolve a response to three or more cowbird eggs, but that the more difficult challenge of distinguishing between two cowbird eggs and one or none of them is also possible.

Similarly, the conditional response of chalk-browed mockingbirds (*Mimus saturninus*) represents another frustrating example that can be used to support either lag or equilibrium. This host is parasitized by shiny cowbirds that lay eggs of two morphs, spotted and immaculate white. It ejects nearly all immaculate cowbird eggs but very few spotted ones, which are not as distinct from its own spotted eggs (Fraga 1985; Mason 1986*a*). An equilibrium advocate would argue that this differential response occurs because the degree of discrimination needed to differentiate between spotted cowbird eggs and mockingbird eggs would lead to recognition errors. On the other hand, mockingbird and spotted cowbird eggs are far from identical in appearance. This could lead an advocate of lag to argue that because humans can easily distinguish virtually all spotted cowbird eggs from mockingbird eggs, the birds should be able to do so too. Indeed some rejecters, even the confamilial brown thrasher (*Toxostoma rufum*, Rothstein 1975*a*) routinely eject cowbird eggs even though their eggs are even more like those of the parasite. Although it has been assumed that these birds can discriminate among egg types at least as easily as can humans, we do not know the visual capabilities of all of these species, which means that both the lag and equilibrium alternatives are viable.

In addition to conditional responses to different situations, Lotem and Nakamura (this vol.) suggest that intraspecific variation in behavior due to individuals with different genotypes also provides a reliable way to measure rejection costs and therefore to assess lag versus equilibrium. The assumption here is that individuals differ only in their responses to parasitic eggs. However, no case of local intraspecific variation in responses is known to be due clearly to different genotypes. But a situation that approaches the conditions that Lotem and

Nakamura propose is Sealy's (1996) discovery that the eastern race of the warbling vireo (*Vireo g. gilvus*) ejects cowbird eggs at nearly 100% of its nests, but pays a small cost of 0.13 vireo eggs per ejection. By contrast, the western race of this species (*V. g. swainsonii*) is heavily parasitized, seems to never reject and loses its entire brood when parasitized (Rothstein et al. 1980; Sealy 1996; Ward and Smith, in press). Although there is some sentiment toward considering these two vireos as separate sibling species (Sealy 1996), they are so similar that their different responses to cowbird eggs seem likely to be due to a few key genetic differences. This further indicates that acceptance by the western form is due to evolutionary lag.

Rejection in warbling vireos indicates that the importance of the costs of puncture ejection (Rohwer and Spaw 1988) may be less than some workers assumed, especially the assumption that small species would pay prohibitively high costs if they ejected. At only 15 g, this rejecter is less than half the 40 g mass of a Bullock's oriole yet its ejection cost is much less than the 0.26 eggs paid by the oriole (Røskaft et al. 1993). Other aspects of Sealy's study can be taken as support for either lag or equilibrium. The western vireos have been exposed to widespread cowbird parasitism for a shorter period (Rothstein 1994), thereby suggesting the importance of lag. On the other hand, because western vireos have slightly smaller bills and masses (12 g versus 15 g), it is possible that ejection costs are too high for them. However, western vireos should still desert parasitized nests because cowbird parasitism means total reproductive failure for them.

The warbling vireo example emphasizes the importance of considering host-specific costs of cowbird parasitism, which can range from zero host young fledging to almost no detectable difference between the productivity of parasitized and unparasitized nests. The most important parameter here is host size, with smaller hosts suffering greater losses. Although Røskaft and Moksnes (this vol.) found only weak support for this relationship, their analysis focussed on a small number of mostly recently studied hosts. A larger number of host species would have shown a stronger relationship (see Robinson et al. 1995*b*; Trine et al., this vol.). In addition, incubation period length is also of major importance (Rothstein 1975*a*) as even relatively large hosts can experience heavy losses if their eggs hatch several days after those of the cowbird. All small vireos and flycatchers suffer the dual disadvantages of small size and relatively long incubation periods and accordingly, nearly always experience total losses when parasitized (Walkinshaw 1961; Rothstein 1975*b*; Goldwasser et al. 1980; Grzybowski et al. 1986; Sedgewick and Knopf 1988).

A number of species that are relatively large and have short incubation periods typically raise all or most of their own young that hatch (e.g., Freeman et al. 1990; Røskaft et al. 1990; Smith and Arcese 1994; Scott and Lemon 1996). The major cost of parasitism for these species are the host eggs removed or damaged by adult cowbirds, but this is a loss that can not be recovered by ejection behavior. (However, some losses seem to be due to the presence of cowbird eggs [Fraga 1985], and these could be avoided by ejection.) Two of these species (song sparrow, *Melospiza melodia* and northern cardinal, *C. cardinalis*) are candidates for recognition costs because their eggs resemble cowbird eggs and may be the most likely examples among North American cowbird hosts for acceptance representing an equilibrium.

The most adaptive option for these species may be to defend their nests against cowbirds to forestall parasitism or egg removal even if parasitism occurs. However, because both species are heavily parasitized (Scott and Lemon 1996; Smith and Myers-Smith, this vol.) even though at least one recognizes and is aggressive towards cowbirds (Robertson and Norman 1976, 1977; Smith et al. 1984), it is either too costly to guard nests at times when parasitism is likely to occur or the failure to do so is an example of lag. Guarding against egg removal by adult cowbirds is even more difficult because cowbirds almost never remove host eggs during their pre-dawn egg-laying visits but do so at other times or on other days (Hann 1941; Friedmann 1963; Mayfield 1963; Sealy 1992).

EVOLUTIONARY LAG VERSUS EQUILIBRIUM AND THE ACCEPTANCE OF PARASITIC YOUNG

Acceptance of parasitic eggs is a riddle because egg recognition occurs in many taxa and is therefore something that has evolved many times. Students of brood parasitism have also addressed the question of acceptance of nonmimetic parasitic young because rejection of such young could allow host species to avoid most of the severe reproductive losses inflicted by brood parasitism. But unlike egg rejection, there is no known example of outright rejection of nonmimetic young. Even systems with mimetic parasitic young do not involve rejection of nonmimetic young, but instead seem to be maintained because mimetic young receive better care. Yet host ejection of nonmimetic young would be adaptive and could even avoid some of the costs of egg rejection. Given its appendages, a hatchling must be a lot easier for a bird to grab in its bill than is an egg. Why then is evidence for nestling recognition so scarce and for nestling rejection totally unknown?

Lotem (1994) and Redondo (1993) have provided important recent attempts to deal with this question. Among the chapters in this book, the question of nestling recognition is dealt with at greatest length by McLean and Maloney (this vol.), who argue that acceptance of parasitic nestlings results from cognitive limitations of birds. Any decision to reject parasitic nestlings could involve severe costs because misimprinting could lead to mistaken rejection of host nestlings, which could result in an evolutionary equilibrium. This constraint would not apply if a host evolved an innate, fail-proof recognition mechanism that safely rejects only parasitic hatchlings (Davies and Brooke 1988). But McLean and Maloney argue that newly hatched altrical birds are very similar across species and that there are no safe rules that could be used to reject only parasitic young, especially given the extent to which nestling appearance changes with time.

While many pairings of host and parasitic species and perhaps some avifaunas have nestlings that are very similar, North American hosts are an exception. Although most hatchlings are similar overall, mouth color, a character that will be seen clearly by feeding parents, is species- and even family-specific. So far as is known, North American passerine families have either red (Emberizidae, Corvidae) or yellow (Tyrannidae, Muscicapidae, Vireonidae, Mimidae) mouths (Ficken 1965). In addition, nearly all species have either white or yellow rictal flanges framing their mouths. While flange color does not seem to vary within species, except for cowbirds (Rothstein 1978*b*), it does vary within families. Thus, any flycatcher, thrush, or vireo that evolved ejection of nestlings with red mouths would never reject its own young but would safely rid itself of cowbird nestlings. Although mouth color would not be a reliable cue for emberizid hosts, flange color would suffice for many species. While McLean and Maloney's argument may not apply to many North American hosts, it may be highly relevant for other avifaunas. Despite the best efforts of many biologists, including the perceptive points raised by McLean and Maloney, no argument in our view provides a wholly convincing explanation for the total lack of nestling rejection among altricial birds.

IS ACCEPTANCE DUE TO AN AVIAN "MAFIA"?

Another argument for equilibrium in parasite–host systems was first advanced by Zahavi (1979), who hypothesized that parasites could make rejection costly by destroying host clutches in nests from which parasitic eggs had been ejected. This "mafia" effect could make acceptance more adaptive than rejection if hosts can raise some of their own young when parasitized. But the mafia effect alone can not make acceptance more adaptive than rejection if parasitism results in complete loss of the host's young, as with most cuckoo and many cowbird hosts. For such hosts, the mafia effect only reduces the selective advantage of rejection in proportion to the rate of nest destruction carried by the parasites. However, if the mafia effect is combined with some other cost, such as recognition errors, it could tip the balance in favor of acceptance.

Rothstein (1982*a*) pointed out that the mafia effect is unlikely to apply to cowbird hosts because many researchers routinely remove cowbird eggs and have reported no resulting increases in nest predation rates. However, a recent experimental study strongly suggests that the great spotted cuckoo may engage in mafia-like behavior (Soler et al. 1995*a*) when its main host, the magpie (*P. pica*), rejects its eggs. In accord with expectations from the mafia model, this is one of the few cuckoos whose nestlings do not always kill off all of the host young. But some other features of this system do not conform to the mafia model, which Zahavi proposed to explain acceptance of nonmimetic eggs. If eggs are mimetic, no special hypothesis is needed to explain their acceptance. Magpies are parasitized by mimetic cuckoo eggs and reject nearly all nonmimetic eggs where sympatric with this cuckoo (Soler and Møller 1990). While it seems likely that great spotted cuckoos have some continuing interest in magpie nests they parasitize (see also Arias-de-Reyna, this vol.), it is not clear that this results in the sort of mafia effect envisaged by Zahavi. Nevertheless, a mafia system would greatly increase the odds of an evolutionary equilibrium.

DOES ACCEPTANCE OCCUR BECAUSE PARASITISM IS BENEFICIAL?

Acceptance of parasitism is an evolutionary enigma because parasitism is assumed to be harmful. In most cases this assumption is based on strong evidence. But there is one well-known case in which a researcher reported that brood parasitism is beneficial under some circumstances. Smith (1968, 1979, 1980) reported that giant cowbirds are involved in a complex set of interactions with their colonial cacique and oropendola hosts and with parasitic flies that can kill nestlings. Nestling cowbirds aid their host nestmates by preening them and removing fly larvae. But these hosts can also gain protection from the flies by nesting near bee or wasp nests, as these colonial insects drive away the flies. Caciques and oropendolas at colonies with no bee or wasp nests accept cowbird eggs because they have no other protection against the flies. But those at colonies with bee or wasp nests, attack adult cowbirds and reject nonmimetic cowbird eggs. This complex system may be limited to Panama where Smith described it, as other researchers have obtained different results in Costa Rica (Webster 1994) and in Peru (Robinson 1988). Smith's work is the only evidence that acceptance of brood parasitism is beneficial.

AN UNMEASURED COST OF PARASITISM

The impact of parasitism on a host's fitness is typically measured by the number of young it fledges relative to comparable conspecifics that are not parasitized (Rothstein 1975*b*; Lotem and Nakamura, this vol., Røskaft and Moksnes, this vol., plus many other studies cited in these chapters). Some workers have carried such analyses one step further by using the mass of nestlings to assess the quality of host young produced from parasitized and unparasitized nests (e.g., Smith 1981). Almost no attention, however, has been paid to the impact that rearing a brood parasite has on an adult host's future reproductive value. Basic life history theory assumes that current reproductive effort reduces future reproductive output (Stearns 1976) through increased adult mortality during or after the breeding season or through a decrease in future reproductive investment. Although some studies have failed to find evidence for a cost of reproductive effort in birds (e.g., DeSteven 1980) others have found that increased effort reduces adult survival (Askemo 1979; Nur 1984) or the likelihood or size of a second brood (Fink et al. 1987). The failure to find such effects in some studies may be due to Type I errors.

No study has yet attempted to factor in the extent to which rearing a brood parasite lowers a host's reproductive value. Thus, hosts that have little in the way of a reduction in the number or possibly even quality of fledglings when they rear a parasite, such as some of the larger cowbird hosts cited above, still have to expend more effort and so are likely to incur a reduction in lifetime fitness. Some authors have assessed the reproductive effort needed to rear parasitic versus host young. Gill (this vol.), for example, determined that rearing a cuckoo is less effort than rearing an average-sized host brood. However, abandoning a cuckoo chick and not rearing anything would require still less effort and therefore should be more adaptive than rearing the parasite. In addition to reductions in the number of host young, this other type of cost of brood parasitism is something that needs to be addressed when assessing the costs of accepting brood parasitism. Similarly, impacts on a host's reproductive value due to the effort needed to renest also must be assessed to determine the benefits of nest desertion.

OVERVIEW OF THE CHAPTERS ON EQUILIBRIUM VERSUS LAG

The three chapters in this part emphasize costs or limitations of host defenses that need to be considered when modeling the coevolutionary process. All three offer some support for the "equilibrium" view, which is not necessarily incompatible with the arms race/evolutionary lag model described at the outset of this introduction. Davies and Brooke (this vol.), for example, argue that some common cuckoo hosts may be at an equilibrium whereas others appear to be engaged in an ongoing arms race. Røskaft and Moksnes suggest that acceptance by many cowbird hosts may be due to lag. The arms race/evolutionary lag model may be particularly relevant in rapidly changing anthropogenic landscapes in which the geographic distributions and populations of both hosts and parasites are in a state of flux. The major challenge in trying to determine whether lag or equilibrium provides a better explanation for the acceptance of parasitic eggs and/or young that seem distinguishable from those of hosts is the difficulty of determining the costs that accepters would pay if they tried to reject parasitism.

OVERVIEW OF PART V: EFFECTS OF PARASITISM ON HOST POPULATION DYNAMICS

Because brood parasitism is often costly for hosts, high frequencies of parasitism could threaten host populations (May and Robinson 1995; Robinson et al. 1995*a*,*b*). Much of the recent attention focused on cowbirds comes from concerns that they may be causing population declines in widespread passerines (Mayfield 1997; Brittingham and Temple 1983; Böhning-Gaese et al. 1993) in addition to the well-documented endangerment of several rare species and subspecies with limited ranges (Post and Wiley 1976; reviewed in Robinson et al. 1995*b*; Cruz et al., this vol.; Trine et al., this vol). As a result, cowbirds have become a major management issue in North America (Robinson et al. 1993; Rothstein and Robinson 1994; Smith et al., in press) leading some to call for as much cowbird eradication as possible (Schram 1994).

Under what circumstances, then, does brood parasitism pose a threat to host populations? This part includes three chapters that review situations in which cowbirds may threaten host populations. Because these chapters deal only with cowbirds, our overview addresses the question of why population-level impacts are more common among cowbird hosts than among the hosts of other brood parasites. Among the chapters in this section, Trine et al. provide a general overview of population level impacts of brown-headed cowbird parasitism, with emphasis on studies in the midwestern U.S. Cruz et al. (this vol.) describe the potential threat posed by the simultaneous invasion of Florida, a formerly cowbird-free area, by shiny cowbirds from the south and brown-headed cowbirds from the north. This situation could pose a severe threat for some host populations in Florida, some of which are small and represent endemic races. Apart from conservation concerns, the recent sympatry between these two cowbirds, the two most generalized brood parasites in the world, offers some exciting research opportunities to study potential competition. The shiny cowbird may pose a particular threat to cavity nesters because of its greater tendency to enter enclosed nests. Lastly, Smith and Myers-Smith (this vol.) provide an insightful analysis of geographic variation in parasitism of one of the brown-headed cowbird's most commonly used hosts, the song sparrow.

POPULATION LEVEL IMPACTS ON HOSTS IN NONCOWBIRD SYSTEMS

Most brood parasites have little impact on host populations (Payne 1977*a*; Wyllie 1981) as a result of coevolutionary processes that have reduced the frequency or impacts of parasitism or because the parasites are uncommon relative to their hosts (Wyllie 1981; May and Robinson 1985; Davies and Brooke 1989*a*,*b*; Rothstein 1990*a*). In particular, most cuckoos and honeyguides are uncommon. Accordingly, European hosts of the common cuckoo usually suffer parasitism frequencies of less than 5%, a stark contrast to those reported in many cowbird–host systems (Fraga, this vol.; Kattan, this vol.; all three chapters in part V). Whether low cuckoo numbers are due to low recruitment as a result of well-defended hosts or to the intrinsic ecology of cuckoos is unclear. By contrast, cowbirds—especially the two most generalized species—are far more common than many, perhaps most of their host species. Furthermore, because individual female cowbirds are not host-specific (Fleischer 1985) unlike female cuckoos, cowbirds can drive rare hosts to extinction. There is no feedback resulting in lower parasitism rates as rare hosts decline because they contribute negligibly to cowbird recruitment. Only declines in the most common hosts or in many host species will cause declines in cowbirds. Cuckoos also have less impact on host populations because they lay 16–25 eggs a year (Payne 1974, 1977*a*; Wyllie 1981), whereas the two generalized cowbirds lay 40 or more (Scott and Ankney 1983; Rothstein et al. 1986; Kattan, this vol.).

Nevertheless, there are situations in which cuckoo parasitism is frequent enough to pose a possible threat to local host populations (Brooke and Davies 1987). In parts of Japan where common cuckoos have undergone recent range expansion, parasitism frequencies often exceed 40% (Higuchi and Sato 1984; Yamagishi and Fujioka 1986; Higuchi, this vol.; Nakamura et al., this vol.). Some cuckoos may be expanding their ranges as a result of human modifications of habitats that create more early successional habitat (Higuchi, this vol.). Great spotted cuckoos also have historically variable geographic ranges in Europe; when they invade an area, they initially parasitize hosts at very high frequencies (Arias-de-Reyna, this vol.; Soler et al., this vol.).

Interestingly, in both of these situations, there is evidence of rapid development of host defenses (Soler et al. 1994; Higuchi, this vol.; Nakamura, this vol.). Cowbird hosts, on the other hand, seem to be much slower in evolving these behaviors (Rothstein 1990*a*). Perhaps the longer history of cuckoo presence in the Old World has resulted in more widespread egg recognition capability in potential hosts (Rothstein 1992). Furthermore, many currently unparasitized Old World birds may retain vestiges of egg recognition from past encounters with parasites, as evidenced by high rejection rates of nonmimetic eggs in two species in Iceland (Davies and Brooke 1989*a*), and may therefore be able to respond quickly once parasitism reoccurs.

Some cuckoo–host systems with high frequencies of parasitism do not seem to have population-level consequences for hosts. Gill (1983, this vol.) found that a host population in New Zealand is affected little by high levels of parasitism because only relatively unproductive second broods are parasitized. Brooker and Brooker (1996) report that splendid fairy-wrens (*Malurus splendens*) in one well-studied Australian population readily renest after raising Horsfield's bronze-cuckoos (*Chrysococcyx basalis*), which enables them to produce, on average, the small number of recruits (0.9/pair/year) needed to keep the population at the carrying capacity. In addition to parasitism, the survival potential of fairy-wren populations is affected by nest predation and fire frequency and smaller populations may not be able to sustain themselves given anthropogenic increases in fire frequency and predation rates (Brooker and Brooker 1994). A likely example of cuckoos exerting a major effect at the population level is the hypothesized extirpation of great reed warblers (*Acrocephalus arundinaceus*) from Hungary after heavy parasitism by common cuckoos (Molnar 1944). Some other specialist host–parasite systems also have high (>20%) frequencies of parasitism (Payne and Payne 1967; Gaston 1976; Jensen and Clinning 1974; Vernon 1984) and may induce local population level effects on hosts.

POPULATION LEVEL IMPACTS ON COWBIRD HOSTS

Consequences of cowbird parasitism for host populations are extensively reviewed in this volume (*M. bonariensis*: Cruz et al., see also Cruz et al. 1985, in press; *M. ater*: Trine et al.; see also Robinson 1995*a,b*). Smith and Myers-Smith (this vol.) further examine the geographic and landscape contexts in which a frequent host is most likely to be parasitized. These essays deal with situations in which human-generated land use changes have enabled cowbirds to expand their geographic ranges and increase their populations. As a result, cowbirds have come into contact with new host populations, which they often parasitize at very high frequencies (see also Post et al. 1979; Trine et al., this vol.). Some of these new hosts have declined almost to the point of complete extirpation over large areas, especially in western North America (Rothstein 1994).

It has often been suggested that new host populations are especially susceptible to cowbird parasitism because they have not evolved resistance (Mayfield 1965; Cruz et al., this vol.). This view is consistent with the observation that some endangered subspecies that are new hosts have conspecific races that are not declining and that have survived long histories of sympatry with cowbirds (Mayfield 1965; Rothstein 1994). However, there is little actual evidence that new hosts have weaker defenses than comparable taxa that are old hosts. For example, among new California hosts, least Bell's vireos (*Vireo bellii pusillus*) often desert parasitized nests (J. Greaves, pers. commun.) just as midwestern conspecifics (*V. b. bellii*) do (Friedmann 1963) and southwestern willow flycatchers (*Empidonax traillii extimus*) desert at least as often as eastern conspecifics (*E. t. traillii*) with much longer histories of contact with cowbirds (Harris 1991). Although both of these taxa are now nearly extinct, they were common in California before cowbirds arrived about 1900. But these near extinctions seem to be due to factors other than a relative lack of defenses. One factor may be especially high rates of parasitism. Perhaps more importantly, these endangered hosts have experienced the loss of about 95% of their riparian habitat since 1900 (Rothstein 1994), which resulted in range contractions and made them especially vulnerable to any other threats such as cowbird-induced declines.

As with these two California taxa, most of the several other hosts that are threatened by cowbird parasitism also have small geographic ranges that are entirely within areas of high cowbird abundance (reviewed in Trine et al., this vol.; Robinson et al. 1995*b*). However, in some cases, such as with the Kirtland's warbler (*Dendroica kirtlandii*), these ranges were small even before the European settlement of North America (Mayfield 1965). Anthropogenic habitat losses are probably just as great a threat to these geographically restricted and endangered taxa as is cowbird parasitism (Robinson et al. 1995*a,b*). Without habitat loss, some or all of these endangered passerines might be able to survive in the face of cowbird parasitism. The importance of habitat changes is further indicated by the fact that two endangered species (black-capped vireo, *Vireo atricapilla* and golden-cheeked warbler, *Dendroica chrysoparia*) now threatened by parasitism occur wholly within the cowbird's ancestral center of abundance (Smith et al., in press).

More geographically widespread hosts should be less vulnerable to cowbird parasitism (May and Robinson 1985) because losses in areas with heavy parasitism may be offset by dispersal from other areas with few cowbirds and high reproductive success (Smith and Myers-Smith, this vol.; Trine et al., this vol.; see also Friedmann 1929, 1963; Hoover and Brittingham 1993). Recent support for this hypothesis comes from studies showing very strong regional gradients in cowbird parasitism in relation to forest cover (Robinson et al. 1995*a*). In mostly forested landscapes, where cowbirds have few feeding areas, parasitism levels are low enough that they pose virtually no threat to hosts. Because these areas also have low nest predation rates (Robinson et al. 1995*a*), they might act as population "sources" (sensu Pulliam 1988) that produce a surplus available to recolonize areas in more fragmented forests in agricultural landscapes where both cowbird parasitism and nest predation rates can be very high (Robinson et al. 1995*a*; see also Robinson 1992; Donovan 1995*a,b*). Such metapopulation and large-scale source-sink dynamics and dispersal apparently occur over hundreds of km in the midwestern United States, where some forest host species have little reproductive success over most of several states yet maintain numerous small populations (Brawn and Robinson 1996). Indeed, the lack of overall population declines in most forest passerines (Peterjohn et al. 1995; James et al., 1996) may be due to high production in a few heavily forested regions of North America. Immigration from areas where cowbirds are absent or rare may also slow the evolution of host defenses against cowbirds in areas where cowbirds have major effects. In addition, some accepter host species are so successful when parasitized because of their size and/or short incubation period that they are unlikely to suffer major net reductions even in the face of heavy parasitism (Arcese et al. 1992; see Smith and Myers-Smith, this vol.).

Nevertheless, several authors and much of the popular literature have suggested that cowbirds may be one of the primary causes of either local (Robinson et al. 1995*a*) or large-scale declines (Mayfield 1977; Brittingham and Temple 1983; Böhning-Gaese et al. 1993). Declines are plausible if there is too much "sink" habitat in which cowbirds are abundant relative to "source" habitat where cowbirds are scarce (modeled by Temple and Cary 1988). Böhning-Gaese et al. (1994) found that the greatest population declines in North America were in the species considered to be the most frequent cowbird hosts.

GAPS IN OUR KNOWLEDGE OF HOST DEMOGRAPHICS

Too little is known about the factors that limit host populations to be certain of the causes of any population declines, especially for birds that may also be limited on their winter grounds (Robinson 1993; Sherry and Holmes 1995). For example, we know little about dispersal distances in songbirds and we have virtually no examples of dispersal on the scale required to recolonize isolated woodlots. We also know very little about the levels of parasitism and nest predation that populations can withstand before they become population "sinks" (discussed in Trine et al., this vol.).

Mortality rates are another major gap in our knowledge of key demographic parameters. These rates are typically estimated from yearly return rates of breeding birds, but because birds may disperse even after one or more breeding attempts (Walkinshaw 1983) our estimates may be incorrect. Thus, we need more extensive demographic studies of color-banded bird populations before we can interpret the viability of host populations.

Considerable uncertainty also exists about the relative roles of nest predation and cowbird parasitism, which tend to be correlated (Gates and Gysel 1978; Temple and Cary 1988; Robinson et al. 1995*a*). Some authors have even argued that nest predation has a much greater effect on host populations than brood parasitism (McGeen 1972; Martin 1992). Parasitism, however, has additional costs that have not been fully considered in most models (e.g., May and Robinson 1985; but see Pease and Grzybowski 1995). Raising cowbirds often consumes an entire nesting season, whereas most temperate region birds renest within a few days of losing nests to predators (e.g., Roth and Johnson 1993; Trine et al., this vol.). For example, a parasitism rate of 10% may have the same impact as a predation rate of 50% because the latter allows for repeated renesting attempts (Rothstein 1990*a*).

MANAGEMENT AND CONTROL OF COWBIRDS

Until we have more data, the most prudent long-term conservation strategy for cowbird hosts is to preserve and restore large blocks of habitat where rates of both cowbird parasitism and nest predation are low (Trine et al., this vol.). Cowbird trapping and removal should be limited to areas and host taxa that have no potential "source" populations and that involve entire host species or races that are endangered (Rothstein and Robinson 1994; see Smith et al., in press for examples). As of this writing, there are cowbird control programs designed to aid at least eight cowbird hosts: Kirtland's warblers in Michigan; two races of Bell's vireos in southern California and Arizona; southwestern willow flycatchers in southern California and Arizona; California gnatcatchers (*Polioptila californica*) in southern California; black-capped vireos in Texas and Oklahoma; golden-cheeked warblers in Texas; and the yellow-shouldered blackbird (*Agelaius xanthomus*) in Puerto Rico.

Because cowbirds are so easily trapped, all of these programs have reduced local parasitism and some have virtually eliminated it. But management efforts must be continued indefinitely as even nearly complete cowbird removal over plots as large as hundreds of square km has no effect on cowbird numbers the next year (Smith et al., in press). We estimate that current control efforts destroy 10 000–20 000 cowbirds per year. The Bell's vireo program in southern California appears to be a great success, with at least a tenfold increase in numbers of adult breeders since it was initiated in 1983 (Griffith and Griffith, in press). But as Trine et al. (this vol.) discuss (see also Rothstein and Robinson 1994; Rothstein and Cook, in press), other control programs have a mixed track record as regards increases in host populations. For example, cowbird control began on the Kirtland's warbler's range in 1972 but the warblers did not begin to increase until 1990 when additional habitat became available (DeCapita, in press). Control may still have been beneficial in keeping the warbler from going extinct between 1972 and 1990 (De Capita, in press) but there is no direct proof of this benefit (Rothstein and Cook, in press).

Many management-oriented people in governmental agencies and in the private businesses that are often hired to do the actual cowbird trapping, have enthusiastically endorsed the perceived benefits of cowbird control. Although many new control programs appear to be starting up, there is no central agency coordinating these efforts. Nor are there established guidelines to justify control programs and carry them out in standardized ways. Some programs are being done with almost no attempt to determine local parasitism rates, which is an important omission (Robinson et al. 1995*b*) because parasitism rates on the same species vary (see next subpart). A recent (October 1997) tabulation estimates that expenditures for cowbird trapping exceed $1 000 000 in California alone (C. Hahn, pers. commun.). Despite such large expenditures, some control programs lack efforts to quantify the pre-control sizes of host populations and have no long-term monitoring efforts to track effects on host numbers. It will be difficult to impossible to assess the efficacy of these programs. Concerns have been raised that cowbird control will become so overused and so poorly applied at times that the well-designed, effective programs may become wrongfully discredited (Rothstein and Robinson 1994).

GEOGRAPHIC VARIATION IN RATES OF COWBIRD PARASITISM

Some cowbird hosts are parasitized at greatly different rates in different regions (Friedmann 1963; Friedmann et al. 1977) and such variation may be a key factor in the ability of some species to withstand cowbird parasitism, as described above. In the most thorough analysis of geographic variation in the parasitism of any cowbird host, Smith and Myers-Smith (this vol.) assess variation in song sparrow parasitism on two geographic scales, across all of North America and within a small part of British Columbia. Because this host is so widespread and so likely to rear cowbirds, it may contribute more overall recruitment to cowbird populations than any other host species. Smith and Myers-Smith use data from both geographic scales to address four competing hypotheses for the determinants of parasitism rates. Both scales suggest that the chief determinant of parasitism rates is simply cowbird abundance and provide little support for other hypotheses such as distinct preferences for particularly good hosts.

Nevertheless, there are significant differences

in parasitism rates among the large number of species that appear to be high-quality cowbird hosts. For example, Smith and Myers-Smith show that wood thrushes are parasitized more heavily than song sparrows everywhere the two species are sympatric. These authors suggest that such a difference could be due to a preference for particular host species or for particular habitats, as the two hosts occur in different habitats. In addition, another determinant of variation in parasitism rates is the degree of temporal overlap between cowbird and host laying seasons. Song sparrows are among the earliest nesting passerines and many of their nests are initiated too early to be parasitized (Smith and Arcese 1994). Wood thrushes breed later and their entire laying season overlaps that of the cowbird (Roth and Johnson 1993). It is unclear whether the timing of cowbird laying seasons is based on the quality and numbers of breeding hosts or on some ecological constraint such as the availability of food for the formation of eggs. Lack (1963) pointed out that the initiation of egg laying in the common cuckoo is probably constrained by food availability, as most regular hosts in Britain begin to lay earlier than the cuckoos. The song sparrow–wood thrush contrast suggests that the same constraint may apply to cowbirds.

OVERVIEW OF PART VI: CONSEQUENCES OF PARASITISM FOR THE MATING SYSTEMS AND LIFE HISTORIES OF BROOD PARASITES

Besides their obvious lack of parental care, brood parasites differ from other birds in several crucial respects. First, their mating systems have evolved without the selective constraints of parental care (e.g., Emlen and Oring 1977). Second, their diets are not constrained by the energetic and nutritional requirements of feeding offspring (e.g., Lack 1968). Third, because they are not tied to a particular nest site, they can range over great distances. And fourth, because they are raised by a different species, the development of their vocalizations and other behaviors is subject to very different constraints from those of most birds (Kroodsma 1982). These features make parasites ideal subjects for testing the generality of models for the evolution of social and mating behavior, foraging behavior, spatial distribution, and vocal development.

Because this part has only a single essay, which focuses on mating systems, we will review briefly the literature on other particularly interesting aspects of the biology of brood parasites, namely their vocal behavior, diet, and spacing systems. Our focus is on ways in which brood parasites both differ from and converge upon the ecology and behavior of parental birds.

BEHAVIOR OF PARASITES

Mating systems

Three striking patterns emerge from Barnard's (this vol.) comprehensive review of the mating systems of brood parasites and these reveal how much more needs to be learned about this subject. First, although some species conform to the prediction that birds freed from parental care will be promiscuous, many species exhibit monogamy and resource-defense-based mating systems. Barnard attributes this to the diversity of behaviors used by parasites to gain access to nests (see parts II and III for examples), to diverse foraging ecologies of the parasites, and to peculiarities of individual host–parasite systems such as unbalanced sex ratios in cowbirds (Rothstein et al. 1986; Yokel 1986, 1989).

Second, there are very few studies of color-marked populations of brood parasites and even studies of the same species sometimes disagree on the classification of the mating system. For example, the major studies of the mating system of the brown-headed cowbird involving marked birds in the field (Darley 1971, 1982; Elliott 1980; Dufty 1982*a*,*b*; Teather and Robertson 1985, 1986; Yokel 1986, 1989; Yokel and Rothstein 1991) and in captivity (Rothstein et al. 1986), suggest that the prevalent mating system is monogamy with a low incidence of polygyny (reviewed in Robinson 1986; Rothstein et al. 1986). However, one field study (Elliott 1980) found no evidence of pair bonds and a prevalence of promiscuity. Whether these differences reflect different ecological conditions (e.g., variable host and/or cowbird densities, varying sex ratios) or different study methods and use of definitions to "cubbyhole" mating behavior remains uncertain (Barnard, this vol.). But some ecological differences probably have importance as Elliott's (1980) study indicating promiscuity and the study with the strongest evidence for monogamy (Yokel 1986) were both based on observed copulations in nature. Despite the lack of extra pair copulations in Yokel's large sample of observed matings, promiscuity or some other mating system besides monogamy might be occurring if extra pair copulations are done under special circumstances that minimize their detection (e.g., in hidden sites). However, a recent DNA fingerprinting study at Delta, Manitoba has also found that monogamy is the prevalent mating system (Alderson, Gibb and Sealy, in prep.).

The third pattern that arises from Barnard's paper is that the relative roles of host and food distribution in the evolution of mating systems remain poorly known for most parasites. Although there is one well-known example of a resource-based mating system in an Asian honeyguide, wherein females trade sex for access to wax (Cronin and Shearman 1977), such behavior is rare or absent in other honeyguides (Short and Horne 1985). Among other parasites, males of some viduine species display from perches that are near concentrations of food and water (Payne 1973*a*, 1985; Payne and Payne 1977; see Barnard, this vol.) suggesting that mate choices are related to both mate quality and to resources.

Song learning and vocal development

It might seem that being raised by a different species would create strong selection for hard-wired genetic programming of song. There is evidence both for and against this view from the brown-headed cowbird, the parasitic bird whose vocal development has been studied most extensively. Most songbirds raised in acoustic isolation develop songs that are obviously abnormal (Kroodsma and Miller 1996). But cowbirds raised in isolation sing normal or nearly normal versions of "perched songs," one of the species' two song types (King et al. 1980). Nevertheless, perched songs are modified greatly by subsequent experience (e.g., West and King 1988; King and West 1990) and these modifications are essential if males are to use these songs with maximal effectiveness.

The other type of song, the "flight whistle," seems to be almost totally learned (O'Loghlen and Rothstein 1993) and, probably as a result, is among the most variable of all bird songs (Rothstein and Fleischer 1987*a*; Rothstein et al. 1988, in press). Although their details vary geographically, perched songs have the same basic structure throughout the species' range (Rothstein et al. 1988) whereas flight whistles occur as distinct dialects in areas usually measuring 10–50 km across. These dialects can range from pure tone whistles to buzzes and trills and dialects only 50 km apart can be utterly confusing to even skilled field workers. Cowbirds have high dispersal rates (Fleischer and Rothstein 1988) and those that disperse from one dialect to another benefit by conforming to their new dialect because flight whistle type is an important mate choice cue for females (O'Loghlen and Rothstein 1995). In contrast to the variable vocalizations of males, females have a distinctive chatter call that varies slightly among individuals, but shows no geographic variation (Burnell and Rothstein 1994). These studies in combination provide a much-cited example of the role of learning and behavioral flexibility despite the expectation of genetically hard-wired behavior.

Some species of viduine parasitic birds learn and incorporate elements of their host's song into their own song (Payne 1973*a*, 1976, 1982, 1985; Payne and Payne 1977). These songs appear to have no interspecific function, but instead are directed at females, which choose males raised by the same hosts (Payne 1973*a*). This song learning has the potential for sympatric speciation by cultural learning, but most host song races also differ geographically which suggests possible allopatric speciation (Payne 1973*a*). Host song learning allows for the exploitation of new hosts by facilitating adoption of a new song type when a new host is used (Payne 1973*b*, 1976). Some authors have suggested that viduines and their estrildine hosts have "cospeciated" in concert because there is such a close one-to-one correspondence between parasitic and host species in this system and because a learned cue, song, that leads to reproductive isolation can be easily acquired (Nicolai 1964). However, patterns of mitochondrial DNA variation argue against the cospeciation model because viduine hosts are an older clade than the viduines themselves (Klein et al. 1993).

The song-based mechanism for assortative mating in viduines is essential for maintenance of adaptations designed to foil host defenses. By contrast, the egg mimicry that occurs in numerous other parasites in response to host rejection behavior is less dependent on assortative mating. If a cuckoo of one gens mates with a male of another gens with a different egg type, the appearance of the female's eggs will be unchanged and will still be suited to her usual host as egg features are determined by the maternal genotype, not by the genotype of the offspring (although this female's daughter may have an inappropriate egg, depending on the inheritance of egg color genes; see Davies and Brooke, this vol.). However, the mimicry that is essential for viduines, gape appearance, is determined by the offspring's genes, which are inherited from both parents. Interbreeding between viduines adapted to different host species will result in poor mimicry in the next generation and is likely to be subject to especially strong selection. Learning host songs is probably a consequence of that selection.

ECOLOGY OF PARASITES

Diet and foraging ecology

Hamilton and Orians (1965) pointed out that brood parasitism is strongly correlated with unusual diets and foraging ecology. Honeyguides consume wax (Friedmann 1955; Friedmann and Kern 1956; Cronin and Sherman 1977), which is very difficult to digest, especially for nestlings, and even prefer wax over honey (Short and Horne 1985). At least two species of honeyguide lead bee predators to hives and eat the wax and larvae that are exposed (Friedmann 1955; Ranger 1955; Fry et al. 1988). The two bee predators most commonly chosen by honeyguides are the ratel (*Mellivora capensis*) and humans and there is much African folklore associated with this remarkable behavior. By carefully mapping guiding sessions, Isack and Reyer (1989) confirmed that honeyguides are already aware of the location of a nearby bee hive when they approach a person. Although cuckoos have diverse diets, most consume hairy caterpillars that would be unpalatable to nestlings (e.g., Wetmore 1968; Rowan 1983; Fry et al. 1988; Gill, this vol.).

Cowbirds often feed on sometimes ephemeral patches of bare ground and short grass created by grazing mammals. Historically, North American cowbirds may have been associated with herds of nomadic ungulates (Mayfield 1965), which led Widman (1907) to suggest that cowbirds evolved parasitism because they could not stay long enough in any one area to take care of their own young. Cowbirds, however, probably evolved parasitism in South America (Friedmann 1929) and early visitors to the plains, such as Custer (1962) reported that small bands of bison were found along water courses much of the time. Ungulate foraging associates, therefore, were present in the plains along riparian strips, precisely where host diversity was greatest. Although there are no quantitative data, field observations indicate that cowbirds have a much more intimate association with bison than cattle (S. Rothstein, pers. obs.). Cowbirds forage on the backs of some large mammals, eat the blood-sucking flies that feed on them, and also benefit by eating insects flushed by the mammals (Friedmann 1929; Robinson 1988). Despite their obvious affinity for associating with ungulates, it is not clear that cowbirds require such associations.

Whether or not unusual foraging ecologies were precursors to the evolution of brood parasitism remains unknown. Such a precursor role is possible in the case of cuckoos because even nonparasitic cuckoos seem to specialize on hairy caterpillars. Alternatively, the exploitation of unusual resources may have been made possible by the evolution of brood parasitism. Disentangling cause and effect here may not be possible.

Habitat selection

At least some brood parasites are extreme habitat generalists. Brown-headed cowbirds, for example, occupy virtually all habitats in North America except areas above treeline and some treeless deserts (Lowther 1993; Robinson et al. 1995*b*). Year after year, they are seen on more breeding bird census plots than any other North American bird species (Rothstein 1994). Their ability to feed and breed in separate areas (below) enables them to exploit hosts in many habitats while remaining relatively specialized in their foraging behavior. Nevertheless, their broad geographic range is at least partly an artifact of anthropogenic changes in landscapes that have created many new feeding opportunities. Other brood parasites may be benefiting from human disturbances that create early successional vegetation favored for foraging (e.g., Glue and Morgan 1972; Arias-de-Reyna, this vol.; Higuchi, this vol.; Nakamura et al., this vol.; Soler et al., this vol.) but habitat selection in highly specialized brood parasites is likely determined by that of their hosts.

Spatial patterns

At least one group of brood parasites, the cowbirds, breed and feed in separate areas (reviewed in Robinson et al. 1995*b*). Radiotelemetry shows that cowbirds routinely commute up to 7 km between feeding and breeding areas (Dufty 1982*a*; Rothstein et al. 1984; Thompson 1994). In some areas, cowbirds roost in a different place from their breeding and feeding areas (Thompson 1994). In the morning, solitary female cowbirds search for nests but in the afternoon, females gather in flocks and search for insects and seeds in pastures, plowed fields, and areas with short-mowed grass. This uncoupling of breeding and feeding areas and the large distances between such areas allow great flexibility in habitat selection and enable cowbirds to parasitize nests over areas as large as 150 km^2, even when such areas contain only one highly preferred feeding site (Rothstein et al. 1984, 1987). The extreme mobility of breeding cowbirds may make local efforts to control their populations by trapping very difficult in landscapes where there are many feeding areas (Rothstein et al. 1987; Robinson et al. 1993). Cowbirds, however, are not locked into using widely separate breeding and feeding areas because they will feed at breeding sites if habitat is suitable (Elliott 1980; Rothstein et al. 1986; Thompson 1994).

Defense of breeding areas against conspecifics is widespread in cuckoos, and apparently in some cowbirds, but mutually exclusive use of breeding areas seems to be more prevalent in cuckoos. Radiotelemetry studies of brown-headed cowbirds and cuckoos (reviewed in Barnard, this vol.) indicate nonoverlapping or only slightly overlapping breeding ranges of females. In some cowbirds, females defend territories (Lowther 1993) in a system of site-based dominance, at least in some areas (Darley 1982; Dufty 1982*a*,*b*; Teather and Robertson 1985, 1986) but females may be only partially successful at excluding other females because of the large sizes of their breeding ranges (Rothstein et al. 1984). High cowbird densities may make defense uneconomical in the Great Plains, where there is little evidence for territoriality (Elliott 1980). Areas where breeding areas overlap may be characterized by high incidences of multiple cowbird or cuckoo parasitism because more than one female may lay in a nest (Elliott 1977, 1978, 1980; Arias-de-Reyna, this vol.; Soler et al., this vol.). But multiple parasitism is far more common in cowbirds than in cuckoos. Perhaps the more clearly defined territoriality in cuckoos, especially as regards mutually exclusive breeding areas, is due more to cuckoos being less common than to intrinsic differences in territorial behavior between cuckoos and cowbirds. In at least one cowbird, however, males and females live together in flocks and show no evidence of territoriality (Robinson 1988).

OVERVIEW OF PART VII: CONSPECIFIC BROOD PARASITISM

The past 15 years have seen a remarkable increase in interest in conspecific brood parasitism (CBP), which occurs when a female lays eggs in nests of conspecifics (reviewed in Yom-Tov 1980; Andersson 1984; Rohwer and Freeman 1989; Petrie and Møller 1991; Power, this vol.). Much of this interest has been generated by the realization, starting with Yom-Tov (1980), that CBP occurs in far more species than was once thought to be the case and that it is often a regularly occurring reproductive strategy rather than the result of rare accidents. As an indication of the rate at which information on CBP is increasing, the number of species in which it has been documented increased from 53 in Yom-Tov's (1980) review to 82 (MacWhirter 1989) and 140 (Rohwer and Freeman 1989) in two reviews only nine years later. CBP is especially widespread in waterfowl and other precocial species (Weller 1959; Rohwer and Freeman 1989; Sorenson 1992, this vol.) and in colonial birds (Brown 1984; Emlen and Wrege 1986; Møller 1987, this vol.; Brown and Brown 1988, 1989; Lank et al. 1989*a*,*b*; Feare 1991; but see Harms et al. 1991; Lyon et al. 1992). Furthermore, CBP probably occurs in many more species but has escaped notice due to the difficulty of detecting it (MacWhirter 1989).

Interest in CBP has also been stirred by new molecular methods that better enable researchers to detect it (reviewed in Power, this vol.). Lastly, CBP is also of interest because of its similarity to extra-pair copulations or EPCs (Petrie and Møller 1991; Power, this vol.), a subject that has seen a similar increase in attention in recent years. However, CBP is costly to both parasitized parents as both raise unrelated young, whereas EPCs are costly only to males. As with interspecific parasitism (Davies et al. 1989), there are major parallels between CBP in insects and birds (Brockmann 1993).

THE ADAPTIVE BASIS OF CONSPECIFIC BROOD PARASITISM

Much of the work on CBP has focused on attempts to determine whether it represents an alternative reproductive mode pursued by individuals who are capable of breeding on their own but who receive increased reproductive output by being parasitic or by pursuing a mixed strategy involving parasitism as well as caring for their own young. Within this framework, it is possible that CBP represents an Evolutionarily Stable Strategy, or ESS, wherein individuals within a species have equal fitnesses regardless of the extent to which they practice CBP. In these views, CBP is an option that birds can choose depending on the net pay-offs they are likely to receive. An alternative possibility is that CBP is pursued by individuals making the best of a bad situation (see Møller, this vol. and Sorenson, this vol., for expanded discussions of this issue). The main distinction between these two alternatives is that parasitism is less profitable than caring for one's own young in the second alternative, but is the only option open to some individuals because they may have lost their own nests during the laying period, they may have been unable to secure a territory or nest site, or they may be too inexperienced to care for young. Yom-Tov's (1980) review indicated that CBP usually results from individuals making the best of a bad situation (for more recent examples see Dhindsa 1983; Lank et al. 1989*a*,*b*; Feare 1991; Weigmann and Lamprecht 1991; Power, this vol.). However, some studies have found little or no CBP despite nest losses during laying (Harms et al. 1991; Lyon et al. 1992) and experiments designed to provoke CBP by destroying nests have met with mixed success (Emlen and Wrege 1986; Feare 1991; Rothstein 1993; Power, this vol.).

Rather than CBP being a best of a bad situation, most recent authors have emphasized that it can be a fitness-enhancing alternative reproductive strategy, even when individuals have the option of taking care of their own young (reviewed by Petrie and Møller 1991). Individuals may be able to increase their reproductive output by spreading the risk among several nests, by reducing competition within their own nests, and by increasing their own fecundity through exploitation of the parental care of other birds (Brown 1984; Møller 1987, this vol.; Brown and Brown 1988; Evans 1988; Jackson 1993, this vol.; Sorenson, this vol.). Among the four chapters in part VII, CBP in Møller's barn swallow (*Hirundo rustica*) study and in Jackson's weaver (*Ploceus taeniopterus*) study appear to be strategies pursued by birds that have the option of laying in their own nests, whereas CBP in Power's European starling (*Sturnus vulgaris*) study appears to be mostly a best of bad situation strategy. In the remaining paper, one of two duck species studied by Sorenson conforms mostly to the former condition, while the other conforms more to the latter.

CONSPECIFIC BROOD PARASITISM AND THE EVOLUTION OF OBLIGATE INTERSPECIFIC PARASITISM

CBP was viewed as a precursor to the evolution of interspecfic brood parasitism by Hamilton and Orians (1965), who suggested that obligate brood parasitism could arise initially from facultative conspecific parasitism by birds that have lost their nests during laying. Experimental tests of the Hamilton–Orians model, however, have yielded mixed results (Stouffer and Power 1991; Rothstein 1993), although many additional species need to be tested.

Yamauchi (1993, 1995) used quantitative genetics models to generalize the Hamilton–Orians model and found: (i) that conspecific brood parasitism is a necessary precursor to interspecific brood parasitism; (ii) that brood parasitism can only pay if the cost of competition within nests is greater than the cost of conspecific parasitism; and (iii) that interspecific brood parasitism can only evolve when the costs of both intra- and interspecific parasitism are small. However, these findings assume that some degree of host defense, such as at least a low rate of egg recognition, will exist as soon as CBP appears, but the large number of species with no hint of egg recognition (part IV) casts doubt on this assumption. Another recent modeling study (Cichon 1996) also indicated that CBP could lead to obligate interspecific parasitism.

Sorenson (this vol.) provided actual examples of ways in which CBP could lead to interspecific parasitism as he documented conditional CBP and interspecific parasitism in two ducks depending upon environmental conditions. One of these species, the redhead (*Aythya americana*) seems much closer to obligate parasitism as most of its parasitism is of other species and appears to be a fitness-enhancing strategy not merely a best of a bad situation strategy.

Ever since the time of Darwin (1920), New World cuckoos in the genus *Coccyzus* have also figured prominently in speculations regarding the evolution of obligate brood parasitism. These birds tend their own nests but also regularly parasitize the nests of congeners and conspecifics, and occasionally even those of passerines (Nolan and Thompson 1975; Potter 1980). They also share a number of characteristics with obligately parasitic cuckoos and are the sort of species likely to evolve obligate interspecific parasitism (Hamilton and Orians 1965). In contrast, Hughes (1996) has suggested that the characters *Coccyzus* shares with parasitic cuckoos are not pre-adaptations for parasitism but are evidence that this genus was once obligately parasitic and has reverted to caring for its own young.

ADAPTATIONS TO FACILITATE AND DEFEAT CONSPECIFIC BROOD PARASITISM

In many respects, CBP is subject to the same coevolutionary forces that shape interspecific parasite–host systems (Petrie and Møller 1991; Power, this vol.). CBP is costly to both sexes of host pairs because it can involve egg removal by the parasites (Emlen and Wrege 1986; Brown and Brown 1988, 1989; Lombardo et al. 1989), the death of some of the host's own young because of overcrowded broods (Jackson, this vol.; Power, this vol.) and the costs of misdirected parental care (Petrie and Møller 1991). These host costs are minimized in precocial species (Lyon and Eadie 1991; Sorenson, this vol.). One species, the ostrich (*Struthio camelus*) may actually benefit from CBP because extra eggs dilute losses of the parent's own eggs to predators (Bertram 1979).

Host defenses against CBP include several striking parallels to defenses described in parts II, III, and IV. Nest guarding has been documented in some hosts (Emlen and Wrege 1986; Gowaty et al. 1989; Møller, this vol.). Møller (this vol.) found that nest defense and CBP were positively density-dependent in colonial barn swallows. Sealy et al. (1989) hypothesized that nest guarding may explain the very low incidence of CBP in yellow warblers, which are territorial. Asynchronous nesting may also decrease the likelihood of CBP (Møller 1987).

One major difference between potential defenses against CBP and interspecific parasitism is that behaviors dependent on egg recognition are unlikely in CBP because parasitic and host eggs have the same features. Species subject to CBP have developed two features to deal with this egg recognition problem. These species remove all eggs, regardless of their appearance, from nests before they start to lay. Evidence that this behavior evolved as a response to CBP comes from the fact that removal of such "premature" eggs is not universal. Accepter species with little or no evidence of CBP accept eggs placed in their nests before they begin to lay (Rothstein 1986, 1993).

The "toss out everything strategy" does not work for CBP that occurs after the host has already begun to lay, but some hosts may have evolved extreme variation in egg appearance among clutches to deal with such parasitism. High variation enables individuals to distinguish between their own eggs and those of conspecifics (Freeman 1988; Møller and Petrie 1990; Jackson, this vol.). Jackson (this vol.) suggests that the intense selection for rejection of conspecific eggs in weavers may secondarily provide an additional, built-in defense against cuckoo eggs. This is contrary to the traditional belief that interpreted cuckoo parasitism as the selection pressure responsible for the evolution of both egg recognition and variable eggs in weavers (Victoria 1972; Rothstein 1990*a*). Nevertheless, it may be worth considering the view that cuckoo parasitism was involved in the initial evolution of egg recognition. The high interclutch variation characteristic of many weaver species would have no value as a host defense in the absence of egg recognition and rejection. So recognition had to precede the appearance of interclutch variation, but recognition evolved in response to CBP would be very difficult if a species had monomorphic eggs. In the latter case, recognition and high interclutch variation would be selected for only if they occurred as a result of new genetic combinations (mutations or recombinants) in the same individual, a highly unlikely event. This problem does not occur if recognition first evolved in response to cuckoo parasitism because initially the cuckoo and weaver eggs were likely to have had some differences. So a host response to interspecific parasitism may have been necessary to initiate the evolution of the current defenses shown by weavers, even if those defenses today have more value against CBP than against cuckoo parasitism.

Regardless of whether weavers evolved their extreme interclutch variation in response to conspecific or interspecific parasitism, it is worthwhile to ask why such variation is so rare among other passerines (although some increase in interclutch variation appears to have evolved in response to common cuckoo parasitism; Øien et al. 1996; Soler and Møller 1996). Nest predation may constrain egg color variation in species with open nests because their eggs are visible, but weavers have enclosed nests that shield their eggs from view. However, unlike most species with enclosed nests (either cavities or domed nests), enough light comes into weaver nests to allow visible discrimination of eggs. So the extreme egg variation of many weaver species may be due to the unusual combination of a nest chamber that conceals the eggs but is still light enough to allow eggs to be seen clearly from the inside.

Except for weavers (Victoria 1972; Jackson, this vol.), rejection of conspecific parasitic eggs has rarely been documented, perhaps because it is hard for researchers to detect but also because it might be difficult for conspecifics to distinguish similar eggs from their own (Rothstein 1990*a*; see also Petrie and Møller 1991).

Evolved responses to CBP may have had a major effect on a critical aspect of avian life history. If CBP is frequent but a bird cannot identify parasitic eggs, it may be adaptive to lay a clutch with a smaller number of eggs than the number of young that can be fed (Power et al. 1989; Power, this vol.). Then if CBP does occur, the host will be able to feed all of the nestlings in its nest and thereby avoid the heavy costs due to starvation of most of the nestlings. This "parasitism insurance hypothesis" could explain why clutch sizes are sometimes smaller than the number of young that can be fed (see also Rothstein 1990*b*).

Documented adaptations that facilitate CBP include stealth and rapid laying by parasites (Brown and Brown 1989; Feare et al. 1989; Power, this vol.) and even the ability to transfer eggs from one nest to another (Brown and Brown 1988). Mated pairs sometimes approach hosts together with the male distracting the hosts and the female quickly laying (Lank et al. 1989*a*,*b*). Because timing is important, parasites must monitor the progress of host nests; this may explain why CBP often involves close neighbors (Møller 1987; Brown and Brown 1989).

EFFECTS OF CONSPECIFIC PARASITISM ON HOST DEMOGRAPHY

A number of authors have considered the possibility that CBP may affect population sizes. May et al. (1991) and Nee and May (1993) argued that CBP could cause population sizes to rise and fall cyclically because it reduces the average fitness. Cichon (1996) argued that CBP may even be an unstable situation unless individuals have defenses. Møller's (this vol.) documentation of positively density-dependent nest guarding and CBP is consistent with these models, although the extent to which the costs of nest guarding have negative population-level consequences remains unknown. Nevertheless, it is likely that CBP is one of the inevitable costs of colonial breeding in some bird species (Hoogland and Sherman 1976; Brown 1984), although it may benefit some of the individuals that parasitize other nests (Brown 1984). Conspecific brood parasitism can even have management consequences; when nest boxes of wood ducks (*Aix sponsa*) are too close together, CBP in dump nests can reduce overall reproductive success of populations (Clawson et al. 1979; Heusmann et al. 1980; Semel and Sherman 1986; Semel et al. 1988; Bellrose and Holm 1994).

REFERENCES

Ali, S. A. (1931). The origin of mimicry in cuckoo's eggs. *J. Bombay Nat. Hist. Soc.*, **34**, 1067–1070.

Andersson, M. (1984). Brood parasitism within species. In: *Producers and scroungers* (ed. C. J. Barnard), pp. 195–228. Chapman & Hall, London.

Ankney, C. D. and D. M. Scott (1980). Changes in nutrient reserves and diet of breeding Brown-headed Cowbirds. *Auk*, **97**, 684–696.

Ankney, D. and D. Scott (1982). On the mating system of brown-headed cowbirds. *Wilson Bull.*, **94**, 260–268.

Appert, O. (1970). Zur Biologie einiger Kua-Arten Madagaskars (Aves: Cuculi). *Zool. Jahrb.*, **97**, 424–453.

Arcese, P., J. N. M. Smith, W. M. Hochachka, C. M. Rogers, and D. Ludwig (1992). Stability, regulation, and the determination of abundance in an insular song sparrow population. *Ecology*, **73**, 805–822.

Arcese, P., J. N. M. Smith, and M. I. Hatch (1996). Nest predation by cowbirds and its consequences for passerine demography. *Proc. Natl. Acad. Sci. USA*, **93**, 4608–4611.

Arias-de-Reyna, L., this vol. Coevolution of the great spotted cuckoo and its hosts.

Askemo, C. (1979). Reproductive effort and the return rate of male pied flycatchers. *Amer. Natur.*, **114**, 748–753.

Baba, R., Y. Nagata, and S. Yamagishi (1990). Brood parasitism and egg robbing among three freshwater fish. *Anim. Behav.*, **40**, 776–778.

Baker, E. C. S. (1942). *Cuckoo problems*. Witherby, London.

Barnard, P., this vol. Variability in the mating systems of parasitic birds.

Barnard, P. and M. B. Markus (1988). Male copulation frequency and female competition for fertilizations in a promiscuous brood parasite, the Pin-tailed Whydah *Vidua macroura*. *Ibis*, **131**, 421–425.

Becking, J. H. (1981). Notes on the breeding of Indian cuckoos. *J. Bombay Nat. Hist. Soc.*, **78**, 201–231.

Bellrose, F. C. and D. J. Holm (1994). *Ecology and management of the Wood Duck*. Stackpole Books, Mechanicsburg, Pennsylvania.

Benson, C. W. and M. P. S. Irwin (1972). The thick-billed cuckoo *Pachycoccyx audeberti* (Schlegel) (Aves: Cuculidae). *Arnoldia (Rhodesia)*, **5**, 1–124.

Bertram, B. C. R. (1979). Ostriches recognize their

own eggs and discard others. *Nature*, **279**, 233–234.

Blankespoor, G. W., J. Oolman, and C. Uthe (1982). Eggshell strength and cowbird parasitism of Red-winged Blackbirds. *Auk*, **99**, 363–365.

Böhning-Gaese, K., M. L. Taper, and J. H. Brown (1993). Are declines in North American insectivorous songbirds due to causes on the breeding range? *Conservation Biol.*, **7**, 76–81.

Braa, A. T., A. Mosknes, and E. Røskaft (1992). Adaptations of Bramblings and Chaffinches towards parasitism by the common cuckoo. *Anim. Behav.*, **43**, 67–78.

Brawn, J. D. and S. K. Robinson (1996). Source-sink population dynamics may complicate the interpretation of large-scale and long-term census trends. *Ecology*, **77**, 3–11.

Briskie, J. V. and S. G. Sealy (1987). Responses of least flycatchers to experimental inter- and intra-specific parasitism. *Condor*, **89**, 899–901.

Briskie, J. V. and S. G. Sealy (1989). Changes in nest defense against a brood parasite over the breeding cycle. *Ethology*, **82**, 61–67.

Briskie, J. V. and S. G. Sealy (1990). Evolution of short incubation periods in the parasitic cowbirds, *Molothrus* spp., *Auk*, **107**, 789–794.

Briskie, J. V., S. G. Sealy, and K. A. Hobson (1992). Behavioral defenses against avian brood parasitism in sympatric and allopatric host populations. *Evolution*, **46**, 334–340.

Brittingham, M. C. and S. A. Temple (1983). Have cowbirds caused forest songbirds to decline? *BioScience*, **33**, 31–35.

Brockmann, H. J. (1993). Parasitizing conspecifics: comparisons between Hymenoptera and birds. *Trends Ecol. Evol.*, **8**, 2–4.

Brooke, M. de L. and N. B. Davies (1987). Recent changes in host usage by cuckoos *Cuculus canorus* in Britain. *J. Anim. Ecol.*, **56**, 873–883.

Brooke, M. de L. and N. B. Davies (1988). Egg mimicry by cuckoos *Cuculus canorus* in relation to discrimination by hosts. *Nature*, **335**, 630–632.

Brooker, L. C. and M. G. Brooker (1990). Why are cuckoos host specific? *Oikos*, **57**, 301–309.

Brooker, L. C. and M. G. Brooker (1994). A model for the effects of fire and fragmentation on the population viability of the splendid fairy-wren. *Pacific Conser. Biol.*, **1**, 344–358.

Brooker, L. C., M. G. Brooker, and A. M. H. Brooker (1990). An alternative population/genetics model for the evolution of egg mimesis and egg crypsis in cuckoos. *J. Theor. Biol.*, **146**, 123–143.

Brooker, M. G. and L. C. Brooker (1989*a*). Cuckoo hosts in Australia. *Australian Zool. Rev.*, **2**, 1–67.

Brooker, M. G. and L. C. Brooker (1989*b*). The comparative breeding behaviour of two sympatric species of cuckoos, Horsfield's bronze-cuckoo *Chrysococcyx basalis* and the shining bronze-cuckoo *C. lucidus* in Western Australia: A new model for the evolution of egg morphology and host specificity in avian brood parasites. *Ibis*, **131**, 528–547.

Brooker, M. G. and L. C. Brooker (1991). Eggshell strength in cuckoos and cowbirds. *Ibis*, **133**, 406–413.

Brooker, M. G. and L. C. Brooker (1992). Evidence for individual female host specificity in two Australian Bronze-cuckoos (*Chrysococcyx* spp.). *Australian J. Zool.*, **40**, 485–493.

Brooker, M. G. and L. C. Brooker (1996). Acceptance by the splendid fairy-wren of parasitism by Horsfields' bronze-cuckoo: further evidence for evolutionary equilibrium in brood parasitism. *Behavioral Ecology*, **7**, 395–407.

Brown, C. R. (1984). Laying eggs in a neighbor's nest: benefit and cost of colonial nesting in swallows. *Science*, **224**, 518–519.

Brown, C. R. and M. B. Brown (1988). A new form of reproductive parasitism in cliff Swallows. *Nature*, **331**, 66–68.

Brown, C. R. and M. B. Brown (1989). Behavioral dynamics of intraspecific brood parasitism in colonial cliff swallows. *Anim. Behav.*, **37**, 777–796.

Brown, R. J., M. N. Brown, M. de L. Brooke, and N. B. Davies (1990). Reactions of parasitized and unparasitized populations of *Acrocephalus* warblers to model cuckoo eggs. *Ibis*, **132**, 109–111.

Burgham, M. C. and J. Picman (1989). Effect of brown-headed cowbirds on the evolution of yellow warbler anti-parasite strategies. *Anim. Behav.*, **38**, 298–308.

Burhans, D. E. Dawn nest arrivals in cowbird hosts: Their role in aggression, cowbird recognition, and response to parasitism. In: *Ecology and management of cowbirds* (eds. J. N. M. Smith, T. Cook, S. K. Robinson, S. I. Rothstein, and S. G. Sealy), Univ. of Texas Press, Austin, Texas (in press).

Burnell, K. and S. I. Rothstein (1994). Variation in the structure of female brown-headed cowbird vocalizations and its relation to vocal function and development. *Condor*, **96**, 703–715.

Carello, C. A. and G. K. Snyder. The effects of host numbers on cowbird parasitism of red-winged blackbirds. In: *Ecology and management of cowbirds* (eds. J. N. M. Smith, T. Cook, S. K. Robinson, S. I. Rothstein, and S. G. Sealy), Univ. of Texas Press, Austin, Texas (in press).

Carey, C. (1986). Possible manipulation of eggshell

conductance of host eggs by brown-headed cowbirds. *Condor*, **88**, 388–390.

Carter, M. D. (1986). The parasitic behavior of the Bronzed Cowbird in south Texas. *Condor*, **88**, 11–25.

Chance, E. (1922). *The cuckoo's secret*. Sedgwick and Jackson, London.

Chance, E. (1940). *The truth about the cuckoo*. Country Life, London.

Cichoń, M. (1996). The evolution of brood parasitism: the role of facultative parasitism. *Behav. Ecol.*, **7**, 137–139.

Clark, K. L. and R. J. Robertson (1979). Spatial and temporal multi-species nesting aggregations in birds as anti-parasite and anti-predator defenses. *Behav. Ecol. Sociobiol.*, **5**, 359–371.

Clark, K. L. and R. J. Robertson (1981). Cowbird parasitism and evolution of anti-parasite strategies in the Yellow Warbler. *Wilson Bull.*, **92**, 244–258.

Clawson, R., G. Hartman, and L. Fredrickson (1979). Dump nesting in a Missouri Wood Duck population. *J. Wildlf. Manag.*, **43**, 347–355.

Clotfelter, E. D. (1995). Courtship displaying and intrasexual competition in the Bronzed Cowbird. *Condor*, **97**, 816–818.

Clotfelter, E. D. and T. Brush (1995). Unusual parasitism by the Bronzed Cowbird. *Condor*, **97**, 814–816.

Courtenay, J. (1967). The juvenile food-begging call of some fledgling cuckoos—vocal mimicry or vocal duplication by natural selection? *Emu*, **67**, 154–157.

Cronin, E. W. and P. W. Sherman (1977). A resource-based mating system: the orange-rumped honeyguide, *Indicator xanthonotus*. *Living Bird*, **15**, 5–37.

Cruz, A. and J. W. Wiley (1989). The decline of an adaptation in the absence of a presumed selection pressure. *Evolution*, **43**, 55–62.

Cruz, A., T. Manolis, and J. W. Wiley (1985). The shiny cowbird: a brood parasite expanding its range in the Caribbean region. Ornithol. Monogr., No. 36, pp. 607–620.

Cruz, A., J. W. Prather, W. Post, and J. W. Wiley. The spread of Shiny and Brown-headed Cowbirds into the Florida region. In: *Ecology and management of cowbirds* (eds. J. N. M. Smith, T. Cook, S. K. Robinson, S. I. Rothstein, and S. G. Sealy), Univ. of Texas Press, Austin, Texas (in press).

Cruz, A., W. Post, J. W. Wiley, C. P. Ortega, T. K. Nakamura, and J. W. Prather, this vol. Potential impacts of Cowbird range expansion in Florida.

Custer, G. A. (1962). *My life on the plains*. University of Oklahoma Press, Norman.

Darley, J. A. (1971). Sex ratio and mortality in the brown-headed cowbird. *Auk*, **88**, 560–566.

Darley, J. A. (1982). Territoriality and mating behavior of the male brown-headed cowbird. *Condor*, **84**, 15–21.

Darley, J. A. (1983). Territorial behavior of the brown-headed cowbird. *Can. J. Zool.*, **61**, 65–69.

Darwin, C. (1920). *The origin of species*. John Murray, London.

Davies, N. B. and M. de L. Brooke (1988). Cuckoos versus reed warblers: adaptations and counteradaptations. *Anim. Behav.*, **36**, 262–284.

Davies, N. B. and M. de L. Brooke (1989*a*). An experimental study of co-evolution between the cuckoo, *Cuculus canorus*, and its hosts. I. Host egg discrimination. *J. Anim. Ecol.*, **58**, 207–224.

Davies, N. B. and M. de L. Brooke (1989*b*). An experimental study of co-evolution between the cuckoo, *Cuculus canorus*, and its hosts. II. Host egg markings, chick discrimination and general discussion. *J. Anim. Ecol.*, **58**, 225–236.

Davies, N. B. and M. de L. Brooke (1991). Coevolution of the cuckoo and its hosts. *Sci. Am.*, **264**, 92–98.

Davies, N. B. and M. de L. Brooke, this vol. Cuckoos versus hosts: experimental evidence for coevolution.

Davies, N. B., A. F. G. Bourke, and M. de L. Brooke (1989). Cuckoos and parasitic ants: interspecific parasitism as an evolutionary arms race. *Trends Ecol. Evol.*, **4**, 274–278.

Davies, N. B., M. de L. Brooke, and A. Kacelnik (1996). Recognition errors and probability of parasitism determine whether reed warblers should accept or reject mimetic cuckoo eggs. *Proc. R. Soc. Lond. B.*, **263**, 925–931.

Davis, D. E. (1940). Social nesting habits of *Guira guira*. *Auk*, **57**, 472–484.

Davis, D. E. (1942). The phylogeny of nesting habits in the Crotophaginae. *Q. Rev. Biol.*, **17**, 115–134.

Dawkins, R. and J. R. Krebs (1979). Arms races between and within species. *Proc. R. Soc. London B. Ser.*, **205**, 489–511.

Dearborn, D. C. (1996). Video documentation of a brown-headed cowbird nestling ejecting an indigo bunting nestling from the nest. *Condor*, **98**, 645–649.

De Geus, D. W. and L. B. Best (1991). Brown-headed cowbirds parasitize loggerhead shrikes: first records for family Laniidae. *Wilson Bull.*, **103**, 504–506.

DeCapita, M. E. Brown-headed cowbird control on Kirtland's warbler nesting areas in Michigan, 1972–1994. In: *Ecology and management of cowbirds* (eds. J. N. M. Smith, T. Cook, S. K. Robinson,

S. I. Rothstein, and S. G. Sealy), Univ. of Texas Press, Austin, Texas (in press).

DeSteven, D. (1980). Clutch size, breeding success and parental survival in the tree swallow (*Tachycineta bicolor*). *Evolution*, **34**, 278–291.

Dhindsa, M. S. (1983). Intraspecific nest parasitism in two species of Indian weaverbirds, *Ploceus benghalensis* and *P. manyar*. *Ibis*, **125**, 243–245.

Donovan, T. M., F. R. Thompson, III, J. Faaborg, and J. R. Probst (1995*a*). Reproductive success of migratory birds in habitat sources and sinks. *Conserv. Biol.*, **9**, 1380–1395.

Donovan, T. M., R. H. Lamberson, A. Kimber, F. R. Thompson, III, and J. Faaborg (1995*b*). Modeling the effects of habitat fragmentation on source and sink demography of neotropical migrant birds. *Conserv. Biol.*, **9**, 1396–1407.

Dow, D. D. (1972). The New Zealand long-tailed cuckoo: nest parasite or predator? *Emu*, **72**, 179–180.

Duckworth, J. W. (1991). Responses of breeding reed warblers *Acrocephalus scirpaceus* to mounts of sparrowhawk *Accipiter nisus*, cuckoo *Cuculus canorus* and jay *Garrulus glandarius*. *Ibis*, **133**, 68–74.

Dufty, A. M., Jr (1982*a*). Movements and activities of radio-tracked Brown-headed Cowbirds. *Auk*, **99**, 316–327.

Dufty, A. M., Jr (1982*b*). Response of Brown-headed Cowbirds to simulated conspecific intruders. *Anim. Behav.*, **30**, 1043–1052.

Dufty, A. M., Jr and J. C. Wingfield (1986). Temporal patterns of circulating LH and steroid hormones in a brood parasite, the brown-headed cowbird. II. Females. *J. Zool.*, **208**, 205–214.

Eastzer, D., P. R. Chu, and A. P. King (1980). The young cowbird: Average or optimal nestling? *Condor*, **82**, 417–425.

Elliott, P. F. (1977). Adaptive significance of cowbird egg distribution. *Auk*, **94**, 590–593.

Elliott, P. F. (1978). Cowbird parasitism in the tallgrass prairie of northeastern Kansas. *Auk*, **95**, 161–167.

Elliott, P. F. (1980). Evolution of promiscuity in the brown-headed cowbird. *Condor*, **82**, 138–141.

Emlen, S. T. and L. W. Oring (1977). Ecology, sexual selection and the evolution of mating systems. *Science*, **197**, 215–223.

Emlen, S. T. and P. H. Wrege (1986). Forced copulations and intraspecific parasitism: two costs of social living in the White-fronted Bee-eater. *Ethology*, **71**, 2–29.

Evans, P. G. H. (1988). Intraspecific nest parasitism in the European starling *Sturnus vulgaris*. *Anim. Behav.*, **36**, 1282–1294.

Facemire, C. F. (1980). Cowbird parasitism of marsh-nesting red-winged blackbirds. *Condor*, **82**, 347–348.

Feare, C. J. (1991). Intraspecific nest parasitism in starlings *Sturnus vulgaris*: effects of disturbance on laying females. *Ibis*, **133**, 75–79.

Ficken, M. S. (1965). Mouth color of nestling passerines and its use in taxonomy. *Wilson Bull.*, **77**, 71–75.

Finch, D. M. (1982). Rejection of cowbird eggs by Crissal Thrashers. *Auk*, **99**, 719–724.

Finch, D. M. (1983). Brood parasitism of the Abert's towhee: timing, frequency and effects. *Condor*, **85**, 355–359.

Fink, M. A., D. J. Milinkovich, and C. F. Thompson (1987). Evolution of clutch size: an experimental test in the house wren (*Troglodytes aedon*). *J. Anim. Ecol.*, **56**, 99–114.

Fleischer, R. C. (1985). A new technique to identify and assess dispersion of eggs of individual brood parasites. *Behav. Ecol. Sociobiol.*, **17**, 91–99.

Fleischer, R. C. and S. I. Rothstein (1988). Known secondary contact and rapid gene flow among subspecies and dialects in the brown-headed cowbird. *Evolution*, **42**, 1146–1158.

Fleischer, R. C. and N. G. Smith (1992). Giant cowbird eggs in the nests of two icterid hosts: The use of morphology and electrophoretic variants to identify individuals and species. *Condor*, **94**, 572–578.

Fleischer, R. C., A. P. Smyth, and S. I. Rothstein (1987). Temporal and age-related variation in the laying rate of the brown-headed cowbird in the eastern Sierra Nevada, CA. *Can. J. Zool.*, **65**, 2724–2730.

Folkers, K. L. and P. E. Lowther (1985). Responses of nesting red-winged blackbirds and Yellow Warblers to brown-headed cowbird. *J. Field Ornith.*, **56**, 175–177.

Fraga, R. M. (1978). The rufous-collared sparrow as a host of the shiny cowbird. *Wilson Bull.*, **90**, 271–284.

Fraga, R. M. (1979). Differences between nestlings and fledglings of screaming and Bay-winged cowbirds. *Wilson Bull.*, **90**, 151–154.

Fraga, R. M. (1983). The eggs of the parasitic screaming cowbird (*Molothrus rufoaxillaris*) and its host, the bay-winged cowbird (*M. badius*): is there evidence for egg mimicry? *J. Ornith.*, **124**, 187–193.

Fraga, R. M. (1985). Host–parasite interactions between chalk-browed mockingbirds and shiny cowbirds. In: *Neotropical ornithology* (eds. P. A. Buckley, M. S. Foster, E. S. Morton, R. S. Ridgely, and F. G. Buckley), Ornithol. Monogr, No. 36, pp. 829–844.

Fraga, R. M. (1996). Further evidence of parasitism of chopi blackbirds (*Gnorimopsar chopi*) by the specialized screaming cowbird (*Molothrus rufoaxillaris*) *Condor* (in press).

Fraga, R. M., this vol. Interactions of the parasitic screaming and shiny cowbirds (*Molothrus rufoaxillaris* and *M. bonariensis*) with a shared host, the bay-winged cowbird (*M. badius*).

Frederick, P. C. and M. A. Shields (1986). Corrections for the underestimation of brood parasitism frequency derived from daily nest inspections. *J. Field Ornith.*, **57**, 224–226.

Freeman, S. (1988). Egg variability and conspecific nest parasitism in *Ploceus* weaverbirds. *Ostrich*, **59**, 49–53.

Freeman, S., D. F. Gori, and S. Rohwer (1990). Red-winged blackbirds and Brown-headed Cowbirds: some aspects of a host-parasite relationship. *Condor*, **92**, 336–340.

Friedmann, H. (1929). *The cowbirds: a study in the biology of social parasitism*. C. C. Thomas, Springfield, IL.

Friedmann, H. (1948). *The parasitic cuckoos of Africa*. Washington Acad. Sci. Monogr. No. 1, Washington, D.C.

Friedmann, H. (1955). The honey-guides. *U.S. National Mus. Bull.*, **208**, 1–292.

Friedmann, H. (1960). The parasitic weaverbirds. *U.S. National Mus. Bull.*, **223**, 1–196.

Friedmann, H. (1963). Host relations of the parasitic cowbirds. *U.S. National Mus. Bull.*, **233**, 1–276.

Friedmann, H. (1964*a*). The history of our knowledge of avian brood parasitism. *Centaurus*, **10**, 282–304.

Friedmann, H. (1964*b*). Evolutionary trends in the avian genus *Clamator*. *Smithson. Misc. Collect.*, **146**, 1–127.

Friedmann, H. (1967). Alloxenia in three African species of *Cuculus*. *Proc. U.S. Nat. Mus.*, **124**, 1–13.

Friedmann, H. (1968). The evolutionary history of the avian genus *Chrysococcyx*. *U.S. National Mus. Bull.*, **265**, 1–137.

Friedmann, H. and J. Kern (1956). The problem of cerophagy or wax eating in the honey guides. *Q. Rev. Biol.*, **31**, 19–30.

Friedmann, H. and L. F. Kiff (1985). The parasitic cowbirds and their hosts. *Proc. West Found. Vert. Zool.*, **2**, 225–304.

Friedmann, H., L. F. Kiff, and S. I. Rothstein (1977). A further contribution to the knowledge of the host relationships of the parasitic cowbirds. Smithson. Contrib. Zool., No. 235.

Fry, C. H., S. Keith, and E. K. Urban (1988). *The birds of Africa, Vol. III*. Academic Press, London.

Gaston, A. J. (1976). Brood parasitism by the pied crested cuckoo *Clamator jacobinus*. *J. Anim. Ecol.*, **45**, 331–348.

Gates, J. E. and N. Giffen (1991). Neotropical migrant birds and edge effects at a forest-steam ecotone. *Wilson Bull.*, **103**, 204–217.

Gates, J. E. and L. W. Gysel (1978). Avian nest dispersion and fledging success in field-forest ecotones. *Ecology*, **59**, 871–883.

Gill, B. J. (1983). Brood parasitism by the shining cuckoo *Chrysococcyx lucidus* at Kaikoura, New Zealand. *Ibis*, **125**, 40–55.

Gill, B. J., this vol. Behavior and ecology of the shining cuckoo, *Chrysococcyx lucidus*.

Gill, S. A. and S. G. Sealy (1996). Nest defence by yellow warblers: recognition of a brood parasite and an avian nest predator. *Behaviour* (in press).

Glue, D. and R. Morgan (1972). Cuckoo hosts in British habitats. *Bird Study*, **19**, 187–192.

Gochfeld, M. (1979*a*). Begging by nestling Shiny Cowbirds: adaptive or maladaptive? *Living Bird*, **17**, 41–50.

Gochfeld, M. (1979*b*). Brood parasite and host coevolution: interactions between shiny cowbirds and two species of meadowlarks. *Amer. Natur.*, **113**, 855–870.

Goguen, C. B. and N. E. Mathews (1996). Nest desertion by blue-gray gnatcatchers in association with cowbird parasitism. *Anim. Behav.*, **52**, 613–619.

Goldwasser, S., D. Gaines, and S. R. Wilbur (1980). The least Bell's vireo in California: a de facto endangered race. *Amer. Birds*, **34**, 742–745.

Gosper, D. (1964). Observations on the breeding of the Koel. *Emu*, **64**, 39–41.

Gowaty, P. A., J. H. Plissner, and T. G. Williams (1989). Behavioural correlates of uncertain parentage: mate guarding and nest guarding by eastern bluebirds, *Sialia sialis*. *Anim. Behav.*, **38**, 272–284.

Graham, D. S. (1988). Responses of five host species to cowbird parasitism. *Condor*, **99**, 588–591.

Griffith, J. T. and J. C. Griffith. Cowbird control and the endangered least Bell's vireo: a management success story. In: *Ecology and management of cowbirds* (eds. J. N. M. Smith, T. Cook, S. K. Robinson, S. I. Rothstein, and S. G. Sealy), Univ. of Texas Press, Austin, Texas (in press).

Grzybowski, J. A., R. B. Clapp, and J. T. Marshall Jr (1986). History and current population status of the black-capped vireo in Oklahoma. *American Birds*, **40**, 1151–1161.

Hahn, D. C. and J. S. Hatfield (1995). Parasitism at the landscape scale: Cowbirds prefer forests. *Conserv. Biol.*, **9**, 1415–1424.

Hamilton, W. J. III and M. E. Hamilton (1965). Breeding characteristics of yellow-billed cuckoos in Arizona. *Proc. Calif. Acad. Sci., 4th Ser.*, **32**, 405–432.

Hamilton, W. J. III and G. H. Orians (1965). Evolution of brood parasitism in altricial birds. *Condor*, **67**, 361–382.

Hahn, H. W. (1941). The cowbird at the nest. *Wilson Bull.*, **53**, 211–221.

Hardcastle, A. (1925). Young cuckoo fed by several birds. *British Birds*, **19**, 100.

Harms, K. E., L. D. Beletsky, and G. H. Orians (1991). Conspecific nest parasitism in three species of New World blackbirds. *Condor*, **93**, 967–974.

Harris, J. H. (1991). Effects of brood parasitism by brown-headed cowbirds on willow flycatcher nesting success along the Kern River, California. *West. Birds*, **22**, 13–26.

Harvey, P. H. and L. Partridge (1988). Of cuckoo clocks and cowbirds. *Nature*, **335**, 586–587.

Haverschmidt, F. (1961). Der Kuckuck *Tapera haevia* und seine Wirte in Surinam. *J. Ornith.*, **102**, 353–359.

Herrick, F. H. (1910). Life and behavior of the cuckoo. *J. Exp. Zool.*, **9**, 169–233.

Heusmann, H. W., R. Bellville, and R. Burrell (1980). Further observations on dump nesting by wood ducks. *J. Wildlf. Manag.*, **44**, 908–915.

Hickman, S. (1990). Evidence of edge species attraction to nature trails within deciduous forest. *Natural Areas Journal*, **10**, 3–5.

Higuchi, H. (1989). Responses of the bush warbler *Cettia diphone* to artificial eggs of *Cuculus canorus* in Japan. *Ibis*, **126**, 398–404.

Higuchi, H., this vol. Host use and egg color of Japanese cuckoos.

Higuchi, H. and S. Sato (1984). An example of character release in host selection and egg colour of cuckoos, *Cuculus* spp. in Japan. *Ibis*, **126**, 398–404.

Hill, D. P. and S. G. Sealy (1994). Desertion of nests parasitized by cowbirds: have clay-colored sparrows evolved an anti-parasite defence? *Anim. Behav.*, **48**, 1063–1070.

Hobson, K. A. and S. G. Sealy (1989). Responses of yellow warblers to the threat of cowbird parasitism. *Anim. Behav.*, **38**, 510–519.

Höhn, E. O. (1959). Prolactin in the cowbird's pituitary in relation to avian brood parasitism. *Nature*, **184**, 2030.

Holford, K. C. and D. D. Roby (1993). Factors limiting fecundity of captive Brown-headed Cowbirds. *Condor*, **95**, 536–545.

Hoogland, J. L. and P. W. Sherman (1976). Advantages and disadvantages of bank swallow (*Riparia riparia*) coloniality. *Ecol. Monogr.*, **46**, 33–58.

Hoover, J. P., M. C. Brittingham, and L. J. Goodrich (1995). Effects of forest patch size on nesting success of wood thrushes. *Auk*, **112**, 146–155.

Hughes, J. M. (1996). Phylogenetic analysis of the Cuculidae (Aves, Cuculiformes) using behavioral and ecological characters. *Auk*, **113**, 10–22.

Hunter, H. C. (1961). Parasitism of the masked weaver *Ploceus velatus arundinaceus*. *Ostrich*, **32**, 55–63.

Isack, H. A. and H.-U. Reyer (1989). Honeyguides and honey gatherers: interspecific communication in a symbiotic relationship. *Science*, **243**, 1343–1346.

Jackson, N. H. and D. D. Roby (1992). Fecundity and egg-laying patterns of captive yearling Brown-headed Cowbirds. *Condor*, **94**, 585–589.

Jackson, W. (1993). Causes of conspecific parasitism in the northern masked weaver. *Behav. Ecol. Sociobiol.*, **32**, 119–126.

Jackson, W. M., this vol. Egg-discrimination and egg-color variability in the northern masked weaver: the importance of conspecific versus interspecific parasitism.

James, F. C. and C. E. McCulloch (1995). The strength of inferences about causes of trends in populations. In: *Ecology and management of neotropical migratory birds* (eds. T. E. Martin and D. M. Finch), pp. 40–54. Oxford University Press, Oxford.

James, F. C., C. E. McCulloch, and D. A. Wiedenfeld (1996). New approaches to the analysis of population trends in land birds. *Ecology*, **77**, 13–27.

Jenner, E. (1788). Observations on the natural history of the cuckoo. *Phil. Trans. R. Soc. London*, **78**, 219–235.

Jensen, R. A. C. and C. F. Clinning (1974). Breeding biology of two cuckoos and their hosts in South West Africa. *Living Bird*, **13**, 5–50.

Jensen, R. A. C. and M. K. Jensen (1969). On the breeding biology of southern African cuckoos. *Ostrich*, **40**, 163–181.

Jensen, R. A. C. and C. J. Vernon (1970). On the biology of the didric cuckoo in southern Africa. *Ostrich*, **41**, 237–246.

Johnsgard, P. A. (1997). *The avian brood parasites: deception at the nest.* Oxford University Press, New York.

Jourdain, F. C. R. (1925). A study of parasitism in the cuckoos. *Proc. Zool. Soc. London*, **1925**, 639–667.

Kattan, G. H. (1993). Reproductive strategy of a generalist brood parasite, the shiny cowbird, in the Cauca Valley, Columbia. Ph.D. dissertation, University of Florida, Gainesville, FL.

Kattan, G. H. (1995). Mechanisms of short incubation period in brood-parasitic cowbirds. *Auk*, **112**, 335–342.

Kattan, G. H. (1996). Growth and provisioning of shiny cowbird and house wren host nestlings. *J. Field Ornith.*, **67**, 434–441.

Kattan, G. H. (1997). Shiny cowbirds follow the "shotgun" strategy of brood parasitism. *Anim. Behav.*, **53**, 647–654.

Kattan, G. H., this vol. Impact of brood parasitism: why do house wrens accept shiny cowbird eggs?

Kelly, C. (1987). A model to explore the rate of spread of mimicry and rejection in hypothetical populations of cuckoos and their hosts. *J. Theor. Biol.*, **125**, 283–299.

Kemal, R. E. and S. I. Rothstein (1988). Mechanisms of avian egg recognition: adaptive responses to eggs with broken shells. *Anim. Behav.*, **36**, 175–183.

King, A. P. and M. J. West (1990). Variation in species-typical behavior: a contemporary issue for comparative psychology. In: *Contemporary issues for comparative psychology* (ed. D. A. Dewsbury), Sinauer Assoc., Sunderland, Mass.

King, A. P., M. J. West, and D. H. Eastzer (1980). Song structure and song development as potential contributors to reproductive isolation in cowbirds (*Molothrus ater*). *J. Comp. Physiol. Psychol.*, **94**, 1028–1039.

Klaas, E. E. (1975). Cowbird parasitism and nesting success in the eastern phoebe. *Occas. Papers Mus. Nat. Hist., Univ. Kansas*, **41**, 1–18.

Klein, N. K. and R. B. Payne. Evolutionary associations of brood parasitic finches (*Vidua*) and their host species: analyses of mitochondrial restriction sites. *Evolution* (in press).

Klein, N. K. and K. V. Rosenberg (1986). Feeding of brown-headed cowbird (*Molothrus ater*) fledglings by more than one "host" species. *Auk*, **103**, 213–214.

Klein, N. K., R. B. Payne, and M. E. D. Nhlane (1993). A molecular genetic perspective on speciation in the brood parasitic *Vidua* finches. Proc. VIII Pan-Afr. Orn. Congr., pp. 29–39.

Kroodsma, D. E. (1982). Learning and the ontogeny of sound signals in birds. In: *Acoustic communication in birds, Vol. 2* (eds. D. E. Kroodsma and E. H. Miller), pp. 1–23. Academic Press, New York.

Kroodsma, D. E. and E. H. Miller (1996). *Ecology and evolution of acoustic communication in birds.* Cornell University Press, Ithaca, New York.

Lack, D. (1963). Cuckoo hosts in England. With an appendix on the cuckoo hosts in Japan, by T. Royama. *Bird Study*, **10**, 185–203.

Lack, D. (1968). *Ecological adaptations for breeding in birds.* Methuen, London.

Lack, D. (1971). *Ecological isolation in birds.* Harvard Univ. Press, Cambridge, Mass.

Lamba, B. S. (1975). The Indian crows: a contribution to their breeding biology, with notes on brood parasitism on them by the Indian koel. *Record. Zool. Surv. India*, **71**, 183–300.

Lank, D. B., E. Cooch, R. F. Rockwell, and F. Cooke, (1989*a*). Environmental and demographic correlates of intraspecific nest parasitism in lesser snow geese, *Chen c. caerulescens*. *J. Anim. Ecol.*, **58**, 29–45.

Lank, D. B., P. Mineau, R. F. Rockwell, and F. Cooke (1989*b*). Intraspecific nest parasitism and extra-pair copulation in lesser snow geese. *Anim. Behav.*, **37**, 74–89.

Lanyon, S. M. (1992). Interspecific brood parasitism in blackbirds (Icterinae): a phylogenetic perspective. *Science*, **255**, 77–79.

Lawes, M. J. and S. Kirkman (1996). Egg recognition and interspecific brood parasitism rates in red bishops (Aves: Ploceidae). *Anim. Behav.*, **52**, 553–563.

Line, L. (1993). Silence of the songbirds. *National Geographic*, **183**, 68–90.

Line, L. (1995). Songbird population losses tied to fragmentation of forest habitat. New York Times "Science Times" April 4, 1995.

Liversidge, R. (1955). Observations on a Piet-My-Vrou (*Cuculus solitarius*) and its host the cape robin (*Cossypha caffra*). *Ostrich*, **26**, 18–27.

Liversidge, R. (1971). The biology of the jacobin cuckoo. *Clamator jacobinus*. Proc. 3rd Pan-Afr. Ornithol. Congr. *Ostrich* (Suppl.), **8**, 117–137.

Lombardo, M. P., H. W. Power, P. C. Stouffer, L. C. Romagnano, and A. S. Hoffenberg (1989). Egg removal and intraspecific brood parasitism in the European starling (*Sturnus vulgaris*). *Behav. Ecol. Sociobiol.*, **24**, 217–223.

Lotem, A. (1993). Learning to recognize nestlings is maladaptive for cuckoo *Cuculus canorus* hosts. *Nature*, **362**, 743–745.

Lotem, A. and H. Nakamura, this vol. Evolutionary equilibria in avian brood parasitism: an alternative to the "arms race-evolutionary lag" concept.

Lotem, A. and S. I. Rothstein (1995). Cuckoo-host coevolution: from snapshots of an arms race to the documentation of microevolution. *Trends Ecol. Evol.*, **10**, 436–437.

Lotem, A., N. Nakamura, and A. Zahavi (1992). Rejection of cuckoo eggs in relation to host age: a possible evolutionary equilibrium. *Behav. Ecol.*, **3**, 128–132.

Lotem, A., H. Nakamura, and A. Zahavi (1995).

Constraints on egg discrimination and cuckoo-host co-evolution. *Anim. Behav.*, **49**, 1185–1209.

Lowther, P. E. (1993). Brown-headed Cowbird (*Molothrus ater*). In: *The birds of North America, No. 47* (eds. A. Poole and F. Gill), The Academy of Natural Sciences, Philadelphia; The American Orthithologists' Union, Washington, D.C.

Lowther, P. E. (1995). Bronzed Cowbird (*Molothrus aeneus*). In: *The birds of North America, No. 144* (eds. A. Poole and F. Gill), The Academy of Natural Sciences, Philadelphia; The American Ornithologists' Union, Washington, D.C.

Lyon, B. E. (1991). Brood parasitism in American coots: avoiding the constraints of parental care. *Acta XX Congressus Internationalis Ornithologici*, pp. 1023–1031.

Lyon, B. E. and J. M. Eadie (1991). Mode of development and interspecific brood parasitism. *Behav. Ecol.*, **2**, 309–318.

Lyon, B. E., L. D. Hamilton, and M. Magrath (1992). The frequency of conspecific brood parasitism and the pattern of laying determinancy in yellow-headed blackbirds. *Condor*, **94**, 590–597.

Macedo, R. H. (1994). Inequities in parental effort and costs of communal breeding in the guira cuckoo. *Ornith. Neotrop.*, **5**, 79–90.

MacWhirter, R. B. (1989). On the rarity of intraspecific brood parasitism. *Condor*, **91**, 485–492.

Madge, S. and H. Burn (1988). *Wildfowl*. Christopher Helm, Bromley, UK.

Marchant, S. (1972). Evolution of the genus *Chrysococcyx*. *Ibis*, **114**, 219–233.

Marchetti, K. (1992). Costs of host defence and the persistence of parasitic cuckoos. *Proc. R. Soc. Lond. Ser. B*, **248**, 41–45.

Mark, D. and B. J. Stutchbury (1994). Response of a forest-interior songbird to the threat of parasitism. *Anim. Behav.*, **47**, 275–280.

Martin, T. E. (1992). Breeding productivity considerations:: What are the appropriate habitat features for management? In: *Ecology and conservation of neotropical migrant landbirds* (eds. J. M. Hagan and D. W. Johnston), pp. 455–493. Smithsonian Inst. Press, Washington, D.C.

Mason, P. (1986*a*). Brood parasitism in a host generalist, the shiny cowbird: I. The quality of different species as hosts. *Auk*, **103**, 52–60.

Mason, P. (1986*b*). Brood parasitism in a host generalist, the shiny cowbird: II. Host selection. *Auk*, **103**, 61–69.

Mason, P. (1987). Pair formation in cowbirds: evidence found for screaming but not shiny cowbirds. *Condor*, **89**, 349–356.

Mason, P. and S. I. Rothstein (1986). Coevolution and avian brood parasitism: Cowbird eggs show evolutionary response to host discrimination. *Evolution*, **40**, 1207–1214.

Mason, P. and S. I. Rothstein (1987). Crypsis versus mimicry and the color of shiny cowbird eggs. *Amer. Natur.*, **130**, 161–167.

May, R. M. and S. K. Robinson (1985). Population dynamics of avian brood parasitism. *Amer. Natur.*, **126**, 475–494.

May, R. M., S. Nee, and C. Watts (1991). Could intraspecific brood parasitism cause population cycles? Proc. XX International Ornithol. Congr., pp. 1012–1022.

Mayfield, H. (1961). Vestiges of propriety interest in nests by the brown-headed cowbird parasitizing the Kirtland's Warbler. *Auk*, **78**, 162–167.

Mayfield, H. F. (1965). The brown-headed cowbird, with old and new hosts. *Living Bird*, **4**, 13–28.

Mayfield, H. F. (1977). Brown-headed cowbird: agent of extermination? *American Birds*, **31**, 107–113.

McGeen, D. S. (1972). Cowbird-host relationships. *Auk*, **89**, 360–380.

McLean, I. (1987). Response to a dangerous enemy: should a brood parasite be mobbed? *Ethology*, **75**, 235–245.

McLean, I. G. and R. F. Maloney, this vol. Brood parasitism, recognition and response: the options.

McLean, I. G. and G. Rhodes (1991). Enemy recognition and response in birds. *Curr. Ornith.*, **8**, 173–211.

McLean, I. G. and J. R. Waas (1987). Do cuckoo chicks mimic the begging calls of their hosts? *Anim. Behav.*, **35**, 1896–1907.

Mermoz, M. E. and J. C. Reboreda (1994). Brood parasitism of the shiny cowbird, *Molothrus bonariensis*, on the brown-and-yellow marshbird, *Pseudoleistes virescens*. *Condor*, **96**, 716–721.

Mermoz, M. E. and J. C. Reboreda (1996). New effective host for the screaming cowbird. *Condor*, **98**, 630–632.

Middleton, A. L. (1977). Effect of cowbird parasitism on American goldfinch nesting. *Auk*, **94**, 304–307.

Moksnes, A. and E. Røskaft (1988). Responses of fieldfares, *Turdus pilaris* and bramblings *Fringilla montifringilla* to experimental parasitism by the cuckoo *Cuculus canorus*. *Ibis*, **130**, 535–539.

Moksnes, A. and E. Røskaft (1989). Adaptations of meadow pipits to parasitism by the common cuckoo. *Behav. Ecol. Sociobiol.*, **24**, 25–30.

Moksnes, A. and E. Røskaft (1992). Responses of some rare cuckoo hosts to mimetic model cuckoo eggs and to foreign conspecific eggs. *Ornis Scand.*, **23**, 17–23.

Moksnes, A., E. Røskaft, and L. Korsnes (1993).

Rejection of cuckoo (*Cuculus canorus*) eggs by meadow pipits (*Anthus pratensis*). *Behav. Ecol.*, **4**, 120–127.

Møller, A. P. (1987). Intraspecific nest parasitism and anti-parasite behaviour in swallows, *Hirundo rustica. Anim. Behav.*, **35**, 247–254.

Møller, A. P., this vol. Density-dependent intraspecific nest parasitism and anti-parasite behavior in the barn swallow *Hirundo rustica*.

Møller, A. P. and M. Petrie (1990). Evolution of intraspecific variability in birds' eggs: is intraspecific nest parasitism the selective agent? *Acta XX Congressus Internationalis Ornithologici*, **II**, 1041–1047.

Molnar, B. (1944). The cuckoo in the Hungarian plain. *Aquila*, **51**, 100–112.

Morel, M.-Y. (1973). Contribution à l'étude dynamique de la population de *Lagonosticta senegala* L. (estrildides) à Richard-Toll (Sénégal). Interrelations avec le parasite *Hypochera chalybeata* (Müller) (viduines). *Mem. Mus. Nat. d'Hist. Nat. Ser. A. Zool.*, **78**, 1–156.

Morgensen, J. (1927). Nota sobre el parasitismo del "Crespin" (*Tapera naevia*). *Hornero*, **4**, 68–70.

Morton, E. S. and S. M. Farabaugh (1979). Infanticide and other adaptations of the nestling striped cuckoo. *Ibis*, **121**, 212–213.

Mundy, P. J. (1973). Vocal mimicry of their hosts by nestlings of the great spotted cuckoo and striped crested cuckoo. *Ibis*, **115**, 602–604.

Nakamura, H. (1990). Brood parasitism of the cuckoo *Cuculus canorus* in Japan and the start of new parasitism on the azure-winged magpie *Cyanopica cyana*. *Jap. J. Ornith.*, **39**, 1–18.

Nakamura, H., S. Kubota and R. Suzuki, this vol. Coevolution between the common cuckoo and its major hosts in Japan: stable versus dynamic specialization on hosts.

Nee, S. and R. M. May (1993). Population-level consequences of conspecific brood parasitism in birds and insects. *J. Theoret. Biol.*, **161**, 95–109.

Neudorf, D. L. and S. G. Sealy (1992). Reactions of four passerine species to threats of predation and cowbird parasitism: enemy recognition or generalized response? *Behaviour*, **123**, 84–105.

Neudorf, D. L. and S. G. Sealy (1994). Sunrise nest attentiveness in cowbird hosts. *Condor*, **96**, 162–169.

Neufeldt, I. (1966). Life history of the Indian cuckoo, *Cuculus micropterus micropterus* Gould, in the Soviet Union. *J. Bombay Nat. Hist. Soc.*, **63**, 399–419.

Nicolai, J. (1964). Der brutparasitismus der viduinae als ethologisches Problem. *Tierpsychol.*, **21**, 127–204.

Nicolai, J. (1974). Mimicry in parasitic birds. *Sci. Am.*, **231**, 92–98.

Nolan, V., Jr (1978). The ecology and behavior of the prairie warbler, *Dendroica discolor. Ornith. Monogr.*, *26*, 1–595.

Nolan, V., Jr and C. F. Thompson (1975). The occurrence and significance of anomalous reproductive activities in two North American non-parasitic cuckoos. *Ibis*, **117**, 496–503.

Norman, R. F. and R. J. Robertson (1975). Nest-searching behavior in the brown-headed cowbird. *Auk*, **92**, 610–611.

Nur, N. (1984). The consequences of brood size for breeding bluetits I. Adult survival, weight change, and the cost of reproduction. *J. Anim. Ecol.*, **53**, 479–496.

O'Loghlen, A. L. (1995). Delayed access to local songs prolongs vocal development in dialect populations of brown-headed cowbirds. *Condor*, **97**, 402–414.

O'Loghlen, A. L. and S. I. Rothstein (1993). An extreme example of delayed vocal development: song learning in a population of wild Brown-headed Cowbirds. *Anim. Behav.*, **46**, 293–304.

O'Loghlen, A. L. and S. I. Rothstein (1995). Culturally correct song dialects are correlated with male age and female song preferences in wild populations of Brown-headed Cowbirds. *Behav. Ecol. Sociobiol.*, **36**, 251–259.

Øien, I. J., A. Moksnes, and E. Røskaft (1995). Evolution of variation in egg color and marking pattern in European passerines: adaptations in a coevolutionary arms race with the cuckoo, *Cuculus canorus*. *Behav. Ecol.*, **6**, 166–174.

Øien, I. J., M. Honza, A. Moksnes, and E. Røskaft (1996). The risk of parasitism in relation to the distance from reed warbler nests to cuckoo perches. *J. Anim. Ecol.*, **65**, 147–153.

Orians, G. H. (1985). *Blackbirds of the Americas*. Univ. of Washington Press, Seattle.

Orians, G. H., E. Røskaft, and L. D. Beletsky (1989). Do Brown-headed Cowbirds lay their eggs at random in the nests of red-winged blackbirds? *Wilson Bull.*, **101**, 599–605.

Ortega, C. P. and A. Cruz (1992). Differential growth patterns of nestling Brown-headed Cowbirds and yellow-headed blackbirds. *Auk*, **109**, 368–376.

Owen, J. H. (1933). The cuckoo in the Felsted district. *Rep. Felsted School Sci. Soc.*, **33**, 25–39.

Payne, R. B. (1967). Interspecific communication signals in parasitic birds. *Amer. Natur.*, **101**, 363–375.

Payne, R. B. (1969). Nest parasitism and display of chestnut sparrows in a colony of grey-capped social weavers. *Ibis*, **111**, 300–307.

Payne, R. B. (1973). Behavior, mimetic songs, and song dialects, and relationships of the parasitic indigobirds (*Vidua*) of Africa. *Ornith. Monogr.*, No. 11.

Payne, R. B. (1973*a*). Vocal mimicry of the paradise whydahs (*Vidua*) and response of female whydahs to the songs of their hosts and their mimics. *Anim. Behav.*, **21**, 762–771.

Payne, R. B. (1973*b*). The breeding season of a parasitic bird, the brown-headed cowbird in central California. *Condor*, **75**, 80–99.

Payne, R. B. (1974). The evolution of clutch size and reproductive rates in parasitic cuckoos. *Evolution*, **28**, 169–181.

Payne, R. B. (1976). Song mimicry and species relationships among the West African pale-winged indigobirds. *Auk*, **93**, 25–38.

Payne, R. B. (1977*a*). The ecology of brood parasitism in birds. *Annu. Rev. Ecol. Syst.*, **8**, 1–28.

Payne, R. B. (1977*b*). Clutch size, egg size, and the consequences of single vs. multiple parasitism in parasitic finches. *Ecology*, **58**, 500–513.

Payne, R. B. (1997*a*). Avian brood parasitism. In: *Host–parasite evolution: general principles and avian models* (eds. D. H. Clayton and J. Moore), pp. 338–369. Oxford University Press, Oxford.

Payne, R. B. (1997*b*). Field identification of the brood-parasitic whydahys *Vidua* and cuckoo finch *Anomalospiza*. *Bull. African Bird Club*, **4**, 18–28.

Payne, R. B. (1982). *Species limits in the indiobirds (Ploceidae, Vidua) of West Africa: mouth mimicry, song mimicry, and description of new species.* Misc. Publ. Mus. Zool. Univ. of Michigan.

Payne, R. B. (1985). The species of parasitic finches in West Africa. *Malimbus*, **7**, 103–113.

Payne, R. B. (1989). Egg size of African honeyguides (Indicatoridae): specialization for brood parasitism? *Tauraco*, **1**, 201–210.

Payne, R. B. (1992). Clutch size, laying periodicity and behaviour in the honeyguides *Indicator indicator* and *I. minor*. Proc. VII Pan-African Ornithological Cong., pp. 537–547.

Payne, R. B. and K. Payne (1967). Cuckoo hosts in southern Africa. *Ostrich*, **38**, 135–143.

Payne, R. B. and K. Payne (1977). Social organization and mating success in local song populations of village indigobirds, *Vidua chalybeata*. *Z. Tierpsych.*, **45**, 113–173.

Payne, R. B. and L. L. Payne, this vol. Nestling eviction and vocal behavior in the Australian glossy cuckoos *Chrysococcyx basalis* and *C. lucidus*.

Payne, R. B., L. L. Payne, and I. Rowley (1985). Splendid wren (*Malurus splendens*) response to cuckoos: an experimental test of social organization in a communal bird. *Behaviour*, **94**, 108–127.

Payne, R. B., L. L. Payne, and J. L. Woods. Song learning in brood parasitic indigobirds *Vidua chalybeata*: song mimicry of their host species. *Anim. Behav.* (in press).

Pease, C. M. and J. A. Grzybowski (1995). Assessing the consequences of brood parasitism and nest predation on seasonal fecundity in passerine birds. *Auk*, **112**, 343–363.

Peck, G. K. and R. D. James (1987). *Breeding birds of Ontario: nidiology and distribution, Volume 2: Passerines*. Royal Ontario Museum, Toronto, Canada.

Peterjohn, B. and J. R. Sauer. Temporal and geographic patterns in population trends of Brown-headed Cowbirds. In: *The ecology and management of cowbirds* (eds. J. N. M. Smith, T. L. Cook, S. K. Robinson, S. I. Rothstein and S. G. Sealy), Univ. of Texas Press, Austin, Texas (in press).

Peterjohn, B. G., J. R. Sauer, and C. S. Robbins (1995). Population trends from the North American Breeding Bird Survey. In: *Ecology and management of neotropical migratory birds* (eds. T. E. Martin and D. M. Finch), pp. 3–39. Oxford University Press, Oxford.

Petit, L. J. (1991). Adaptive tolerance of cowbird parasitism by prothonotary warblers: a consequence of nest-site limitation? *Anim. Behav.*, **41**, 425–432.

Petrie, M. and A. P. Møller (1991). Laying eggs in others' nests: intraspecific brood parasitism in birds. *Trends Ecol. Evol.*, **6**, 315–320.

Phillips, W. W. A. (1948). Cuckoo problems of Ceylon. *Spolia Zeylanica*, **25**, 45–60.

Picman, J. (1989). Mechanisms of increased puncture resistance of eggs of brown-headed cowbirds. *Auk*, **106**, 577–583.

Piper, W. H. (1994). Courtship, copulation, nesting behavior and brood parasitism in the Venezuelan stripe-backed wren. *Condor*, **96**, 654–671.

Poiani, A. and M. A. Elgar (1994). Cooperative breeding in the Australian avifauna and brood parasitism by cuckoos (*Cuculidae*). *Anim. Behav.*, **47**, 697–706.

Post, W. and J. W. Wiley (1976). The Yellow-shouldered Blackbird present and future. *Amer. Birds*, **30**, 13–20.

Post, W., T. K. Nakamura, and A. Cruz (1990). Patterns of Shiny Cowbird parasitism in St Lucia and southwestern Puerto Rico. *Condor*, **92**, 461–492.

Potter, E. F. (1980). Notes on nesting yellow-billed cuckoos. *J. Field Ornith.*, **51**, 17–29.

Power, H. W., this vol. Quality control and the important questions in avian conspecific parasitism.

Power, H. W., E. D. Kennedy, L. C. Romagnano, M. P. Lombardo, A. S. Hoffenberg (1989). The parasitism assurance hypothesis: why starlings leave space for parasitic eggs. *Condor*, **91**, 753–765.

Preble, N. A. (1957). Nesting habits of the yellow-billed cuckoo. *Amer. Midl. Natur.*, **57**, 474–482.

Prescott, K. (1947). Unusual behavior of a cowbird and a scarlet tanager at a red-eyed vireo nest. *Wilson Bull.*, **59**, 210.

Pulliam, H. R. (1988). Sources, sinks, and population regulation. *Amer. Natur.*, **137**, 550–566.

Rahn, H., L. Curran-Everett, and D. T. Booth (1988). Eggshell differences between parasitic and non-parasitic Icteridae. *Condor*, **90**, 962–964.

Ralph, C. P. (1975). Life style of *Coccyzus pumilus*, a tropical cuckoo. *Condor*, **77**, 73–83.

Ranger, G. A. (1955). On three species of honey-guide. The greater (*Indicator indicator*), the lesser (*I. minor*) and the scaly-throated (*I. variegatus*). *Ostrich*, **36**, 70–87.

Redondo, T. (1993). Exploitation of host mechanisms for parental care by avian brood parasites. *Etología*, **3**, 235–297.

Redondo, T. and L. Arias-de-Reyna (1988). Vocal mimicry of hosts by great spotted cuckoo *Clamator glandarius*: further evidence. *Ibis*, **130**, 540–544.

Reed, R. A. (1968). Studies of the diederik cuckoo *Chrysococcyx caprius* in the Transvaal. *Ibis*, **110**, 321–331.

Rice, J. (1974). Social and competitive interactions between two species of vireos (Aves: Vireonidae). Ph.D. dissertation, University of Toronto, Canada.

Rich, A. C., D. S. Dobkin, and L. J. Niles (1994). Defining forest fragmentation by corridor width: The influence of narrow forest-dividing corridors on forest-nesting birds in southern New Jersey. *Conserv. Biol.*, **8**, 1109–1121.

Ricklefs, R. E. (1969). An analysis of nesting mortality in birds. *Smithson. Contrib. Zool.*, **9**, 1–48.

Robbins, M. B. and D. A. Easterla (1981). Range expansion of the bronzed cowbird with the first Missouri record. *Condor*, **83**, 270–272.

Robertson, R. J. and R. F. Norman (1976). Behavioral defenses to brood-parasitism by potential hosts of the brown-headed cowbird. *Condor*, **78**, 166–173.

Robertson, R. J. and R. F. Norman (1977). The function and evolution of aggressive host behavior towards the brown-headed cowbird (*Molothrus ater*). *Can. J. Zool.*, **55**, 508–518.

Robinson, S. K. (1986). The evolution of social behavior and mating systems in the blackbirds (Icterinae). In: *Ecological aspects of social evolution* (eds. D. I. Rubenstein and R. W. Wrangham), pp. 175–200. Princeton Univ. Press, Princeton, N.J.

Robinson, S. K. (1988). Foraging ecology and host relationships of Giant Cowbirds in southeastern Peru. *Wilson Bull.*, **100**, 224–235.

Robinson, S. K. (1992). Population dynamics of breeding neotropical migrants in a fragmented landscape. In: *Ecology and conservation of neotropical migrant landbirds* (eds. J. M. Hagan and D. W. Johnston), pp. 408–418. Smithsonian Institution Press, Washington, D.C.

Robinson, S. K. (1993). Conservation problems of neotropical migrant land birds. *Trans. N. Amer. Wildlf. Natur. Res. Conf.*, **58**, 379–389.

Robinson, S. K., J. A. Grzybowski, Jr, S. I. Rothstein, L. J. Petit, M. C. Brittingham, and F. R. Thompson, III (1993). Management implications of cowbird parasitism on neotropical migratory birds. In: *Status and management of neotropical migratory birds* (eds. D. M. Finch and P. W. Stangel), pp. 93–102. USDA-USFS Gen. Tech. Rep. RM-229, Fort Collins, Colorado.

Robinson, S. K., J. P. Hoover, J. Herkert, and R. Jack (1996). Cowbird parasitism in fragmented landscapes: effects of tract size, habitat, and abundance of hosts. In: *The ecology and management of cowbirds* (eds. J. N. M. Smith, T. L. Cook, S. K. Robinson, S. I. Rothstein, and S. G. Sealy), Univ. of Texas Press, Austin, Texas.

Robinson, S. K., F. R. Thompson III, T. M. Donovan, D. R. Whitehead, and J. Faaborg (1995*a*). Regional forest fragmentation and the nesting success of migratory birds. *Science*, **267**, 1987–1990.

Robinson, S. K., S. I. Rothstein, M. C. Brittingham, L. J. Petit, and J. A. Grzybowski (1995*b*). Ecology and behavior of cowbirds and their impact on host populations. In: *Ecology and management of neotropical migratory birds* (eds. T. E. Martin and D. M. Finch), pp. 428–460. Oxford University Press, New York.

Rohwer, F. C. and S. Freeman (1989). The distribution of conspecific nest parasitism in birds. *Can. J. Zool.*, **67**, 239–253.

Rohwer, S. and C. D. Spaw (1988). Evolutionary lag versus bill-size constraints: a comparative study of the acceptance of cowbird eggs by old hosts. *Evol. Ecol.*, **2**, 27–36.

Rohwer, S., C. D. Spaw, and E. Røskaft (1989). Costs to northern orioles of puncture-ejecting parasitic cowbird eggs from their nests. *Auk*, **106**, 734–738.

Røskaft, E. and A. Moksnes, this vol. Coevolution between brood parasites and their hosts: an optimality theory approach.

Røskaft, E., G. H. Orians, and L. D. Beletsky (1990). Why do Red-winged Blackbirds accept eggs of Brown-headed Cowbirds? *Evol. Ecol.*, **4**, 35–42.

Røskaft, E., S. Rohwer, and C. D. Spaw (1993). Cost of puncture ejection compared with costs of rearing cowbird chicks for Northern Orioles. *Ornis Scandinivica*, **24**, 28–32.

Roth, R. R. and R. K. Johnson (1993). Long-term dynamics of a wood thrush population breeding in a forest fragment. *Auk*, **110**, 37–48.

Rothstein, S. I. (1974). Mechanisms of avian egg recognition: possible learned and innate factors. *Auk*, **91**, 796–807.

Rothstein, S. I. (1975*a*). An experimental and teleonomic investigation of avian brood parasitism. *Condor*, **77**, 250–271.

Rothstein, S. I. (1975*b*). Evolutionary rates and host defenses against avian brood parasitism. *Amer. Natur.*, **109**, 161–176.

Rothstein, S. I. (1976*a*). Cowbird parasitism of the cedar waxwing and its evolutionary implications. *Auk*, **93**, 498–509.

Rothstein, S. I. (1976*b*). Experiments on defenses cedar waxwings use against cowbird parasites. *Auk*, **93**, 675–691.

Rothstein, S. I. (1977). Cowbird parasitism and egg recognition of the northern oriole. *Wilson Bull.*, **89**, 21–32.

Rothstein, S. I. (1978*a*). Mechanisms of avian egg-recognition: additional evidence for learned components. *Anim. Behav.*, **26**, 671–677.

Rothstein, S. I. (1978*b*). Geographical variation in the nestling colorations of parasitic cowbirds. *Auk*, **95**, 152–160.

Rothstein, S. I. (1982*a*). Successes and failures in avian egg and nestling recognition with comments on the utility of optimality reasoning. *Amer. Zool.*, **22**, 547–560.

Rothstein, S. I. (1982*b*). Mechanisms of avian egg recognition: which egg parameters elicit responses by rejector species? *Behav. Ecol. Sociobiol.*, **11**, 229–239.

Rothstein, S. I. (1986). A test of optimality: egg recognition in the eastern phoebe. *Anim. Behav.*, **34**, 1109–1119.

Rothstein, S. I. (1990*a*). A model system for coevolution: avian brood parasitism. *Annu. Rev. Ecol. Syst.*, **21**, 481–508.

Rothstein, S. I. (1990*b*). Brood parasitism and clutch-size determination in birds. *Trends Ecol. Evol.*, **5**, 101–102.

Rothstein, S. I. (1992). Brood parasitism, the importance of experiments and host defences of avifaunas on different continents. Proc. VII Pan-Afr. Ornithol. Congr., pp. 521–535.

Rothstein, S. I. (1993). An experimental test of the Hamilton–Orians hypothesis for the origin of avian brood parasitism. *Condor*, **95**, 1000–1005.

Rothstein, S. I. (1994). The cowbird's invasion of the Far West: History, causes and consequences experienced by host species. In: *Studies in avian biology, No. 15. A century of avifaunal change in western North America* (eds. J. R. Jehl, Jr, and N. K. Johnson), pp. 301–315. Cooper Ornithological Society.

Rothstein, S. I. and T. Cook. Overview and synthesis: Cowbird management, host population regulation and efforts to save endangered species. In: *Ecology and management of cowbirds* (eds. J. N. M. Smith, T. Cook, S. K. Robinson, S. I. Rothstein, and S. G. Sealy), Univ. of Texas Press, Austin, Texas (in press).

Rothstein, S. I. and R. C. Fleischer (1987*a*). Vocal dialects and their possible relation to honest status signalling in the brown-headed cowbird. *Condor*, **89**, 1–23.

Rothstein, S. I. and R. C. Fleischer (1987*b*). Brown-headed cowbirds learn flight whistles after the juvenile period. *Auk*, **104**, 512–516.

Rothstein, S. I. and S. K. Robinson (1994). Conservation and coevolutionary implications of brood parasitism by cowbirds. *Trends Ecol. Evol.*, **9**, 162–164.

Rothstein, S. I., J. Verner, and E. Stevens (1980). Range expansion and diurnal changes in dispersion of the brown-headed cowbird in the Sierra Nevada. *Auk*, **97**, 253–267.

Rothstein, S. I., J. Verner, and E. Stevens (1984). Radio-tracing confirms a unique diurnal pattern of spatial occurrence in the brown-headed cowbird. *Ecology*, **65**, 77–88.

Rothstein, S. I., D. A. Yokel, and R. C. Fleischer (1986). Mating and spacing systems, female fecundity and vocal dialects in captive and free-ranging brown-headed cowbirds. *Curr. Ornith.*, **3**, 127–185.

Rothstein, S. I., J. Verner, E. Stevens, and L. V. Ritter (1987). Behavioral differences among sex and age classes of the brown-headed cowbird and their relation to the efficacy of a control program. *Wilson Bull.*, **99**, 322–337.

Rothstein, S. I., D. A. Yokel, and R. C. Fleischer (1988). The agonistic and sexual functions of vocalizations of male brown-headed cowbirds, *Molothrus ater. Anim. Behav.*, **36**, 73–86.

Rothstein, S. I., J. Verner, and C. Farmer. Cowbird vocalizations: an overview and the use of playbacks to enhance cowbird detectability. In: *Ecology and management of cowbirds* (eds. J. N. M. Smith, T. L. Cook, S. K. Robinson, S. I. Rothstein, and S. G. Sealy), Univ. of Texas Press, Austin, Texas (in press).

Rowan, M. K. (1983). *The doves, parrots, louries and cuckoos of southern Africa.* David Philip, Cape Town.

Salvador, G. A. (1982). Estudio de parasitismo del Crespin, *Tapera naevia. Hist. Natur.*, **2**, 65–70.

Sato, T. (1986). A brood parasitic catfish of mouthbreeding cichlid fishes in Lake Tanganyika. *Nature*, **323**, 58–59.

Schram, B. (1994). An open solicitation for cowbird recipes. *Birding*, August, 254–256.

Scott, D. M. (1977). Cowbird parasitism on the gray catbird at London, Ontario. *Auk*, **94**, 18–27.

Scott, D. M. (1991). The time of day of egg laying by the brown-headed cowbird and other Icterines. *Can. J. Zool.*, **69**, 2093–2099.

Scott, D. M. and C. D. Ankney (1983). The laying cycle of brown-headed cowbirds: passerine chickens? *Auk*, **100**, 583–593.

Scott, D. M. and R. E. Lemon (1996). Differential reproductive success of brown-headed cowbirds with northern cardinals and three other hosts. *Condor*, **98**, 259–271.

Scott, D. M., P. J. Weatherhead, and C. D. Ankney (1992). Egg-eating by female Brown-headed Cowbirds. *Condor*, **94**, 579–584.

Scott, P. E. and B. R. McKinney (1994). Brown-headed cowbird removes Blue-grey Gnatcatcher nestlings. *J. Field Ornith.*, **65**, 363–364.

Sealy, S. G. (1992). Removal of yellow warbler eggs in association with cowbird parasitism. *Condor*, **94**, 40–54.

Sealy, S. G. (1995). Burial of cowbird eggs by parasitized yellow warblers: an empirical and experimental study. *Anim. Behav.*, **49**, 877–889.

Sealy, S. G. (1996). Evolution of host defenses against brood parasitism: implications of puncture-ejection by a small passerine. *Auk*, **113**, 346–355.

Sealy, S. G. and D. L. Neudorf (1995). Male northern orioles eject cowbird eggs: implications for the evolution of rejection behavior. *Condor*, **97**, 369–373.

Sealy, S. G., E. A. Hobson, and J. V. Briskie (1989). Responses of yellow warblers to experimental intraspecific brood parasitism. *J. Field Ornith.*, **60**, 224–229.

Sealy, S. G., D. L. Neudorf, and D. P. Hill (1995). Rapid laying in brown-headed cowbirds *Molothrus ater* and other parasitic birds. *Ibis*, **137**, 76–84.

Sealy, S. G., D. L. Neudorf, K. A. Hobson, and S. A. Gill, this vol. Nest defense by potential hosts of the brown-headed cowbird: methodological approaches, benefits of defense and coevolution.

Sedgewick, J. A. and F. L. Knopf (1988). A high incidence of brown-headed cowbird parasitism of willow flycatchers. *Condor*, **90**, 253–256.

Semel, B. and P. W. Sherman (1986). Dynamics of nest parasitism in wood ducks. *Auk*, **103**, 813–816.

Semel, B., P. W. Sherman, and S. M. Byers (1988). Effects of brood parasitism and nest box placement on wood duck breeding biology. *Condor*, **90**, 920–930.

Seppä, J. (1969). The cuckoo's ability to find a nest where it can lay an egg. *Ornis Fennica*, **46**, 78–79.

Shaw, P. (1984). The social behavior of the pin-tailed whydah *Vidua macroura* in northern Ghana. *Ibis*, **126**, 463–473.

Sherry, T. W. and R. T. Holmes (1995). Summer versus winter limitation of populations: What are the issues and what is the evidence? In: *Ecology and management of neotropical migratory birds* (eds. T. E. Martin and D. M. Finch), pp. 85–120. Oxford University Press, Oxford.

Short, L. L. and J. F. M. Horne (1985). Behavioral notes on the nest-parasitic Afrotropical honeyguides (Aves: Indicatoridae). *Am. Mus. Novitates*, **2825**, 1–46.

Short, L. L. and J. F. M. Horne (1992). Honeyguide-host interactions. Proc. VII Pan-Afr. Ornith. Congr., pp. 549–552.

Sibley, C. G. and J. E. Ahlquist (1990). *Phylogeny and classification of birds: A study in molecular evolution.* Yale Univ. Press, New Haven, CT.

Sick, H. (1985). *Ornithologia brasileira, uma introducao.* Universidade de Brasilia, Brasilia, Brazil.

Sick, H. (1993). *Birds in Brazil.* Princeton Univ. Press, Princeton, N.J.

Skutch, A. F. (1961). Helpers among birds. *Condor*, **63**, 198–226.

Slack, R. D. (1976). Nest guarding behavior by male gray catbirds. *Auk*, **93**, 292–300.

Smith, J. N. M. (1981). Cowbird parasitism, host fitness, and age of the host female in an island song sparrow population. *Condor*, **83**, 152–161.

Smith, J. N. M. (1994). Cowbirds, conservation villains or convenient scapegoats? *Birding*, **August 1994**, 257–259.

Smith, J. N. M. and P. Arcese (1994). Brown-headed Cowbirds and an island population of song sparrows: a 16-year study. *Condor*, **96**, 916–934.

Smith, J. N. M. and I. H. Myers-Smith, this vol. Spatial variation in parasitism of song sparrows by brown-headed cowbirds.

Smith, J. N. M., P. Arcese, and I. G. McLean (1984).

Age, experience, and enemy recognition by wild song sparrows. *Behav. Ecol. Sociobiol.*, **14**, 101–106.

Smith, J. N. M., T. Cook, S. K. Robinson, S. I. Rothstein, and S. G. Sealy (eds.). *Ecology and management of cowbirds.* Univ. of Texas Press, Austin, Texas (in press).

Smith, N. G. (1968). The advantage of being parasitized. *Nature*, **219**, 690–694.

Smith, N. G. (1979). Alternate responses by hosts to parasites which may be helpful or harmful. In: *Host–parasite interfaces* (ed. B. B. Nickel), pp. 7–15. Academic Press, N.Y.

Smith, N. G. (1980). Some evolutionary, ecological, and behavioral correlates of communal nesting by birds with wasps or bees. *Acta XVII Congressus Internationalis Ornithologici*, **2**, 1199–1205.

Soler, J. J. and A. P. Møller (1996). A comparative analysis of the evolution of variation in appearance of eggs of European passerines in relation to brood parasitism. *Behav. Ecol.*, **7**, 89–94.

Soler, M. and A. P. Møller (1990). Duration of sympatry and coevolution between the great spotted cuckoo and its magpie host. *Nature*, **343**, 748–750.

Soler, M., J. J. Soler, J. G. Martinez, and A. P. Møller (1994). Microevolutionary change in host response to a brood parasite. *Behav. Ecol. Sociobiol.*, **35**, 295–301.

Soler, M., J. J. Soler, J. G. Martinez, and A. P. Møller (1995*a*). Host manipulation by great spotted cuckoos: evidence for an avian mafia? *Evolution*, **49**, 770–775.

Soler, M., J. J. Palomino, J. G. Martinez, and J. J. Soler (1995*b*) Communal parental care by monogamous magpie hosts of fledgling Great Spotted Cuckoos. *Condor*, **97**, 804–810.

Soler, M., J. J. Soler, and J. G. Martinez, this vol. Duration of sympatry and coevolution between the great spotted cuckoo (*Clamator glandarius*) and its primary host, the magpie (*Pica pica*).

Sorenson, M. (1992). Comment: why is conspecific nest parasitism more frequent in waterfowl than in other birds? *Can. J. Zool.*, **70**, 1856–1858.

Sorenson, M. D., this vol. Patterns of parasitic egg laying and typical nesting in redhead and canvasback ducks.

Southern, W. E. (1958). Nesting of the red-eyed vireo in the Douglas Lake region, Michigan. *Jack-Pine Warbler*, **36**, 105–130, 185.

Spaw, C. D. and S. Rohwer (1987). A comparative study of eggshell thickness in cowbirds and other passerines. *Condor*, **89**, 307–318.

Spencer, O. R. (1943). Nesting habits of the black-billed cuckoo. *Wilson Bull.*, **55**, 11–22.

Stearns, S. C. (1976). Life history tactics: a review of the data. *Q. Rev. Biol.*, **51**, 3–47.

Stevens, E. (1982). Variation in the oral flange color of brown-headed cowbird nestlings: an adaptation for brood parasitism? M.A. Thesis, Univ. California, Santa Barbara.

Steyn, P. (1973). Some notes on the breeding biology of the striped cuckoo. *Ostrich*, **44**, 163–169.

Stouffer, P. C. and H. W. Power (1991). Brood parasitism by starlings experimentally forced to desert their nests. *Anim. Behav.*, **41**, 537–539.

Strahl, S. D. (1988). The social organization and behavior of the hoatzin *Opisthocomus hoatzin* in Central Venezuela. *Ibis*, **130**, 483–502.

Swynnerton, C. F. M. (1916). On the coloration of the mouths and eggs of birds II. On the coloration of eggs. *Ibis*, **4**, 529–606.

Swynnerton, C. F. M. (1918). Rejection by birds of eggs unlike their own: with remarks on some of the cuckoo problems. *Ibis*, **6**, 127–154.

Takasu, F., K. Kawasaki, H. Nakamura, J. E. Cohen, and N. Shigesada (1993). Modeling the population dynamics of a cuckoo–host association and the evolution of host defenses. *Amer. Natur.*, **142**, 819–839.

Teather, K. L. and R. J. Robertson (1985). Female spacing patterns in brown-headed cowbirds. *Can. J. Zool.*, **63**, 218–222.

Teather, K. L. and R. J. Robertson (1986). Pair bonds and factors influencing the diversity of mating systems in brown-headed cowbirds. *Condor*, **88**, 63–69.

Temple, S. A. and J. R. Cary (1988). Modeling dynamics of habitat-interior bird populations in fragmented landscapes. *Conserv. Biol.*, **2**, 340–347.

Terborgh, J., S. K. Robinson, T. A. Parker, III, C. A. Munn, and N. Pierpont (1990). Structure and organization of an Amazonian forest bird community. *Ecol. Monogr.*, **60**, 213–238.

Thompson, F. R., III (1994). Temporal and spatial patterns of breeding brown-headed cowbirds in the midwestern United States. *Auk*, **111**, 979–990.

Thompson, F. R., III and W. D. Dijak. Differences in movements, home range, and habitat preferences of female Brown-headed Cowbirds in three midwestern landscapes. In: *The ecology and management of cowbirds* (eds. J. N. M. Smith, T. L. Cook, S. K. Robinson, S. I. Rothstein, and S. G. Sealy), Univ. of Texas Press, Austin, Texas (in press).

Thompson, F. R., III, J. R. Probst, and M. G. Raphael (1995). Impacts of silviculture: overview and management. In: *Ecology and management of neotropical migratory birds* (eds. T. E. Martin and

D. M. Finch), pp. 201–219. Oxford University Press, Oxford.

Trine, C. L., W. D. Robinson, and S. K. Robinson, this vol. Consequences of brown-headed cowbird brood parasitism for host population dynamics.

Uyehara, J. C. and P. M. Narins (1995). Nest defense by willow flycatchers to brood-parasitic intruders. *Condor*, **97**, 361–368.

Uyehara, J. C. and M. J. Whitfield. Vegetative cover and parasitism in territories of the southwestern willow flycatcher (*Empidonax traillii extimus*). In: *Ecology and management of cowbirds* (eds. J. N. M. Smith, T. Cook, S. K. Robinson, S. I. Rothstein, and S. G. Sealy), Univ. of Texas Press, Austin, Texas (in press).

Vehrencamp, S. L. (1978). The adaptive significance of communcal nesting in groove-billed anis (*Crotophaga sulcirostris*). *Behav. Ecol. Sociobiol.*, **4**, 1–33.

Vernon, C. J. (1964). The breeding of the cuckoo-weaver (*Anomalospiza imberbis* (Cabanis)) in southern Rhodesia. *Ostrich*, **35**, 260–263.

Vernon, C. J. (1984). The breeding biology of the thick-billed cuckoo. Proc. 5th Pan-Afr. Ornith. Congr., pp. 825–840.

Vernon, C. J. (1987). Bill hooks of *Protodiscus* nestlings. *Ostrich*, **58**, 187.

Victoria, J. K. (1972). Clutch characteristics and egg discriminative ability of the African village weaverbird (*Ploceus cucullatus*). *Ibis*, **114**, 367–376.

Von Haartman, L. (1981). Coevolution of the cuckoo *Cuculus canorus* and a regular cuckoo host. *Ornis Fenn.*, **58**, 1–10.

Walkinshaw, L. H. (1961). The effect of parasitism by the brown-headed cowbird on *Empidonax* flycatchers in Michigan. *Auk*, **78**, 266–268.

Walkinshaw, L. H. (1983). *Kirtland's Warbler: the natural history of an endangered species*. Bulletin 58, Cranbrook Inst. Sci., Bloomfield Hills, Michigan.

Ward, D. and J. N. M. Smith. Inter-habitat differences in parasitism frequencies by brown-headed cowbirds in the Okanagon Valley, British Columbia. In: *Ecology and management of cowbirds* (eds. J. N. M. Smith, T. Cook, S. K. Robinson, S. I. Rothstein, and S. G. Sealy), Univ. of Texas Press, Austin, Texas (in press).

Ward, D., A. K. Lindholm, and J. N. M. Smith (1996). Multiple parasitism of the red-winged blackbird: further experimental evidence of evolutionary lag in a common host of the brown-headed cowbird. *Auk*, **113**, 408–413.

Weatherhead, P. J. (1989). Sex ratios, host-specific reproductive success, and impact of brown-headed cowbird parasitism. *Auk*, **106**, 358–366.

Webster, M. S. (1994). Interspecific brood parasitism of Montezuma oropendolas by giant cowbirds: parasitism or mutualism? *Condor*, **96**, 794–798.

Weigmann, C. and J. Lamprecht (1991). Intraspecific nest parasitism in bar-headed geese, *Anser indicus*. *Anim. Behav.*, **41**, 677–688.

Weller, M. W. (1959). Parasitic laying in the redhead (*Aythya americana*) and other North American Anatidae. *Ecol. Monogr.*, **29**, 333–365.

Weller, M. W. (1968). The breeding biology of the parasitic black-headed duck. *Living Bird*, **7**, 169–207.

West, M. J. and A. P. King (1988). Female visual displays affect the development of male song in the cowbird. *Nature*, **334**, 244–246.

West, M. J., A. P. King, and D. H. Eastzer (1981*a*). The cowbird: reflections on development from an unlikely source. *Amer. Sci.*, **69**, 56–66.

West, M. J., A. P. King, and D. H. Eastzer (1981*b*). Validating the female bioassay of cowbird song: relating differences in song potency to mating success. *Anim. Behav.*, **29**, 490–501.

Wetmore, A. (1968). *Birds of the Republic of Panama. Part 2 – Columbidae (Pigeons) to Picidae (Woodpeckers)*. Smiths. Misc. Coll. 150, part 2.

Whitcomb, R. F., C. S. Robbins, J. F. Lynch, B. L. Whitcomb, M. K. Klimkiewicz, and D. Bystrak (1981). Effects of forest fragmentation on the avifauna of the eastern deciduous forest. In: *Forest island dynamics in man-dominated landscapes* (eds. R. L. Burgess and D. M. Sharpe), pp. 125–205. Springer-Verlag, New York.

Widman, O. (1907). A preliminary catalog of the birds of Missouri. *Trans. Acad. Sci. St. Louis*, **17**, 1–288.

Wiedenfeld, D. A. Cowbird population changes and their relationship to changes in some host species. In: *The ecology and management of cowbirds* (eds. J. N. M. Smith, T. L. Cook, S. K. Robinson, S. I. Rothstein, and S. G. Sealy), Univ. of Texas Press, Austin, Texas (in press).

Wiley, J. W. (1986). Growth of shiny cowbird and host chicks. *Wilson Bull.*, **98**, 126–131.

Wiley, J. W. (1988). Host selection by the shiny cowbird. *Condor*, **90**, 289–303.

Wiley, R. H. and M. S. Wiley (1980). Spacing and timing in the nesting ecology of a tropical blackbird: comparison of populations in different environments. *Ecol. Monogr.*, **50**, 153–178.

Winker, K., J. H. Rappole, and M. A. Ramos (1990). Population dynamics of the wood thrush in southern Veracruz, Mexico. *Condor*, **92**, 444–460.

Woodward, P. W. (1983). Behavioral ecology of fledgling Brown-headed cowbirds and their hosts. *Condor*, **85**, 151–163.

Woodward, P. W. and J. C. Woodward (1979). Brown-headed cowbird parasitism on eastern bluebird. *Wilson Bull.*, **91**, 321–322.

Wyllie, I. (1981). *The cuckoo*. Universe Books, N.Y.

Yamagishi, S. and M. Fujioka (1986). Heavy brood parasitism by the common cuckoo *Cuculus canorus* on the azure-winged magpie *Cyanopica cyana*. *Tori*, **34**, 91–96.

Yamauchi, A. (1993). Theory of intraspecific nest parasitism in birds. *Anim. Behav.*, **46**, 335–345.

Yamauchi, A. (1995). Theory of evolution of nest parasitism in birds. *Amer. Natur.*, **145**, 434–456.

Yokel, D. A. (1986). Monogamy and brood parasitism: an unlikely pair. *Anim. Behav.*, **34**, 1348–1358.

Yokel, D. A. (1989). Intrasexual aggression and the mating behavior of brown-headed cowbirds: their relation to population densities and sex ratios. *Condor*, **91**, 43–51.

Yokel, D. A. and S. I. Rothstein (1991). The basis for female choice in an avian brood parasite. *Behav. Ecol. Sociobiol.*, **29**, 39–45.

Yom-Tov, Y. (1980). Intraspecific nest parasitism in birds. *Biol. Rev.*, **55**, 93–108.

Zahavi, A. (1979). Parasitism and nest predation in parasitic cuckoos. *Amer. Natur.*, **113**, 157–159.

Zimmerman, J. L. (1983). Cowbird parasitism of dickcissels in different habitats and at different nest densities. *Wilson Bull.*, **95**, 7–22.

Zuñiga, J. M. and T. Redondo (1992). No evidence for variable duration of sympatry between the great spotted cuckoo and its magpie host. *Nature*, **359**, 410–411.

II

COEVOLUTION BETWEEN CUCKOOS AND THEIR HOSTS

This part of the book includes seven chapters detailing the complex coevolutionary interactions between cuckoos and their hosts. Cuckoos are the most diverse, and probably the oldest, lineage of brood parasites. Cuckoo–host systems have provided some of the classic examples of coevolutionary arms races expected in specialized systems that involve one parasite and one, or a few hosts. These chapters include examples of genetically programmed and learned adaptations of hosts and parasites (see also chapter 14), microevolutionary processes, and interspecific competition for hosts (possible character displacement). For a comprehensive overview of the rich literature on coevolution in cuckoo–host systems, see pages 10–16 in chapter 1.

2

Cuckoos versus Hosts

Experimental Evidence for Coevolution

N. B. DAVIES, M. de L. BROOKE

In this chapter we review work on the interactions between the cuckoo *Cuculus canorus* and its hosts, to show how simple field experiments combined with comparative studies can be used to dissect the stages of coevolution between two parties. Living organisms bear the scars of their evolutionary history and comparative studies can be used to help reconstruct past events. We begin by assessing the evidence that *Cuculus canorus* is divided into distinct host-specific races. Next, we describe the behavior involved in parasitizing hosts. Then we summarize experimental evidence for cuckoo–host coevolution, showing first how the cuckoo has responded to selection pressures from the hosts and second how the hosts, in turn, have responded to selection from cuckoos. Finally, we speculate on the time course of the evolution of the various adaptations and counteradaptations.

CUCULUS CANORUS AND ITS HOSTS

Host-specific races of cuckoo (gentes)?

In Britain, the cuckoo parasitizes five main hosts; the meadow pipit *Anthus pratensis* in moorland, the reed warbler *Acrocephalus scirpaceus* in marshland, the dunnock *Prunella modularis* and robin *Erithacus rubecula* in woodland and farmland, and the pied wagtail *Motacilla alba* in open country. The cuckoo is not a particularly common bird, with only about 21 000 females laying eggs in Britain each year (Brooke and Davies 1987), so the frequency of parasitism of even the favorite hosts is not high, on average 5% or much less (table 2.1).

Detailed observations have shown that individual female cuckoos tend to favor one particular host species and lay distinctive eggs, which match to varying degrees the eggs of their respective hosts (Chance 1922, 1940; Jourdain 1925; Baker 1942; Wyllie 1981). Thus it has long been supposed (e.g., Newton 1893) that *Cuculus canorus* is divided into genetically distinct strains, referred to as "gentes" (singular "gens"). We have measured the color and darkness of the cuckoo eggs from the various gentes in Britain and have shown that there are, indeed, significant differences among them (Brooke and Davies 1988). Meadow pipit-cuckoos lay brownish eggs, reed warber-cuckoos greenish eggs and pied wagtail-cuckoos pale, grayish-white eggs. These three gentes mimic the eggs of their hosts (fig. 2.1). By contrast, although the dunnock-cuckoo lays a distinctive egg (Brooke and Davies 1988), there is clearly no color-mimicry; the pale spotted cuckoo egg contrasts sharply with the uniform blue egg of this host (fig. 2.1). Based on a survey of museum collections, Moksnes and Røskaft (1995) suggest that there are at least 15 different cuckoo egg-morphs in Europe, with the major hosts being small passerines which nest in low vegetation (warblers, chats, shrikes, finches) or on the ground (pipits, wagtails, buntings).

Table 2.1 The five main hosts of the cuckoo in Britain, with their frequency of parasitism as measured by the proportion of parasitized nests recorded from 1939–1982 in the nest record scheme of the British Trust for Ornithology

		Nest record cards (B.T.O.)		
Host	Breeding pairs in Britain and Ireland	Total	With cuckoo	Nests parasitized (%)
Meadow pipit	3×10^6	5331	142	2.66
Reed warbler	60 000	6927	384	5.54
Dunnock	5×10^6	23 352	453	1.94
Pied wagtail	500 000	4945	21	0.42
Robin	5×10^6	12 917	38	0.29

Source: Brooke and Davies (1987) and Davies and Brooke (1989*b*).

These different cuckoo gentes would remain distinct if three conditions held, namely: (i) hosts rejected badly-matching eggs (for which there is good evidence—see later); (ii) daughters inherited their egg color from their mother (no evidence, but this seems likely); and (iii) females came to parasitize the same species of host that reared them, perhaps learning the host characteristics through imprinting. Thus, for example, a female cuckoo raised by reed warblers would, when she was adult, prefer to parasitize reed warblers. If she, like her mother, laid a greenish egg, then the match between cuckoo and host egg would be maintained over successive cuckoo generations.

There would be two possible outcomes of such host-specificity by cuckoos:

Figure 2.1 The left-hand column shows the eggs of four main host species parasitized by the cuckoo in Britain. From top to bottom these are: (1) meadow pipit, (2) reed warbler, (3) dunnock and (4) pied wagtail. The central column is a typical example of a cuckoo egg laid in the corresponding host nest. The right-hand column shows an example of a model egg painted to mimic the corresponding cuckoo egg, for testing host-discrimination. In the case of the dunnock (row 3), where there is no color mimicry by the cuckoo, the model egg was painted pale blue, the same color as the host egg, so that responses to blue mimetic eggs could also be tested. Photograph from Davies (1992), with permission from Oxford University Press.

1. Cryptic species

Host faithfulness by female cuckoos and assortative mating between male and female cuckoos reared by the same host would both be necessary to maintain the integrity of the gentes if both parents influenced their daughter's egg characteristics. Although the males do not play any part in egg laying itself, they defend territories containing host nests (Dröscher 1988; Nakamura and Miyazawa 1990) and their choice of which territory and hosts to defend could provide a mechanism for assortative mating. If both sexes remained faithful to the host species that reared them, *Cuculus canorus* would be divided into several genetically distinct species. Although there is no morphologic evidence to support this view, the cuckoo gentes could represent cryptic or incipient species, similar to those in certain phytophagous insects (Feder et al. 1988; Thompson 1994).

2. Female-specific host races

The alternative is that only females remain host specific and so only female lines are distinct (Jourdain 1925; Baker 1942). This last possibility is a real one if heterogametic females alone influence their daughter's egg type, perhaps because the w (sex) chromosome codes for egg characteristics (Sheppard 1958; for an analogous case in butterfly mimicry, see Clarke and Sheppard 1962). Males would then be free to interbreed with various female types without disrupting the maintenance of egg mimicry and this would maintain *Cuculus canorus* as one species.

Maintenance of gentes: imprinting?

We tested whether host imprinting occurred by laboratory experiments with seven cuckoos originating from reed warbler nests. Two were transferred soon after hatching to robin nests, while the other five were raised by reed warblers, their natural host. When about 12 days old, the nestling cuckoos and their foster parents were transferred from the wild to aviaries, where the fosterers continued to raise the cuckoos to independence. The fosterers were then released, and the cuckoos retained and presented, at one and two years of age, with a behavioral choice between robins and reed warblers. Each cuckoo was tested in a long cage with adjacent cages at one end containing a reed warbler and at the other end a robin. We then scored the proportion of time the cuckoo spent near either host, assuming that proximity was an indicator of preference (Bateson 1966). None of the cuckoos (five females and two males) displayed any consistent preference for the host that reared them even, in the case of females, after estradiol implantation (Brooke and Davies 1991).

With negative evidence, it is difficult to know whether we have done an inadequate experiment with regard to rearing or preference testing, or whether cuckoos really do not imprint on their hosts. It is possible, for example, that natal philopatry or habitat imprinting might be sufficient to maintain the gentes because, in Britain at least, there is reasonable geographic separation between them. Even with a crude separation of the gentes based on habitat differences, most female cuckoos could end up favoring the nests of their foster-parent species. In Britain, meadow pipits in moorland, reed warblers in reed beds and dunnocks in farmland are among the commonest birds with readily found nests in their respective habitats and would, therefore, receive a major share of the eggs of a randomly searching female cuckoo (Brooke and Davies 1991). Alternatively, cuckoos may select a particular host nest type (Moksnes and Røskaft 1995). In support of this idea, only 44% of host clutches in museum collections contained cuckoo eggs of the morph most closely matching that host, whereas 77% of the cuckoo eggs had been laid in nest sites characteristic of the best-matching host (Moksnes and Røskaft 1995).

Detailed field studies are now needed of the choices made by female cuckoos in relation to the host nests available. Do females select any host within a particular habitat, any host with a particular nest site, or do they have favorite host species and simply choose alternative hosts if there are no suitable nests of their main host, as suggested long ago by Chance (1940)?

Genetic differentiation of gentes?

If cuckoo gentes consist only of female-specific host races, then there should be significant differences between gentes in mitochondrial DNA (transmitted through females only) but not in nuclear DNA (which segregates through both sexes). If, on the other hand, cuckoo gentes represent cryptic species then there should be coincident variation in both nuclear and mitochondrial DNA markers. Together with Lisle Gibbs we recently conducted the first survey of variation in both the mitochondrial and nuclear genomes of cuckoo chicks raised by the three most common hosts in Britain, namely the meadow pipit, the reed warbler and the dunnock. We focused on the control region of mtDNA, which in birds generally shows the most rapid evolution and hence the highest levels of intraspecific variation.

We found no significant differences among cuckoos raised by the three hosts in the number of repeats in the control region of the mtDNA, nor in base sequence of another rapidly evolving mtDNA gene (ND6), nor in the allele frequencies of three microsatellite loci in the nuclear genome (Gibbs et al. 1996). Given that cuckoos parasitizing these three hosts do lay different eggs (see above) and that individual female cuckoos lay an egg of a constant type, it seems likely that there must, nevertheless, be genetic differences between the gentes. Our failure to detect them could arise for two reasons:

(a) There may be frequent gene flow between cuckoos exploiting the three hosts. In support of this idea is Moksnes and Røskaft's finding (1995) that cuckoos often laid in nests of hosts other than those for which their egg was the best match. Thus the samples we analyzed, in which race-type was assigned on the basis of host nest in which the cuckoo chick was found, may actually consist of a heterogeneous collection of "races" as defined by egg morphology alone. A study based on gentes as defined by egg morphs may reveal significant differences.
(b) Gens evolution may be so rapid that even rapidly evolving markers cannot track it. Cuckoos and their hosts recolonized Britain in the last 14 000 years following the last glacial retreat and, furthermore, there is evidence that cuckoos may change their hosts with time causing rapid evolution of host defenses and cuckoo egg-types, perhaps within a few hundred or a few thousand years (see later). Thus adaptive evolution at the loci determining cuckoo egg color may have occurred exceptionally quickly with little change in the (presumably) neutral region of the genome containing the repetitive DNA markers (see Gibbs et al. 1996 for further discussion).

Having set the scene with a discussion of cuckoo gentes, we now turn to a description of cuckoo behavior and evidence for coevolution with hosts.

THE CUCKOO'S PROCEDURE FOR PARASITIZING HOSTS

The parasitic habits of the cuckoo were known to Aristotle, writing some 2300 years ago: "it lays its eggs in the nests of smaller birds" (Peck 1970, p. 271); "they . . . do not sit, nor hatch, nor bring up their young" (Hett 1936, p. 241). However, it was not until the 1920s that the female's egg-laying behavior was described in detail in the pioneering study of meadow pipit-cuckoos by Edgar P. Chance (Chance 1922). He showed that the female cuckoo usually finds nests by watching the hosts build. Then she waits until the hosts have begun their clutch and parasitizes the nest one afternoon during the host's laying period, laying just one egg per host nest. Before laying, she remains quietly on a perch nearby, sometimes for an hour or more. Then she glides down to the nest. She removes one host egg, sometimes more, lays her own directly into the nest and flies off carrying the host egg in her bill, which she later swallows whole. The total time spent at the host nest is less than 10 seconds and it is difficult to believe that she can have laid in such a short time. Yet a visit to the nest will reveal the cuckoo egg, lying among the host clutch, usually matching the host eggs well in color and pattern and sometimes identifiable only by its slightly larger size, more rounded shape and more glossy shell. This remarkable performance has since been shown to be the same for cuckoos exploiting other hosts, such as reed warblers (Wyllie 1981) and marsh warblers *Acrocephalus palustris* (Gärtner 1981, 1982).

The cuckoo lays on average eight eggs a year, with a maximum of 25 recorded for one of Chance's meadow pipit-cuckoos (Chance 1940) and also for a reed warbler-cuckoo studied by Bayliss (1988). Cuckoos lay on alternate days', usually in two batches separated by several days rest. If nests of the favored host are in short supply, then the female may turn to alternative hosts in her territory. For example, reed warbler-cuckoos may sometimes lay in the nests of sedge warblers *Acrocephalus schoenobaenus* and meadow pipit-cuckoos may lay in the nests of tree pipits *Anthus trivialis* or yellowhammers *Emberiza citrinella.* These cases may provide the starting-point for new gentes if the young cuckoos imprint on the hosts that rear them (see above).

The female cuckoo plunders nests that she finds at the incubation or chick stage, which are too advanced for parasitism, swallowing whole clutches of eggs or broods of young chicks (Gärtner 1982). This induces the hosts to start a replacement clutch and so increases the availability of suitable nests to parasitize. On our study site for reed warbler-cuckoos, the cuckoo was a major predator of reed warbler nests. We obtained indirect evidence that the cuckoo remembered the location of nests she had parasitized and left these untouched because only 22% of parasitized nests suffered predation at the egg stage compared with 41% of unparasitized nests—a statistically significant difference (Davies and Brooke 1988).

Having laid her egg, the cuckoo then abandons it, leaving all subsequent care to the hosts. The cuckoo chick usually hatches first, after an incubation period of just 11–12 days (Wyllie 1981), benefiting from a head start when it is laid because of development inside the female's oviduct prior to laying (Perrins 1967). Then follows the most remarkable event of all. Just a few hours old, and still naked and blind, the cuckoo chick balances the host eggs one by one on its back and ejects them from the nest. Any newly hatched young suffer the same gruesome fate and so the cuckoo chick becomes the sole occupant of the nest. This extraordinary behavior was known to Aristotle, "when the young bird is born it casts out of the nest those with whom it has so far lived" (Hett 1936, p. 241), but the first detailed description was by Edward Jenner in 1788, who noted that the cuckoo chick had a depression in its back which helped it to balance the host eggs. The ejection behavior is strongest from about 8 to 36 hours after hatching, and thereafter declines. If any host young remain after this time they are usually smothered by the faster-growing cuckoo chick and die anyway. Thus, the hosts lose all their reproductive success from a successfully parasitized nest. Even while the foster parents brood it, the young cuckoo may continue to eject eggs and, incredibly, the host will stand aside to enable the cuckoo to accomplish its task, looking on while the parasite completes the destruction of the host's own reproduction.

The hosts then slave away, feeding the young cuckoo in the nest for about 20 days, and then for a further two weeks after fledging, until it becomes independent. This presents an extraordinary spectacle. When the nestling cuckoo is full grown it overflows the tiny host nest, which sometimes disintegrates or at best is flattened into a little platform. The hosts seem to risk being devoured themselves as they bow deep into the enormous gape with food, and once the chick has fledged they often have to perch on the cuckoo's back to reach the mouth of a fledgling eight times their own weight! Nevertheless, throughout chick-rearing the hosts behave as if nothing was amiss and as if they regarded the monster as one of their own young.

These interactions between the cuckoo and its hosts are clearly cases where we might expect coevolution. Selection should favor host abilities to decrease cuckoo success. This, in turn, should select for improved trickery by the cuckoo, leading perhaps to ever more intricate adaptations and counteradaptations. We now discuss our experiments to test how each party has evolved in response to the other.

HAVE CUCKOOS EVOLVED IN RESPONSE TO SELECTION FROM HOSTS?

The design of the cuckoo's egg-laying procedure: experiments with reed warblers

We would have good evidence that the cuckoo has evolved in relation to selective pressures exerted by the host if we could show that the cuckoo's egg-laying tactics, described above, were specifically designed to circumvent host defenses. To test this, we followed the experimental technique pioneered by Stephen Rothstein (1975*a*) and behaved as cuckoos ourselves, parasitizing host nests with model cuckoo eggs. The models, of the exact size and weight of real cuckoo eggs, were made of resin and painted to resemble the different eggs laid by various cuckoo gentes. Our idea was to vary the different aspects of the cuckoo's procedure to see how the hosts responded.

The study was done during 1985 and 1986 on reed warblers in the fenland around Cambridge. Of a total of 274 nests monitored daily during laying, 44 (16%) were naturally parasitized by cuckoos. Two of these clutches were destroyed by predators soon after laying. At eight of the remaining 42 nests (19%), the reed warblers rejected the cuckoo egg, four by desertion and four by ejection. Thus although the hosts sometimes rebelled, at the majority of the nests (81%) the cuckoo egg was accepted. The aim of our experiments was to discover the functional significance of each part of the cuckoo's egg-laying tactics to see whether they aided deception of hosts and, if so, how (Davies and Brooke 1988). We consider each part in turn.

Host egg mimicry

We first copied the procedure of real cuckoos by removing an egg from a reed warbler nest one

Table 2.2 Summary of experiments on reed warblers, using model cuckoo eggs to examine the host responses to variation in the cuckoo's egg-laying procedure

Experimental parasitism		% nests (n) where model cuckoo egg rejected
(a) *Host egg mimicry varied*: afternoon parasitism during host laying period.		
Nonmimetic eggs	pied wagtail gens	81.3 (16)
	redstart gens	69.6 (23)
	meadow pipit gens	44.4 (9)
Mimetic egg	reed warbler gens*	3.3 (30)
(b) *Stage of parasitism varied*		
Mimetic egg before host lays		100 (6)
(c) *Time of laying varied*		
Mimetic egg at dawn		46.2 (13)
(d) *Speed of laying varied*		
Mimetic egg after stuffed cuckoo on nest		39.1 (23)
(c) *Size of egg varied*		
Giant mimetic egg		40.0 (15)
(f) *Removal of a host egg varied*		
Mimetic egg	with one host egg removed	8.3 (12)
	no host egg removed	0 (11)
Nonmimetic egg	with one host egg removed	64.0 (25)
	no host egg removed	65.0 (20)

Note: *Totals for green and brown spotted models, the ones which most closely resembled the reed warbler's own eggs.

Statistical analysis (χ^2 tests, 1 df): In (a) all three nonmimetic eggs more likely to be rejected than mimetic eggs ($P < 0.001$ for first two nonmimetic types, $P < 0.01$ for the third). Mimetic eggs in (b) $P < 0.001$, (c) $P < 0.01$, (d) $P < 0.01$ and (e) $P < 0.01$, all more likely to be rejected compared with mimetic eggs in (a). No significant effects of host egg removal in (f) for either mimetic or nonmimetic egg rejection.

Source: Davies and Brooke (1988).

afternoon during the laying period and substituting a model cuckoo egg. The warblers accepted almost all the model eggs that resembled their own in color (painted like the reed warbler gens of cuckoo) but rejected two-thirds of those that did not (painted like three other gentes; table 2.2), either by ejecting them from the nest or by desertion. Clearly then, host discrimination has selected for a mimetic cuckoo egg, as first suggested by Baker (1913); see also Higuchi (1989) and Lotem et al. (1995).

This may seem to be an obvious and expected result but there are two alternative explanations for the mimicry. Alfred Russel Wallace (1889) interpreted bird egg colors as an example of protective coloration and suggested that cuckoo eggs have come to resemble host eggs so the clutch is not made conspicuous to predators; "if each bird's eggs are to some extent protected by their harmony of colour with their surroundings, the presence of a larger and very differently coloured egg in the nest might be dangerous and lead to the destruction of the whole set. Those cuckoos, therefore, which most frequently placed their eggs among the kinds which they resembled would in the long run leave most progeny, and thus the very frequent accord in colour might have been brought about" (Wallace, op. cit. p. 216). This same suggestion was also made more recently by Harrison (1968). However, selective predation is unlikely to be an important pressure favoring cuckoo egg mimicry. There was no tendency for reed warbler nests with one of our nonmimetic model eggs to be more likely to get depredated (Davies and Brooke 1988) and two other experimental studies have also shown that clutches with an odd egg do not suffer greater predation (Mason and Rothstein 1987; Davies 1992).

Another hypothesis stems from the observation that a proportion of parasitized nests are later parasitized again, by a second cuckoo (six out of the 44 parasitized nests in our study of reed warbler-cuckoos). As described above, the cuckoo first removes a host egg before laying her own. If the nest had already been parasitized, it would clearly pay the second cuckoo to pick out the first cuckoo's egg for removal, because there is only room for one cuckoo in the nest; if two hatch out, then one ejects the other and it is the chick from the second laid cuckoo egg which is likely to hatch last and so, being smaller, suffer ejection. Host egg mimicry may evolve, therefore, because it reduces the chance that second cuckoos will be able to discriminate, and remove, a cuckoo egg from the clutch (Davies and Brooke 1988). Some cuckoos visited nests in which we had placed a model egg and there was, indeed, a tendency (though not quite significant statistically on our small sample size) for the cuckoo to remove a model if it was unlike the host eggs in coloration (Davies and Brooke 1988). Discrimination by cuckoos themselves may have therefore played some part in the evolution of egg mimicry. However, it is likely to be a much weaker selective force than host discrimination, at least for most European hosts. For example, on our fenland study sites, 14% of parasitized reed warbler nests were visited by a second cuckoo and our model egg experiments showed that cuckoos removed a nonmimetic egg in 60% of cases. So egg replacement by cuckoos led to rejection of nonmimetic eggs in 8% cases (0.14×0.6), well below that arising from host rejection (69%; 33 of 48 cases in table 2.2).

Nevertheless, this last hypothesis emphasizes the importance of testing host responses to cuckoo eggs experimentally to investigate coevolution. In some cases competition among cuckoos themselves, in the complete absence of any host defenses, could lead to evolutionary change in cuckoo egg coloration (Brooker and Brooker 1989; Brooker et al. 1990). In the same way, competition among predators for limited prey can lead to changes in predator lineages even in the absence of any evolutionary response by prey.

Stage of parasitism

To test whether other features of the cuckoo's behavior were also tailored to beat host defenses, we used mimetic model eggs (reed warbler gens) and varied each part of the procedure in turn. When models were placed in nests before the hosts themselves began to lay, even these mimetic eggs were all rejected (table 2.2). Thus the hosts seemed to adopt the rule "any egg appearing in the nest before I start to lay cannot be mine, so reject it." This explains why cuckoos wait until hosts begin to lay before they parasitize a nest.

Once the host had begun its clutch, the number of host eggs had no effect on rejection frequency. Nevertheless, real cuckoos preferred to lay early on in the host laying period with significantly fewer reed warbler nests parasitized at the four-egg stage, or later, than expected from the proportion of nests that would be vulnerable at this stage (Davies and Brooke 1988). Most reed warblers have clutches of four eggs, but because incubation usually begins at the three- or even two-egg stage, cuckoos that laid at the four-egg stage would be at a disadvantage as their chicks would be less likely to hatch before the host chicks, and host chicks are probably more difficult to eject than host eggs. Thus, selection has favored cuckoos which parasitize host nests not too early and not too late.

Time of laying

When a mimetic model was placed in a nest at dawn, the reed warblers were more likely to reject it than an afternoon parasitism (table 2.2). Afternoon laying is therefore an important part of the cuckoo's trickery, perhaps because later in the day the warblers are less likely to be in attendance at the nest (they lay their own eggs at dawn). In a study of intra-specific brood parasitism in starlings *Sturnus vulgaris* Feare et al. (1982) noted that parasitic eggs were more often laid in the afternoon, compared with the normal early morning laying. Afternoon laying, however, is not characteristic of all brood parasites. Brown-headed cowbirds *Molothrus ater* lay at dawn (Scott 1991) as do some parasitic cuckoos (*Clamator jacobinus*; Gaston 1976).

Speed of laying

The remarkable speed of laying is also important. When we placed a stuffed cuckoo on the warblers' nest, to simulate a female that was slow to lay, the warblers mobbed it vigorously and were subsequently more likely to reject even well-matching models (table 2.2). This response may, in part, have been a result of increased general excitement rather than a specific reaction to the cuckoo itself, because in our experiments a stuffed jackdaw *Corvus monedula* (a nest predator) also stimulated increased rejection of model eggs. However, there was a tendency (though not significant statistically) for the stuffed cuckoo to elicit stronger rejection of eggs (Davies and Brooke 1988). Of course, even if the increased rejection is due to a generalized response to any nest intruder it would still select for rapid laying by cuckoos. In similar experiments, Moksnes and Røskaft (1989) found that the presentation of a stuffed cuckoo also stimulated increased rejection of model cuckoo eggs by meadow pipits. By contrast, a stuffed willow grouse *Lagopus lagopus*, which was not mobbed, had no effect. These results support the view that there are costs of rejection, which the hosts are more willing to incur if they have better evidence that they have been parasitized (Davies and Brooke 1988; Davies et al. 1996).

Size of the cuckoo's egg

Parasitic cuckoos lay smaller eggs than nesting cuckoos of the same body size (Payne 1974). The egg of *Cuculus canorus* weighs on average only 3.4 g (Wyllie 1981), whereas a nonparasitic cuckoo of the same body size (100 g) would be expected to lay a 10 g egg. When we placed such giant eggs in reed warbler nests, they were more likely to be rejected even when painted to match the color of the warbler's eggs (table 2.2). Still, host discrimination may not be the only selective pressure favoring a small cuckoo egg—large eggs may simply be difficult for small hosts to incubate.

Removal of a host egg

All aspects of the cuckoo's strategy discussed so far are adapted to increase the chances that the host will accept the parasitic egg. One feature, however, apparently has nothing to do with host defenses, namely the cuckoo's habit of removing an egg before she lays. Our experiments showed that the warblers accepted almost all mimetic models, and nonmimetic models stood the same chance of rejection, regardless of whether a host egg was removed or not (table 2.2). Thus hosts are not alerted to the presence of a parasitic egg by counting an extra egg in their clutch. Moksnes and Røskaft (1989) also found that meadow pipits were just as likely to accept a model cuckoo egg whether one of their own eggs was removed or not (see Sealy 1992 for the same result with a cowbird host).

Our experiments did suggest, however, that egg removal might increase incubation efficiency. There was a greater incidence of unhatched reed warbler eggs in nests where we did not remove a host egg to make room for a model than at nests where we removed a host egg (Davies and Brooke 1988). A second advantage of egg removal is that the female cuckoo eats the host egg and so gains a free meal. Why, then, does she not remove all the host eggs and so gain even more nourishment? Our experiments showed that host responses again provided the answer. When we reduced the warbler's final clutch size to three eggs, they never deserted but a reduction to two eggs caused 20% desertion and a reduction to one egg caused 87% desertion (Davies and Brooke 1988). Most reed warblers (63% clutches) lay a clutch of four, though clutches of three (18%) and five (15%) are also common (Bibby 1978). Therefore the maximum the cuckoo can safely take from a clutch of three is one (she replaces this with her own egg, so the final clutch is three), and from a clutch of four the maximum is two (two removed then her own egg added to give a clutch of three). This argument predicts quite well the observed behavior of cuckoos, which is to remove usually one but sometimes two host eggs, rarely more (Wyllie 1981). It is worth noting here that desertion in response to partial clutch reduction appears to be universal in passerines so it is unlikely to have evolved in response to brood parasitism *per se* (Rothstein 1982*a*).

Ejection by cuckoo chick

Although reed warblers nearly always deserted a single egg, they never deserted a single chick (either their own or a cuckoo). This explains very neatly why it is the cuckoo chick which takes on the task of ejecting the remainder of the nest contents rather than its mother, earlier on during the egg stage. Although there is a limit to the number of host eggs the laying female cuckoo can remove, later on when the cuckoo chick hatches it can eject the entire nest contents without penalty.

Absence of chick mimicry

Reed warblers quickly note and reject strange eggs but show surprising tolerance of the cuckoo chick. Indeed, there is no evidence that any host of *Cuculus canorus* ever rejects cuckoo chicks. Yet the newly hatched cuckoo's pink body and vivid orange gape is quite unlike the warbler chick's black skin and yellow gape with two conspicuous black tongue spots. Given that the reed warblers can detect these kinds of color and pattern differences among eggs why can they not do so among chicks? Any reed warbler with the rule "feed only chicks with tongue spots" would gain immediate protection against cuckoos.

One possibility is that hosts tolerate cuckoo chicks because they have nothing to compare them with; the cuckoo chick sees to that by hatching first and doing away with the rest of the brood. We tested this by giving warblers the opportunity to compare a cuckoo chick with their own young. We strapped two nests side by side, one holding a cuckoo chick, the other holding a reed warbler chick. Faced with this choice, the hosts fed the contents of both nests indiscriminately, simply feeding whichever chick was begging when they returned with food (Davies and Brooke 1988).

Dawkins and Krebs (1979) have suggested that whereas the cuckoo relies on deception to get its egg accepted, it relies on a different trick, manipulation, at the chick stage. They suggest that the hosts may not be able to resist a cuckoo chick any more "than the junkie can resist his fix." However, we found that reed warblers and other hosts would accept chicks of other species in among their own brood, raising them as their own, despite their different appearance (Davies and Brooke 1988, 1989*a*). This shows that the cuckoo chick does not possess any special super-stimuli necessary to elicit feeding. Rather, it seems as if the hosts simply fail to show any discrimination at the chick stage and will feed any begging mouth in their nest. We found that cuckoo chicks raised by reed warblers were fed on the same types of invertebrate prey that the hosts brought to a brood of their own young. The cuckoo chick was also provisioned in the nest at about the same rate as an average warbler brood (three or four young), suggesting that its development is adapted to a feeding rate the hosts can normally sustain (Brooke and Davies 1989).

The failure to discriminate cuckoo chicks cannot, therefore, be attributed either to the host's lack of opportunity for comparison with their own young, nor to the possession of special manipulative signals by cuckoos. If no hosts of brood parasites showed rejection of nonmimetic young, we could claim that there may be some design constraint which prevents chick discrimination from evolving in birds. However, in some cases where the brood parasite is raised alongside the host's own young, there is good chick mimicry, apparently selected for by host discrimination against young unlike their own. For example, near perfect nestling mimicry occurs between the parasitic screaming cowbird *Molothrus rufoaxillaris* and its host the bay-winged cowbird *M. badius* (Fraga, this vol.) which is unlikely to be simply attributable to common descent (Lack 1968, p. 94; Lanyon 1992), and nestlings of parasitic viduines mimic the intricate mouth patterns of their particular species of estrildid finch host, which preferentially feed nestlings with matching mouth markings (Nicolai 1964; Payne 1982).

We suggested that the contrast between the fine discrimination of cuckoo eggs and the blind acceptance of cuckoo chicks by hosts of *Cuculus canorus* is the outcome of two selective pressures (Davies and Brooke 1988). First, there will be stronger selection for rejection of cuckoo eggs because the earlier the hosts spot the parasitism the better. Not only are they then more likely to save their current brood (the cuckoo chick ejects the host eggs soon after it hatches), but even if they are unable to do this, for example because they reject by desertion, rejection early, at the egg stage, is more likely to give them the chance to raise a second brood that year. By the time the chicks hatch this may be too late. Second, discrimination of strange chicks may be a more difficult task than discrimination of strange eggs. An egg looks the same throughout incubation whereas a chick changes in appearance as it grows. In addition, many hosts, like reed warblers, have brood hierarchies resulting from asynchronous hatching. There will, therefore, be considerable variation in appearance even within the host's own brood and to spot a cuckoo chick among this variability may not be easy. Mistakes in chick recognition will also be more costly than mistakes in egg recognition; if a host mistakenly rejects one of its own chicks it loses more investment than if it mistakenly rejects one of its own eggs. But these explanations are not wholly satisfactory for, while they suggest why egg rejection may be more advantageous than chick rejection, they do not explain the apparent absence of chick rejection.

Recently, Lotem (1993) and Redondo (1993) have pointed out that a consideration of the mechanisms of rejection may explain exactly why discrimination of chicks may be more costly than that of eggs. Experiments have shown that hosts come to reject parasite eggs by learning the characteristics of their own eggs during their first breeding attempt and then rejecting eggs which are different from this learned set (Rothstein 1974, 1982*b*; Lotem et al. 1992). The cost of this "learning rule" is that of misimprinting on a parasite egg if the host is parasitized in its first attempt, in which case it will then be doomed always to accept parasite eggs. Lotem (1993) showed that the learning rule is nevertheless adaptive at the egg stage because the benefits of correct rejection exceed the misimprinting cost. However, at the chick stage the misimprinting cost is substantial if the parasite chick ejects the host eggs. Then, if the host is parasitized the first time it breeds, it imprints only on the parasite chick and so will reject its own young in future breeding attempts. In this case, the rule "accept all chicks in my nest" does better than the learning rule and

Lotem suggests that this explains why hosts of *Cuculus* and other ejecting parasites discriminate at the egg stage but not the chick stage. He points out that the learning rule is adaptive in cases where parasite and host young are raised together, which could explain why these are the cases with better nestling mimicry by the parasite (Davies and Brooke 1988).

Redondo (1993) emphasizes that hosts use simple rules to allocate food among their brood, such as "feed the largest or most strongly begging chick" (Redondo and Castro 1992; Kacelnik et al. 1995). Parasites can exploit these rules, which may be costly for hosts to change because of their advantage in unparasitized nests. In 95% of reed warbler nests, for example, feeding a large, strongly begging chick is an adaptive response by the parents and only in the 5% of cases they are parasitized does the rule mis-fire. Furthermore, parasite chicks may exaggerate some features, such as begging intensity, to compensate for their odd appearance. Experiments are needed to investigate these ideas.

Degree of host egg mimicry in different cuckoo gentes

The results above show that much of the cuckoo's egg-laying procedure is designed to circumvent host defenses, which is the first piece of evidence that the cuckoo has responded to selective pressures from the hosts. The second piece of evidence comes from comparing the degree of egg mimicry shown by the different cuckoo gentes. Our experiments with meadow pipits and pied wagtails revealed that, like reed warblers, they tended to reject model eggs unlike their own and were more likely to accept a well-matching model representing the egg type laid by their own cuckoo gentes (Brooke and Davies 1988). Thus, there is evidence that the mimetic eggs of pipit and wagtail cuckoos have been selected through host discrimination.

By contrast, dunnocks, for whom the cuckoo does not lay a mimetic egg, showed no discrimination and accepted model eggs of any color. We wondered if dunnocks had poor color vision, or found it difficult to discern egg color in their nests which are built in dense cover. Perhaps in the dark nest their own blue eggs and the various models would all appear a similar shade of gray. We therefore did further experiments with white or black model eggs, which should have been easily detected as different in shade, and nine out of ten were accepted (one was deserted). Even presentation of a stuffed cuckoo failed to elicit rejection (8/8 pipit-type cuckoo eggs accepted). Moreover, dunnocks accepted whole clutches of model eggs unlike their own, and then accepted their own eggs back again after three days, so their acceptance of a single model egg was not simply because they regarded it as a harmless lump of resin (Davies and Brooke 1989*a*)!

In conclusion, the degree of mimicry exhibited by the various cuckoo gentes in Britain reflects the degree of discrimination shown by their respective hosts, indicating that cuckoos clearly evolve in response to selection from hosts. A cuckoo gens in eastern Europe lays an immaculate blue egg for its host, the redstart *Phoenicurus phoenicurus*, whose eggs are blue like those of the dunnock (von Haartman 1981; Moksnes et al. 1995). Thus cuckoos have the potential to mimic dunnock eggs; they do not simply because dunnocks have not required them to do so. This still leaves unresolved the problem of why the dunnock-cuckoo nevertheless lays a distinctive egg-type, different in darkness from the other gentes and no more variable (fig. 2.1). One possibility is that secondary hosts have selected for the cuckoo egg-type; many woodland and farmland passerines have pale mottled eggs, and so the dunnock-cuckoo egg may be a generalized match for a variety of hosts. Alternatively, the dunnock-cuckoo egg may be cryptic, blending in with the nest lining, and so may reduce the chance of removal by second cuckoos, as suggested by the Brookers for some Australian cuckoos which lay a specific cryptic egg type despite the complete absence of any host discrimination (Brooker and Brooker 1989).

HAVE HOSTS EVOLVED IN RESPONSE TO SELECTION FROM CUCKOOS?

The two kinds of evidence above show that cuckoos have evolved in response to selection from hosts. Have hosts, in turn, responded to selection from cuckoos? Three sources of evidence suggest that indeed they have, so that the interaction involves true coevolution, namely each party evolving in relation to selective pressure from the other (Janzen 1980).

Responses to cuckoo eggs by species unsuitable as hosts

If the rejection of eggs shown by hosts of the various mimicking cuckoo gentes cited above has evolved specifically in response to cuckoo parasitism, then species which have never interacted with cuckoos should show no rejection of eggs unlike their own. There are two groups of birds which we can be sure have not been affected by cuckoos. The first are seed-eaters (e.g., finches, Fringillidae) which are unsuitable as hosts because the young cuckoo needs an invertebrate diet for successful rearing. The second may have a suitable diet but nest in small holes inaccessible to the laying female cuckoo (tits Paridae, pied flycatcher *Ficedula hypoleuca*, wheatear *Oenanthe oenanthe*, starling *Sturnus vulgaris*, house sparrow *Passer domesticus*, swift *Apus apus*). To this list can be added the swallow *Hirundo rustica*, which feeds its young by regurgitation (all cuckoo hosts feed their young from the bill).

Figure 2.2(a) shows that "unsuitable" hosts exhibited little or no rejection of eggs unlike their own, compared with the various degrees of rejection shown by "suitable" hosts, namely passerines with open nests which feed their young on invertebrates. It could be argued that demonstrating that hole nesters are accepters is trivial because they cannot see their eggs anyway. However, Mason and Rothstein (1986) have shown that a "dark nesting" host, the hornero *Furnarius rufus*, has sensitive egg recognition based exclusively on egg size, so in principle it seems as if hole nesters could evolve counteradaptations to cuckoos if they were ever exploited.

Some comparisons between closely related species in fig. 2.2(a) are particularly interesting. Thus in the tribe Muscicapini, the spotted flycatcher *Muscicapa striata*, whose open nest is exploitable by cuckoos, showed strong rejection of eggs unlike its own, whereas the pied flycatcher, which nests in holes and so is inaccessible to cuckoos, showed no rejection at all. In the finch family, Fringillidae, the one species that feeds its young predominantly on invertebrates, and is therefore suitable as a cuckoo host (the chaffinch *Fringilla coelebs*), showed strong rejection (see also Braa et al. 1992), while four species that feed their young mostly on seeds (the greenfinch *Carduelis chloris*, the linnet *Acanthis cannabinna*, the redpoll *A. flammea* and the bullfinch *Pyrrhula pyrrhula*) showed no rejection. These results provide evidence that rejection is not strongly constrained by taxonomy. Nevertheless, there are problems in assuming that species provide independent data points for comparative analysis (Harvey and Pagel 1991). Figure 2.2(b) presents a more conservative analysis using independent contrasts and shows that the difference in rejection between unsuitable and suitable hosts is still highly significant.

Response to eggs by populations of favorite host species isolated from cuckoos

These comparative data suggest that before cuckoos began to parasitize meadow pipits, pied wagtails and other current favorite hosts, these species would have shown no rejection of eggs unlike their own. We cannot go back in time to test this, but we could do the next best experiment. *Cuculus canorus* breeds from western Europe to Japan but not in Iceland, which does however have isolated populations of meadow pipits and white wagtails *Motacilla alba alba* (of which the pied wagtail is a separate subspecies *M. alba yarrellii*). We therefore took our model eggs to Iceland with exciting results (table 2.3). Both pipits and wagtails showed much less discrimination against eggs unlike their own than did members of the parasitized populations of these same species in Britain. They did, however, reject some of our pure blue model eggs of the redstart gens and so, unlike the unsuitable host species, they were not completely naive. Possibly the Icelandic populations were derived from parasitized populations from other parts of Europe and still have a legacy of some of their ancestors' egg discrimination. Further work is needed to test to what extent the greater acceptance by Icelandic hosts is caused by genetic differences or represents phenotypic plasticity, with hosts simply being less likely to reject if they assess parasitism to be absent, for example because they do not encounter cuckoos.

Moksnes et al. (1991*a*) have shown since that Norwegian meadow pipits, which are commonly parasitized (Moksnes and Røskaft 1987), rejected (by desertion) only 8% of nonmimetic models, much less than in Britain and even less than our value for the Icelandic population. This may partly reflect the different types of model egg used, but Moksnes et al. (1993) suggest that a shorter breeding season in higher latitudes, with less opportunity for re-nesting, may favor less host desertion. This idea is an interesting alternative hypothesis for explaining the lower rejection in Iceland than in Britain. However, contrary to its prediction, Moksnes et al. (1993) found no tendency for the Norwegian pipits to reject more earlier in the season, when re-nesting was possible.

Adaptive responses to adult cuckoos

Moksnes et al. (1991*a*) have shown, using experiments with model eggs and a dummy cuckoo, that there is a correlation between a species' rejection of unlike eggs and its aggressive response to the

(a) SUITABLE HOSTS (N=24)

Blackbird
Reed w.* Pied wag.*
Chaffinch Gt. reed w.*
Wren* Icter. w. Brambling*
Dunnock* Sp. Fly. Willow w.
Fieldfare Redstart* S. Thrush Reed bunt.
Sedge w.* M. pipit* Blackcap Chiffchaff
Robin* Redwing Bluethroat Yell. wag. Yellowham.

0 20 40 60 80 100

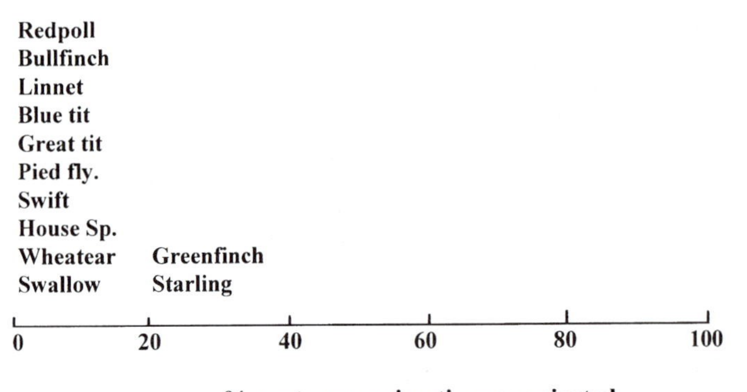

Figure 2.2 (a) Percentages of nests where a model egg unlike the host's own eggs was rejected. Species suitable as cuckoo hosts show varying degrees of rejection (mean $\pm$ 1SE = 59 $\pm$ 7%, N = 24), while species that are unsuitable as hosts, and which therefore have no history of coevolution with the cuckoo, show largely no rejection (5 $\pm$ 3%, N = 12; $P < 0.001$, Mann–Whitney U-test, 2-tailed). Within suitable species, some rarely used hosts (unstarred) show stronger rejection (mean 71 $\pm$ 7%, N = 14) than the most commonly used hosts (starred; 42 $\pm$ 12%, N = 10; $P < 0.10$, Mann–Whitney U-test, two-tailed). Data are means of the two studies by Davies and Brooke (1989*a*) and Moksnes et al. (1991*a*), considering species with at least five nests tested per study. There was good agreement between these two studies; comparing % rejection for all 16 species tested in both studies, $r_s = 0.818$, $P < 0.01$; for nine suitable hosts $r_s = 0.713$, $P < 0.05$. (b) Analysis of independent contrasts from the data in (a) using the method of Purvis (1991) and the phylogeny of Sibley and Ahlquist (1990). Comparing rejection frequency of suitable (R and C) and unsuitable (U) hosts; all seven contrasts (1 to 7) showed greater rejection (+) in suitable hosts (mean $\pm$ 1SE contrast score = 18.6 $\pm$ 3.9, significantly different from zero, $P < 0.01$). Within suitable hosts, there were six contrasts (A to F) between common (C) and rare (R) hosts. Four showed greater rejection (+) in rare hosts and two showed less (−). (Mean contrast = 4.5 $\pm$ 5.0, not significantly different from zero.)

cuckoo. Species unsuitable as hosts, such as hole-nesting titmice and the pied flycatcher, and the seed-eating greenfinch, tended to show less aggression to the cuckoo than did suitable hosts, namely open-nesting species with an invertebrate diet. This suggests that aggression to the cuckoo is a specific response to parasitism.

The way in which reed warblers mob cuckoos compared with other predators also supports the view that hosts display specific aggressive responses suited to the costs of parasitism (Duckworth 1991). Cuckoos are a threat to the nest contents but not to the host parents and adult reed warblers will readily attack a cuckoo near their

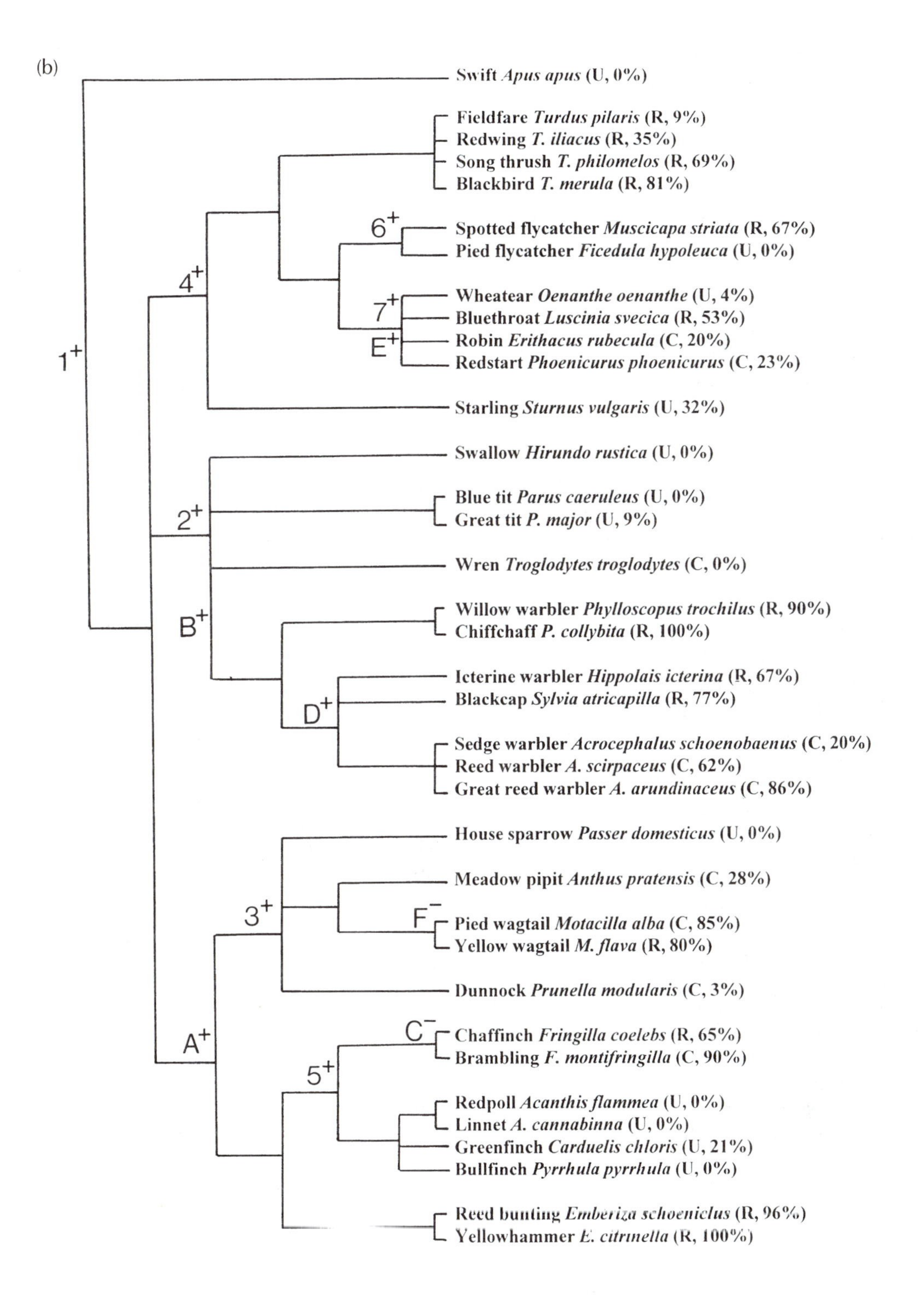

Figure 2.2 (*continued*)

Table 2.3 Rejection frequencies of model cuckoo eggs by meadow pipits and pied/white wagtails in parasitized populations (Britain) and unparasitized populations, isolated from cuckoos (Iceland)

		% nests (n) where model rejected		
Host	Model cuckoo egg-type	Britain (parasitized)		Iceland (unparasitized)
Meadow pipit	Pied wagtail gens	36.0 (25)	*	0 (13)
(*Anthus pratensis*)	Redstart gens	83.3 (18)	*	36.4 (11)
Pied/white wagtail	Meadow pipit gens	66.7 (18)	***	0 (15)
(*Motacilla alba*)	Redstart gens	76.9 (13)	N.S.	50.0 (10)

χ^2 tests, 1 df: *P <0.05; *** P <0.001. N.S., not significant.

Source: Davies and Brooke (1989*a*).

nest, pulling feathers from a cuckoo, whether alive or stuffed. Indeed, speedy intervention by the experimenter is needed if a stuffed cuckoo is to avoid destruction! Great reed warblers *Acrocephalus arundinaceus* have even been seen to knock cuckoos into the water under their nest, where they sometimes drown (Molnar 1944). Rapid laying by cuckoos may therefore reduce damage from host attacks. By contrast, reed warblers are less ready to approach a jay *Garrulus glandarius* and even less likely to approach a sparrowhawk *Accipiter nisus*; the former is a danger not only to the nest but also to the adult, while the latter is a danger to the adults only.

Further evidence of adaptive responses to cuckoos is the fact that reactions to a cuckoo more or less cease once the young fledge (the cuckoo is no threat to fledglings) whereas alarm to a jay or sparrowhawk continues (both are capable of killing fledglings; Duckworth 1991). Clearly the host's different responses to these three predators are adapted to their respective dangers to the nest and the adults themselves.

It would be interesting to study to what extent learned and innate components are involved in host recognition of cuckoos. Studies of mobbing by cowbird hosts (Robertson and Norman 1977; Smith et al. 1984) and of enemy recognition by fish (Magurran 1990, Huntingford and Wright 1992) suggest that both are likely to be involved, with a genetic predisposition to respond to experience that varies between host species and populations depending on their history of parasitism. The clearest case for some component of cuckoo recognition being due to a genetic predisposition is the Snares Island subspecies of the black tit *Petroica macrocephala dannefaerdi*, which responds to long-tailed cuckoo *Eudynamys taitenis* mounts even though it is not sympatric with any brood parasite (McLean and Maloney, this vol.).

SPECULATIONS ON THE STAGES AND TIME COURSE OF CUCKOO–HOST COEVOLUTION

Hypothesis for a coevolutionary sequence

The observations and experiments in the previous two sections show that both parties have evolved in response to selection from the other. We conclude with some speculations about the time course and outcome of cuckoo–host coevolution. Our results led us to suggest the following sequence (Davies and Brooke 1989*b*), summarized below incorporating modifications by Rothstein (1990).

1. Initially, before it becomes a victim of the cuckoo, the host shows no rejection of eggs unlike its own.

2. Once cuckoos begin to parasitize, selection favors host discrimination. Assuming that rejecter genes are already present at a low frequency in the host population, the time taken for discrimination to spread will depend on the frequency of parasitism (fig. 2.3). With parasitism levels of 5% or less (table 2.1) rejection could take many hundred, perhaps several thousand, generations to spread through the host population because most hosts never encounter cuckoos, so rejection has only a small selective advantage. Even when rejection has reached a detectable frequency, say 10% (lower curve in fig. 2.3), it still takes many hundred generations to spread to a high frequency (90%,

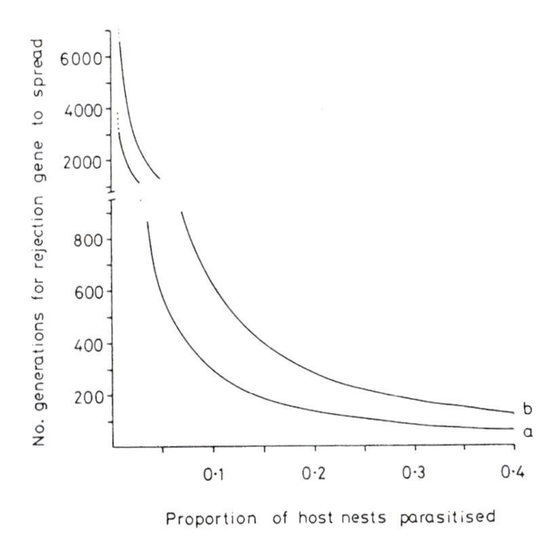

Figure 2.3 The number of generations for a hypothetical dominant "rejecter gene" to spread through the host population, calculated using Haldane's formula, depends on the proportion of nests parasitized by cuckoos. With the low rates of parasitism experienced by most hosts in Britain (table 2.1) the spread is slow. Number of generations for the gene to increase from a frequency of 10^{-6} to 0.1 (curve a) and 0.9 (curve b). From Davies and Brooke (1989*b*).

upper curve). Rothstein (1975*b*, 1990) argues that evolutionary lag in host responses might also occur because it takes time for the appearance of suitable genetic variation that enables rejection.

3. Once hosts begin to reject unlike eggs, selection will favor a mimetic cuckoo egg. Figure 2.4 shows the likely sequence of events, based on the simulation model by Kelly (1987). The selective advantage of host rejection depends on the frequency of parasitism, while that of parasite egg mimicry depends on the frequency of host rejection. The figure shows three stages in the arms race: (i) Initially host rejection increases more rapidly than does parasite egg mimicry, until the frequency of host rejection exceeds the frequency of parasitism (Rothstein 1990). (ii) Thereafter mimicry increases more rapidly and is likely to become fixed in the cuckoo population before rejection has spread through the host population. There may be a general reason for expecting the cuckoo to be one step ahead at this stage of the arms race, which Dawkins and Krebs (1979) call the "rare enemy effect," namely every cuckoo encounters a host but not every host encounters a cuckoo. In fig. 2.4, for example, when 50% of the hosts are rejecting unlike eggs selection is very strong for a mimetic cuckoo egg. But still only 5% or less of the hosts are encountering cuckoos, so the spread of rejection still proceeds slowly. As fig. 2.4 shows, there is therefore a stage where the cuckoo lays a mimetic egg but not all hosts in the population reject unlike eggs. (iii) In the final stage, there is an unexpected twist to the sequence. As mimicry becomes more common, the increase in host rejection slows down and may even stall completely if mimicry reaches fixation (Kelly 1987; Rothstein 1990). This arises because rejecter individuals that encounter mimetic eggs will not detect that they have been parasitized, and so will have the same fitness as accepters. Thus the spread of mimicry in the parasite population lowers the selective advantage of host rejection.

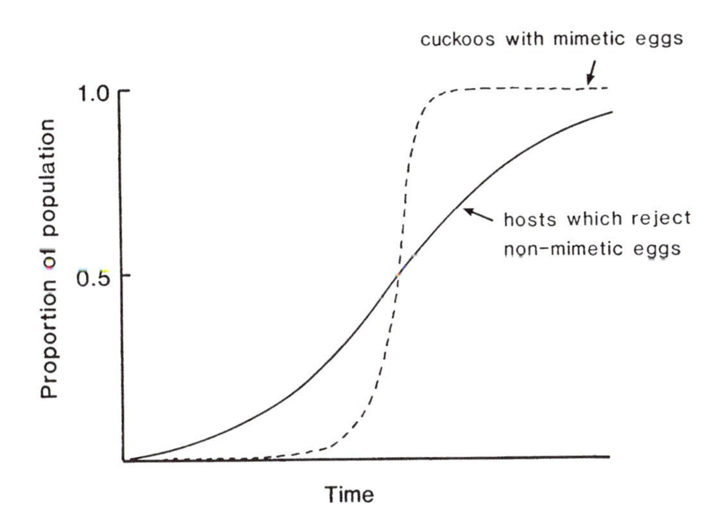

Figure 2.4 Results of the simulation model of Kelly (1987), showing the spread of egg mimicry in the cuckoo population and rejection of unlike eggs in the host population. The time course of the spread depends on the rate of parasitism (fig. 2.3).

4. Under high levels of parasitism, as occur in some hosts of cowbirds, hosts may be driven to extinction before they have time to evolve rejection (May and Robinson 1985). If, however, host and parasite populations coexist for a sufficient length of time then we can imagine two outcomes. (i) If there are other unexploited, naive accepter species available, then as hosts evolve rejection it may pay the cuckoo to turn to alternative hosts. (ii) Once all suitable hosts exhibit rejection of unlike eggs, then the only possibility for the cuckoo is to evolve better mimicry of one host's eggs. The result would be several gentes of specialist cuckoos. If parasitism frequency is low, and if there are costs of rejection, then an equilibrium may be established where the cuckoo lays a mimetic egg and the hosts accept it (Lotem et al. 1995; Davies et al. 1996).

As the cuckoo evolves a mimetic egg, hosts could escape parasitism by means of more distinctive markings on their eggs to signify "this is my egg" (Swynnerton 1918). This would select for forgery by the cuckoo to indicate "and so is this," leading perhaps to more distinctive signatures and better forgeries. We predicted, therefore, that in response to parasitism hosts will evolve less variation within clutches and greater variation between clutches, two characteristics which would make it easier for individuals to spot a parasitic egg. Our limited comparative survey of egg markings, comparing species which have interacted with cuckoos and those which have not, did not provide any evidence for this signature hypothesis (Davies and Brooke 1989*b*) but more recently, Øien et al. (1995) and Soler and Møller (1996) analyzed a larger data set and found evidence for both predictions.

There are, of course, other possible host responses to parasitism, including better nest concealment or defense and variation in the timing of breeding such as synchrony to swamp parasites or unpredictability to escape their attention.

5. Once freed from parasitism, because its cuckoo gens has gone extinct or changed to exploit other species, the host may revert back to become an accepter of unlike eggs. The speed of this transition will depend on the costs of rejection (see below). With high rejection costs reversion to acceptance may be fast and so could maintain a pool of accepter species available for subsequent recycling through the coevolutionary sequence (Marchetti 1992). If rejection had little costs then old hosts might remain as strong rejecters for a long time, perhaps never losing their "ghost of adaptation past." Indeed, Rothstein (1990) suggests that some hosts may be descended from lineages which have retained egg rejection even through speciation events.

Evidence for the proposed sequence

Two sources of evidence have been used to test this proposed coevolutionary sequence.

(a) Phylogenetic analysis

Before parasitism evolves, no hosts show rejection (stage 1) so the first parasite should be a generalist to maximize its encounter rate with hosts. As more hosts evolve defenses, we predict that parasites will be forced to specialize (stage 4).

In an important paper, Lanyon (1992) used a molecular phylogeny of the parasitic cowbirds and related New World blackbirds (Icterinae) to test how parasites may change their host-specificity with time. The most parsimonious cladistic tree showed that brood parasitism evolved just once in this group and that species which are currently the most specialized in their use of hosts are the most ancient, with the two most generalist parasite species being the most recently derived. Lanyon concluded, therefore, that specialist parasites evolve into generalists, the opposite of what we would predict. However, "number of hosts" is unlikely to be a stable species characteristic. It could equally well be argued that because the generalist parasites are younger, then this represents the starting condition and that the older species have had more time to interact with hosts and to evolve specialization. Their current specialist behavior may not reflect their ancestral state.

(b) Observations and experiments on behavior

The results of several recent studies can be interpreted using the framework in the previous section. Brown et al. (1990) found that unparasitized species of *Acrocephalus* warblers showed no rejection of nonmimetic eggs, unlike parasitized populations of reed warblers *A. scirpaceus* (stage 1). Stage (1) is also supported by the work of Soler and Møller (1990) on magpies *Pica pica*, hosts of the great spotted cuckoo *Clamator glandarius*. They found that magpies geographically isolated from the parasite showed no rejection of eggs unlike their own, whereas in areas of parasitism they did reject. Furthermore, rejection was stronger in areas of ancient than in recent sympatry and Soler et al. (1994) documented an increase in host rejection over the period 1982–1992 in an area of recent sympatry in southern Spain (stage 2). (See Zuniga and Redondo 1992 for an alternative interpretation.)

Cruz and Wiley (1989) have shown that a population of the village weaver *Ploceus cucullatus*, introduced into Hispaniola from Africa in the eighteenth century, shows significantly weaker rejection of nonmimetic eggs than stock recently derived from Africa (Victoria 1972). This provides an example of stage (5), where rejection has declined presumably because of the absence of selection from the African parasite, the didric cuckoo *Chrysococcyx caprius*.

Finally, the idea that brood parasites may change their use of hosts with time, so promoting arms races with new hosts, is supported by observations of the cuckoos *Clamator glandarius* in Spain (Arias-de-Reyna and Hidalgo 1982) and *Cuculus canorus* in Japan (Yamagishi and Fujioka 1986; Nakamura 1990), both of which have recently begun to parasitize the same new host, the azure-winged magpie *Cyanopica cyana*. In Spain, this host shows much less rejection of nonmimetic eggs than the common magpie *Pica pica*, which is the regular and presumably much older host. Especially exciting is Nakamura's study (1990; this vol.) which has documented a recent increase in the frequency of defenses against parasitism by this new host. Nakamura (1990) has also suggested that the decline in parasitism of the Siberian meadow bunting *Emberiza cioides*, which used to be a major host of *Cuculus canorus* in Japan but is now rarely parasitized, might be linked to the strong discrimination shown by this host.

These observations of the dynamic nature of cuckoo–host interactions raise the possibility that the different degrees of rejection of nonmimetic eggs shown by suitable hosts in fig. 2.2 might represent snapshots of different stages of a continuing arms race (Davies and Brooke 1989*b*, 1991; Davies et al. 1989; Moksnes et al. 1991*a*). On this view dunnocks would be relatively recent victims, still lagging behind in their counter-adaptations to a new selective pressure (early in stage (2)). Meadow pipits, pied wagtails and reed warblers, also current favorite hosts, would be at stage (3) where the cuckoo gentes lay mimetic eggs but rejection of unlike eggs by the hosts is well below 100%. Likewise the redstart-cuckoo gens in Finland lays perfect mimetic eggs (immaculate blue) even though redstarts rejected eggs unlike their own in only about 35% of the experiments by von Haartman (1981) and Järvinen (1984).

Other suitable, but rarely used hosts, such as chaffinch, spotted flycatcher and reed bunting, might be at stage (5), having been common hosts in the past and evolving such strong rejection that the cuckoo was forced to change to new hosts. There is a hint in fig. 2.2 that rarely used hosts on average do indeed show stronger rejection than current favorite hosts though the difference is not significant either using species as independent data points (fig. 2.2(a)) or independent contrasts (fig. 2.2(b)).

Continuing arms race or equilibrium?

The alternative hypothesis for the interspecific variation in rejection of nonmimetic eggs is that the systems are all at equilibrium with the degree of rejection now shown by each species under stabilizing selection. Our experiments with reed warblers (Davies and Brooke 1988) showed that there are two costs of rejecting parasitic eggs, first an "ejection" cost (damage of own eggs while rejecting the parasite egg) and second "recognition" costs (mistakes in identifying the parasitic egg—see also Lotem et al. 1995). In theory, acceptance may be a better option for hosts if the costs of rejection are high and the rates of parasitism low (fig. 1 in Davies and Brooke 1989*b*; see also Rohwer and Spaw 1988; Petit 1991; Lotem et al. 1995).

While our calculations support the idea that, because of recognition costs, it pays to accept *mimetic* cuckoo eggs when parasitism frequency is low (Davies et al. 1996), it is hard to believe that this argument can explain all cases of acceptance of *nonmimetic* eggs. We could find no evidence from current costs and benefits of rejection to support the equilibrium hypothesis for variation between species in acceptance of nonmimetic eggs (Davies and Brooke 1989*a*). Thus, there was no relationship betrween the frequency of rejection of nonmimetic eggs and frequency of cuckoo parasitism; for example, dunnocks suffer similar rates of parasitism to the meadow pipit and a much greater rate than the pied wagtail (table 2.1), both strong rejecters. There was also no obvious relationship between rejection frequency of nonmimetic eggs and costs of rejection. For example, we found that small-billed hosts were more likely to reject by desertion, probably because they were unable to eject cuckoo eggs from their nest (see also Moksnes et al. 1991*b*). Desertion is a more costly method of rejection because the host loses its current clutch. Furthermore, smaller-billed hosts were more likely to suffer the cost of cracking their own eggs whenever they tried to reject by ejection. Nevertheless, smaller-billed hosts were not more likely to accept, so rejection costs influenced the method of rejection but not rejection frequency.

However, these data on *current* selective pressures may not be an ideal test of the equilibrium hypothesis. As shown by Takasu et al. (1993), a model which combines the evolution of host rejection together with population dynamics produces various outcomes. Below a critical parasitism rate, which depends also on the costs of rejection, rejection does not spread. Even when this threshold is exceeded, rejection may not spread to fixation but can reach an equilibrium at less than 100%. Furthermore, parasitism rates are likely to vary over time in response to host rejection. Thus, for example, although current parasitism rates of dunnocks and meadow pipits are similar, it is possible that the rejection behavior of the pipits evolved under higher parasitism levels in the past. As the pipits evolved rejection, this may have driven down the population of their cuckoo gens to a lower level and hence resulted in a lower current parasitism rate. Perhaps the dunnock parsitism rate, by contrast, has always been at the low current level and below the threshold for the evolution of rejection? According to this view, the acceptance by dunnocks could be at equilibrium.

In conclusion, it seems likely that species and population differences in rejection reflect a mix of a continuing arms race and systems at different equilibrium points. On the one hand, there is good evidence for the dynamic nature of parasite–host interactions, with at least some hosts currently evolving increased or reduced rejection in response to changing parasitism (see previous section and also Rothstein 1990, Briskie et al. 1992). On the other hand, it is also likely that some mechanisms of egg recognition might preclude hosts from ever showing 100% rejection, and different species might stabilize at different degrees of rejection depending on mechanisms of egg discrimination and recognition costs (see Lotem et al. 1992, 1995, this vol.). For example, hosts with greater variation between clutches (more distinctive individual egg markings) have stronger rejection of eggs unlike their own (Øien et al. 1995; Soler and Møller 1996). Therefore, irrespective of whether egg markings evolve in response to parasitism (see above), current egg patterns may influence the ability of hosts to recognize parasite eggs.

Further work on the mechanisms used by different species to recognize their own and parasite eggs is likely to reveal a far richer coevolutionary story than the one suggested here, with hosts responding to increased parasitism by evolving ever more intricate mechanisms for learning egg characteristics. Hopefully, advances in molecular techniques will allow improved studies of genetic differences between cuckoo gentes. If the magnitude of these differences could be used to estimate the time for which they have been isolated, then we would have a better idea about the time course over which these various parasite and host adaptations and counteradaptations have evolved.

ACKNOWLEDGMENTS We thank Steve Rothstein and Eivin Røskaft for helpful reviews, Arnon Lotem, Hiroshi Nakamura and Amotz Zahavi for discussion, and the Natural Environment Research Council, the Science and Engineering Research Council and the Royal Society for supporting our work. We particularly thank Hiroshi Nakamura for hosting the meeting on brood parasitism at Shinshu University, Nagano, Japan, in August 1991, where the discussions helped us with the writing of this chapter.

REFERENCES

Arias-de-Reyna, L. and S. J. Hidalgo (1982). An investigation into egg acceptance by azure-winged magpies and host recognition of great spotted cuckoo chicks. *Anim. Behav.*, **30**, 819–823.

Baker, E. C. S. (1913). The evolution of adaptation in parasitic cuckoos' eggs. *Ibis*, **1913**, 384–398.

Baker, E. C. S. (1942). *Cuckoo problems*. Witherby, London.

Bateson, P. P. G. (1966). The characteristics and context of imprinting. *Biol. Rev.*, **41**, 177–220.

Bayliss, M. (1988). Cuckoo X breaks records. *B.T.O. News*, **159**, 7.

Bibby, C. J. (1978). Some breeding statistics of reed and sedge warblers. *Bird Study*, **25**, 207–222.

Braa, A. T., A. Moksnes, and E. Røskaft (1992). Adaptations of bramblings and chaffinches towards parasitism by the common cuckoo. *Anim. Behav.*, **43**, 67–78.

Briskie, J. V., S. G. Sealy, and K. A. Hobson (1992). Behavioural defences against avian brood parasitism in sympatric and allopatric host populations. *Evolution*, **46**, 334–340.

Brooke, M. de L. and N. B. Davies (1987). Recent changes in host usage by cuckoos *Cuculus canorus* in Britain. *J. Anim. Ecol.*, **56**, 873–883.

Brooke, M. de L. and N. B. Davies (1988). Egg mimicry by cuckoos *Cuculus canorus* in relation to discrimination by hosts. *Nature*, **335**, 630–632.

Brooke, M. de L. and N. B. Davies (1989). Provisioning of nestling cuckoos *Cuculus canorus* by reed warbler *Acrocephalus scirpaceus* hosts. *Ibis*, **131**, 250–256.

Brooke, M. de L. and N. B. Davies (1991). A failure to demonstrate host imprinting in the cuckoo *Cuculus canorus* and alternative hypotheses for the maintenance of egg mimicry. *Ethology*, **89**, 154–166.

Brooker, L. C., M. G. Brooker, and A. M. H. Brooker (1990). An alternative population/genetics model for the evolution of egg mimesis and egg crypsis in cuckoos. *J. Theor. Biol.*, **146**, 123–143.

Brooker, M. G. and L. C. Brooker (1989). The comparative breeding behaviour of two sympatric cuckoos, Horsfield's bronze-cuckoo *Chrysococcyx basalis* and the shining bronze-cuckoo *C. lucidus*, in western Australia: a new model for the evolution of egg morphology and host specificity in avian brood parasites. *Ibis*, **131**, 528–547.

Brown, R. J., M. N. Brown, M. de L. Brooke, and N. B. Davies (1990). Reactions of parasitized and unparasitized populations of *Acrocephalus* warblers to model cuckoo eggs. *Ibis*, **132**, 109–111.

Chance, E. P. (1922). *The cuckoo's secret*. Sidgwick & Jackson, London.

Chance, E. P. (1940). *The truth about the cuckoo*. Country Life, London.

Clarke, C. A. and P. M. Sheppard (1962). The genetics of the mimetic butterfly *Papilio glaucus*. *Ecology*, **43**, 159–161.

Cruz, A. and J. W. Wiley (1989). The decline of an adaptation in the absence of a presumed selection pressure. *Evolution*, **43**, 55–62.

Davies, N. B. (1992). *Dunnock behaviour and social evolution*. Oxford University Press.

Davies, N. B. and M. de L. Brooke (1988). Cuckoos versus reed warblers: adaptations and counteradaptations. *Anim. Behav.*, **36**, 262–284.

Davies, N. B. and M. de L. Brooke (1989*a*). An experimental study of co-evolution between the cuckoo, *Cuculus canorus*, and its hosts. I. Host egg discrimination. *J. Anim. Ecol.*, **58**, 207–224.

Davies, N. B. and M. de L. Brooke (1989*b*). An experimental study of co-evolution between the cuckoo, *Cuculus canorus*, and its hosts. II. Host egg markings, chick discrimination and general discussion. *J. Anim. Ecol.*, **58**, 225–236.

Davies, N. B. and M. de L. Brooke (1991). Coevolution of the cuckoo and its hosts. *Sci. Am.*, **264**(1), 92–98.

Davies, N. B., A. F. G. Bourke, and M. de L. Brooke (1989). Cuckoos and parasitic ants: interspecific brood parasitism as an evolutionary arms race. *Trends Ecol. Evol.*, **4**, 274–278.

Davies, N. B., M. de L. Brooke, and A. Kacelnik (1996). Recognition errors and probability of parasitism determine whether reed warblers should accept or reject mimetic cuckoo eggs. *Proc. R. Soc. Lond. B.*, **263**, 925–931.

Dawkins, R. and J. R. Krebs (1979). Arms races between and within species. *Proc. R. Soc. Lond. B.*, **205**, 489–511.

Dröscher, L. (1988). A study on radio-tracking of the European cuckoo (*Cuculus canorus canorus*). *Proc. Int. 100 Do-G. meeting*, pp. 187–193.

Duckworth, J. W. (1991). Responses of breeding reed warblers *Acrocephalus scirpaceus* to mounts of sparrowhawk *Accipiter nisus*, cuckoo *Cuculus canorus* and jay *Garrulus glandarius*. *Ibis*, **133**, 68–74.

Feare, C. J., P. L. Spencer, and D. A. T. Constantine (1982). Time of egg-laying of starlings *Sturnus vulgaris*. *Ibis*, **124**, 174–178.

Feder, J. L., C. A. Chilcote, and G. L. Bush (1988). Genetic differentiation between sympatric host-races of the apple maggot fly, *Rhagoletis pomonella*. *Nature*, **336**, 61–64.

Gärtner, K. (1981). Das Wegnehmen von Wirtsvogeleiern durch den kuckuck (*Cuculus canorus*). *Ornithol. Mitt.*, **33**, 115–131.

Gärtner, K. (1982). Zur Ablehnung von Eiern und Jungen des kuckucks (*Cuculus canorus*) durch die Wirtsvögel: Beobachtungen und experimentelle Untersuchungen an Sumpfrohrsänger. *Vogelwelt*, **103**, 201–224.

Gaston, A. J. (1976). Brood parasitism by the pied crested cuckoo *Clamator jacobinus*. *J. Anim. Ecol.*, **45**, 331–345.

Gibbs, H. L., M. de L. Brooke, and N. B. Davies (1996). Analysis of genetic differentiation of host races of the common cuckoo *Cuculus canorus* using mitochondrial and microsatellite DNA variation. *Proc. R. Soc. Lond. B.*, **263**, 89–96.

Harrison, C. J. O. (1968). Egg mimicry in British cuckoos. *Bird Study*, **15**, 22–28.

Harvey, P. H. and M. D. Pagel (1991). *The comparative method in evolutionary biology*. Oxford University Press, Oxford.

Hett, W. S. (1936). *Aristotle: Minor works. On marvellous things heard*. Heinemann, London.

Higuchi, H. (1989). Responses of the bush warbler *Cettia diphone* to artificial eggs of *Cuculus* cuckoos in Japan. *Ibis*, **131**, 94–98.

Huntingford, F. A. and P. J. Wright (1992). Inherited population differences in avoidance conditioning in three-spined sticklebacks, *Gasterosteus aculeatus*. *Behaviour*, **122**, 264–273.

Janzen, D. H. (1980). When is it co-evolution? *Evolution*, **34**, 611–612.

Järvinen, A. (1984). Relationship between the common cuckoo *Cuculus canorus* and its host the redstart *Phoenicurus phoenicurus*. *Ornis Fenn.*, **61**, 84–88.

Jenner, E. (1788). Observations on the natural history of the cuckoo. *Phil. Trans. R. Soc. Lond.*, **78**, 219–237.

Jourdain, F. C. R. (1925). A study of parasitism in the cuckoos. *Proc. Zool. Soc. Lond.*, **1925**, 639–667.

Kacelnik, A., P. A. Cotton, L. Stirling, and J. Wright (1995). Food allocation among nestling starlings: sibling competition and the scope of parental choice. *Proc. R. Soc. Lond. B.*, **259**, 259–263.

Kelly, C. (1987). A model to explore the rate of spread of mimicry and rejection in hypothetical populations of cuckoos and their hosts. *J. Theor. Biol.*, **125**, 283–299.

Lack, D. (1968). *Ecological adaptations for breeding in birds.* Methuen, London.

Lanyon, S. M. (1992). Interspecific brood parasitism in blackbirds (Icterinae): a phylogenetic perspective. *Science*, **255**, 77–79.

Lotem, A. (1993). Learning to recognize nestlings is maladaptive for cuckoo *Cuculus canorus* hosts. *Nature*, **362**, 743–745.

Lotem, A., H. Nakamura, and A. Zahavi (1992). Rejection of cuckoo eggs in relation to host age: a possible evolutionary equilibrium. *Behav. Ecol.*, **3**, 128–132.

Lotem, A., H. Nakamura, and A. Zahavi (1995). Constraints on egg discrimination and cuckoo-host co-evolution. *Anim. Behav.*, **49**, 1185–1209.

Magurran, A. E. (1990). The inheritance and development of minnow anti-predator behaviour. *Anim. Behav.*, **39**, 834–842.

Marchetti, K. (1992). Cost to host defence and the persistence of parasitic cuckoos. *Proc. R. Soc. Lond. B.*, **248**, 41-45.

Mason, P. and S. I. Rothstein (1986). Coevolution and avian brood parasitism: cowbird eggs show evolutionary response to host discrimination. *Evolution*, **40**, 1207–1214.

Mason, P. and S. I. Rothstein (1987). Crypsis versus mimicry and the color of shiny cowbird eggs. *Amer. Natur.*, **130**, 161–167.

May, R. M. and S. K. Robinson (1985). Population dynamics of avian brood parasitism. *Amer. Natur.*, **126**, 475–494.

Moksnes, A. and E. Røskaft (1987). Cuckoo host interactions in Norwegian mountain areas. *Ornis Scand.*, **18**, 168–172.

Moksnes, A. and E. Røskaft (1989). Adaptations of meadow pipits to parasitism by the common cuckoo. *Behav. Ecol. Sociobiol.*, **24**, 25–30.

Moksnes, A. and E. Røskaft (1995). Egg morphs and host preference in the common cuckoo (*Cuculus canorus*): an analysis of cuckoo and host eggs from European museum collections. *J. Zool. Lond.*, **236**, 625–648.

Moksnes, A., E. Røskaft, A. T. Braa, L. Korsnes, H. M. Lampe, and H. Ch. Pedersen (1991*a*). Behavioural responses of potential hosts towards artificial cuckoo eggs and dummies. *Behaviour*, **116**, 64–89.

Moksnes, A., E. Røskaft, and A. T. Braa (1991*b*). Rejection behaviour by common cuckoo hosts towards artificial brood parasite eggs. *Auk*, **108**, 348–354.

Moksnes, A., E. Røskaft, and L. Korsnes (1993). Rejection of cuckoo (*Cuculus canorus*) eggs by meadow pipits (*Anthus pratensis*). *Behav. Ecol.*, **4**, 120–127.

Moksnes, A., E. Røskaft, and T. Tysse (1995). On the evolution of blue cuckoo eggs in Europe. *J. Avian Biol.*, **26**, 13–19.

Molnar, B. (1944). The cuckoo in the Hungarian plain. *Aquila*, **51**, 100–112.

Nakamura, H. (1990). Brood parasitism by the cuckoo *Cuculus canorus* in Japan and the start of new parasitism on the azure-winged magpie *Cyanopica cyana*. *Jap. J. Ornithol.*, **39**, 1–18.

Nakamura, H. and Y. Miyazawa (1990). Social system among cuckoo males *Cuculus canorus* at Kayanodaira heights, central Japan. *Bull. Inst. Nature Educ.*, Shiga Heights, Shinshu Univ., **27**, 17–27.

Newton, A. (1893). A dictionary of birds. Part I. A. & C. Black, London.

Nicolai, J. (1964). Der Brutparasitismus der Viduinae als ethologisches Problem. *Z. Tierpsychol.*, **21**, 127–204.

Øien, I. J., A. Moksnes, and E. Røskaft (1995). Evolution of variation in egg colour and marking pattern in European passerines: adaptations in a coevolutionary arms race with the cuckoo *Cuculus canorus*. *Behav. Ecol.*, **6**, 166–174.

Payne, R. B. (1974). The evolution of clutch size and reproductive rates in parasitic cuckoos. *Evolution*, **28**, 169–181.

Payne, R. B. (1982). Species limits in the indigobirds (Ploceidae, *Vidua*) of West Africa: mouth mimicry, song mimicry, and description of new species. *Misc. Publ. Univ. Mich. Museum Zool.*, 102.

Peck, A. L. (1970). *Aristotle: Historia animalium* Vol. II. Heinemann, London.

Perrins, C. M. (1967). The short apparent incubation period of the cuckoo. *Brit. Birds*, **60**, 51–52.

Petit, L. J. (1991). Adaptive tolerance of cowbird parasitism by prothonotary warblers: a consequence of nest-site limitation? *Anim. Behav.*, **41**, 425–432.

Purvis, A. (1991). Comparative analysis by independent contrast. Version 1.2, Users guide. University of Oxford.

Redondo, T. (1993). Exploitation of host mechanisms for parental care by avian brood parasites. *Etologia*, **3**, 235–297.

Redondo, T. and F. Castro (1992). Signalling of nutritional need by magpie nestlings. *Ethology*, **92**, 193–204.

Robertson, R. J. and R. F. Norman (1977). The function and evolution of aggressive host behaviour towards the brown-headed cowbird (Molothrus ater). *Can. J. Zool.*, **55**, 508–518.

Rohwer, S. and C. D. Spaw (1988). Evolutionary lag versus bill-size constraints: a comparative study of the acceptance of cowbird eggs by old hosts. *Evol. Ecol.*, **2**, 27–36.

Rothstein, S. I. (1974). Mechanisms of avian egg recognition: possible learned and innate factors. *Auk*, **91**, 796–807.

Rothstein, S. I. (1975*a*). An experimental and teleonomic investigation of avian brood parasitism. *Condor*, **77**, 250–271.

Rothstein, S. I. (1975*b*). Evolutionary rates and host defenses against avian brood parasitism. *Amer. Natur.*, **109**, 161–176.

Rothstein, S. I. (1982*a*). Successes and failures in avian egg recognition with comments on the utility of optimality reasoning. *Amer. Zool.*, **22**, 547–560.

Rothstein, S. I. (1982*b*). Mechanisms of avian egg recognition: which egg parameters elicit responses by rejector species? *Behav. Ecol. Sociobiol.*, **11**, 229–239.

Rothstein, S. I. (1990). A model system for coevolution: avian brood parasitism. *Annu. Rev. Ecol. Syst.*, **21**, 481–508.

Scott, D. M. (1991). The time of day of egg laying by the brown-headed cowbird and other icterines. *Can. J. Zool.*, **69**, 2093–2099.

Sealy, S. G. (1992). Removal of yellow warbler eggs in association with cowbird parasitism. *Condor*, **94**, 40–54.

Sheppard, P. M. (1958). *Natural selection and heredity*. Hutchinson, London.

Sibley, C. G. and J. E. Ahlquist (1990). *Phylogeny and classification of birds: A study in molecular evolution*. Yale University Press, New Haven.

Smith, J. N. M., P. Arcese, and I. G. McLean (1984). Age, experience and enemy recognition by wild song sparrows. *Behav. Ecol. Sociobiol.*, **14**, 101–106.

Soler, J. J. and A. P. Møller (1996). A comparative analysis of the evolution of variation in appearance of eggs of European passerines in relation to brood parasitism. *Behav. Ecol.*, **7**, 89–94.

Soler, M. and A. P. Møller (1990). Duration of sympatry and coevolution between the great spotted cuckoo and its magpie host. *Nature*, **343**, 748–750.

Soler, M., J. J. Soler, J. G. Martinez, and A. P. Møller (1994). Micro-evolutionary change in host response to a brood parasite. *Behav. Ecol. Sociobiol.*, **35**, 295–301.

Swynnerton, C. F. M. (1918). Rejections by birds of eggs unlike their own: with remarks on some of the cuckoo problems. *Ibis*, **1918**, 127–154.

Takasu, F., K. Kawasaki, H. Nakamura, J. E. Cohen, and N. Shigesada (1993). Modeling the population dynamics of a cuckoo-host association and the evolution of host defenses. *Amer. Natur.*, **142**, 819–839.

Thompson, J. N. (1994). *The coevolutionary process.* University of Chicago Press, Chicago, London.

Victoria, J. K. (1972). Clutch characteristics and egg discriminative ability of the African village weaverbird *Ploceus cucullatus*. *Ibis*, **114**, 367–376.

von Haartman, L. (1981). Coevolution of the cuckoo *Cuculus canorus* and a regular cuckoo host. *Ornis Fenn.*, **58**, 1–10.

Wallace, A. R. (1889). *Darwinism: An exposition of the theory of natural selection with some of its applications*. Macmillan, London.

Wyllie, I. (1981). *The cuckoo*. Batsford, London.

Yamagishi, S. and M. Fujioka (1986). Heavy brood parasitism by the common cuckoo *Cuculus canorus* on the azure-winged magpie *Cyanopica cyana*. *Tori*, **34**, 91–96.

Zuniga, J. M. and T. Redondo (1992). No evidence for variable duration of sympatry between the great spotted cuckoo and its magpie host. *Nature*, **359**, 410–411.

3

Host Use and Egg Color of Japanese Cuckoos

HIROYOSHI HIGUCHI

Different species of parasitic cuckoos deposit their eggs in nests of different host species, although a few incidental hosts are used in common by two or more species of cuckoo (e.g., Baker 1927, 1942; Kobayashi and Ishizawa 1940; Friedmann 1964, 1967*a*; Kiyosu 1965; Jensen and Jensen 1969; Lack 1971; Gaston 1976; Payne 1977). Friedmann (1967*b*) proposed the term *alloxenia* to describe the situation in which related species of parasites tend to use different species of hosts. Host segregation probably evolved as a result of interspecific competition (Lack 1971; Wyllie 1981), but the process or mechanism of its evolution has not been proven.

Different species of cuckoos also tend to lay eggs similar in color to the eggs of their main hosts. Egg mimicry can be found in many species of cuckoos (e.g., Kobayashi and Ishizawa 1940; Baker 1942; Kiyosu 1965; Payne 1977; Wyllie 1981; Tarboton 1986; Higuchi and Sato 1984). There are three hypotheses for egg mimicry in cuckoos: (i) discriminating ability of host species; (ii) convergence on cryptic patterns through general predation; and (iii) selection through predation by other cuckoos. Many studies showed that host species tend to reject eggs dissimilar in color to their own and accept more similar eggs (e.g., Rensch 1924; Southern 1954; Lack 1968; Alvarez et al. 1976; von Haartman 1981; Davies and Brooke 1988; Higuchi 1989). These studies therefore strongly suggest that host discrimination is the reason for egg mimicry in cuckoos. The second hypothesis postulates that cuckoo eggs resemble host eggs so that the clutch is not more conspicuous to predators (Wallace 1889; Harrison 1968; von Haartman 1981). Bird eggs are to some extent cryptic with their surroundings, so the addition of a differently colored egg to the nest may lead to the predation of the whole clutch. However, there is little direct support for this hypothesis (Mason and Rothstein 1987; Davies and Brooke 1988). The third hypothesis stems from the observation that some parasitized nests are later parasitized by a second cuckoo (Wyllie 1981; Davies and Brooke 1988). It would clearly pay the second cuckoo to remove the first cuckoo's egg because only one cuckoo is ever reared per nest and the earlier-laid egg is likely to hatch first. Egg mimicry may therefore evolve because it reduces the chance that second cuckoos will be able to discriminate the first cuckoo's egg from the host's eggs (Davies and Brooke 1988; Brooker and Brooker 1989, 1990; Brooker et al. 1990). This process is plausible, but field data are scarce to test the hypothesis.

Many earlier studies of host selection in communities with several cuckoo species may have misclassified host relationships because cuckoo eggs were collected immediately upon being found, and eggs were identified based mainly on egg characters such as color and markings (e.g., Kobayashi and Ishizawa 1940; Baker 1942). This undoubtedly led to errors in the past, and to date some or many examples of host

Figure 3.1 The Japanese Islands, showing the study areas (•) and the northern limit (----) of the range of little cuckoos.

species and egg colors reported may be wrong.

In Japan, four species of *Cuculus* cuckoos breed sympatrically: common cuckoo, *C. canorus*; Himalayan cuckoo, *C. saturatus*; little cuckoo *C. poliocephalus*; and Horsfield's hawk cuckoo, *C. fugax*. The number and species composition of cuckoos are different among the Japanese Islands, a situation that provides an excellent opportunity to study interrelationships among different species of cuckoos. I studied host use and egg color of Japanese cuckoos by field experiments and by a reliable observation method based on the appearance of nestling cuckoos (Higuchi and Payne 1986). In this chapter, I first present data on host use and egg color of Japanese cuckoos. I then refer to the responses of host species to differently colored artificial eggs, and finally discuss the evolution and mechanisms of host use by cuckoos. It is shown that host use changes in the presence or absence of closely related cuckoo species and that the color of cuckoo eggs may also vary with changes in host species with different egg color preferences. I conclude that interspecific competition is important for the evolution of host use and egg color in cuckoos.

MATERIALS AND METHODS

Field observations

Field data on host selection and egg color were collected in 1980–85 and 1988–90 from Mt Hakone, Mt Fuji, and the surrounding areas of central Honshu (the largest main island), Miyake Island of the Izu Islands off the Pacific coast of central Honshu, and Asahikawa, Wakkanai and the surrounding areas of central and northern Hokkaido (the northernmost and second largest main island; see fig. 3.1). All four cuckoo species breed in central Honshu (WBSJ 1980), whereas only little cuckoos breed in the Izu Islands (Higuchi 1973). Common, Himalayan, and Horsfield's hawk cuckoos breed in central and northern Hokkaido, though Horsfield's hawk cuckoos are sporadically distributed. The northern limit of the breeding distribution of little cuckoos is at latitude 42°30′ in the Oshima Peninsula of southern Hokkaido (OSJ 1974; WBSJ 1980; Higuchi and Sato 1984).

Extensive nest searches were conducted in various kinds of habitats in the study areas, without focusing on the nests of particular host species. When a parasitized nest was found, the size and weight of the cuckoo and host eggs were measured with a caliper and a balance, respectively, and notes taken on the egg colors. The cuckoo eggs were distinguished from the host eggs by their size (cuckoo eggs are usually larger) and color differences. Some cuckoo and host eggs were photographed. The contents of parasitized nests were checked again at least twice on later days. The species identification of cuckoo eggs was determined by the plumage and body size of nestlings 10 or more days old. This is the safest method to determine species identification, as the plumage and body size of the young of the four species of cuckoos are quite different.

The plumages of the nestling cuckoos aged more than nine days are summarized as follows. The upper plumages of young Himalayan and little cuckoos are similar: dark blackish brown in the former and slatey black or dark blackish brown in the latter. The throat and breast are dark blackish brown in young Himalayan cuckoos, whereas they are banded with black and white in young little cuckoos. The rest of the lower parts are banded with black and white in both species, but the bars are obscure and much darker in young Himalayan cuckoos (Higuchi and Payne 1986). The upper plumage of young common cuckoos is dark brown, barred with rufous, each feather having white margins that are absent in the other three species. One or two conspicuous spots are present on the head, though these are missing from the other three species. The throat, breast and other lower parts are barred white and black (Yamashina 1941; Higuchi and Sato 1984). Young Horsfield's hawk cuckoos are the most distinctive of all cuckoo nestlings; their underparts are not barred but are instead dirty white streaked with dark blackish brown. The upper plumage is blackish brown with a thick white bar around the collar. There are two large spots on both sides of the throat (Yamashina 1941).

Little cuckoo nestlings are less than half the size of other nestling cuckoos. The typical body weight of nestlings aged at least nine days is 25–30 g in little cuckoos, 55–75 g in common and Himalayan cuckoos, and 60–80 g in Horsfield's hawk cuckoos (Higuchi and Payne 1986; H. Higuchi, pers. obs.). The differences in body size of the young cuckoos correspond to those of the adult cuckoos. The mean body weight of adults is, respectively, 54.4 g in little cuckoos, 120.0 g in common cuckoos, 101.2 g in Himalayan cuckoos, and 123.7 g in Horsfield's hawk cuckoos (Kobayashi 1939; Enomoto 1941).

Because nest contents were not checked systematically, it was often difficult to know the reason for losses of cuckoo eggs or young during the incubation and nestling periods. Thus, the frequencies of egg ejection, nest desertion, or predation among these nests will not be discussed. Cases in which nestling or fledgling cuckoos were found but the eggs from which they hatched were not seen, were also excluded.

Field data from Mt. Fuji and Mt. Hakone on Honshu were grouped together, as were data from Asahikawa and Wakkanai on Hokkaido. This was because data were insufficient to compare these pairs of sites and because there were no obvious differences in habitats and the species composition of cuckoos and major hosts between sites on the same island.

Experiments on egg color

This study was carried out on Miyake Island, 1982–1985, its purpose being to determine the responses of bush warblers to different egg colors to elucidate the mechanisms by which the reddish eggs of little and Himalayan cuckoos have evolved (Higuchi 1989).

Experiments were performed at nests on which incubation had proceeded for four days or less and were thus close to the laying stage when the cuckoos usually deposit their eggs. Cuckoos also sometimes (not "usually" as described in Higuchi 1989) lay during the host's early incubation period. One egg model was added to each of 47 nests, and two to four egg models were introduced into each of the 55 other nests. In the latter cases, one egg model was added on the first day and was exchanged for another artificial egg on the next day. Hence, only one egg was in a nest at a time. One real egg was removed from each bush warbler nest when the first artificial egg was added. No birds were tested more than once with the same egg type.

Plastic egg models of six colors were used in the experiments: red (R), orange (O), pink (P), gray spotted with orange (GO), gray (G), and white (W). The spots on a GO egg covered 40–50% of the egg's surface area. The five colors used in painting were specified by the Munsell code. In the Munsell color system, each color occupies a particular site in the three dimensions of Hue, Value, and Chroma (Munsell 1929), and the closer the two colors, the more similar they are. In this system, the difference in color between bush warbler eggs and the plastic experimental eggs increases in the following order: R, O, P, G, and W. For controls, real bush warbler eggs were also used. The size of plastic eggs (mean size: 19.2×12.8 mm and 1.2 g) was similar to that of bush warbler eggs (18.3×13.6 mm and 1.8 g).

Each nest was checked 24–30 hours after the introduction of the egg. The responses of bush warblers were classified as : (i) acceptance; (ii) desertion of the entire clutch; and (iii) ejection of the introduced egg. However, because ejection was rare, ejection and desertion were grouped together into rejection for analysis. Ejection was probably rare because the plastic eggs used were too hard for bush warblers to peck holes if they tried to eject by spiking. Bush warblers also often deserted nests when real eggs of dissimilar colors were introduced (H. Higuchi, unpublished results).

HOST USE AND EGG COLOR

Honshu

A total of 139 parasitized nests were found in central Honshu. There was one cuckoo egg in each of 133 nests and two in each of the other six. In 51 of the nests with a single cuckoo egg and all nests with two cuckoo eggs, the cuckoo eggs or recently hatched nestlings were lost, usually with all host eggs and nestlings, which precluded identification of cuckoo eggs. Host use based on the other 82 nests is shown in fig. 3.2. This figure does not necessarily show the real frequencies of parasitism by the cuckoos, because of eggs which were not identified to species. Rather, this figure indicates levels of successful parasitism by the cuckoos.

All but one of the 28 young little cuckoos on Honshu were raised by bush warblers, *Cettia diphone*; the other was raised by cavity-nesting wrens, *Troglodytes troglodytes*. All the eggs from which the young cuckoos hatched were chocolate brown (fig. 3.3), a very similar color to that of bush warbler eggs. Wren eggs, however, are plain white.

All but one of the 12 young Himalayan cuckoos were fostered by crowned willow warblers, *Phylloscopus coronatus*. The other successful host was the Narcissus flycatcher, *Ficedula narcissina*. The eggs from which the young cuckoos hatched were white (seven) or very light brown (five), wreathed with small dark brown markings, particularly near the large end where the markings occurred in an irregular ring (see fig. 3.5, left egg). The egg colors were somewhat similar to those of Narcissus flycatchers but were not similar to those of crowned willow warblers, which were plain white.

Half of the 16 young Horsfield's hawk cuckoos were raised by Siberian blue robins, *Erithacus cyane*, and five by blue-and-white flycatchers, *Cyanoptila cyanomelana*. The other successful hosts were Siberian bluechats *Tarsiger cyanurus* and Narcissus flycatchers. The eggs from which the young cuckoos hatched were plain blue (14) or blue with some dark blue markings near the large end (two). The egg colors were similar to those of Siberian blue robins, but not to those of the other hosts, which were whitish with small brownish markings.

About one-third of the 26 young common cuckoos were fostered by bull-headed shrikes, *Lanius bucephalus*, and another one-third by great reed warblers, *Acrocephalus arundinaceus*. The other successful hosts were meadow buntings (*Emberiza cioides*), gray wagtails (*Motacilla cinerea*), and black-faced buntings (*Emberiza spodocephala*). The eggs from which the young cuckoos hatched varied in color and markings. Ground color ranged from grayish white to very light brown. All eggs were marked with streaks, fine spots, lines, or irregular markings in many parts, particularly near the large end. The color of such markings ranged from light brown to blackish brown. Some eggs were somewhat similar in coloration to those of the main hosts such as bull-headed shrikes, great reed warblers, and meadow buntings, but others were not.

Two of the eight cuckoo eggs in the nests of bull-headed shrikes, another two of the eight cuckoo eggs in the nests of great reed warblers, and one of the four cuckoo eggs in the nests of meadow buntings were very similar in color and markings to the eggs of the respective host species (fig. 3.4). Eggs that looked like those of bull-headed shrikes had irregular spots and markings, particularly near the large end in an irregular ring. Eggs that looked like those of great reed warblers had many irregular spots and markings throughout the whole surface area. Eggs similar to those of meadow buntings had line markings, particularly near the large end.

Egg size was largest in Horsfield's hawk cuckoos and smallest in Himalayan cuckoos (table 3.1). The egg sizes were significantly different

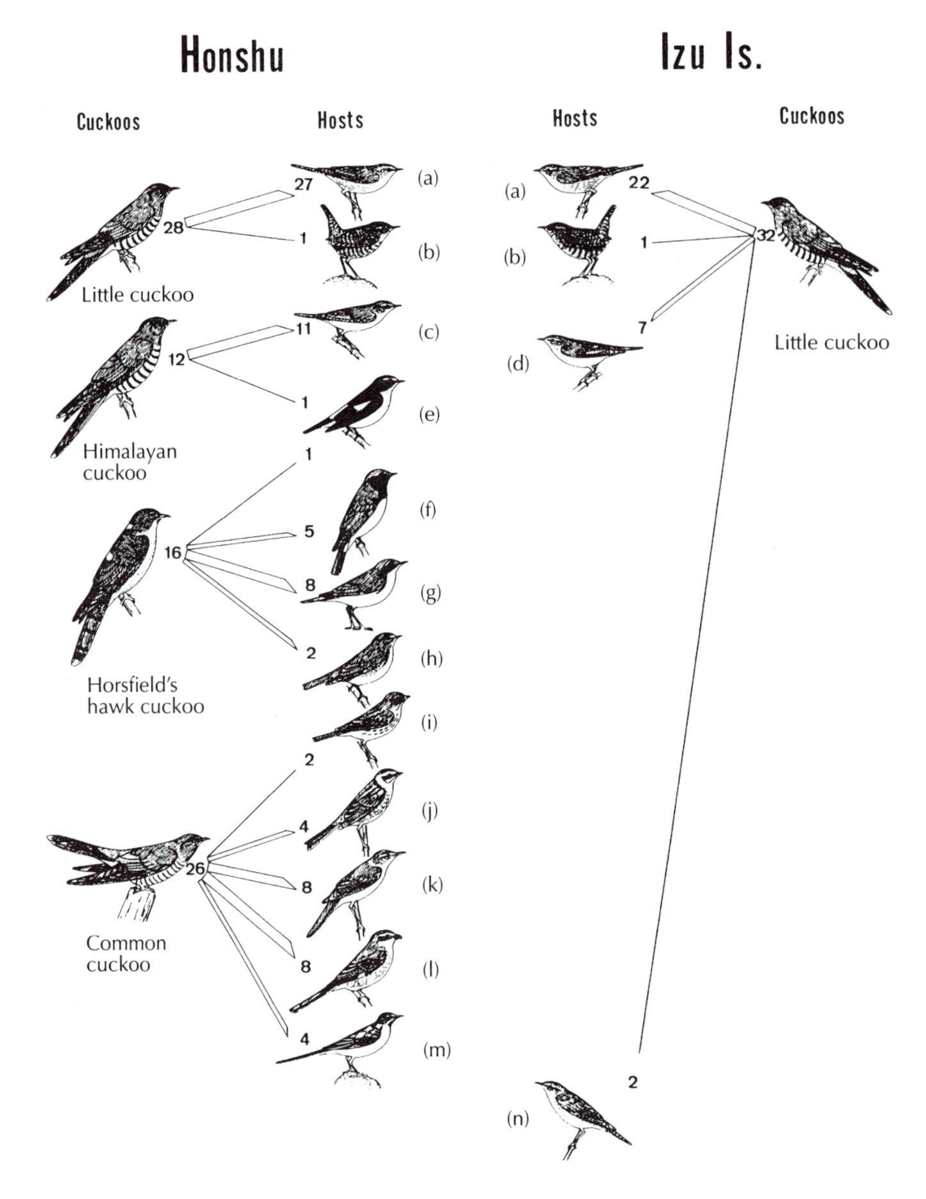

Figure 3.2 Host species and frequency of parasitism in the four species of Japanese cuckoos. The species of hosts shown are as follows: (a) bush warblers; (b) wren; (c) crowned willow warblers; (d) Ijima's willow warblers; (e) narcissus flycatcher; (f) blue-and-white flycatcher; (g) Siberian blue robin; (h) Siberian bluechat; (i) black-faced bunting; (j) meadow bunting; (k) great reed warbler; (l) bull-headed shrike; (m) gray wagtail; (n) Middendorff's grasshopper warbler. Numerals next to cuckoos represent the number of cases of parasitism in which the cuckoo species was determined by the appearance of the young. Numerals next to host species represent the number of the nests with the young cuckoos.

between all pairs of cuckoo species ($P < 0.01$ for each of the three characters of length, width, and weight, *t*-tests, two-tailed). It is notable that Himalayan cuckoos lay smaller eggs than little cuckoos despite their much larger body size (table 3.1). This is probably because most Himalayan cuckoos parasitize crowned willow warblers which lay very small eggs.

The following species have also been reported as hosts for Honshu cuckoos, when the cuckoo species could be reliably identified by the appearance of the young either reported in the litera-

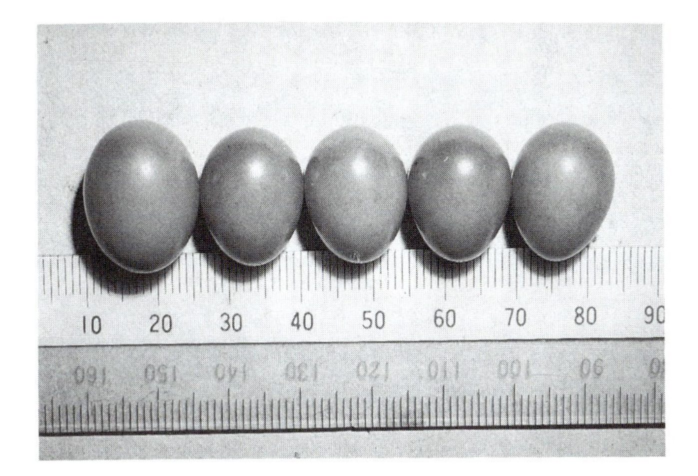

Figure 3.3 The eggs of little cuckoos (left) and bush warblers.

ture or from personal observations including photographs: short-tailed bush warblers *Cettia squameiceps* for Horsfield's hawk cuckoos; azure-winged magpies *Cyanopica cyana* (Yamagishi and Fujioka 1986; Nakamura 1990); black-browed reed warblers *Acrocephalus bistrigiceps*; brown shrikes *Lanius cristatus*; stonechat *Saxicola torquata*; Japanese accentor *Prunella rubida*; brown thrush *Turdus chrysolaus* (Hota, A., Imanishi, T., and Yoshino, T. pers. comm.); and gray thrush *T. cardis* (Hayashi 1978) for common cuckoos. These records are all from central Honshu (see also Nakamura et al., this vol.).

The Izu Islands

All 58 parasitized nests observed on Miyake Island had only one cuckoo egg. The species identification of the eggs was determined in 32 cases by the appearance of the young, all of which were little cuckoos. Twenty-two of the 32 young were raised by bush warblers, and seven by Ijima's willow warblers *Phylloscopus ijimae* (fig. 3.2). The other successful hosts were Middendorff's grasshopper warblers *Locustella ochotensis* and wrens. It is notable that willow warblers (crowned willow warblers) that were parasitized by Himalayan cuckoos in Honshu were occasionally parasitized by little cuckoos on this island.

Ijima's and crowned willow warblers are closely related allopatric species, and are sometimes considered different subspecies of the same species (OSJ 1942). Data in fig. 3.2 show that bush warblers made up a significantly smaller proportion of the successful little cuckoo hosts on Miyake Island than on the mainland of Honshu (22 of 32 versus 27 of 28, $P = 0.005$, Fisher's exact probability test, two-tailed). The range of host selection by little cuckoos seems to have expanded in the absence of other cuckoo species on Miyake Island.

Because little cuckoos are the only cuckoo species breeding on Miyake Island, we can assume that the eggs lost during the incubation and early nestling periods were also little cuckoo eggs. Thus, three more species were used as hosts: meadow buntings; bull-headed shrikes; and white-eyes (table 3.2). There was a positive significant correlation between % parasitism and the % of cuckoo eggs that were successful (Spearman correlation $r_s = 0.785$, $P < 0.05$, $N = 7$), which suggests avoidance of low-quality hosts.

All the eggs of little cuckoos on Miyake Island were chocolate-brown. The egg color was very similar to that of bush warblers, but different from those of all other host species. Ijima's willow warblers and wrens lay plain white eggs, and meadow buntings, bull-headed shrikes, and Middendorff's grasshopper warblers lay whitish eggs with brownish or reddish markings. It is not known how often these host species accept the chocolate brown eggs of little cuckoos.

The egg sizes of little cuckoos on Miyake Island and Honshu were not significantly different ($P > 0.05$ for each of the three characters of length, width, and weight (t-tests, two-tailed, table 3.1).

Figure 3.4 Good examples of common cuckoo egg mimicry in the nests of (a) meadow buntings, (b) great reed warblers, and (c) bull-headed shrikes. Eggs shown by arrows are cuckoo eggs. Although these pictures were not taken in the study areas but in Nobeyama, Nagano, central Honshu, the color and markings are very similar to those of the study areas.

Table 3.1 Egg sizes of Japanese cuckoos and two of the main host species

Species (Mean bodyweight*)		Egg measurements: Honshu Length (mm)	Width (mm)	Weight (g)	Miyake Island Length (mm)	Width (mm)	Weight (g)	Hokkaido Length (mm)	Width (mm)	Weight (g)
Little cuckoo	*n*	21	21	14	21	21	19			
Cuculus poliocephalus	x̄	21.67	15.74	2.82	21.70	15.69	2.76			
(54.4 g)	S.E.	0.16	0.08	0.05	0.14	0.05	0.05			
Himalayan cuckoo	*n*	9	9	9				8	8	8
C. saturatus	x̄	19.52	14.39	2.31				22.07	15.77	2.85
(101.2 g)	S.E.	0.22	0.11	0.06				0.21	0.12	0.06
Horsfield's hawk cuckoo	*n*	12	12	12				5	5	5
C. fugax	x̄	28.13	19.85	5.59				28.24	19.91	5.75
(123.7 g)	S.E.	0.24	0.09	0.11				0.13	0.04	0.13
Common cuckoo	*n*	18	18	18				16	16	16
C. canorus	x̄	23.39	17.66	3.67				23.37	17.61	3.69
(120.0 g)	S.E.	0.19	0.09	0.07				0.21	0.09	0.07
Bush warbler	*n*	46	46	46	35	35	35	28	28	28
Cettia diphone	x̄	18.10	13.89	1.82	18.31	14.05	1.82	18.17	13.93	1.84
(17.6 g)	S.E.	0.11	0.07	0.02	0.10	0.08	0.02	0.12	0.10	0.03
Crowned willow warbler	*n*	21	21	21						
Phylloscopus coronatus	x̄	15.81	12.64	1.15						
(10.5 g)	S.E.	0.16	0.08	0.04						

*From Enomoto (1941).

Table 3.2 Parasitism frequencies of little cuckoos on Miyake Island of the Izu Islands

Species	No. of nests found	No. of nests parasitized	% parasitized	No. of nests parasitized successfully	% of nests parasitized successfully
Bush warbler (*Cettia diphone*)	98	32	32.7	22	68.8
Wren (*Troglodytes troglodytes*)	25	4	16.0	1	25.0
Ijima's willow warbler (*Phylloscopus ijimae*)	64	12	18.8	7	58.3
Meadow bunting (*Emberiza cioides*)	12	2	16.7	0	0.0
Bull headed shrike (*Lanius bucephalus*)	6	1	16.7	0	0.0
Middendorf's grasshopper warbler (*Locustella ochotensis*)	18	5	27.8	2	40.0
Japanese white-eye (*Zosterops japonica*)	42	2	4.8	0	0.0

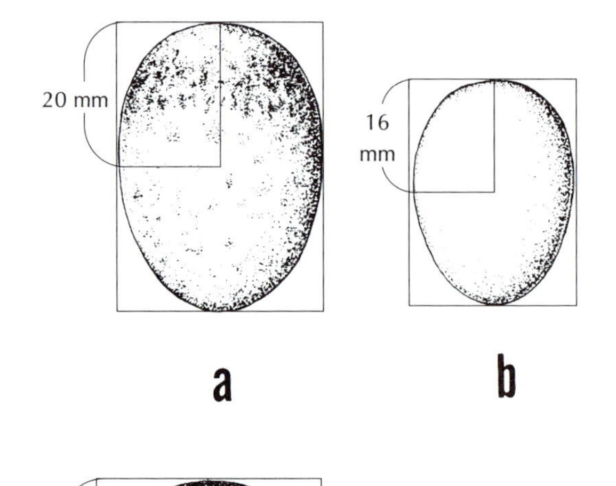

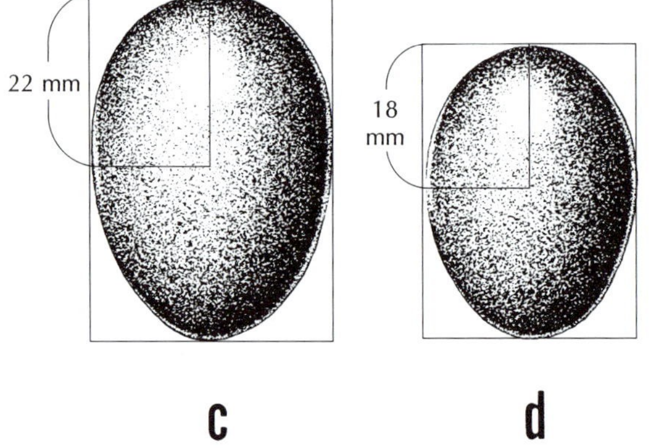

Figure 3.5 The eggs of (a) Himalayan cuckoos and (b) crowned willow warblers on Honshu, and of (c) Himalayan cuckoos and (d) bush warblers on Hokkaido. Shown in natural sizes.

Hokkaido

A total of 56 nests with cuckoo eggs were found in central and northern Hokkaido, each with only one cuckoo egg. Twenty-three of the 36 cuckoo young that I was able to identify were common cuckoos, eight were Himalayan cuckoos, and the other five were Horsfield's hawk cuckoos.

All the eight young Himalayan cuckoos were raised by bush warblers. The eggs from which the young cuckoos hatched were chocolate-brown (five eggs) and orange-brown (three) in color. These eggs were completely different in color from those of the same species in Honshu (fig. 3.5). The chocolate-brown eggs were very similar to bush warbler eggs and to little cuckoo eggs from Honshu and Miyake Island, and the orange-brown eggs were somewhat paler than bush warbler and little cuckoo eggs. Apparently, Himalayan cuckoos on Hokkaido lay reddish eggs in the nests of bush warblers in the absence of little cuckoos. Parasitism on bush warblers was observed in both Asahikawa (seven cases) and Wakkanai (one case).

Six of the 23 young common cuckoos were fostered by black-browed reed warblers, another six by white wagtails *Motacilla alba*, four by great reed warblers, and another four by bull-headed shrikes. The other three were raised by black-faced buntings (two cases) and gray-headed buntings *Emberiza fucata* (one case). Three (black-browed reed warblers, white wagtails and gray-headed buntings) of the six species were not recorded as hosts in my study on Honshu (black-browed reed warblers have been recorded by other observers), but all the host species including these three belong to one of the four groups of birds that were often parasitized on Honshu: reed warblers, shrikes, buntings, and wagtails. Statistical tests culd not be made because of insufficient sample sizes, but host selection by the common cuckoo appears to be basically the same on both Honshu and Hokkaido, unlike the case for the Himalayan cuckoo.

The eggs from which common cuckoos hatched were similar in color and markings to those of the same species on Honshu, although no eggs occurred with line markings similar to those of meadow buntings on Honshu. Ground color varied from grayish white to very light brown. Streaks, spots, and irregular markings were light brown to blackish brown in color. As on Honshu, some common cuckoo eggs were similar in color to the host eggs with which they were found, but most were not.

Three of the five young Horsfield's hawk cuckoos were raised by Siberian blue robins, and the other two by blue-and-white flycatchers. The eggs from which the young cuckoos hatched were plain blue. As on Honshu, egg color was similar to that of Siberian blue robins, but not to those of blue-and-white flycatchers. These two host species were also the main hosts of the cuckoo on Honshu.

The eggs of common and Horsfield's hawk cuckoos were not significantly different in size from those of the same species on Honshu, respectively ($P > 0.05$ for the three characters, *t*-test; table 3.1). The eggs of Himalayan cuckoos were significantly larger in all dimensions than those of the same species on Honshu ($P < 0.01$), and did not differ significantly from those of little cuckoos on Honshu and Miyake Island ($P > 0.05$). Himalayan cuckoos on Hokkaido probably evolved the same egg size as little cuckoos on Honshu and Miyake Island because they parasitize the same host species, bush warblers. The egg size of bush warblers is not significantly different among Honshu, Miyake Island, and Hokkaido populations ($P > 0.05$; table 3.1). These facts suggest that bush warblers have a strong egg size preference as a cuckoo host. Further studies using plastic eggs of different sizes would clarify bush warbler egg size preferences.

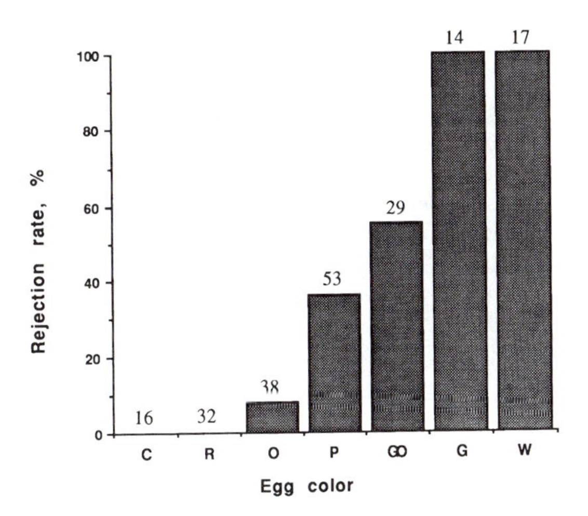

Figure 3.6 The rejection rates of bush warblers to eggs of different colors. C: control eggs of bush warblers; R: red; O: orange; P: pink; GO: gray with orange spots; G: gray; W: white. Numerals above bars represent the number of tests.

The following species are also known as hosts for Hokkaido cuckoos, based on the appearance of the young: willow warblers (species unknown) for Himalayan cuckoos (Y. Iijima, pers. comm.) and Siberian rubythroats *Erithacus calliope* (S. Takano, pers. comm.). These records were from central (Tokachi) and eastern (Nemuro) Hokkaido, respectively.

Table 3.3 summarizes the main host species and egg colors of Japanese cuckoos.

RESPONSES OF HOST SPECIES TO DIFFERENT EGG MODELS

The order of introduction of different egg models did not influence the rejection rates (Higuchi 1989). The rejection rates of the seven egg models ranged from 0% of C and R to 100% of G and W (fig 3.6). There were no significant differences among responses to C, R, and O, between P and GO, or between G and W ($P > 0.05$, Fisher's exact probability tests, two-tailed), but there were significant differences among other combinations of colors ($P < 0.01$, Fisher's exact probability tests, two-tailed). Because both rejection rate (fig. 3.6) and the difference in color from the bush warbler's egg increase in the order R, O, P, G, and W (see "Materials and methods"), it is apparent that bush warblers are most likely to reject those eggs that are least like their own. It is remarkable that the rejection rate of GO was much smaller than that of G. Presumably, the orange spots of GO contributed to its similarity to bush warbler eggs and to its acceptance.

These results suggest that the chocolate-brown eggs of little cuckoos on Honshu and of Himalayan cuckoos on Hokkaido evolved in response to bush warbler rejection of eggs unlike their own. Another explanation instead of mimicry of host eggs is that cuckoos might have evolved the same egg color because, by laying in the same nest as the hosts, they are subject to the same selection pressures for crypsis related to nest structure and predation (Harrison 1968; von Haartman 1981; Mason and Rothstein 1987). Bush warblers build nests with a side entrance in dense bamboo thickets. The same type of nest is built by

Table 3.3 Summary of host use and egg appearance in Japanese cuckoos

	Species			
Host and egg appearance	Little cuckoo	Himalayan cuckoo	Horsfield's hawk cuckoo	Common cuckoo
Honshu				
Major hosts	Bush warbler	Crowned willow warbler	Siberian blue robin Blue-and-white flycatcher	Shrikes, reed warblers, buntings, wagtails
Egg coloration	Chocolate-brown	White to light brown wreathed with markings	Plain blue or blue with spots	Pale ground color with variable markings
Eggs similar to hosts?	Yes	No	Yes for Siberian blue robin	Usually not
Miyake Island		Absent	Absent	Absent
Major hosts	Bush warbler, Ijima's willow warbler	—	—	—
Egg coloration	Chocolate-brown	—	—	—
Eggs similar to hosts?	Yes for bush warbler; No for willow warbler	—	—	—
Northern Hokkaido	Absent			
Major hosts	—	Bush warbler	Siberian blue robin Blue-and-white flycatcher	Shrikes, reed warblers, buntings, wagtails
Egg coloration	—	Chocolate to orange-brown	Plain blue	Pale ground color with variable markings
Eggs similar to hosts	—	Yes	Yes for Siberian blue robin	Usually not

Ijima's willow warblers in similar habitat on Miyake Island, but that species lays plain white eggs. Therefore, it seems unlikely that the domed nest of bush warblers has influenced the similarity in egg color of the cuckoos and bush warblers. Generally, the cryptic explanation for egg mimicry may not be relevant to species with domed nests because the eggs are hidden from the view of predators.

Selection by second cuckoos could not be evaluated because of the lack of data on cuckoo predation. However, if selection by cuckoos themselves occurs, I predict that it would be less strong than the demonstrable discriminative ability of bush warblers.

EVOLUTION AND MECHANISMS OF HOST SELECTION BY CUCKOOS

Host use described in this paper does not mean real host selection, because only successful host species are treated here, and there may be some other host species that ejected cuckoo eggs and were not noticed to have been parasitized. For example, it is possible that Himalayan cuckoos on Honshu frequently parasitize bush warblers, but that the warbers remove such nonmimetic eggs so quickly that people almost never see them. Therefore, even though Himalayan cuckoos show a shift in egg color between Honshu and Hokkaido, they may not show a comparable shift in host selection. However, in this case, bush warblers usually do not eject dissimilar eggs but desert the whole clutch (see "Experiments on egg color"), and I have never found any deserted bush warbler nest with eggs of cuckoos other than little cuckoos on Honshu and Himalayan cuckoos on Hokkaido (H. Higuchi, pers. observ.). Cuckoos may parasitize other host species than reported in this paper, but the frequency of parasitism would be low (see Rothstein 1971, 1977 for more details of this "unseen" parasitism problem).

The main host species are clearly segregated among different species of cuckoos. Each cuckoo species has one or more main hosts, and tends to lay eggs similar in color to those of their main hosts, although there are some exceptions. Each cuckoo species therefore specializes on particular host species. Specialization on particular hosts increases breeding success of individual cuckoos. If different species of cuckoos randomly laid their eggs in the nests of hosts, the frequency of nests with more than one cuckoo egg would increase. However, usually only one cuckoo young can survive in a nest because young *Cuculus* cuckoos expel all other eggs and nestlings from the nest (Nibe 1979; Wyllie 1981; Higuchi 1985).

Egg similarity or egg mimicry in cuckoos probably evolved through discriminative ability of host species and their intolerance to dissimilar eggs, as shown in the above experiments and other studies (e.g., Victoria 1972; Alvarez et al. 1976; von Haartman 1981; Davies and Brooke 1988; see Davies and Brooke 1988; Brooker and Brooker 1989, 1990; Brooker et al. 1990 for alternative explanations related to second cuckoo predation). This process is a typical example of an evolutionary arms race or coevolution (Krebs and Davies 1987; Davies and Brooke 1988; Davies et al. 1989; Rothstein 1990).

It is therefore probable that host segregation has evolved through interspecific competition. The idea is supported by latter examples of host changes in the absence of closely related species of cuckoos. On Miyake Island of the Izu Islands, the range of host selection by little cuckoos appears to have expanded in the absence of other cuckoo species. However, the expansion seems to be limited to host species that may accept the chocolate-brown eggs of the cuckoo. In central and northern Hokkaido where little cuckoos do not breed, Himalayan cuckoos parasitize bush warblers and lay reddish eggs, as little cuckoos do on Honshu and the Izu Islands. Himalayan cuckoos also parasitize willow warblers in Hokkaido (Y. Ijima, pers. comm.) as on Honshu, though the egg color of the cuckoos has not been observed. In the case of Hokkaido, we may say that a "resource" (bush warblers and their nests) is being exploited by Himalayan cuckoos in the absence of little cuckoos, and that a morphologic character (egg color) has evolved to facilitate the ecologic shift. This kind of shift in host selection and egg color seems to exemplify niche release and character release (Grant 1972) in the absence of closely related species.

What determines the main hosts for each cuckoo species? Each cuckoo species prefers common host species in its habitat. For example, crowned willow warblers are one of the dominant species in well-developed deciduous or mixed forests, which are preferred habitats for Himalayan cuckoos. The main hosts of common cuckoos, such as great reed warblers, bull-headed shrikes, and meadow buntings, are common in such open habitats as reedbeds, scattered open forests, and forest edges, which are also inhabited by common cuckoos. Choosing common species as hosts increases the opportunities for parasitism by cuckoos. Host segregation is therefore also related to habitat segregation of cuckoos. Little cuckoos prefer deciduous or mixed forests with bamboo thickets. Horsfield's hawk cuckoos occur in dense forests at more than 1000 m above sea level on Honshu, though they are also seen in lower forests on Hokkaido. Habitat segregation, however, is not necessarily complete in cuckoos, and different species of cuckoos may be seen in the same kind of habitat. Little, Himalayan and Horsfield's hawk cuckoos, for example, co-occur in forests of around 1000 m above sea level in central Honshu (H. Higuchi, pers. observ.). Moreover, cuckoos have wide home ranges, and often fly over several types of habitats. Nevertheless, even in such cases, different species of cuckoos parasitize different species of hosts. Therefore, host segregation may be partially independent from habitat segregation.

Host segregation or preference is also associated with the nest types of hosts. Common cuckoos parasitize birds such as shrikes, reed warblers, and buntings that build cup-shaped nests among twigs, whereas Horsfield's hawk cuckoos prefer robins and flycatchers that build cup-shaped nests in depressions and shelves at ground level. Bush warblers, the main host of little cuckoos, build dome-shaped nests with a side entrance among bamboo, and crowned willow warblers, the main host of Himalayan cuckoos on Honshu, build similar nests on the ground. This nest type preference may be the reason why Himalayan cuckoos, not common or Horsfield's hawk cuckoos, parasitize bush warblers in the absence of little cuckoos on Hokkaido.

Egg similarity or egg mimicry is a product of each cuckoo species or population parasitizing particular host species. The degree of egg mimicry is affected by several factors. First, it is affected by the degree of host discrimination of dissimilar eggs (Wyllie 1981), as shown in its difference between bush warblers and willow warblers as hosts of Himalayan cuckoos. Second, the length of the history of parasitism may affect both the degree of egg mimicry and the discriminative ability of hosts (Baker 1942; Wyllie 1981; Davies and Brooke 1988; Nakamura 1990). Himalayan cuckoos on Hokkaido sometimes lay orange-brown eggs, which do not match accurately the chocolate-brown eggs of bush warblers. This may imply that Himalayan cuckoos have parasitized bush warblers for less time than little cuckoos, which always lay the same color eggs as bush warblers. Third, the degree of egg mimicry is affected by the amount of spatial and therefore reproductive isolation of each host-specific group of individuals within one species of cuckoo (Wyllie 1981). This applies to common cuckoos that may have different host races or gentes. On Honshu and Hokkaido, different types of common cuckoo eggs can be found in the same area, and some good examples of egg mimicry can be seen, which suggests that some gentes exist. The frequency of occurrence of egg mimicry, however, is low in spite of the high discriminative ability of major hosts such as meadow buntings, great reed warblers, and bull-headed shrikes (Nakamura 1990 and pers. oberv.), and it appears that differentiation into gentes is poor. Poor differentiation may be the results of recent habitat destruction or fragmentation in cuckoo habitats such as reedbeds and open forests. It may be difficult to develop and maintain egg mimicry for a particular host in such a dynamic situation. In Europe, common cuckoos show good egg mimicry and well-defined gentes in large areas of homogeneous habitat, such as the forests of eastern Europe and Scandinavia, or the reedbeds of Hungary. However, gentes do not seem to have remained separate and egg mimicry is relatively poor in Europe where the landscape has been broken up by cultivation (Southern 1954; Wyllie 1981).

ACKNOWLEDGMENTS I am indebted to N. Ise, M. Hasegawa, H. Saito, T. Sadaoka, Y. Takeuchi, and M. Yamamoto for field assistance; to R. Kakizawa, K. Kobayashi, A. Sasagawa, K. Shichinohe, and K. Yoshikane for laboratory work; and to A. Hotta, Y. Iijima, S. Imanishi, and T. Yoshino for their unpublished field data. My thanks are also due to S. I. Rothstein, S. K. Robinson, C. A. Haas and H. Nakamura for reviewing the draft and to H. Sano and T. Yoshino for providing illustrations or photographs used in this chapter. This work was supported in part by the Japan Ministry of Education, Science and Culture.

REFERENCES

Alvarez, F., L. A. De Reyna, and M. Segura (1976). Experimental brood parasitism of the Magpie (*Pica pica*). *Anim. Behav.*, **24**, 907–916.

Baker, E. C. S. (1927). *The fauna of British India, Bird, IV.* Taylor and Francis, London.

Baker, E. C. S. (1942). *Cuckoo problems.* Witherby, London.

Brooker, L. C. and M. G. Brooker (1990). Why are cuckoos host specific? *Oikos*, **57**, 301–309.

Brooker, L. C., M. G. Brooker, and A. M. H. Brooker (1990). An alternative population/genetics model for the evolution of egg mimesis and egg crypsis in cuckoos. *J. Theor. Biol.*, **146**, 123–143.

Brooker, M. G. and L. C. Brooker (1989). The comparative breeding behaviour of two sympatric cuckoos, Horsfield's Bronze-Cuckoo *Chrysococcyx basalis* and the Shining Bronze-Cuckoo *C. lucidus*, in western Australia: a new model for the evolution of egg morphology and host specificity in avian brood parasites. *Ibis*, **131**, 528–547.

Davies, N. B. and M. de L. Brooke (1988). Cuckoos versus reed warblers: adaptations and counteradaptations. *Anim. Behav.*, **36**, 262–284.

Davies, N. B., A. F. G. Brooke, and M. de L. Brooke (1989). Cuckoos and parasitic ants: interspecific brood parasitism as an evolutionary arms race. *Trends Ecol. Evol.*, **4**, 274–278.

Enomoto, K. (1941). *A guide to wild birds, Vol. 2.* Wild Bird Society of Japan Osaka Chapter, Osaka (in Japanese).

Friedmann, H. (1964). Evolutionary trends in the avian genus *Clamator. Smithsonian Misx. Coll.*, **146**, 4.

Friedmann, H. (1967*a*). Alloxenia in three sympatric African species of *Cuculus. Proc. U.S. Nat. Mus.*, **124**, 3633.

Friedmann, H. (1967*b*). Evolutionary terms for parasitic species. *Syst. Zool.*, **16**, 175.

Gaston, A. J. (1976). Brood parasitism by the Pied Crested Cuckoo *Clamator jacobinus. J. Anim. Ecol.*, **45**, 331–348.

Grant, P. R. (1972). Convergent and divergent character displacement. *Biol. J. Linn. Soc.*, **4**, 39–68.

Harrison, C. J. O. (1968). Egg mimicry in British Cuckoos. *Bird Study*, **15**, 22–28.

Hayashi, T. (1978). A young Common Cuckoo reared by Gray Thrushes. *Tancho*, **5**, 2–4 (in Japanese).

Higuchi, H. (1973). Birds of the Izu Islands. (1) Distribution and habitat of the breeding land and freshwater birds. *Tori*, **22**, 14–24 (in Japanese with English summary).

Higuchi, H. (1985). *The secret of reddish eggs.* Shisakusha, Tokyo (in Japanese).

Higuchi, H. (1989). Responses of the Bush Warbler *Cettia diphone* to artificial eggs of *Cuculus* cuckoos in Japan. *Ibis*, **131**, 94–98.

Higuchi, H. and R. B. Payne (1986). Nestling and

fledgling plumages of *Cuculus saturatus horsfieldi* and *C. poliocephalus poliocephalus* in Japan. *Jap. J. Ornithol.*, **35**, 61–65.

Higuchi, H. and S. Sato (1984). An example of character release in host selection and egg colour of cuckoos *Cuculus* spp. in Japan. *Ibis*, **126**, 398–404.

Jensen, R. A. C. and M. K. Jensen (1969). On the breeding biology of southern African cuckoos. *Ostrich*, **40**, 163–181.

Kiyosu, Y. (1965). *The birds of Japan, II.* Kodansha, Tokyo (in Japanese).

Kobayashi, K. (1939). The eggs and hosts of cuckoos. *Hyogoken Hakubutsugakkai Kaiho*, **18**, 21–25 (in Japanese).

Kobayashi, K. and T. Ishizawa (1940). *The eggs of Japanese birds, II.* Published by the authors, Kobe.

Krebs, J. R. and N. B. Davies (1987). *An introduction to behavioral ecology*, 2nd ed., Sinauer Associates, Sunderland, Mass.

Lack, D. (1968). *Ecological adaptations for breeding in birds.* Methuen, London.

Lack, D. (1971). *Ecological isolation in birds.* Blackwell, Oxford.

Mason, P. and S. I. Rothstein (1987). Crypsis versus mimicry and the color of Shiny Cowbird eggs. *Amer. Natur.*, **130**, 161–167.

Munsell, A. H. (1929). *Munsell book of color.* Munsell Color Company, Baltimore.

Nakamura, H. (1990). Brood parasitism by the Cuckoo *Cuculus canorus* in Japan and the start of new parasitism on the Azure-winged Magpie *Cyanopica cyana. Jap. J. Ornithol.*, **39**, 1–18.

Nibe, T. (1979). *The ecology of wild birds, I.* Taishukan, Tokyo (in Japanese).

OSJ (Ornithological Society of Japan) (1942). *A hand-list of the Japanese birds*, 3rd and revised ed., Ornithological Society of Japan, Tokyo.

OSJ (1974). *Check-list of Japanese birds*, 5th and revised ed., Gakken, Tokyo.

Payne, R. B. (1977). The ecology of brood parasitism in birds. *Annu Rev. Ecol. Syst.*, **8**, 1–28.

Rensch, B. (1924). Zur Entstehung der Mimikry der Kuckuckseier, *J. Ornith.*, **72**, 461–472.

Rothstein, S. I. (1971). Observation and experiment in the analysis of interactions between brood parasites and their hosts. *Amer. Natur.*, **105**, 71–74.

Rothstein, S. I. (1977) Cowbird parasitism and egg recognition of the northern oriole. *Wilson Bull.*, **89**, 21–32.

Rothstein, S. I. (1990). A model system for coevolution: avian brood parasitism. *Annu. Rev. Ecol. Syst.*, **21**, 481–508.

Southern, H. N. (1954). Mimicry in cuckoo's egg. In *Evolution as a process* (ed. J. Huxley), pp. 219–232. George Allen & Unwin, London.

Tarboton, W. (1986). African cuckoo: the agony and ecstasy of being a parasite. *Bokmakierie*, **38**, 109–111.

Victoria, J. K. (1972). Clutch characteristics and egg discriminative ability of the African Village Weaverbird *Ploceus cucullatus. Ibis*, **114**, 367–376.

von Haartman, L. (1981). Co-evolution of the Cuckoo *Cuculus canorus* and a regular Cuckoo host. *Ornis Fennica*, **58**, 1–10.

Wallace, A. R. (1889). *Darwinism: An exposition of the theory of natural selection with some of its applications*. Macmillan, London.

WBSJ (Wild Bird Society of Japan) (1980). *The breeding bird survey in Japan 1978.* Wild Bird Society of Japan, Tokyo (in Japanese).

Wyllie, I. (1981). *The cuckoo.* Batsford, London.

Yamagishi, S. and M. Fujioka (1986). Heavy brood parasitism by the Common Cuckoo *Cuculus canorus* on the Azure-winged Magpie *Cyanopica cyana. Tori*, **34**, 91–96.

Yamashina, Y. (1941). *A Natural History of Japanese Birds, Vol. 2.* Iwanami-shoten, Tokyo (in Japanese).

4

Coevolution between the Common Cuckoo and Its Major Hosts in Japan

Stable versus Dynamic Specialization on Hosts

HIROSHI NAKAMURA, SATOSHI KUBOTA, REIKO SUZUKI

The common cuckoo *Cuculus canorus* is a brood parasite known to have laid its eggs in the nest of more than 125 species of passerine birds (Wyllie 1981). But many of these are rare or accidental hosts. Cuckoo populations in each region concentrate on only a few host species and individual female cuckoos are thought to specialize on just one host species and to mimic the eggs of that host species. The egg mimicry is believed to have evolved in relation to egg discrimination by host species (Chance 1940; Baker 1942; Wyllie 1981; Brooke and Davies 1988). In the parasitic relations between the cuckoo and its hosts, the latter gain no reproductive output at all due to the parasitic nestling's eviction behavior (Davies and Brooke, this vol.). Therefore, selection should favor host abilities to decrease cuckoo parasitism which in turn will lead to improved or more elaborate trickery by the cuckoo (Brooke and Davies 1987). Such coevolution between parasites and hosts is expected to lead to intricate adaptations and counteradaptations. It has recently attracted considerable attention because it may be described by a general coevolutionary model (Rothstein 1975*a*, 1990; May and Robinson 1985; Kelly 1987; Harvey and Partridge 1988; Davies and Brooke 1989*a*,*b*).

However, there are few data confirming coevolutionary responses by parasitic birds and their hosts because no one has fully documented evolutionary changes during the course of their research (Lotem and Rothstein 1995). Instead, such changes are inferred to have happened by various comparative studies. Furthermore there are alternative hypotheses concerning some aspects of parasite–host coevolution. For example, the perplexing acceptance many North American bird species show towards brood parasitism by the brown-headed cowbird *Molothrus ater* has been attributed to evolutionary lag (Rothstein 1975*a*,*b*, 1982*a*). Similarly, a study of cuckoos in Europe suggested that the varying degrees of egg discrimination shown among suitable host species represent snap shots in evolutionary time of different stages of a continuing arms race between cuckoos and hosts (Davies and Brooke 1989*b*). By contrast, Rohwer and Spaw (1988) and Lotem et al. (1995) suggest that host species that accept parasitism are at an evolutionary equilibrium due to costs of rejection such as breakage of host eggs during attempts to remove parasitic eggs and mistaken removal of host eggs.

The parasitic relations between the common cuckoo and its hosts in central Japan are different from those in Europe. Common cuckoos in Japan have a larger number of major hosts within a single area (six species as opposed to one or two in Europe), a high frequency of parasitism (more than 20% as opposed to rates generally below 10%), and recent changes in host usage in the past 60 years (Nakamura 1990). The most dramatic change is the start of parasitism on the azure-winged magpie *Cyanopica cyana* (Yamagishi and Fujioka 1986; Nakamura 1990).

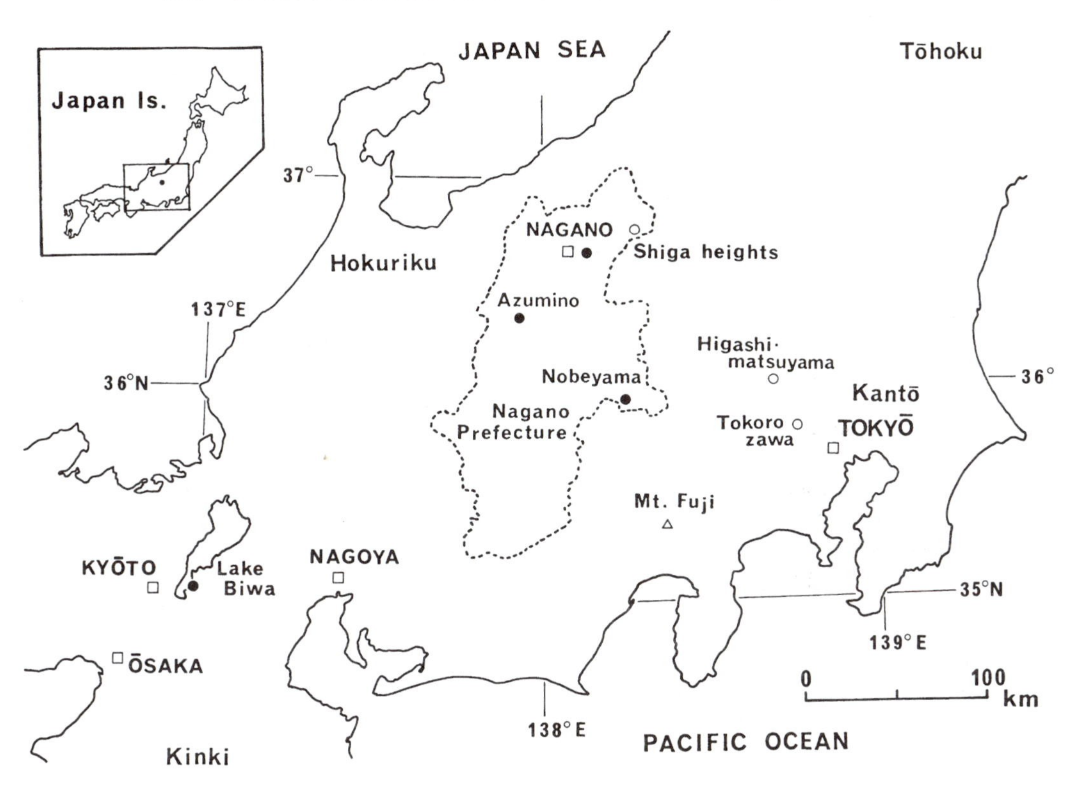

Figure 4.1 Map of central Japan, showing places mentioned in the text.

This new host–parasite relationship spread quickly throughout the host's breeding range within 20 years (Nakamura 1990).

In this paper we report on general cuckoo–host interactions in central Japan, concentrating on the host's responses to parasitic eggs and the degree of egg mimicry by cuckoos. The most novel aspect of our study is documentation of a rapid increase in egg discrimination by newly parasitized populations of the magpie. This increase has been so rapid that it is difficult to explain only by genetic changes. We also discuss the possibility that cuckoo gentes are somewhat flexible in host use and therefore capable of host shifts. A comparison of egg markings between the cuckoo and its hosts in central Japan indicates two possible coevolutionary pathways. One is adaptation in the direction of stable specialization involving parasitism and egg mimicry of the same small set of host species for a long time. The other is dynamic specialization in which cuckoos with poorly mimetic eggs parasitize a frequently changing set of host species through adaptation and counteradaptation.

STUDY AREAS AND METHODS

Major study areas

Observations and experiments were done in six breeding seasons, 1984 to 1989, mainly at the Chikuma River and its surrounding areas, Nagano City, Nagano Prefecture, central Japan (fig. 4.1). Extensive areas of waste land, reed beds, willow bushes, and acacia forests occur along the river and in its bed. Rice fields, apple orchards, and villages lie around the river. Many host species breed along this river. Major hosts in this area are the azure-winged magpie, great reed warbler *Acrocephalus arundinaceus* and bull-headed shrike *Lanius bucephalus* (Nakamura 1990). The egg discrimination of magpies at Nagano was compared with discrimination at two other sites in Nagano Prefecture. Azumino is 50 km southwest of Nagano City at the foot of the Japanese Alps. Nobeyama is a highland of Mt. Yatugatake 84 km south of Nagano City (see fig. 4.1). The magpie populations in these three areas had been exposed to cuckoo parasitism for different periods when this study was done: Azumino — about 20 years, Nagano — about 15 years, and Nobeyama — about 10 years (Nakamura 1990).

General observations and natural parasitism

We found great reed warbler and bull-headed shrike nests by searching through reedbeds or bushes and magpie nests mainly by observing behavior of the birds. Searches for nests were done throughout the breeding season; however, our searching efforts differed somewhat according to hosts, seasons, and years. We did not search intensively for magpie nests in the first year (1984). Nests found during the building and egg-laying stages were visited daily or every few days. Cuckoo eggs found in the nests were measured (length and breadth) and weighed. Each cuckoo egg was placed on a gray cloth and four or five close-up photographs (Kodak Ektachrome 100 HC color slide film) were taken from various directions under daylight. Thereafter the nests were visited daily or every few days to check the responses of hosts to cuckoo eggs. Some eggs of great reed warblers, bull-headed shrikes and magpies were also measured, weighed and photographed after clutch completion.

Experimental parasitism

To test the egg discrimination ability of hosts, experiments were conducted with model cuckoo eggs and real shrike eggs. The model cuckoo eggs were made from paper-clay and measured $23.2 \pm 0.9 \times 17.6 \pm 0.7$ mm (mean ± SD of 20 model eggs) which was close to the size of real cuckoo eggs in this area (mean of $142 = 23.0 \pm 1.1$ mm, 17.4 ± 0.5 mm). The weight of the model eggs (3.8 ± 0.6 g, $n = 20$) was adjusted to that of real cuckoo eggs (3.8 ± 0.35 g, $n = 142$) by putting two shotgun balls into the center of each model egg. The model cuckoo eggs were painted with water colors to match the egg patterns of real cuckoo eggs. The most typical egg markings (lines, dots, and blotches, see below) of local cuckoos were copied exactly on the model eggs, which were then covered with dull gloss varnish.

Real shrike eggs were used in experiments in addition to the artificial eggs because shrike eggs are similar to cuckoo eggs in size, weight, egg color, and markings, except that shrike eggs have no line markings. The artificial eggs and real shrike eggs were warmed with a hot-water bottle or by hand before being put in host nests, to match the temperature of real host eggs. Experiments were done following the procedure pioneered by Rothstein (1975*a*). All the model eggs were placed in nests in the afternoon during the laying period or early incubation of the hosts. One host egg was removed at the same time. The response of hosts to the experimental parasitism was checked every day. Pecks by hosts left detectable marks on the paper-clay cuckoo eggs and eggs with the peck marks were regarded as rejections. The experiment was done once at each nest. These experiments were done mainly in 1985–87 at the Chikuma River, Nagano City.

Egg discrimination of magpies was tested by the same methods at Azumino and Nobeyama in 1987–89. In addition to these three areas in Nagano Prefecture, the same experiments were done 130 km to the southeast at Higashimatuyama, Saitama Prefecture (fig. 4.1) in 1989 using model cuckoo eggs and real shrike eggs. In addition, the egg discrimination ability of great reed warblers at Lake Biwa (a non-parasitized population about 270 km southwest of Nagano; see fig. 4.1) was examined in 1987 with model cuckoo eggs and real shrike eggs.

Analyses of egg markings from photographs

Analyses of cuckoo egg markings were done from the photographs. Each cuckoo egg was classified as one of eight egg types (A–H) based on combinations of three types of markings: lines, dots (spots less than 1 mm in diameter), and blotches (spots more than 1 mm). Figure 4.2 shows examples of eggs A–D, the most common types. In addition to this egg type analysis, each cuckoo egg was analyzed for the amount of lines, dots, and blotches. The amount of each marking was estimated as the percentage of an egg covered by marks and classified into 10 grades. The assignment of grades was done by comparison of pictures which showed the exact amount of markings in each grade. Besides the amount and types of markings, these cuckoo eggs varied in ground color and in the color of the markings. But these color differences were not extreme and their representation in the photos was influenced slightly by lighting conditions. Therefore, color differences were ignored in this paper.

RESULTS

The sizes and shapes of cuckoo and host eggs

Table 4.1 shows sizes and shapes of cuckoo and host eggs collected at Nagano City. As is clear from fig. 4.2, azure-winged magpie eggs were

Table 4.1 Egg size and shape of cuckoos and their hosts at Nagano City

	n	Weight (g)	Size (mm) (Mean ± S.D.) Length	Breadth	Shape (length/breadth)
Host eggs					
Azure-winged magpie	81	5.39 ± 0.55	26.5 ± 1.2	19.7 ± 0.7	1.35 ± 0.12
Bull-headed shrike	95	3.83 ± 0.38	23.6 ± 0.5	17.8 ± 0.6	1.32 ± 0.09
Great reed warbler	215	2.79 ± 0.28	22.1 ± 1.1	15.9 ± 0.5	1.39 ± 0.10
Cuckoo eggs					
From a-w magpie nests	63	3.81 ± 0.33	23.3 ± 0.9	17.5 ± 0.5	1.33 ± 0.11
From b-h shrike nests	20	3.81 ± 0.33	23.0 ± 1.0	17.6 ± 0.5	1.30 ± 0.13
From g. r. warbler nests	59	3.78 ± 0.23	22.7 ± 0.7	17.3 ± 0.5	1.31 ± 0.11

larger than those of the two other major hosts in this region, bull-headed shrikes and great reed warblers, with a mass 1.41 times that of cuckoo eggs laid in their nests. Warbler eggs were the smallest, with a mass 0.74 times that of cuckoo eggs laid in warbler nests. On the other hand, shrikes had eggs almost the same mass as cuckoo eggs (table 4.1). The egg shape (length/breadth) of the shrike was the most rounded (1.32) and the shape of the warbler the most narrow (1.39). The eggs of these three different hosts had statistically significant differences (t-tests, $P < 0.01$) from each other in weight, egg length, egg breadth, and egg shape.

Although cuckoo eggs laid in these different host nests were fairly similar in size and shape there were some significant differences among them. Cuckoo eggs laid in great reed warbler and magpie nests differed as regards weight ($t = 2.1$, $P < 0.05$), length ($t = 2.6$, $P < 0.01$) and shape ($t = 2.0$, $P < 0.05$) but not breadth (table 4.1). In addition, cuckoo eggs in reed warbler nests

Figure 4.2 Eggs of the three major host species discussed in this study. Upper row, left to right are eggs of azure-winged magpie, bull-headed shrike and great reed warbler (two eggs). The two eggs shown for the warbler are an example of some of the variants that occur within this species. Bottom row, left to right are cuckoo eggs of types A–D. Note that all three hosts have type D eggs.

Figure 4.3 Some examples of cuckoo eggs laid in great reed warbler nests, except for the two eggs on the left end of the bottom row, which were laid by the warblers.

had smaller breadths than those in shrike nests ($t = 2.1$, $P < 0.05$) and cuckoo eggs in magpie nests were more rounded than those in shrike nests ($t = 2.3$, $P \leqslant 0.05$).

The size differences among cuckoo eggs from nests of the three hosts paralleled size differences among the host eggs. The longest cuckoo eggs were laid in magpie nests and the shortest in great reed warbler nests. These correspondences indicate the existence of egg mimicry in size, but probably not in shape as one of the two significant shape differences was in the wrong direction (cuckoo eggs from warbler versus magpie nests).

Types of markings on cuckoo eggs

Markings on cuckoo eggs were variable and this was somewhat true even for eggs found in the nests of a single host species (fig. 4.3). Table 4.2 shows the variation in the markings of cuckoo eggs laid in different host nests. Cuckoo eggs laid in both magpie nests and shrike nests were of four types (A, B, C, and D). However, type A cuckoo eggs were not found in the great reed warbler nests. Type C cuckoo eggs with lines, dots and blotches were the most common type for all three hosts and greatly outnumbered all other types put together for cuckoo eggs from

Table 4.2 The types of markings on cuckoo eggs laid in different host nests at the Chikuma River, Nagano City

		Egg types								
		A	B	C	D	E	F	G	H	
	Lines	•	•	•	—	—	—	•	—	
	Dots	—	•	•	•	—	•	—	—	
Hosts	Blotches	—	—	•	•	•	—	•	—	Total
Azure-winged magpie		15	2	28	10	—	—	—	—	55
Bull-headed shrike		1	1	14	3	—	—	—	—	19
Great reed warbler		—	5	69	10	—	—	—	—	84

Note: These data were collected from 1984 to 1987. The sample size of cuckoo eggs is smaller than in table 4.1 because all the eggs collected outside our main study area were not included in this table.

Table 4.3 Local differences in cuckoo egg types laid in the nests of the azure-winged magpie

		Egg types								
		A	B	C	D	E	F	G	H	
	Lines	•	•	•	—	—	—	•	—	
Local	Dots	—	•	•	•	—	•	—	—	
populations	Blotches	—	—	•	•	•	—	•	—	Total
Azumino		—	2	43	6	—	—	—	—	51
Nagano		15	2	28	10	—	—	—	—	55
Nobeyama		2	2	28	4	—	—	4	—	40

Note: These data were collected from 1984 to 1987.

shrike and warbler nests. The proportion of type C cuckoo eggs was 77.4% in great reed warbler nests, 73.7% for the shrike but only 50.9% (28/55) for the magpie. The difference in the proportion of type C eggs is significant for the magpie versus the warbler ($\chi^2 = 13.9$, $P < 0.001$, d.f. = 1 for this and all subsequent χ^2 tests) but not for the magpie versus the shrike ($\chi^2 = 2.98$, $P > 0.05$). The second most common cuckoo egg type in magpie nests, type A at 27.3%, was significantly less common in warbler ($\chi^2 = 22.9$, $P < 0.001$) and shrike ($\chi^2 = 4.04$, $P < 0.05$) nests. Egg-type variation was the smallest in cuckoo eggs in great reed warbler nests.

There were local differences in the types of cuckoo eggs collected from different populations of the magpie (table 4.3). Type A cuckoo eggs were not collected at Azumino, where the magpies had the longest history of cuckoo parasitism (about 20 years). Egg-type variation was also smallest at Azumino and most (84.3%) cuckoo eggs there were type C. Nobeyama had the shortest history of cuckoo parasitism on magpies (about 10 years) and the most variable cuckoo eggs (table 4.3). Type G cuckoo eggs were observed only in this area. The major difference among these cuckoo eggs from different areas was the significantly higher proportion of type A eggs at Nagano relative to Azumino ($\chi^2 = 14.03$, $P < 0.001$).

Egg discrimination by hosts

Table 4.4 shows host responses to naturally deposited cuckoo eggs at the Chikuma River, Nagano City. Three types of host rejection were observed: ejection, nest desertion and "building over." Rejection rate was highest (44.7%) for the reed warbler and lowest (34.7%) for the magpie, but this difference was not significant ($\chi^2 = 0.63$, $P < 0.1$). Ejection was the most frequent rejection mode used by reed warblers whereas the other two species most often rejected by nest desertion but none of the differences in rejection mode was significant statistically.

Because observed responses to naturally deposited parasitic eggs may be a biased representation of a host's actual egg discrimination behavior, experiments were done at the Chikuma River by placing model cuckoo eggs and real shrike or warbler eggs into nests (table 4.5). Rejection by building over was not observed in

Table 4.4 Responses of hosts to cuckoo eggs laid at the Chikuma River, Nagano City

	Number of cuckoo eggs					
	Rejected by					
Hosts	Ejection	Desertion	Building over	Accepted	Rejections/Total	%
Azure-winged magpie	8	9	—	32	17/49	34.7
Bull-headed shrike	1	4	2	10	7/17	41.2
Great reed warbler	13	7	1	26	21/47	44.7

Table 4.5 Responses of hosts to experimentally placed model cuckoo eggs and real shrike or warbler eggs at the Chikuma River, Nagano City

		Number of eggs					
		Rejected by					
Hosts	Experimental egg type	Ejection	Desertion	Building over	Accepted	Rejections/Total	%
Azure-winged magpie	Model cuckoo egg	13	7	—	14	20/34	58.9
	Real shrike egg	4	—	—	5	4/9	44.4
Bull-headed shrike	Model cuckoo egg	2	2	—	5	4/9	44.4
	Real warbler egg	4	—	—	5	4/9	44.4
Great reed warbler	Model cuckoo egg	20	6	—	9	26/35	74.3
	Real shrike egg	17	2	—	5	19/24	79.2

these experiments. Ejection, the rejection mode most difficult to detect in responses to naturally laid eggs, was the most frequent type of rejection for all three species. All three species rejected experimentally placed eggs more frequently than naturally deposited eggs but the difference was significant only for great reed warbler responses to experimentally placed shrike versus naturally laid cuckoo eggs ($\chi^2 = 6.34$, $P < 0.05$). The great reed warbler showed the highest rejection rates in these experiments but there were no significant differences in rejection rates among the three hosts.

Table 4.6 shows local differences in magpie responses to cuckoo eggs. Rejections by egg ejection and nest desertion were observed at both Azumino and Nagano. Magpies at Azumino rejected at a higher rate than those at Nagano (41.7% versus 34.7%) but this difference was not significant ($\chi^2 = 1.38$, $P = 0.17$). However, all the cuckoo eggs found in magpie nests at Nobeyama were accepted, in spite of the wide variation in cuckoo eggs at this locality (table 4.3). Rejection was significantly less frequent at Nobeyama than at Azumino ($\chi^2 = 18.1$, $P < 0.001$) and Nagano ($\chi^2 = 12.0$, $P < 0.001$).

Table 4.7 shows local differences in magpie egg discrimination as elucidated by experimental parasitism. All three populations in Nagano Prefecture showed higher rejection rates to the model cuckoo eggs than to the real shrike eggs, but none of these differences was significant. The Azumino population had the highest rejection rate for both model cuckoo eggs and real shrike eggs while the Nobeyama population had the lowest rates. Thus, these results were almost the same as the trends observed in real cuckoo parasitism. When data for model cuckoo and real shrike eggs were combined, magpies at Nobeyama had a significantly lower rejection rate (10%) than at Azumino (61.5%, $P < 0.025$ Fisher exact test) and at Nagano (55.8%, $\chi^2 = 5.12$, $P < 0.05$). Because the magpies at Nobeyama have been parasitized for the shortest period of time, these results show

Table 4.6 Local differences in magpie responses to cuckoo eggs laid at three localities in Nagano Prefecture

	Number of cuckoo eggs					
	Rejected by					
Local populations	Ejection	Desertion	Building over	Accepted	Rejection/Total	%
Azumino	4	1	—	7	5/12	41.7
Nagano	8	9	—	32	17/49	34.7
Nobeyama[1]	—	—	—	32	0/32	0.0

[1]Most data from Nobeyama were provided by S. Imanishi.

Table 4.7 Local differences in magpie responses to experimentally placed model cuckoo eggs and real shrike eggs in central Japan

Local populations	Egg type	Number of eggs: Rejected by: Ejection	Desertion	Building over	Accepted	Rejections/Total	%
Nagano Prefecture							
Azumino	Model cuckoo egg	3	2	—	2	5/7	71.4
	Real shrike egg	1	2	—	3	3/6	50.0
Nagano	Model cuckoo egg	13	7	—	14	20/34	58.8
	Real shrike egg	4	—	—	5	4/9	44.4
Nobeyama	Model cuckoo egg	—	1	—	4	1/5	20.0
	Real shrike egg	—	—	—	5	0/5	0.0
Saitama Prefecture							
Higashi-Matuyama	Model cuckoo egg	7	1	—	5	8/13	61.5

that magpie egg discrimination is closely related to the history of cuckoo parasitism. The slightly higher overall rejection rate at Azumino was not significantly different from the rate at Nagano.

None of 14 magpie nests found in the egg-laying or early incubation stages at Higashi-matuyama in Saitama Prefecture was parasitized. However, magpies there had about as high a rejection rate (61.5%) of model cuckoo eggs as magpies at Azumino and Nagano (table 4.7). The frequency of parasitism on magpie populations living in the Kanto Plain (includes Higashi-matuyama) is very low even though cuckoos are common there (Nakamura 1990). Thus, magpies at Higashi-matuyama retain strong egg discrimination despite the low parasitism rate that they experience.

Lake Biwa is in the lowland of the Kinki district, Honshu, where cuckoos do not occur (Ezaki 1981, 1987). In spite of the absence of parasitism, the reed warbler population in this area showed higher rejection rates of both model cuckoo eggs and real shrike eggs than did the parasitized population at Nagano City (table 4.8) but none of the differences was significant (to model cuckoo eggs $\chi^2 = 1.10$, $P = 0.13$; to real shrike egg $\chi^2 = 0.32$, $P = 0.59$). The rejection rate of the reed warbler population at Nagano changed with the advance of the breeding season because yearlings bred later and showed less discrimination than older birds (Lotem et al. 1992). The experiments at Lake Biwa were done early in the breeding season (early June). During the same period at Nagano, the frequency of rejection of model cuckoo eggs was 88.2% (15/17) and of real shrike eggs, 81.0% (17/21). These rejection rates are close to those from Lake Biwa and show that the reed warbler population there has about the same level of egg discrimination as the Nagano population, in spite of their lack of parasitism.

Table 4.8 Great reed warbler responses to experimentally placed eggs in a parasitized population (Chikuma River at Nagano City) and a population isolated from cuckoos (Lake Biwa)

Local populations	Experimental egg type	Number of eggs: Rejected by: Ejection	Desertion	Building over	Accepted	Rejections/Total	%
Chikuma River	Model cuckoo egg	20	6	—	9	26/35	74.3
	Real shrike egg	17	2	—	5	19/24	79.2
Lake Biwa	Model cuckoo egg	17	1	—	2	18/20	90.0
	Real shrike egg	12	—	—	1	12/13	92.3

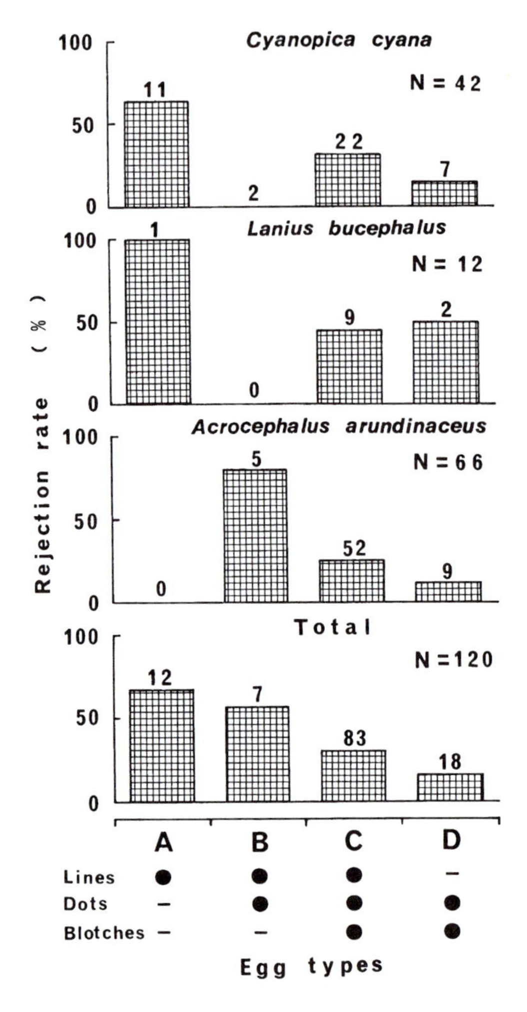

Figure 4.4 Responses of the three major hosts at the Chikuma River, Nagano City to naturally laid cuckoo eggs of different types. The graph at the bottom, labeled Total, is for all three species combined. Egg types were based on the markings on eggs, shown at the bottom of the figure (see table 4.2). Note that only types A–D occurred in cuckoo eggs and that all three hosts have type D eggs.

Cuckoo egg types and their rejection rates

The rejection rates of different types of cuckoo eggs (table 4.2) are shown in fig. 4.4. All three hosts have eggs without lines but with dots and blotches (i.e., type D eggs; see fig. 4.2). Seven of 11 (63.6%) cuckoo eggs with only lines (type A) laid in magpie nests were rejected whereas magpies rejected only seven of 22 (31.8%) and one of seven (14.3%) of C- and D-type cuckoo eggs, respectively (fig. 4.4). The magpie rejection rate of A eggs is significantly higher than that of D eggs ($\chi^2 = 4.22$, $P < 0.05$). Thus, magpies were more likely to accept cuckoo eggs the more similar these were to their own eggs. The same trend was observed in great reed warblers, which rejected four of five (80.0%) type B cuckoo eggs, but only 13 of 52 (25%) type C and one of nine (11.1%) type D eggs. The warbler rejection rate of B eggs is significantly higher than for D (Fisher test, $P < 0.025$) and C eggs ($\chi^2 = 4.23$, $P < 0.05$). Sample sizes of cuckoo eggs from shrike nests were too small to confirm any trends. Figure 4.4 also shows rejection rates of each cuckoo egg type for all three hosts combined. Note that the rejection rate of A eggs was the highest (69.2%) and that the rate decreased in the order of types B (57.1%), C (28.9%), and D (16.7%). The over-

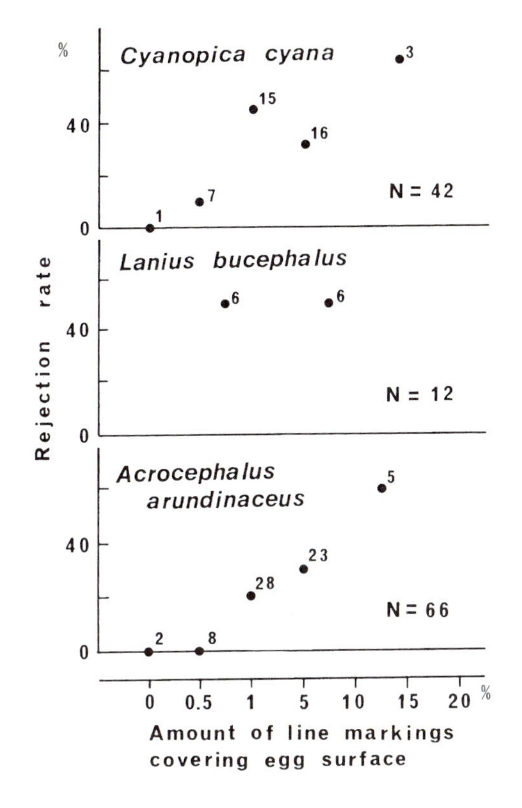

Figure 4.5 Rejection rate of major hosts at Nagano City as a function of the extent of line markings on naturally laid cuckoo eggs. All three hosts have no lone markings on their eggs. The numbers next to each data point indicate sample size.

all rejection rate was significantly higher for A eggs than for C ($\chi^2 = 6.6$, $P < 0.01$) and D eggs ($\chi^2 = 7.75$, $P < 0.01$), which again shows selection against eggs with lines.

Cuckoo eggs with many lines were especially likely to be rejected by both magpies and by great reed warblers (fig. 4.5). Magpies rejected one of eight (12.5%) cuckoo eggs with $\leq 0.5\%$ of their surfaces covered with lines and 20 of 34 (58.8%) with $\geq 1\%$ lines, a significant difference ($\chi^2 = 3.86$, $P < 0.05$). Similarly, great reed warbler nests rejected none of ten cuckoo eggs with $\leq 0.5\%$ lines but 16 of 56 with $\geq 1\%$ lines ($\chi^2 = 7.23$, $P < 0.05$). Thus, selection pressures due to host egg discrimination are acting against cuckoo eggs wth line markings.

Egg types of the overall host community in central Japan

The overall similarity between cuckoo and host eggs can be assessed by considering all host species. Six major and 14 other host species have been recorded in central Japan (Nakamura 1990). These 20 hosts have seven of the possible eight egg types (table 4.9). Twelve species have only one egg type and eight have two egg types. In most species with two egg types, one type was predominant. The most common egg type, D, occurred in 16 species.

Type C was the most common among cuckoo eggs (70.3% of 158 cuckoo eggs; table 4.2) but occurred in only three host species, black-faced bunting *Emberiza spodocephala*, Japanese yellow bunting *E. sulphurata*, and black-browed reed warbler *Acrocephalus bistrigiceps.* Furthermore, the major egg type for these three species was type D and there is only one record (Kobayashi and Ishizawa 1940) of a type C egg of the black-faced bunting egg (from the foot of Mt. Fuji outside of Nagano Prefecture). Although type D eggs were the most prevalent kind among hosts (table 4.9), they were the second most common type among cuckoo eggs and made up only 14.6% (23 or 158; table 4.2) cuckoo eggs. The third most prevalent type of cuckoo eggs, type A

Table 4.9 Egg types of cuckoo hosts in central Japan

		Egg types							
		A	B	C	D	E	F	G	H
	Lines	•	•	•	—	—	—	•	—
	Dots	—	•	•	•	—	•	—	—
Species	Blotches	—	—	•	•	•	—	•	—
*Azure-winged magpie									
Cyanopica cyana					•				
Japanese grosbeak									
Euphona personata		•						•	
*Black-faced bunting									
Emberiza spodocephala				•[1]	•				
Gray-headed bunting									
Emberiza fucata					•				
Siberian meadow bunting									
Emberiza cioides		•						•	
Japanese yellow bunting									
Emberiza sulphurata				•	•				
*Great reed warbler									
Acrocephalus arundinaceus					•				
Black-browed reed warbler									
Acrocephalus bistrigiceps				•	•				
Gray thrush									
Turdus cardis					•				
Brown thrush									
Turdus chrysolaus					•				
Stonechat									
Saxicola torquata					•				
Japanese accentor									
Prunella rubida									•
*Bull-headed shrike									
Lanius bucephalus					•				
*Brown shrike									
Lanius cristatus					•				
Thick-billed shrike									
Lanius tigrinus					•				
Brown-eared bulbul									
Hypsipetes amaurotis					•	•			
Gray wagtail									
Motacilla cinerea					•				
Japanese wagtail									
Motacilla grandis					•		•		
*Indian tree pipit									
Anthus hodgsoni							•		
Skylark									
Alauda arvensis					•		•		

*Major host in central Japan. [1]Only one case of a C-type egg is known and this was from outside of Nagano Prefecture.

(10.1%, 16 of 158), occurred in only two host species, Japanese grosbeak (*Euphona personata*) and Siberian meadow bunting. Because the most common kind of cuckoo egg, type C, is rare among hosts, cuckoo egg types do not closely match those of the current hosts, which is consistent with recent host shifts.

DISCUSSION

Our major findings suggest that the common cuckoo and its hosts in central Japan are more poorly adapted to one another than is the case for the same cuckoo in Europe (Davies and Brooke this vol.). In particular, the degree of egg mimicry is relatively poor and one major host, the azure-winged magpie, shows a high incidence of acceptance of nonmimetic eggs. Nevertheless, the latter host shows spatial variation in responses to nonmimetic eggs and this finding suggests that egg recognition is rapidly changing in response to cuckoo parasitism. In contrast to the relatively low rejection rates of parasitized azure-winged magpies, populations of certain other species that are not currently parasitized show highly developed egg recognition behavior. These and other findings raise a series of questions, which we address in the first several subsections of this Discussion.

How did a new host, the azure-winged magpie, develop egg discrimination ability so rapidly?

Cuckoo parasitism on magpies in central Japan began about 20 years before our study and spread rapidly throughout the magpie's range as a result of spatial overlap due to range expansion by both species (Nakamura 1990). Some magpie populations developed rejection of cuckoo eggs within 10 years and the level of discrimination is positively correlated with the duration of sympatry with cuckoos (table 4.6 and 4.7). Similarly, a rapid increase in rejection rate has been documented even within a single population (H. Nakamura, in prep.)

How could the magpie populations develop egg discrimination so quickly? The parasitism rate on the magpie is high (more than 50%) and magpies that accept a cuckoo egg are unlikely to fledge their own offspring. Therefore rejection is clearly advantageous. As first modelled explicitly by Rothstein (1975*a*,*b*), natural selection is the process usually thought to be responsible for an increase in rejection behavior. Rothstein (1975*a*) found that many host species of the brown-headed cowbird *Molothrus ater* nearly always accepted cowbird eggs. He suggested that these host species have remained accepters because suitable mutations or recombinations that could trigger rejection have not yet appeared (Rothstein 1975*b*, 1982*a*). However, in the case of the magpies, it is likely that genetic variation coding for rejection was present before the start of cuckoo parasitism because some magpie populations developed rejection behavior immediately after the start of parasitism. Magpies are highly sedentary (S. Yamagishi and S. Imanishi, pers. comm.) so it is unlikely that rejection behavior was present in only a small area and spread rapidly due to gene flow, that is, dispersal of magpies.

Rothstein (1975*b*, 1982*a*) suggested that, once rejecter alleles occur, cowbird hosts will change from acceptors to rejecters within as little as 20 to 100 years. Davies and Brooke (1989*b*) suggested that the change may take longer for cuckoo hosts because of the relatively low rates of parasitism that they experience. Thus, it is unlikely that natural selection alone can explain the rapid increase in rejection rate by magpies over a period of 10–20 years. Furthermore, because the average life span (S. Yamagishi, pers. comm.) of the magpie is 3.3 years (male) and 2.3 years (female), magpie populations turn over slowly. Rather than a replacement of accepter genotypes by rejecter genotypes, it is more parsimonious to explain the rapid increase in rejection by phenotypic responses to the start of heavy parasitism by cuckoos.

The sight of a cuckoo on a host's nest stimulates greater rejection of cuckoo eggs by both reed warblers *Acrocephalus scirpaceus* (Davies and Brooke 1988) and meadow pipits *Anthus pratensis* (Moksnes and Roskaft 1989). A similar stimulation of rejection may have occurred in the establishment of egg discrimination by magpies as cuckoos clearly made many visits to their nests. Multiple parasitism was common (Nakamura 1990) and cuckoos also made frequent visits to magpie nests before laying in them. In addition, as this bird is highly social (Hosono 1981, 1983), individuals may have learned to eject eggs by watching other magpies. Although the recent changes may be largely phenotypic, it is unlikely that they would have occurred if the magpie lacked a genetic predisposition to acquire rejection via phenotypic responses.

Why do unparasitized populations have high levels of egg discrimination ability?

The magpie population at Higashi-matuyama in Saitama Prefecture had as high a rate of egg rejection as populations subjected to more than 15 years of parasitism at Nagano (table 4.7). Despite this highly developed egg discrimination, the frequency of cuckoo parasitism on the magpies in the Kanto plain, including Higashi-matuyama, is low (Nakamura 1990). At Tokorozawa, 30 km from Higashi-matuyama, the rate was 3.8% (3/80) in 1977–78 and there has been no increase in parasitism since then (Nakamura 1990). Why has the dramatic increase of parasitism of magpies observed in recent years in Nagano Prefecture

(Nakamura 1990) failed to take place in the Kanto plain, around the Tokyo metropolitan area? Magpies have lived in this plain for at least 60 years (Kiyosu 1952), which predates the start of their range expansion about 30–40 years ago from Kanto to Tohoku (Nakamura 1990).

Members of the Tokyo branch of the Wild Bird Society of Japan have established that cuckoos have occurred in suburbs of Tokyo (Kunitachi, Kokubunji, Irima City, etc.) for more than 50 years (Kawachi 1979). Thus, magpies and cuckoos have lived together in the Kanto Plain–Tokyo area for a long time and before the start of the magpie's range expansion. The only record of parasitism in the Tokyo area occurred at Kokubunji in 1976 (Kawachi 1979) but it is possible that magpies in this region experienced heavy cuckoo parasitism in the past and developed egg discrimination ability during that period. This egg discrimination may have made it adaptive for the cuckoo to switch to other hosts. According to this explanation, the magpie population retains a high level of rejection behavior, thereby making it still adaptive for cuckoos to parasitize other hosts. To test these ideas further it is necessary to examine past records on cuckoo parasitism and to do comparative studies of cuckoo parasitism and egg discrimination of magpies over wide areas around the Kanto plain.

The magpie is not the only species in Japan that shows a high level of rejection in the absence of parasitism. Great reed warblers at Lake Biwa had rejection rates as high as those of a parasitized population at the Chikuma River (table 4.8) although cuckoos do not breed in the lowlands of western Japan, including Lake Biwa (WBSJ 1980). One possible explanation for the rejection behavior at Lake Biwa is gene flow from other populations where these behaviors are selected for. If rejection at Lake Biwa is maintained by gene flow, it should occur at a lower rate than in parasitized populations but that was not the case (table 4.8). Thus, the most likely explanation for the strong egg discrimination of great reed warblers at Lake Biwa is that this population was parasitized by cuckoos in the past and has retained the egg discrimination it developed in that period even though it is no longer sympatric with cuckoos. A detailed examination of this problem may be made in a future paper dealing with the past distribution of the cuckoo in Japan and with host aggression to cuckoo mounts.

These discoveries of strong egg discrimination in the absence of parasitism, differ from previously described results on other cuckoo–host systems. For example, meadow pipits and white wagtails in Iceland are isolated from cuckoos and showed significantly less discrimination against eggs unlike their own than did conspecific populations in Britain, which are parasitized (Davies and Brooke 1989*a*). Soler and Møller (1990) demonstrated that rejection of eggs of the parasitic great spotted cuckoo *Clamator glandarius* by the magpie *Pica pica* varies among populations. In a population of ancient sympatry between the cuckoo and magpie, magpies showed high rejection. Magpies in an area of recent sympatry showed less rejection and those in a population isolated from great spotted cuckoos show no rejection. These differing results may exist because the parasite–host interactions of the common cuckoo in Japan are more dynamic than are interactions involving this and other species of cuckoos in other regions. This possibility will be addressed later in this chapter.

Why are cuckoo eggs laid in magpie nests so variable?

Cuckoos have much more variable eggs (tables 4.2 and 4.3) than do their hosts, most of which have only one egg type (table 4.9). Data from Nagano show that cuckoo eggs laid in nests of a new host, the magpie, are more variable than those laid in nests of an old host, the great reed warbler (table 4.2). Similarly, Nobeyama, one of the newest areas of cuckoo parasitism on magpies, had more variable cuckoo eggs than areas where parasitism of magpies has a longer history (table 4.3).

Individual female cuckoos are thought to specialize on one host and to be separated into gentes, each of which has a single egg type (Chance 1940; Baker 1942; Wyllie 1981; Davies and Brooke 1989*a*,*b*). Female cuckoos, however, are known to turn to other hosts when nests of a favorite host are not available (Chance 1940). The recent rise in parasitism of magpies in Japan is an example of such a host shift that has become permanent (Nakamura 1990).

If this shift to the magpie was due to a single female cuckoo or even a small number of females, today's descendants of the first female(s) to parasitize magpies should have a limited range of variation. However, variation was larger among cuckoo eggs from magpie nests than among eggs from older hosts and variation was greatest in the newest areas of cuckoo parasitism (tables 4.2 and 4.3). Thus it is likely that many female cuckoos are responsible for the host shift to the magpie.

Cuckoo parasitism on magpies in Nobeyama began after magpies invaded this area and became sympatric with cuckoos that were already present. Besides the azure-winged magpie, there are four other major hosts (parasitism frequency more than 20%) in Nobeyama: great reed warbler, bull-headed shrike, black-faced bunting, and brown shrike (*Lanius cristatus*) (Nakamura 1990). The especially high variation among cuckoo eggs in magpie nests in this area may exist because cuckoos parasitizing two or more of these other major hosts made the shift to the magpie simultaneously. A shift to parasitism of the magpie may have been facilitated by this host's lack of aggressiveness towards cuckoos and by its habit of leaving its nest unattended for long periods during the egg-laying stage (Yamagishi and Fujioka 1990; H. Nakamura and M. Nakamura, in prep.). As with the discussion (above) of host egg rejection in the absence of parasitism, this consideration of the large amount of variation in cuckoo eggs laid in magpie nests indicates that the cuckoo's host relations in Japan are relatively dynamic. It also suggests that cuckoo gentes are not as clearly delimited or as strongly entrained to a single host species (or a small set of species with similar eggs), as is thought to be the case in Europe (Davies and Brooke, this vol.).

Are cuckoo egg markings adapted to host egg markings?

If gentes in Japan are clearly delimited and if each is adapted to a different host, then there should be a close correspondence between cuckoo eggs and the host eggs with which they are usually found. However, the most common type of cuckoo egg in the nests of all three major hosts in central Japan was type C (lines, dots and blotches; table 4.2) even though these hosts have type D eggs (fig. 4.2) as do most hosts in central Japan (table 4.9). In fact there are no hosts that typically lay type C eggs in central Japan, so cuckoo egg markings did not correspond closely to host egg markings. Thus the degree of egg mimicry in Japan is much weaker than this same cuckoo shows in Europe.

Nevertheless, there is evidence that egg mimicry is of adaptive value. Magpie populations parasitized by cuckoos for 10 years or more, are exerting heavy selection pressures due to their well-developed egg rejection behavior (table 4.6). Selection seems to be strongest against cuckoo eggs with lines (figs. 4.4 and 4.5). Cuckoos parasitizing magpies may have disadvantageous line patterns at the present time because they are descended from cuckoos that specialized on some of the few potential host species with lined egg markings. Two possible hosts in this category are the Siberian meadow bunting and Japanese grosbeak (table 4.9). The former was one of the major hosts in central Japan more than 60 years ago but is now a rare host in this area (Nakamura 1990). Cuckoo eggs with remarkable lines, which are similar to those on bunting eggs, were in fact a speciality of this district (Ishizawa 1930). The bunting is presently one of the most numerous species in Nagano Prefecture (Nakamura 1990) so the cuckoo's shift to other hosts is not due to a scarcity of buntings. Instead, we suggest that the cuckoo shifted to other hosts because the bunting developed strong egg discrimination. Currently, buntings have the strongest egg discrimination ability among potential cuckoo host species in Japan (H. Nakamura, S. Imanishi and R. Suzuki, unpublished). Thus, line markings observed on the present-day cuckoo eggs in magpie nests may be remnants from a gens that mimicked the bunting eggs. Currently, cuckoos that parasitize magpies may be in a transition stage from having eggs with many lines (egg types A and B) to eggs with fewer or no lines (types C and D) due to the effects of natural selection.

Host shifts such as that responsible for parasitism of the magpie, may reduce reproductive isolation among lineages of cuckoos specializing on different host species, thereby leading to the current situation in which most cuckoo eggs have all three of the possible types of egg markings (table 4.2). Besides being due to past host shifts, the high frequency of cuckoo eggs with all types of markings may facilitate additional host shifts because they result in the cuckoo population retaining a large range of genetic variation for egg markings. This may allow cuckoos that shift to a new host to begin quickly to develop egg mimicry as soon as the host develops egg rejection because features of the host's eggs can be duplicated from the range of genetic variation present in the cuckoo. It is unlikely that cuckoo eggs with all types of markings, especially lines, have adaptive value in the short term. Species with such eggs are not only few in central Japan (table 4.9), but eggs with lines only or with dots, blotches, and lines are in fact rare among Northern Temperate birds in general. In Europe, lines are found only in some bunting species (*Emberiza*; Harrison 1975) and in North America they occur only in some blackbirds (icterines) and a few other species (Harrison 1978). Egg markings of passerines are thought to have evolved to maximize concealment (Lack 1958) and eggs with all three types of markings may be rare because only one or two kinds of markings are sufficient for concealment.

Coevolution between cuckoos and their hosts

Hosts of parasitic birds sometimes lack egg discrimination, as is the case with many cowbird hosts in North America, which may not have yet had the right genetic variation needed for discrimination to develop (Rothstein 1975b). Thus, a likely scenario for an association between a cuckoo and a new host species is as follows: in the initial stages after cuckoos begin to parasitize a host species with few or no defenses – as may have been the case with cuckoo parasitism of the magpie in Japan – parasitism is highly successful. Both the number of cuckoos in the new gens and the parasitism frequency on the new host increase rapidly, as observed between cuckoos and magpies at Nobeyama (Nakamura 1990). The success of the new gens, however, does not continue at such a high level because the intense parasitism results in a shortage of hosts and increased competition among members of the gens. This competition leads to an increase in the frequency of multiply parasitized nests. As many as four or five female cuckoos have laid in some magpie nests (Yamagish and Fujioka 1986; Nakamura 1990). Multiple parasitism leads to a decline in the per capita success rate of the new cuckoo gens because only one cuckoo will survive in multiply parasitized nests. In addition, the development of host defenses may also contribute to a decline in the gens' success rate. This sequence of events may produce dynamic relationships among changes in breeding success, numbers of parasites and hosts, and rates of parasitism. If the host does not develop defenses and parasitism levels remain high, a host population may even be driven to extinction with the cuckoo gens itself also going extinct.

Regardless of the causes, it is likely that a high level of parasitism of a particular species is an unstable situation. If the rate of parasitism declines, two alternative pathways may occur. One pathway is towards stable specialization, in which a parasite victimizes the same single to several host species for a long time. The other is towards dynamic specialization and involves the parasite using a small set of host species whose make-up changes frequently. The number of hosts used at any one time and in a single region is similar in both pathways or slightly smaller in stable specialization. But in dynamic specialization, cuckoos shift away from a host when it develops a high frequency of efficient defenses and hosts and parasites go through never-ending cycles of adaptation and counteradaptation.

Stable specialization

For specialization to be stable on the same set of host species, accurate egg mimicry needs to reach a high frequency in the cuckoo population. Highly developed egg mimicry may cause the egg rejection behavior of hosts to become less adaptive for two reasons. First, if egg mimicry is highly accurate, then the host's egg discrimination will have to be very sensitive to detect cuckoo eggs. But highly sensitive discrimination may lead to recognition errors in which unparasitized rejecter hosts incur recognition costs by occasionally rejecting their own eggs if these are slightly atypical in appearance (Lotem et al. 1992, 1995). Secondly, as cuckoo egg mimicry improves, hosts have to become more intolerant of eggs that deviate from the norm to detect parasitic eggs successfully. There is evidence that the sensitivity of host egg discrimination is modulated in this way because the degree of egg discrimination in North American rejecters of cowbird eggs seems to be correlated with the amount of difference between their eggs and cowbird eggs (Rothstein 1982*b*). But increased intolerance of even slightly unusual eggs may lead to increased recognition costs. If these costs are sufficiently high because of accurate egg mimicry, selection may favor a decrease in discrimination sensitivity which will lead to some rejecter individuals accepting highly mimetic cuckoo eggs. This will lessen the selection differential between rejecter and accepter individuals because some of the former will also accept cuckoo eggs (Kelly 1987). Whether selection ceases to favor increased intolerance by the host depends in large part on the frequency of parasitism. If parasitism is rare, paying recognition costs may be less adaptive than the small likelihood that a host will have to pay the cost of being parasitized. Thus, for these various reasons, the net benefits and costs of rejection behavior may become equal to those for acceptance behavior, especially if the parasite becomes uncommon relative to the host (Rothstein 1990). After that the host and parasite relation may be stable for a long time, as shown in the upper part of fig. 4.6. Furthermore, if host rejection declines or ceases, the increase in cuckoo egg mimicry will slow down or cease altogether because selection of cuckoo eggs operates through the egg discrimination by hosts (but see Brooker and Brooker 1989).

Stable specialization does not seem to be the case in central Japan, but a possible example of it occurs in Britain where cuckoos exploit five main hosts: meadow pipits *Anthus pratensis*;

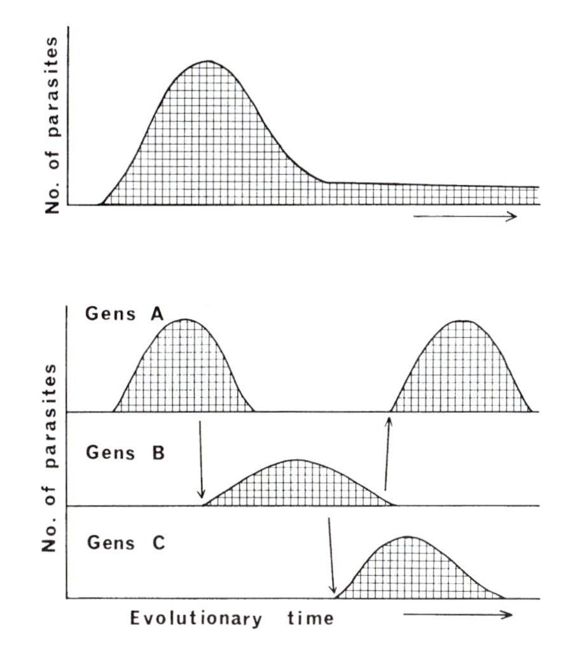

Figure 4.6 The number of cuckoos over evolutionary time in two alternative coevolutionary pathways. The upper graph represents stable specialization and shows the course of population change of a gens parasitic on a single host. The lower graphs represent dynamic specialization and show the development of new gentes. Note that cuckoo numbers are permanently high only in the lower graphs. In dynamic specialization, gens A eventually shifts to another host species and becomes gens B because of the development of highly refined egg recognition by host A. Host B eventually also develops highly refined egg recognition, which results in a shift back to host A and/or to new host C.

reed warblers *Acrocephalus scirpaceus*; dunnocks *Prunella modularis*; robins *Erithacus rubecula*; and pied wagtails *Motacilla alba*. There is a generalized match between parasitic eggs and host eggs, with the exception of the dunnock (Brooke and Davies 1988) and egg mimicry is better than in Japan (this chapter; H. Nakamura and S. Kubota, unpublished).

The parasitism rates of these four main hosts in Britain whose eggs are mimicked, meadow pipits, reed warblers, robins, and pied wagtails, were 2.21%, 7.29%, 0.13% and 0.21%, respectively (Brooke and Davies 1987). Stable specialization seems also to occur in Norway, where meadow pipits are one of the most frequent hosts and cuckoos lay mimetic eggs; the parasitism rate is 6.4% (Moksnes and Roskaft 1987). Despite its small size, there are six major cuckoo hosts in Nagano Prefecture. This relatively high number of major hosts may be related to the large range of habitats that occur in this region, from lowland areas to high elevations in the Japanese Alps (Nakamura 1990). The parasitism rates of major hosts in Nagano Prefecture are all more than 20% and are thus higher than in Europe, which is one of the major differences between cuckoo parasitism in Europe and in central Japan (Nakamura 1990). Thus, the relatively high rates of parasitism in Japan may forestall stable specialization because selection continues to favor host rejection regardless of the degree of egg mimicry by the cuckoo. Ever-increasing rates of host rejection may favor hosts shifts, such as away from buntings and towards magpies, as has been documented in Japan.

These facts suggest that most of the parasitic relationships between the common cuckoo and its hosts in Europe are closer to an evolutionary equilibrium state than those in central Japan. Although Brooke and Davies (1987) showed that there have been significant changes in the parasitism rates of five cuckoo hosts in Britain over the past 40 years, these changes occurred within low parasitism rates (less than 3%), except reed warblers which changed from 1.80% to 7.29%. Thus these changes do not alter the general picture of relative stability in Britain.

In addition to the relatively high abundance of

a parasite, another factor that may make a parasite–host system unstable is disturbance of habitats by human activities, especially in this century. Under primeval conditions, some local Japanese populations of the Siberian meadow bunting gens cuckoos and the great reed warbler gens cuckoos might have represented stable specialization. Because of heavy rainfall, Japan was once mostly forested and the common cuckoo, which does not live in forests, must have been much rarer and more local than it is today. This may have facilitated the development of well-defined gentes because of reproductive isolation among cuckoos specializing on different host species. But when forests were cleared, cuckoos increased in numbers and different gentes may have mixed. This, in addition to increasing numbers of cuckoos, would also tend to destabilize parasite–host interactions. So it is possible that the dynamic parasite–host interactions in Japan are due to anthropogenic habitat disturbances which greatly increased cuckoo abundance.

The view that habitat disturbance may destabilize parasite–host interactions is supported by three congeners of the common cuckoo that also breed in central Japan. These three species (Horsfield's hawk cuckoo *C. fugax*; Himalayan cuckoo *C. saturatus*; and little cuckoo *C. poliocephalus*) all occur within forests, which were the original habitats throughout much of Japan. They segregate their woodland passerine hosts among each other and all except the Himalayan cuckoo mimic the eggs of their respective hosts (Royama 1963; Higuchi 1986). The densities of these woodland cuckoos are low in general and parasitism rates are nearly always lower than for the common cuckoos in Japan. Thus, all three woodland cuckoos are relatively uncommon specialists whose parasite–host relations may represent stable equilibrium states. Nevertheless, these specialist woodland cuckoos still have the potential for evolutionary change (Higuchi and Sato 1984; Higuchi, this vol.).

Dynamic specialization

If a host develops a high level of defense, such as finely tuned egg discrimination and especially strong aggressiveness to cuckoos, a parasite–host interaction may become unstable and other pathways can occur. The cuckoo gens may decline and be driven to extinction. Alternatively, another way for the cuckoo gens to survive is to shift to a new host, as has happened with the common cuckoo and the magpie in Japan. This shift seems to be successful and the relation between the markings on cuckoo eggs and the local duration of parasitism suggests that a new cuckoo gens is developing, as shown in fig. 4.6 (lower).

Once a host population is emancipated from parasitism, its egg discrimination behavior may eventually be lost. However, if the behavior has strong genetic determinants, it may be retained for a long time, as may be the case for the magpie population in the Kanto Plain and the unparasitized great reed warbler population at Lake Biwa. Still though, if birds reject their own eggs in the absence of parasitism, the host should slowly revert back to become an accepter host (Davies and Brooke 1989*b*). A possible example of such a decline in host defenses is the village weaver *Ploceus cucullatus.* It rejects dissimilar eggs in its native Africa where it is parasitized by cuckoos (Victoria 1972). But weavers introduced sometime before 1797 to Hispaniola, where no parasites occurred, have lost most of their rejection behavior (Cruz and Wiley 1989). If a host population that has lost its defense again becomes parasitized, a new cuckoo gens may start and the same sequence of events can be repeated (fig. 4.6 lower), resulting in dynamic specialization.

Unstable interactions may block the development of egg mimicry as discussed previously in this chapter. Furthermore, this instability may result in most cuckoos having a generalized egg such as type C with markings (dots, blotches, and lines) that match those of the full range of potential hosts.

ACKNOWLEDGMENTS We thank Yoshitomo Miyazawa, Hiroshi Kobayashi, Masao Nakamura, Masahiro Ogawa, Toshiuki Yoshino and Hiroshi Uchida who helped with our field work and/or experiments, and Sadao Imanishi who allowed us to cite his unpublished data and offered useful information. The first half of this work was supported in part by a Grant-in-Aid for Special Project Research on Biological Aspects of Optimal Strategy and Social Structure from the Ministry of Education, Science and Culture and the second half by a Grant-in-Aid for Scientific Research (C) from The Ministry of Education, Science and Culture. We particularly thank Dr. Stephen Rothstein (University of California, Santa Barbara), who provided useful literature and made helpful comments to a draft of this paper. We thank also Drs. Amotz Zahavi and Arnon Lotem (Tel Aviv University) for their useful comments to the development of ideas.

REFERENCES

Baker, E. C. S. (1942). *Cuckoo problems*. Witherby, London.

Brooke, M. de L. and N. B. Davies (1987). Recent changes in host usage by cuckoos *Cuculus canorus* in Britain. *J. Anim. Ecol.*, **56**, 873–883.

Brooke, M. de L. and N. B. Davies (1988). Egg mimicry by cuckoos *Cuculus canorus* in relation to discrimination by hosts. *Nature*, **335**, 630–632.

Brooker, M. G. and L. C. Brooker (1989). The comparative breeding behaviour of two sympatric species of cuckoos, Horsfield's bronze-cuckoo *Chrysococcyx basalis* and the shining bronze-cuckoo *C. lucidus* in Western Australia: a new model for the evolution of egg morphology and host specificity in avian brood parasites. *Ibis*, **131**, 528–547.

Chance, E. (1940). *The truth about the cuckoo*. Country Life, London.

Cruz, A. and J. W. Wiley (1989). The decline of an adaptation in the absence of a presumed selection pressure. *Evolution*, **43**, 55–62.

Davies, N. B. and M. de L. Brooke (1988). Cuckoos versus reed warblers: adaptations and counteradaptations. *Anim. Behav.*, **36**, 262–284.

Davies, N. B. and M. de L. Brooke (1989*a*). An experimental study of co-evolution between the cuckoo, *Cuculus canorus*, and its hosts, I Host egg discrimination. *J. Anim. Ecol.*, **58**, 207–224.

Davies, N. B. and M. de L. Brooke (1989*b*). An experimental study of co-evolution between the cuckoo, *Cuculus canorus*, and its hosts, II Host egg markings, chick discrimination and general discussion. *J. Anim. Ecol.*, **58**, 225–236.

Ezaki, Y. (1981). Female behavior and pair relation of the polygynous great reed warbler *Acrocephalus arundinaceus* (Aves: Sylviinae). *Physiol. Ecol. Japan*, **18**, 77–91.

Ezaki, Y. (1987). Male time budgets and recovery of singing rate after pairing in polygamous great reed warbler. *Jap. J. Ornith.*, **36**, 1–11.

Harrison, C. (1975). *A field guide to the nests, eggs and nestlings of European birds*. Collins, London.

Harrison, C. (1978). *A field guide to the nests, eggs and nestlings of North American birds*. Collins, New York.

Harvey, P. H. L. and L. Partridge (1988). Evolutionary biology of cuckoo clocks and cowbirds. *Nature*, **335**, 586–587.

Higuchi, H. (1986). Adaptations for brood parasitism in *Cuculus* cuckoos. In: *Breeding strategies in birds* (ed. S. Yamagishi), Tokaidaigaku-shuppankai, Tokyo (in Japanese).

Higuchi, H. and S. Sato (1984). An example of character release in host selection and egg colour of Cuckoos *Cuculus* spp. in Japan. *Ibis*, **126**, 398–404.

Hosono, T. (1981). *The life of the Azure-winged Magpie*. Shinmaishuppanbu, Nagano (in Japanese).

Hosono, T. (1983). A study of life history of Blue Magpie (11). Breeding helpers and nest-parasitism by Cuckoo. *Misc. Rep. Yamashina Inst. Ornithol.*, **15**, 63–71 (in Japanese with English summary).

Ishizawa, T. (1930). Some new studies about the breeding of Cuculidae in Japan. *Tori*, **6**, 382–408 (in Japanese).

Kawachi, H. (1979). Breeding records of the cuckoo *Cuculus canorus* in the Tokyo Metropolitan area. *Bul. Tokyo branch, Wild bird society of Japan*, **43**, 1–2 (in Japanese).

Kelly, C. (1987). A model to explore the rate of spread of mimicry and rejection in hypothetical populations of cuckoos and their hosts. *J. Theor. Biol.*, **125**, 283–299.

Kiyosu, Y. (1952). *The birds of Japan, II*. Kodansha, Tokyo (in Japanese).

Kobayashi, K. and T. Ishizawa (1940). *The eggs of Japanese birds*. Kobayashi and Ishizawa, Kobe.

Lack, D. (1958). The significance of the colour of Turdine eggs. *Ibis*, **100**, 145–166.

Lotem, A. and S. I. Rothstein (1995). Cuckoo–host coevolution: from snapshots of an arms race to the documentation of microevolution. *Trends Ecol. Evol.*, **10**, 436–437.

Lotem, A., N. Nakamura, and A. Zahavi (1992). Rejection of cuckoo eggs in relation to host age: a possible evolutionary equilibrium. *Behav. Ecol.*, **3**, 128–132.

Lotem, A., H. Nakamura, and A. Zahavi (1995). Constraints on egg discrimination and cuckoo–host co-evolution. *Anim. Behav.*, **49**, 1185–209.

May, R. M. and S. K. Robinson (1985). Population dynamic of avian brood parasitism. *Amer. Natur.*, **126**, 475–493.

Moksnes, A. and E. Roskaft (1987). Cuckoo host interactions in Norwegian mountain areas. *Ornis Scand.*, **18**, 168–172.

Nakamura, H. (1990). Brood parasitism by the cuckoo *Cuculus canorus* in Japan and the start of new parasitism on the azure-winged magpie *Cyanopica cyana. Jap. J. Ornith.*, **39**, 1–18.

Rohwer, S. and C. D. Spaw (1988). Evolutionary lag versus bill size constraints: a comparative study of the acceptance of cowbird egg by old host. *Evol. Ecol.*, **2**, 27–36.

Rothstein, S. I. (1975*a*). Evolutionary rates and host defenses against avian brood parasitism. *Amer. Natur.*, **109**, 161–176.

Rothstein, S. I. (1975*b*). An experimental and teleonomic investigation of avian brood parasitism. *Condor*, **77**, 250–271.

Rothstein, S. I. (1982*a*). Successes and failures in avian egg and nestling recognition with comments on the utility of optimality reasoning. *Amer. Zool.*, **22**, 547–560.

Rothstein, S. I. (1982*b*). Mechanisms of avian egg recognition: which egg parameters elicit responses by rejecter species? *Behav. Ecol. Sociobiol.*, **11**, 229–239.

Rothstein, S. I. (1990). A model system for coevolution: avian brood parasitism. *Annu. Rev. Ecol. Syst.*, **21**, 481–508.

Royama, T. (1963). Cuckoo hosts in Japan. *Bird Study*, **10**, 201–202.

Soler, M. and A. P. Moller (1990). Duration of sympatry and coevolution between the great spotted cuckoo and its magpie host. *Nature*, **343**, 748–750.

Victoria, J. K. (1972). Clutch characteristics and egg discriminative ability of the African village weaverbird (*Ploceus cucullatus*). *Ibis*, **114**, 367–376.

WBSJ (Wild Bird Society of Japan) (1980). *The breeding bird survey in Japan 1978*. Wild Bird Society of Japan, Tokyo (in Japanese).

Wyllie, I. (1981). *The cuckoo.* Batsford, London.

Yamagishi, S. and M. Fujioka (1986). Heavy brood parasitism by the common cuckoo *Cuculus canorus* on the azure-winged magpie *Cyanopica cyana. Tori*, **34**, 91–96.

5

Duration of Sympatry and Coevolution between the Great Spotted Cuckoo (*Clamator glandarius*) and Its Primary Host, the Magpie (*Pica pica*)

M. SOLER, J. J. SOLER, J. G. MARTINEZ

Avian brood parasitism is an excellent system for studies of coevolution, as the interacting species are few, and often only two (Rothstein 1990). Parasites exploit hosts by laying eggs in their nests, and leaving parental care of the parasitic offspring to the host. The fitness cost of parasitism to hosts is often high because: (1) parasite females, before laying, usually remove or damage at least one host egg; (ii) the parasite young ejects all host offspring or outcompetes most of them for food during the nestling period; and (iii) the host often provides extensive parental care for considerably longer than when hosts provide care for conspecific young (Payne 1977; Rothstein 1990). Thus, strong selection pressures may favor hosts that evolve discriminatory abilities which prevent parasitism or reduce its costs. The scene is subsequently set for an evolutionary arms race between the brood parasite and its hosts (Dawkins and Krebs 1979; Harvey and Partridge 1988), and this can result in rapid coevolution.

The assumption that brood parasitism should result in coevolution between parasites and hosts has been supported by three pieces of information: (1) the laying procedure of parasitic cuckoos matches host defences (e.g., rapid laying and laying in the afternoon) (Davies and Brooke 1988); (2) egg mimicry of cuckoos parallels the discriminatory ability of the host species (Brooke and Davies 1988); and (3) actual and potential host species have stronger anti-parasite responses than species unsuitable as hosts (Davies and Brooke 1988).

The great spotted cuckoo (*Clamator glandarius*) and its magpie (*Pica pica*) host are excellent subjects for studies of coevolution, because this cuckoo is a specialist brood parasite which, in the Palearctic, only occasionally parasitizes other species of the corvid family different from the magpie (Veillard 1973; Cramp 1985; Soler 1990). This cuckoo occurs throughout Africa, the Mediterranean basin and Asia Minor (Rowan 1983; Fry et al. 1988). Great spotted cuckoo Palearctic populations are migratory and those breeding in Europe are believed to winter in Africa south of the Sahara (Cramp 1985). Individuals usually arrive in southern Spain in early March or late February (Soler 1990) and adults leave the breeding areas at the beginning of July with a peak of departures in August (Cramp 1985).

Over most of its range, the great spotted cuckoo is primarily parasitic on Corvidae. In Europe, mainly the magpie is parasitized, but other corvids are used occasionally (Soler 1990). In Africa, the most frequent hosts are *Corvus albus* and *Corvus capensis*; however, this cuckoo also makes use of open and hole-nesting Sturnidae (Jensen and Jensen 1969; Rowan 1983).

There are important differences in adaptations for brood parasitism displayed by the two species of European brood parasites: the great spotted cuckoo and the European cuckoo (*Cuculus canorus*). Most important, in contrast to the European cuckoo, the great spotted cuckoo has only one egg type (i.e., it does not possess gentes with eggs adapted to different host species), and the nestling does not eject host eggs or young. A previous paper (Soler 1990) provided information about the differences in parasitic adaptations between these two cuckoos: (1) Unlike the European cuckoo, adult great spotted cuckoos do not remove host eggs; they only damage some while laying (pecking or crushing), and the host ejects damaged eggs as is done by virtually all bird species (Kemal and Rothstein 1988). (2) Crushed eggs are more frequent in nests parasitized by the great spotted cuckoo than in those parasitized by other cuckoo species; this method of damaging host eggs is probably more advantageous in the genus *Clamator.* (3) Davies and Brooke (1988) have provided some evidence that the European cuckoo is a major predator on host eggs; however, there is no evidence that the great spotted cuckoo depredates host nests (unless its egg is rejected by the host; see below). (4) Although all regular hosts of the European cuckoo are smaller than it, the great spotted cuckoo nearly always parasitizes hosts larger than itself. Thus, it is advantageous first, that the male and female cooperate at laying to overcome host aggression and second, that the female lays its egg more quickly (3 s; Arias-de-Reyna et al. 1982) than the European cuckoo (10 s; Wyllie 1981) and other brood parasites. (5) In the great spotted cuckoo, chicks compete with host chicks and therefore this cuckoo has evolved the following adaptations: (i) parasite eggs hatch considerably earlier than those of the host, because of a shorter incubation period; and (ii) parasite chicks develop faster than host chicks (Soler and Soler 1991). However, both cuckoo species have these two adaptations.

The great spotted cuckoo has increased its geographic range considerably during the twentieth century (Cramp 1985) and has therfore coexisted with its corvid hosts for different lengths of time. This cuckoo colonized Spain's Hoya de Guadix region only recently (Soler 1990; Soler and Møller 1990), possibly due to the fact that the region is 900–1100 m above sea level (a.s.l.) and it is unusual to find this species above 500 m a.s.l. (Cramp 1985).

In this chapter we analyze the relationship between the great spotted cuckoo and its magpie host in a recently colonized area (Hoya de Guadix) and in an area of longstanding sympatry (Santa Fe). We predict that, if coevolutionary arms races occur between the parasitic cuckoo and its host, the counteradaptations of the host populations in areas of recent sympatry would be less developed. Consequently, the parasitic cuckoo would have greater reproductive success than it does in areas of ancient sympatry, and the rate of ejection of parasitic eggs should increase as duration of exposure to the brood parasite increases.

MATERIALS AND METHODS

Field work

Field work was done between 1982 and 1994, but most data were collected in 1982, 1983, and 1990–1994. Nests were visited at least once a week. Usual breeding parameters, such as laying date (date on which the first egg was laid; in all analyses, day 1 = 1 April), clutch size, hatching date, number of eggs hatched and number of fledglings were recorded for each magpie nest whenever possible. Furthermore, a number of different measurements of great spotted cuckoo parasitism were recorded, mainly: (i) the percentage of magpie nests parasitized in a particular area or year; (ii) the percentage of parasitized magpie nests with more than one cuckoo egg; (iii) the percentage of parasitized magpie nests with cuckoo eggs laid by more than one female. Multiple parasitism by more than a single female great spotted cuckoo was determined on the basis of differences in the appearance of cuckoo eggs (color and shape) and laying dates; (iv) the number of fledgling cuckoos per magpie nest, which was estimated as the number of cuckoo nestlings present on the last visit to a parasitized magpie nest on day 14, before the presumed date of fledging (day 15–19 in different nests). The responses of magpies to both experimental and natural parasitism were classified as either acceptance (the cuckoo egg remained in the nest), or rejection (the cuckoo egg disappeared between two nest visits). During this long-term study, we performed a variety of experiments. When necessary, only nonexperimental nests have been considered in the analyses.

All measurements of distances and areas in the Hoya de Guadix area were obtained from aerial photographs taken at a scale of 1:8000 in 1990.

Experimental procedure

Most of the magpie nests found during the egg-laying stage were used for an experimental test of rejection behavior. Nests were randomly assigned to two groups: (i) magpie nests to which a mimetic model egg resembling a great spotted cuckoo egg

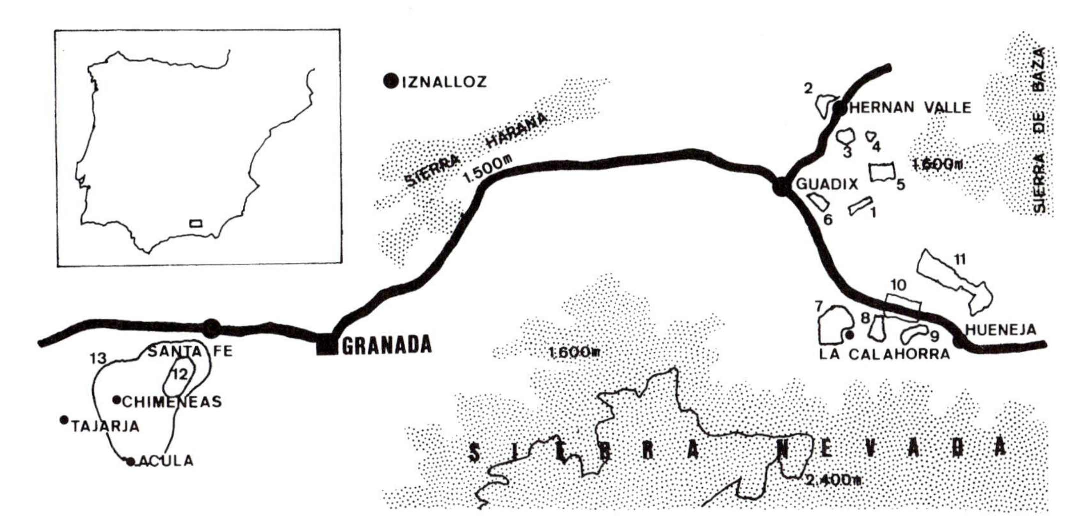

Figure 5.1 The two study areas (Santa Fe and Hoya de Guadix) and the main route (Santa Fe — Granada — Guadix) between them. Shaded areas are mountains and the contour lines are shown. The 11 study plots in the Hoya de Guadix area are as follows: 1 = Zaragüil; 2 = Hernan Valle; 3 = Fuente Alamo; 4 = La Fuente; 5 = Ladihonda; 6 = Via; 7 = La Calahorra; 8 = Ferreira; 9 = Dolar; 10 = Carretera; 11 = Hueneja. In addition, note areas 12 (Santa Fe; our study area) and 13 (Chimeneas; Zuñiga and Redondo 1992 study area). The three groups of plots considered in some analyses are as follows: Guadix: plots 1 to 6; La Calahorra: plots 7 to 9; Hueneja: plots 10 and 11.

was added; and (ii) magpie nests to which a nonmimetic model egg was added. Mimetic model cuckoo eggs were made of plaster of Paris mixed with glue and painted with acrylic paints to mimic real eggs. These eggs resembled cuckoo eggs in both size and mass (Soler and Møller 1990). Nonmimetic eggs were quail *Coturnix coturnix* eggs painted red which differed markedly in appearance from mimetic models and real magpie eggs, both with respect to ground color and markings, and were therefore highly nonmimetic. (Fig. 1 in Soler and Møller 1990). The magpie nests involved in the experiment were matched with respect to laying date and study plot because of random assignment to treatments.

All magpie nests in the experiment were revisited after five days, at which time magpies were classified as rejecters if the model egg was missing, or accepters if it was present.

Statistical analyses

The statistical methods of Sokal and Rohlf (1979) were followed. When nonparametric statistics were needed, the methods described by Siegel and Castellan (1988) were employed. Values given are mean ± SE. When a variable was discontinuous, data were square-root transformed (Sokal and Rohlf 1979) and/or nonparametric tests were used. All tests were two-tailed. Unless otherwise stated, data used in the analyses were collected in the Hoya de Guadix.

Analysis of hatching, fledging, and breeding success were made by calculating the success rate for each nest, and individual nests were considered as independent data.

The temporal change in ejection rate was tested in a model I linear regression with the annual estimates of ejection as the dependent variable and year as the independent variable. The estimates of ejection rate were square-root arcsine-transformed in order to fulfil the requirement of a normal frequency distribution of the dependent variable in a model I linear regression. All tests were two-tailed.

STUDY AREAS

This work was carried out in the Hoya de Guadix and Santa Fe, two areas near Granada in southern Spain (fig. 5.1).

The Hoya de Guadix, a high-altitude plateau at about 1000 m a.s.l., was apparently colonized recently by the great spotted cuckoo (see below). A total of 11 study plots (0.57–4.15 km^2 in area) within the Hoya de Guadix, separated by cultivated land with few or no potential nest sites for the magpie, were included in this study (fig. 5.1). The distance between neighboring study plots was 0.5–8 km, and the two most distant study plots were 25 km apart. The ejection rate of natural eggs was studied by considering each study plot as statistically independent but most reproductive parameters were pooled across all the study plots. Parasitism rates and the numbers of great spotted cuckoo eggs per nest were partitioned into three groups with different study plots grouped according to proximity, the type of habitat, and tree density.

1. The Guadix area, containing six different plots (see fig. 5.1), is a cereal-producing plain at 900–1100 m a.s.l. It has many gullies with clay cliffs rich in crevices and holes used for nesting by jackdaws (*Corvus monedula*), ravens (*Corvus corax*), and choughs (*Pyrrhocorax pyrrhocorax*). Magpies and carrion crows (*Corvus corone*) mostly nest in almond trees (*Prunus dulcis*) and holm oaks (*Quercus rotundifolia*). Except for the raven, all corvids have been parasitized by great spotted cuckoos in this area (Soler 1990).
2. The Hueneja area, comprised of two different plots (fig. 5.1) is about 30 km east of Guadix, at 800–900 m a.s.l. It consists mainly of almond groves and cereal fields. There were 30–40 pairs of magpies and three to five of carrion crows, both species being parasitized.
3. La Calahorra, having three different plots (fig. 5.1), is about 15 km south of Guadix, at 900–1000 m a.s.l. It consists mainly of almond groves and cereal fields, with some parts irrigated. Only magpies breed in this area. These three study areas are included in the Hoya de Guadix region, the presumably recently colonized area.

Santa Fe, 14 km west of Granada and 700–800 m a.s.l. is a cereal-producing plain with a small tamarisk (*Tamarix gallica*) forest and some holm oak trees. Only magpies nest in this area, in which breeding parameters have values similar to those in the literature. This is a presumed area of longstanding sympatry between the great spotted cuckoo and its magpie host.

RESULTS AND DISCUSSION

Incidence of parasitism

In the Hoya de Guadix there are five corvid species. The raven was never parasitized (0 of 26 nests), the jackdaw and the chough were parasitized only sporadically (6/290 and 8/162 respectively) and the carrion crow was parasitized rarely (4/47) (Solver 1990). Thus, these corvids are secondary hosts, while the magpie is parasitized frequently (54.8%; table 5.1) and can be considered to be the only primary host.

Table 5.1 Frequency of great spotted cuckoo parasitism of the magpie host according to study area and year

	Guadix			Hueneja			La Calahorra			Hoya de Guadix (total)			Sante Fe		
Study year	nf	np	%	nf	np	%	nf	np	%	nf	np	%	nf	np	%
1982	6	5	83.3	6	5	83.3	—	—	—	12	10	83.3	4	2	50.0
1983	28	9	32.1	10	3	30.0	—	—	—	38	12	31.6	8	4	50.0
1987–88	—	—	—	7	2	28.6	—	—	—	7	2	28.6	—	—	—
1989	21	5	23.8	13	6	46.2	—	—	—	34	11	32.4	26	5	19.2
1990	18	9	50.0	18	13	72.2	35	11	31.4	71	33	46.5		—	—
1991	19	10	52.6	40	31	77.5	52	24	46.2	111	65	58.6	—	—	—
1992	30	18	60.0	39	29	74.4	93	64	68.8	162	111	68.5	—	—	—
1993	33	23	69.7	40	30	75.0	111	58	52.3	184	111	60.3	—	—	—
1994	17	11	64.7	35	19	54.3	95	46	48.4	147	76	51.7	—	—	—
Total	172	90	52.3	208	138	66.3	386	203	52.6	766	431	54.8	38	11	28.9

nf = number of nests found; np = number of nests parasitized by the great spotted cuckoo; % = percentage of nests parasitized.

Altogether, 804 magpie nests were found, 442 of which (55.0%) were parasitized (table 5.1). The incidence of parasitism in the magpie ranged from 23.8% in Guadix in 1989 to 83.3% in Guadix and Hueneja in 1982 (table 5.1). To be valid, comparisons of parasitism rates among areas must not be confounded by comparisons among years. Considering only the data for 1990–1994, years when all sites in Hoya de Guadix were studied, Hueneja suffered a higher rate of nest parasitism (70.9%, $n = 172$) than did Guadix (60.7%, $n = 119$; $\chi^2 = 3.3$, $P = 0.07$) or La Calahorra (52.6%, $n = 386$; $\chi^2 = 16.4$, $P = 0.00001$). These results are independent of the laying date, since, during the same four years magpies laid at a similar time in the three areas of the Hoya de Guadix: Guadix ($\bar{x} = 29.3 \pm 1.61$, $n = 78$), Hueneja ($\bar{x} = 27.3 \pm 1.22$, $n = 127$) and La Calahorra ($\bar{x} = 26.1 \pm 0.65$, $n = 278$) ($F_{2,480} = 2.15$, $P = 0.12$). The great spotted cuckoo started to breed at the same time as the magpie (fig. 5.2). The rate of parasitism varied significantly among years ($\chi^2 = 13.48$, d.f. = 4, $P = 0.007$).

The percentage of parasitized nests in Hoya de Guadix, the area of recent sympatry (54.8%) was higher than in Santa Fe, the area of longstanding sympatry (28.9%; table 5.1) and than the rates of parasitism reported in the literature: 6% (1/17) (Valverde 1971), 2.4% (4/164) (Alvarez and Arias-de-Reyna 1974), and 22.2% (14/63) and 19.0% (16/84), respectively, in two different years (Arias-de-Reyna et al. 1982). Statistical comparisons were not made because they may be confounded by temporal variation. However, the differences are strong enough to suggest a higher rate of parasitism in the area of recent sympatry, which could be due to less developed host defenses, e.g., rejection of parasitic eggs (see below) and direct attack against adult parasites.

Parasitism in relation to breeding chronology

The magpie's laying period is from early April to early June, and, as the breeding season proceeds, the frequency of parasitism increases ($r_s = 0.93$, $P < 0.0005$, $n = 8$; fig. 5.2A), perhaps because late in the season host nests are very scarce. The number of cuckoo eggs per parasitized nest showed a similar tendency ($r_s = 0.83$, $P = 0.01$, $n = 8$; fig. 5.2B), also because late-season host nests are very scarce. The total number of cuckoo eggs laid per 10-day period showed a normal distribution peaking at the end of April, exactly following the tendency of magpie nests availability (fig. 5.2C).

Egg-laying behavior

Brood parasites usually remove or peck at least one host egg (Payne 1977; Wyllie 1981; Brooker et al. 1988). Mountford and Ferguson-Lees (1961) and Valverde (1971) found host-egg damage or removal in nests parasitized by the great spotted cuckoo. However, in other studies (Alvarez and Arias-de-Reyna 1974; Arias-de-Reyna et al. 1982), pecking or removal of eggs was not observed. In our study, 62.2% of the parasitized nests ($n = 360$) had damaged host eggs, whereas egg damage was never found in nonparasitized nests; 36.2% of the nests with damaged host eggs ($n = 224$) had one damaged egg, but some nests had two (55), three (41), four (27), five (11), six (5), or more (4) damaged eggs. These eggs were pecked, crushed or cracked (Soler 1990). In 136 parasitized nests, there were no damaged eggs, but in some cases this could have been because magpies removed damaged eggs before we saw them (Soler 1990). For this reason, the host clutches in parasitized nests were smaller than those in unparasitized nests (Soler 1990). The damaging of host eggs may be advantageous for the parasite by decreasing the number of future competitors of the parasitic chick in the nest.

Parasitic birds frequently remove a host egg, usually shortly before they lay their own egg (Payne 1977; Brooker et al. 1988); however, we have no evidence that great spotted cuckoos remove host eggs, nor do other authors (Arias-de-Reyna et al. 1982). The pied crested cuckoo (*Clamator jacobinus*) also does not remove host eggs (Liversidge 1971; Gaston 1976). This absence of egg removal in *Clamator* may be caused by the use of large host species, since removing large host eggs can be too costly in terms of the time involved. It may be important for *Clamator* cuckoos to minimize the time spent in the host nest, since the large hosts may attack and injure a parasite caught by surprise.

Crushed host eggs are probably the consequence of the impact of the parasite's egg, since the cuckoo lays from the rim of the nest (Arias-de-Reyna et al. 1982). It has been suggested that parasitic eggs may also be damaged in the process (Gaston 1976; Arias-de-Reyna et al. 1982), but we have never found a crushed great spotted cuckoo egg together with crushed host eggs in a nest with less than six cuckoo eggs ($n = 186$). This is to be expected because, as Lack (1968) stated, this laying behavior selects for eggs with thick shells (Brichambaut 1973). The laying of eggs from the rim of the nest may thus be a strategy to reduce the

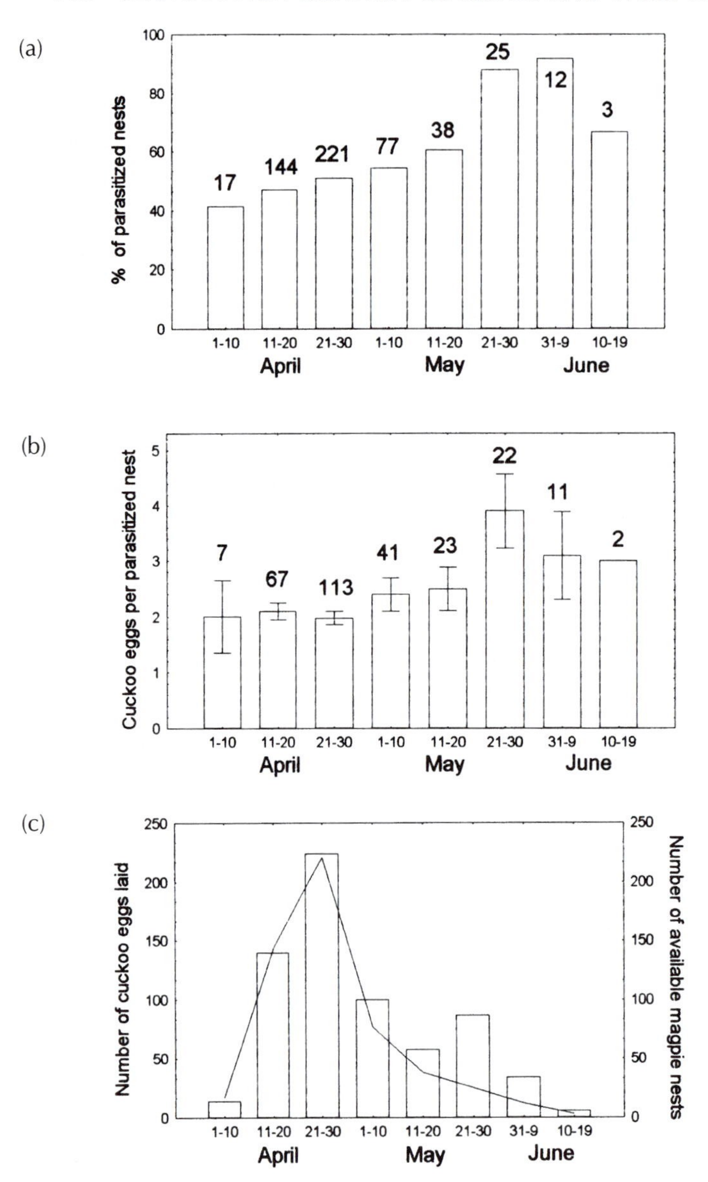

Figure 5.2 Measures of cuckoo parasitism of magpies at the Hoya de Guadix in 10-day periods. (A) Percentage of parasitized nests. (B) Number of parasitic eggs ($\bar{x} \pm$ SE) per parasitized nest. (C) Number of cuckoo eggs laid (bars) in relation to magpie nest availability (lines). Numbers above bars in (A) and (B) are sample sizes. Sample sizes for each 10-day period in (A) represent relative nest availability because search effort was equal across all 10-day periods.

number of competing host nestlings, and has also been recorded in the pied crested cuckoo (Gaston 1976). This laying strategy is used in the genus *Clamator* because, besides crushing some host eggs, it shortens the duration of laying. According to Arias-de-Reyna et al. (1982), laying takes less than 3 s, while in the European cuckoo it requires about 10 s (Wyllie 1981). Fast laying is important for *Clamator* cuckoos because of the already mentioned risk of being attacked by their large hosts.

If host eggs are damaged as a result of parasitic laying, the number of damaged host eggs should increase as the number of parasitic eggs per nest increases. Since magpies remove damaged eggs, the number of parasite eggs per nest should increase as the number of undamaged host eggs

Table 5.2 Number of great spotted cuckoo eggs per parasitized nest according to study area and year

	Number of parasitic eggs per parasitized nest														
Study year	1	2	3	4	5	6	7	8	9	10	11	12	n	x̄	S.E.
1982, 83, 87–88	6	8	7	—	1	—	—	—	—	—	—	—	22	2.2	0.21
1989	7	3	1	—	—	—	—	—	—	—	—	—	11	1.5	0.21
1990	11	6	2	2	1	4	—	1	2	—	1	1	31	3.7	0.58
1991	27	17	11	4	1	1	1	—	—	—	—	—	61	2.0	0.17
1992	39	34	12	8	9	3	1	1	—	1	—	—	108	2.4	0.17
1993	34	32	17	9	8	2	3	—	—	—	—	—	105	2.5	0.15
1994	28	25	8	9	1	1	—	—	1	—	—	—	73	2.2	0.17
Study area															
Guadix	34	30	7	6	2	—	2	1	2	1	—	—	85	2.3	0.21
Hueneja	43	29	31	12	4	6	2	1	1	—	1	1	131	2.7	0.17
Calahorra	75	66	20	13	15	5	1	—	—	—	—	—	195	2.2	0.10
Hoya de Guadix (Total)	152	125	58	31	21	11	5	2	3	1	1	1	411	2.4	0.09
Sante Fe (Total)	7	1	2	—	—	—	—	—	—	—	—	—	10	1.5	0.27

The section giving data according to year includes only Hoya de Guadix. Sample sizes are smaller than in table 5.1 because we could not determine the number of cuckoo eggs in some nests.

decreases, which is in fact the case ($r_s = -0.27$, $P < 0.0001$, $n = 353$).

Number of great spotted cuckoo eggs per nest

In Hoya de Guadix and Santa Fe, 63.0% ($n = 411$) and 30% ($n = 10$) of the parasitized nests, respectively, had two or more parasitic eggs per nest (table 5.2). In the great spotted cuckoo, as in other *Clamator* species, females may lay more than once in a given host nest (Rowan 1983). Great spotted cuckoo females lay an average of 1.4 eggs per nest (SE = 0.06, $n = 101$; see references in Soler 1990). In contrast, brood parasites in general usually lay one egg per host nest (Payne 1977). This difference may be due to the fact that hatchling *Clamator* take no direct action against other nestlings, but this, in turn, may be because the hosts of *Clamator* species are larger than the parasite and can probably rear more than one parasite chick.

The number of parasitic eggs per parasitized nest varied significantly when the three areas from the Hoya de Guadix and the five main years (1990–1994) were considered together (two-way ANOVA, $F(8, 363) = 3.66$, $P = 0.0004$). Significant differences were found among years ($F(4, 363) = 2.75$, $P = 0.03$), as a consequence of the larger number of parasite eggs per nest in 1990 ($\bar{x} = 3.7$; table 5.2). Differences among areas were also significant ($F(2, 363) = 7.36$, $P = 0.0007$) because most of the multiparasitized nests were in Hueneja. A post-hoc comparison showed that only data from Hueneja in 1990 were significantly different from the rest of the years and areas (Tukey HSD test; $P < 0.05$). Nothing is known about the movements and population dynamics in the great spotted cuckoo, but differences between neighboring areas in the same region and between consecutive years suggest that varying numbers of individuals may arrive in the different years (perhaps due to differences in recruitment or in overwinter surivival in Africa), and that they prefer certain areas. In Hueneja, the rate of nest parasitism was also higher than in the other areas (see above).

In Santa Fe, the area of longstanding sympatry, there were 1.5 cuckoo eggs per parasitized nest (S.E. = 0.27, $n = 10$), very close to that found in the literature (see above). In Hoya de Guadix, the area of recent sympatry, the mean was 2.4 (S.E. = 0.09, $n = 411$) and was significantly higher than in Santa Fe ($t = 3.34$, $P < 0.05$) and than in data reported in the literature ($t = 33.3$, $P < 0.0001$; references in Soler, 1990). However, these comparisons are not completely reliable because of the extensive year-to-year variation.

Because magpies can rear more than one parasitic chick, there was a significant positive

Table 5.3 Number of great spotted cuckoo females laying in the same nest

	Number of females laying in the same nest								
Study years	1	2	3	4	5	6	n	$\bar{x}$	S.E.
1982–83	13	5	1	—	—	—	19	1.37	0.14
1990	1	3	1	2	2	—	9	3.11	0.48
1991	0	11	1	—	—	—	12	2.08	0.08
1992	11	26	18	5	1	1	62	2.39	0.13
1993	19	23	11	1	1	—	55	1.95	0.12
1994	9	10	3	1	—	—	23	1.83	0.17
Total	53	78	35	9	4	1	180	2.09	0.07

Only Hoya de Guadix nests with more than one cuckoo egg have been considered.

correlation between the number of parasitic eggs laid and hatched per nest ($r_s = 0.78$, $P < 0.00001$, $n = 213$) and also between the number of eggs and chicks fledged per nest ($r_s = 0.66$, $P < 00001$, $n = 213$). However, this does not necessarily mean that multiple parasitism is advantageous. In order to address this question, we correlated the number of cuckoo eggs per nest with the number of eggs hatched and with the number of chicks fledged for each egg that was laid. A significant negative correlation was found in both cases ($r_s = -0.25$, $P < 0.00001$, $n = 213$ and $r_s = -0.41$, $P < 0.00001$, $n = 213$, respectively). This suggests that multiple parasitism may be disadvantageous, both with respect to hatching and fledging success.

Number of great spotted cuckoo females laying in the same nest

Using egg characteristics and laying dates, we determined that more than one female cuckoo laid in 127 of 180 nests with more than one cuckoo egg (table 5.3). Mountford and Ferguson-Lees (1961) and Arias-de-Reyna et al. (1982) also found nests with eggs from more than one female, and the latter explained this by: (i) territorial overlap; (ii) a lack of host nests with egg laying in progress; and (iii) a scarcity of host nests late in the season. According to the first explanation, since the great spotted cuckoo is assumed to be territorial (Arias-de-Reyna et al. 1987), few nests should have eggs from more than one female, and the same prediction applies to the second explanation because in all the considered 10-day periods, there were available host nests with egg laying in progress (fig. 5.2A). Thus, according to the first and second explanations, few nests should have eggs from more than one female, as in other studies (see Arias-de-Reyna et al. 1982). However, in the Hoya de Guadix region, 38.7% of the parasitized nests ($n = 328$) and 70.6% of the parasitized nests with more than one parasitic egg ($n = 180$; table 5.3) had eggs from more than one female. Thus, the first and second explanations can not account completely for the Hoya de Gaudix area. The third explanation seems the most likely, even though there is no direct support for it.

Parasitism in relation to host breeding cycle

Parasitic eggs were usually deposited during the laying period of the host (70.6% of 494 cuckoo eggs). However, 55 cuckoo eggs were laid in 22 nests shortly before the host started laying (table 5.4). In nests with more than one cuckoo egg, the second and third eggs were usually laid after the host already had at least four eggs and often after clutches were completed (table 5.4). Of nine eggs for which Arias-de-Reyna et al. (1982) knew the timing of laying, one was laid before the host started laying and another on the day after clutch completion. Among nests with multiple parasitism in our study, 11 eggs were deposited five or more days after the host completed its clutch (table 5.4).

Effect of the number of parasitic eggs per nest on the breeding success of the parasite and the host

Hatching success, fledging success and breeding success of the parasite in nests with more than four parasitic eggs were lower than in nests with one to three parasitic eggs (Table 5.5; Mann–Whitney U-test, $Z = 2.38$, $P = 0.02$; $Z = 3.59$, $P = 0.0003$; $Z = 3.92$, $P < 0.00001$, respectively). This is be-

Table 5.4 The laying stages at which magpie nests were parasitized

No. of eggs already laid by host	Sequence of cuckoo eggs laid					
	First	Second	Third	Fourth	Fifth and later	Total
0	22	(15)	(9)	(6)	(3)	22
1–3	44	13	8	3	1	69
4–7	41	31	18	5	10	105
Total laying period	167	89	50	22	21	349
Clutch completed	45	36	19	12	11	123

Numbers in parentheses indicate cases in which magpies may have laid eggs which were destroyed by the parasite and afterwards ejected by the magpie before being noted by an observer. For this reason we considered a total of only 22 eggs laid before magpies started laying. The total laying period includes all cuckoo eggs deposited during the magpie laying period. The sample sizes do not add up because the total row includes eggs in two previous rows, as well as those cuckoo eggs deposited on undetermined days during the magpie laying period.

cause unhatched eggs and chick mortality were considerably higher in nests with more than four cuckoo eggs.

The number of parasitic eggs per nest considerably affected the breeding success of the magpie (table 5.5). There was a significant negative correlation between number of parasitic eggs per nest and magpie eggs hatched ($r_s = -0.39$, $P < 0.01$, $n = 190$), and magpie chicks fledged ($r_s = -0.34$, $P < 0.01$, $n = 212$). In fact only one magpie fledged from the 38 nests with four or more cuckoo eggs that fledged any young.

Effect of the host breeding cycle on the breeding success of the parasite and the host

The stage in the breeding cycle at which parasitism occurred was important for parasite and host breeding success (table 5.6). In the 22 nests that were parasitized before the magpies started laying,

Table 5.5 Breeding success of great spotted cuckoos and magpies in relation to the number of parasite eggs per nest (in nests that fledged at least one parasitic or host young)

No. of parasitic eggs per nest	Eggs hatched ($\bar{x}$ ± S.E., n)	Chicks fledged ($\bar{x}$ ± S.E., n)	Hatching success (% ± S.E., n)	Fledging success (% ± S.E., n)	Breeding success (% ± S.E., n)
Clamator glandarius					
1	0.8 ± 0.0, 81	0.8 ± 0.0, 81	81.5 ± 4.3, 81	100 ± 0.0, 66	81.5 ± 4.3, 81
2	1.8 ± 0.1, 60	1.6 ± 0.1, 60	88.3 ± 3.6, 60	92.7 ± 2.4, 55	80.8 ± 4.0, 60
3	2.1 ± 0.2, 32	1.8 ± 0.2, 32	68.8 ± 5.4, 32	92.5 ± 3.7, 31	61.5 ± 5.2, 32
4	3.3 ± 0.2, 18	2.6 ± 0.2, 18	83.3 ± 5.7, 18	80.4 ± 4.8, 17	63.9 ± 5.4, 18
5	3.8 ± 0.4, 11	2.6 ± 0.5, 11	76.4 ± 8.0, 11	75.8 ± 9.7, 10	52.7 ± 9.8, 11
6	4.5 ± 1.5, 4	2.0 ± 0.7, 5	75.0 ± 25.0, 4	38.9 ± 14.7, 3	33.3 ± 11.8, 5
7–12	5.9 ± 0.9, 7	3.8 ± 0.3, 6	66.4 ± 7.3, 7	77.6 ± 8.6, 6	46.9 ± 6.0, 6
Pica pica					
1	2.0 ± 0.3, 68	1.1 ± 0.2, 81	33.8 ± 4.2, 68	53.2 ± 6.3, 41	17.6 ± 2.7, 81
2	1.4 ± 0.3, 52	0.5 ± 0.2, 60	26.5 ± 4.4, 52	31.0 ± 8.3, 27	8.0 ± 2.5, 58
3	0.9 ± 0.3, 29	0.3 ± 0.2, 31	21.1 ± 6.4, 27	24.0 ± 10.4, 10	6.1 ± 3.4, 28
4	0.4 ± 0.2, 18	0.0 ± 0.0, 18	9.4 ± 4.5, 18	0.0 ± 0.0, 4	0.0 ± 0.0, 18
5	0.0 ± 0.0, 9	0.0 ± 0.0, 9	0.0 ± 0.0, 9	—	0.0 ± 0.0, 9
6	0.2 ± 0.2, 5	0.2 ± 0.2, 5	4.2 ± 4.2, 3	100.0 ± 0.0, 1	4.2 ± 4.2, 3
7–12	0.0 ± 0.0, 7	0.0 ± 0.0, 6	0.0 ± 0.0, 4	—	0.0 ± 0.0, 3

n = number of nests. Only data from the Hoya de Guadix area have been included. Only successful nests have been considered. Nests where the cuckoo egg was rejected have not been considered. Hatching success is mean percentage of eggs that hatched, fledging success is mean percentage of hatchlings that fledged, and breeding success is mean percentage of eggs that produced fledglings.

Table 5.6 Breeding success of great spotted cuckoos and magpies in relation to time of parasitism during the breeding cycle

	No. of eggs already laid by host	Hatching success			Fledging success			Breeding success		
		n	$\bar{x}$ (%)	S.E.	n	$\bar{x}$ (%)	S.E.	n	$\bar{x}$ (%)	S.E.
	1–3	7	100.0	0.0	7	100.0	0.0	7	100.0	0.0
	4–7	9	100.0	0.0	9	100.0	0.0	9	100.0	0.0
Clamator glandarius	Total laying period	31	100.0	0.0	31	100.0	0.0	31	100.0	0.0
	Clutch completed	23	60.9	10.4	14	100.0	0.0	23	60.9	10.4
	1–3	6	17.1	14.0	2	0.0	0.0	7	0.0	0.0
	4–7	8	20.5	6.5	5	30.0	13.3	9	10.4	4.6
Pica pica	Total laying period	25	22.0	6.0	12	31.9	11.3	31	8.7	2.7
	Clutch completed	22	42.5	7.4	17	76.3	7.3	23	30.2	5.6

Only nests that received one parasitic egg in only one period of the breeding cycle have been considered.

the cuckoo eggs suffered different forms of rejection (see below). Cuckoo eggs laid during the host's laying period had a higher hatching success than did those laid after clutch completion (100% versus 60.9%, Table 5.6; Mann–Whitney U-test, $Z = 3.8$, $P = 0.0002$). Because every cuckoo hatchling fledged, the identical contrast also occurred for breeding success (table 5.6).

Magpie breeding success was lower (0.0% versus 10.4%) when the nest received a cuckoo egg early in the laying sequence (one to three magpie eggs) than late (four to seven eggs) (Mann–Whitney U-test, $Z = 1.94$, $P = 0.05$). When a cuckoo egg was laid after clutch completion, all three indices of magpie breeding success (table 5.6) were significantly higher than when the parasitic egg was laid during the laying period (hatching success: Mann–Whitney U-test, $Z = 2.06$, $P = 0.04$; fledging success: $Z = 2.8$, $P = 0.005$ and breeding success: $Z = 3.34$, $P = 0.0008$).

Breeding success of the parasite

An average of 1.55 great spotted cuckoo chicks fledged from each parasitized magpie nest (table 5.7). Egg losses per nest were higher ($\bar{x} = 0.44 \pm 0.06$) than chick losses ($\bar{x} = 0.25 \pm 0.05$) and 92.17% of parasitic hatchlings fledged (table 5.7).

There are few other data on great spotted cuckoo breeding success. Arias-de-Reyna et al. (1982) presented data similar to those of this study, but hatching success was lower in their study (56.0% and 42.1% in two different years), because of host rejection (see above). The cuckoo's fledging success was poorer when parasitizing the carrion crow (its secondary host) than when parasitizing the magpie (primary host), as the carrion crow's large size gives it clear advantages in nestling competition (Soler 1990).

Analysis of the success of each parasitic egg according to the order of its appearance in a nest (table 5.8) showed a significant negative correlation between laying sequence and hatching success ($r = -0.95$, $P < 0.0001$, $n = 8$) and also between laying sequence and fledging success ($r = -0.98$, $P < 0.0001$, $n = 8$). The sixth and subsequent cuckoo eggs in the laying sequence never produced fledgings because of starvation resulting from competition with parasitic chicks that hatched earlier.

Table 5.7 Measures of breeding success of the great spotted cuckoo in magpie nests

	n	$\bar{x}$	S.E.
Eggs hatched	220	1.87	0.10
Eggs unhatched	212	0.44	0.06
Chicks fledged	229	1.55	0.07
Chicks dying	214	0.25	0.05
Hatching success	213	80.77%	2.28
Fledging success	195	92.17%	1.34
Breeding success	213	73.21%	2.40

n = number of nests. Only successful nests have been considered.

Table 5.8 Success of parasitic eggs in relation to their laying sequence

		Hatching success		Fledging success		Breeding success	
Laying sequence	Eggs laid	*n*	%	*n*	%	*n*	%
1	217	196	90.3	196	100.0	196	90.3
2	135	111	82.2	96	86.5	96	71.1
3	71	44	62.0	31	70.5	31	43.7
4	38	26	68.4	12	46.2	12	31.6
5	21	11	52.4	3	27.3	3	14.3
6	10	6	60.0	0	0.0	0	0.0
7	7	3	42.9	0	0.0	0	0.0
8–12	9	1	11.1	0	0.0	0	0.0

Here, each parasitic egg has been considered as an independent datum.

Breeding success of the cuckoo females according to the sequence of laying

As shown earlier, more than one female cuckoo often laid in the same magpie nest. Breeding success of the first, second and third females laying in the same nest has been analyzed by using only nests where each female laid only one egg. Hatching, fledging, and breeding success of the first female were 100% in each category ($n = 22$). Hatching success, fledging success, and breeding success of the second female (72.7%, 75.0%, and 54.6% respectively, $n = 22$) were significantly lower than those of the first female (Fisher exact test, $P = 0.01$, $P = 0.03$, and $P = 0.003$, respectively). Hatching, fledging, and breeding success of the third female were each 0% ($n = 5$), and were each significantly lower than for the first female (Fisher exact test, $P = 0.0001$). Also, hatching and breeding success of these third females was significantly lower than those of the second female (Fisher exact test, $P = 0.006$ and $P = 0.05$, respectively).

Breeding success of the host

All parameters relating to magpie breeding success were significantly different between parasitized and nonparasitized nests (U-test; table 5.9). Magpie breeding success in parasitized nests was low. Only 22.9% of 227 successful nests (those fledging at least one parasitic or host nestling) fledged some magpie chicks, resulting in an average of 0.61 fledged chicks per nest, which is significantly less than that of nonparasitized nests ($X = 3.47$; table 5.9). The number of eggs hatched was lower in parasitized than in nonparasitized nests as a consequence of egg destruction by the parasite. The number of magpie chicks that died per nest was significantly lower in the parasitized nests (table 5.9), but this occurred simply because there were more host chicks in unparasitized nests.

Table 5.9 Breeding success of the magpie in parasitized and unparasitized nests

	Parasitized			Nonparasitized		
	n	$\bar{x}$	S.E.	*n*	$\bar{x}$	S.E.
Eggs hatched	197	1.26	0.13	112	4.98	0.19**
Eggs unhatched	182	3.89	0.16	108	1.81	0.18**
Chicks fledged	227	0.61	0.09	131	3.47	0.13**
Chicks dying	190	0.72	0.09	108	1.56	0.14**
Hatching success	185	23.95%	2.35	109	73.33%	2.57**
Fledging success	84	39.97%	4.57	108	73.43%	2.25**
Breeding success	204	10.37%	1.45	122	50.47%	1.89**

Comparisons have been made by Mann–Whitney U-tests. Probability level: ** = $P < 0.0001$. n = number of nests. Only successful nests have been considered.

The latter nests had a lower mortality rate for nestlings (26.6% versus 60.0%, from table 5.9).

The data in table 5.9 do not show one additional way in which parasitism is likely to reduce magpie breeding success. Magpies that fledged from unparasitized nests weighed more than those from parasitized nests (Soler and Soler 1991) and probably had higher survival rates.

Nest predation suffered by parasitized and nonparasitized nests

The predation rate of nonparasitized nests was higher than that of parasitized ones in all the study years (table 5.10). In no year were the differences significant (Fisher exact tests, χ^2, ns); however, when data from the five years were pooled, nonparasitized nests had a significantly higher predation rate (31.9%, $n = 232$) than parasitized nests (20.2%, $n = 326$; $\chi^2 = 9.79$, $P = 0.002$).

This higher predation rate in nonparasitized nests may be due to two reasons. First, cuckoos may prey selectively on nonparasitized host nests. Nest destruction by cuckoos could result from: (i) cuckoos forcing magpies to raise parasitic offspring by destroying or preying upon host eggs or nestlings following host ejection of parasite offspring (see Zahavi 1979; Soler et al. 1995*a*) or (ii) cuckoos preying upon host nestlings to increase the relaying frequency of the host (Soler et al., 1995*a*). Second, parasitized nests may be safer against predation because: (i) cuckoos protect parasitized nests from predators (Soler et al., manuscript submitted); and/or (ii) cuckoos may preferentially parasitize safer nests (in fact it has recently been demonstrated experimentally that great spottted cuckoos selectively parasitize high-quality magpie pairs (Soler et al. 1995*b*). However, at present, we have no evidence to choose one hypothesis over the other.

Table 5.10 Predation rate in parasitized and nonparasitized nests

	Parasitized		Nonparasitized	
Year	%	*n*	%	*n*
1990	25.0	16	53.8	13
1991	32.1	53	40.0	30
1992	12.8	86	17.6	34
1993	14.4	90	22.6	53
1994	20.8	48	32.7	52
Total	20.2	326	31.9	232

Only nests which remained unmanipulated (control nests) each year have been used.

Host responses

Egg rejection is the most important host defense, and it appears that avian egg recognition evolved in response to brood parasitism (Rothstein 1990). Previous works on egg recognition indicate that many species may be categorized as "rejecters" or "accepters" (Rothstein 1975*a*,*b*, 1990; Davies and Brooke 1989). Magpies are considered rejecters (Alvarez et al. 1976). We performed experimental egg manipulations in 85 nests of the five corvid species in our study area (31 magpie, 14 carrion crow, 22 jackdaw, 7 raven and 11 chough nests), and rejection was found only among magpies (see Soler 1990 for more detailed information). Therefore, the other four species can be categorized as accepters (see also Yom-Tov 1976).

Magpies in Hoya de Guadix rejected 78.3% of 23 quail eggs painted with bright colors, but only 12.5% of eight chicken eggs. These values were lower than expected, considering that in other areas magpies reject all nonmimetic eggs (Arias-de-Reyna and Hidalgo 1982). Ejection was the only defensive response to experimental eggs. Magpies never ejected one of their own eggs.

In accordance with the results obtained in the experimental egg manipulations, jackdaws ($n = 6$), carrion crows ($n = 4$) and choughs ($n = 8$) never rejected naturally deposited cuckoo eggs (Soler 1990). In the Santa Fe area, magpies ejected the great spotted cuckoo egg in two of ten nests. In one nest lacking parasite eggs, there was one pecked egg, indicating that the nest had been parasitized, the parasite egg having been ejected afterwards. This suggests a rejection rate of 27.3%. These data from the Santa Fe area agree with those of Arias-de-Reyna et al. (1982), who found a rejection rate of 25% (2/8) in Central Sierra Morena, which is more than 150 km from Santa Fe and 200 km from Guadix.

In Hoya de Guadix, the area of presumably recent sympatry, magpies showed rejection at only 24 of 458 (5.2%) parasitized nests. This rejection rate was lower than the 27.3% rate in the area of presumably longstanding sympatry ($\chi^2 = 9.61$, $P = 0.002$) and lower than data previously reported by Arias-de-Reyna et al. (1982) ($\chi^2 = 2.68$, $P = 0.02$). Ejection was the only type of host rejection that occurred when the nests were parasitized during the host laying period. However, other host responses (Davies and

Brooke 1988; Graham 1988; Rothstein 1990) occurred in 22 nests in which the parasitic egg or eggs were deposited before the host started laying: ten nests were deserted; in three nests, magpies buried the parasite eggs with nesting material and laid a replacement clutch on the new nest floor; in two, magpies ejected one and two parasite eggs, respectively; in one, magpies pecked two parasite eggs. In these latter three nests, magpies started laying in the same nest shortly after ejecting or pecking the parasitic eggs. The magpies accepted the cuckoo eggs in the remaining six of these 22 nests.

The evolution of egg discrimination by the hosts

Because changes in host defenses select for changes in parasitic counterdefenses, it has been suggested that there is a coevolutionary arms race between cuckoos and their hosts (Dawkins and Krebs 1979; Mason and Rothstein 1987; Brooke and Davies 1988). There is strong evidence suggesting that parasitism is the factor responsible for the evolution of egg rejection by hosts (see review in Rothstein 1990). Why, then, did jackdaws, choughs, and carrion crows not reject nonmimetic eggs? One explanation which could be valid for the first two species is that they are unusual and recent hosts. They were only recently first recorded as hosts of the great spotted cuckoo (Zuñiga et al. 1983), and therefore may have not yet had time to evolve counteradaptations. However, the carrion crow requires another explanation, as it has long been parasitized (Valverde 1953). Given that rejection of parasitic eggs involves costs to the host expressed as recognition errors (Davies and Brooke 1988, 1989; Rothstein 1990), and that parasitism by the great spotted cuckoo has little impact on the carrion crow (Soler 1990), discrimination and ejection may be more costly than accepting the parasite egg. Thus, the lack of egg rejection in the carrion crow in our study area could be due to a state of equilibrium (Lotem, this vol.). However, acceptance in the jackdaw and the chough may be the result of an evolutionary lag, since the minor impact of the parasitism on the carrion crow is a consequence of large size of the host (Soler 1990). These two other corvids are much smaller and for them the impact of parasitism could be stronger.

A related question is why the rejection rate in the area of recent sympatry was lower than in the area of older sympatry and elsewhere (see earlier). We predicted that great spotted cuckoo parasitism influences magpie discrimination. This prediction was tested (Soler and Møller 1990) by analyzing magpie responses to parasitic eggs by introducing a painted mimetic model of great spotted cuckoo eggs, a nonmimetic egg (quail egg painted red), or both, into magpie nests in three different areas: Hoya de Guadix, an area of recent colonization by the cuckoos; Santa Fe, an area of longstanding sympatry; and Uppsala (Sweden), where the great spotted cuckoo is absent.

Magpie responses to the introduction of both kinds of eggs differed significantly among sites. Mimetic eggs had the highest ejection rate, 78% ($n = 9$), in the Santa Fe area of longstanding sympatry, with low values of 14% ($n = 14$) in the Guadix area of recent sympatry and no ejection ($n = 12$) in the Uppsala area (fig. 5.3; $G^2 = 18.93$, $P < 0.001$). Nonmimetic eggs had an ejection rate of 100% ($n = 9$) in Santa Fe and 71% ($n = 14$) in Guadix, with no ejection ($n = 12$) in Uppsala (fig. 5.3; $G^2 = 31.47$, $P < 0.001$). There was a highly significant difference in ejection rate between mimetic (14%) and nonmimetic (71%) eggs in the Guadix area of recent sympatry (fig. 5.3; $G^2 = 10.01$, $P < 0.005$). Host discrimination against mimetic cuckoo eggs was therefore most pronounced in the area of longstanding sympatry, whereas nonmimetic eggs were largely discriminated against in both areas of sympatry (Soler and Møller 1990).

These experimental results demonstrated that the rejection of great spotted cuckoo eggs by the magpie hosts varies among populations, depending on cuckoo parasitism. Such differences in egg recognition among host populations affect the spatial population dynamics of parasites and their hosts (Soler and Møller 1990).

Is Guadix actually a recently colonized area?

Zuñiga and Redondo (1992) argued against our interpretation of the different magpie responses between Guadix and Santa Fe. Observations of parasitism from Santa Fe and Guadix were used to suggest that there was no evidence for variable duration of sympatry between the great spotted cuckoo and its magpie host.

Our case for recent sympatry in the Guadix area was supported by the first great spotted cuckoo having been shot there in 1962 (Soler 1990). The cuckoo was shown by the hunter to his peers, but nobody recognized the species. Given that the great spotted cuckoo is a conspicuous bird with a very loud and characteristic call (Cramp 1985) which cannot easily be missed, and that

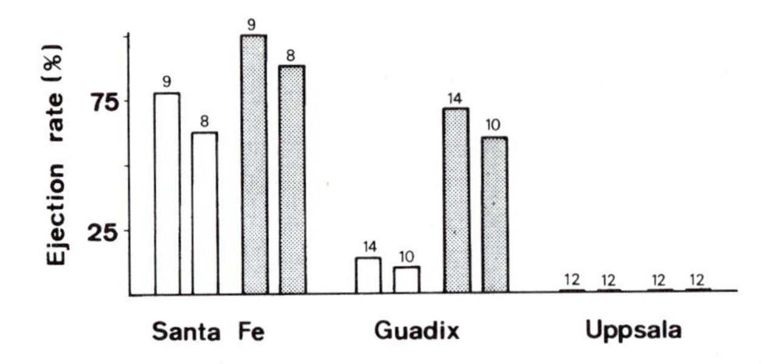

Figure 5.3 Ejection rate (%) of mimetic and nonmimetic eggs from magpie nests in the three areas of different duration of sympatry (see text for details of areas). The two unshaded columns for each study site represent ejection rates of mimetic eggs added singly (left column) or simultaneously with a nonmimetic egg (right column). The two shaded columns for each study site represent ejection rates of nonmimetic eggs added singly (left column) or simultaneously with a mimetic egg (right column). The number of magpie nests used in the experiments is indicated. (After Soler and Møller 1990.)

several of the hunters had an intimate knowledge of the Guadix fauna since the 1940s, it seems unlikely that this cuckoo was present but went unnoticed prior to the early 1960s. Zuñiga and Redondo (1992) cited a record of great spotted cuckoos at Iznalloz (42 km from Guadix) in 1885 as evidence of an early occurrence in the Guadix area. However, Iznalloz is only 30 km from Santa Fe and is separated from Hoya de Guadix by the Sierra Harana mountain (see fig. 5.1), and great spotted cuckoos rarely breed in mountains (Soler 1990).

We demonstrated that rejection of great spotted cuckoo eggs by the magpie hosts varies between populations, depending on the duration of sympatry with cuckoos. This conclusion was challenged by Zuñiga and Redondo (1992) who showed that differences in parasitism rates between Guadix and Santa Fe appeared to contradict those previously reported by Soler and Møller (1990). This conclusion may be unwaranted for three reasons.

First, Zuñiga and Redondo (1992) did not study magpies only in the Santa Fe area, as stated in their paper, but rather in Chimeneas (T. Redondo, pers. commun.) which is approximately 16 km from Santa Fe (see fig 5.1). The habitat in Chimeneas (*Olea europea* and mainly *Prunus dulcis*) differs from that in our study site in Santa Fe (*Tamarix gallica* and *Quercus rotundifolia*), and this may affect both the ability of cuckoos to parasitize magpies and the ability of magpies to defend their nests against parasites. Second, Zuñiga and Redondo (1992) compared observational data from Guadix during 1982–1984 with observational data from Chimeneas during 1985 and 1989–1991. However, the frequency of parasitism in different areas cannot be compared without considering temporal effects, which we have shown to be significant (Soler et al. 1994). Third, while Soler and Møller (1990) compared experimental data from three areas during the same year, Zuñiga and Redondo (1992) compared observational data from different years. Evaluation of temporal changes in host responses to parasites based on observational data is suspect, primarily because some cases of natural parasitism may go unnoticed because of egg ejection by hosts before the contents of host nests have been checked (see also Soler et al. 1994). It is imperative to check host responses using an experimental approach.

Further evidence has recently been provided which supports the idea that Guadix is a recently colonized area. We (Soler et al. 1994) tested the prediction that if Guadix is a recently colonized area, the rate of ejection of experimentally introduced egg models should increase as the duration of exposure to the brood parasite increases. The rejection rate of mimetic model eggs increased from 14% (n = 14 nests) in 1989 to 27% (n = 11 nests) in 1992 to 33% (n = 21 nests) in 1993. This increase of 4.72% (S.E. = 0.40) per year is marginally significant (F = 139.45, d.f. = 1.1, r^2 = 0.99, P = 0.054). The ejection rate of nonmimetic model eggs, which was 61% (n = 31 nests) in 1983 and 1984, increased to 71% (n = 14 nests) in 1989, 73% (n = 11 nests) in 1992, and 89% (n = 18 nests) in 1993. This increase of 2.27% (S.E = 0.63) per year was statistically significant (F = 12.78, d.f. = 1.3, r^2 = 0.81, P = 0.037). These recent increases in rejection rates in the Guadix area support the idea that Guadix is a recently colonized area.

Our experiments on egg ejection by magpies have shown an increasing tendency to discriminate against alien eggs, even when the potentially confusing effects of other variables were controlled experimentally. Only two potential cases of recent changes in the response of hosts to brood parasitism have been documented however; these have not been verified experimentally (Soler et al. 1994). The rapid increase in host response to cuckoo eggs could be due either to a genetic change or to some sort of learning.

ACKNOWLEDGMENTS We are most grateful to Anders P. Møller, Juan Moreno and Stephen I. Rothstein for valuable comments on the manuscript. We are grateful to Javier Minguela, Mariano Paracuellos, Jesús Sánchez (Lechu) and all the friends that helped us with the field work. We are indebted to Eduardo Ramirez for preparing fig. 5.1. We are specially indebted to Anders Pape Møller, not only for his helpful comments, but also for his collaboration in obtaining most of the information presented in this paper. Financial support was given by the Commission of the European Communities (SCI*-CT92-0772), DGICYT PB91-0084-CO3-02 research project, and Junta de Andalucía (grupo 4104: Comportamiento y Ecología Animal).

REFERENCES

Alvarez, F. and L. Arias-de-Reyna (1974). Mecanismos de parasitación por *Clamator glandarius* y defensa por *Pica pica. Doñana, Acta Vert*, **1**, 43–66.

Alvarez, F., L. Arias-de-Reyna, and M. Segura (1976). Experimental brood parasitism of the magpie (*Pica pica*). *Anim. Behav*., **24**, 907–916.

Arias-de-Reyna, L. and S. Hidalgo (1982). An investigation into egg-acceptance by azure-winged magpies and host-recognition by great spotted cuckoo chicks. *Anim. Behav*., **30**, 819–823.

Arias-de-Reyna, L., P. Recuerda, M. Corvillo, and I. Aguilar (1982). Reproducción del Críalo (*Clamator glandarius*) en S. Morena Central. *Doñana, Acta Vert*., **9**, 177–193.

Arias-de-Reyna, L., P. Recuerda, M. Corvillo, and A. Cruz (1987). Territory in the great spotted cuckoo (*Clamator glandarius*). *Journal für Ornithologie*, **128**, 231–239.

Brichambaut, J. (1973). Contribution de l'oologie á la connaissance de la biologie du coucou-geai *Clamator glandarius. Alauda*, **41**, 353–361.

Brooke, M. de L. and N. B. Davies (1988). Eggs mimicry by cuckoos *Cuculus canorus* in relation to discrimination by host. *Nature*, **355**, 630–632.

Brooker, M. G., L. C. Brooker, and I. Rowley (1988). Egg deposition by the bronze cuckoos *Chrysococcyx basalis* and *Ch. lucidus*. *Emu*, **88**, 107–108.

Calvo, F. (1979). *Estadística aplicada.* Deusto S.A., Bilbao.

Cramp, S. (ed.) (1985). *The birds of the Western Palearctic Vol. IV*. Oxford University Press, Oxford.

Davies, N. B. and M. de L. Brooke (1988). Cuckoos versus reed warblers, Adaptations and counteradaptations. *Anim. Behav*., **36**, 262–284.

Davies, N. B. and M. de L. Brooke (1989). An experimental study of coevolution between the cuckoo, *Cuculus canorus* and its host. I. Host egg discrimination. *J. Anim. Ecol*., **58**, 207–224.

Dawkins, R. and J. R. Krebs (1979). Arms race between and within species. *Proc. R. Soc. Lond. B*., **205**, 489–511.

Fry, C. H., S. Keith, and E. K. Urban (1988). *The birds of Africa, Vol. III.* Academic Press, London.

Gaston, A. J. (1976). Brood parasitism by the pied crested cuckoo (*Clamator jacobinus*). *J. Anim. Ecol*., **45**, 331–348.

Graham, D. S. (1988). Responses of five host species to cowbird parasitism. *Condor*, **90**, 588–591.

Harvey, P. H. and L. Partridge (1988). Of cuckoo clocks and cowbirds. *Nature*, **335**, 586–587.

Jensen, R. A. C. and M. K. Jensen (1969). On the breeding biology of Southern African cuckoos. *Ostrich*, **40**, 163–181.

Kemal, R. E. and S. I. Rothstein (1988). Mechanisms of avian egg recognition: adaptive responses to eggs with broken shells. *Anim. Behav*., **36**, 175–183.

Lack, D. (1968). *Ecological adaptations for breeding in birds.* Chapman & Hall, London.

Liversidge, R. (1971). The biology of the Jacobin Cuckoo (*Clamator jacobinus*). Proceedings 32nd Ornithological Congress. *Ostrich Supplement*, **8**, 177–237.

Mason, P. and S. I. Rothstein (1987). Crypsis versus mimicry and the color of shiny cowbird eggs. *Amer. Natur*., **130**, 161–167.

Mountford, G. and I. Ferguson-Lees (1961). The bird of the Coto Doñana. *Ibis*, **103a**, 86–109.

Payne, R. B. (1977). The ecology of brood parasitism in birds. *Annu. Rev. Ecol. Sys*., **8**, 1–28.

Rothstein, S. I. (1975*a*). An experimental and teleonomic investigation of avian brood parasitism. *Condor*, **77**, 250–271.

Rothstein, S. I. (1975*b*). Evolutionary rates and host defenses against avian brood parasitism. *Amer. Natur.*, **109**, 161–176.

Rothstein, S. I. (1990). A model system for coevolution: avian brood parasitism. *Annu. Rev. Ecol. Sys.*, **21**, 481–508.

Rowan, M. K. (1983). *The doves, parrots, louries and cuckoos of Southern Africa.* David Philip, Cape Town.

Siegel, S. and N. V. Castellan, Jr (1988). *Nonparametric statistics for the behavioral sciences.* McGraw-Hill, New York.

Sokal, R. R. and F. J. Rohlf (1979). *Biometría, Principios y métodos estadísticos en la investigación biológica*. Blume, Madrid.

Soler, J. J., M. Soler, A. P. Møller, and J. G. Martinez (1995*b*). Does the Great Spotted Cuckoo choose Magpie hosts according to their parenting ability? *Behav. Ecol. Sociobiol.*, **36**, 201–206.

Soler, M. (1990). Relationships between the Great Spotted Cuckoo, *Clamator glandarius*, and its corvid hosts in a recently colonized area. *Ornis Scandinavica*, **21**, 212–223.

Soler, M. and A. P. Møller (1990). Duration of sympatry and coevolution between the Great Spotted Cuckoo and its Magpie host. *Nature*, **343**, 748–750.

Soler, M. and J. J. Soler (1991). Growth and development of Great Spotted Cuckoos and their Magpie host. *Condor*, **93**, 49–54.

Soler, M., J. J. Soler, J. G. Martinez and A. P. Møller (1994). Micro-evolutionary change in host response to a brood parasite. *Behav. Ecol. Sociobiol.*, **35**, 295–301.

Soler, M., J. J. Soler, J. G. Martinez, and A. P. Møller (1995*a*). Magpie host manipulation by great spotted cuckoos: evidence for an avian mafia? *Evolution*, **49**, 770–775.

Valverde, J. A. (1953). Contribution a la biologie du coucou-geai, *Clamator glandarius* L. I. Notes sur le coucou-geai en Castille. *Oiseaux*, **23**, 288–296. **49**, 770–775.

Valverde, J. A. (1971). Notas sobre la biología reproductora del Críalo (*Clamator glandarius*). *Ardeola (volumen especial)*, 591–647.

Veillard, J. (1973). Remarques sur l'adaptation du parasitisme chez le Coucou-geai. *Alauda*, **41**, 362–364.

Wyllie, I. (1981). *The Cuckoo*. Batsford, London.

Yom-Tov, Y. (1976). Recognition of eggs and young by the carrion crow (*Corvus corone*). *Behaviour*, **59**, 247–251.

Zahavi, A. (1979). Parasitism and nest predation in parasitic cuckoos. *Amer. Natur.*, **113**, 157–159.

Zuñiga, J. M. and T. Redondo (1992). No evidence for variable duration of sympatry between the Great Spotted Cuckoo and its Magpie host. *Nature*, **359**, 410–411.

Zuñiga, J. M., M. Soler and I. Camacho (1983). Nota sobre nuevas especies parasitadas por el críalo (*Clamator glandarius*) en España. *Doñana, Acta Vert.*, **10**, 207–209.

6

Coevolution of the Great Spotted Cuckoo and Its Hosts

LUIS ARIAS-de-REYNA

In the past decade, the great spotted cuckoo (*Clamator glandarius*) has come to be one of the most well-studied parasitic birds (Redondo and Arias-de-Reyna 1989; Soler 1990; Soler and Møller 1990; Soler and Soler 1991; Zuñiga and Redondo 1992*a*,*b*; Redondo 1993; Arias-de-Reyna 1994; Soler et al. 1994). It differs from the two other most well-studied parasites, the common cuckoo (*Cuculus canorus*) and the brown-headed cowbird (*Molothrus ater*), in several critical ways. These two latter species show little geographic variation in the number of host species they use. In each region where it occurs, the common cuckoo specializes on several host species while the brown-headed cowbird acts as an extreme generalist and parasitizes numerous passerine species with which it co-occurs. In contrast, the great spotted cuckoo specializes on a single host species in some regions but uses a wide range of host species in other areas. In addition, some aspects of the great spotted cuckoo's host interactions suggest that it is a more highly coevolved parasite than either of the two other well-known parasites. For example, mated pairs parasitize host nests in a coordinated fashion that negates host nest defenses. There is also some evidence that great spotted cuckoos protect host nests from predators. In this chapter, I review these and other adaptations that great spotted cuckoos have for parasitism and concentrate on the extent to which this species can be considered a specialist or a generalist in terms of the number of host species it uses. Most of the data I review are from Spain, where all intensive studies of this parasite have been carried out. But I begin with an overview of this species' entire geographic range.

Cramp (1985) and Fry et al. (1988) divided the great spotted cuckoo into four largely disjunct populations (fig. 6.1). In southern Africa, it parasitizes at least 10 species in the Sturnidae and about six in the Corvidae. Each of the three other populations, in central Africa and in the eastern and western Mediterranean, specialize on one to a few corvid species (fig. 6.1). The intensively studied great spotted cuckoo in Spain specializes on a single corvid, the magpie (*Pica pica*) with the carrion crow (*Corvus corone*) being an important secondary host and several other corvids such as jackdaw (*Corvus monedula*), red-billed chough (*Pyrrhocorax pyrrhocorax*), azure-winged magpie (*Cyanopica cyanus*) and raven (*Corvus corax*) reported as occasional hosts (Valverde 1953; Lévêque 1968; Araujo and Landín 1970; Zuñiga et al. 1983; Soler, 1990). There are a few other rare host species such as another corvid, the jay (*Garrulus glandarius*) and common kestrel (*Falco tinnunculus*) (Valverde 1971). In addition, I once observed a woodchat shrike (*Lanius senator*) repeatedly feed a great spotted cuckoo fledgling for 15 minutes, but this sort of observation is not definite evidence of parasitism (Zuñiga and Redondo, 1992*a*; Redondo, 1993).

Both Palearctic subpopulations of great

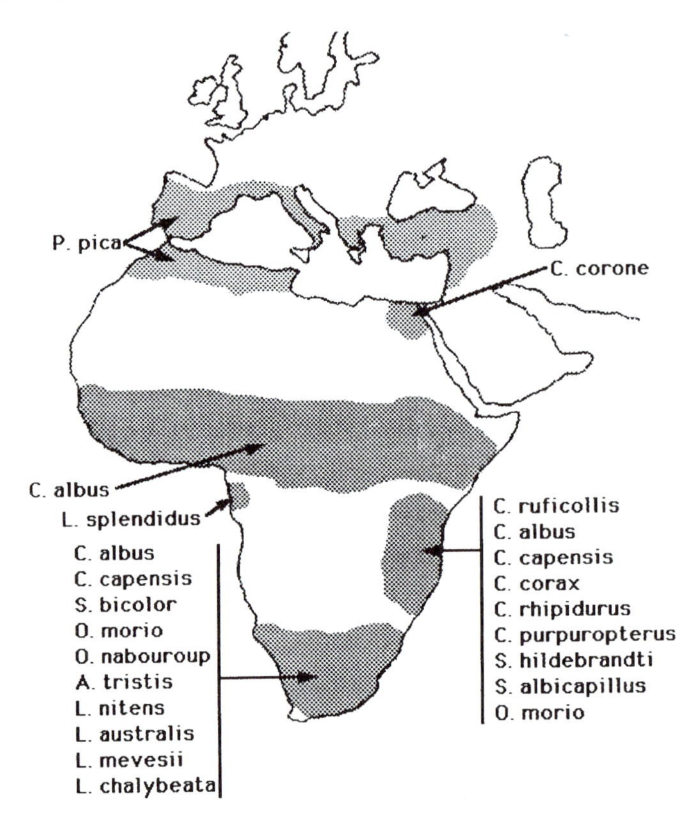

Figure 6.1 Distribution of the great spotted cuckoo. Each shaded area represents a different subpopulation of the parasitic species. The usual hosts for each area are listed. Host genera are abbreviated as follows: *A.* = *Acridotheres*; *C.* = *Corvus*; *L.* = *Lamprotornis*; *O.* = *Onychognathus*; *P.* = *Pica*; *S.* = *Spreo.*

spotted cuckoos are colonizing new areas. The western Mediterranean population entered Europe via the Iberian peninsula. Early reports of parasitism of the magpie by the great spotted cuckoo in Spain date from 1869 or earlier in Aranjuez and 1885 in Granada [Saunders (1869), in Friedmann (1948) and Sánchez García (1885), in Zuñiga and Redondo 1992*b*)], but the great spotted cuckoo probably colonized Spain much earlier. The western subpopulation reached France at the end of the last century but the cuckoo was regarded as an accidental species there until 1936. However, by the middle of this century, hundreds of pairs were reproducing throughout southwestern France (Lévêque 1968; Yeatman 1974). A similar pattern of range extension was observed in the eastern Mediterranean subpopulation, the size of which is currently increasing steadily (Lévêque 1968; Di Carlo 1969, 1971; Demartis 1971). It entered Europe via the Middle East and has moved westwards to Italy and also, possibly, France.

The great spottted cuckoo should have more hosts in its area of origin, which is probably South Africa, than in newly colonized areas because a parasitic cuckoo with tendencies towards specialization may parasitize increasing numbers of host species the longer it is in an area (Arias-de-Reyna 1985; Cramp 1985). Its range has expanded mainly thanks to corvids; however, based on the species it parasitizes in South Africa, sturnids were probably the first to be parasitized. Sturnids nest in holes, which may have led the great spotted cuckoo to use different strategies from those employed to parasitize species not nesting in holes. But this possibility has not previously been explored.

Unfortunately, there is little detailed information on host species in Africa and the Eastern Mediterranean region. The impact of parasitism in these regions is poorly known, as are rates of parasitism. This uncertainty precludes classification of the hosts as usual, occasional or rare, and hinders reliable estimation of their numbers in each geographic area. In any case, new species are certainly being parasitized as this cuckoo's range expands, so an analysis of the mechanisms involved is therefore in order (Arias-de-Reyna 1985).

STUDY AREAS AND METHODS

In Europe, the great spotted cuckoo has been thoroughly studied at only a few Spanish locations, of which the following are worth special mention.

Central Sierra Morena (Córdoba and Jaén) — This habitat is characterized by an open, virtually monospecific woodland of oaks (*Quercus rotundifolia*). Trees grow in a nearly uniform distribution (oak savannah = "monte adehesado") and there are gentle slopes. As regards potential hosts, three corvid species breed in this region, the magpie, the azure-winged magpie and the jackdaw, but in different places. Carrion crow and red-billed chough also occur in the area, though in small numbers. Azure-winged magpies use both oak and pine (*Pinus pinaster*) woodland, in addition to olive groves. Studies on various magpie and azure-winged magpie colonies have been carried out in the area by monitoring nests of both species every 2–7 days, depending on the nature of the particular study.

Santa Fe (Granada) — This region has cereal crops, olive groves, and scattered oaks on very gentle slopes. There are linear groves of taray (*Tamarix gallica*) in two gullies where magpies, the only corvid in this area, nest.

Guadix (Granada) — This area has two main habitats: (i) flat ground (80% of the overall area) with cereal crops and very scant pasture or oak savannah and scattered almond (*Prunus amigdalus*) orchards; (ii) streams between 3 and 7 km long and 40 to 300 m wide, where retamo (*Retama* spp.) is the major species. Five corvids coexist sympatrically (magpie, carrion crow, jackdaw, raven, and red-billed chough).

The Guadix and Santa Fe areas (the latter is ca. 60 km southwest of the former) have been studied by Soler, Redondo and Zuñiga, both individually and with other coworkers since 1978. The Sierra Morena area has been surveyed virtually uninterruptedly by the author and his coworkers since 1978. Details of the data collection and experimental methodologies are presented in the results.

RESULTS AND DISCUSSION

As with other parasitic species, the major mechanisms whereby the great spotted cuckoo enhances its reproductive success (see Rothstein 1990) are as follows: (i) special laying tactics; (ii) increased egg laying; (iii) rapid chick development; (iv) a short incubation period; (v) decreased nestling competition; (vi) egg mimicry and (vii) parasitization of new host species. These mechanisms and major aspects of parasitism in each of the above study areas are summarized in table 6.1 and are discussed below.

Parasite laying tactics

Laying more than one egg per nest may result in damage to great spotted cuckoo eggs and decrease competition between parasitic chicks as well as between parasitic and host chicks, particularly when the host is large (e.g., the carrion crow). When more than one female cuckoo lays in a host nest, the hatching and breeding success rates of the second female (six eggs laid, none hatched) are significantly lower than those of the first female to lay (40 eggs laid, 36 hatched, and 30 fledged) (Soler 1990). Also, scattering eggs among host nests, as opposed to laying many eggs in one or a few nests decreases the likelihood that all of a female's eggs will be lost to predation (Payne 1974; 1977*a*). For these reasons, the cuckoos should avoid multiple parasitism.

The average number of eggs laid by the great spotted cuckoo in nests of its usual host in areas where only the magpie breeds varies from 1.13 at Sierra Morena to 2.44 at Sante Fe (table 6.1). In areas with more than one host, the average number of eggs per nest is similar and ranges from 1.20 to 2.2.

To distribute eggs adaptively, a mated cuckoo pair can use the following tactics, among others: host recognition; special laying procedures; parasite–host laying synchronization; territoriality; and the provision of benefits to the host. Each tactic is discussed in detail below.

Recognition of hosts

To check whether great spotted cuckoo chicks learn to recognize their hosts through some sort of

Table 6.1 Some data on the great spotted cuckoo and its hosts from the more studied locations in Spain

Location	Host	Distance (m) between nests*	Percentage of nests parasitized	Number of parasitic eggs/nest#	Fledgling great spotted cuckoos/nest#	Source
Sierra Morena	*Pica pica*	111.4 ±55.3 (63)	19.0 (84)–	1.13 ±0.12 (30)	0.86 ±0.77 (14)	Arias-de-Reyna et al.
Santa Fe	*P. pica*	271.1 ± 13.03 (11-	22.2 (63)	1.8 ±0.49 (5)	—	(1982)
		6)	50 (12)			Soler and Møller (1990) in
						Zúñiga and Redondo
	P. pica			2.44 ±0.12 (143)	1.34 ±0.11 (127)	(1992*b*)
		—	60.7 (242)			Zúñiga and Redondo
Gaudix	*P. pica*			2.2 ±0.21 (23)	—	(1992*b*)
		—	42 (57)			Soler and Møller, 1990, in
						Zúñiga and Redondo
	P. pica			1.20 ±0.45 (5)	—	(1992*b*)
		221.5 ± 154.3 (71)	—	1.33 ±0.5 (6)	—	J. Zúñiga, unpublished
	P. pica			2.02 ±0.17 (40)	0.88 ±0.15 (41)	results
		—	30.8 (130)			Zúñiga and Redondo
	Corvus corone			1.45 ±0.49 (31)	—	(1992*b*)
		473.1 ± 190 (86)	—			J. Zúñiga, unpublished
	C. monedula			1.17 ±0.17 (6)	—	results
	P. pyrrhocorax	—	2.1 (290)	2.25 ±0.75 (4)	—	Soler (1990)
Guadix + Hueneja	*C. corone*	—	4.9 (162)	1.25 ±0.25 (4)	0.5 ±0.29 (4)	Soler (1990)
		—	8.5 (47)			Soler (1990

*Values ± S.E. Sample size in parentheses.
#Data are for parasitized nests only.

imprinting process (Payne 1977*a*), we transplanted single, newly hatched great spotted cuckoo chicks from magpie nests into other magpie and azure-winged magpie nests containing host chicks of the same size (Arias-de-Reyna and Hidalgo 1982). The great spotted cuckoo chicks reared by both species were captured as fledglings when about 2 weeks old and hand-reared until 4 weeks old. In tests with stuffed specimens of magpie and azured-winged magpie, the great spotted cuckoo fledglings reacted favorably (with approach behavior) to the species into whose nest they were transplanted and unfavorably (with avoidance behavior) to the other species. This experiment demonstrated that 1-month-old great spotted cuckoo chicks learn to recognize their host parents.

There is also evidence of vocal mimicry by great spotted cuckoo chicks being reared by magpies or carrion crows (Redondo and Arias-de-Reyna 1988) as previously observed by Mundy (1973) in the great spotted and striped crested cuckoos (*C. levaillantii*). It is not known whether these differential vocalizations by great spotted cuckoo young also occur during adulthood and if they are used for mate selection by great spotted cuckoos (Arias-de-Reyna 1985; Brooke and Davies 1991; Davies and Brooke 1991). One study (Redondo and Arias-de-Reyna 1988) has shown that the vocal mimicry is due to learning, and not to genetic differences among the parasites.

Special laying procedures

Before laying, a mated pair of great spotted cuckoos exhibits two distinct behaviors to identify the reproductive stage of its potential host. After flying to a fairly long distance from the host nest, one or both members remain motionless for long periods (over 1 hour in some cases) watching the hosts and presumably gathering information about host breeding stages. They repeat this behavior near several nests. The second strategy involves approaching the host nest while both members vocalize. The cuckoos are usually attacked and chased by the hosts and the intensity of the attack may allow the cuckoos to assess the hosts' nest stage.

The second type of assessment behavior resembles the sequence of events in the "distraction strategy" used for egg laying. The latter draws magpies away from their nests, allowing easy access by the female great spotted cuckoo (Alvarez and Arias-de-Reyna 1974*a*; Alvarez et al. 1976; Arias-de-Reyna et al. 1982). Distraction begins when the male great spotted cuckoo flies around a magpie nest singing loudly. Both magpie nest owners attack and follow him. Meanwhile, the female great spotted cuckoo approaches the magpie nest silently and inconspicuously as soon as the nest owners leave and lays her egg, in only 2–3 s, from the edge of the nest (Arias-de-Reyna et al. 1982). Then, she flies away, also singing loudly. When the magpies note the parasitic female, they return to the nest, permitting both parasitic parents to escape. This strategy is also used by the great spotted cuckoo to parasitize pied crows (*Corvus albus*) (Mundy and Cook 1977), and is also seemingly employed by Jacobin cuckoo (*Clamator jacobinbus*) (Liversidge 1971; Gaston 1976) and striped crested cuckoo (Steyn 1973), as well as most African Cuculinae whose laying tactics are known [e.g., African cuckoo (*Cuculus gularis*) and emerald cuckoo (*Chrysococcyx caprius*); see Fry et al. 1988]. All observations of laying by great spotted cuckoos occurred during the three first hours after dawn (Mundy and Crook 1977; Arias-de-Reyna et al. 1984*a*), which coincides with the laying time for Jacobin cuckoo (Liversidge 1961, 1971; Gaston 1976), but not with the afternoon laying of the common cuckoo (Davies and Brooke 1988).

Parasite–host laying synchronization

Host and parasite laying is synchronized (Alvarez et al. 1976, Arias-de-Reyna et al. 1982), as in Jacobin cuckoo (Liversidge 1971; Gaston, 1976) and common cuckoo (Seel 1973; Davies and Brooke 1988). For example, there was a statistically significant correlation ($r = 0.85$, $P < 0.001$) between the laying dates of magpies and great spotted cuckoos in the Sierra Morena during 1978 and 1979 (Arias-de-Reyna et al. 1982). In fact, the parasite and the host started and stopped laying on the same dates (fig. 6.2) (Arias-de-Reyna et al. 1982; see also Soler 1990).

Territoriality

If laying one egg per nest is adaptive, as argued above, cuckoos would benefit from territorial behavior, which would minimize the number of females that lay in a single nest. Territoriality is exhibited by the great spotted cuckoo (Mundy and Cook 1977; Arias-de-Reyna et al. 1982, 1987), striped crested cuckoo (Liversidge, 1961, 1971) and Jacobin cuckoo (Gaston 1976). Territory size per great spotted cuckoo pair varies widely, from

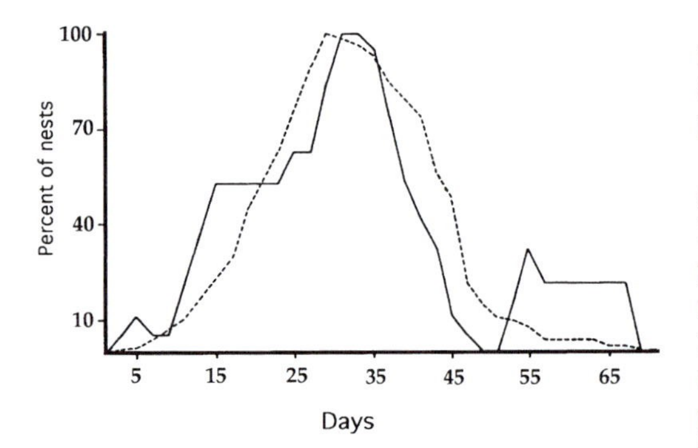

Figure 6.2 Percent of all parasitized and unparasitized magpie nests with eggs (in laying or incubation stages) that were active over the course of the laying season. The continuous and dotted lines show percentages for parasitized and unparasitized nests, respectively. Day 1 represents the laying of the first magpie and cuckoo eggs of the season, both of which occurred on 5 April. Note that the peaks for parasitized and unparasitized nests differ by only about three days.

1 km^2 (Valverde 1971) and 2.5 km^2 (Mundy and Cook 1977) to 8–9 km^2 (Mountford and Ferguson-Lees 1961).

To study territoriality, we used six simultaneous lookouts at fixed observation spots with a clear view of the study area (9 km^2) on alternate days from mid-February to mid-July. Each lookout used a 1:1000 map to draw flight routes for each great spotted cuckoo sighted; the behavior observed in each place was also recorded. Territories delineated by this approach were about 2.9 km^2 in size and contained about 40 magpie nests (see fig. 6.3).

Territories were defended and held until no parasitizable magpie nests remained, after which the cuckoos occurred throughout the study area and laid their eggs in magpie nests that had already been parasitized by another female or in nests with eggs that were hatching (Arias-de-Reyna et al., 1987). Territories were defended even when host nests were scarce (10–15 magpie nests only) and nearly all nests were multiply parasitized (T. Redondo, pers. comm.).

Providing some benefit to the host

Brood parasites do not incur the energetic and time expenses needed to care for the young. These savings may allow them to defend the nests they parasitize against predation. Such defense may also allow maintenance or even expansion of the potential host population.

Our results do indeed suggest that great spotted cuckoos provide hosts with some benefits. In Sierra Morena in 1978 and 1979, the 7.7% predation rate on 26 parasitized nests was significantly lower than the 40.0% predation rate on 136 unparasitized nests within cuckoo territories (Fisher test, two-sided, $P = 0.001$). Furthermore, parasitized and unparasitized nests did not differ in physical characteristics such as nest height, nest to trunk distance, cover of the nest dome, trunk perimeter, orientation and slope of the hillside, cover within 10 m around the nest tree (L. Arias-de-Reyna et al., unpublished results). Nevertheless, parasitized nests may indeed have been chosen on account of some special feature that was not measured by our research group.

As regards the decreased extent of predation of the parasitized population, a comparison of predation on magpie nests within great spotted cuckoo territories versus those in adjacent areas with no parasitism revealed significantly less predation in the territorial areas (40.0% of 50 versus 67.4% of 46 nests; Fisher test, two-sided, $P = 0.0084$). However, we cannot unequivocally ascribe this result to the great spotted cuckoo since territories may also be established in places with smaller numbers of predators. Based on the homogeneity of the study area, the predator population, essentially consisting of rodents (*Mus*, *Apodemus*) and lizards (*Lacerta lepida*), did not seem to be different, even though this assertion cannot be confirmed owing to a lack of data on predator abundance.

Increased egg laying

Data for parasitic cuckoos, finches, and cowbirds show that these birds lay more eggs than do females of related nonparasitic species (Payne 1965, 1973, 1977*a,b*; Scott and Ankney 1979, 1980; Ankney 1985). In captivity, the great spotted cuckoo lays 15–16 eggs, in three series of five or six eggs each (Frisch 1969). Histologic techniques showed that birds in nature lay on alternate days in series of three to six eggs, with an intervening rest period between series of about six days, up to a maximum of 18 eggs (Payne 1974, 1977*a,b*). In a field study (Arias-de-Reyna et al. 1982), we used an egg shape ratio (length/width) to differentiate between the eggs of the two female cuckoos that were using a particular study area. One female laid 18 eggs on alternate days in three series of six eggs,

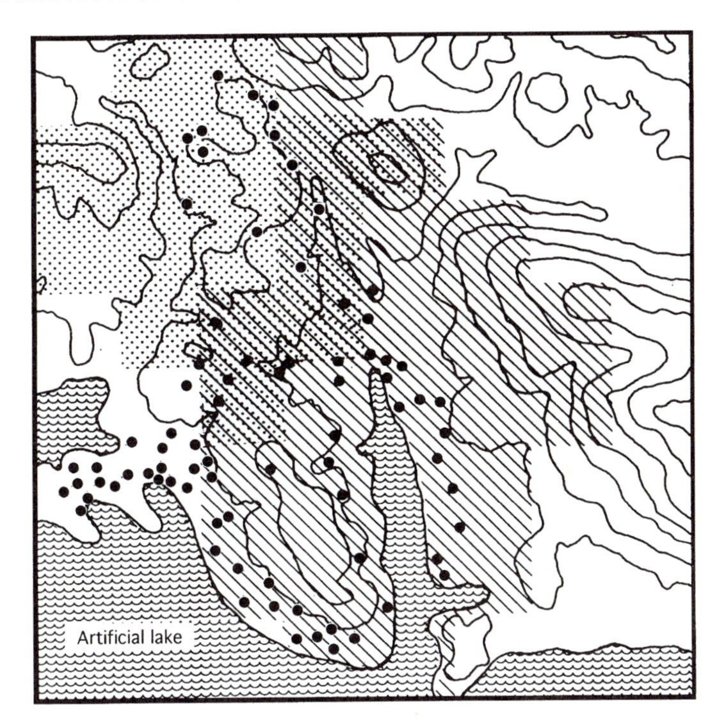

Figure 6.3 Representation of two great spotted cuckoo territories on map of the study area in 1979. The northwest territory (denoted with dots) extended beyond study area and overlapped the second territory (denoted with diagonal lines) near the center of the figure. Filled circles represent magpie nests. The area shown is 3 km by 3 km.

each of which was separated by four and five days. Most nonparasitic cuculiform species also lay in series of 4–6 eggs (Fry et al. 1988). Thus, brood parasitism appears to shorten the interval between series, and not the number of eggs in a series.

Rapid chick development

Parasitic nestlings receive more food and experience decreased nestling competition if they have a shorter incubation period, grow faster, and are more competitive than host chicks (Payne 1977*b*; Marvil and Cruz 1989). These advantages can even allow a parasitic female to be successful if she lays while the host is already in the incubation period (Arias-de-Reyna 1985). As with other parasites, development in great spotted cuckoo chicks is faster than in their magpie and carrion crow hosts (Valverde 1971; Mundy and Cook 1977; Alvarez and Arias-de-Reyna 1974*a*; Arias-de-Reyna et al. 1982; Soler and Soler 1991).

Soler and Soler (1991) reported that great spotted cuckoo chicks in magpie nests weigh 7.8 g at hatching. The chicks rapidly gained weight and the inflection point of the growth curve was at T50 = 8.12 days, with an asymptote at A = 132.5 g. Chick growth was steady from hatching to the 14th day, after which it leveled off (Soler and Soler 1991). The nestling period of parasitic chicks was 20.3 days (SE = 0.34, range = 17–27 days, n = 29). Also, they grew faster than host chicks in unparasitized nests (Soler and Soler 1991).

Experiments showed that great spotted cuckoo nestlings could survive even when magpie nestlings were 8 days older. They were able to leave the nest at 15 days of age to accompany an older magpie chick and were still fed by their host parents (Alvarez and Arias-de-Reyna 1974*a*). When the host is larger than the great spotted cuckoo (e.g., the carrion crow), the parasite's growth can be similar to that with the magpie (Mundy and Cook 1977; Soler and Soler 1991). However, competition may be quite strong and some great spotted cuckoos may starve (three of five carrion crow nests) in the presence of larger host chicks (Soler 1990).

Table 6.2 Incubation period of great spotted cuckoo and some hosts

Species	Incubation period (days)	Source
Clamator glandarius	13–14	Alvarez and Arias-de-Reyna (1974*a*)
	12.8	Arias-de-Reyna et al. (1982)
	13.4	J. Zúñiga, unpublished results
Pica pica	18.3	J. Zúñiga, unpublished results
	21–22	Birkhead (1991)
	19.38	F. Castro, unpublished results
	17.1	Arias-de-Reyna et al. (1984*b*)
	18.0	Alvarez and Arias-de-Reyna (1974*b*)
Corvus corone	20.7	Soler (1984)
Corvus capensis	18–19	Skead (1952)
Corvus monedula	21	Soler (1984)
Corvus albus	18–19	Mundy and Cook (1977)
Corvus ruficollis	20–22	Goodwin (1976)
P. pyrrhocorax	21.3	J. Zúñiga, unpublished results

A short incubation period

Table 6.2 shows that the incubation period of the great spotted cuckoo is less than that of its hosts, as has been shown for brood parasites in general (Liversidge 1961; Payne 1965; Perrins 1967; Lack 1968; Seel 1973; Scott and Ankney 1979). Because parasitic cuckoos lay on alternate days, their eggs begin to develop in the oviduct.

Parasitic cuckoos lay smaller eggs than nesting cuckoos (Payne 1974) and because egg size and incubation period are positively correlated (Rahn and Ar 1974), the parasites should have a shorter hatching period. However, incubation time for a given egg size does not differ between parasitic and nonparasitic cuckoos (Davies and Brooke 1988). Thus, short incubation periods in parasitic cuckoos appear to be a consequence of small egg size rather than a feature evolved in direct response to parasitism. However, the great spotted cuckoo has large eggs for its body size with respect to other parasitic cuckoos and eggs similar in size to those of nesting cuckoos (Payne 1974). So, its shorter incubation period could have been evolved in response to parasitism.

Decreased nest competition

The great spotted cuckoo minimizes competition with host chicks by destroying host eggs and by the behavior of its own nestlings. Great spotted cuckoo females lay from the nest edge and because their eggshells are thicker than those of the magpie, as is true of other brood parasites (Lack 1968; Liversidge 1971; Brooker and Brooker 1991), one or two magpie eggs are often broken during laying (Alvarez and Arias-de-Reyna 1974*a*; Alvarez et al. 1976; Arias-de-Reyna et al. 1982; Soler, 1990). These broken eggs are subsequently ejected by the nest owner (Arias-de-Reyna et al. 1982; Soler, 1990). Thus, the mean number of host eggs left in parasitized host nests is considerably smaller than in unparasitized nests (4.0 versus 5.1 for carrion crow, 3.7 versus 5.2 for jackdaw, and 3.5 versus 4.2 for red-billed chough; Soler 1990). This strategy is common among other parasitic cuckoos (Jensen and Vernon 1970; Brosset 1976; Gaston 1976; Gill 1983). However, Brooker and Brooker (1991) suggest that the advantage of a strong eggshell is to protect the parasitic egg from damage caused by other parasitic eggs.

The second strategy used to decrease competition in the nest is increased begging by the parasitic chick which drives the magpie chicks to starvation. Begging bouts by great spotted cuckoo chicks were much longer than those of host chicks; the former quavered unceasingly until they were supplied with food, so they monopolized most of the food resources supplied by the nest owners (Redondo and Arias-de-Reyna 1988) and excluded competitors. Only 19.2% of the parasitized nests fledged any magpie host chicks (Soler 1990) in Granada in 1983 and 1984; and none of the host chicks in parasitized nests in Doñana and Central Sierra Morena survived (Alvarez and Arias-de-Reyna 1974*a*; Arias-de-Reyna et al. 1982).

The situation is seemingly different when the host is a larger corvid, such as the carrion crow. In every reported case, some host chicks survived (Soler 1990) and parasitic nestlings experienced strong competition from the carrion crow nestlings whose size rapidly surpassed that of the great spotted cuckoos. The proportion of great spotted cuckoos that successfully fledged was higher with the magpie (82.9%, n = 35) than with the carrion crow (40%, n = 5) (Soler 1990). Thus, brood reduction (Clark and Wilson 1981) occurs among the cuckoo nestlings as a result of starvation. Differential fledging success with large- and small-sized hosts has also been reported for other species of *Clamator* (Gramet 1970; Gaston 1976).

Because multiple parasitism is so common in the great spotted cuckoo, parasitic nestlings often compete with one another. When there is more than one cuckoo chick in a magpie nest, some may starve (Soler 1990; Soler and Soler 1991). However, all of the great spotted cuckoos sometimes survive (Valverde 1953; Gramet 1970; Arias-de-Reyna et al. 1982).

Egg mimicry

The most effective host defense mechanism is ejection of parasitic eggs, which probably elicits the evolution of parasitic mimicry of host eggs as a counterdefense (Baker 1942; Lack 1968; Rothstein 1975*b*; 1976; 1982*a*; 1990; Payne 1977*a*; Scott 1977; Finch 1982). Thus, parasites and their hosts are engaged in a coevolutionary arms race (Davies and Brooke 1988, 1989*a*,*b*; Rothstein 1990; Redondo 1993).

Many cuckoo species lay polymorphic eggs, each type (gens) being adapted to a particular host (Cramp 1985; Fry et al. 1988). The existence of gentes may require reproductive isolation, because egg color probably has an important genetic component. The selection for egg mimicry can be quite strong because hosts may change from accepters to rejecters very quickly (Rothstein 1975*a*,*b*; Takasu et al. 1993).

Great spotted cuckoo eggs resemble magpie eggs (Valverde 1971, Brinchambaut 1973; Alvarez and Arias-de-Reyna 1974*a*). In experiments with model eggs of various shapes, sizes and colors, Alvarez et al. (1976) showed that magpies ejected all nonmimetic eggs in Doñana, whereas real or model great spotted cuckoo eggs resembling magpie eggs were accepted. In Guadix, Soler (1990) carried out experiments on egg-recognition in the five potential corvid host species (carrion crow, jackdaw, raven, magpie, and red-billed chough) by using painted quail and domestic hen eggs; all of the corvids, except magpies, were accepters as also found by Zuñiga et al. (1983) for the jackdaw, chough and carrion crow.

As noted by Arias-de-Reyna and Hidalgo (1982) and Davies and Brooke (1989*b*), a species must accept at least some parasitic eggs to be considered a potential host of a specialist parasite. However, once they become a usual host, such species may evolve the ability to recognize their own eggs (Rothstein 1982*b*; Moksnes 1992) and therefore behave as rejecters of nonmimetic eggs. Host rejection could spread within a few hundred years (Kelly 1987; Davies and Brooke 1989b). Thus, if a new host species evolves rejection behavior, the parasite has a fairly short time to develop mimetic eggs (Takasu et al. 1993); otherwise, parasitic success with the host will be nil. Also, specialist brood parasitic species are likely to exhibit a high potential to evolve new egg coloration, thereby allowing them to develop mimetic eggs; any specialist parasites that have lacked this potential are likely to have gone extinct.

Soler and Møller (1990) used model eggs that mimicked magpie eggs, nonmimetic quail eggs, and real magpie eggs to carry out experiments on magpies from two populations in Granada (Guadix and Santa Fe) and one in Uppsala (Sweden). They found that the level of rejection covaried positively with the duration of sympatry, which was longest at Sante Fe and nonexistent at Uppsala since the great spotted cuckoo does not occur in Sweden (Soler et al., this vol.). However, the relatively low rate of rejection at Guadix does not necessarily imply that the host population only recently has become parasitized (Takasu et al. 1993). I agree with Zuñiga and Redondo (1992*b*) who suggested that the differences in rejection rates between Guadix and Santa Fe may reflect lower densities of great spotted cuckoos at Guadix and therefore lower contact rates with adult parasites. Contact with adult parasites can increase the likelihood that a host will reject a foreign egg (see Davies and Brooke 1988; Moksnes and Røskaft 1989).

New host species

The great spotted cuckoo is currently expanding its range in Europe, where its populations are steadily increasing (Cramp 1985). The increases in range and population size are probably due to high rates of breeding success in combination with dispersal due to territoriality. Increased cuckoo

Table 6.3 Numbers of magpie and great spotted cuckoo fledglings

Location	Fledgling magpies/nest	Fledgling *Clamator*/ parasitic female(*)	Source
Parasitized populations			
Doñana	2.13	—	Alvarez and Arias-de-Reyna (1974*b*)
Doñana	2.00	—	Valverde (1971)
Sierra Morena	1.56	8	Arias-de-Reyna et al. (1982)
Granada	1.26	—	J. Zúñiga, unpublished results
Guadix + Hueneja + Santa Fe	—	9	Soler (1990)
Santa Fe	—	10	Zúñiga and Redondo (1992*b*)
Guadix	—	6	Zúñiga and Redondo (1992*b*)
Unparasitized populations			
Doñana (#)	2.10		Castro (1993)
Holanda	0.98		Baeyens (1981*a*)
Leicester (U.K.)	3.18		Holyoak (1974)
Bristol (U.K.)	0.82		Vines (1981)
Sheffield (U.K.)	1.81		Birkhead (1991)

(*)Fledgling *Clamator* = (number of *Clamator* eggs × % success)/100.
Parasitic females = number of *Clamator* eggs/18.
(#) No parasitism recorded in the study year.

See Birkhead (1991) for breeding success of magpies in other locations.

populations coupled with a possible decrease in host populations should favor the use of new host species.

Parasite versus host breeding success

If a specialized parasite is to develop a stable interaction with its host, it must have features that keep it from decimating the population of its usual host, i.e., its impact on the host should allow for the reproductive potential of the parasitized population to be maintained. May and Robinson (1985) suggest that specialist parasites rarely have an impact on their host populations.

The breeding success of great spotted cuckoos parasitizing magpies or carrion crows in Sierra Morena and both Granada areas was between 6 and 10 fledglings per female (see table 6.3). Therefore, unless mortality until reproductive age is very high, the great spotted cuckoo enjoys a much higher rate of reproductive success than is typical for nesting birds.

Within populations, the numbers of fledged magpie chicks are typically smaller for parasitized than for unparasitized nests (Arias-de-Reyna et al. 1982; Soler 1990). The overall number of fledged magpie chicks per nest in both parasitized (1.26 to 2.13) and unparasitized (0.8 to 3.18) populations varies widely (see table 6.3), and is always much smaller than that typically achieved by individual great spotted cuckoos in one season. Nevertheless, the great spotted cuckoo probably exerts little pressure on magpie populations as the number of fledglings in parasitized populations is essentially the same as in unparasitized populations (see table 6.3).

The azure-winged magpie as a new host

There is little evidence of parasitism of the azure-winged magpie by the great spotted cuckoo (Valverde 1953; Friedmann 1964). But it is often parasitized by the common cuckoo in Japan (Hosono 1983; Nakamura 1990; Nakamura et al., this vol.; Takasu et al. 1993). By placing great spotted cuckoo eggs in azure-winged magpie nests at a Sierra Morena site near Córdoba, we found that 25% of azure-winged magpies are rejecters (Arias-de-Reyna and Hidalgo 1982). Eggs of the latter are much smaller and are colored differently than those of the cuckoo and the magpie. From 1991 to 1994, our group studied azure-winged magpies at two other sites in the Sierra Morena, Andújar and Montoro. At the former, the great spotted cuckoo parasitized a low-density magpie population (similar to that in Santa Fe). Andújar also had a large population of azure-winged magpies (above five nests per ha). On three occasions we observed the great spotted cuckoo's whole distraction process among trees where azure-winged magpie nests were located. However, none of the nests was parasitized. There are no great spotted cuckoos in the Montoro area. The three azure-winged magpie colonies we have studied are linearly arranged over a distance of 40 km between Córdoba and Montoro.

The reproductive success of a parasitic bird in azure-winged magpie nests might be different than in magpie nests owing to the smaller chick size at hatching and growth asymptotic values in the former (5.5 g and 70 g versus 8.5 g and 180 g, respectively). The small size of the azure-winged magpie could limit its potential to feed great spotted cuckoo chicks. However, the great spotted cuckoo had a high breeding success in artificially parasitized azure-winged magpie nests (Redondo and Arias-de-Reyna 1989). So it appears that cuckoo success is potentially similar with both magpies. The presence of helpers (Hosono 1983), which can supply additional food (Wilkinson and Brown 1984), may allow the azure-winged magpie to raise cuckoos despite its small size. However, the lack of frequent parasitism of the azure-winged magpie could be due to its cooperative breeding which results in both communal defense against brood parasites and possibly in rejection of parasitic eggs. Such communal defense has been demonstrated in hosts of other cuckoo species (Ferguson 1994).

Aggression towards adult great spotted cuckoos and egg rejection should be weak in populations that are allopatric with the parasite (e.g., the Montoro population). In the Andújar colony, in sympatry with the great spotted cuckoo, the number of pairs that attacked adult dummies of the parasite during laying and hatching was higher than in the Montoro colony ($\bar{x} = 5.56$, S.E. = 0.51 for Andújar; $\bar{x} = 3.64$, S.E. = 0.46 for Montoro; ANOVA, $F = 7.67$, $P < 0.013$; Arias-de-Reyna et al., unpublished results). This difference is not due to a generalized stronger response to all nest intruders at Andújar because we found no significant differences between the responses the Andújar and Montoro azure-winged magpies showed to aerial and mammalian predators (L. Arias-de-Reyna et al., unpubl. results).

In contrast to azure-winged magpies near Córdoba, those at Andújar and Montoro show high rates of rejection towards foreign eggs. But azure-winged magpies at the latter two sites rejected magpie eggs (which mimic great spotted cuckoo eggs) at different intensities. The Andújar colony ejected 100% and the Montoro colony 70% of the magpie eggs ($c^2 = 4.357$, df = 1, two-sample test, $P = 0.037$; Arias-de-Reyna et al., unpubl. results). The time magpie eggs remained in azure-winged magpie nests was significantly shorter in the Andújar ($\bar{x} = 3.6$ days, S.E. = 2.40, $n = 10$) than in the Montoro colony ($\bar{x} = 8.3$ days, S.E. = 2.82, $n = 10$) (ANOVA, $F = 5.01$, $P < 0.045$; Arias-de-Reyna et al., unpublished results).

The distance over which azure-winged magpie young disperse from their natal colony is unknown; however, because the Andújar and Montoro colonies are fairly close, some individuals may disperse between them, so ejection at Montoro may be induced by prior experience with parasitism by the great spotted cuckoo. Therefore, the differences between the Córdoba population, consisting mostly of egg accepters, and the Andújar and Montoro populations, which are egg rejecters to differing extents, may be ascribed to learning and subsequent dispersal between populations. However, it is clear that egg rejection can have a nonlearned (i.e., genetic) basis (Moksnes et al. 1990; Braa et al. 1992; Briskie et al. 1992; Mark and Stutchbury 1994), and the host defenses at Montoro could be due solely to gene flow (see Davies and Brooke 1989*a*; Briskie et al. 1992).

ACKNOWLEDGMENTS Our field studies cited in this article were made possible by the close cooperation and discussions with many coworkers since 1977. I wish to thank Pilar Recuerda, without whom this article would never have materialized. I am also indebted to Jesús Marín, Alberto Redondo-Villa, Paqui Castro and Tomás Redondo, and especially to Jesús Zuñiga, for the use of unpublished results. Finally, I wish to thank S. Rothstein and an anonymous reviewer for their helpful comments on an earlier manuscript draft.

REFERENCES

Alvarez, F. and L. Arias-de-Reyna (1974*a*). Mecanismos de parasitación por *Clamator glandarius* y defensa por *Pica pica*. *Doñana, Acta Vert.*, **1**, 43–65.

Alvarez, F. and L. Arias-de-Reyna (1974*b*). Reproducción de la urraca (*P. pica*) en Doñana. *Doñana, Acta Vert.*, **1**, 77–95.

Alvarez, F., L. Arias-de-Reyna, and M. Segura (1976). Experimental brood parasitism in the magpie (*P. pica*). *Anim. Behav.*, **24**, 907–916.

Ankney, C. D. (1985). Variation in weight and composition of brown-headed cowbird eggs. *Condor*, **87**, 296–299.

Araujo, J. and A. Landin (1970). *Corvus corax* hospedador de *Clamator glandarius. Ardeola*, **16**, 267–268.

Arias-de-Reyna, L. (1985). Parasitismo de incubación en el críalo (*Clamator glandarius*). *Misc. Zool.*, **9**, 419–425.

Arias-de-Reyna, L. (1994). Parasitismo de incubación a nuevos hospedadores en cuculidae del viejo mundo: estrategias para maximizar el proceso. *Actas del V Congreso Nacional de Etología y II Iberoamericano* (in press).

Arias-de-Reyna, L. and S. J. Hidalgo (1982). An investigation into egg-acceptance by azure-winged magpies and host-recognition by great spotted cuckoo chicks. *Anim. Behav.*, **30**, 819–823.

Arias-de-Reyna, L., P. Recuerda, M. Corvillo, and I. Aguilar (1982). Reproducción del críalo (*Clamator glandarius*) en Sierra Morena Central. *Doñana, Acta Vert.*, **9**, 177–193.

Arias-de-Reyna, L., J. Trujillo, P. Recuerda, M. Corvillo and A. Cruz (1984*a*). Ritmo de actividad del Críalo (*Clamator glandarius*) en periodo de reproducción (Aves, Cuculidae). *Hist. Nat.*, **27**, 237–242.

Arias-de-Reyna, L., P. Recuerda, M. Corvillo, and A Cruz (1984*b*). Reproducción de la urraca (*Pica pica*) en Sierra Morena (Andalucía). *Doñana, Acta Vert*, **11**(1), 79–22.

Arias-de-Reyna, L., P. Recuerda, J. Trujillo, M. Corvillo, and A. Cruz (1987). Territory in the great spotted cuckoo (*Clamator glandarius*). *J. Ornith.*, **128**, 231–239.

Baeyens, G. (1982). The role of the sexes in territory defence in the magpie (*Pica pica*). *Ardea*, **69**, 69–82.

Baker, E. C. S. (1942). *Cuckoo problems.* Witherby, London.

Birkhead, T. R. (1991). *The magpies*. T. & A. D. Poyser, London.

Braa, A. T., A. Moksnes and E. Røskaft (1992). Adaptations of bramblings and chaffinches towards parasitism by the common cuckoo. *Anim. Behav.*, **43**, 67–78.

Brinchambaut, J. (1973). Contribution de l'oologie á la connaissance de la biologie du coucou-geai *Clamator glandarius. Alauda*, **41**, 353–361.

Briskie, J. V., S. Sealy, and K. A. Hobson (1992). Behavioral defenses against avian brood parasitism in sympatric and allopatric host populations. *Evolution*, **46**, 334–340.

Brooke, M. de L. and N. B. Davies (1991). A failure to demonstrate host imprinting in the cuckoo (*Cuculus canorus*) and alternative hypotheses for the maintenance of egg mimicry. *Ethology*, **89**, 154–166.

Brooker, M. G. and L. C. Brooker (1991). Eggshell strength in cuckoos and cowbirds. *Ibis*, **133**, 406–413.

Brosset, A. (1976). Observations sur le parasitisme de la reproduction du coucou Emeraude *Chrysococcyx cupreus* au Gabon. *L'oiseau et R.F.O.*, **46**, 201–208.

Castro, F. (1993). Evolución de los sistemas comunicativos fiables en las relaciones paternofiliales de aves altriciales. Tesis Doctoral. Univ. Córdoba.

Clark, A. B. and D. S. Wilson (1981). Avian breeding adaptations: hatching asynchrony, brood reduction and nest failure. *Q. Rev. Biol.*, **56**, 253–277.

Cramp, S. (ed.) (1985). *The birds of the Western Palearctic. Vol. IV.* Oxford University Press, Oxford.

Davies, N. B. and M. D. Brooke (1988). Cuckoos versus reed warblers. Adaptations and counteradaptations. *Anim. Behav.*, **36**, 262–284.

Davies, N. B. and M. D. Brooke (1989*a*). An experimental study of coevolution between the Cuckoo, *Cuculus canorus*, and its host. I. Host egg discrimination. *J. Anim. Ecol.*, **58**, 207–224.

Davies, N. B. and M. de L. Brooke (1989*b*). An experimental study of coevolution between the Cuckoo, *Cuculus canorus*, and its host. II. Host egg markings, chick discrimination and general discussion. *J. Anim. Ecol.*, **58**, 225–236.

Davies, N. B. and M. de L. Brooke (1991). Coevolution of the cuckoo and its hosts. *Sci. Amer.*, **264**, 66–73.

Demartis, A. M. (1971). Il cuculo dal ciuffo – *Clamator glandarius* (L) in Sardegna. *Rev. Ital. Ornith.*, **41**, 373–389.

Di Carlo, E. A. (1969). Ulteriori notizie sul Cuculo dal ciuffo (Clamator glandarius). *Rev. Ital. Ornith.*, **39**, 3–6.

Di Carlo, E. A. (1971). Appunti sulla biologia del cuculu dal ciuffo (*Clamator glandarius*). *Rev. Ital. Ornith.*, **41**, 86–107.

Ferguson, J. W. H. (1994). The importance of low host densities for successful parasitism of Diederik Cuckoos on Red-Bishop birds. *S. Afr. J. Zool.*, **29**, 70–73.

Finch, D. M. (1982). Ejection of cowbird eggs by crissal thrashers. *Auk*, **99**, 719–721.

Friedmann, H. (1948). *The parasitic cuckoos of Africa. Monogr. No. 1*, Washington Acad. Sci., Washington, D.C.

Friedmann, H. (1964). Evolutionary trends in the avian genus *Clamator. Smithson. Mis. Coll.*, **146**, 1–127.

Frisch, O. von (1969). Die Entwicklung des Häherkuckucks (*Clamator glandarius*) im nest der Wirtsvögel und seine Nachzucht in gefangenschaft. *Z. Tierpsychol.*, **26**, 641–650.

Fry, C. H., S. Keith and K. Urban (1988). *The birds of Africa. Vol. III. Parrots to Woodpeckers.* Academic Press, London.

Gaston, A. J. (1976). Brood parasitism by the pied crested cuckoo (*Clamator jacobinus*). *J. Anim. Ecol.*, **45**, 331–345.

Gill, B. J. (1983). Brood-parasitism by the shining cuckoo *Chrysococcyx lucidus* at Kaikoura, New Zealand. *Ibis*, **125**, 40–55.

Goodwin, D. (1976). *Crows of the world.* Brit. Museum (Natural History), London.

Gramet, H. (1970). Contribution a l'etude des comportements parentaux chez quelques corvides pendant la periode de reproduction. Ph. D. Thesis, University of Rennes.

Holyoak, D. (1974). Territorial and feeding behaviour of the magpie. *Bird Study*, **21**, 117–128.

Hosono, T. (1983). A study of the life history of Blue Magpie (II): Breeding helpers and nest parasitism by Cuckoo. *J. Yamashina's Inst. Ornith.*, **15**, 63–71.

Jensen, R. A. C. and C. J. Vernon (1970). On the biology of the Diedrik Cuckoo *Chrysococcyx caprius* in southern Africa. *Ostrich*, **41**, 237–246.

Kelly, C. (1987). A model to explore the rate of spread of mimicry and rejection in hypothetical populations of cuckoos and their hosts. *J. Theor. Biol.*, **125**, 283–299.

Lack, D. (1968). *Ecological adaptation for breeding in birds.* Methuen, London.

Lévêque, R. (1968). Über Verbreitung, Bestandesvermehrung und Zug des Häherkuckucks, *Clamator glandarius* L. in West Europe. *Orn. Beob.*, **65**, 43–71.

Liversidge, R. (1961). Pre-incubation development of *Clamator jacobinus. Ibis*, **103**, 624.

Liversidge, R. (1971). The biology of the Jacobin Cuckoo (*Clamator jacobinus*). *Proc. 3rd Pan-Afr. Ornith. Congr., Ostrich Suppl.* **8**, 117–137.

Mark, D. and B. J. Stutchbury (1994). Response of a forest-interior songbird to the threat of cowbird parasitism. *Anim. Behav.*, **47**, 275–280.

Marvil, R. E. and A. Cruz (1989). Impact of Brown-headed Cowbird parasitism on the reproductive success of the Solitary Vireo. *Auk*, **106**, 476–480.

May, R. M. and S. K. Robinson (1985). Population dynamics of avian brood parasitism. *Amer. Natur.*, **126**, 475–494.

Moksnes, A. (1992). Egg recognition in chaffinches and bramblings. *Anim. Behav.*, **43**, 993–995.

Mountford, G. and I. Ferguson-Lees (1961). The birds of the Coto Doñana. *Ibis*, **103a**, 86–109.

Mundy, P. J. (1973). Vocal mimicry of their host by nestlings of the Great Spotted Cuckoo and Striped Crested Cuckoo. *Ibis*, **115**, 602–604.

Mundy, P. J. and A. W. Cook (1977). Observations on the breeding of the pied crow and great spotted cuckoo in Northern Nigeria. *Ostrich*, **48**, 72–84.

Nakamura, H. (1990). Brood parasitism by the cuckoo *Cuculus canorus* in Japan and the start of new parasitism of the Azure-winged Magpie *Cyanopica cyana. Jap. J. Ornith.*, **39**, 1–18.

Payne, R. B. (1965). Clutch size and numbers of egg laid by Brown-headed Cowbirds. *Condor*, **67**, 44–60.

Payne, R. B. (1973). Individual laying histories and the clutch size and number of eggs in parasitic cuckoos. *Condor*, **75**, 414–438.

Payne, R. B. (1974). The evolution of clutch size and reproductive rates in parasitic cuckoos. *Evolution*, **28**, 169–181.

Payne, R. B. (1977*a*). The ecology of brood parasitism in birds. *Annu. Rev. Ecol. Syst.*, **8**, 1–28.

Payne, R. B. (1977*b*). Clutch size, egg size, and the consequences of single vs. multiple parasitism in parasitic finches. *Ecology*, **58**, 500–513.

Perrins, C. M. (1967). The short apparent incubation period of the cuckoo. *Brit. Birds.*, **60**, 51–52.

Rahn, H. and A. Ar (1974). The avian egg: incubation time and water loss. *Condor*, **76**, 147–152.

Redondo, T. (1993). Exploitation of host mechanisms for parental care by avian brood parasites. *Etología*, **3**, 235–297.

Redondo, T. and L. Arias-de-Reyna (1988). Vocal mimicry of hosts by Great Spotted Cuckoo *Clamator glandarius*: further evidence. *Ibis*, **130**, 540–544.

Redondo, T. and L. Arias-de-Reyna (1989). High breeding success in experimentally parasitized broods of azure-winged magpies (*Cyanopica cyana*). *Le Gerfaut*, **77**, 149–152.

Rothstein, S. I. (1975*a*). Evolutionary rates and host defenses against avian brood parasitism. *Amer. Natur.*, **109**, 161–176.

Rothstein, S. I. (1975*b*). An experimental and teleonomic investigation of avian brood parasitism. *Condor*, **77**, 250–271.

Rothstein, S. I. (1976). Experiments on defenses Cedar Waxwings use against Cowbird parasitism. *Auk*, **93**, 675–691.

Rothstein, S. I. (1982*a*). Successes and failures in avian egg and nestling recognition with comments on the utility of optimality reasoning. *Amer. Zool.*, **22**, 547–560.

Rothstein, S. I. (1982*b*). Mechanism of avian egg recognition: which egg parameters elicit responses by rejecter species? *Behav. Ecol. Sociobiol.*, **11**, 229–239.

Rothstein, S. I. (1990). A model system for coevolution: avian brood parasitism. *Annu. Rev. Ecol. Syst.*, **21**, 481–508.

Sánchez García, J. (1885). *Catálogo de los mamíferos y aves observados en la provincia de Granada*, Real Sociedad Económica de Granada, Granada.

Scott, D. M. (1977). Cowbird parasitism on the Gray Catbird at London, Ontario. *Auk*, **94**, 18–27.

Scott, D. M. and C. D. Ankney (1979). Evolution of a method for estimating the laying rate of brown-headed cowbird. *Auk*, **96**, 483–488.

Scott, D. M. and C. D. Ankney (1980). Fecundity of the brown-headed cowbird in southern Ontario. *Auk*, **97**, 677–683.

Seel, D. C. (1973). Egg-laying by the cuckoo. *Brit. Birds*, **66**, 528–535.

Skead, C. J. (1952). The constancy of incubation. *Willson Bull.*, **74**, 115–152.

Soler, M. (1984). Biometría y biología de la grajilla (*Corvus monedula*). Doctoral Thesis, University Granada.

Soler, M. (1990). Relationships between the Great Spotted Cuckoo *Clamator glandarius* and its corvid hosts in a recently colonized area. *Ornis Scand.*, **21**, 212–223.

Soler, M. and A. P. Møller (1990). Duration of sympatry and coevolution between the great spotted cuckoo and its magpie host. *Nature*, **343**, 748–750.

Soler, M. and J. J. Soler (1991). Growth and development of Great Spotted Cuckoos and their magpie host. *Condor*, **93**, 49–54.

Soler, M., J. J. Soler, J. G. Martinez and A. P. Møller (1994). Micro-evolutionary change in host response to a brood parasite. *Behav. Ecol. Sociobiol.*, **35**, 295–301.

Steyn, P. (1973). Some notes on the breeding biology of the Striped Cuckoo. *Ostrich*, **44**, 163–169.

Takasu, F., K. Kawasaki, H. Nakamura, J. E. Cohen and N. Shigesada (1993). Modeling the population-dynamics of a cuckoo–host association and the evolution of host defenses. *Amer. Natur.*, **142**, 819–839.

Valverde, J. A. (1953). Contribution a la biologie du coucou-geai, *Clamator glandarius* L. I. Notes sur le coucou-geai en Castille. *Oiseaux et R.F.O.*, **23**, 288–296.

Valverde, J. A. (1971). Notas sobre la biología reproductora del Críalo (*Clamator glandarius*). *Ardeola* (vol. especial), 591–647.

Vines, G. (1981). A socio-ecology of Magpies *Pica pica*. *Ibis*, **123**, 190–202.

Wilkinson, R. and A. E. Brown (1984). Effect of helpers on the feeding rates of nestlings in the Chesnut-bellied Starling *Spreo pulcher*. *J. Anim. Ecol.*, **53**, 301–310.

Yeatman, L. (1974). *Vie sexualle des oiseaux*. Leson Ed, Paris.

Zuñiga, J. M. and T. Redondo (1992*a*). Adoption of Great Spotted Cuckoo *Clamator glandarius* fledglings by Magpies *Pica pica*. *Bird Study*, **39**, 200–202.

Zuñiga, J. M. and T. Redondo (1992*b*). No evidence for variable duration of sympatry between the great spotted cuckoo and its magpie host. *Nature*, **359**, 410–411.

Zuñiga, J. M., M. Soler, and I. Camacho (1983). Nota sobre nuevas especies parasitadas por el Críalo (*Clamator glandarius*) en España. *Doñana, Acta Vert.*, **10**, 207–209.

7

Behavior and Ecology of the Shining Cuckoo *Chrysococcyx lucidus*

B. J. GILL

Chrysococcyx lucidus (Gmelin 1788), called the shining cuckoo in New Zealand and shining bronze cuckoo in Australia, is one of the small "glossy cuckoos" monographed by Friedmann (1968). In this review of the shining cuckoo's behavior and ecology, I focus mainly on New Zealand data (Gill 1982*c*, 1983*a*), but make comparisons and contrasts with the situation in Australia (Brooker and Brooker 1989*a*,*b*) and New Caledonia (Hannecart and Letocart 1980) where possible. These references are the main sources of the data given, and for simplicity and readability they may not be cited at every point.

THE HOSTS

There is no evidence that shining cuckoos parasitize any host other than the gray warbler *Gerygone igata* on the main islands of New Zealand (Gill 1983*a*). On the Chatham Islands east of New Zealand the host is *G. albofrontata* (Dennison et al. 1984). On Norfolk Island the cuckoo parasitizes *G. modesta* (Schodde et al. 1983), but it probably does not breed on Lord Howe Island where the local warbler (*G. insularis*) is extinct. In New Caledonia and Vanuatu (and probably on Rennell and Bellona Islands) the host is *G. flavolateralis* (Mayr 1932; Hannecart and Letocart 1980).

The major biologic (i.e., regularly parasitized and successful) hosts in Australia are six species of *Acanthiza*, the most important of which are *A. chrysorrhoa* on the mainland and *A. pusilla* in Tasmania (Brooker and Brooker 1989*a*). Thus, all biologic hosts of the shining cuckoo throughout its entire range belong to one family (Pardalotidae). The species of *Gerygone* and *Acanthiza* are broadly similar in appearance, weigh 6–10 g, eat small arthropods, build enclosed nests, and lay their eggs at 48-h intervals.

Gray warblers glean invertebrate prey from foliage and are common throughout New Zealand. They average 6.4 g, about one-fourth the mass of the parasite. Gray warblers are sedentary, territorial, solitary nesting, and monogamous (Gill 1982*a*). The enclosed nest is built in or near foliage at any height from 30 cm to 20 m up in the forest canopy, and has a circular entrance about 30 mm wide. The usual clutch size is four, with 48-h intervals between successive eggs. Incubation, by the female alone, starts with the laying of the last egg. The incubation period is 17–21 days and the nestling period 15–19 days.

THE PARASITE AND ITS BREEDING ADAPTATIONS

The weights of 40 adult shining cuckoos from New Zealand (thawed after deep freezing) ranged from 17.0–27.5 g [mean = 22.8 ± 2.71 g (S.D.)] with no significant difference between the sexes as determined at autopsy (B. J. Gill, unpublished data). An additional female with an egg in the oviduct weighed 28.6 g. Australian birds averaged 23.4 g (n = 101, S.E. = 3.5, range = 16.0–35.0 g; Brooker and Brooker 1989*a*). The plumage of the upper parts is dark green with an iridescent copper sheen. The underparts are white with dark green transverse bars. The sexes are alike in plumage. The coloration is cryptic as the birds are difficult to see when perched among foliage.

In the temperate zone, shining cuckoos breed during the austral summer in three discrete areas: southwest Australia, southeast Australia (eastern New South Wales, Victoria, and Tasmania), and New Zealand (see references in Gill 1983*c*). Australian populations are usually regarded as belonging to subspecies *plagosus* and New Zealand birds to the nominate race. The laying seasons are August to December in southern Australia and October to December in New Zealand. Cuckoos vacate these areas during the southern winter, at which time there is an influx of the species in tropical areas to the north: the Lesser Sunda Islands, New Guinea, New Britain, and the Solomon Islands. There is no evidence of breeding north of about 10°S latitude.

In eastern Australia between about 10° and 39°S (Queensland and New South Wales), shining cuckoos tend to occur all year. They breed as far north as about 28°S (close to Brisbane), breeding records further north being uncertain (Brooker and Brooker 1989*a*). The subspecies *layardi* in New Caledonia, Vanuatu, and nearby islands (10–23°S) is a sedentary breeding population (Mayr 1932; Hannecart and Letocart 1980). A subspecies *harterti* on Rennell and Bellona Islands (south of the Solomons at about 12°S) is also said to be sedentary (Mayr 1932).

Morphologic traits, especially bill widths, of museum specimens suggest that New Zealand shining cuckoos winter in the archipelago from New Britain east to the Solomon Islands (Mayr 1932; Gill 1983*c*). It has been believed that they take a direct transoceanic route — a return journey of 6000 km with landfall possible only on Norfolk and Lord Howe Islands — which would be one of the longest migrations across sea by any land bird.

However, evidence (Gill 1983*c*) suggests that New Zealand birds take a longer route via New South Wales and Queensland that involves much shorter transoceanic distances. It appears that New Zealand and Australian birds — supposedly separate subspecies — intermingle in eastern Australia, particularly in September and October (when land birds are breeding), but not in November. Significantly, November is the peak month of laying by New Zealand birds so it is unlikely that any breed in Australia. Presumably an instinct to return to natal areas and seek out the species of host that raised them, drives New Zealand birds to transit Australia without breeding, and so maintains the genetic isolation of the populations.

In bill width — the main subspecific character — New Zealand and Western Australian birds are well separated, but the distinction between the Tasmanian population and the other two is not as great. The plumages of these populations are very similar whereas the race *layardi* has distinctive plumage. There are therefore grounds for synonymizing *plagosus* with *lucidus* (Gill 1983*c*). However, there are differences in appearance of the nestlings in Australia and New Zealand (see subsection, Nestling mimicry) that might argue for subspecies.

Few shining cuckoos have been banded. In New Zealand, one cuckoo banded as a nestling and another caught in a mist-net, were recorded in the same areas one or two years later (see Gill 1983*a*).

Shining Cuckoos in New Zealand eat insects (mainly caterpillars and beetles) and spiders (Gill 1980, and unpublished data). Most of the prey are less than 20 mm long, with a significant number of items only 1–5 mm long. Many of the caterpillars have defensive hairs (e.g., larvae of the magpie moth *Nyctemera annulata*, Arctiidae) and many of the beetles are ladybirds (Coccinellidae) which are supposed to have defensive fluids toxic to vertebrates.

Shining cuckoos have behavioral postures that may enhance their crypticity while they watch their hosts nesting. Gill (1982*c*) saw an adult cuckoo of unknown sex suddenly sleek its feathers and stiffen, head in line with body at 45° to the horizontal, as two nesting warblers foraged to within a few meters. The cuckoo remained still except for a slight rotation of its head, as if to keep the warblers in view.

Shining cuckoos lay directly into the host's enclosed nest by entering it and emerging backwards (Brooker et al. 1988). After she lays, the female cuckoo carries off in her bill one host egg. Thus the clutch size does not increase. Shining cuckoos have been seen swallowing the contents of eggs and discarding the shells (see examples in Macdonald and Gill 1991). Warbler-like eggshell fragments have twice been found in the stomachs of female cuckoos collected in November (B. Gill, unpublished).

Individually, shining cuckoos lay one egg per host nest. Cases of double parasitism arising from laying by two shining cuckoos in the same nest are unknown in New Zealand. In Australia 4% of 833 host nests held two shining cuckoo eggs and another 37 held a shining cuckoo egg plus the egg of another species of cuckoo (Brooker and Brooker 1989*a*).

Shining cuckoos may defend territories during laying. In New Zealand, a color-banded cuckoo was seen six times during November 1978 (peak month for laying) over an area of forest of at least 5 ha. It had one companion and no other cuckoos were noted in the same area. A theoretical estimate of home range size in this habitat was 20 ha, based on a model developed for cuckoos that parasitize only one host species in a given region and lay one egg per parasitized nest (Gill 1983*a*). In the model, C, the minimum area needed to support laying by one female cuckoo, equals Hn/p where H is the average area of the host's territory, n is the number of eggs the female cuckoo lays in one season, and p is the frequency of parasitism. The number of eggs that shining cuckoos lay in one season is unknown, but can be estimated assuming they lay at a similar rate to a southern African congener (*Ch. caprius*) which produces 1.3–1.8 eggs per week over the laying season (Payne 1973).

The shining cuckoo's egg (mean 18.7 × 12.6 mm, 1.9 g; Gill 1983*a*) is only slightly larger than that of the gray warbler (mean 17.1 × 12.1 mm, 1.5 g; Gill 1982*a*). Egg weight as a percentage of adult weight is 7–8% for the cuckoo (Gill 1983*a*; Brooker and Brooker 1989*a*) compared with 23% for the warbler (Gill 1982*a*). The shining cuckoo's egg (unspotted olive-green) does not mimic the gray warbler's (white with reddish-brown speckles). The same egg colors apply for hosts and parasite in Australia and New Caledonia (Hannecart and Letocart 1980; Brooker and Brooker 1989*b*), except that the eggs of some Australian hosts are unspotted white.

In Western Australia, four shining cuckoo eggs in yellow-rumped thornbill nests took 13.8–14.8 days to hatch (Brooker and Brooker 1989*b*). Measuring incubation period in the same way, from the start of incubation (rather than from the day the cuckoo laid as in Gill 1983*a*), the data for three eggs in gray warbler nests in New Zealand were 13.5, 15, and 16 days (mean 14.8 days). Thus, the incubation period seems to be similar and less than for the respective hosts in both countries. Eviction of the host's eggs or young by nestling shining cuckoos is universal. The cuckoo's nestling period is 18–23 days (pooling data for Australia and New Zealand), slightly longer than for the hosts' own young.

MIMICRY

Lack of egg mimicry

Shining cuckoos display many of the breeding adaptations typical of brood-parasitic cuckoos, the most notable exception being production of a nonmimetic egg radically different in color and pattern from that of the hosts. The lack of egg mimicry implies a failure of the hosts to recognize and/or reject foreign eggs. Various model eggs experimentally introduced to nests of yellow-rumped and western thornbills were accepted in every case (Brooker and Brooker 1989*b*).

There are at least three possible explanations of the lack of rejection behavior.

1. *The hosts do not see the cuckoo's egg clearly enough.* Brooker et al. (1990) suggested that there is insufficient light in the dimly lit enclosed nest for the host to respond to the foreign dark-colored egg and to develop rejection behavior (see paragraph (B) below).
2. *The hosts recognize shining cuckoo eggs as foreign, but are physically incapable of removing or puncturing whole eggs with their short thin bills.* They could desert a nest in which they detected a cuckoo's egg, but the cost of renesting is high (see sub-section, Population effects). Cuckoo eggs occasionally fail to hatch, or survive after hatching, and this may be another factor that weakens selection in favor of deserting nests.
3. *The host–parasite association is too recent for rejection behavior to have spread through the host populations.* This is a possible explanation for the acceptance of non-mimetic cuckoo (*Cuculus canorus*) eggs by dunnocks *Prunella modularis* in Britain (Brooke and Davies 1988). Rejection is expected to spread slowly if the frequency of parasitism is low (Kelly 1987). These arguments do not seem to fit shining cuckoos because frequencies of parasitism tend to be high and the hosts are unlikely to be recent (Brooker et al. 1990). As an endemic species in a nonendemic genus, the gray warbler has probably had ancestors in New Zealand since the Pleistocene, perhaps as long as a million years (Fleming

1962). New Zealand shining cuckoos are an endemic subspecies, which implies approximate origins as far back as 15 000 years. Adjacent tropical winter ranges of Australian and New Zealand shining cuckoo populations give way to adjacent summer ranges covering vast areas in which are parasitized only covered-nesting species of *Gerygone* and *Acanthiza.* A host–parasite relationship so recent as to have precluded the development and spread of rejection behavior by the hosts is unlikely to be so consistent over so large an area.

The dark color *per se* of the shining cuckoo's egg is a characteristic for which at least three possible explanations have been suggested.

(A) The color is of no adaptive significance and succeeds because the hosts do not discriminate against it. It may have been fixed in an ancestral population and subsequently not selected against because it incurs no cost.

(B) The color is cryptic in the dark enclosed nest and deceives the hosts [see paragraph (1) above]. Harrison (1968) suggested that in cuckoos parasitizing hosts with covered nests, natural selection might induce cuckoos to produce dark eggs that would be less visible than the host's (usually white) eggs and less likely to provoke rejection. Referring specifically to shining cuckoos parasitizing yellow-rumped thornbills, Marchant (1972) suggested that it may have been advantageous for the cuckoo to retain or develop dark eggs "so that the host could see them less easily in the dark interior of the nest."

(C) The color is cryptic and deceives other cuckoos. Brooker and Brooker (1989*b*) suggested that the cryptic egg color of shining cuckoos evolved in the first instance, not to deceive the hosts (which are probably all "accepters" anyway), but to deceive other female cuckoos (conspecifics or not) which may compete for nests. In cases of double parasitism of a nest, a cryptic egg would help to ensure that a second female did not remove the original cuckoo's egg when laying her own. If the first cuckoo's egg remains, it will hatch first and evict the second cuckoo's egg or chick.

New Zealand shining cuckoos compete for hosts with no other cuckoo species: the only other New Zealand cuckoo, the long-tailed cuckoo *Eudynamys taitensis*, is much larger and parasitizes larger hosts. As for intraspecific competition, nests containing two shining cuckoo eggs have not been reported in New Zealand, and there is circumstantial evidence that the cuckoos may be territorial during the peak laying period (see subsection on Breeding adaptations). Thus, present circumstances in New Zealand do not support hypothesis (C) above, though this could still be the correct explanation for an ancestral population.

Nestling mimicry

Nestlings of most parasitic cuckoos (subfamily Cuculinae) are naked at hatching. The only reported exceptions appear to be three species of Australasian *Chrysococcyx*: *Ch. russatus* ("four white tufts on the rear of the crown"; McGill and Goddard 1979); *Ch. minutillus* ("pale yellowish plumules on the crown and along the back"; McGill and Goddard 1979); and *Ch. lucidus.* In the latter, New Zealand nestlings have long but sparse, white trichoptiles on the head and back (see Fig. 2 of Gill 1982*c*) whereas Western Australian nestlings have short, whitish, easily overlooked trichoptiles on the head only (see Fig. 2 of Brooker and Brooker 1986). A single nestling from southeast Australia was described as naked (Courtney and Marchant 1971), but sparse down could have been overlooked.

In New Zealand, the newly hatched shining cuckoo seems to mimic the gray warbler chick in appearance: neonates of both species have white natal down and white rictal flanges that become yellow. Western Australian shining cuckoos at hatching are slightly different (table 7.1), but they too resemble their hosts in the color of the mouth lining and rictal flanges.

McGill and Goddard (1979) reported that nestlings of *Chrysococcyx russatus* and their host in north Queensland (*Gerygone magnirostris*) both have black skin and white natal down. They also stated that the pale pink skin and pale yellowish natal down of the nestling of *Ch. minutillus* "to a remarkable degree simulates that of the nestlings of *G. olivacea*," the host in northeastern New South Wales. Although they were most interested in the implications of this for the systematics of the two cuckoos, which some had considered to be conspecific, they stated that the coloring could be "localised adaptation by the cuckoos to different foster-parents." Both hosts have covered nests like those of the gray warbler.

Table 7.1 Coloration of newly hatched nestlings of the shining cuckoo and its hosts in New Zealand (Gill 1982*c*, 1983*b*) and Western Australia (Brooker and Brooker 1986, 1987)

	Down	Rictal flanges	Mouth lining
New Zealand			
Cuckoo	Sparse, white	White, later yellow	Pink
Gray warbler	Abundant, white	White, later yellow	?
Western Australia			
Cuckoo	Very sparse, whitish	Bright yellow	Yellow
Yellow-rumped thornbill *Acanthiza chrysorrhoa*	Abundant, gray	Cream	Yellow
Western thornbill *A. inornata*	Abundant, dark gray	Yellow	Orange-yellow

These observations suggest a degree of nestling mimicry in three Australian *Chrysococcyx.* If the resemblance of newly hatched shining cuckoos to their hosts' neonates is mimicry rather than coincidence, it may be because adult hosts can recognize a foreign nestling among their own. However, discrimination against foreign nestlings would be puzzling in a species that accepts a strange egg. It could be merely that nestlings with the wrong colors in the dimly lit, enclosed nest would be overlooked at feeding (M. G. Brooker, pers. comm.). This needs to be explored by experiment.

Reed warblers *Acrocephalus scirpaceus* reject foreign eggs (and have induced a gens of European cuckoo *Cuculus canorus* to mimic their eggs), but accept a nestling of a different species placed with their brood (Davies and Brooke 1988). Chick discriminatioin in response to brood parasitism apparently does not occur in hosts of *Cuculus canorus* (Davies and Brooke 1989). However, what applies for open, well-lit nests may not be reievant to the covered nests parasitized by *Chrysococcyx* cuckoos.

Instead of mimicry, the greater development of natal down on New Zealand, relative to Australian, shining cuckoos could be a thermoregulatory adaptation to a cooler climate (R. B. Payne, pers. comm.). However, the down is probably too sparse to counter heat loss, and significant down occurs on *Chrysococcyx* cuckoos from northeast New South Wales and Queensland, as cited above, where it is warmer than in New Zealand.

The begging calls of the nestling shining cuckoo mimic those of the young gray warbler (McLean and Waas 1987), but experiments have shown that the adult warblers can discriminate between the two types of call (McLean and Rhodes 1991).

INTERSPECIFIC COMPETITION FOR FOOD IN NESTLINGS

Laying by New Zealand shining cuckoos relative to the gray warbler's incubation cycle was determined at eight nests (Gill 1983*a*). At only one nest, possibly two, did the cuckoo lay before incubation started, thus ensuring that her egg began developing as early as the hosts' eggs did. In five cases the cuckoo laid one, two, two, seven and seven to ten days after the start of incubation. The late laying in the latter nests resulted in the cuckoo eggs hatching several days after all of the host eggs. In the eighth case, a cuckoo parasitized an abandoned clutch of one. The timing of parasitism was thus inappropriate or seriously suboptimal in about one-third of cases.

There were only three nests at which both the day of laying by the cuckoo and the circumstances at eviction by the cuckoo nestling were noted. Nest C12 typified efficient parasitism. The female cuckoo laid on the day of the warblers' penultimate egg of four. The young cuckoo hatched two days before any warblers, was heavier than the warblers when they hatched, and evicted them on their second or third days in the nest.

The other two nests were parasitized less efficiently. Nest C3 was parasitized two days after the clutch was complete and incubation began. The cuckoo hatched on the same day as one warbler, but a day after the other two. For its first four days the cuckoo had to compete for food with warbler chicks, all of which grew (fig. 7.1). The cuckoo's mass on the first four days was about average, but the mass rose to above average thereafter (6.5 g vs. 5.3 g, 7.6 g vs. 6.6 g; average masses given in Gill 1982*c*), suggesting that the cuckoo should have been heavier while it shared the nest with the warbler chicks. At nest A5 the cuckoo laid seven

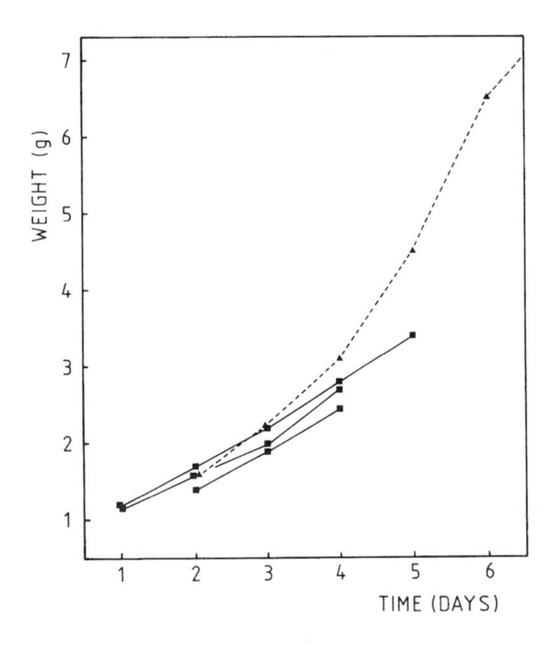

Figure 7.1 Interspecific competition for food between three gray warbler nestlings (■, —) and a shining cuckoo brood parasite (▲, ----) as indicated by growth in weight. Two warblers hatched on the first day; the cuckoo and the third warbler hatched the following day. The cuckoo evicted each warbler during the 24 hours following the last day on which its weight is shown. The nest was at Kaikoura, New Zealand.

days after incubation started and this led to seven days of interspecific competition for food between nestlings before eviction took place. This greatly reduced the nestling cuckoo's growth rate (see Fig. 3 of Gill 1983*a*).

By the time nestling cuckoos were old enough to perform eviction (3–7 days old, $n = 8$), host nestlings were present in seven of eight nests (not exactly the same sample as indicated at the start of this section). At four of five nests at which nestlings were weighed (including the three nests named above), gray warbler nestmates coexisted with a cuckoo for at least two days and increased in weight before eviction. This shows that interspecific competition for food was frequent. This tendency must extend the period for which parasitized nests hold nestlings, increasing their vulnerability to predators, compared with situations where the cuckoo evicts eggs and receives all food that the hosts bring.

The risk of predation, and the risk that a cuckoo nestling may be physically too small to evict its nestmates, are likely to select in favor of early laying by cuckoos (before the start of incubation by the host), and early eviction by nestling cuckoos. But such selection may have been weaker for shining cuckoos in New Zealand's predator-free past when mammalian predation was not a factor (see subsection, Impact of predation).

The modal week of the second bout of laying by gray warblers follows six weeks after the peak week of "arrival" of shining cuckoos in New Zealand (that during which cuckoos are most frequently first heard calling; see Fig. 1 of Gill 1983*a*). Cuckoos and hosts continue laying for another five weeks. It therefore seems unlikely that the cuckoos are constrained to lay inefficiently by a shortage of time.

Could the mimicry seemingly exhibited by shining cuckoo nestlings be liniked to their frequent cohabitation with host nestlings, and be the basis for comparison that this must afford the parent warblers? Perhaps in the absence of mammalian predators, any tendency in gray warblers to reject cuckoo nestlings was countered by the evolution of mimicry rather than by improving the timing of laying to remove host young before the warblers could compare them with the cuckoo chick. Mimicry of host nestlings is exhibited by nestlings of several parasitic cuckoos that do not evict host young and are reared with them (see references in Davies and Brooke 1988). New Zealand shining cuckoos seem to be intermediate in that host and cuckoo nestlings often coexist for longer than is usual in evicting species.

Shining cuckoos breeding in Australia have long had to cope with predation by both snakes and mammals. In Western Australia, laying was well synchronized with the start of incubation and nestling cuckoos evicted host nestlings earlier than did New Zealand birds (Brooker and Brooker 1989*b*). These behaviors could be adaptations to shorten the life of the nest during the vulnerable nestling phase, but Brooker and Brooker (1989*b*) suggested that evicting nest contents from gray warbler nests may be physically more difficult for the cuckoo than from Australian thornbill nests. I doubt that the latter is so because gray warbler nests have a simple construction with an unconcealed entrance (Gill 1983*b*). Among Australian hosts, nests of the yellow-rumped thornbill are complicated conglomerate structures with hidden entrances, and viewing the contents is difficult, even with a dental mirror and torch (Marchant 1972; Brooker and Brooker 1989*b*). The entrance diameter is about the same (around 30 mm) in gray warblers, yellow-rumped thornbills and western thornbills.

EFFECT OF PARASITISM ON THE HOST

Individual effects on the adult host

In New Zealand, single shining cuckoo nestlings never weighed as much as the combined weight of a brood of four gray warblers of the same age, and never weighed as much as broods of three except for their last few days in the nest (Gill 1982*b*). Cuckoos were brooded more on average than three or four warblers of the same age, for a greater part of their longer nestling period, and in longer attentive periods. This may be because the nestling cuckoo develops endothermy later than warblers do. On the other hand, the single nestling cuckoo was visited with food less often than broods of three or four warblers, and had fewer fecal sacs removed. Thus, on balance it seems that raising a cuckoo to fledging involves the foster-parents in less effort than a brood of their own young, particularly for male warblers, which join their mate in feeding nestlings but do not brood.

Population effects

At Kaikoura (42°S latitude), the only site where breeding by the gray warbler has been studied in depth, laying by the shining cuckoo (October to December) coincides with a second ("late") period of laying by the warbler (Gill 1983*a*). The warbler's "early" clutches are laid from the end of August to the beginning of October, but escape parasitism because the migratory cuckoo is not then either present or ready to breed. The modal time of arrival of cuckoos in New Zealand, as indicated by first reports of their calls, is late September to early October (Cunningham 1955).

Laying clutches before shining cuckoos arrive is not likely to be an adaptation to minimize the rate of parasitism (as suggested by May and Robinson 1985). Other insectivorous songbirds in the same study area breed as early as warblers and for as long (Gill et al. 1983) but lack brood parasites. A universal explanation of when these birds start breeding, probably involving food supply, is more likely. However, producing clutches before the brood parasite can lay preadapts gray warblers to withstand the impact of brood parasitism (Diamond 1984).

There was no significant difference between the size of early and late warbler clutches (Gill 1982*a*), but late nests were less successful. Early nests each produced 2.0 fledgling warblers on average, whereas unparasitized late nests yielded only 1.1, probably because mammalian predation was more intense later in the season when the predators had bred and were more numerous. Mammalian predation seemed to be the main factor depressing the warbler's reproductive success, accounting for the loss of at least 15% of active nests (including parasitized nests) throughout the season. There were many other failed nests at which predation or disturbance by mammals may have been the cause, but it was impossible to tell.

The incidence of parasitism of warbler nests initiated in mid-October or later, that is, those available for parasitism, was 55% (n = 40). Late nests were built more quickly than early ones, were harder to find and were under-represented in the study (Gill 1982*a*), but the same number of warblers probably nested in both early and late periods. Therefore, if the rate of parasitism is to be expressed over the whole breeding season of the host, the best approximation is probably 55% or 28%. In Western Australia, shining cuckoos parasitized 26% of yellow-rumped thornbill nests (Brooker and Brooker 1989*b*). In this case the cuckoo is present during the host's entire laying season, and a second host is parasitized (8% of western thornbill nests).

Of 53 gray warbler eggs laid in 16 nests that became parasitized, only one egg (2%) survived to produce a fledgling. The associated cuckoo egg did not hatch. Of 70 warbler eggs laid in 20 unparasitized late nests, 23 (33%) produced fledglings. Therefore, the effect of parasitism on parasitized nests was to prevent 33 − 2% (=31%) of all warbler eggs from yielding fledglings.

The average number of warbler fledglings was 2.0 from early nests, 1.1 from unparasitized late nests, and 0.06 from parasitized nests (one fledgling from 16 nests). An estimate of average warbler productivity in the absence of parasitism is 2.0 + 1.1 = 3.1 fledglings/pair/year. With parasitism, estimated average warbler productivity falls 18% to 2.0 + 0.45 (1.1) + 0.55 (0.06) = 2.53.

This is a much smaller loss to the warbler compared with losses from predation. Parasitism caused the loss of 21% of eggs that did not hatch and 17% of nestlings that did not fledge. The respective figures for predation were up to 52% and up to 57%, there being some doubt as to whether predation had occurred in some cases. However, the 18% loss cited in the previous paragraph is rather misleading because it is the loss to parasitism after predation has occurred, and would be higher if predation were absent. Unfortunately, predation cannot be removed from consideration, because of the many failed nests at which disturbance by mammals may or may not have occurred.

Nevertheless, an 18% loss to parasitism is a potentially strong selection pressure. Also, parasitism can be a stronger selection pressure than predation even if it is less common than predation, because parasitism diverts the hosts' effort for an entire nesting cycle whereas birds readily renest after predation (Rothstein 1990). However, the latter argument may not apply to gray warblers because their cost of re-nesting is higher than for typical north-temperate songbirds. Gray warblers take at least a week to build the complex nest. It then takes seven days to lay the clutch of four because of the 48-h laying interval. On average, the clutch is 93% of the female's body weight. In the four cases where gray warblers lost a clutch or brood and their next nest was found, there were intervals of 11, 12, 22, and 32 days before the first egg of the new clutch was laid. No warblers were known to make more than three nesting attempts in a season.

EXTERNAL INFLUENCES ON THE HOST–PARASITE INTERACTION

Geography and conservation

Gray warblers occur throughout New Zealand to 1400–1500 m above sea level, in all types of forest, and any other habitat offering at least low shrubs for nesting. Shining cuckoos are similarly widespread (see Bull et al. 1985), and their summer range is therefore about 228 000 km^2 — the area of New Zealand less an allowance for lakes and the alpine zone. The winter range of New Zealand shining cuckoos (the Solomon Islands, Bogainville, New Britain, New Ireland, and nearby smaller islands) is about 87 000 km^2.

During the past 100 years much native forest has been cleared in New Zealand, but this has probably not greatly affected total numbers of gray warblers because native forest has been replaced by other habitats broadly suitable for warblers. Deforestation in the shining cuckoo's winter range, which is likely to increase, could reduce cuckoo numbers in the future.

Impact of predation

New Zealand has always been free of land snakes. During the past 200 years, European contact has seen the introduction of ship rats (*Rattus rattus*), Norway rats (*R. norvegicus*), three species of *Mustela*, and feral cats (*Felis catus*). New Zealand was previously free of mammalian land predators, excepting *Rattus exulans* introduced by Polynesians 1000 years ago. Predation by mammals on adults and nests is therefore a major new factor affecting the survival of birds in New Zealand. Since it applies to both host and parasite, it is not likely to have greatly disturbed their coevolution.

The major factor causing loss of eggs and nestlings of both host and parasite at Kaikoura seemed to be predation by small mammals, but in absolute terms, mammalian predation threatens the long-term survival of neither species. Some 52% of shining cuckoo eggs survived to produce fledglings, and given that each female probably laid about 16 eggs per season (see reasoning in Gill 1983*a*), the recruitment rate seems healthy, even for a migratory species. On average, gray warbler pairs produced 2.53 fledglings per year, despite both parasitism and predation (see subsection on Population effects). This is healthy productivity in a sedentary species in which the annual survival rate of adults is at least 82% (Gill 1982*a*).

ACKNOWLEDGMENTS I thank M. G. Brooker, I. Jamieson, I. G. McLean, R. B. Payne, and S. I. Rothstein for helpful comments on a draft of this chapter.

REFERENCES

Brooke, M. de L. and N. B. Davies (1988). Egg mimicry by Cuckoos *Cuculus canorus* in relation to discrimination by hosts. *Nature*, **335**, 630–632.

Brooker, L. C., M. G. Brooker and A. M. H. Brooker (1990). An alternative population/genetics model for the evolution of egg mimesis and egg crypsis in cuckoos. *J. Theor. Biol.*, **146**, 123–143.

Brooker, M. and L. Brooker (1986). Identification and development of the nestling cuckoos, *Chrysococcyx basalis* and *C. lucidus plagosus*, in Western Australia. *Aust. Wildlf. Res.*, **13**, 197–202.

Brooker, M. G. and L. C. Brooker (1987). Description of some neonatal passerines in Western Australia. *Corella*, **11**, 116–118.

Brooker, M. G. and L. C. Brooker (1989*a*). Cuckoo hosts in Australia. *Aust. Zool. Rev.*, **2**, 1–67.

Brooker, M. G. and L. C. Brooker (1989*b*). The comparative breeding behaviour of two sympatric cuckoos, Horsfield's Bronze-Cuckoo *Chrysococcyx basalis* and the Shining Bronze-Cuckoo *C. lucidus*, in Western Australia: a new model for the evolution of egg morphology and host specificity in avian brood parasites. *Ibis*, **131**, 528–547.

Brooker, M. G., L. C. Brooker, and I. Rowley (1988). Egg deposition by the bronze-cuckoos *Chrysococcyx basalis* and *Ch. lucidus. Emu*, **88**, 107–109.

Bull, P. C., P. D. Gaze, and C. J. R. Robertson (1985). *The Atlas of Bird Distribution in New Zealand.* Ornithological Society, Wellington.

Courtney, J. and S. Marchant (1971). Breeding details of some common birds in south-eastern Australia. *Emu*, **71**, 121–133.

Cunningham, J. M. (1955). The dates of arrival of the Shining Cuckoo in New Zealand in 1953. *Notornis*, **6**, 121–130.

Davies, N. B. and M. de L. Brooke (1988). Cuckoos versus Reed Warblers: adaptations and counteradaptations. *Anim. Behav.*, **36**, 262–284.

Davies, N. B. and M. de L. Brooke (1989). An experimental study of co-evolution between the Cuckoo, *Cuculus canorus*, and its hosts. II. Host egg markings, chick discrimination and general discussion. *J. Anim. Ecol.*, **58**, 225–236.

Dennison, M. D., H. A. Robertson, and D. Crouchley (1984). Breeding of the Chatham Island Warbler (*Gerygone albofrontata*). *Notornis*, **31**, 97–105.

Diamond, J. M. (1984). Evolution of stable exploiter–victim systems. *Nature*, **310**, 632.

Fleming, C. A. (1962). History of the New Zealand land bird fauna. *Notornis*, **9**, 270–274.

Friedmann, H. (1968). The evolutionary history of the avian genus *Chrysococcyx. U.S. Nat. Mus. Bull.*, **265**, 1–137.

Gill, B. J. (1980). Foods of the Shining Cuckoo (*Chrysococcyx lucidus.* Aves: Cuculidae) in New Zealand, *N.Z. J. Ecol.*, **3**, 138–140.

Gill, B. J. (1982*a*). Breeding of the Grey Warbler *Gerygone igata* at Kaikoura, New Zealand. *Ibis*, **124**, 123–147.

Gill, B. J. (1982*b*). The Grey Warbler's care of nestlings: a comparison between unparasitised broods and those comprising a Shining Bronze-cuckoo. *Emu*, **82**, 177–181

Gill, B. J. (1982*c*). Notes on the Shining Cuckoo (*Chrysococcyx lucidus*) in New Zealand. *Notornis*, **29**, 215–227.

Gill, B. J. (1983*a*). Brood-parasitism by the Shining Cuckoo *Chrysococcyx lucidus* at Kaikoura, New Zealand. *Ibis*, **125**, 40–55.

Gill, B. J. (1983*b*). Breeding habits of the Grey Warbler (*Gerygone igata*). *Notornis*, **30**, 137–165.

Gill, B. J. (1983*c*). Morphology and migration of *Chrysococcyx lucidus*, an Australasian cuckoo. *N.Z. J. Zool.*, **10**, 371–381.

Gill, B. J., R. G. Powlesland, and M. H. Powlesland (1983). Laying seasons of three insectivorous song-birds at Kowhai Bush, Kaikoura. *Notornis*, **30**, 81–85.

Hannecart, F. and Y. Letocart (1980). *Oiseaux de Nouvelle Calédonie et des Loyautes. Tome 1.* Les Editions Cardinalis, Nouméa.

Harrison, C. J. O. (1968). Egg mimicry in British cuckoos. *Bird Study*, **15**, 22–28.

Kelly, C. (1987). A model to explore the rate of spread of mimicry and rejection in hypothetical populations of cuckoos and their hosts. *J. Theor. Biol.*, **125**, 283–299.

Marchant, S. (1972). Evolution of the genus *Chrysococcyx. Ibis*, **114**, 219–233.

May, R. M. and S. K. Robinson (1985). Population dynamics of avian brood parasitism. *Amer. Natur.*, **126**, 475–494.

Mayr, E. (1932). Birds collected during the Whitney south sea expedition. Part 19. Notes on the bronze cuckoo *Chalcites lucidus* and its subspecies. *Amer. Mus. Nov.*, **520**, 1–9.

Mcdonald, C. and B. J. Gill (1991). Shining Cuckoo eating an egg. *Notornis*, **38**, 250–251.

McGill, A. R. and M. T. Goddard (1979). the little bronze cuckoo in New South Wales. *Australian Birds*, **14**, 23–24.

McLean, I. G. and G. Rhodes (1991). Enemy recognition and response in birds. *Curr. Ornith.*, **8**, 173–211.

McLean, I. G. and J. R. Waas (1987). Do cuckoo chicks mimic the begging calls of their hosts? *Anim. Behav.*, **35**, 1896–1907.

Payne, R. B. (1973) Individual laying histories and the clutch size and number of eggs of parasitic cuckoos. *Condor*, **75**, 414–438.

Rothstein, S. I. (1990). A model system for coevolution: avian brood parasitism. *Annu. Rev. Ecol. Syst.*, **21**, 481–508.

Schodde, R., P. Fullagar and N. Hermes (1983). *A review of Norfolk Island birds: past and present.* Australian National Parks and Wildlife Service Special Publication, no. 8, pp. 1–119.

8

Nestling Eviction and Vocal Begging Behaviors in the Australian Glossy Cuckoos *Chrysococcyx basalis* and *C. lucidus*

ROBERT B. PAYNE, LAURA L. PAYNE

In most species of brood parasitic cuckoos, nestlings kill their foster nestmates by evicting them from the nest (Jenner 1788; Wyllie 1981; Davies and Brooke 1988; Payne 1997a). Nestling cuckoos of species in the genera *Cuculus*, glossy cuckoos *Chrysococcyx*, drongo cuckoos *Surniculus*, and koels *Eudynamys* (in Australia only) evict the eggs and nestlings of their host species (Chance 1922; Gosper 1964; Friedmann 1968; Becking 1981; Rowan 1983; Cramp 1985; Crouther 1985), whereas in the *Clamator* cuckoos, koels in India, and channel-billed cuckoos (*Scythrops novaehollandiae*) the nestling cuckoos tolerate the foster young or other cuckoos in the nest and are reared together with them (Friedmann 1964; Ali and Ripley 1969; Goddard and Marchant 1983). By attacking its nestmates the evicting young cuckoo monopolizes parental care and receives all the food that would go to the fosterer's own brood.

The young cuckoo also controls the behavior of its foster parents with loud and persistent begging. Begging calls that are loud and persistent are sufficient to attract the care of the foster parent in most species of birds. Nevertheless, the calls of certain cuckoos resemble the begging calls of the foster young and these cuckoos may mimic the young of their host species (Cramp 1985; McLean and Griffin 1991; Redondo 1993). The mechanisms in eviction, the behavior of the foster parent, and the variation in begging calls within and among species of parasitic cuckoos are of interest to studies both of parasite–host coevolution (Payne 1977, 1997b; Rothstein 1990) and of sibling rivalry in conflict for parental care in nonparasitic birds (Godfray 1995).

While the ultimate explanation of evicting behavior is the removal of rivals for parental care, the proximate mechanisms leading to eviction and to the gain of parental care from the foster parents are not well known and lead to several questions. What conditions of time, temperature, and behavior are involved in eviction of the foster eggs and nestlings? How does nestling behavior differ in parasitic cuckoos and in birds with parental care by their own species? What is the origin of the resemblance of begging calls of the parasitic cuckoos to the begging calls of their host species?

Here, we report our field observations of brood parasitism by the Australian Horsfield's bronze cuckoo (*Chrysococcyx basalis*) on the cooperatively breeding splendid fairy-wren (*Malurus splendens*). Our information suggests that evicting behavior is controlled by the environmental and body temperatures of the nestling cuckoo and the daily cycle of brooding and feeding by the foster parents, as well as by tactile stimulation of the eggs and nestlings of the host. Nestlings and fledglings of this cuckoo had different begging calls when reared by fairy-wrens than when reared by other host species. A second Australian species, the shining cuckoo (*C. lucidus*), did not vary its begging calls with its host species.

METHODS

Study area and population

We observed cuckoos at Gooseberry Hill in Western Australia, 31°57′S, 116°05′E, in 1983, 1985, and 1986. We tape recorded the adults, nestlings, and fledgling cuckoos in the care of their foster parents, and the songs, adult calls, and begging calls of their host species. We observed evicting behavior of the nestling cuckoos in relation to age and body temperature. The natural history of the cuckoos was described by Brooker and Brooker (1986, 1989*a*,*b*, 1992). Observations were coordinated with studies of the social behavior and breeding success of the fairy-wrens (Payne et al. 1985, 1988*a*,*b*; Russell and Rowley 1988; Rowley and Russell 1989; Rowley et al. 1989). *C. basalis* parasitized splendid fairy-wrens, the cooperatively breeding western thornbills (*Acanthiza inornata*), and pairs of scarlet robins (*Petroica multicolor*). On the same hillside, *C. lucidus* parasitized western thornbills and yellow-rumped thornbills (*A. chrysorrhoa*).

Our identifications were based on birds that we first observed as eggs or nestlings and were compared with museum specimens. Eggs are white with brownish spots in *C. basalis*, and unspotted light brown in *C. lucidus* (Beruldsen 1980). *C. basalis* nestlings have a naked dark skin and white rictal flanges until day 10, and *C. lucidus* nestlings have a naked paler skin and yellow rictal flanges. In fledgling *C. basalis* the throat is gray, the side of the face is dark, often with a dark streak behind the eye, and the rectrices are rufous on the outer and inner webs. In fledgling *C. lucidus* the throat is whitish, the side of the face is pale, and the outer webs of the rectrices lack rufous. Most fledged cuckoos at Gooseberry Hill were photographed with a 500-mm lens. Several young cuckoos were color-banded and photographed as nestlings and were observed in the care of their foster parents after they left the nest.

Eviction behavior and temperature of nestling cuckoos

Eviction by nestling cuckoos was observed directly in 1985 and 1986 and was inferred in all years by finding the eggs or nestlings of the host species outside the nest and the young cuckoo alone in the nest. Six nestling *C. basalis* were intensively observed and tested during 22 periods in 1986. Their temperatures before and during eviction, their behavior and movements, and their vocalizations were recorded.

Observations were made before light, at dawn, and at sunrise when temperatures were cool, and during warmer daytime hours. A digital thermometer (Jenco 7000H) with a small thermocouple (Jenco J4L, time constant 0.3 s) was used to record ambient air temperature T_A, nestling temperature T_B, and temperature in the fairy-wren's nest chamber T_N. T_A was recorded in shade within 0.5 m of ground; fairy-wren nests were usually less than 1 m above the ground in a bush. T_B was taken with the thermocouple inserted 3 mm into the cloaca and recorded when the temperature stabilized within 0.3°C for at least 2 s. T_N was taken with the thermocouple on the floor of the nest chamber.

To facilitate observation we removed a nestling cuckoo from the fairy-wren nest and placed it in a styrofoam box in the shape of a nest cup. To test the eviction response we placed a finger on its back; the touch was comparable with a mass of about 1–2 g, similar to a fairy-wren egg or day-old nestling. We used the movements of the tarsi and bracing of the head and neck to assess eviction and noted the cumulative time (sec) the cuckoo held an evicting posture in 60 s. The cuckoo was allowed to rest and cool for 2–4 min; we then recorded its T_B and T_A, and touched it again to test its eviction. These procedures were repeated until the cuckoo no longer evicted when touched or until its T_B approached T_A. Between eviction trials we counted leg kicks and body movements like those seen when the cuckoo was actively evicting, and noted its begging behavior. We cooled the cuckoo by exposing it to ambient air temperature. When the cuckoo no longer responded, we warmed it in the sun or against our bodies, and we then noted its behavior as its T_B returned to the T_B at the beginning of the test period.

We recognized two phases of eviction behavior. First, the nestling cuckoo moved about actively in a "searching" phase, which corresponded to its activity when it moved about the nest and came into contact with an egg or nestling. Second, the cuckoo shifted its body under the egg, pushed up, and lifted the egg, in a "push-up" phase of eviction when it actually removed the egg from the host nest.

Begging calls of nestling and fledgling cuckoos

We tape recorded the calls of the undisturbed nestling and fledged cuckoos and the young of the host species when they were fed by the (foster) parents, when the parent approached and entered

Table 8.1 Hatching date and mass and the age at eviction by nestling *Chrysococcyx basalis* in nests of *Malurus splendens*

	Cuckoo			
Wren nest	Hatching	Eviction	Mass (g)	Age at eviction (days)[a]
L85	27 Oct	30 Oct	—	2–3
U85	11 Nov	14 Nov	4.4	3[b]
Q85	13 Nov	13 Nov	—	0[c]
NN85	14 Nov	16 Nov	2.6	2
EE85	20 Nov	21 Nov	—	1
D85	26 Nov	28 Nov	—	2
A86	13 Oct	14 Oct	3.1[d]	1
I86	22 Oct	24 Oct	3.7[d]	2
LL86	3 Nov	5 Nov	2.3	2
W86	8 Nov	10 Nov	2.6	2
N86	10 Nov	12 Nov	2.5	2
R86	12 Nov	14 Nov	2.2	2
M86	8 Dec	10 Dec	3.0	2

[a]All evictions were of host eggs, except in five nests (Q85, U85, NN85, N86, M86) where they were of host nestlings, and in another (W86) the cuckoo first evicted a developed host egg then later in the day a day-old host nestling.
[b]Foster young hatched two days earlier than cuckoo.
[c]Top of nest was removed during incubation.
[d]Nestling mass was determined later in the day.

the nest, or when we lightly tapped the nest. Fledged birds were recorded while the parent was foraging and when it approached and fed the young.

Recordings were made with a Uher 4200 recorder and Sennheiser MKH 804 microphone, or a Sony TCM-5000 recorder and Sennheiser ME-40 microphone mounted in a 30-cm parabolic reflector. A Uniscan II and dot matrix painter and a Kay Elemetrics DSP Sona-Graph 5500 and printer 5510 were used to visualize the calls and print the spectrograms. Time and frequency measurements were taken from spectrograms with a transform size (band width) of 300 Hz. The sample included more than 4000 calls recorded from 28 foster broods and 33 young cuckoos, some of which were recorded repeatedly as both nestlings and fledglings.

RESULTS

Eviction

Size, age, eviction times, and eviction behavior

The adult host fairy-wren averaged 10 g (Rowley 1981). Adult cuckoos *C. basalis* at Gooseberry Hill ranged from 20.8 to 21.8 g (mean 21.2 g, $n = 4$); one *C. lucidus* was 21.1 g. Fairy-wren eggs were 1.0–1.2 g ($n = 2$) and at hatching two nestlings were 0.8 and 1.0 g before they were fed. Cuckoo eggs were the same size as the eggs of their hosts (Beruldsen 1980). Nestling cuckoos *C. basalis* were 1.1–1.5 g in the afternoon of hatching ($n = 4$). Two days later in early morning before they were fed they were 2.2–2.6 g ($n = 5$). By day 7 the nestling cuckoo was heavier than the adult fairy-wren. Nestling cuckoos were feathered from 12 days and reached adult size 2–3 days before they fledged at 18 or 19 days after hatching ($n = 4$).

Incubation periods of cuckoo eggs are shorter than in the host species. Our observations from laying to hatching were 11 days ($n = 3$) and 12 days ($n = 2$) for *C. basalis* and 12 days ($n = 1$) for *C. lucidus.* Gill (1983, this vol.) and Brooker and Brooker (1989*a*) also found short incubation periods in these cuckoos.

The domed structure of the nests of fairy-wrens and thornbills usually delayed the actual eviction of the host eggs until the cuckoo had grown for a day or two (table 8.1). The side entrance of the nest is about 10 mm above the floor of the nest cavity. At hatching and on the next day a cuckoo is too small to lift the foster eggs or nestlings to the

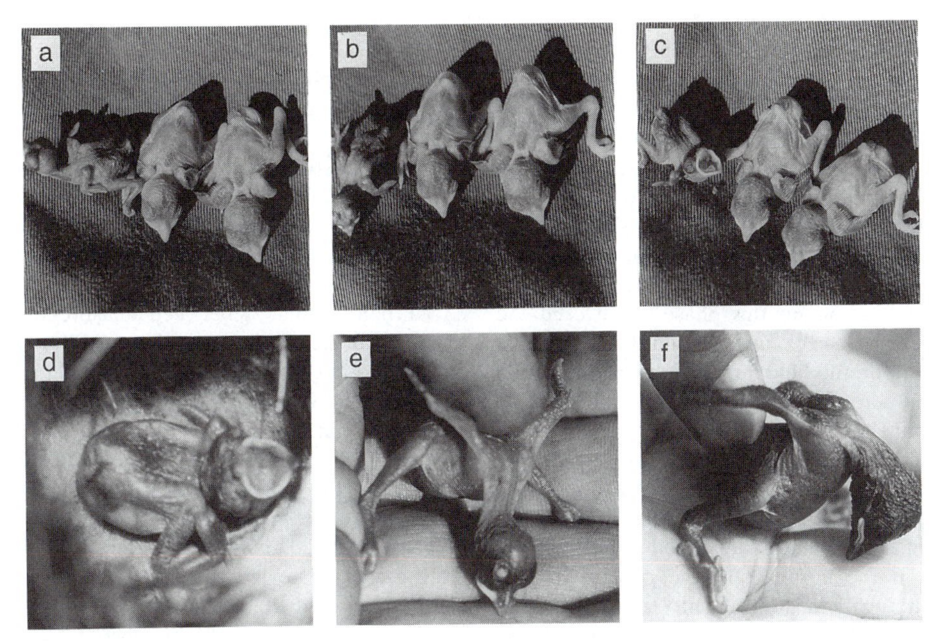

Figure 8.1 (a–c) A 2-day-old *C. basalis* (on the left) with its foster siblings, two 4-day-old *M. splendens*. Note the active begging and movement of the cuckoo within 60 s on a cool morning (the three nestlings were aligned 20 s before the first photo). (d) A 1-day-old nestling cuckoo *Chrysococcyx basalis* alone in nest of *Malurus splendens* after evicting the nestling fairy-wrens. (e) A 3-day-old *C. basalis* in a "push-up" posture, "evicting" a finger touching its back; note the position of the legs, wings, and head and neck. (f) A 5-day nestling *C. basalis* "evicting" a finger; note the use of head and neck as a brace.

rim of the nest entrance. In six cases when we visited a fairy-wren nest in early morning, *C. basalis* evicted within an hour of sunrise. In two nests we observed eviction at midday. One cuckoo hatched in a nest where the roof and side had been removed by an unsuccessful predator; it evicted its day-old nestmate in early afternoon on the day the cuckoo hatched. The other cuckoo evicted on day 3; it hatched a day after its foster siblings and it was too small to evict on the morning of day 2.

Nestling cuckoos sometimes evict later if they are too small to evict by the second day. In one nest, *C. basalis* hatched when the fairy-wrens were three days old. The two-day cuckoo evicted one nestling when the fairy-wren was five-days old. The other fairy-wren weighed 6.7 g and was identical in size and appearance to the evicted nestling, while the cuckoo was 2.6 g; so a cuckoo can evict a nestling more than twice its own size. The remaining fairy-wren disappeared from the nest the next day. The large size of the older host nestlings can sometimes save them from eviction, as in a nest where a cuckoo hatched when the fairy-wren was five days old, the cuckoo disappeared and the fairy-wren survived to fledge.

Nestling cuckoos evict other nestling cuckoos in the rare instance where two cuckoo eggs are laid and hatch in the same nest. In 1983 we found a nest of a western thornbill with one egg of the thornbill, one of *C. basalis* and one of *C. lucidus*. The cuckoos hatched 12 days later. By the second day the thornbill egg was evicted (it caught in the spines of the nest shrub) and the two cuckoo nestlings struggled to evict each other. By the fourth day only the nestling *C. lucidus* remained in the nest.

Elicitation of the eviction response

Eviction was recorded in detail in the cuckoos we observed either in the hand or in a test box. All unfeathered nestling *C. lucidus* and *C. basalis* had the same eviction behavior.

Cuckoos moved about pushing with their legs in the initial "searching" phase of eviction when they were not in contact with an egg or our finger, both in the host nest and test box (fig. 8.1). The cuckoo moved backwards until it came into contact with the edge of the nest or test box, or it rotated by pushing with one leg then the other.

In a nest these movements brought the nestling cuckoo into contact with an egg or nestling of the host, and in our test it came into contact with a finger. The movements were like those we saw when it was actively evicting in the host nest and the cuckoo shifted its position and orientation in the nest.

When the cuckoo came into contact with an egg or finger on its back, it shifted into the "push-up" phase of eviction. In this phase, it lowered the head, flexed the neck, and pressed the bill against the nest floor and held a rigid, tonic position (fig. 8.1e,f). When a finger was on its back, the cuckoo moved forward and shifted the touch from thorax to pelvic region. When a finger was on a side, the cuckoo moved in that direction and positioned its back under it, then took on the push-up posture. The behavior was the same as that observed of a cuckoo in the nest, where the cuckoo backed under the egg and lifted it onto the back, wings outstretched, balanced the egg as it pushed backwards and upwards, and expelled the egg onto the rim of the nest entrance where the egg fell from the nest.

Several morphologic features of the nestling cuckoos were associated with eviction. Nestlings had a flat back (fig. 8.1d). Two dorsolateral pelvic ridges, one on each side of the midline, protruded 1–2 mm above the midline and stabilized the host's egg during eviction. The large unfeathered wing with its large first digit or alula (noted also by Mellor 1917) held the cuckoo nestling in place as it pushed against the side of the nest. The rump moved upwards with an egg (finger, in our tests) balanced on its back until the cuckoo reached the opening of the nest.

Activity, eviction, and body temperature

The nestling cuckoo was active before it actually evicted, in the initial "searching" phase of eviction. We counted leg kicks in six cuckoos during 57 periods each of 30 s between eviction tests, and found an average of 5.9 kicks, per 30 s. T_B during this activity ranged from 35.9° to 23.7°C, and one cuckoo was active at 17.8°C. We also noted undirected movements that we termed "thrashings"; during 11 periods of 8 min each, cuckoos had a mean thrashing activity for 8.1 s per min. The cuckoo became quite active at low temperatures T_B and usually rotated its body halfway around during each min (fig. 8.1a–c).

Nestling cuckoos evicted persistently in a "push-up" posture when their T_B dropped 2–3°C and they were touched on the back (figs. 8.2 and 8.3). They maintained the posture while T_B was above 20°C. They increased "searching" activity as they cooled and the touch was removed, and they repeatedly returned to the "push-up" posture when touched on the back. Cooling was rapid and went to a lower T_B in early morning when T_A ranged from 10° to 20°C. Small nestlings (1.9 to 3.0 g) cooled rapidly, with T_B dropping from 5.7° to 11.7°C below initial T_B between measurements. T_B changes during midday ranged from 0.6° to 4.6°C. Larger nestlings in early morning cooled from 5.4° to 12.0°C below their initial T_B, and during midday their change in T_B ranged from 2.3° to 5.2°C. When a nestling was touched for more than 100 s, it abandoned the eviction posture, then begged or rested. Both nestlings whose T_B dropped to 20°C or below stopped giving the eviction response.

The cuckoo's eviction activities did not maintain a high body temperature, inasmuch as T_B decreased during the activity (figs. 8.2 and 8.3) and no shivering or other behavioral evidence of active thermoregulation was observed. The nestling cuckoo cooled about as rapidly as a nestling fairy-wren. In the brood where both were measured, at a T_A of 12.7°C, T_B of an inactive hatchling fairy-wren (0.9 g) dropped to 21.3°C in the nest by the time T_B of the active cuckoo (2.6 g) dropped from 29.5° to 21.2°C. The cuckoo evicted the hatchling fairy-wren later in the morning.

Eviction varied with age. Of the four 0–1-day cuckoos that we monitored for temperature and eviction behavior, three (75%) assumed the eviction posture. All six cuckoos at two days of age evicted for at least 20 s. Duration of the eviction posture dropped after five days, though at seven days one cuckoo evicted when its T_B was lowered from 28° to 24°C. By 11 days the cuckoos were partly feathered and maintained a constant and elevated T_B (38.3–40.2°C) for 10 min or longer (T_A ranged from 28° to 32°C) and the cuckoos did not evict.

Eviction was elicited up to six times in 30 min when T_B was between 22° and 38°C. Most nestlings evicted when T_B was lower than 30°C, but below 20°C they no longer evicted. When we warmed them, they evicted again. They evicted when T_B was greater than 34°C, but not above 38°C.

T_N was recorded infrequently. When the female fairy-wren flushed ($n = 5$) at our approach, the nest chamber was warmer than T_A. Early morning T_A was as low as 10°C, while T_N was usually 30–32°C. Within 1 to 2 min after the female left the nest, T_N fell within 3°C or less ($n = 3$) of T_A. T_N in nests where the female was absent when we ap-

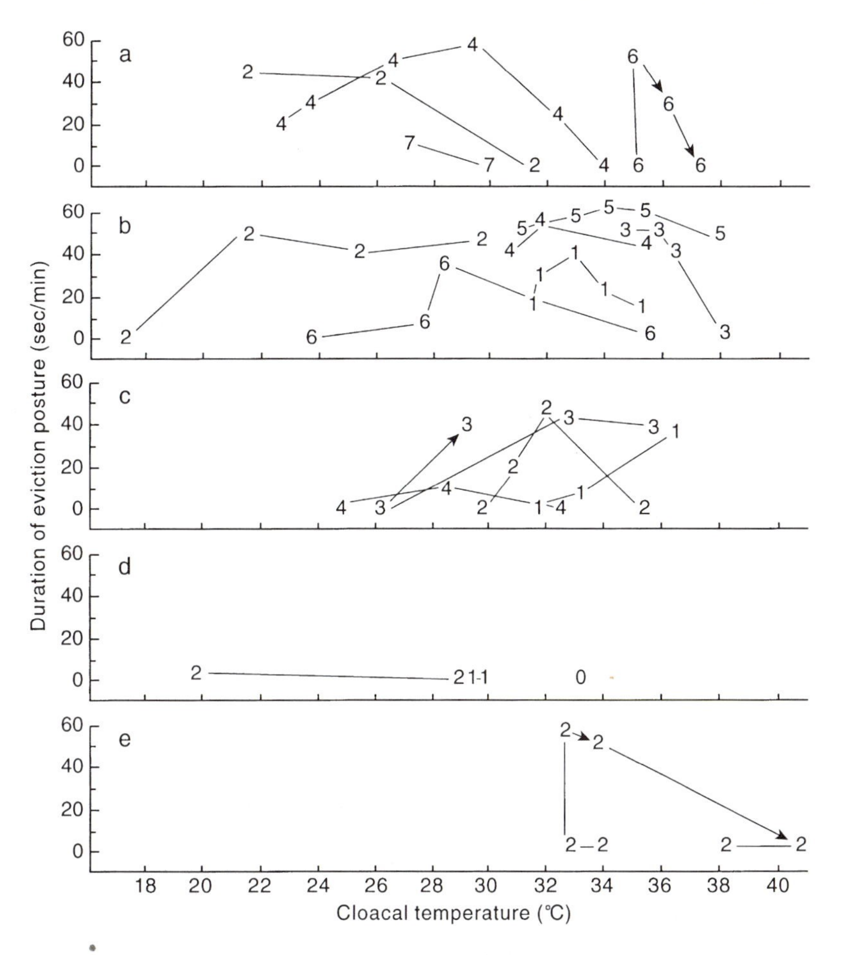

Figure 8.2 Duration of eviction posture of nestling *Chrysococcyx basalis* in relation to age and cloacal temperature T_B. The vertical axis indicates the time (s) the cuckoo maintained an eviction posture at T_B. The figure includes all cases where a cuckoo was tested at least twice and its T_B was recorded. Letters identify individual cuckoos, and the numbers indicate their age (days). Lines connect consecutive tests of an individual on a day. Except for cases with an arrow, nestlings in this figure were tested at progressively cooler temperatures.

proached was similar to T_A, and the T_B of the nestling cuckoo was falling.

In contrast to the nestling cuckoos, the nestlings of the host fairy-wren of the same age (0–5 days) observed at T_A between 22° and 30°C remained motionless while chilled until they were warmed to 30°C.

The form and context of the behavior in these observations indicate that the nestling cuckoo's movements as it cooled were associated with eviction, and the time of eviction by the cuckoos was related to the temporal pattern of behavior in the host fairy-wren. During 28 days of early morning observations in 1986, we watched the fairy-wrens for an hour before sunrise. Before dawn, the males sang continuously from high perches for about 20 min. At the end of the singing period about 20 min before sunrise, the brooding female left the nest and fed with the males in her social group. This was the time of day when air temperatures were coolest and when we recorded the low initial T_B and the greatest decreases in T_B of the cuckoo nestling. This was also the time when the cuckoo evicted the contents of the nest of the host. Our main observation on the temperature of eviction is that cuckoo nestlings are active and can operate in cool temperatures, and not that cooling is the necessary initial stimulus to evict.

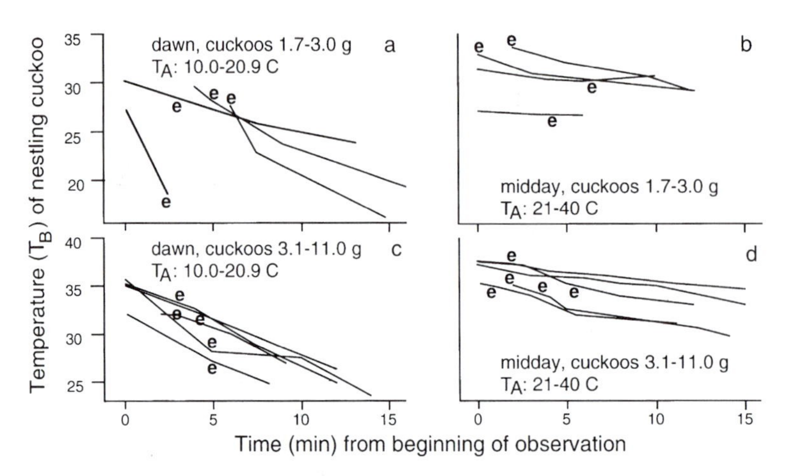

Figure 8.3 Cooling of body temperature T_B and the elapsed time to simulated eviction by nestling *Chrysococcyx basalis* after removal from the nests of *Malurus splendens.* Except for the newly hatched cuckoo, all birds had evicted the contents of their nest. Cloacal T_B was measured at 3-min intervals after the cuckoo was removed from the nest. Note the rapid cooling of T_B at dawn, when temperatures were cool, and the cooling and lower terminal T_B in the nestlings smaller than 3.0 g, the size at which they evict the contents of the nest. The initial eviction response for the day is indicated by **e**.

Begging calls

Begging calls of the host species

The begging calls of the young were distinctive among the four host species (table 8.2; fig. 8.4). Yellow-rumped thornbills (*Acanthiza chrysorrhoa*) gave monotonic whines with two concentrations of energy at 7 and 8 kHz and a duration of about 0.25 s; the whines were repeated at intervals of 0.1 to 0.2 s. Western thornbills (*A. inornata*) gave buzzy calls that rose in pitch at the beginning of a call, especially in the nestlings. The calls of nestlings had a duration of 0.2–0.3 s and calls of fledglings were longer at 0.4 s; both had a prominent frequency band at 7–8 kHz. The begging calls of fledged robins *Petroica multicolor* were whistles that descended from 8 to 2.5 kHz and ended with a buzz. Small nestling fairy-wrens *Malurus splendens* gave soft "peep" or "ti" notes at a frequency of 2–3 kHz. These calls develop into longer buzzy notes that were higher in pitch at 6–8 kHz by fledging. Fledged fairy-wrens gave series of short (<0.2 s) notes at intervals of < 0.1 s,

Table 8.2 Begging calls of four host species in Western Australia

Host species	Age (days)	No. of broods	Begging call[a]					
			peep	sit	ze-ba	whine	buzz	U
A. chrysorrhoa	fledgling	4				4		
A. inornata	nestling (7, 8)	2					2	
	fledgling	14					14	
P. multicolor	fledgling	1		1[b]				
M. splendens	nestling (4–8)	3	1[b]					2
	fledgling	5				2[b]		5

[a]Numbers indicate the numbers of broods recorded for each call type.
[b]"peep" and "whine" not identical with those of the cuckoo.

See text and fig. 8.4 for descriptions and spectrograms of calls.

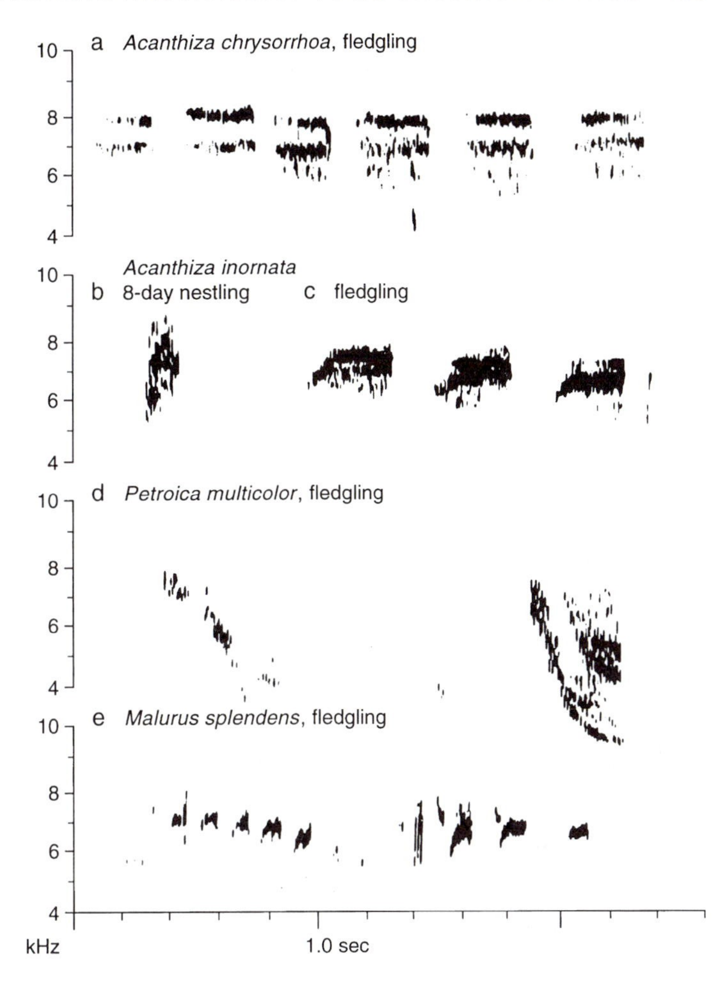

Figure 8.4 Begging calls of four host species. (a) *Acanthiza chrysorrhoa*, fledgling. (b) *Acanthiza inornata*, 8-day nestling. (c) *Acanthiza inornata*, fledgling. (d) *Petroica multicolor*, fledgling. (e) *Malurus splendens*, 37-day fledgling.

phrased in a series of notes that fell in pitch, then rose and fell again, in a pattern suggestive of adult song (Payne et al. 1988*a*) and described as a "gushing reel" (Slater 1975). The note itself rose then descended in pitch, forming an inverted "U" in spectrograms (fig. 8.4e; see also fig 8.7d). Occasional U-shaped notes were prolonged into a short whine or buzz. The calls became more complex with age and in newly independent fairy-wrens the series appeared to develop directly into early song.

Begging calls of C. lucidus

The main begging call of shining cuckoos was a "whine" (table 8.3), a single or double band of sound at about 8 kHz. The call was maintained at a constant pitch or decreased slightly and lasted 0.3–0.4 s. A fledgling cuckoo called persistently as its foster parents foraged nearby. The call was similar in cuckoos fed by *A. chrysorrhoa* and by *A. inornata* (fig. 8.5). "Whine" of Australian *C. lucidus* was similar to "whine" of a nestling *C. lucidus* in New Zealand in the care of gray warblers (*Gerygone igata*) (McLean and Waas 1987). The calls were longer when the foster parent was present. Calls did not vary with cuckoo age from 13 to 49 days, when a color-banded fledgling in the care of its *A. inornata* foster parents (fig. 8.5g) actively elicited feedings from them by pursuing and calling until it was fed; it also caught and ate insects by itself while the foster thornbills foraged nearby.

Table 8.3 Begging calls of two *Chrysococcyx* cuckoos in Western Australia

Cuckoo	Host species	Age	*n*	Begging call[a]					
				peep	sit	ze-ba	whine	buzz	U
C. lucidus	*A. chrysorrhoa*	nestling	2	1			2		
		fledgling	5			3	5		
	A. inornata	nestling	1			1	1		
		fledgling	5			3	5		
C. basalis	*A. inornata*	nestling	2	1	1	1	2		
		fledgling	3	1	1	2	3		
	P. multicolor	fledgling	5			2	5		
	M. splendens	nestling	12	8	5	4			5
		fledgling	6		6	6			6

[a]Numbers indicate the numbers of cuckoos recorded for each call type.

See text and figs. 8.5–8.7 for descriptions and spectrograms of calls.

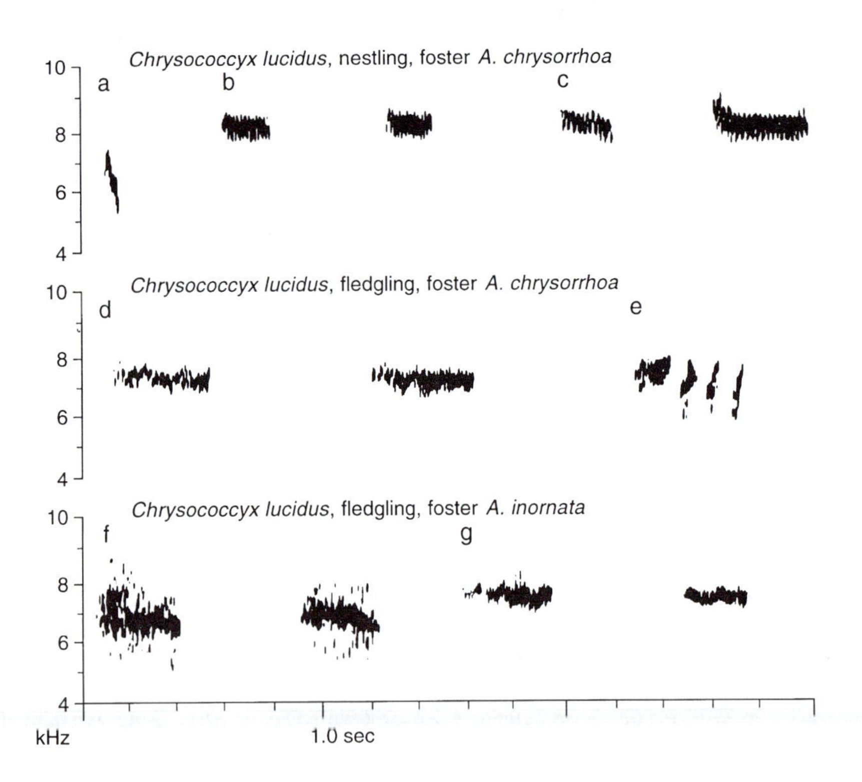

Figure 8.5 Begging calls of cuckoos *Chrysococcyx lucidus* in the care of the host *A. chrysorrhoa.* (a) A 4-day nestling, "peep". (b) A 7-day nestling. (c) A 13-day nestling. (d) Fledgling whine. (e) Fledgling "ze-ba" call given when the foster parent was away and foraging; the long whine calls were given when it returned and the cuckoo begged; same bird and calling bout as (d). (f) Whines of 30-day fledgling banded as a nestling in *A. inornata* nest. (g) A 49-day fledgling, same bird as (f).

Another begging call was "ze-ba," a buzz of 0.1–0.2 s ("ze") followed by short (0.01–0.02-s), rising calls ("ba") (fig 8.5e). It was given repeatedly by a fledgling cuckoo while the foster thornbill foraged in another tree out of sight of the cuckoo. The fledgling also gave "ze-ba" after a feed while the thornbill remained within 1 m. Occasionally during a feed, the introductory "ze" was omitted, and the call was a series of up to six "bas."

Young nestling cuckoos gave "peep" calls that attracted the foster parent. "Peep" was short (0.05–0.07 s), rising then descending in pitch, and louder on the descending part (fig 8.5a). "Peep" was heard from hatching to five days after hatching.

Begging calls of C. basalis

Horsfield's bronze cuckoos had five call types related to foster parental care.

Fledgling cuckoos gave a "whine" when begging from a foster thornbill *A. inornata* or robin *P. multicolor* (table 8.3). "Whines" changed irregularly in pitch but usually rose slightly at the beginning and fell at the end, were wavy in form, and lasted 0.3 s or longer (fig. 8.6). Cuckoos fed by thornbills had shorter modulations and more initial rise, and fledglings fed by robins had longer modulations, but these were minor variations. In contrast, the begging calls of cuckoos fed by fairy-wrens *M. splendens* consisted of inverted "U" notes, each about 0.2 s and delivered in a reeling series (figs. 8.6 and 8.7).

Three other call types were given by *C. basalis* reared by all three host species (table 8.3). "Sit" was a short descending note given by nestlings and young fledged within four days. It sometimes preceded feeds. "Ze-ba" calls involved an introductory "ze" element usually longer than 0.1 s, and "ba" elements similar to "sit" (fig. 8.6d). "Ze-bas" were given by a 13-day nestling cuckoo as its foster parent approached, and fledglings gave "ze-bas" between feeds when the foster parent was not nearby. The elements are longer in *C. basalis* than in *C. lucidus*, and they descend rather than ascend in pitch. Finally, "peep" (fig. 8.7b) was given as early as hatching when the nestling cuckoo had a cap of eggshell, and through the next 10 days as the female fairy-wren approached the nest and brooded the nestling cuckoo, and when the cuckoo was alone in the nest. A variant "peep" that descended more than 2 kHz was given by a nestling in distress in the hand. A fledgling reared by *A. inornata* gave a similar call when a brown goshawk (*Accipiter fasciatus*) circled overhead.

The sequence of begging calls in a fledgling cuckoo accompanied by an adult robin changed during the feeding bouts. At first the cuckoo gave persistent "whines" while the foster parent approached, then it called with higher pitch, variable duration, and shorter intervals. After a feed, the calls changed to "ze-ba" and "whines." Similar patterns were recorded with the other host species, with the inverted "U" begging calls leading to "ze-ba" calls in cuckoos begging from fairy-wrens.

Cuckoos gave inverted "U" notes while still in the nest of fairy-wrens. The calls appear to be derived from nestling "peeps" (fig. 8.7), and nestlings at five to seven days gave "peeps" intermediate in form between the earlier "peep" and the "U" of the fledged cuckoos. They were given singly with more than 0.2 s between calls. On day 10, two nestlings gave "U" notes and one delivered "U"s in a series. By day 14 the form and delivery of calls were similar to fledgling cuckoos' (figs. 8.6e and 8.7e). Fledglings gave the calls through day 32, 14 days after they left the nest and the latest that we saw a cuckoo attended by a fairy-wren. Calls were given in rapid series with successive notes higher then lower in pitch. Nestling and fledgling cuckoos gave more than 80 notes in a series of "reels."

Begging calls of young cuckoos resemble the calls of the foster young within and among species (tables 8.2 and 8.3). The rising buzz of young thornbills was not matched exactly, nor was the descending buzzy note of robins, and some cuckoo calls ("ze-ba," "sit") did not closely match the calls of the foster young. However, the begging calls of cuckoo and host species were similar, both in *C. lucidus* reared by thornbills *A. chrysorrhoa*, and in the "U"s in *C. basalis* reared by fairy-wrens.

Summary of begging calls of C. basalis *fledglings reared by different host species*

Overall, three *C. basalis* fledglings reared by thornbills *A. inornata* and five fledglings reared by robins *P. multicolor* gave only "whines" when begging, whereas each of six *C. basalis* fledglings reared by fairy-wrens *M. splendens* gave only "U" calls. The latter differed significantly from cuckoos reared by the two other host species in the occurrence of individuals that gave "U" calls ($P \leq 0.05$, two-tailed Fisher exact test on 6 of 6 versus 0 of 3, and $P < 0.02$ for 6 of 6 versus 0 of 5).

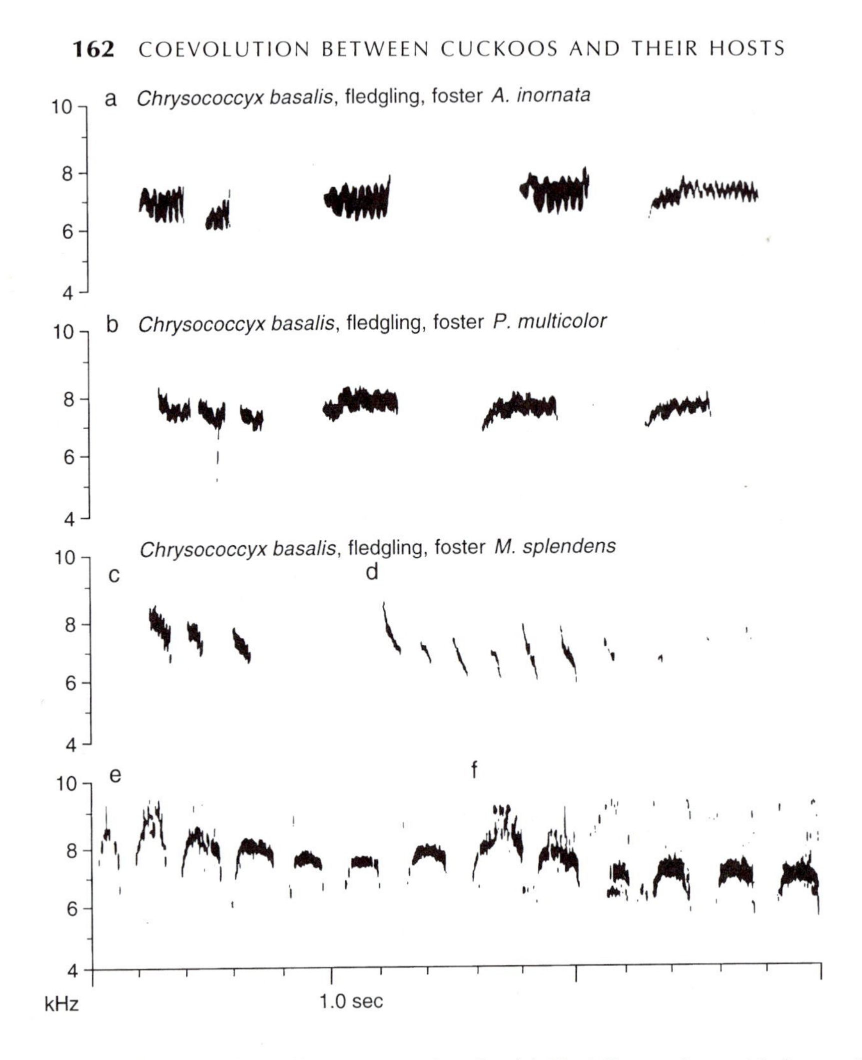

Figure 8.6 Begging calls of cuckoos *Chrysococcyx basalis.* (a) Fledgling cuckoo with foster *Acanthiza inornata.* (b) Fledgling with foster *Petroica multicolor.* (c) 17-day nestling cuckoo "ze-ba" with foster *Malurus splendens.* (d) "Sit" calls, 23-day cuckoo fledged from nest of *M. splendens.* The cuckoo gave these calls also as a nestling, and was fed while giving these calls on its first two days out of the nest. (e, f) Two series of beggings, U-note sequences, same fledgling cuckoo at 33 days.

Attraction of foster parents to the begging call

Cuckoo fledglings actively associated with their foster parents by flying towards them, calling, and begging with fluffed plumage, slightly drooped wings, and open mouth. Both species of cuckoos were loud and persistent in calling when the foster parents were out of sight and when they were nearby. The cuckoos were more conspicuous and louder than the nestlings or fledglings of the host species at the same age.

We broadcast recordings of begging calls when the foster parent flew into the area near their cuckoo young. In one test the calls of a 15-day-old nestling cuckoo *C. basalis* in a nest of thornbill *A. inornata* were played two days after the cuckoo fledged. The pair fed the cuckoo repeatedly for an hour before playback, on playback they approached the speaker within 2 m, then in 10 s the pair flew to the calling cuckoo and fed it. In a second test, the same calls were broadcast to a fairy-wren that attended a young *C. basalis* fledged from her nest six days earlier. She ignored the speaker and fed her cuckoo as it gave "sits." These first two tests used playbacks of a descending "whine." In a third test, "U" calls of a 16-day nestling *C. basalis* were played to its foster fairy-

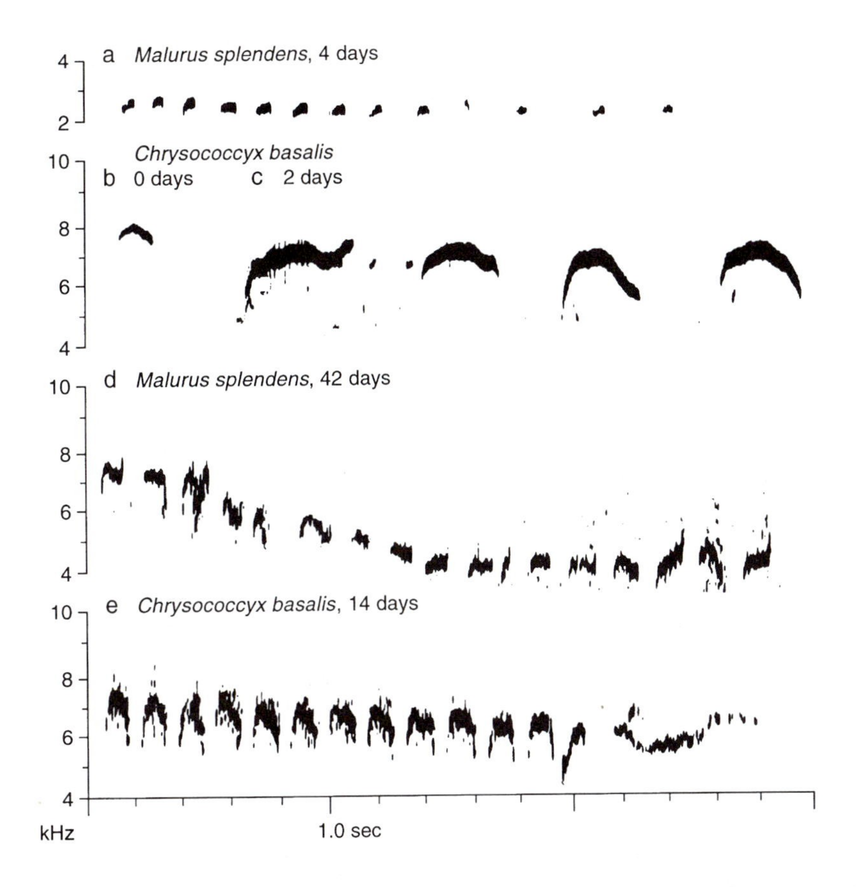

Figure 8.7 Development of begging calls in host *Malurus splendens* and cuckoo *Chrysococcyx basalis.* (a) *Malurus splendens* 4-day nestling calls at 2 kHz. (b) *C. basalis* hatching "peep" recorded within 2 h of hatching. (c) 2-day *C. basalis* in the same nest and same day as (a) where the fairy-wrens hatched two days earlier, their calls had no obvious effect on development of calls of the cuckoo. (d) 42-day *M. splendens* fledgling. (e) 14-day nestling *C. basalis*; the "whine" at the end occurred at a transfer of food by the foster mother. Note the inverted U-notes of *M. splendens*, their increase in pitch with age, and the variation in form of the calls within a begging sequence.

wren as she brought food to the cuckoo, which had fledged seven days earlier from her nest; within 10 s she approached the speaker and not the cuckoo. Further tests are necessary to determine whether the attraction of a foster parent varies with its experience in rearing cuckoos with different begging calls.

DISCUSSION

Evicting behavior

The young glossy cuckoos, like all other parasitic cuckoos, are altricial and ectothermic. They depend on the foster parent to provide food and to brood them and maintain a warm T_B for growth and development. With the size difference between parasite and foster young, the young cuckoo uses as much food as the entire brood of the host (Rowley 1981). Elimination of the competition for resources by the hosts' young is the function of eviction by the cuckoo nestling.

Evicting behavior of nestling *Chrysococcyx basalis* and *C. lucidus* is similar to that of other cuckoo species (common cuckoo, *Cuculus canorus*; Chance 1922; Makatsch 1955; Wyllie 1981; Maljchevski 1987; Yoshino 1988; Matsuda and Uchida 1990: red-chested cuckoo, *C. solitarius*; Liversidge 1955: koels in Australia; Gosper 1964). The coordinated use of the bill and head, wings with large alula, and tarsi as a brace is similar to eviction in *Cuculus* cuckoos (Molnár 1947; Cramp 1985). Like other species of *Chrysococcyx* glossy cuckoos (Jensen and Vernon 1970; Jensen and Clinning 1974), young *C. basalis* and *C. lucidus* have a flatter back than the more hollow back of the *Cuculus* cuckoos (*C. canorus, C. solitarius, C. saturatus*, Horsfield's cuckoo *C. horsfieldi*, and lesser cuckoo *C. poliocephalus*; Chance 1922; Liversidge 1955; Wyllie 1981; Gill 1982; Higuchi 1986). Like other species of glossy cuckoos and common cuckoos (Skead 1952; Wyllie 1981; Cramp 1985), the Australian glossy cuckoos evict from hatching day to day 2 depending on the shape of the nest and the size of the host eggs and nestlings, and they can evict through day 4 or 5. The eviction posture of the nestling cuckoos is similar to that of a hatching cuckoo, and the eviction movements may be derived from the movements of hatching (von Frisch 1969).

Our observations suggest an advantage for cuckoo nestlings that evict over a prolonged period because of variation in laying schedules. When a cuckoo egg is laid several days after the host eggs and the cuckoo hatches later than the host young, the cuckoo nestling may need a few days to grow large enough to evict its more slowly growing nestmates (see also Gill, this vol.).

We suggest that the nestling cuckoo remains inactive while it is brooded and kept warm, and it evicts as it cools when the foster female leaves the nest to feed. The female fairy-wren remains on the nest while their males sang before dawn, then joins the males and feeds with them as it becomes light from 20 min before to 15 min after sunrise. After she leaves the nest, the cuckoo cools and moves about the nest, comes into contact with other objects, and then evicts, as we found some nestling cuckoos evicting the host egg from the nest before sunrise. A female fairy-wren at 10 g is larger than a small nestling cuckoo at 1–2.6 g, and when she is on the nest her mass might inhibit eviction; however, in our tests we did not notice a decrease in intensity or duration of eviction when we increased our finger force on the cuckoo's back. Alternatively, a nestling cuckoo may rapidly warm by conduction of heat from the brooding foster mother, and an increased body or skin temperature may terminate a bout of evicting behavior.

The absence of the foster female fairy-wren while the nestling cuckoo evicts its nestmates might prevent her interference in the struggle within her brood. It is unknown whether parental interference would occur; it has not been observed in fairy-wrens. It is also unlikely in hosts of common cuckoo, as a nestling can evict both when the foster mother is away and when she is on the nest; the foster parent lifts its own body while the cuckoo evicts the host's own egg from the nest (Chance 1922; Wyllie 1981; Maljchevski 1987; Yoshino 1988).

The activity of cuckoo nestlings at low temperatures that we observed in the glossy cuckoos in Western Australia is similar to that in common cuckoos in northern Russia as described by Promptov and Loukina (1940). "Temperature changes greatly affect the activity of the Cuckoo, which is most active and vigorous at the relatively low temperature of 15° [C], when the young of the host (Willow Warbler [= *Phylloscopus trochilus*]) may be rigid with cold" (Dement'ev and Gladkov 1966:497). Promptov and Loukina (1940:89–90) reported that the cuckoo evicts better at lower temperatures, and noted that the cuckoo's skin is cold when it is active. They suggested this behavior is adaptive: in natural conditions the cuckoo evicts the host young when the host female is absent, when the nest is cold and host chicks are relatively inactive, while the cuckoo lies quietly on the nest floor when the host female warms the nest. Further work by Maljchevski (1987:86–87) showed that nestling cuckoos like passerines were inactive at temperatures lower than 15°C, and were most active when T_B was 35–37°C, as when the nest is brooded. This remarkable facultative active behavior of the altricial nestlings of parasitic cuckoos at cool temperatures appears to have been overlooked in recent accounts of the thermal ecology of young birds (O'Connor 1984; Haftorn 1988; Ricklefs 1988; Whittow and Tazawa 1991).

The activity of young parasitic cuckoo nestlings at low temperature is associated with brood parasitism and not simply with being a cuckoo. In Michigan, we have observed shivering and maintenance of elevated T_B and a lack of response to the touch test in a nonparasitic nestling black-billed cuckoo (*Coccyzus erythropthalmus*) at five days; unlike the parasitic cuckoos, it was partly feathered by this age. Other altricial songbird nestlings that we have observed are inactive at low temperatures, as are the fairy-wrens and willow warblers.

The termination of evicting responses in older nestling cuckoos may involve developmental changes such as endothermy and maintenance of high T_B including insulation of the plumage and thermal inertia with increasing body size, as in nestlings of other altricial birds (Dunn 1975; O'Connor 1984; Whittow and Tazawa 1991). Common cuckoos are altricial and their T_B is low and variable until about day 9 when they are feathered and begin to maintain a constant high T_B (Hund and Prinzinger 1980). In another brood parasitic bird, the brown-headed cowbird (*Molothrus ater*), nestling endothermy does not develop earlier than in the host species or in other nesting species in the family Icteridae (Dunn 1975). The tactile stimulus leading to eviction may vary among cuckoo species and even within a species: nestling *C. lucidus* in Australia are naked whereas nestlings in New Zealand are covered with down, yet both evict (our observations; Gill 1982, 1983, this vol.). Several physiological questions would be of interest to pursue with these altricial nestlings that are active at low temperatures.

Begging calls

Unattended fledgling *C. lucidus* use short calls to maintain social contact with the foraging foster parent. When the foster parent approaches, the fledgling increases its rate and duration of calls. No differences were observed in the calls given by young cuckoos reared by the two thornbill species. These calls are similar to the begging calls of the two host species.

In contrast, the begging calls of young *C. basalis* vary with the host species that rear them. Nestling and fledgling cuckoos give intense, rapidly repeated inverted "U"s when their *M. splendens* foster parents arrive with food. Cuckoos in the care of *A. inornata* or *P. multicolor* give "whines." "Sit" and "ze-ba" calls of the nestling and fledgling cuckoos do not vary with host species.

Four explanations of the differences in the begging calls of *C. basalis* with different host species are (1) misidentification of the cuckoo, (2) the cuckoo begged from a bird other than its foster parent, (3) developmental responses of the cuckoo vary with the social or acoustic environment of the host species, and (4) innate differences occur among host races of the cuckoo.

(1) Mistaken identification seems unlikely. Many of these cuckoos were examined and photographed in the hand, and they were color banded and observed repeatedly. *C. basalis* is well known as a brood parasite of splendid fairy-wrens on Gooseberry Hill (Rowley 1981; Brooker and Brooker 1989*a*). One *C. basalis* reared by a pair of color-banded western thornbills was photographed before it fledged, as were three *C. lucidus*, and all four cuckoos and their thornbill foster parents were identified by their color bands after fledging. Three color-banded fledglings of *C. basalis* reared by western thornbills and three unbanded *C. basalis* attended and fed by scarlet robins were photographed, and each had rufous outer tail, dark face, and no barring on the breast. The scarlet robin is a common and successful fosterer of *C. basalis* and it is not known to rear *C. lucidus* (Beruldsen 1980; Brooker and Brooker 1989*b*, 1992).

(2) Six cuckoos color-banded as nestlings were observed and tape recorded for an hour or longer after fledging. Each cuckoo was fed only by its own foster parents and their helpers in the same social group. In no case did it attract other individuals or other species. Other fledged cuckoos that we watched for shorter periods likewise were fed only by their foster group. Cuckoos sometimes attract adults of other species and more than one species may feed a cuckoo (Reed 1968; Smith 1989), but adoption is unusual and was not seen in this study.

(3) Begging calls of nestling cuckoos developed without the cuckoo ever hearing the foster nestlings, when it had evicted the host eggs. The begging calls did not closely resemble the calls of the foster parents, though the song of the splendid fairy-wren is somewhat like the call of the fairy-wren when disturbed and like the begging calls of the cuckoo. Call imitation during nestling life would be without a known parallel in other birds, which generally do not give learned vocalizations until several weeks or months of age. Experiments have shown no effect on early vocal development when songbirds are given different auditory experiences before hatching or as young nestlings (Thielcke-Poltz and Thielcke 1960; Konishi and Nottebohm 1969). There are similarities between begging calls of a cuckoo and contact notes or parts of song of its host species, and these might be related to the cuckoo hearing its foster parents. Nevertheless, the lack of difference in the calls of cuckoos reared by thornbills and by robins, host species with dissimilar songs and calls, and the difference between these cuckoos and the cuckoos reared by the fairy-wrens, lead us to question whether the variation in their calls is due to vocal imitation.

Another kind of vocal learning, without imitation, might involve an initial variation in the begging calls of a nestling cuckoo followed by the development of the variants that the foster parents respond to most strongly. In vocal learning in another brood parasite, the brown-headed cowbird (*Molothrus ater*), naive yearling males tend to repeat the song types to which a female gives a visual response (West and King 1988). An attrition of innate calls that were not associated with parental care by the foster parent would parallel the selective attrition of endogenous song elements recorded in immature sparrows (Marler and Peters 1982). In our observations the cuckoo calls were distinct by the time of fledging, nestling cuckoos of the fairy-wrens gave the inverted "U" calls as did the fledglings (fig. 8.7), and we detected no greater variability in the calls of nestlings than in the fledged cuckoos.

(4) Races that are specialized for certain host species are suggested by the polymorphism in egg color and pattern in some species of cuckoos. Egg polymorphism is unknown in *C. basalis* (Beruldsen 1980; Brooker and Brooker 1989*a*), and we noted no differences in the appearance of nestling or fledgling cuckoos reared by different host species, or in the songs of adult cuckoos on Gooseberry Hill and elsewhere in Australia. Differences in cuckoos reared by fairy-wrens and by its other host species nevertheless could have a genetic basis, with the offspring of certain females marked by their begging calls. Such races could be maintained by the same behaviors proposed for the maintenance of "gentes" or egg-color and pattern races (Southern 1954), such as imprinting on the foster parent or attraction to the habitat where the young cuckoos were reared. Habitat imprinting seems unlikely in the cuckoo *C. basalis*, insofar as both splendid fairy-wrens and western thornbills commonly build their nests in the same kind of vegetation: both host species nested in grasstrees (*Xanthorrhoea preissii*), bottlebush (*Calothamnus quadrifidus*, *C. sanguineus*), spiny hakea (*Hakea incrassata*, *H. petiolaris*), and prickly moses (*Acacia pulchella*), and all three species nested in marri (*Eucalyptus calophylla*) (names after Seabrook 1983). Fairy-wrens and thornbills both build a covered domed nest; the robins build an open cup. Some average differences in nest site were observed (thornbills often nest higher than fairy-wrens, and scarlet robins nest higher than both in the marri trees, as they do elsewhere; Robinson 1990), but all three fosters nested in the same patches of habitat, heath shrub on Gooseberry Hill.

The difference in begging calls in young *C. basalis* reared by fairy-wrens and by thornbills is due to the presence of two distinct races of the cuckoo, each specialized in its begging calls. Robins are less common as hosts (Brooker and Brooker 1989a) and were apparently parasitized by the thornbill race of *C. basalis*.

The source of variation of begging calls in *C. basalis* might be determined in future studies. Recordings of begging calls of cuckoos from the same host species in different areas are needed to test whether the distinctive calls are widespread behaviors. Cross-fostering eggs and young are needed to test whether the calls differentiate in response to the host environment or their development is determined by the specificity of a female cuckoo. Individual host specificity by a female cuckoo is suggested by the lack of parasitism in the 64 nests of fairy-wrens observed on Gooseberry Hill in 1983; some female *C. basalis* cuckoos successfully parasitized western thornbills in the same year. All successful parasitism of scarlet robins was observed in 1985, though robins were parasitized in other years as well (Brooker and Brooker 1992). Friedmann (1968:63) mentioned other variations in the incidence of parasitism and suggested they result from host specificity of individual female glossy cuckoos. In the common cuckoo, an individual female is known to be species-specific in her choice of a host species (Chance 1922; Wyllie 1981; Riddiford 1987; Dröscher 1990).

In addition to their advantage over the host species in size, loudness, and persistence in begging (Wyllie 1981; Crouther 1985; Davies and Brooke 1988), the young cuckoo may gain parental care with a begging call that is similar to the begging call of the host species. The variation in calls between cuckoo species suggests vocal mimicry. In Nigeria, the begging calls of nestling great spotted cuckoos (*Clamator glandarius*) were more like the begging calls of its host pied crow (*Corvus albus*) than like the begging calls of the congeneric African crested cuckoo (*C. levaillantii*) (Mundy 1973). In New Zealand, McLean and Waas (1987) found different begging calls in two cuckoo species, shining cuckoo and long-tailed cuckoo (*Eudynamys taitensis*), each with a call like its host species. In playback experiments the foster parents were attracted to the calls of their cuckoo chick (McLean and Rhodes 1991), as they were in our playbacks in Australia. McLean and Griffin (1991) argue that since begging calls do not vary with habitat among songbird species, the similar begging calls of cuckoos and their fosters suggest mimicry.

Other studies have reported differences in begging calls within a cuckoo species when reared by different host species, including another glossy cuckoo, the African diederik cuckoo (*Chrysococcyx caprius*) (Reed 1968). Morton and Farabaugh (1979) hand-reared a young American striped cuckoo (*Tapera naevia*) in Panama and found its call somewhat like the young of the host species, and different from the call of a young cuckoo in the nest of another host species as described by Haverschmidt (1961). Redondo and Arias-de-Reyna (1988) found differences in calls of *Clamator glandarius* reared by two species of corvids in Spain. Calls of the nestling corvids changed with age (Redondo and Exposito 1990); the development of calls in the young cuckoos did not track these changes. Mundy (1973) and Redondo and Arias-de-Reyna (1988) suggested that young cuckoos might learn the calls of their foster parents, which resemble the begging calls of the foster young. It is unknown whether the individual female cuckoos are species-specific in their host selection and have call races, or the calls are learned. Courtenay (1967) suggested that cuckoo begging calls evolved by natural selection to match the calls of host species, but did not consider the variation within a species of parasitic cuckoo. Host-specific mimicry of begging calls is unknown in the parasitic cowbirds (Gochfeld 1979; Woodward 1983; Broughton et al. 1987), except for the most specialized cowbird species (*Molothrus rufoaxillaris*; Fraga, this vol.). It is also unknown in nestlings of the closely related species of host-specific parasitic *Vidua* finches (Payne 1982). More observations and tape recordings are needed to describe the begging calls of brood parasites reared by different species of hosts.

ACKNOWLEDGMENTS We thank I. Rowley for the opportunity to work at Gooseberry Hill, the CSIRO Division of Wildlife and Rangelands Research for facilities, and M. Brooker, G. Chapman, and J. Leone for finding nests. The Western Australian Museum and the American Museum of Natural History loaned specimens. M. Brooker, H. Higuchi, R. Liversidge, E. S. Morton, H. Nakamura, and E. Røskarft commented on their observations of cuckoos. For translations of Russian and Japanese publications we thank L. R. Skrynnikova and T. Yuri. For comments on the manuscript we thank J.-C. Belles-Isles, H. Higuchi, J. M. Olson, S. I. Rothstein, and I. Rowley. M. van Bolt completed the illustrations. Fieldwork was supported by a research grant from the National Geographic Society and by NSF grant BNS8415753.

REFERENCES

Ali, S. and S. D. Ripley (1969). *Handbook of the birds of India and Pakistan. Vol. 3.* Oxford University Press, Bombay.

Becking, J. H. (1981). Notes on the breeding of Indian cuckoos. *J. Bombay Nat. Hist. Soc.*, **78**, 201–231.

Beruldsen, G. (1980). *A field guide to nests and eggs of Australian birds.* Rigby, Adelaide.

Brooker, M. and L. Brooker (1992). Evidence for individual female host specificity in two Australian Bronze-cuckoos (*Chrysococcyx* spp.). *Aust. J. Zool.*, **40**, 485–493.

Brooker, M. and L. Brooker (1986). Identification and development of the nestling cuckoos, *Chrysococcyx basalis* and *C. lucidus plagosus*, in Western Australia. *Aust. Wildlf. Res.*, **13**, 197–202.

Brooker, M. G. and L. C. Brooker (1989*a*). The comparative breeding behaviour of two sympatric cuckoos, Horsfield's Bronze-cuckoo *Chrysococcyx basalis* and the Shining Bronze-cuckoo *C. lucidus*, in Western Australia: a new model for the evolution of egg morphology and host specificity in avian brood parasites. *Ibis*, **131**, 528–547.

Brooker, M. G. and L. C. Brooker (1989*b*). Cuckoo hosts in Australia. *Aust. Zool. Rev.*, **2**, 1–67.

Broughton, K. E., A. L. A. Middleton, and E. D. Bailey (1987). Early vocalizations of the Brown-headed Cowbird and three host species. *Bird Behaviour*, **7**, 27–30.

Chance, E. (1922). *The Cuckoo's secret.* Sidgwick and Jackson, London.

Courtenay, J. (1967). The juvenile food-begging call of some fledgling cuckoos – vocal mimicry or vocal duplication by natural selection? *Emu*, **67**, 154–157.

Cramp, S. A. (ed.) (1985). *Handbook of the birds of Europe, the Middle East and North Africa, Vol. 4.* Oxford University Press, Oxford.

Crouther, M. M. (1985). Some breeding records of the Common Koel *Eudynamis scolopacea. Australian Bird Watcher*, **11**, 49–56.

Davies, N. B. and M. de L. Brooke (1988). Cuckoos versus Reed Warblers: adaptations and counter-adaptations. *Anim. Behav.* **36**, 262–284.

Dement'ev, G. P. and N. A. Gladkov (eds) (1966). *Birds of the Soviet Union, vol. 1.* (Translation of: *Ptitsy Sovetskogo Soyuza*, Moskow, 1951). Israel Program for Scientific Translations, Jerusalem.

Dröscher, L. (1990). A study on radio-tracking of the European Cuckoo (*Cuculus canorus canorus*). In: *Current topics in avian biology* (eds R. van den Elzen, K.-L. Schuchmann, and K. Schmidt-Koenig), pp. 187–193. Deutschen Ornithologen-Gesellschaft, Bonn.

Dunn, E. H. (1975). The timing of endothermy in the development of altricial birds. *Condor*, **77**, 288–293.

Friedmann, H. (1956). Further data on African parasitic cuckoos. *Proc. U.S. Natl Mus.*, **106**, 377–408.

Friedmann, H. (1964). Evolutionary trends in the avian genus *Clamator. Smithsonian Misc. Coll.*, **146**(4).

Friedmann, H. (1968). The evolutionary history of the avian genus *Chrysococcyx. U.S. Nat. Museum Bull.*, **265**.

Gill, B. J. (1982). Notes on the Shining Cuckoo (*Chrysococcyx lucidus*) in New Zealand. *Notornis*, **9**, 215–227.

Gill, B. J. (1983). Brood parasitism by the Shining Cuckoo *Chrysococcyx lucidus* at Kaikoura, New Zealand. *Ibis*, **125**, 40–55.

Gochfeld, M. (1979). Begging by nestling Shiny Cowbirds: adaptive or maladaptive. *Living Bird*, **17**, 41–50.

Goddard, M. T. and S. Marchant (1983). The parasitic habits of the Channel-billed Cuckoo *Scythrops novaehollandiae* in Australia. *Australian Birds*, **17**, 65–72.

Godfray, H. C. J. (1995). Signaling of need between parents and young: parent–offspring conflict and sibling rivalry. *Amer. Natur.*, **146**, 1–24.

Gosper, D. (1964). Observations on the breeding of the Koel. *Emu*, **64**, 39–41.

Haftorn, S. (1988). Incubating female passerines do not let the egg temperature fall below the "physiological zero temperature" during their absences from the nest. *Ornis Scand.*, **21**, 255–264.

Haverschmidt, F. (1961). Der Kuckuck *Tapera naevia* und seine Wirte in Surinam. *J. Ornith*, **102**, 353–359.

Higuchi, H. (1986). Adaptations for brood parasitism in cuckoos. In: *Breeding strategies in birds, vol. 2* (ed. S. Yamagishi), pp. 1–31. Tokaidaigaku Shuppan, Tokyo (in Japanese).

Hund, K. and R. Prinzinger (1980). Zur Jugendentwicklung der Körpertemperature und des Körpergewichtes beim Kuckuck *Cuculus canorus. Ökol. Vogel.*, **2**, 130–131.

Jenner, E. (1788). Observations on the natural history of the Cuckoo. *Phil. Trans. Roy. Soc. London*, **78**, 219–235.

Jensen, R. A. C. and C. F. Clinning (1974). Breeding biology of two cuckoos and their hosts in South West Africa. *Living Bird*, **13**, 5–50.

Jensen, R. A. C. and C. J. Vernon (1970). On the biology of the Didric Cuckoo in southern Africa. *Ostrich*, **41**, 237–246.

Konishi, M. and F. Nottebohm (1969). Experimental studies in the ontogeny of avian vocalizations. In: *Bird vocalizations* (ed. R. A. Hinde), pp. 29–48. Cambridge University Press, Cambridge.

Liversidge, R. (1955). Observations on a Piet-my-Vrou (*Cuculus solitarius*) and its host the Cape Robin (*Cossypha caffra*). *Ostrich*, **26**, 18–27.

Makatsch, W. (1955). *Der Brutparasitismus in der Vogelwelt.* Neumann Verlag, Radebeul and Berlin.

Maljchevski, A. C. (1987). *Cuckoo and its hosts.* Leningrad University, Leningrad (in Russian).

Marler, P. and S. Peters (1982). Developmental overproduction and selective attrition: new processes in the epigenesis of birdsong. *Dev. Psychobiol.*, **15**, 369–378.

Matsuda, T. and H. Uchida (1990). *Rearing strategies of cuckoos.* Color Nature Series 68. Akane-shobo, Tokyo (in Japanese).

McLean, I. G. and J. M. Griffin (1991). Structure, function, and mimicry in begging calls of passerines and cuckoos. *Acta XX Congr. Internat. Ornithol.*, **2**, 1273–1284.

McLean, I. G. and G. Rhodes (1991). Enemy recognition and response in birds. *Curr. Ornith.*, **8**, 173–211.

McLean, I. G. and J. R. Waas (1987). Do cuckoo chicks mimic the begging calls of their hosts? *Anim. Behav.*, **35**, 1896–1898.

Mellor, J. W. (1917). Notes on hatching of cuckoo and wren. *S. Australian Ornith.*, **3**, 18.

Molnár, B. (1947). The Cuckoo in the Hungarian plain. *Aquila*, **51–54**, 100–112.

Morton, E. S. and S. M. Farabaugh (1979). Infanticide and other adaptations of the nestling Striped Cuckoo *Tapera naevia. Ibis*, **121**, 212–213.

Mundy, P. J. (1973). Vocal mimicry of their hosts by nestlings of the Great Spotted Cuckoo and Striped Crested Cuckoo. *Ibis*, **115**, 602–604.

O'Connor, R. J. (1984). *The growth and development of birds.* Wiley-Interscience, John Wiley, New York.

Payne, R. B. (1977). The ecology of brood parasitism in birds. *Annu. Rev. Ecol. Syst.*, **8**, 1–28.

Payne, R. B. (1982). Species limits in the indigobirds (Ploceidae, *Vidua*) of West Africa: mouth mimicry, song mimicry, and description of new species. Misc. Publ. University of Michigan Museum of Zoology 162.

Payne, R. B. (1997a). Family Cuculidae (cuckoos). In: *Handbook of the birds of the world, vol. 7* (eds J. del Hoyo, A. Elliott and J. Sargatal), pp. 508–607. Birdlife/Lynx Edicions, Barcelona.

Payne, R. B. (1997b). Avian brood parasitism. In: *Host–parasite evolution, general principles and avian models* (eds D. H. Clayton and J. Moore), pp. 338–369. Oxford University Press, Oxford.

Payne, R. B., L. L. Payne, and I. Rowley (1985).

Splendid Wren *Malurus splendens* response to cuckoos: an experimental test of helping and social organization in a communal bird. *Behaviour*, **94**, 108–127.

Payne, R. B., L. L. Payne, and I. Rowley (1988*a*). Kin and social relationships in Splendid Fairy-wrens: recognition by song in a cooperative bird. *Anim. Behav.*, **36**, 1341–1351.

Payne, R. B., L. L. Payne, and I. Rowley (1988*b*). Kinship and nest defence in cooperative birds: Splendid Fairy-wrens *Malurus splendens. Anim. Behav.*, **36**, 939–941.

Promptov, A. N. and E. W. Loukina (1940). Sur les rapports biologiques réciproques du coucou (*Cuculus canorus* L.) avec certain espèces d'oiseaux qui l'élèvent ses petits. *Bull. Soc. Nat. Moscou Sect. Biol.*, **49**(5–6), 82–96 (in Russian, French résumé).

Redondo, T. (1993). Exploitation of host mechanisms for parental care by avian brood parasites. *Etologia*, **3**, 235–297.

Redondo, T. and L. Arias-de-Reyna (1988). Vocal mimicry of hosts by Great Spotted Cuckoo *Clamator glandarius:* further evidence. *Ibis*, **130**, 540–544.

Redondo, T. and F. Exposito (1990). Structural variations in the begging calls of nestling Magpies *Pica pica* and their role in the development of adult voice. *Ethology*, **84**, 307–318.

Reed, R. A. (1968). Studies of the Diederik Cuckoo *Chrysococcyx caprius* in the Transvaal. *Ibis*, **110**, 321–331.

Ricklefs, R. E. (1988). Adaptations to cold in bird chicks. In: *Physiology of cold adaptation in birds* (eds C. Bech and R. E. Reinertsen), pp. 329–338. NATO Advances Sciences Institute Series A, vol. 173, Plenum Press, New York.

Riddiford, N. (1987). Why do Cuckoos *Cuculus canorus* use so many species of hosts? *Bird Study*, **33**, 1–5.

Robinson, D. (1990). The nesting ecology of sympatic Scarlet Robin *Petroica multicolor* and Flame Robin *P. phoenicea* populations in open eucalypt forest. *Emu*, **90**, 40–52.

Rothstein, S. I. (1990). A model system for coevolution: avian brood parasitism. *Annu. Rev. Ecol. Syst.*, **21**, 481–508.

Rowan, M. K. (1983). *The doves, parrots, louries and cuckoos of southern Africa.* David Philip, Cape Town.

Rowley, I. (1981). The communal way of life in the Splendid Wren, *Malurus splendens. Z. Tierpsychol.*, **55**, 228–267.

Rowley, I. and E. Russell (1989). Splendid Fairy-wren. In: *Lifetime reproduction in birds* (ed. I. Newton), pp. 233–252. Academic Press, London.

Rowley, I., E. Russell, R. B. Payne, and L. L. Payne (1989). Plural breeding in the Splendid Fairy-wren, *Malurus splendens* (Aves: Maluridae), a cooperative breeder. *Ethology*, **83**, 229–247.

Russell, E. M. and I. Rowley (1988). Helper contributions to reproductive success in the Splendid Fairy-wren (*Malurus splendens*). *Behav. Ecol. Sociobiol.*, **22**, 131–140.

Seabrook, J. (1983). *List of the determined flora of the specified areas of the valley of the Helena River, slopes and associated creeks.* CSIRO Division of Wildlife Research, Helena Valley, Western Australia.

Skead, C. J. (1952). Cuckoo studies on a South African farm (Part II). *Ostrich*, **23**, 1–15.

Slater, P. (1975). *A field guide to Australian birds. Vol. 2, Passerines.* Scottish Academic Press Ltd, Edinburgh.

Smith, L. H. (1989). Feeding of young Pallid Cuckoo by four passerine species. *Australian Bird Watcher*, **13**, 99–100.

Southern, H. N. (1954). Mimicry in cuckoos' eggs. In: *Evolution as a process* (eds J. Huxley, A. C. Hardy, and E. B. Ford), pp. 257–270. Collier, New York.

Thielcke-Poltz, H. and G. Thielcke (1960). Akustisches Lernen verschieden alter schallisolierter Amsel (*Turdus merula* L.) und die Entwicklung erlernter Motiv ohne und mit künstlichem Einfluss von Testosteron. *Z. Tierpsychol.*, **17**, 211–244.

von Frisch, O. (1969). Die Entwicklung des Häherkuckucks (*Clamator glandarius*) im Nest der Wirtsvögel und seine Nachtzuch in Gefangenschaft. *Z. Tierpsychol.*, **26**, 641–650.

West, M. J. and A. P. King (1988). Female visual displays affect the development of male song in the cowbird. *Nature*, **334**, 244–246.

Whittow, G. C. and H. Tazawa (1991). The early development of thermoregulation in birds. *Physiol. Zool.*, **64**, 1371–1390.

Woodward, P. W. (1983). Behavioral ecology of fledgling Brown-headed Cowbirds and their hosts. *Condor*, **85**, 151–163.

Wyllie, I. (1981). *The cuckoo.* Batsford, London.

Yoshino, T. (1988). *Cuckoo* [*Kara Shizes Shirizu*]. Kaisei-Sha, Ichigaya, Tokyo (in Japanese).

III

COEVOLUTION BETWEEN COWBIRDS AND THEIR HOSTS

This part includes chapters on coevolutionary processes in the cowbirds, and reflects the extreme range (from 3 to 221) in the number of hosts that occurs within this group. One cowbird–host system (chapter 9) rivals or exceeds the complexity of coevolutionary processes in some of the cuckoo–host systems described in part I. Others, in contrast, involve the most generalized of all brood parasites in which coevolutionary interactions are poorly developed. For a comprehensive overview of cowbird–host coevolution, see pages 16–23 in chapter 1.

9

Interactions of the Parasitic Screaming and Shiny Cowbirds (*Molothrus rufoaxillaris* and *M. bonariensis*) with a Shared Host, the Bay-Winged Cowbird (*M. badius*)

ROSENDO M. FRAGA

This chapter deals with the triangular interactions of three avian species currently classified in the genus *Molothrus*, a system consisting of two brood parasites sharing a host. The host species is the bay-winged cowbird (*Molothrus badius*), quite different from the other cowbirds because it is non-parasitic, has extensive parental care, and lives in permanent groups that show cooperative breeding (Fraga 1972, 1991, 1992). The screaming cowbird (*Molothrus rufoaxillaris*) is one of the most specialized avian brood parasites and has *M. badius* as its only host over most of its range. Lastly, the shiny cowbird (*M. bonariensis*) is a generalist avian parasite with hundreds of hosts (Friedmann and Kiff 1985) that regularly parasitizes *M. badius*. A recent phylogeny based on mitochondrial DNA (Lanyon 1992) suggests that *M. badius* is not closely related to the parasitic cowbirds. The DNA data show that the two parasitic cowbirds *M. rufoaxillaris* and *M. bonariensis* are members of a monophyletic clade, with *bonariensis* being the most recent species.

This three-sided host-parasite system seems ideally suited for coevolutionary studies (Rothstein 1990). A higher degree of coevolution is expected between the *badius* host and its old and specialized parasite (*rufoaxillaris*); coevolution between the host and the recent generalist parasite (*bonariensis*) should be weak. The best-known adaptation to parasitism in *rufoaxillaris* is the remarkable resemblance of its chicks to those of the *badius* host (Hudson 1920; Lack 1968; Fraga 1979), which seems to be a case of mimicry (Lanyon 1992). Nevertheless, the selective value of this resemblance has not been investigated. *M. bonariensis* chicks do not resemble host chicks closely (Hudson 1920; Friedmann 1929; Fraga 1983, 1986) and their interaction with *M. badius* provides a natural experiment on the advantages of mimicry.

If parasites evolve towards diminished virulence during evolutionary time, then *rufoaxillaris* parasitism should be less severe than *bonariensis* parasitism. This hypothesis of coevolutionary amelioration has been criticized (Anderson and May 1982; Ewald 1983; Rothstein 1990) and could be tested with the three-sided system explored here. Alternatively, over evolutionary time host and parasites may be involved in escalated conflicts (the "arms race" hypothesis; Davies and Brooke 1989*a*,*b*) or may reach some sort of equilibrium. From a demographic viewpoint, populations of specialized parasites depend on host density, and the association could show oscillations or stability (May and Robinson 1985).

In this chapter I provide an updated outline of the natural history of the parasites of *M. badius*, with an emphasis on the lesser-known species (*M. rufoaxillaris*). I compare the reproductive success of the generalist and specialist, and the costs of both for host reproduction. The adaptations of both parasites are contrasted within the framework of coevolutionary models. I also inves-

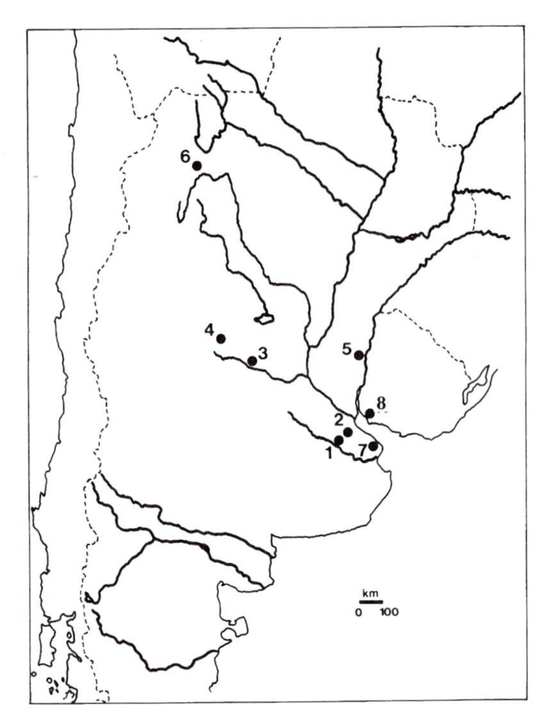

Figure 9.1 Location of study area and other sites in Argentina and Uruguay: 1: study area, 2 = La Armonia; 3 = Villa María; 4 = El Rosario; 5 = Parque Nacional El Palmar; 6 = Valle de Lerma (Hoy and Ottow 1964); 7 = Estancias El Destino and El Talar (Mason 1980, 1987); 8 = Piedra de los Indios, Uruguay.

tigate the possible selective factors responsible for host specialization in *M. rufoaxillaris.*

STUDY AREAS AND METHODS

My main study site (study area 1 in fig. 9.1) was Estancia La Candelaria, Buenos Aires Province, Argentina (35° 15′ S, 59° 13′ W), mostly between 1973 and 1979. Unless otherwise stated, all data in this chapter were obtained there. The study area was rectangular, measuring 1200 × 500 m (60 ha) and contained a large sector of woodland surrounded by pastures and smaller woodlots (Fraga 1985). Host nests and records of banded screaming cowbirds were located with the help of an aerial photograph scaled 1:4800. Distances exceeding 100 m were calculated to the nearest meter with this photograph. I resided at the study area for 120–130 days per year.

At the study area, about one-third of *badius* nests were located in nest boxes (Fraga 1988). Eggs found in nests of *badius* were marked with waterproof ink for identification, and scored for several variables. Numerical variables included laying date, maximum length and width, and mass; categorical ones described ground color and pattern of marking (Fraga 1983). Color photographs were taken of most parasitized clutches. The in-

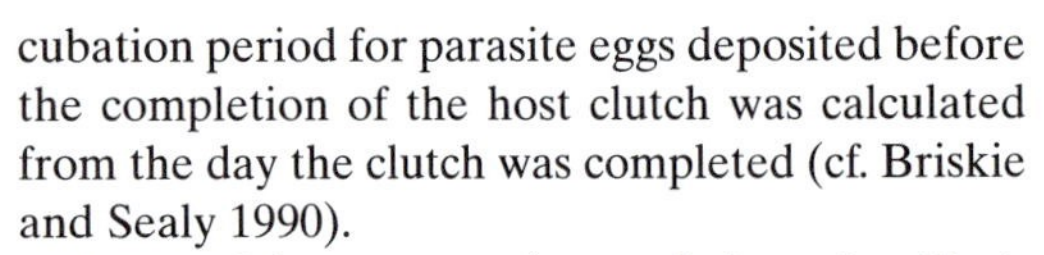

cubation period for parasite eggs deposited before the completion of the host clutch was calculated from the day the clutch was completed (cf. Briskie and Sealy 1990).

Eggs of *bonariensis* (generalist), *rufoaxillaris* (specialist), and their common *badius* host have been described elsewhere (Fraga 1979, 1983, 1988). Nestlings of the host and its two parasites were identified with diagnostic features (Fraga 1979) in the color of the skin and bill. Nestlings were banded with numbered aluminum bands and colored bands, and were weighed at regular intervals (1–2 days).

Behavioral interactions between host and parasites were recorded at 25 nests before egg hatching and 12 nests after hatching; observation periods lasted 45–75 min, between 7:30 and 13:00, and from 16:30 until sunset. The nests were monitored from a car, at distances not exceeding 25 m. Vocalizations were recorded with a Uher 4000 IC tape recorder plus an Electrovoice DL42 ultradirectional microphone.

Additional data on *M. rufoaxillaris* were obtained at other sites (fig. 9.1) in Argentina and Uruguay. In Buenos Aires Province host–parasite interactions were observed at Estancia La Armonia (study area 2: 34° 55′S, 59° 02′W) during 10 days (November 1983–January 1984). Sites in Córdoba Province were located in the lowlands near

Villa María (study area 3: 32° 25′S, 63° 15′W) and at Estancia El Rosario in the Sierra Grande (study area 4: 31° 16′S, 64° 28′W, altitude 700–750 m). In total, 59 days of observations were spent in Córdoba province, between November 1984 and January 1985. Descriptions of all the additional sites can be found in Fraga (1986, 1988). Lastly, some experiments were carried out with chick acceptance at Piedra de los Indios (site 5 in fig. 9.1) near Colonia, Uruguay.

Experiments on egg acceptance were conducted at all sites by introducing 31 eggs in nests of *badius*, at different stages of the breeding cycle. Introduced eggs were natural (n = 22) or artificial (n = 9). Natural eggs were obtained from deserted host nests. The artificial eggs were made of plaster of Paris and coated with acrylic paint, resembling the color and marks (if any) of real eggs. For evaluating the "grasp index" of Rohwer and Spaw (1988), eight study skins of *M. badius* at the bird collection of the Estación Biológica de Doñana, Sevilla were used.

To experiment with nestling acceptance, nestling shiny cowbirds were introduced into three nests found in study sites 1 and 2. A trial at the last site also showed that *badius* adults would feed nestlings placed just outside their nest. Therefore, I placed nestlings and fledglings of the host, as well as of shiny and screaming cowbirds, in small cages placed within 30–40 cm of host nests at sites 3 and 5. This allowed the food items provisioned to juveniles of the three species, and the number of host adults feeding or approaching them to be counted. Two juveniles were presented, in separate cages, at each experiment. The relative position of the young was changed every 50 min during the experiments.

Table 9.1 Major morphological, ecological, and behavioral differences between *Molothrus rufoaxillaris* and *M. bonariensis* in Central Argentina

	Species	
Character	*rufoaxillaris*	*bonariensis*
Sexual dimorphism		
Male color	jet black	glossy black
Female color	jet black	brown[a]
Size	0.81	0.81
Male song	simple	complex
Seasonal movements	no	yes
Follows grazing mammals	yes	yes
Scale of flock size	10s	10–1000 s
Mating	pairs	variable
Host species	1 (3)**	200+
Relative egg mass (%)	8.4	8.7
(Ns)	(117, 26)	(33, 17)
Chick plumage	host-like	female-like

Sexual size dimorphism is expressed as the ratio of female/male mean mass. Weight data for *rufoaxillaris* (n = 58) taken from Mason (1980); data for *bonariensis* (n = 31) are my own, from study area 1. Relative egg mass is the ratio (in %) of mean fresh egg mass/mean female mass. Data on egg masses taken from Fraga (1983, and unpubl.); female weights from sources mentioned above.

[a] A blackish female morph ("melanogyna") exists in some *bonariensis* populations (Hellmayr 1937).

**Two other hosts known from areas outside the present study area.

RESULTS

Natural history data

The host *M. badius* is a 40–50 g icterine blackbird, mostly pale brown with rufous wing feathers and blackish areas around the eyes; the juvenile plumage is like that of the adult. *M. badius* is sexually monomorphic in plumage and slightly dimorphic in mass (Fraga 1992). As most cooperative breeding birds, it is sedentary and lives in permanent groups (Fraga 1991). I have never seen *badius* following grazing mammals, and it is far more arboreal than the parasitic cowbirds. *M. badius* is sometimes regarded as a nest parasite (Payne 1977; Lanyon 1992), but it breeds in a variety of covered nest sites (Fraga 1989), and in study area 1 was basically a hole (cavity) nester.

The two parasitic cowbirds *M. rufoaxillaris* and *bonariensis* share a glossy black plumage in males, a similar size (50–60 g) and the habit of foraging with grazing mammals, and differ mostly in breeding ecology (table 9.1). *M. rufoaxillaris* was described as a species in 1866 by Cassin; previously it was confused with the better-known *M. bonariensis*. Hudson (1920) discovered that *M. rufoaxillaris* was parasitic on *M. badius*. and noticed the remarkable similarity between juveniles of this parasite and its host at the end of his ornithological studies in Argentina (in 1872). Evidence that *M. bonariensis* was also parasitic on *badius* was found in 1910–1920 in northwestern Argentina (Miller 1917; Friedmann 1929).

The size difference between the sexes of *M.*

rufoaxillaris, similar to that of *bonariensis* (table 9.1), is detectable in the field. Behavior and concomitant vocalizations also help to separate the sexes. The loudest vocalization of male *M. rufoaxillaris* (cf. Wetmore 1926; Friedmann 1929), produced during the song spread display (Orians and Christman 1968) is a burst of sound covering a wide frequency range, immediately followed by a whistle, the whole often audible at hundreds of meters. The sexual and agonistic contexts of this vocalization indicate that it is the species' "song." According to Hudson (1920) male *rufoaxillaris* have specially enlarged tracheas to amplify sound production, but I have not seen a confirmation of this statement.

Nonbreeding *rufoaxillaris* usually occur in mixed flocks with other icterines (cf. Belton 1985). I have seen mixed flocks of screaming cowbirds with four other species of icterines, particularly with their *badius* host. Both species commonly shared roosting sites in trees. *Badius* flocks without *rufoaxillaris* occurred mostly in wooded country. Counts in 30 mixed host–parasite flocks in open country (fall and winter, 1975 and 1976) may give an idea of the relative abundance of both species: I observed a total of 396 *badius* and 85 *rufoaxillaris* (ratio 3.7 : 1).

Screaming cowbirds in mixed flocks benefited from the alarm calls of *badius*. In 12 days of observations during the fall of 1977, I saw 34 attacks (all unsuccessful) by Aplomado falcons (*Falco femoralis*), an important avian predator, on mixed flocks feeding on stubble fields. *Badius* individuals detecting the incoming falcons, which mostly attack flying birds, gave hawk-alarm calls that immobilized the entire flock on the ground for several minutes. Screaming cowbirds did not produce alarm vocalizations.

All the literature on *rufoaxillaris* indicates the existence of consorting or pairing behavior (e.g., Hudson 1920; Friedmann 1929; Hoy and Ottow 1964; Willis and Oniki 1985; Mason 1987; Narosky and Yzurieta 1987). Mason (1987) provided a statistical test for consorting in trapped screaming cowbirds. I have seen screaming cowbirds mostly in pairs or flocks of pairs. Paired females usually respond to male songs with a screech, in a simple duet. Male–female duets occur in other icterines (e.g., *Amblycercus holosericeus*, Stiles and Skutch 1989; and even *Molothrus ater*, S. Rothstein, pers. comm.). In my single-banded *rufoaxillaris* pair the male–female association lasted for a full breeding season: both male and female were captured in the same mist net (11 November 1977), and were seen together in 13 subsequent observations until 13 March 1978. I never saw the male apart from the female. Duets involving the two individuals were frequently heard.

Host–parasite interactions

Host specificity of M. rufoaxillaris

The geographic ranges of *M. rufoaxillaris* and its host *M. badius* overlap extensively (fig. 9.2) and both are expanding their range in southeastern Brazil (Willis and Oniki 1984). Except for northeastern Brazil (subspecies *M. b. fringillarius*), both species are almost everywhere sympatric and syntopic. The generalist *M. bonariensis* overlaps with both species everywhere.

For a long time, *M. rufoaxillaris* was thought to have *badius* as its only host (Friedmann 1929, 1963). Hudson (1870) provided a record of parasitism on the brown-and-yellow marshbird (*Pseudoleistes virescens*, another icterine) based on juveniles, not on eggs. The marshbird has recently been confirmed as an effective (and rare) host by Mermoz and Reboreda (1996) who found *P. virescens* successfully rearing a *rufoaxillaris* chick (in a sample of 338 nests). Sick (1985) found evidence indicating that a third icterine, the Chopi blackbird (*Gnorimopsar chopi*) is a regular host of *M. rufoaxillaris* in Parana, southeast Brazil. In three *Gnorimopsar* nests found in 1973, Sick observed five *rufoaxillaris* chicks. Similarly, Fraga (1996) found that *Gnorimopsar* was parasitized in northeastern Argentina. *Gnorimopsar* resembles *badius* in breeding season and choice of nest sites (holes or the crowns of palm trees; De la Peña 1987 and pers. obs.), which would facilitate host shifts.

The geographic range of *rufoaxillaris* indicates that *badius* is its major host: *badius* is the only nesting icterine found everywhere within the range of *rufoaxillaris* (Fraga 1986), particularly in the semiarid western half of Argentina. Parasitism of *Pseudoleistes virescens* by *rufoaxillaris* in eastern Argentina is infrequent (Mermoz and Reboreda, in press). Sick (1985) suggested that brood parasitism of *Gnorimopsar* by *rufoaxillaris* is recent, possibly the result of range expansion by the cowbird.

I saw *rufoaxillaris* visiting nests of other birds, but most visits were directed to nest sites that could be used by *badius* (table 9.2). In my study site I found eggs or nestlings of *rufoaxillaris* only in nests of *badius*, although I examined about 300 nests of 30 other passerine species (including the species in table 9.2). All *rufoaxillaris* fledglings I

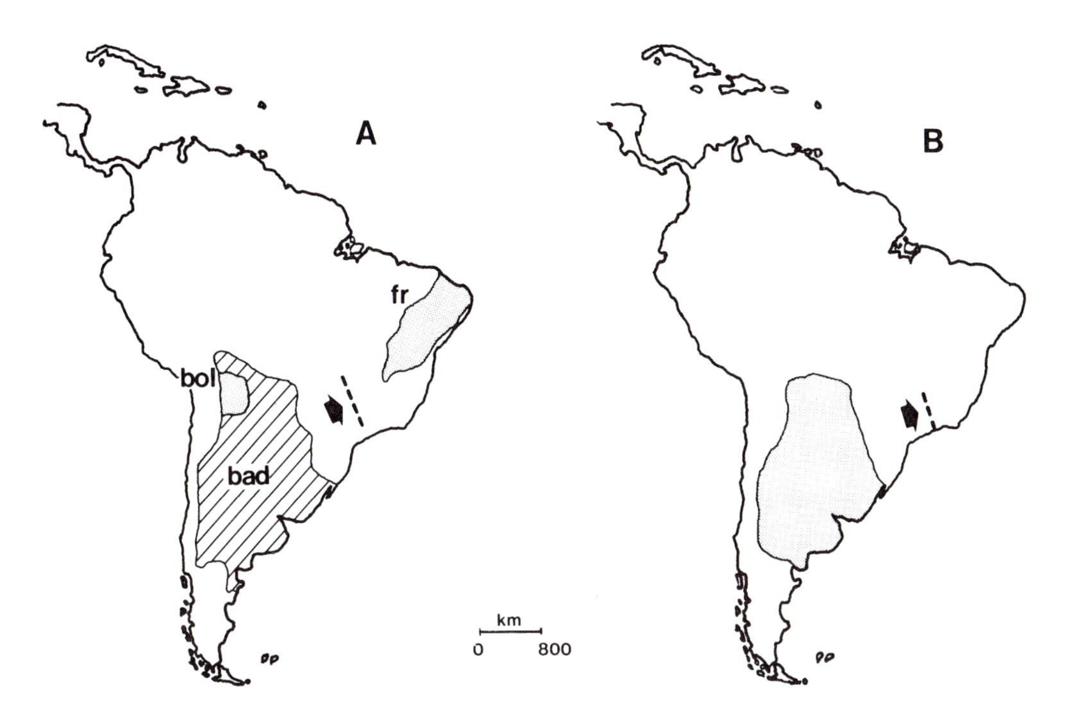

Figure 9.2 Geographic distribution of *M. badius* (A) and *M. rufoaxillaris* (B) in South America (sources in Fraga 1986). The subspecies of *M. badius* are: bad = *badius*, bol = *bolivianus*, and fr = *fringillarius*. The present distribution of *fringillarius* is quite patchy (Studer and Vieillard 1988). Arrows indicate range expansions in southeastern Brazil.

Table 9.2 Species of birds (other than *M. badius*) with records of nest visits by *M. rufoaxillaris*, at the four study areas in Argentina

Species			Nest used by *M. badius*?
Nonpasserines			
Picui dove	*Columbina picui*	(O)	No
Monk parakeet	*Myopsitta monacha*	(D)	Yes*
Guira cuckoo	*Guira guira*	(O)	No
Green flicker	*Colaptes melanochloros*	(H)	Yes*
Passerines			
Rufous hornero	*Furnarius rufus*	(D)	Yes*
Lenatero	*Annumbius annumbi*	(D)	Yes*
Cacholote	*Pseudoseisura lophotes*	(D)	Yes*
Great kiskadee	*Pitangus sulphuratus*	(D)	Yes*
Cattle tyrant	*Machetornis rixosus*	(D)	Yes
House wren	*Troglodytes aedon*	(H)	Yes
Saffron finch	*Sicalis flaveola*	(H, D)	Yes*

Nest types for these species categorized as H (hole), D (domed or covered) and O (open). Use of nest sites by *badius* is based on my own records. Species with more than five recorded visits by screaming cowbirds are marked with *.

have seen in Argentina (over 200) were escorted or reared by *badius*.

Breeding seasons

At study area 1, egg-laying in *badius* occurred from 27 October to 10 March (Fraga 1988). By extrapolation from fledging records, during the 1976–1977 season *badius* eggs were laid up to the end of March. Eggs of *rufoaxillaris* (n = 189 with known deposition date) occurred from 18 November to 15 March. Only two *badius* clutches were completed before 15 November. At study areas 3 and 4, the first *badius* clutches were deposited after the first rainstorms in November. In Northern Argentina the breeding season for *badius* and *rufoaxillaris* is the rainy season (November to April) (Hoy and Ottow 1964).

The breeding season of the generalist parasite *bonariensis* (all its hosts combined) at study area 1 was earlier, from 26 September to 7 February. From mid-February to mid-March, *bonariensis* was seldom seen (apparently the species is migratory, cf. Lucero 1982). At study area 1 about 20–35% of the *badius* clutches (depending on the year) were produced too late for *bonariensis* parasitism.

Copulation in *rufoaxillaris* was observed only from December to March. All but one of the eight copulations occurred in the morning, between 07:23 and 10:47, at distances of 15 to 25 m from a nest site of *badius* (not always an active nest). The exception occurred near a roost site, at 19:51. Males gave song spread displays just before the females assumed the solicitation posture, similar to that of other female icterines (Orians and Christman 1968).

Behavioral aspects of the host–parasite interaction

The start of nest-searching behavior in the parasite did not parallel its egg-laying chronology. Pairs or small groups of *rufoaxillaris* visited some traditional host nest sites in advance of the *badius* breeding season, in late fall and winter. My earliest record for an off-season visit is from 28 May 1978. Visits to empty nests, with abundant song spread displays and concomitant vocalizations, were perhaps related to pair formation in the parasite. In addition, domed nests or holes that could be used by *badius* were also visited (table 9.2) even when occupied by birds other than *badius*. The nonhost species usually attacked the visiting *rufoaxillaris* (over 30 records, mostly for *Furnarius rufus* and *Pitangus sulphuratus*).

During the breeding season, pairs or small groups of *rufoaxillaris* visited host nests with a mean frequency of 2.33 visits/hour (range 0 to 9 visits/hour, n = 63 watches at 21 nests). The rates apply to host nests watched from the nest-building stage to the first half of the incubation period. Visitation rates by *rufoaxillaris* were apparently higher during morning watches (3.14 visits/h before 13.00, and 2.2 afterwards), but the difference was nonsignificant and nests were visited up to sunset. The data indicate that most nests were visited 20 times (or more) per day by *rufoaxillaris*. Group size of *rufoaxillaris* varied between one and four pairs per visit (five pairs at site 3 in Córdoba). Groups larger than one pair were transitory, as pairs often departed independently in different directions. The behavioral data suggest nonexclusive use of host nests; up to five *rufoaxillaris* eggs were deposited in a nest the same day. Gregariousness during the breeding season also occurred at feeding and roosting sites.

The number of *badius* nests visited by each pair of *rufoaxillaris* every season is not well known. As a minimal estimate, a banded pair of parasites was seen in 12 different nests from November to March. Two host nests located 860 m apart were visited the same day, indicating considerable daily movements.

Most *rufoaxillaris* visits were noisy and conspicuous (cf. Mason 1987), particularly when more than one pair was involved. Both sexes of *rufoaxillaris* produced audible wing flaps just before landing and on departure. Parasites usually landed 10 m or more from the nest. Subsequently, males either gave song spreads from exposed branches, or tail-flicked while uttering series of monosyllabic notes. These behaviors also occurred in other sites or contexts, such as roosting sites. In most cases the parasites departed after 1 or 2 min, apparently because the male *badius* was guarding, or the *badius* female was seen in or near the nest. The longest visit I recorded took 9 min.

Male *badius* guarded the nest for 66% of the daytime, alternating in nest attendance with their mates (Fraga 1992). Nest-guarding by *badius* started well in advance of egg laying (Fraga 1992). Host guardians chased or attacked the visiting *rufoaxillaris* in only 34 of 141 visits (23%) during watches. Attacks occurred only when the *rufoaxillaris* approached the nest; otherwise the parasites apparently were ignored. Host attacks sometimes involved physical contact, including knocking down a parasite. More commonly I saw chases or supplantings. All but two of the attacks were directed at parasite females. Female *rufoaxillaris* were able to enter unguarded, empty nests in only two of the 141 visits (1.4%); both parasite females briefly sat in the (empty) nest cups. Two other attempts at entering nests were prevented at the last minute by an incoming host guardian. I did not observe male *rufoaxillaris* entering a *badius* nest.

At site 4 in Córdoba I experimented with the nest-guarding of a nesting *badius* group by placing mounted specimens of male and female *rufoaxillaris* at two shrubs, 6 and 10 m away from the nest. The male *rufoaxillaris* was mounted in a song spread position. In 10 trials the *badius* attacked the nearest specimen first, although I alternated the position of the male and female mounts. Attacks on *rufoaxillaris* mounts involved pecking at the head. In a different nest, the mounts were attacked when placed as far as 25 m, but this occurred only once during 2 h of exposure. Both nests contained eggs.

During the watches at site 1, I observed seven single *bonariensis* females perched at distances of less than 25 m of a *badius* nest. Once a female was attacked by the nesting *badius*.

Table 9.3 Mean linear dimensions (mm), volume (length × width^2/2), in cm^3 and shape (width/length) of incubated (accepted) and rejected (ejected plus deserted) eggs of *M. rufoaxillaris*, vs. those of its host *M. badius*

	Length	Width	Volume	Shape
Parasite eggs				
Accepted (n = 57)	23.1	17.9	3.72	0.773
Rejected (n = 104)	23.2	17.8	3.69	0.771
Host eggs				
All eggs (n = 226)	23.7	17.7	3.78	0.744

t-tests between accepted and rejected parasite eggs show no significant differences in dimensions, volume and shape between these outcomes (t values of 0.047, 0.593, 0.457, 0.403, d.f. = 159 for each variable, $P > 0.5$ in all cases).

Eggs of parasite and host

Eggs of *rufoaxillaris* and *badius* in Central Argentina are spotted and/or blotched, but otherwise quite variable in color and markings (Fraga 1983; table 9.3). Data on linear dimensions, shape, estimated volume, and relative mass of eggs can be fund in table 9.3 and Fraga (1983). Relative egg mass for the host is 9.2% (*badius* egg weights from Fraga 1983; 14 host female weights from my own data).

Most primary sources mentioned that eggshells of *rufoaxillaris* are heavier or thicker than those of *badius* (e.g., Hoy and Ottow 1964). Earlier sources may wrongly include eggs of *bonariensis* in the samples, but De la Peña (1987) was aware of this problem. His data on eggshell mass, controlled for egg size, show that *rufoaxillaris* has a mean eggshell mass 18% larger than that of *badius*. This difference in eggshell mass probably reflects eggshell thickness.

Frequency and intensity of parasitism

Eggs of *rufoaxillaris* were found in more than 80% of *badius* nests in all study areas (table 9.4). Hoy and Ottow (1964) reported a 92.9% incidence of *rufoaxillaris* parasitism in 14 *badius* nests, and

Table 9.4 Frequency and intensity of *M. rufoaxillaris* parasitism at the four study areas in Argentina. Figures for *M. rufoaxillaris* eggs per nest include only the nests parasitized by this cowbird

Locality	No. of nests	Percent parasitized	Parasitic eggs per nest: Mean	Parasitic eggs per nest: S.D.
Study area 1	79	87.3	3.54	2.42
Study area 2	6	100	2.83	1.47
Study area 3	6	100	5.83	3.76
Study area 4	6	83.3	2.80	1.04
Total	97	88.7	3.60	2.47

The four areas do not significantly differ in numbers of *M. rufoaxillaris* eggs per nest (Kruskal–Wallis = 2.79, d.f. = 3, P = 0.423).

Mason (1980) an incidence of 100% in 15 *badius* nests. The figures are high, but similar parasitism levels have been reported in a few *bonariensis* hosts (e.g., Fraga 1985). More remarkable was the high number of *rufoaxillaris* eggs per parasitized nest (table 9.4). Hoy and Ottow (1964) reported from 1 to 15 *rufoaxillaris* eggs per nest (mean = 4.92); Mason (1980) reported 1 to 19 eggs (mean 8.07 eggs). Figures for my four sites do not differ significantly from each other (table 9.4); at study site 1, nest boxes and natural sites were equally parasitized (Fraga 1988).

Multiple parasitism was the prevalent trend among screaming cowbirds: 81% of the parasitized nests at study area 1 received two or more parasitic eggs (fig. 9.3), with a mean of 3.09 eggs (n = all 79 nests). These figures possibly underestimate the intensity of parasitism, as some parasitic eggs could have been ejected and not found. The observed frequency of host nests with 0, 1, 2 ... n *rufoaxillaris* eggs did not fit a Poisson distribution (χ^2 goodness of fit, with five categories = 11.63, df = 4, P = 0.02). The variance/mean ratio (2.1) indicate a contagious distribution.

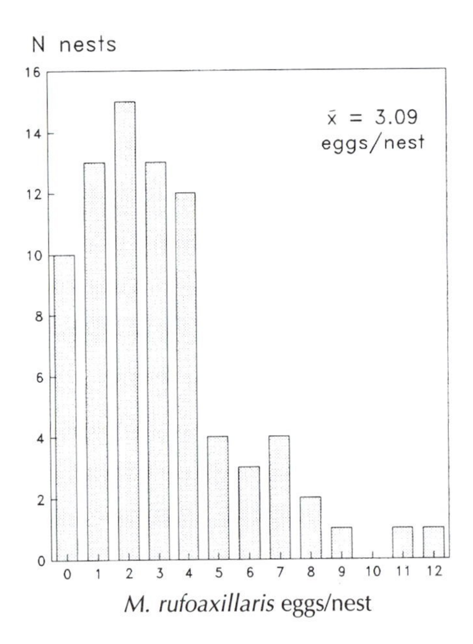

Figure 9.3 Numbers of screaming cowbird eggs deposited in all *M. badius* nests (n = 79). Numbers of nests with 0, 1, 2, ... n eggs significantly differ from those expected under a Poisson distribution (χ^2 goodness of fit, with five categories = 11.63, d.f. = 4, P = 0.02). The ratio variance/mean is 2.1, indicating a clumped (contagious) distribution of eggs.

I observed *M. bonariensis* parasitism on *badius* at study areas 1 (24.1% of 79 nests) and 3 (16.7% of six nests). At study area 1, the egg-laying season of *M. bonariensis* started earlier than that of *badius*, covering the first 100 days of this host's breeding season. One *bonariensis* egg was found per nest, except for one nest with two eggs at study area 1. Overall, only 20 of 550 eggs (3.6%) found in *badius* nests at study area 1 were of *bonariensis*; these data indicate that *badius* was a secondary host for this parasite.

Misplaced parasitic eggs, laying synchrony, and egg acceptance

Hoy and Ottow (1964) and Mason (1980) described cases of *rufoaxillaris* laying eggs in abandoned nests. At study area 1, I found six nests containing eggs of *rufoaxillaris* (n = 19) that perhaps never held host eggs. All these nests had been previously used by *badius*. In other cases, *rufoaxillaris* laid two eggs in the wrong nest cup, when *badius* placed their nest cup in the middle of the 90-cm entrance tunnel of a huge nest of sticks built by the furnariid *Anumbius annumbi*, rather than in the original brood chamber at the bottom. The parasitic eggs were laid in the chamber and not in the tunnel.

Hoy and Ottow (1964) described the following features of the *badius–rufoaxillaris* interaction: (i) screaming cowbirds often lay eggs before the *badius* laying period, and these eggs are ejected. The ejected eggs can be found intact below the nest; (ii) The host accepts and incubates parasitic eggs laid after the start of its own clutch, but all the eggs may be ejected from the nest if the parasitized clutch becomes too large. Nests with large clutches may be also deserted. Hoy also observed egg puncturing in the system (Hoy, in Friedmann 1963).

Ideally, brood parasites should deposit their eggs during the host's laying period. Parastic cowbirds often synchronize egg laying with that of their hosts by monitoring nest-building activities (e.g., Hann 1941; Norman and Robertson 1975; Fraga 1978). *M. badius* is a difficult host in this respect because it breeds in closed, dark spaces, and the nest itself is rarely visible from the outside (Fraga 1988). The amount of nest building by the host is variable (Fraga 1992), and in some types of sites (e.g., domed structures of grass built by some tyrant flycatchers) the female *badius* simply added a few items to the lining before starting the clutch.

A variable period of female inactivity (pre-laying period) precedes the deposition of the first host egg. In a sample of 21 nests with complete records, 6.2 days (range 3–12 days) elapsed on average between the apparent completion of the nest and the laying of the first host egg. The host guarded the nest throughout this period (Fraga 1991).

I obtained data on laying synchrony from 49 parasitized *badius* sets, finding that 28 of 189 (14.3%) *rufoaxillaris* eggs were premature (fig. 9.4). Mason (1980) estimated that 87% of *rufoaxillaris* eggs were deposited before the start of the host clutch, and were then ejected or deserted. He believed that laying asynchrony was the main cause of failure for the parasite. The difference between the two studies may reflect my finding that not all ejected eggs were premature, or even belonged to *M. rufoaxillaris.*

A total of 62 *rufoaxillaris* eggs in the sample (32.8%; fig 9.4) were deposited after the completion of the host clutch. Those deposited five days or more after the last host egg (17/189 or 9%) were too late for successful incubation. As an extreme, two *rufoaxillaris* eggs were laid in nests containing small nestlings.

Among 11 parasitic *bonariensis* eggs with known deposition dates, seven were laid with the host clutch, two eggs were premature (and ejected), and two eggs were late; the two parasites did not differ in synchronization ($\chi^2 = 1.022$, df = 2, $P = 0.60$).

Experiments to test the effect of laying synchrony on egg acceptance involved the introduction of eggs into *M. badius* nests at different stages. At study area 1, I introduced two natural *rufoaxillaris* eggs into two empty nests. One was ejected out of the nest cup, the other was partially buried with nest material. Egg burial was not observed again, but the case should probably be considered as rejection. Six artificial eggs designed to mimic *rufoaxillaris* eggs placed in six nest boxes before the start of laying (three at study area 2, three at study area 4) were ejected within 48 h, in one case within 9–10 h. Five eggs were found just outside the nest cup, but inside the nest box; the other egg was on the ground below the box.

On the other hand, eggs introduced during or after the host laying period were accepted or incubated, even those strikingly different from host eggs (such as unmarked white eggs, natural or artificial). I placed artificial eggs at study sites 2 ($n = 2$) and 3 ($n = 1$) and these were equally accepted (i.e., incubated). At study site 1, I introduced a total of 20 natural eggs into 19 nests: none was ejected, and 16 actually hatched. These

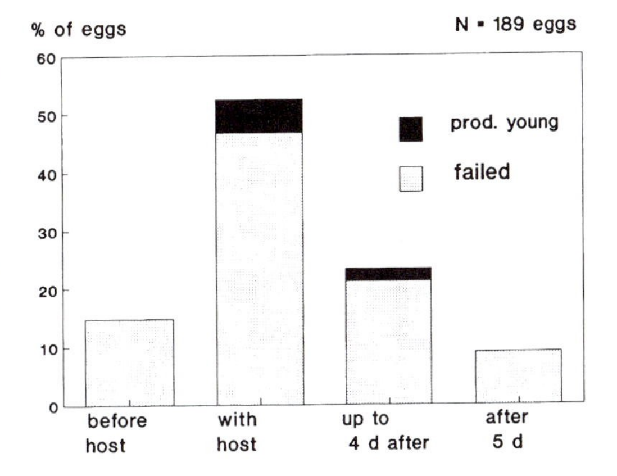

Figure 9.4 Percent of *M. rufoaxillaris* eggs deposited at different stages of the host nesting cycle (49 nests), with data on success (ratio fledglings/eggs) for different levels of synchronization. Eggs deposited before the host eggs, or five (or more) days afterwards produced no young.

results agree with those reported in the literature (Hoy and Ottow 1964; Mason 1980), showing that acceptance of parasitic eggs by *badius* depends primarily on synchrony.

Further support for this hypothesis is the lack of significant differences in measurements, shape, and volume of incubated and rejected eggs (table 9.3). Although color and maculation of eggs could not be scored as objectively, eggs of *rufoaxillaris* are predominantly pinkish, those of *badius* mostly bluish-gray (Fraga 1983). I found no evidence of differential ejection of pinkish (reddish to buff) *rufoaxillaris* eggs (52 of 143 accepted) vs. blue-gray *rufoaxillaris* eggs (6 of 24 accepted) ($\chi^2 = 1.17$; df = 1, $P > 0.25$).

Destruction of host eggs

I obtained several lines of evidence of host egg destruction in this host–parasite system. First, in some nests I found yolk-stained eggs or nest-lining in my first check, indicating previous egg damage (all these clutches also had less than three host eggs). Second, I observed 18 punctured *badius* eggs before they were removed from the nest, presumably by the host. Third, 44 more marked host eggs disappeared from parasitized nests between first and last checks. The missing eggs were presumably damaged and then removed. The difference in host clutch between nonparasitized and parasitized nests is significant (table 9.5: using the clutch sizes found in last checks for parasitized nests, Mann–Whitney z = 6.599, $P = 0.01$). I did not observe damaged eggs in nonparasitized nests.

Data in table 9.5 indicate that both parasites damaged and/or removed host eggs. Shiny cowbirds *M. bonariensis* are well known to puncture eggs (e.g., Fraga 1979, 1985), but this behavior has not been reported or quantified for *M. rufoaxillaris.* Final host clutches did not significantly differ between the three categories of parasitized nests in table 9.5 (Kruskal–Wallis = 0.375, $P = 0.829$), suggesting that both parasites inflicted similar levels of egg damage to their common host.

I had direct observations of egg damage by *rufoaxillaris.* Female *rufoaxillaris* entered three abandoned host nests which I had just examined. Immediate checks showed punctured eggs. Female *rufoaxillaris* were seen destroying eggs of saffron finches (*Sicalis flaveola*) and picui doves (*Columbina picui*). In the last case the eggs were eaten. Lastly, in 1979 I placed two nest boxes, each containing five abandoned eggs of three species of passerines (including four eggs of *rufoaxillaris*) within 15 m of active *badius* nests. Both experimental nests were visited within 30 min by *rufoaxillaris*, and the parasite females entered the boxes. Immediate checks showed one and two punctured eggs (not of *rufoaxillaris*) at each box. In *badius* nests, however, *rufoaxillaris* eggs (10.2% of 245) were punctured and/or removed between nest checks (implying intraspecific puncturing).

Table 9.5 Clutch size (mean ± S.D.) of *M. badius* in parasitized and nonparasitized nests at study area 1

	Host clutch (mean ± S.D.)	
Nest category	Initial	Final
Parasitized only by *M. rufoaxillaris* ($n = 56$)	2.85 ± 1.35	1.97 ± 1.47
Parasitized only by *M. bonariensis* ($n = 6$)	2.83 ± 0.98	2.16 ± 1.17
Parasitized by both ($n = 13$)	2.46 ± 1.48	1.92 ± 1.71
All parasitized nests ($n = 75$)	2.79 ± 1.48	1.97 ± 1.47
Nonparasitized nests ($n = 4$)	3.75 ± 0.50	3.75 ± 0.50

In 37 parasitized nests host clutch size decreased from first to last checks, so an initial and a final value are given. The difference between these two values reflects egg damage by the brood parasites.

Ejection and desertion of entire clutches

Clutch size also affects host behavior, regardless of synchrony. In my study, parasitized clutches accepted by *badius* contained a range of three to eight eggs (mean, SD = 5.0, 1.48, $n = 30$), and smaller ($n = 5$) or larger ($n = 4$) clutches were ejected or deserted by the host. According to Hoy and Ottow (1964), nesting *badius* receiving "too many" eggs of *rufoaxillaris* will attempt to eject them from the nest cup, or else desert the nest. Hoy and Ottow did not state what was the maximum number of eggs that *badius* would incubate. However, they mention a parasitized clutch of 12 eggs incubated "for a short time." and another of eight eggs incubated for a longer time.

Clutch ejection is the removal of all the eggs present in the nest cup, followed by the laying of a replacement *badius* clutch in the same site. I include here only 13 cases where the replacement clutch was of the same host female. A total of 64 parasitic eggs (at least 55 of *rufoaxillaris*) and seven host eggs were removed from the nest cup. Clutch ejection was also observed at study areas 2 and 3.

Egg removal was indiscriminate as to species. I experimented with clutch ejection by putting back two rejected host eggs in two empty nest cups; both eggs were out of the nest cup within 2 h. Hoy and Ottow (1964) also reported *badius* removing its own eggs.

In nest boxes ($n = 10$) and in a nest cavity the ejected eggs remained within the box outside the nest cup, but otherwise intact. In nests of furnariids the ejected eggs remained in the entrance tunnel (cf. Hoy and Ottow 1964), but eventually some fell down to the ground. It is quite plausible that some cases of egg ejection attributed by Mason (1980) to premature laying by the parasites were actually cases of clutch ejection.

Clutch ejection was apparently done by *badius* females. In the most striking case 12 eggs with a combined total mass of 47 g were rolled out of a nest cup in a box within 4–8 h. Only a banded female *badius* was within the nest box during three partial watches (total 1 h) in the first 4 h. The female was perched outside the nest cup, with its head inside it, but I could not see how it ejected the eggs.

Rothstein (1975, 1977) and Rohwer and Spaw (1988) have documented two techniques for ejecting the parasitic eggs of brown-headed cow-

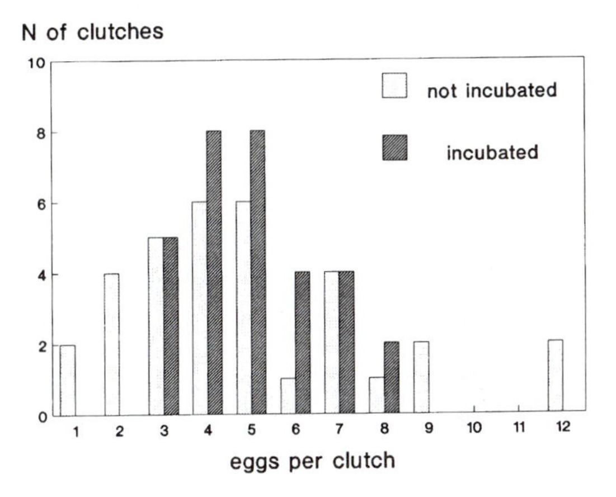

Figure 9.5 Numbers of eggs (total clutch size) remaining in nonincubated vs. incubated parasitized nests: means were 4.94 and 5.00 eggs, respectively. Clutch size does not differ significantly between categories (Mann–Whitney $U = 454$, $P = 0.430$).

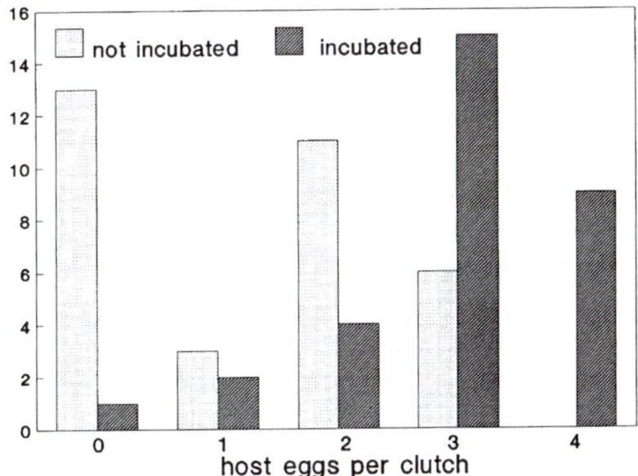

Figure 9.6 Numbers of host eggs remaining in non-incubated vs. incubated parasitized clutches: means were 1.30 and 2.94 eggs, respectively. Both categories of clutches differ in number of host eggs (Mann–Whitney $U = 155$, $P < 0.001$).

birds: egg grasping and egg puncturing. Puncture ejectors pierce the shell and carry the ejected egg spiked in the bill; grasp ejectors carry the egg in the open gape and may leave the egg intact. Natural eggs undoubtedly ejected by *M. badius* were not punctured, including eggs remaining in nest cavities or boxes and those that fell to the ground, but could be recovered intact. Similar results were reported by Hoy and Ottow (1964).

M. badius grasp index (sensu Rohwer and Spaw 1988, i.e., the tomial length multiplied by the comisural breadth of the bill) may not be high enough to reject *rufoaxillaris* eggs by grasping them. The grasp index for *badius* (study skins) is 141 mm^2. The minimum grasp index for ejecting *M. ater* eggs is around 200 mm^2 (Rohwer and Spaw 1988). Egg width is the critical parameter for estimating the minimum grasp index, and eggs of *rufoaxillaris* (8% wider than those of *ater*; Fraga 1983) would require a grasp index of 217 mm^2, substantially higher than the value I estimated for *badius*. I suspect that *badius* ejects by rolling eggs over the nest walls. Nest cups of *M. badius* were 3–5 cm deep.

Do egg numbers and composition of parasitized clutches influence host behavior? Host behavior in this section is treated as a dichotomy: acceptance equals incubation, and rejection equals nonincubation, i.e., it includes clutch ejection and desertion. Parasitized nest sites that were deserted ($n = 20$) contained a total of 30 host eggs and 56 parasite eggs (55 of *rufoaxillaris*).

Mean clutch sizes (host plus parasite eggs) did not differ significantly between outcomes, although the variation was smaller for incubated clutches (fig. 9.5). Accepted clutches had the higher number of host eggs, and ejected clutches the smaller (fig. 9.6; means of 2.93 and 0.61, respectively). Number of parasite eggs per clutch also showed significant differences between outcomes (fig. 9.7), and reversed the trend for host eggs: fewer parasite eggs (mean 2.06) were found in accepted clutches, and larger numbers (mean 4.92) in ejected clutches. These trends persist even when clutches smaller than three eggs or larger than eight eggs are excluded from the tests. As a result, the composition of accepted clutches was strongly biased towards the host. Only one clutch without host eggs was accepted and incubated.

Incubation period and egg hatching

The incubation period for *badius* was 13 days in most nests where three to four host eggs hatched. Hatching asynchrony for host eggs only once exceeded 48 h. The incubation period for single *rufoaxillaris* eggs was 12 days, but, under equal conditions, parasite eggs hatched up to 8 h before any host egg. Up to six eggs hatched in parasitized broods, and because of late laying by *rufoaxillaris*,

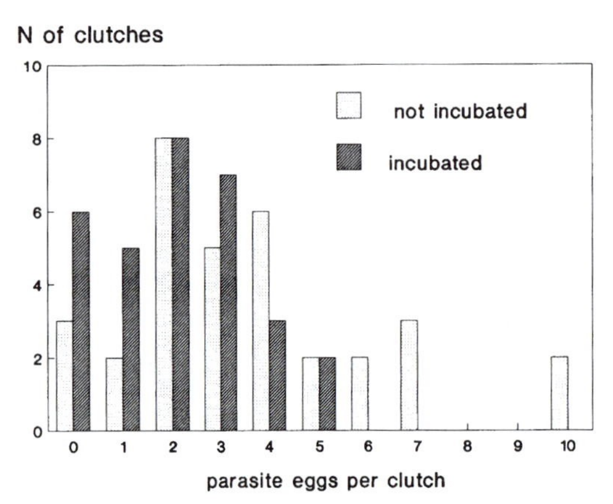

Figure 9.7 Numbers of *M. rufoaxillaris* eggs remaining in nonincubated vs. incubated parasitized clutches: means were 3.64 and 2.06 eggs, respectively. Both categories of clutches differ in number of *rufoaxillaris* eggs (Mann–Whitney $U = 701$, $P = 0.01$).

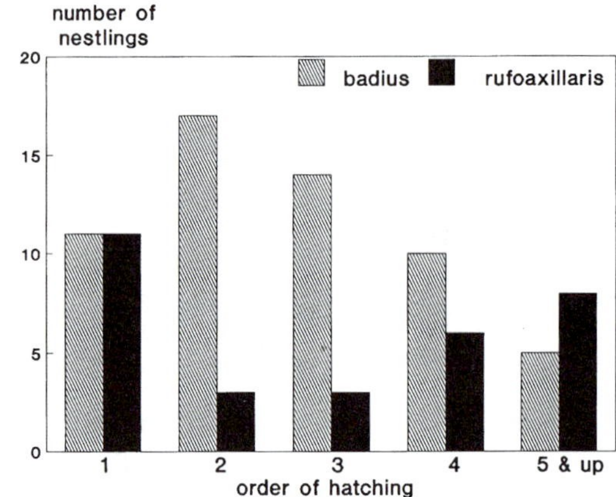

Figure 9.8 Relative order of hatching for *M. badius* and *M. rufoaxillaris* nestlings in mixed broods. Both species differ in hatching order ($G = 9.77$, d.f. = 4, $P < 0.05$).

hatching asynchronies of up to five days were observed in the largest broods.

The parasite's advantage of a shorter incubation period was often lost because of poorly synchronized laying. *M. rufoaxillaris* nestlings hatched with a bimodal pattern (fig. 9.8; i.e., usually first or last in mixed broods), which strongly affected nestling survival (see below). Of 59 *rufoaxillaris* eggs, 40.7% failed to hatch compared with only 9% of 86 *badius* eggs ($\chi^2 = 18.81$, df = 2, $P < 0.005$). The high incidence of hatching failures in the parasite partially results from late laying. Of 21 unhatched eggs with known laying date, 13 were laid after the host laying period. No *rufoaxillaris* egg hatched more than 72 h after the last host egg. Poor synchronization and late laying is a major cause of hatching failure in another specialized avian parasite, *Vidua chalybeata* (Morel 1973). Nine parasitized clutches (out of 28 that were incubated) produced only host nestlings as a result of hatching failures in *rufoaxillaris.*

Even considering properly synchronized eggs, significantly more parasite eggs (8 of 25) than host eggs (8 of 86) failed to hatch ($\chi^2 = 6.38$, df = 2, $P < 0.01$). Two failed *rufoaxillaris* eggs were runt, and five other eggs I opened were sterile or lacked yolk. I used t-tests to assess size differences between unhatched and hatched *rufoaxillaris* eggs laid with the host clutch: volume of unhatched eggs was significantly smaller (means of 3.9 cm^3 vs. 4.2 cm^3; t = 2.23, $P = 0.03$).

Interactions between host and parasite adults during the nestling period

Screaming cowbirds continued to visit host nests during the chick stage, but less often than before. I recorded a total of eight visits to eight nests during 1310 min of observation, during the morning, giving a rate of 0.35 visits/h. Nest-guarding *badius* attacked *rufoaxillaris* during visits (Fraga 1991). Even in the cases where no nest guardians were present, visits by parasites lasted a maximum of 2 min.

I never saw any parental behavior from the visiting *rufoaxillaris.* Although other parasitic cowbirds may prey on nestlings (Briskie and Sealy 1989), screaming cowbirds did not appear to harm host or parasite nestlings. The significance of visits at this stage is not clear, but, as reported before, a few eggs were laid during the first days of the nestling period.

Nestling growth and interactions

Recently hatched chicks of the three species (*badius*, *rufoaxillaris*, and *bonariensis*) differ mostly in skin color (pinkish in *rufoaxillaris*, yellowish-orange in the other two) and bill color (presence of a dark spot in the bill tip of *badius*). The resemblance even extends to the vocalizations (see figs. 9.9 and 9.10).

Growth data for *badius* and *rufoaxillaris* chicks (table 9.6) were fitted to the logistic model (Rick-

Table 9.6 Growth parameters for *Molothrus badius* and *M. rufoaxillaris* chicks, using data from nestlings hatched first or second in the brood

		Species	
Parameters		*M. rufoaxillaris* (n = 11)	*M. badius* (n = 33)
Asymptote A (g)	Mean	47.1	34.9
	Range	40.4–54.6	31.0–40.9
Growth constant K	Mean	0.503	0.485
	Range	0.434–0.635	0.399–0.582
t_{10-90} (days)		8.7	9.1
Asymptote as % of adult weight		87%	76%

Host and parasite differ in asymptotic value A (Mann–Whitney $U = 326$, $P < 0.001$), but not in growth constant K (Mann–Whitney $U = 241$, $P = 0.745$).

Data were fitted to the logistic model (Ricklefs 1986) with a nonlinear procedure. n is the number of chicks used for parameter estimation. t_{10-90} is the time (in days) to complete 90% of growth, in this model the ratio 4.4/K.

lefs 1983). The asymptotic mass value A is 24% higher in parasite than in host chicks. In the field, semi-independent *rufoaxillaris* fledglings often looked bigger than *badius* adults. Asymptotes for parasites are also closer to the mean adult weight, indicating that parasite nestlings leave the nest at a more advanced state of growth, and possibly that postfledging dependence should be shorter in the parasite.

Although helpers occurred at most host nests (Fraga 1991), I never observed more than five chicks fledged from a *badius* nest, although up to seven eggs hatched in parasitized nests. In terms of biomass, the largest fledged brood consisted of three host and two *rufoaxillaris* nestlings. Nests with five or more chicks suffered brood reduction. Brood reduction by starvation was inferred to occur if: (i) only nestlings hatched in positions 5 or later died; (ii) there were no indications of partial predation, disease or high ectoparasite loads; and (iii) nestlings that died or were missing showed a decline in weight. Most affected nestlings died within four days of hatching. In all, 11.1% of eight *badius* and 29% of 31 *rufoaxillaris* nestlings died from starvation, reflecting the late hatching of parasite chicks.

One additional host chick (aged eight days) was probably pushed out of the nest by two older *rufoaxillaris* nestlings. The host nestling had a normal weight, and was found on the ground below the nest (a ruined *Furnarius* nest with a large hole on the top). I placed it back in the nest, but by the next day it was in the same site and dead. The two *rufoaxillaris* filled completely the nest cup.

Postfledging development of rufoaxillaris *chicks*

The remarkable similarity in plumage and vocalizations between chicks of *badius* and *rufoaxillaris* (figs. 9.9 and 9.10) does not extend to behavioral traits related to the host's social system. Host chicks frequently solicit preening from adults or from other chicks with a head-up posture (Selander 1964) but I never saw *rufoaxillaris* chicks soliciting or performing allopreening. Nevertheless, adult *badius* preened *rufoaxillaris* fledglings in the first two weeks. Host young may engage in mutual allopreening, but I never saw *rufoaxillaris* chicks doing this. When threatened by *badius* adults, host chicks assume submissive postures (head withdrawal) apparently absent in young *rufoaxillaris.*

The host-like plumage of *rufoaxillaris* chicks subsists only during the postfledgling period of dependence. The distinctive adult black plumage is acquired in patches (Hudson 1920). A chick from study area 3 reared in captivity showed the first black feathers 38 days after hatching. Judging by weight, this individual was a female. The first black blotches appeared in the scapulars and in the back, and the tail feathers simultaneously became black. Similarly, eight free-living banded *rufoaxillaris* chicks at study site 1, which were seen weekly, started molting into the black adult plumage 33–44 days after hatching

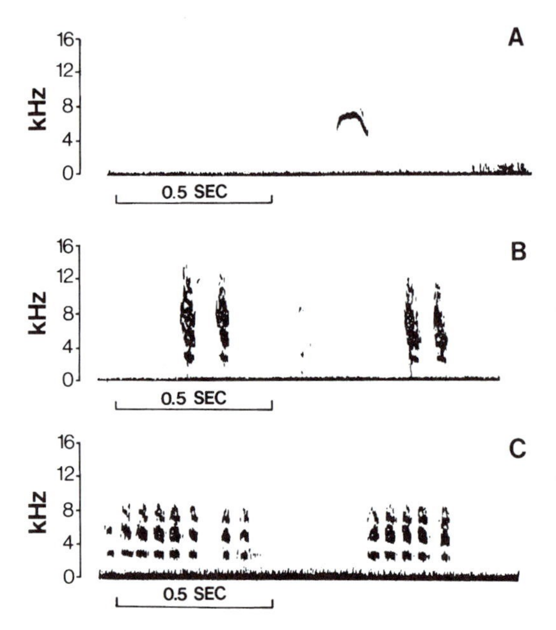

Figure 9.9 Vocalizations of *M. badius* juveniles. A: 1-day-old nestling; B: contact call, fledgling (age 14 days); C: begging call, fledgling (age 31 days).

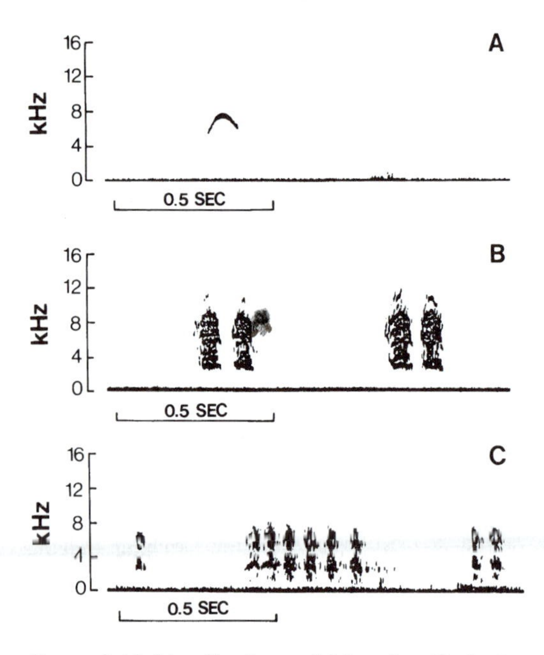

Figure 9.10 Vocalizations of *M. rufoaxillaris* juveniles. A: 1-day-old nestling; B: contact call, fledgling (age 15 days); C: begging call, fledgling (age 28 days).

(mean ± S.D. = 39.0 ± 3.4 days). The first blotches appeared in the flanks, scapulars and back. This molting stage coincided with the transition to nutritional independence. Three young *rufoaxillaris* starting the molt produced recognizable versions of calls of conspecific adults. As a result of dispersal and/or mortality, data for subsequent stages of the molt were more difficult to obtain. One individual was half black, with rufescent wing feathers and brownish belly, at 74 days of age. Another juvenile was completely black, adult-like, 133 days after hatching.

Chicks of *M. bonariensis* differ from those of *badius* and *rufoaxillaris* mostly in having uniform brownish color, in lacking rufous wing feathers, in having either yellow or white rictal flanges (invariably white in the other two), and in having whistled, tremulous begging vocalizations (fig. 9.11). In my experience, begging calls of *bonariensis* chicks reared by different hosts (10 species) throughout Argentina were basically similar.

Breeding success of M. rufoaxillaris *and* bonariensis

As the nesting cycle progresses, the proportion of parasitic *rufoaxillaris* propagules (eggs, nestlings, fledglings) decreases; fig. 9.12 summarizes these changes. At fledging time the ratio among young was 3.3 host per parasite (62/19), closely resembling the ratio found among adults in mixed flocks (3.7 : 1). The overall ratio of nestlings to eggs for *rufoaxillaris* was 31/257 (0.119) and the ratio of fledglings to eggs was 19/257 (0.074). Success was higher for eggs deposited during the host laying period (fledglings/eggs ratio of 0.111; see fig. 9.5). The ratio nestlings/eggs for *bonariensis* was 3/20 (0.15) and the ratio fledglings/eggs 1/20 (0.05).

Negative effect of brood parasitism

Data in table 9.5 show that in nests parasitized only by *rufoaxillaris*, the host clutch decreased on average from 3.75 to 1.97 eggs (47.5%); this damage can be attributed to *rufoaxillaris* parasitism. As 87% of *badius* nests were parasitized by *rufoaxillaris*, the average *badius* clutch was reduced by 41.3% (0.475 × 0.87) by this parasite. Similar calculations applied to nests parasitized only by *bonariensis* (table 9.5) show a 3.2% reduction in the average *badius* clutch. The two figures reflect the benefits of avoiding parasitism.

If parasitism nevertheless occurs, the benefits of ejecting all *rufoaxillaris* eggs could be calcu-

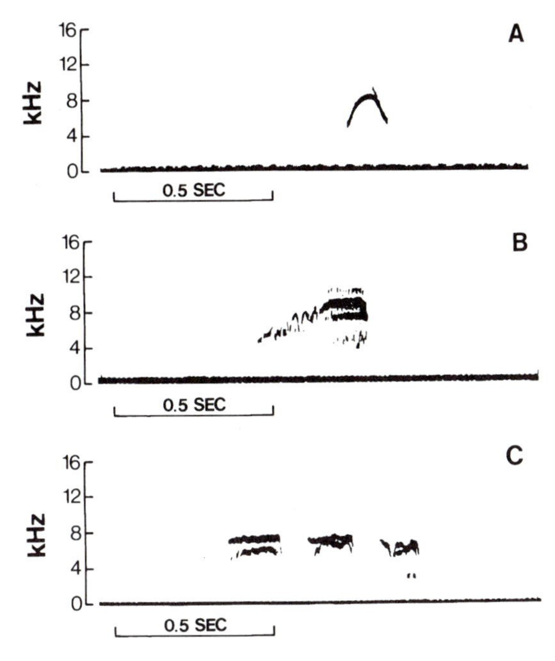

Figure 9.11 Vocalizations of *M. bonariensis* juveniles. A: 1-day-old nestling; B: begging call (12-day-old chick); C: begging call, fledgling (age about 30 days).

lated as follows. The main benefit would be the avoidance of *badius* nestling mortality that could be attributed to the presence of older and bigger parasite nestlings. Only eight of 67 host nestlings (0.119) died from starvation or crowding in nests parasitized solely by *M. rufoaxillaris*; this ratio multiplied by the frequency of parasitism ($0.119 \times 0.87 = 0.103$) provides a minimal estimate of the advantage of rejecting *rufoaxillaris* eggs. Other benefits to hypothetical egg-ejecting

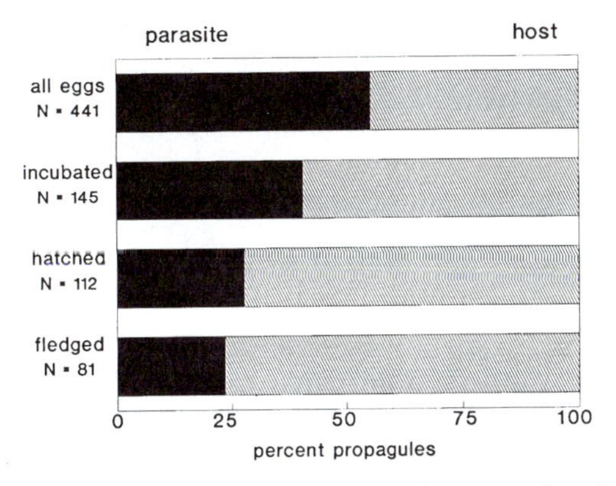

Figure 9.12 Proportion of *M. badius/M. rufoaxillaris* propagules (eggs, nestlings or fledglings) (in %) at different stages of the nesting cycle.

badius parents are more difficult to estimate. Parasite nestlings are bigger and probably require more food than host nestlings, but 35% of the provisioning to nestlings was carried out by helpers (Fraga 1991, 1992), which reduced the cost for nonejecting parents.

Chick mimicry

M. bonariensis chicks are unlike those of *badius* and can be used to test whether or not the host will rear nonmimetic parasite chicks. Three *bonariensis* nestlings hatched naturally in three *badius* nests at study site 1. One hatched late in a nest with five chicks and died in four days, after reaching only 50% of the mean weight of *bonariensis* nestlings. One other nestling grew poorly under unusual circumstances (Fraga 1986). The third *bonariensis* nestling hatched with two host nestlings and was normally fledged. I paid particular attention to this juvenile: during the first day it moved only 25 m from the nest, vs. 80 m for the host young. The *bonariensis* chick was fed only twice during 6 h of observation, never attracted more than a single *badius* adult, and was left alone for periods of up to 40 min. By contrast, recently fledged *badius* chicks may attract up to six adults (Fraga 1991). I could not find the *bonariensis* fledgling after 24 h, although the two host young survived. I have not seen any other fledgling *bonariensis* reared or provisioned by *badius* in nature.

These observations were complemented with experiments involving transplant of four *bonariensis* chicks into three *badius* nests (two nests at study area 1, one at study area 2). The *bonariensis* chicks (two aged one day, two aged six days) were obtained from nests of *Zonotrichia capensis* and *Mimus saturninus*. Three transplanted chicks were the oldest chicks in the experimental nests. All the transplanted nestlings grew normally (i.e., reaching weights like those reared by the natural hosts), and were fledged. Again none of the chicks survived more than 24 h out of the nest.

Further evidence that nonmimetic *bonariensis* chicks are provisioned by *badius* was obtained in experiments at site 5 by placing chicks outside a *badius* nest containing recently hatched nestlings. When I presented simultaneously a *bonariensis* and a *badius* chick (aged 7–8 days) the first received 23 items, the second 14 items; the ratio did not differ significantly from equality ($\chi^2 = 0.67$, df = 2, $P = < 0.1$). Begging activity during the experiment was greater in the *bonariensis* chick, and lower in the *badius* chick. Only two *badius* adults were provisioning the nest, and the parent that brooded the chicks inside (probably the female, Fraga 1991) never fed the outside chicks.

In the first experiment at site 3, the chicks located outside the nest were a *rufoaxillaris* and a *bonariensis* (aged 9 and 12 days), first presented separately, and later simultaneously, for 2 h. The results were different: the *rufoaxillaris* chick was fed eight times, the *bonariensis* never, even when presented alone. The chicks inside the nest (three *badius*, one *rufoaxillaris*, aged 11–13 days) were provisioned 13 times. The provisioners were a *badius* trio. The *bonariensis* chick was begging loudly throughout the experiment. In a second experiment at the same site, I presented the same chicks to a second group of four *badius* adults feeding three host nestlings, aged about 10 days. In 3 h the nestlings inside were provisioned five times, the *rufoaxillaris* outside two times, and the *bonariensis* was not fed. In a third experiment I presented a *badius* and a *bonariensis* chick (both aged 11 days) to a group of nine adults associated with young fledglings. The experimental nestlings were placed in cages 18 m apart and 12 m from a shrub where the fledglings were hidden. Both experimental nestlings vocalized during 40 min. During this time I counted the number of adults attracted to the experimental nestlings: only one adult briefly visited the *bonariensis* chick, whereas up to six adults surrounded the *badius* nestling, and attempted to lure the chick away, with repeated flights towards the shrub. In nature I have seen similar behavior directed to premature fledglings that cannot fly well. Results at study site 3 suggest that, at fledging time, *bonariensis* chicks may receive less care than mimetic or host chicks, but more data are needed.

DISCUSSION

Host specialization in *M. rufoaxillaris*

Why is *M. rufoaxillaris* more specialized than congeneric species? The percentage of *rufoaxillaris* eggs that produce fledglings is small (7.4%) but still higher than that of *bonariensis* with two common hosts (Fraga 1985, 1986). Specialization may be favored if *badius* provides *rufoaxillaris* with a higher mean reproductive success than other hosts. No natural parasitism of *rufoaxillaris* on other hosts occurred in my study area, and only experiments can test this hypothesis. I introduced 10 eggs and two nestlings of *rufoaxillaris* into nests of four other passerines (*Furnarius rufus*, *Mimus saturninus*, *Troglodytes aedon*, *Zonotrichia capensis*), none of which fledged. *Furnarius* ejected the eggs, but the other species accepted the eggs, and at least one nestling. Clearly, the sample is too small to test the hypothesis, and the presence of over 200 species of potential alternative hosts will make further tests difficult.

Specialized or narrow niche breadths could be the result of interspecific competition (Futuyma and Moreno 1988). If experiments show the existence of good alternative hosts for *rufoaxillaris*, there is a possibility that sympatric brood parasites (*M. bonariensis*, also the cuckoo *Tapera naevia*) are excluding *rufoaxillaris* from these hosts, perhaps through egg or chick damage (for a similar argument in cuckoos, see Brooker and Broker 1990). *M. bonariensis* parasitized 40% of the passerine species at study area 1 (Fraga 1985).

Coevolutionary adaptations

Table 9.7 outlines the most plausible traits that could qualify as defenses and counterdefenses in this system, including those suggested in the literature.

Nest guarding

Aggressive nest guarding may not be a coevolved defense, because it is also directed against predators. Nevertheless, screaming cowbirds were the most frequent heterospecific visitors to *badius* nests, and may have selected for increased nest vigilance and aggressiveness in the host. Nest defense against *rufoaxillaris* possibly reduces the intensity of parasitism and helps to disrupt the

Table 9.7 List of possible coevolved defenses and counterdefenses related to brood parasitism of *M. rufoaxillaris* on *M. badius* that are discussed in the text.

Host defense	Parasite counterdefense
(a) Nest guarding	Pairing, mate cooperation, and gregarious visits*
(b) Cryptic nesting behavior*	High visitation rate
(c) Ejection of asynchronous eggs	
(d) Discrimination of dissimilar young*	Mimicry of host young*

Those traits marked with * have also been suggested as coevolved adaptations in the literature (Lack 1968; Mason 1980, 1987).

laying synchrony of the parasites. The fact that nests are aggressively guarded well before egg laying (Fraga 1991) suggests an antiparasitic defense. A comparison of the time allocated to nest guarding by male *badius* in northeastern Brazil, in the absence of *rufoaxillaris*, could be revealing.

Consorting male *rufoaxillaris* could facilitate parasitism by females by attacking or distracting the guarding host individuals, a form of parental care. I have not seen male *rufoaxillaris* attacking nest-guarding *badius*, although they are 20% heavier than host males (ratio of mean weights). The hypothesis of male distraction has been advanced for *Clamator* cuckoos (Alvarez and Arias-de-Reyna 1974), and for giant cowbirds *Scaphidura oryzivora* (Robinson 1988). Males of these brood parasites often approach or display conspicuously around host nests, and are consequently chased by host guardians (pers. obs.). Mason (1980, 1987) extended the host-distraction hypothesis to male *rufoaxillaris*, but I observed that when females attempted to enter *badius* nests, males invariably remained at some distance. Male *rufoaxillaris* visiting host nests were noisy, but no more than at roosting or feeding sites. Nevertheless, because female *rufoaxillaris* were never solitary during the breeding season, the hypothesis of distraction is hard to disprove. Female *bonariensis* were solitary, but still parasitized *badius.* Even if it occurs, male distraction of the host by itself cannot be the only factor selecting for pairing or monogamy among *rufoaxillaris.* Male distraction could be shared between females, as it happened in giant cowbirds *Scaphidura oryzivora* in Panama (pers. obs.). In Panama, most giant cowbird groups visiting oropendola colonies consisted of one male and two to four females, and the females were often ignored while the male was chased by a few host individuals around the colony trees (see also Robinson 1988).

Gregariousness in *rufoaxillaris* may counteract the aggressive nest vigilance of *badius* (Mason 1987; Rothstein 1990). I saw a few episodes where male *badius*, in the presence of two visiting pairs of parasites, had difficulties in keeping the parasitic females out of the nest. Gregariousness in the host had no significant effect on the incidence and intensity of parasitism (Fraga 1986).

Cryptic breeding behavior

The potentially deleterious effects of *rufoaxillaris* parasitism were diluted by the poor timing of its egg deposition. My data show that 48% of the screaming cowbird eggs were deposited before or after the host clutch, with a concomitant decrease in hatching rates and nestling survival. In this section I consider all the host's behavioral traits that may affect the parasite's laying synchrony under the name "cryptic breeding behavior." This term includes all the host behaviors prior to the start of the clutch, i.e., from nest building to pre-laying period.

The use of covered nest sites by *badius* has been reported for the allopatric *fringillarius* population (Ihering 1914; Coelho 1979), and may not be a coevolved antiparasitic defense. Plausibly the host nest-building behavior is confusing to *rufoaxillaris* females, but it does not necessarily qualify as a defense. The long and variable pre-laying period of *badius* females may induce premature deposition of parasitic eggs (Mason 1980) but it may be related to food availability for clutch formation (Fraga 1991).

Overall, the data suggest that the cryptic breeding behavior of *badius* may have preceded parasitism by *rufoaxillaris*, which raises the question of how parasitism of this difficult host evolved. On the other hand, it is quite plausible that *rufoaxillaris* females require high visitation rates to host nests for a better timing of egg deposition.

Egg ejection

Ejection of parasitic eggs, even if limited to prematurely deposited eggs, qualifies as an evolved defense against brood parasitism (Rothstein 1990). Bird species recently exposed to brood parasitism usually accept eggs deposited before their own (e.g., Rothstein 1986; Davies and Brooke 1989*a*). Ejection of premature eggs is the defense with the lowest cost (no possible mistakes, no damage to the host's eggs) and with the highest benefit (no parasitic chick hatching first). Nest desertion at this stage has the cost of finding a new nest site, which is high in a cavity nester such as *badius.* A problem with ejection of premature eggs as a defense against heterospecific parasitism is that it also occurred in rare situations of conspecific parasitism (nest sharing by female *badius*; Fraga 1991). Nest sharing was rare at study area 1 (one or two cases in 79 nests), but it may be more frequent in areas where good nest sites are scarce.

Ejection of whole clutches by *badius* occurred only in the context of brood parasitism, and is probably a coevolved defense. Clutch ejection may have evolved after ejection of premature eggs, because the behavioral repertoire of movements involved in both contexts is the same. This defense has a cost, because host eggs were also ejected. But still clutch ejection is less costly than nest desertion, because it allows the reuse of good nest sites.

Absence of egg recognition

M. badius accepted and incubated eggs with color and maculation very unlike their own. With an incidence of 87% of parasitism, and the deleterious effects of parasitism, true egg recognition and ejection of all parasitic eggs were perhaps expected. What factors may have prevented the evolution of this defense? A list of hypotheses to explain the absence of true egg rejection in *badius* includes: (i) mutualism between host and parasite (i.e., a system such as that described by Smith 1968); (ii) constraints of a small bill size (Rothstein 1975; Rohwer and Spaw 1988); (iii) vindictive parasites (Zahavi 1979); and (iv) absence of true egg recognition because of nesting in cavities and other dark places, where eggs cannot be seen (Rothstein 1990). The first explanation can be rejected, because acceptance of parasitic eggs involves a selective cost to *badius*. The second hypothesis is plausible because *badius* may have too small a bill for true grasp-ejection, and puncture ejection is absent. Nevertheless, *badius* ejects eggs at some stages of the nesting cycle, though perhaps with considerable difficulty. Moksnes et al. (1991) have shown that one cuckoo host with a grasp index similar to that of *badius* was able to grasp-eject cuckoo eggs.

Vindictive parasitism (Zahavi 1979; Guilford and Read 1990) could occur if brood parasites revisit parasitized nests, and "punish" the hosts that ejected their eggs by destroying the remaining clutch. *M. rufoaxillaris* revist host nests (data from two banded pairs) and can destroy clutches. Nevertheless, this explanation requires strong philopatry and territoriality in the parasite, qualities that *rufoaxillaris* may not have. If this mechanism is operating, it is hard to understand why vindictive parasitism has not stopped the ejection of premature parasitic eggs by *M. badius*.

The fourth explanation is that egg recognition is difficult in the dark interior of most *badius* nests. Comparative data suggest that many hosts of brood parasites nesting in domed sites or cavities lack egg recognition and ejection (e.g., Gill 1983; Brooker and Brooker 1989; but see Mason and Rothstein 1986). *M. badius*, however, is not an obligate hole nester (Fraga 1988). A proportion of *badius* nest sites appear well illuminated, such as the domed structures of twigs or grass built by furnariids or tyrant flycatchers. Such structures were the commonest nest sites at study sites 3 and 4 (Fraga 1986, 1988). Furthermore, nesting *badius* often opened gaps in the walls of sticks of furnariid structures, allowing more light to enter.

I favor an integrated explanation that includes costs and benefits of egg ejection at different nesting stages (cf. Rothstein 1976; Davies and Brooke 1988). A major cost of egg ejection is the mistaken ejection of the host's own eggs. Such mistakes are quite plausible: on first sight I was sometimes in doubt about the identity of 5–10% of the *rufoaxillaris* eggs. Once I originally classified as *rufoaxillaris* a *badius* egg that hatched, and vice-versa (two errors in 137 hatched eggs). The chances of mistaken identifications are probably higher for the host birds than for myself (I used magnifying glasses in extreme cases). Here the poor light conditions of some nests would aggravate the problem. Small bill dimensions (grasp index) may create another cost of ejection: the accidental damage to the host's own eggs (see above). As calculated above, the advantage of ejecting *rufoaxillaris* eggs may not exceed 10–15%; perhaps the cost matched or exceeded this benefit during most of the nesting cycle. This model explains the pattern of egg ejection seen in *M. badius*. Ejection occurs when the costs (due to mistakes or accidental damages) are 0 (premature parasitic eggs), or as a last resort, to avoid nest site desertion. And ejection is always indiscriminate (or unselective, Moksnes et al. 1991).

Chick recognition

Mimicry between young of hosts and brood parasites has rarely evolved (Rothstein 1990); the present case, and that of viduines (Morel 1973) are the best known (but see also Redondo and Arias-de-Reyna 1988). Chick mimicry occurs mostly when host and parasite are reared together (Lotem 1992) as in the system reported here. My data suggest that close resemblance to *badius* chicks is not needed within the nest, but may have an adaptive value for fledged young. The shared signals (visual and vocal) between *rufoaxillaris* and *badius* may then constitute mimicry, particularly because parasite and host are not closely related (Lanyon 1992). Mimicry may have evolved because *badius* young are quite mobile, and leave the territory as soon as they fly efficiently (Fraga

1991, 1992). The rufescent wing color shared between *badius* and *rufoaxillaris* chicks is a conspicuous signal in flight (pers. obs.). The shared chick vocalizations cover a wide frequency range and are loud and repetitive (figs. 9.9 and 9.10), providing easy clues about the location of highly mobile young. The lack of these characters in *bonariensis* chicks may account for their poor postfledging survival with *badius* as host.

Coevolution in the system

From an ecological viewpoint, the host–parasite system outlined in this chapter seemed relatively stable: the populations of host and parasite were reasonably constant throughout my study. In the simplified model of one-host–one-parasite explored by May and Robinson (1985), stability depends on host longevity, and on the reduction of host productivity in parasitized nests. Annual survival for *badius* breeders was 76.2% (Fraga 1991) and host productivity in parasitized nests was reduced by less than 50%. Stability, or no effects of parasitism on host numbers, are thus possible. Host populations, as those of other communal breeding birds, were age-stratified (Fraga 1991) with a large number of nonbreeders. This means that the number of breeders is not directly dependent on the reproductive output and parasitism rate of the previous breeding season.

From an evolutionary viewpoint, are *badius* and *rufoaxillaris* involved in an escalated evolutionary conflict, as expected under the "arms race" model? The major coevolutionary pressures on the host appear to be avoidance of parasitism, with the parasite winning. The incidence of parasitism in the system was higher, but the effect of parasitism milder, than in *Cuculus* cuckoos. Selective pressures on *badius* for egg ejection seem moderate. The main obstacle for a more successful and virulent parasitism by *rufoaxillaris* is the poor synchronization of its egg deposition with the host's clutch formation. Perhaps an evolutionary equilibrium has been reached. The amount of damage inflicted to the host by *bonariensis* and *rufoaxillaris* is similar, and thus I found no evidence of "coevolutionary amelioration" in the older and more specialized parasite.

How did the specialized parasitism of *rufoaxillaris* evolve? Phylogenetic data (Lanyon 1992) suggest that *M. rufoaxillaris* is an older species than the generalist cowbirds *ater* and *bonariensis*, but this is no proof that specialization is the ancestral character state of parasitic cowbirds. The acquisition of a new icterine host by *M. rufoaxillaris* in Brazil shows that host specificity is not a fixed behavioral trait in this cowbird. I favor the hypothesis that host specificity in *rufoaxillaris* evolved from a more generalized parasitism. Specialization in the ancestral *rufoaxillaris* may have required the evolution of juvenile mimicry in the parasite, plausibly in geographic areas (like western Argentina) where *badius* was a numerically important host.

ACKNOWLEDGMENTS My research in Córdoba province was financed by NSF grant DEB 8214999 to S. I. Rothstein. P. Mason provided the artificial eggs used in experiments. S. and L. Salvador greatly helped me at Villa María. J. Cabot allowed me the use of the bird collection in his care. T. Redondo, S. I. Rothstein, and S. K. Robinson commented on earlier drafts of this chapter. The Centre d'Estudis Avançats (CSIC, Blanes), J. Vilá, and R. Banchs gave logistical support for printing the manuscript.

REFERENCES

Alvarez, F. and L. Arias-de-Reyna (1974). Mecanismos de parasitización por *Clamator glandarius* y defensa por *Pica pica*. *Doñana Acta Vert.*, **1**, 43–65.

Anderson, R. M. and R. M. May (1982). Coevolution of hosts and parasites. *Parasitology*, **85**, 411–426.

Ankney, C. D. and D. M. Scott (1982). On the mating system of Brown-headed Cowbirds. *Wilson Bull.*, **94**, 260–268.

Belton, W. (1985). Birds of Rio Grande do Sul, Brazil. Part 2. Formicariidae through Corvidae. *Bull. Amer. Mus. Nat. Hist.*, **180**, 1–242.

Briskie, J. V. and S. G. Sealy (1989). Changes in nest defense against a brood parasite over the breeding cycle. *Ethology*, **82**, 61–67.

Briskie, J. V. and S. G. Sealy (1990). Evolution of short incubation periods in the parasitic cowbirds, *Molothrus* sp. *Auk*, **107**, 789–794

Brooker, I. C. and M. G. Brooker (1990). Why are cuckoos host specific? *Oikos*, **57**, 301–309.

Brooker, M. G. and I. C. Brooker (1989). The comparative behaviour of two sympatric cuckoos, Horsfield's Bronze-Cuckoo *Chrysococcyx basalis* and the Shining Bronze-Cuckoo *C. lucidus*, in Western Australia, a new model for the evolution of egg morphology and host specificity in avian brood parasites. *Ibis*, **131**, 528–547.

Coelho, A. G. M. (1979). As aves da estaçao ecológica de Tapacurá, Pernambuco. *Notulae Biologicae (Recife, Brazil)*, **2**, 1–18.

Davies, N. B. and M. L. De Brooke (1988). Cuckoo versus reed warblers: adaptations and counteradaptations. *Anim. Behav.*, **36**, 262–284.

Davies, N. B. and M. L. De Brooke (1989*a*). An experimental study of co-evolution between the cuckoo, *Cuculus canorus*, and its hosts. I. Host egg discrimination. *J. Anim. Ecol.*, **58**, 207–224.

Davies, N. B. and M. L. De Brooke (1989*b*). An experimental study of co-evolution between the cuckoo, *Cuculus canorus*, and its hosts. II. Host egg markings, chick discrimination and general discussion. *J. Anim. Ecol.*, **58**, 225–236.

De la Peña, M. R. (1987). *Nidos y huevos de aves Argentinas.* Imprenta Lux, Santa Fe, Argentina.

Ewald, P. W. (1983). Host–parasite relations, vectors, and the evolution of disease severity. *Annu. Rev. Ecol. Syst.*, **14**, 465–485.

Fraga, R. M. (1972). Cooperative breeding and a case of successive polyandry in the Bay-winged Cowbird. *Auk*, **89**, 447–449.

Fraga, R. M. (1978). The Rufous-collared Sparrow as a host of the Shiny Cowbird. *Wilson Bull.*, **90**, 271–284.

Fraga, R. M. (1979). Differences between nestlings and fledglings of Screaming and Bay-winged Cowbirds. *Wilson Bull.*, **90**, 151–154.

Fraga, R. M. (1983). The eggs of the parasitic Screaming Cowbird (*Molothrus rufoaxillaris*) and its host, the Bay-winged Cowbird (*M. badius*), is there evidence for mimicry? *J. Ornith.*, **124**, 187–194.

Fraga, R. M. (1985). Host–parasite interactions between Chalk-browed Mockingbirds and Shiny Cowbirds. Neotropical Ornithology, Ornithological Monographs No. 36, pp. 829–844.

Fraga, R. M. (1986). The Bay-winged Cowbird (*Molothrus badius*) and its brood parasites, interactions, coevolution and comparative efficiency. Ph.D. Thesis. University of California, Santa Barbara.

Fraga, R. M. (1988). Nest sites and breeding success of Bay-winged Cowbirds (*Molothrus badius*). *J. Ornith.*, **129**, 175–183.

Fraga, R. M. (1991). The social system of a communal breeder, the bay-winged cowbird *Molothrus badius. Ethology*, **89**, 195–210.

Fraga, R. M. (1992). Biparental care in bay-winged cowbirds *Molothrus badius*. *Ardea*, **80**, 389–393.

Fraga, R. M. (1996). Further evidence of parasitism of chopi blackbirds (*Gnorimopsar chopi*) by the specialized screaming cowbird (*Molothrus rufoaxillaris*). *Condor*, **98**, 866–867.

Friedmann, H. (1929). *The cowbirds.* C. C. Thomas, Springfield, Illinois, USA.

Friedmann, H. (1963). Host relations of the parasitic cowbirds. *U.S. Natl. Mus. Bull.*, **223**, 1–276.

Friedmann, H. and L. Kiff (1985). The parasitic cowbirds and their hosts. *Proc. West. Found. Vertebr. Zool.*, **2**, 1–302.

Futuyma, D. J. and G. Moreno (1988). The evolution of ecological specialization. *Annu. Rev. Ecol. Syst.*, **19**, 207–233.

Gill, B. J. (1983). Brood-parasitism by the Shining Cuckoo *Chrysococcyx lucidus* at Kaikoura, New Zealand. *Ibis*, **125**, 40–55.

Guilford, T. and A. F. Read (1990). Zahavian cuckoos and the evolution of nestling discrimination by hosts. *Anim. Behav.*, **39**, 600–601.

Hann, H. W. (1941). The cowbird at the nest. *Wilson Bull.*, **53**, 211–221.

Hellmayr, C. E. (1937). Catalogue of the birds of the Americas. *Field Mus. Nat. Hist., Zool Ser.*, **13**, Part 10.

Hoy, G. and H. Ottow (1964). Biological studies on the molothrini cowbirds (Icteridae) of Argentina. *Auk*, **81**, 186–203.

Hudson, W. H. (1870). *Letters on the ornithology of Buenos Ayres.* Re-edited by D. E. Dewar. Cornell University Press.

Hudson, W. H. (1920) *Birds of La Plata.* J. M. Dent, London.

Ihering, H. Von (1914). Biologia e clasificaçao das cuculidas brazileiras. *Rev. Mus. Paul.*, **9**, 371–390.

Lack, D. (1971). Ecological adaptions for breeding in birds. Methuen, London.

Lanyon, S. M. (1992). Interspecific brood parasitism in blackbirds (Icterinae): a phylogenetic perspective. *Science*, **255**, 77–79.

Lotem, A. (1992). Learning to recognize nestlings is maladaptive for cuckoo *Cuculus canorus* hosts. *Nature*, **362**, 743–745.

Lucero, M. M. (1982). *El anillado de aves en la República Argentina.* Miscelánea No. 74, Fundación M. Lillo, Tucumán, Argentina.

Mason, P. (1980). Ecological and evolutionary aspects of host selection in cowbirds. Ph.D. Thesis. University of Texas at Austin.

Mason, P. (1987). Pair formation in cowbirds, evidence found for screaming but not for shiny cowbirds. *Condor*, **89**, 349–356.

Mason, P. and S. I. Rothstein (1986). Coevolution and avian brood parasitism: cowbird eggs show evolutionary response to host discrimination. *Evolution*, **40**, 1207–1214.

May, R. M. and S. I. Robinson (1985). Population dynamics of avian brood parasitism. *Amer. Natur.*, **126**, 475–494.

Mermoz, M. and J. C. Reboreda (1996). New effective host for the Screaming Cowbird, *Molothrus rufoaxillaris. Condor*, **98**, 630–632.

Miller, L. (1917). Field notes on *Molothrus. Bull. Amer. Mus. Nat. Hist.*, **1917**, 579–592.

Moksnes, A., E. Roskaft, and A. T. Braa (1991). Rejection behavior by common cuckoo hosts towards artificial brood parasite eggs. *Auk*, **108**, 348–354.

Morel, M. Y. (1973). Contribution a l'etude dinamique de de la population de *Lagonosticta senegala* L. (Estrildides) a Richard-Toll (Senegal). Interrelations avec le parasite *Hypochoera chalybeata* (Muller) (Viduines). *Mem. Museum Nat. Hist. Nat. Paris Ser. A*, **37**, 3–156.

Narosky, T. and D. Yzurieta (1987). *Guia para la identificación de las aves de Argentina y Uruguay*. Vázquez Mazzini Editores, Buenos Aires.

Norman, R. F. and R. J. Robertson (1975). Nest-searching behavior in the brown-headed cowbird. *Auk*, **92**, 610–611.

Orians, G. H. and G. M. Christman (1968). A comparative study of the behavior of Red-winged, Tricolored and Yellow-headed Blackbirds. *Univ. Calif. Publ. Zool.*, **84**, 1–81.

Payne, R. B. (1977). The ecology of brood parasitism in birds. *Annu. Rev. Ecol. Syst.*, **8**, 1–28.

Redondo, T. and L. Arias-de-Reyna (1988). Vocal mimicry of hosts by Great Spotted Cuckoo *Clamator glandarius*, further evidence. *Ibis*, **130**, 540–544.

Ricklefs, R. E. (1983). Avian postnatal development. In: *Avian biology, vol. 7* (eds D. S. Farner, J. R. King, and K. C. Parkes), pp. 1–83. Academic Press, New York.

Robinson, S. K. (1988). Foraging ecology and host relationships of Giant Cowbirds in Southeastern Peru. *Wilson Bull.*, **100**, 224–235.

Rohwer, S. and C. D. Spaw (1988). Evolutionary lag versus bill size constraints, a comparative study of the acceptance of cowbird eggs by old hosts. *Evol. Ecol.*, **2**, 27–36.

Rothstein, S. I. (1975). An experimental and teleonomic investigation of avian brood parasitism. *Condor*, **77**, 250–271.

Rothstein, S. I. (1976). Experiments on defenses cedar waxwings use against cowbird parasites. *Auk*, **93**, 675–691.

Rothstein, S. I. (1977). Cowbird parasitism and egg recognition in the northern oriole. *Wilson Bull.*, **89**, 21–32.

Rothstein, S. I. (1986). A test of optimality: egg recognition in the eastern phoebe. *Anim. Behav.*, **34**, 1109–1119.

Rothstein, S. I. (1990). A model system for coevolution, avian brood parasitism. *Annu. Rev. Ecol. Syst.*, **21**, 481–508.

Selander, R. K. (1964). Behavior of captive South American cowbirds. *Auk*, **81**, 394–402.

Sick, H. (1985). *Ornitologia brasileira, uma introducao.* Universidade de Brasilia, Brasilia, Brazil.

Smith, N. G. (1968). The advantage of being parasitized. *Nature*, **219**, 690–694.

Stiles, F. G. and A. F. Skutch (1989). *A guide to the birds of Costa Rica.* Comstock, Ithaca, New York.

Studer, A. and J. Veilliard (1988). Premières données étho-ecologiques sur L'Icteridé brésilien *Curaeus Forbesi* (Sclater 1986) (Aves, Passeriformes). *Revue Suisse Zool.*, **95**, 1063–1077.

Wetmore, A. (1926). Notes on the birds of Argentina, Paraguay, Uruguay and Chile. *Bull. U.S. Natur. Mus.*, **133**, 383–388.

Willis, E. and Y. Oniki (1985). Bird specimens new for the state of Sao Paulo, Brazil. *Rev. Bras. Biol.*, **45**, 105–108.

Zahavi, A. (1979). Parasitism and nest predation in parasitic cuckoos. *Amer. Natur.*, **113**, 157–159.

10

Nest Defense by Potential Hosts of the Brown-Headed Cowbird

Methodological Approaches, Benefits of Defense, and Coevolution

SPENCER G. SEALY, DIANE L. NEUDORF, KEITH A. HOBSON, SHARON A. GILL

Brood parasites have evolved many adaptations that enhance the likelihood that their parasitic life style will be successful, but success for a parasite usually means reduced reproductive output for the hosts (Rothstein 1975, 1990; Payne 1977; May and Robinson 1985; Davies and Brooke 1988; Sealy 1992). Not surprisingly, many hosts have evolved ways of reducing the impact of parasitism on them. Some hosts eject parasitic eggs or desert parasitized clutches, but these behaviors may not recover all of the losses due to parasitism as hosts may already have lost an egg to the laying parasite. Furthermore, these responses may incur costs as hosts may damage or mistakenly eject some of their own eggs, or take additional time, and energy may be expended re-lining nests or building new ones before laying new clutches (Rothstein 1975, 1976; Clark and Robertson 1981; Rohwer et al. 1989; Røskaft et al. 1990; Sealy 1992, 1995; Hill and Sealy 1994). These anti-parasite strategies are initiated after hosts have been parasitized. Especially in species that accept parasitism, the best protection hosts might have against brood parasitism is to avoid being parasitized in the first place. Hosts are thought to achieve this by constructing inconspicuous nests, behaving furtively while building nests and when near them or, perhaps most importantly, through nest defense.

Nest defense is an important aspect of parental care in birds. Although brood parasitism may threaten a bird's reproductive effort, it involves little risk to a host's immediate survival. To maximize benefits, parents should gauge their defense according to the value of the nest contents (e.g., number, age, or quality of offspring; see Regelmann and Curio 1983) and the type of enemy involved (e.g., brood parasite or predator; Neudorf and Sealy 1992). Recognition of different enemies permits parents to behave optimally when defending their nests (Patterson et al. 1980).

In the arms race (Dawkins and Krebs 1979) between avian brood parasites and their hosts, a host feature can properly be called a defense if it both reduces the negative impact of parasitism on the host and has evolved in response to, or is currently maintained by, selection pressures arising from brood parasitism. Similarly, adaptations by the parasite should be called counteradaptations only if they evolved in response to host defenses (Rothstein 1990). Many anti-predator adaptations have coevolved with predator hunting strategies (e.g., Gilbert and Raven 1975), but few examples exist of hosts defending their nests against brood parasites directly in response to the threat the parasites pose, rather than defense being just a generalized response to any intruder near the nest. We suggest that nest defense behaviors by some hosts of avian brood parasites are adaptations that reduce the negative impact of parasitism on the hosts.

In many host–parasite systems, encounters between the parasite and hosts are seldom observed by researchers. This is especially true with parasitic brown-headed cowbirds (*Molothrus ater*). Workers interested in the dynamics of parasite–host interactions have simulated encounters between cowbirds and potential hosts using models of the parasite placed near host nests and quantifying host responses to them. Robertson and Norman (1975, 1976) were the first to study quantitatively the responses of hosts to models of the female brown-headed cowbird and to controls. Since the publication of their papers, several researchers have used models to simulate interactions between cowbirds and other parasitic species and hosts at host nests.

In this chapter, we synthesize the results of studies of host responses to the threat of brood parasitism, concentrating mainly on studies of brown-headed cowbirds because interactions between cowbirds and their hosts have received the most attention. Drawing upon our work on this subject conducted in a host community at Delta Marsh, Manitoba, we review progress made in studies of behavioral defenses against the threat of brood parasitism by assessing methodologic approaches, addressing analytical problems, and examining the defensive strategies of hosts that accept or reject cowbird eggs, and those that are parasitized by cowbirds at different frequencies. We also assess the effectiveness of defense in deterring brood parasitism and whether nest defense can properly be called an anti-parasite adaptation. We conclude by examining features of the laying behavior of cowbirds and other brood parasites that suggest behavior evolved in response to host nest defense and, hence, is coevolutionary.

USE OF MODELS TO ELICIT BEHAVIORAL RESPONSES

Presentation

Although model testing (table 10.1) probably cannot fully reveal the true nature of responses by hosts to live cowbirds, an advantage of this approach is that standardized models allow equivalent sets of stimulus testing, which facilitates making comparisons (Robertson and Norman 1976, 1977; but see below). Differences in the way models have been presented, however, have prevented detailed comparisons among studies (Hobson and Sealy 1989). For this reason, we urge researchers to agree on an approach to model presentations. Most workers investigating host responses to brood parasites have used freeze-dried models or specimens mounted taxidermically in life-like poses (Edwards et al. 1949, 1950; Robertson and Norman 1976, 1977; Smith et al. 1984; Briskie and Sealy 1989; Hobson and Sealy 1989; Cruz et al. 1990; Moksnes et al. 1990; Ortega and Cruz 1991; Neudorf and Sealy 1992; Bazin and Sealy 1993; Gill and Sealy 1996), although others have used museum study skins (Folkers 1982; Folkers and Lowther 1985). Live birds may elicit more realistic responses from hosts (e.g., Knight and Temple 1986*a*; Burgham and Picman 1989), but cages or tethers used to restrain them may

Table 10.1 Recommended protocol for experiments that test host responses to brood parasites at the nest

Experimental variable	Recommendation
Number of different models presented at each nest	Three.
Parasitic female model	
Control model	Species similar in size and shape to the brood parasite. This may or may not be a novel stimulus but should pose no threat to host or nest.
Predator model	Egg and nestling predator that does not prey on adult birds.
Distance of models from nest	0.5 m
Time between model presentations	5 to 20 min depending on the species. Successive days are appropriate if only two models are presented.
Length of model presentation	5 min
Testing at multiple stages	Naive pairs should be used at each stage to avoid carry-over aggression or habituation (see Knight and Temple 1986*a*,*b*).
Quantifying host responses	Quantify each behavioral response separately (e.g., display, close pass) rather than use an overall subjective index.

affect their behavior. Models seem to be the most feasible and best way to facilitate duplication of experimental conditions among trials.

In their analyses of studies using models of predators to test nest-defense behavior, Knight and Temple (1986*a*,*b*) suggested that inconsistencies among studies may be attributed, in part, to the fact that some researchers were themselves visible before they placed models near nests, or even flushed birds from nests before testing them (e.g., Roëll and Bossema 1982). Sometimes this is unavoidable, particularly when hosts are incubating. In a study of the responses of whiteheads (*Mohoua albicilla*) to a model long-tailed cuckoo (*Eudynamys taitensis*), McLean (1987) ensured that hosts were not disturbed during tests by waiting until birds left their nests before presenting models. In our studies of host responses to predator and cowbird models (e.g., Hobson et al. 1988; Hobson and Sealy 1989; Neudorf and Sealy 1992), we occasionally flushed incubating females but found that the response of the returning birds often changed dramatically once they discovered the model near their nests. Duckworth (1991), on the other hand, suggested that model presentations that involve flushing birds mimic natural situations where birds are flushed by a large herbivore. Nevertheless, researchers should be cautious when flushing birds from their nests. Responses of birds to an observer during trials can be reduced by using a blind or by observing from a considerable distance from the nest (e.g., Ortega and Cruz 1991). Blinds and ladders used to place models should be set up near nests well in advance of tests.

Several workers have presented models at nests using poles (e.g., Robertson and Norman 1976, 1977; Folkers 1982; Smith et al. 1984; Hobson et al. 1988). Poles, which are particularly useful for testing high nests, may be painted to match the vegetation and placed near the nest prior to tests to allow hosts to habituate to these objects (Neudorf and Sealy 1992). However, in most studies, models were clipped directly to the available vegetation (Hobson and Sealy 1989; Bazin and Sealy 1993; see also Mark and Stutchbury 1994; Gill and Sealy 1996), which may be easier and preferable for low nests. For studies to be comparable, the posture of models must be consistent. Knight and Temple (1986*b*) found that the direct stare of a predator inhibited aggressive responses by birds. As live birds adopt different postures, their use in studies of nest defense may affect results from one trial to another.

Mark and Stutchbury (1994) presented female cowbird models simultaneously with playback of the "chatter" call. The playback may attract the host's attention toward its nest sooner, although female cowbirds usually do not vocalize when they parasitize nests (Neudorf and Sealy 1994). Gill et al. (1997*a*) showed that yellow warblers respond to female cowbird vocalizations as they would a cowbird model near their nest. Playbacks of only cowbird vocalizations near nests, therefore, are useful for studying recognition cues of hosts.

Most workers who have tested host responses to cowbird models have presented consecutively more than one model per nest. Robertson and Norman (1976) determined experimentally that habituation did not alter responses to subsequent models if models were presented with at least 5 minutes between trials, to a maximum of three consecutive trials. The tendency for habituation of behavior between models, however, may be species-specific and workers should perform their own trials before establishing a model-testing protocol. In our studies, Hobson and Sealy (1989) waited 20 minutes between model presentations, whereas Neudorf and Sealy (1992) and Bazin and Sealy (1993) used 5 minutes. Other workers have presented parasite and control models on successive days (McLean 1987; Mark and Stutchbury 1994). In cases where the second model is presented too soon after the first, a potential problem is carry-over aggression, which may result in the bird giving responses elicited by the first model to the second. Gill (1995) found that carry-over aggression of yellow warblers from models of cowbirds to other models occurred occasionally. Hobson and Sealy (1989) described the "seet" vocalization by yellow warblers, elicited directly in response to models of female cowbirds. In Gill's (1995) study, however, yellow warblers were more likely to give seet calls to other models if the cowbird model was presented first, although usually individuals switched responses within seconds when presented with the next model. Thus, carry-over aggression can be important in nest-defense studies. Clearly, a model-testing protocol must strike a balance between reducing habituation and carry-over aggression between models and expediency. Randomizing the order in which models are presented should reduce the influence of carry-over aggression across treatments, particularly in situations where little time is available between presentations. One way around multiple presentations is to present control and cowbird models simultaneously, equidistant from the nest, thus providing birds with a choice (Mark and Stutchbury 1994). In this method, however, it is difficult to distinguish between host responses to cowbirds and to controls, and such situations are unlikely to occur in nature.

Knight and Temple (1986*a*,*b*) argued that it may be inappropriate to test nests with models more than once over the nesting cycle. They suggested that repeated visits to, or tests at the same nest, might gradually modify the bird's behavior by positive reinforcement. Studies concerned with behavioral responses of hosts toward cowbirds typically test birds during the laying or early incubation stages when hosts are most susceptible to being parasitized. Workers studying changes in host response over the season should attempt to test only naïve pairs at all stages (e.g., Hobson and Sealy 1989).

Controls and experimental design

To distinguish host responses to cowbirds from generalized responses to nest intruders, control models of putatively innocuous species have been used, in addition to a third model representing a nest predator (see also below). Several workers have reported that host species react more intensely toward cowbird models than controls and suggested that hosts recognize the cowbird as a threat (e.g., Briskie and Sealy 1989; Burgham and Picman 1989; Ortega and Cruz 1991; Mark and Stutchbury 1994). Various species of sparrows and finches have been used as controls (e.g., Smith et al. 1984; Folkers and Lowther 1985; Burgham and Picman 1989; Hobson and Sealy 1989). Some of these species were considerably smaller than the cowbird model (e.g., Smith et al. 1984). In our studies, we used a fox sparrow (*Passerella iliaca*) model as a control because it and the female cowbird are similar in size, and the fox sparrow occurs in our study site only as a migrant (Hobson and Sealy 1989), thus removing confounding influences due to differences in prior experience with the species (see also Bazin and Sealy 1993). Mark and Stutchbury (1994) used a veery (*Catharus fuscescens*) as a control because it is common on their study area and did not threaten the host species under investigation, the hooded warbler (*Wilsonia citrina*). The veery, too, is about the same size and shape as a female cowbird. Although the choice of an appropriate control species in these experiments has not been consistent among the studies, this apparently does not matter. Robertson and Norman (1977) found no differences in responses of hosts to three species of sparrow used as controls. We found no significant differences in responses by yellow warblers to three species (red-winged blackbird, fox sparrow, song sparrow) used as controls (Gill et al., unpublished data).

A potentially insidious problem in studies using models, and one that may be linked to high variance in individual response, is the general absence of a "multiple stimuli" approach to hypothesis testing (Kroodsma 1989). Using a single cowbird model represents only a single stimulus to test a distribution of possible responses. Ideally, several different cowbird, predator, and control models should be used to safeguard against pseudoreplication in sampling design (Hurlbert 1984).

A phenomenon common to studies of host responses to cowbird models, and nest-defense behavior in general, is a high degree of variation in the responses among individuals (e.g., Roëll and Bossema 1982; Regelmann and Curio 1983; McLean et al. 1986; Hobson and Sealy 1989). This variation may reflect intrinsic differences among individuals or the unavoidable failure to control all variables that affect nest defense (see Montgomerie and Weatherhead 1988). Variation among individuals in response to a particular threat could make it more difficult for the parasite or predator to predict the response, thus benefiting nest owners. We do not suggest, however, that this variation has been selected for by the pressures of parasitism and/or predation. This variation underlines the need for careful experimental design, such as a randomized block design that involves large sample sizes.

Quantifying response

Robertson and Norman (1976) used an aggressive score index to rate and compare host responses to models. Their method is attractive because it renders a complex pattern of behavioral responses into a single ordinal value and, understandably, several workers have adopted this approach (e.g., Folkers 1982; Folkers and Lowther 1984; Burgham and Picman 1989; Ortega and Cruz 1991). However, based on arguments by Moran et al. (1981), Smith et al. (1984) pointed out that the aggressive score index has two main drawbacks: the details of each response are not known and, hence, the method cannot be repeated by others, and combining responses that mix motor patterns and proximity (distance) measures can lead to problems of interpretation. Distinct categories of responses should be analyzed separately. Montgomerie and Weatherhead (1988) pointed out further that even closely related species often use very different nest-defense behaviors and cautioned against assessing the intensity of nest defense on a common scale. Indeed, a scale describes behavior *a priori*; if a species is being tested for the first time, its range of responses will not be known prior to testing. By recording as many behaviors as possible, researchers will not bias their results against responses that are not on a pre-ordinaed scale. Even if a species has been tested with models, other models may elicit new or previously unreported behavior.

Another problem with the use of an aggressive score index is that the index presupposes that aggression is the primary defense mechanism used by hosts. We caution against this *a priori* approach. Aggressive behavior may not deter brood parasites or predators as effectively as distraction displays and alarm calling. For example, many shorebird species rely on distraction displays rather than aggressive responses when defending their nests against predators (Sordahl 1986). Alarm calling and distraction displays are not aggressive behavior *per se*; thus, using them in a generalized score confuses the nature and possible evolutionary consequences of a host's response.

Using a finer-scale analysis of behavior is more intricate, but we believe it permits better comparisons within and between species. Yellow warbler females, for example, responded to female cowbird models by rushing to and sitting in their nests ("nest-protection behavior"; see Hobson and Sealy 1989; Gill and Sealy 1996). This behavior may represent an effective defense, but it would not be classified as aggressive and, therefore, it would be ranked low, if it was ranked at all. Also, yellow warblers utter seet calls primarily in response to cowbirds, but give "chip" calls to predators (Hobson et al. 1988; Hobson and Sealy 1989; Gill and Sealy 1996). Categorizing all vocalizations under "alarm calls" would obscure this difference. Also, different alarm calls may function differently in nest defense (Gill 1995). Thus, incorporating the data into a generalized score would obscure the importance of behavior that may have evolved in response to parasitism by cowbirds (see also Neudorf and Sealy 1992; Gill 1995). As a compromise, Maloney and McLean (1995) presented both an intensity index and test statistics of individual response behaviors on which the index was based. The development of their index seems to get around the problem of combining distance and other behaviors. Maloney and McLean ranked each behavior separately based on clearly defined criteria.

CHARACTERISTICS OF RESPONSES

Few workers have compared the responses of different host species to cowbirds (Folkers 1982; Neudorf and Sealy 1992) and only Robertson and Norman (1976, 1977) compared such responses over a host community. The latter authors tested host populations in areas of ancient (Manitoba) and recent (Ontario) sympatry with cowbirds, but their sample sizes for most species tested were too small. Since then, additional studies (Briskie and Sealy 1989; Hobson and Sealy 1989; Bazin and Sealy 1993; Grieef 1995; Gill and Sealy 1996; this study) have allowed comparison of the responses of 11 species in a host community at Delta Marsh, Manitoba (table 10.2). Some of these species accept cowbird eggs, some reject them, and two species show both rejection and acceptance. Rothstein (1975) regarded as "rejecter species" those that removed, damaged or buried eggs, or deserted the nests, within five days of the experimental introduction of cowbird eggs. He regarded species that rarely show these behaviors as "accepter species." In the following section, we compare responses of accepter and rejecter species to cowbird models.

The response of 11 hosts to the threat of cowbird parasitism is compared by reviewing our previous work conducted at Delta Marsh and reporting original results on three more host species tested in this study. Execution of all studies followed the same general methodology as outlined earlier, thereby allowing direct comparisons among studies. First, we present new data on the responses of three large hosts tested in this study and then compare results obtained from all hosts tested so far. We examine responses of accepter and rejecter species to female cowbirds. We address the question of whether hosts recognize cowbirds as a unique threat or simply respond to them in a generalized manner as they would to any intruder at the nest.

Responses of three large hosts to threat of cowbird parasitism

American robins (*Turdus migratorius*) reject cowbird eggs (Rothstein 1982; Briskie et al. 1992), whereas yellow-headed blackbirds (*Xanthocephalus xanthocephalus*) and common grackles usually accept cowbird eggs (references to accepter/rejecter status of species in table 10.2). At Delta Marsh, these potential host species are rarely parasitized (Neudorf and Sealy 1994). We tested the responses of these species to cowbirds

Table 10.2 Number of close passes (mean ± S.E.) directed toward a female cowbird, control, and predator model over 5-min periods during the laying stage[a]

Host and status[b]	Parasitism frequency (*n*)	Body mass (g)	Model: Control	Model: Cowbird	Model: Predator	*n*[c]	*P*[d]	Source
Accepters								
Least flycatcher	2.9 (463)	10.3	0	1.9 ± 0.4	—	40	0.008	Briskie and Sealy (1989)
Yellow warbler[e]	19.1 (2163)	9.2	0	0.1 ± 0.1	0	35	N.S.	Gill and Sealy (1996)
Clay-colored sparrow	10.8 (204)	12	0	0.5 ± 0.4	0		N.S.	Grieef (1995)
Yellow-headed blackbird	1.5 (67)	49.3	0*	0.7 ± 0.3**	1.4 ± 0.6**	19	0.01	This study
Red-winged blackbird	19.7 (213)	41.5	0.9 ± 0.2*	4.1 ± 0.9**	6.1 ± 1.5**	47	0.0001	Neudorf and Sealy (1992)
Common grackle[e]	2.3 (44)	100	0.2 ± 0.1*	0.9 ± 0.5**	4.3 ± 1.4***	19	0.001	This study
Rejecters								
Eastern kingbird	0.01 (402)	43.5	5.0 ± 2.0*	5.2 ± 1.5*	18.5 ± 5.3**		0.001	Bazin and Sealy (1993)
American robin	4.4 (92)	77.3	0.1 ± 0.1*	0.1 ± 0.1*	3.9 ± 1.9**	18	0.005	This study
Gray catbird	5.0 (101)	36.9	0.2 ± 0.1*	0.3 ± 0.1*	0.9 ± 0.4**	56	0.006	Neudorf and Sealy (1992)
Cedar waxwing	0 (14)	33	0	0.1 ± 0.1	0.3 ± 0.2	15	N.S.	Neudorf and Sealy (1992)
Baltimore oriole	3.3 (153)	33.3	0.7 ± 0.4*	0.8 ± 0.3*	7.6 ± 2.2**	25	0.0001	Neudorf and Sealy (1992)

[a]Responses of least flycatchers are at nest-building, laying, incubation, and nestling stages combined.

[b]Host status based on experimental manipulations conducted by Rothstein 1975, 1976, 1977, 1982; Briskie and Sealy 1987; Ortega and Cruz 1988; Peer 1993; Hill and Sealy 1994; Sealy 1995; Sealy and Bazin 1995.

[c]Number of nests tested with each model.

[d]Results of the Friedmann test (Wilcoxon signed-rank test for least flycatcher).

[e]Although considered by Rothstein (1975) to be accepters, just over 10% of common grackles reject cowbird eggs (Peer 1993) and yellow warblers bury most cowbird eggs laid early in their laying stage (Sealy 1995).

*, **, ***Results of multiple comparisons. Means with different superscripts are significantly different ($P<0.05$). N.S., not significant.

by presenting taxidermic mounts of a female cowbird, fox sparrow and predator (common grackle or blue jay) during the laying stage of the nesting cycle. Field work was conducted at the University of Manitoba Field Station (Delta Marsh) between May and July, 1991. Models were presented level with the nest rim and facing it, on poles or clipped to vegetation, 0.5 m from the nest. Each of the three models was presented for 5 min at each nest with at least a 10-min rest period between presentations. After the model was placed near the nest, the observer retreated to a blind and waited for a host to respond. Responses of males and females were combined to obtain the total level of defense at each nest. Defense variables included: time spent within 2 m of the model, sitting in the nest, threat displays, alarm calls, close passes, and contacts. The first three categories were quantified as the number of 10-s intervals within which these behaviors occurred, whereas the remaining variables were quantified as the actual number of times they occurred over the 5 min. Female yellow-headed blackbirds sometimes uttered harsh calls while responding to the model. We referred to these as squawks. American robins and common grackles uttered high-pitched alarm calls at models, that we referrred to as screams. Data were analyzed using the Friedmann test and associated multiple comparisons (see Neudorf and Sealy 1992 for details of analyses).

Yellow-headed blackbirds and common grackles responded more intensively to the cowbird than the sparrow model (tables 10.3 and

Table 10.3 Responses (mean ± S.E.) of yellow-headed blackbirds to models of a fox sparrow, brown-headed cowbird and common grackle placed near their nests during the laying stage (n = 19 nests)

	Model			
Response[a]	Sparrow	Cowbird	Grackle	P[b]
Time within 2 m	23.3 ± 2.4	23.6 ± 2.3	23.2 ± 2.4	N.S.
Squawk	10.6 ± 4.7*	37.0 ± 17.1*,**	57.6 ± 18.6**	0.01
Close pass	0.0*	0.7 ± 0.3**	1.4 ± 0.6**	0.01
Contact	0.2 ± 0.1	1.4 ± 0.9	1.7 ± 1.0	N.S.
Threat display	0.0*	1.3 ± 1.2**	1.7 ± 0.9**	0.05
Time female spent in nest	7.1 ± 2.5	5.1 ± 4.4	1.4 ± 1.1	N.S.

[a]Time within 2 m, threat displays and time in the nest were quantified as the number of 10-s intervals that birds engaged in these behaviors during the 5-min trial (maximum of 30), whereas the remaining behaviors were quantified as the actual number of times they occurred during the 5-min trial.

[b]Results of Friedmann test among models.

*, **Multiple comparisons. Means with different superscripts are significantly different ($P < 0.05$).

10.4) by giving more close passes (table 10.3). By contrast, American robins responded similarly to the cowbird and sparrow models (table 10.5). All three hosts responded most intensely to the predator model (tables 10.3–10.5). American robins and common grackles responded to the predator model with significantly more alarm calls, close passes, and contacts than to the cowbird or sparrow model.

Despite being rarely parasitized by cowbirds (but see Dufty 1994), yellow-headed blackbirds and common grackles apparently recognize cowbirds as a threat to their nests. Because the presence of cowbird nestlings in nests of large hosts has little negative effect on host nestling survival (Ortega and Cruz 1988; see also Weatherhead 1989), the main cost to these hosts of cowbird parasitism may be from egg removal by female cowbirds. Cowbirds have been observed removing eggs of another large host at Delta Marsh, the American robin (Sealy 1994). Yellow-headed blackbirds and common grackles are large enough to eject cowbird eggs, but rejection behavior does not occur or does so at a low level in these species at Delta Marsh. Individuals of other populations of these species, however, exhibit low levels of rejection (Rothstein 1975; Peer 1993; Dufty 1994).

Table 10.4 Responses (mean ± S.E.) of common grackles to models of a fox sparrow, brown-headed cowbird and a blue jay placed near their nests during the laying stage (n = 19 nests)

	Model			
Response[a]	Sparrow	Cowbird	Jay	P[b]
Time within 2 m	15.7 ± 2.5*	18.6 ± 2.8*	25.6 ± 2.2**	0.001
Scream	6.1 ± 3.2*	2.7 ± 0.6*	36.2 ± 11.0**	0.001
Close pass	0.2 ± 0.1*	0.9 ± 0.5**	4.3 ± 1.4***	0.001
Contact	0.3 ± 0.2*	1.0 ± 0.5*	16.7 ± 5.7**	0.001
Threat display	0.0*	0.0*	2.4 ± 0.9**	0.001
Time female spent in nest	6.0 ± 2.3	7.4 ± 2.5	6.0 ± 2.0	N.S.

[a]Time within 2 m, threat displays and time in the nest were quantified as the number of 10-s intervals that birds engaged in these behaviors during the 5-min trial (maximum of 30), whereas the remaining behaviors were quantified as the actual number of times they occurred during the 5-min trial.

[b]Results of Friedmann test among models.

*, **, ***Results of multiple comparisons. Means with different superscripts are significantly different ($P < 0.05$). N.S., not significant.

Table 10.5 Responses (mean ± S.E.) of American robins to models of a fox sparrow, brown-headed cowbird and a common grackle placed near their nests during the laying stage (n = 18 nests)

	Model			
Responses[a]	Sparrow	Cowbird	Grackle	P^b
Time within 2 m	9.4 ± 3.1	9.6 ± 3.1	11.3 ± 2.7	N.S.
Scream	19.6 ± 11.1*	16.6 ± 6.4*	61.9 ± 17.0**	0.001
Close pass	0.1 ± 0.1*	0.1 ± 0.1*	3.9 ± 1.9**	0.005
Contact	0.0*	0.3 ± 0.2*	5.0 ± 1.8**	0.001
Time female spent in nest	6.8 ± 2.9	8.0 ± 3.1	1.2 ± 1.2	N.S.

[a]Time within 2 m and time in the nest were quantified as the number of 10-s intervals that birds engaged in these behaviors during the 5-min trial (maximum of 30), whereas the remaining behaviors were quantified as the actual number of times they occurred during the 5-min trial.

[b]Results of Friedmann test among models.

*, **Results of multiple caomparisons. Means with different superscripts are significantly different ($P < 0.05$). N.S., not significant.

Comparison among hosts at Delta Marsh

Robertson and Norman (1976) suggested that accepter species should defend their nests more intensely than rejecter species because they incur higher costs if parasitized. There is support for this hypothesis (Folkers 1982; Neudorf and Sealy 1992). The cost of parasitism, however, varies among hosts depending on the frequency of parasitism and whether the hosts accept or reject cowbird eggs. We expected hosts to defend their nests with an intensity proportional to the costs they incur when parasitized, but only if the defense prevents or reduces parasitism (Neudorf and Sealy 1992).

An alternative hypothesis suggests that rejecters should defend their nests despite their use of egg rejection to eliminate parasitism, and that this defense should be stronger than in accepters. One reason rejecters may be more likely to show stronger nest defense than accepters, in addition to their generally larger size, is that they have evolved at least one defense against parasitism (egg discrimination). During the same period of evolution, additional defenses against cowbirds, such as nest defense, also may have been favored. Indeed, when parasitized, rejecters may lose an egg to the female cowbird. The combined development of two defenses may have resulted in rejecters being more likely also to exhibit nest defense. However, there is little support for this suggestion.

On the other hand, accepters may not have had enough time to develop any effective defenses, assuming that acceptance is not a result of their small size (see Rohwer and Spaw 1988; see also Sealy 1996). Nevertheless, responses that are specific to cowbirds and responses that are more intense at laying, when the threat of parasitism is greatest, suggest that accepters do have defenses. In the followng section, we compare responses of accepter and rejecter species to cowbird models.

Responses of 11 hosts to cowbirds are outlined in table 10.2. During the laying period, all hosts were tested with female cowbird and fox sparrow models. With the exception of the least flycatcher (*Empidonax minimus*), hosts were tested with an egg and nestling predator (either a common grackle or blue jay model). For all hosts, many variables of defense were examined (e.g., Hobson and Sealy 1989; Neudorf and Sealy 1992), but to facilitate comparison, close passes were chosen because they were used by all hosts in nest defense and they were considered to be one of the most intensive responses to cowbirds.

Accepter species were not more likely to give close passes in response to the cowbird model than rejecter species. In fact, the overall mean number of close passes by accepters (1.4) is almost identical to that of rejecters (1.3). If relative responses to the cowbird compared with the responses to the fox sparrow models are examined, however, four of the six accepter species gave more close passes to the cowbird than the fox sparrow, whereas all five rejecter species responded similarly to the cowbird and fox sparrow in the number of close passes. Accepters and rejecters differed significantly in their defensive responses (Fisher exact test, $P = 0.045$). Rejecters used a more generalized response to all models, whereas accepters recognized the cowbird as a unique threat. Both types of defense response may deter cowbirds. Low levels of nest defense in some species, however, may be partly attributable to the fact that an immobile model is not enough to elicit a full response. For example, American robins and yellow-headed blackbirds were defensive toward us when we inspected their nests, but their responses to models sometimes were weak. Alternatively, yellow warblers may respond more intensely to model cowbirds than to live cowbirds, possibly because they cannot move a model away from their nest (Gill 1995). However, during five acts of parasitism witnessed by us at yellow warbler nests, adult females pecked the cowbird as she laid an egg (Sealy et al., in press; see also Sealy et al. 1995).

Level of aggression by hosts toward the cowbird, measured as the number of close passes, was not significantly related to host size (Spearman Rank Correlation, $r = 0.29$, d.f. = 11, $P = 0.35$) or aggression directed at the fox sparrow model ($r = 0.52$, d.f. = 11, $P = 0.10$). However, the relationship between body mass and close passes directed at the predator model approached significance ($r = 0.60$, d.f. = 10, $P = 0.07$). Smaller hosts may be more at risk when they attack a predator than larger hosts. In support of this idea (Gill and Sealy 1996), yellow warblers did not approach the grackle model as closely as they did the cowbird model.

Our results support Robertson and Norman's (1977) hypothesis. Accepter species clearly recognized female cowbirds as a greater threat than the sparrow, regardless of the frequency of parasitism they experience. Rejecter species, on the other hand, did not distinguish between cowbird and control. Rejecters, however, may still deter cowbirds because their responses were aggressive, but not oriented specifically toward cowbirds. In the following section, we discuss in greater detail the recognition displayed by different hosts.

Evolved recognition or generalized response?

The previous section showed that accepters seemed to recognize cowbirds as a threat, whereas rejecters responded to cowbirds and control with similar levels of defense. Of the accepter species we tested, yellow warblers provide the most powerful evidence for recognition of the cowbird as a unique threat. Yellow warblers responded uniquely to the cowbird model: the cowbird model primarily elicited seet calls and nest-protection behavior – behaviors that were most pronounced at the laying stage (Hobson and Sealy 1989; Gill and Sealy 1996). Yellow warblers gave other structurally distinct calls (primarily chip calls, but also "metallic chips", "warbles") in response mainly to nest predators (Gill 1995). Burgham and Picman (1989) also observed female yellow warblers sitting in the nest in response to a cowbird model, but they interpreted this behavior as a resumption of incubation. At least one other warbler species, the American redstart (*Setophaga ruticilla*), responded in this manner to live cowbirds but not to other species that came near its nest (Benson 1939:119–124). Further evidence for the idea of unique recognition of the cowbird is shown by the decrease in nest-defense behavior later in the nesting cycle when cowbirds are less threatening (Hobson and Sealy 1989; Neudorf and Sealy 1992; Gill and Sealy 1996), and infrequent defensive behavior in a population of yellow warblers allopatric with cowbirds (Briskie et al. 1992; Gill 1995).

Red-winged blackbirds responded more defensively to the cowbird model at laying than to either the control or predator model, and they became less defensive toward cowbirds later in the nesting cycle (Neudorf and Sealy 1992). By contrast, defenses toward the predator model increased over the nesting cycle, whereas responses to the fox sparrow remained about the same. This suggests that red-winged blackbirds recognized the specific threat cowbirds pose. Conversely, least flycatchers responded similarly to cowbird models over the entire nesting cycle. This suggests that at later nest stages least flycatchers perceived the cowbird as a predator (Briskie and Sealy 1989) if, indeed, they recognize a cowbird as a distinct threat. Cowbirds have been reported taking nestling passerine birds (e.g., Scott et al. 1992; Scott and McKinney 1994), and at one site where the host breeding season is protracted, evidence suggests that cowbirds employ nest predation as a strategy to force hosts to renest, thus creating new parasitism opportunities (Arcese et al. 1996).

There was no correlation between the frequency of parasitism at our site among accepters and defensive behavior as measured by close passes at the cowbird model (Spearman Rank Correlation, $r = 0.08$, df = 6, $P = 0.85$). Both yellow warblers and clay-colored sparrows were parasitized frequently but each gave low levels of aggressive behavior (table 10.2). Nevertheless, both of these species seem to recognize the cowbird (Grieef 1995; Gill and Sealy 1996). All of the remaining accepter species tested made more close passes at the cowbird than to the fox sparrow model.

Previous studies (Robertson and Norman 1977; Neudorf and Sealy 1992; Bazin and Sealy 1993) have shown that not all rejecters behave defensively toward cowbird models. Cedar waxwings and gray catbirds responded similarly, at low levels, toward cowbird and control models (Neudorf and Sealy 1992), which suggests that neither species perceived the cowbird as a threat. American robins also responded with low intensity to the cowbird and control models (table 10.5) and eastern kingbirds responded slightly more aggressively to the cowbird model than to the control model (Bazin and Sealy 1993). On the other hand, Robertson and Norman (1977) found that gray catbirds and American robins were significantly more aggressive toward a female cowbird model than to the control. In nature, gray catbirds and American robins have been observed occasionally to attack female cowbirds violently and drive them off (Leathers 1956; Friedmann 1963; Scott 1977). Different previous experience with cowbirds on the part of hosts or different methodologies used in the two studies may account for these contradictory results. Indeed, rejecter species may be parasitized more frequently in areas of recent contact, such as Ontario, where many host species in Robertson and Norman's (1977) study were tested (see Scott 1977; Sealy and Bazin 1995; Sealy et al., unpubl. data).

If cowbird recognition is learned (see Smith et al. 1984), rejecters may have shown little aggression because they rarely experience cowbirds near their nests. If recognition is innate, in some or all species, it is difficult to see why natural selection would not favor recognition of cowbirds in rejecters, unless rejecters are never parasitized. Rejecters can still suffer the loss of eggs removed by female cowbirds and given their usually large size, aggression by them should be effective and should incur little risk (but see Friedmann 1929; Leathers 1956). Egg ejection can be costly, too. When attempting to eject a cowbird egg by puncture-ejection, Baltimore orioles (*Icterus galbula*) and warbling vireos (*Vireo gilvus*) occasionally damage their own eggs, which they then remove (Rothstein 1977; Rohwer et al. 1989; Sealy and Neudorf 1995; Sealy 1996). Neudorf and Sealy (1992) interpreted the slightly more defensive responses of Baltimore orioles to the cowbird than control model as due to occasional damage suffered by their own eggs during puncture-ejection. Regardless of its own level of nest guarding, a species should drive away a cowbird that it discovers near its nest, especially if the species is large and is likely to be able to do so, as is the case with most rejecters.

With the exception of cedar waxwings, all species we studied (table 10.2) responded defensively to the common grackle model (see also Neudorf and Sealy 1992). This demonstrates the importance of presenting a model of a nonparasitic enemy in addition to cowbird and control models. Most studies have not included a predator (but see Burgham and Picman 1989; Duckworth 1991; Bazin and Sealy 1993; Gill and Sealy 1996) and, thus, when hosts showed little response it could not be determined whether they were inherently nonaggressive (like cedar waxwings), or simply do not recognize the cowbird.

Arguably, hosts do not need to recognize female cowbirds to receive the benefits of nest defense. They could simply respond defensively to an intruder of similar shape and size. Egg ejecters do not recognize cowbird eggs *per se* but instead eject any egg that differs substantially from their own eggs (Rothstein 1978, 1982; Sealy and Bazin 1995). A similar mechanism may operate in nest defense whereby hosts react defensively toward a range of intruders (S. I. Rothstein, pers. commun.), although available evidence reveals that this is not generally the case. A generalized response given to any intruder near the nest may be costly to nest owners in time and energy, especially responses to species that are not threatening. Some hosts recognize female cowbirds as specific threats and the responses drop off after they have laid their own eggs and the threat of parasitism has declined (Hobson and Sealy 1989; Neudorf and Sealy 1992; Gill and Sealy 1996). This is evidence that nest defense behaviors in these species are innate, although learning may be used to refine recognition skills and adjust the nature of the responses (see Maloney and McLean 1995). Hobson and Sealy (1989) found that yearling yellow warblers responded slightly less intensely to the threat of parasitism than did older or previously parasitized individuals. Hosts could hone their responses to cowbirds as enemies because they have found cowbirds at their nests. Cowbirds damage nests, not the nest owners themselves.

McLean et al. (1996) reported that captive-reared, predator-naïve rufous hare-wallabies (*Lagorchestes hirsutus*) reacted cautiously to mammalian predators in captivity, which suggests either some genetic recognition abilities for a generalized mammalian predator, or perhaps that hare-wallabies are simply generally cautious in the presence of an unknown animal. Wallabies became even more cautious after conditioning techniques were used to teach them to associate fright with a fox or cat (McLean et al. 1996; see also Maloney and McLean 1995).

More testing is required to distinguish between genetic and learned abilities of hosts of brood parasites. Opportunities by hosts to experience cowbirds at their nests seem to be variable, and depend on the species. We recorded few visits to nests by brown-headed cowbirds during hundreds of hours of observation of passerine nests at all nesting stages (Neudorf and Sealy 1994; Sealy et al., in press; Neudorf et al., unpubl. data). This is in contrast with apparently frequent nest visits recorded in studies of other parasitic cowbirds (e.g., Post and Wiley 1977; Carter 1986; Webster 1994).

DOES NEST DEFENSE REDUCE PARASITISM FREQUENCY?

Determination of the effectiveness of nest defense has proven difficult, partly because it is not known what the frequencies of parasitism might have been in the absence of nest defense or at which nests adults may have thwarted parasitism. Nests at which a cowbird egg has already been ejected are recorded only as unparasitized or parasitized. Nevertheless, attacks on female cowbirds by large hosts have been observed in nature (e.g., Friedmann 1929; Leathers 1956; see also Sealy et al. 1995), but female cowbirds also have been observed being driven away from nests that were later parasitized (e.g., Scott 1977; Briskie and Sealy 1987; Hill and Sealy 1994). As long as host fitness is increased by nest defense behavior, even a small amount, and the defense has no detectable costs (probably true for most hosts, especially large ones), the behavior should be selected.

Attempts have been made to determine whether nest defense by potential hosts deters cowbirds or predators by comparing the responses of birds whose nests were parasitized or depredated with nests that were not parasitized or were successful. Some studies have shown that defensive nest owners were more likely to fledge more offspring (e.g., Andersson et al. 1980; Greig-Smith 1980; Blancher and Robertson 1982; Folkers and Lowther 1985; Knight and Temple 1986*a*), but others have recorded the opposite result (Seppä 1969; Robertson and Norman 1976, 1977; Searcy 1979; Röell and Bossema 1982; Smith et al. 1984; McLean et al. 1986; Onnebrink and Curio 1991). The latter finding has led some of these authors to suggest that brood parasites and predators possibly use nest defense responses to locate nests. In addition to gaining information about nest location, brood parasites may use the reactions of nest owners to their intrusions to assess the age and experience of prospective hosts (Robertson and Norman 1977; Smith 1981; see also Gill et al., 1997*b*). Smith (1981) reported that aggression by song sparrows seems to facilitate parasitism, not prevent it. This suggests that recognition of cowbirds by song sparrows is not an evolved behavior.

Examining the effectiveness of defense by comparing responses of birds to models at parasitized and unparasitized nests, however, is problematic. Robertson and Norman (1977) suggested that host aggression is an effective anti-parasite strategy where hosts, such as red-winged blackbirds, nest at high densities (where they recorded the greatest aggression). However, caution is needed when interpreting differences in parasitism frequencies on red-winged blackbirds because dense colonies often are situated in marshes away from perches that cowbirds use to find nests and, hence, low frequencies of parasitism may merely reflect the infrequent attempts by cowbirds to parasitize these nests (see Freeman et al. 1990).

Using the methods outlined in Neudorf and Sealy (1992) and Gill and Sealy (1996), we experimentally tested two species (yellow warblers, red-winged blackbirds) that are frequently parasitized at Delta Marsh to determine whether parasitized nest owners respond to the threat of parasitism more or less intensely than unparasitized birds. Yellow warblers accept cowbird eggs laid late in the laying period (Sealy 1995) and red-winged blackbirds accept them any time in the nesting cycle (Ortega and Cruz 1988). We hypothesized that defense at nests would deter cowbird parasitism and predicted that parasitized nest owners would react less intensely in response to the cowbird model than unparasitized nest owners. Models were presented at naturally parasitized nests of both species either before or after the nests had been parasitized. Because of the small sample of parasitized nests, the data could not be partitioned into nests tested before and nests tested after parasitism. The results were equivocal. Generally, parasitized and unparasitized yellow warblers (table 10.6) and red-winged blackbirds (table 10.7) responded similarly to the models. Unparasitized yellow warblers uttered more chip calls toward the cowbird model (Wilcoxon two-sample tests, $t_R = -1.82$, $P = 0.0343$) and gave fewer seet calls

Table 10.6 Responses (mean ± S.E.) of nine parasitized and 26 unparasitized female yellow warblers at the egg-laying stage

	Model					
	Sparrow		Cowbird		Grackle	
Response[a]	Parasitized	Unparasitized	Parasitized	Unparasitized	Parasitized	Unparasitized
Time within 2 m	20.3 ± 3.8	18.5 ± 2.3	27.3 ± 1.8	27.6 ± 1.1	5.9 ± 3.5	10.7 ± 2.0
Seet	44.8 ± 19.9*	7.4 ± 3.2*	89.6 ± 44.7	22.5 ± 4.0	35.8 ± 35.0	1.6 ± 0.5
Chip	45.2 ± 31.3	15.0 ± 4.4	0.0	3.2 ± 1.8	81.1 ± 30.2	56.3 ± 15.7
Contact	0.0	0.0	0.1 ± 0.1	0.3 ± 0.2	0.0	0.0
Distraction display	0.6 ± 0.6	0.2 ± 0.2	1.7 ± 0.6	1.0 ± 0.9	2.6 ± 1.5	6.0 ± 1.9
Nest-protection behavior	9.6 ± 4.6	5.2 ± 1.7	19.4 ± 4.2	18.4 ± 2.6	3.2 ± 3.2	1.5 ± 1.2

[a]Time within 2 m, distraction display and nest-protection behavior were quantified as the number of 10-s intervals that birds engaged in these behaviors during the 5-min trial (maximum 30), whereas the remaining behaviors were quantified as the actual number of times they occurred during the 5-min trial.
*$P < 0.05$ (Wilcoxon two-sample tests).

to the sparrow model ($t_R = 2.46$, $P = 0.0070$) than did parasitized nest owners. None of the other comparisons was significant ($|t_R| < 1.82$, n.s.).

Unparasitized red-winged blackbirds approached the sparrow model significantly more closely than parasitized individuals ($t_R = 2.26$, $P = 0.0285$; see table 10.7), but no other behavior at parasitized and unparasitized nests differed significantly ($|t_R| < 1.64$, n.s.; see Neudorf 1991). Responses by parasitized and unparasitized hosts did not differ, which provides evidence for innate recognition of cowbirds. Because we do not know which, if any, individuals were present at the time the cowbird came to lay, we cannot draw conclusions. Effectiveness of defense against cowbirds may be determined only through watching host nests at times when cowbirds come to lay.

Vigilance at the nest by hosts when parasitism is most likely to occur should be the key to successful defense against parasitism or predation. Nest owners often are assumed to be at their nests when predation or parasitism occurs (Redondo and Carranza 1989; but see Scott 1991); however, nest attentiveness by adults at times when these events are most likely to occur is not known for most species. Acts of parasitism (Sealy et al. 1995) and nest predation (Pettingill 1976; Sealy 1994) occur within seconds or minutes, respectively; thus, nest owners on or near their nests when these

Table 10.7 Responses (mean ± S.E.) of 16 parasitized and 31 unparasitized red-winged blackbirds at the egg-laying stage

	Model					
	Sparrow		Cowbird		Grackle	
Response[a]	Parasitized	Unparasitized	Parasitized	Unparasitized	Parasitized	Unparasitized
Time within 2 m	15.2 ± 3.9*	24.9 ± 2.3*	36.4 ± 4.1	37.1 ± 2.2	34.3 ± 4.3	35.1 ± 2.8
Check	51.1 ± 12.7	41.8 ± 8.2	100.8 ± 18.9	63.8 ± 9.5	156.6 ± 23.8	113.8 ± 17.7
Growl	0.0	0.0	0.6 ± 0.5	0.5 ± 0.3	0.8 ± 0.4	1.8 ± 1.1
Scream	0.0	4.4 ± 4.2	15.6 ± 12.9	26.7 ± 17.1	1.9 ± 1.3	19.5 ± 10.4
Close pass	0.8 ± 0.4	0.9 ± 0.3	4.0 ± 1.3	4.2 ± 1.2	7.1 ± 2.7	5.2 ± 1.7
Contact	0.3 ± 0.2	0.9 ± 0.3	12.8 ± 3.5	16.0 ± 3.8	5.8 ± 3.0	8.8 ± 4.6

[a]Time within 2 m was quantified as the number of 10-s intervals (maximum 30) that birds engaged in this behavior during the 5-min trial, whereas the remaining behaviors were quantified as the actual number of times they occurred during the 5-min trial.
*$P < 0.05$ (Wilcoxon two-sample tests).

events happen should be in the best position to prevent the event (Sealy and Neudorf 1994; also see below). Parasitism occurs consistently within a few minutes before sunrise (Scott 1991; Neudorf and Sealy 1994) and, hence, it may be more predictable to hosts, but predation may occur throughout the day (Pettingill 1976; Sealy 1994) and nest vigilance in response to this threat may have to be a trade-off between it and other necessary activities, such as foraging and territorial maintenance.

Neudorf and Sealy (1994) quantified the attentiveness at sunrise of 10 potential cowbird hosts, including yellow warblers and red-winged blackbirds, on the morning the host females laid their second eggs. As cowbirds lay just before dawn (Scott 1991; Neudorf and Sealy 1994; Sealy et al., in press), nest owners were predicted to be attentive at this time (Neudorf and Sealy 1994). There was no relationship, however, between nest attentiveness and the frequency of parasitism among the hosts studied. One-fourth of female red-winged blackbirds roosted on their nests during the early laying period (Neudorf and Sealy 1994), whereas more than half of the female yellow warblers roosted on their nests until after the threat of parasitism had passed that morning (Neudorf and Sealy 1994). Most cowbirds, however, parasitize yellow warblers before the warblers lay their second eggs (Sealy 1992), when most females have not begun to roost on or near their nests (Sealy et al., in press). However, individuals of all species that roosted were much more likely to be present at the nest when cowbirds arrive to lay (Neudorf and Sealy 1994).

Roosting may be a manifestation of some other activity such as the onset incubation of behavior. For example, in the yellow warbler, we have shown that features of pre-sunrise nest attentiveness do not conform to the expectation that the roosting behavior evolved in response to cowbird parasitism (Sealy et al., in press). Female yellow warblers spent little time at their nests before sunrise prior to clutch initiation and early laying (Sealy 1992, 1995). If nest attentiveness in yellow warblers is a defense against brood parasitism, it should be at a high level on the days the first and second warbler eggs are laid. Instead, attentiveness increased each day to a plateau about one day before clutches were complete, and incubation had begun (Hébert and Sealy 1992). This pattern of roosting seems to be linked to the gradual onset of incubation rather than being an anti-parasite strategy, although the host may benefit if it is on or at its nest when a cowbird comes along.

At other times durng the day, cowbirds visit nests to inspect them (Mayfield 1961; Neudorf et al., unpubl. data) and to remove host eggs (Sealy 1992). To deter cowbirds at these times, hosts must also be vigilant then and, hence, balance vigilance with other activities. Song sparrows provisioned with food were more vigilant at their nests and were parasitized less frequently (Arcese and Smith 1988). Barn swallows (*Hirundo rustica*) that guarded their nests more were less frequently parasitized by conspecifics (Møller 1989). These studies suggest that hosts can prevent parasitism if they are at their nests at the time parasitism is likely to occur.

DO INTERACTIONS INVOLVED IN NEST DEFENSE RESULT FROM COEVOLUTION?

Coevolution is *reciprocal* (italics ours) change in interacting species (Thompson 1982; see also Janzen 1980). Avian brood parasitism provides some of the most persuasive examples of coevolution, especially ejection of nonmimetic eggs by some hosts and egg mimicry by some parasites (Payne 1977; Rothstein 1990). Not all traits of brood parasites and hosts, however, have necessarily coevolved. Recognition of the female cowbird and nest defense directed toward the cowbird are among the features of hosts that Rothstein (1990) considered unlikely always to be coevolved adaptations. For coevolution to apply, hosts must have: (i) responded to the parasite or presence of a parasitic egg in their nests as threats to their fitness; (ii) developed effective defenses against the parasite or parasitism; and (iii) had these defenses in turn countered by the parasite.

Interactions between many cuckoo species and their hosts have certainly involved reciprocal change, for example, the evolution of egg ejection behavior by hosts and egg mimicry by the parasite (Rothstein 1990). Although some cowbird hosts have evolved ejection behavior (e.g., Rothstein 1975, 1982; Rohwer and Spaw 1988; Sealy and Bazin 1995; Sealy 1996), the generalist cowbirds have not countered this defense by evolving egg mimicry (Rothstein 1990; Fraga 1983; also see Smith 1968). Rather, cowbirds may have shifted from some hosts that developed ejection behavior to hosts that accept parasitism (see Davies and Brooke 1989*a*,*b*; Sealy and Bazin 1995), a situation that Thompson (1992) called "coevolutionary alteration."

Some host species recognize the appearance and calls of brown-headed cowbirds (Gill and

Sealy 1996; Gill et al., 1997*a*). Indeed, two responses (nest-protection behavior and seet calls) are given by yellow warblers primarily in response to the appearance or calls of a female cowbird at the nest (Gill and Sealy 1996; Gill et al., 1997*a*). These behaviors are rarely given in areas where there is no cowbird parasitism (Briskie et al. 1992; Gill 1995), which suggests that the traits evolved due to cowbird parasitism. As shown in the previous section, direct evidence is lacking to indicate that these responses sometimes thwart parasitism (see also Sealy et al., in press). Some level of deterrence of parasitism by hosts and counteraction of it by cowbirds would be necessary before these interactions can be considered coevolved. Indeed, any bird is likely to react defensively toward another bird (or other animal) that enters its nest, especially if the intruder is not a threat to the parents. This would likely be the case regardless of the threat of brood parasitism and, therefore, this behavior is not enough evidence to conclude that the responses evolved directly in response to the threat of parasitism. The key here, however, is that the responses of some host species are directed selectively toward a threat before that threat reaches the nest. This suggests recognition of the threat, and not merely a generalized response. As emphasized above, yellow warblers respond to cowbirds differently than they do to avian and mammalian predators. This is some of the most convincing evidence of coevolution involving nest defense behavior, although questions remain regarding ways in which cowbirds have countered this behavior.

The selective pressures to which parasites have responded also must be identified, whether these are to host defense or the need to reduce the chance of the parasite being detected by hosts that may be more likely to reject parasitic eggs if they first see the parasite at their nest. These pressures are difficult to identify because they are not mutually exclusive. It is noteworthy that parasites behave in several ways that apparently lessen the chance that a host will see or intercept them at the nest. Brown-headed cowbirds lay their eggs around sunrise, when some hosts are inattentive and they, as well as many other parasites, have dull or cryptic plumages, behave furtively near host nests, and lay their eggs within a few seconds (Payne 1967; Feare et al. 1982; Davies and Brooke 1988; Scott 1991; Neudorf and Sealy 1994; Sealy et al. 1994; Sealy et al., unpubl. data). These features of parasites certainly suggest that responses to host behavior have evolved, but they may reflect only constraints placed upon parasites by other activities, such as predator avoidance. Recent tests demonstrated that rejection did not increase in two cowbird hosts presented with a model cowbird on the nest followed by experimental parasitism (Hill and Sealy 1994; Sealy 1995), although rejection has been shown to increase in some cuckoo hosts tested simiarly (Davies and Brooke 1988; Moksnes et al. 1993).

We have shown that accepter species recognize cowbirds as threats to their nests regardless of the frequency of parasitism, which supports Robertson and Norman's (1977) hypothesis. Evidence suggests that this behavior is largely innate, but may be refined through experience with cowbirds at the nest. That cowbirds have adaptations that minimize detection at host nests suggests a coevolutionary response. Hosts may be lagging behind in counter-defenses as they largely are not vigilant at the time cowbirds are most likely to parasitize nests. We suggest further research be done on the effects of experience on recognition of different threats. Testing hosts that experience different frequencies of exposure to cowbirds, including unparasitized populations, should provide a better understanding of the evolution of differential recognition of stimuli and the role recognition plays in the breeding biology of these birds.

ACKNOWLEDGMENTS We thank the staff of the University of Manitoba Field Station (Delta Marsh) for providing support and comfortable living conditions during the field work. The officers of the Portage Country Club and Delta Waterfowl and Wetlands Research Station permitted us to conduct some of the field work on their properties. We gratefully acknowledge the individuals who assisted us in the field over the years. We are grateful to Scott K. Robinson and Stephen I. Rothstein for their constructive criticism of several drafts of the manuscript. Financial support for the work reported on here was provided by Manitoba Department of Natural Resources (Wildlife Branch), Natural Sciences and Engineering Research Council of Canada, and University of Manitoba Research Grants Program.

REFERENCES

Andersson, M., C. G. Wiklund, and H. Rundgren (1980). Parental defence of offspring: a model and an example. *Anim. Behav.*, **28**, 536–542.

Arcese, P. and J. N. M. Smith (1988). Effects of population density and supplemental food on reproduction in Song Sparrows. *J. Anim. Ecol.*, **57**, 119–136.

Arcese, P., J. N. M. Smith, and M. I. Hatch (1996). Nest predation by cowbirds and its consequences for passerine demography. *Proc. Natl Acad. Sci. USA*, **93**, 4608–4611.

Bazin, R. C. and S. G. Sealy (1993). Experiments on the responses of a rejector species to threats of predation and cowbird parasitism. *Ethology*, **94**, 326–338.

Benson, M. H (1939). A study of the American Redstart (*Setohaga ruticilla* Swainson). M.S. Thesis. Cornell University, Ithaca, NY.

Blancher, P. J. and R. J. Robertson (1982). Kingbird aggression: does it deter predation?*Anim. Behav.*, **30**, 929–930.

Briskie, J. V. and S. G. Sealy (1987). Responses of Least Flycatchers to experimental inter- and intraspecific brood parasitism. *Condor*, **89**, 899–901.

Briskie, J. V. and S. G. Sealy (1989). Changes in nest defense against a brood parasite over the breeding cycle. *Ethology*, **82**, 61–67.

Briskie, J. V., S. G. Sealy, and K. A. Hobson (1992). Behavioral defenses against avian brood parasitism in sympatric and allopatric host populations. *Evolution*, **46**, 334–340.

Burgham, M. C. J. and J. Picman (1989). Effect of Brown-headed Cowbirds on the evolution of Yellow Warbler anti-parasite strategies. *Anim. Behav.*, **38**, 298–308.

Carter, M. D. (1986). The parasitic behavior of the Bronzed Cowbird in south Texas. *Condor*, **88**, 11–25.

Clark, K. L. and R. J. Robertson (1979). Spatial and temporal multi-species nesting aggregations in birds as anti-parasite and anti-predator strategies. *Behav. Ecol. Sociobiol.*, **5**, 359–371.

Clark, K. L. and R. J. Robertson (1981). Cowbird parasitism and evolution of anti-parasite strategies in the Yellow Warbler. *Wilson Bull.*, **93**, 249–258.

Cruz, A., T. D. Manolis, and R. W. Andrew (1990) Reproductive interactions of the Shiny Cowbird *Molothrus bonariensis* and the Yellow-shouldered Blackbird *Agelaius icterocephalus* in Trinidad. *Ibis*, **132**, 436–444.

Davies, N. B. and M. de L. Brooke (1988). Cuckoos versus Reed Warblers: adaptations and counteradaptations. *Anim. Behav.*, **36**, 262–284.

Davies, N. B. and M. de L. Brooke (1989*a*). An experimental study of co-evolution between the Cuckoo, *Cuculus canorus*, and its hosts. I. Host egg discrimination. *J. Anim. Ecol.*, **58**, 207–224.

Davies, N. B. and M. de L. Brooke (1989*b*). An experimental study of co-evolution between the Cuckoo, *Cuculus canorus*, and its hosts. II. Host egg markings, chick discrimination and general discussion. *J. Anim. Ecol.*, **58**, 225–236.

Dawkins, R. and J. R. Krebs (1979). Arms races between and within species. *Proc. R. Soc. Lond. B*, **205**, 489–511.

Duckworth, J. W. (1991). Responses of breeding Reed Warblers *Acrocephalus scirpaceus* to mounts of Sparrowhawk *Accipiter nisus*, Cuckoo *Cuculus canorus* and Jay *Garrulus glandarius. Ibis*, **133**, 68–74.

Dufty, A. M., Jr (1994). Rejection of foreign eggs by Yellow-headed Blackbirds. *Condor*, **56**, 799–801.

Edwards, G., E. Hosking, and S. Smith (1949). Reactions of some passerine birds to a stuffed Cuckoo. *British Birds*, **42**, 13–19.

Edwards, G., E. Hosking, and S. Smith (1950). Reactions of some passerine birds to a stuffed Cuckoo. II. A detailed study of the Willow-Warbler. *Brit. Birds*, **43**, 144–150.

Feare, C. J., P. L. Spencer, and D. A. T. Constantine (1982). Time of egg-laying of Starlings *Sturnus vulgaris. Ibis*, **124**, 174–178.

Folkers, K. L. (1982). Host behavioral defenses to cowbird parasitism. *Bull. Kansas Ornith. Soc.*, **33**, 32–34.

Folkers, K. L. and P. E. Lowther (1985). Responses of nesting Red-winged Blackbirds and Yellow Warblers to Brown-headed Cowbirds. *J. Field Ornith.*, **56**, 175–177.

Fraga, R. M. (1983). The eggs of the parasitic Screaming Cowbird (*Molothrus rufoaxillaris*) and its host, the Bay-winged Cowbird (*M. badius*): is there evidence for mimicry? *J. Ornith.*, **124**, 187–193.

Freeman, S., D. F. Gori, and S. Rohwer (1990). Red-winged Blackbirds and Brown-headed Cowbirds: some aspects of a host–parasite relationship. *Condor*, **92**, 336–340.

Friedmann, H. (1929). *The cowbirds: a study in the biology of social parasitism.* C. C. Thomas, Springfield, IL.

Friedmann, H. (1963). Host relations of the parasitic cowbirds. *U.S. National Mus. Bull.*, **233**, 1–276.

Gilbert, L. E. and P. E. Raven (eds) (1975). *Coevolution of animals and plants.* University of Texas Press, Austin, Texas.

Gill, S. A. (1995). Information transfer, function, and evolution of Yellow Warbler alarm calls. M.Sc. Thesis. University of Manitoba, Winnipeg, Manitoba.

Gill, S. A. and S. G. Sealy (1996). Nest defence by Yellow Warblers: Recognition of a brood parasite and an avian nest predator. *Behaviour*, **133**, 263–282.

Gill, S. A., D. L. Neudorf, and S. G. Sealy (1997*a*). Host responses to cowbirds near the nest: cues for recogniton. *Anim. Behav.*, **53**, 1287–1293.

Gill, S. A., P. M. Grieef, L. M. Staib, and S. G. Sealy (1997*b*). Does nest defense deter or facilitate cowbird parasitism? A test of the nesting-cue hypothesis. *Ethology*, **103**, 56–71.

Greig-Smith, P. W. (1980). Parental investment in nest defence by Stonechats (*Saxicola torquata*). *Anim. Behav.*, **28**, 604–619.

Grieef, P. M. (1995). Cues used by brood parasites and predators to locate nests. M.Sc. Thesis. University of Manitoba, Winnipeg, Manitoba.

Hébert, P. N. and S. G. Sealy (1992). Onset of incubation in Yellow Warblers: a test of the hormonal hypothesis. *Auk*, **109**, 249–255.

Hill, D. P. and S. G. Sealy (1994). Desertion of nests parasitized by cowbirds: have Clay-coloured Sparrows evolved an anti-parasite strategy? *Anim. Behav.*, **48**, 1063–1070.

Hobson, K. A., M. L. Bouchart, and S. G. Sealy (1988). Responses of naive Yellow Warblers to a novel nest predator. *Anim. Behav.*, **36**, 1823–1830.

Hobson, K. A. and S. G. Sealy (1989). Responses of Yellow Warblers to the threat of cowbird parasitism. *Anim. Behav.*, **38**, 510–519.

Hurlbert, S. H. (1984). Pseudoreplication and the design of ecological field experiments. *Ecol. Monogr.*, **54**, 187–211.

Janzen, D. H. (1980). When is it coevolution? *Evolution*, **34**, 611–612.

Knight, R. L. and S. A. Temple (1986*a*). Why does intensity of avian nest defense increase during the nesting cycle? *Auk*, **103**, 318–327.

Knight, R. L. and S. A. Temple (1986*b*). Methodological problems in studies of avian nest defense. *Anim. Behav.*, **34**, 561–566.

Kroodsma, D. E. (1989). Suggested experimental design for song playbacks. *Anim. Behav.*, **37**, 600–609.

Leathers, C. L. (1956). Incubating American Robin repels female Brown-headed Cowbird. *Wilson Bull.*, **68**, 68.

Maloney, R. F. and I. G. McLean (1995). Historical and experimental learned predator recognition in free-living New Zealand Robins. *Anim. Behav.*, **50**, 1193–1201.

Mark, D. and B. J. Stutchbury (1994). Response of a forest-interior songbird to the threat of cowbird parasitism. *Anim. Behav.*, **47**, 275–280.

May, R. M. and S. K. Robinson (1985). Population dynamics of avian brood parasitism. *Amer. Natur.*, **126**, 475–494.

Mayfield, H. F. (1961). Vestiges of a proprietary interest in nests by the Brown-headed Cowbird parasitizing the Kirtland's Warbler. *Auk*, **78**, 162–167.

McLean, I. (1987). Responses to a dangerous enemy: should a brood parasite be mobbed? *Ethology*, **75**, 235–245.

McLean, I. G., G. Lundie-Jenkins, and P. J. Jarman (1996). Teaching an endangered mammal to recognise predators. *Biol. Conserv.*, **75**, 51–62.

McLean, I. G., J. N. M. Smith, and K. G. Stewart (1986). Mobbing behaviour, nest exposure, and breeding success in the American Robin. *Behaviour*, **96**, 171–186.

Moksnes, A., E. Røskaft, A. T. Braa, L. Korsnes, H. M. Lampe, and H. C. Pedersen (1990). Behavioral responses of potential hosts towards artificial cuckoo eggs and dummies. *Behaviour*, **116**, 64–89.

Moksnes, A., E. Røskaft, and L. Korsnes (1993). Rejection of Cuckoo (*Cuculus canorus*) eggs by Meadow Pipits (*Anthus pratensis*). *Behav. Ecol.*, **4**, 120–127.

Møller, A. P. (1989). Intraspecific nest parasitism in the Swallow *Hirundo rustica*: the importance of neighbors. *Behav. Ecol. Sociobiol.*, **25**, 33–38.

Montgomerie, R. D. and P. J. Weatherhead (1988). Risks and rewards of nest defense by parent birds. *Q. Rev. Biol.*, **63**, 167–187.

Moran, G., J. C. Fentress, and I. Golani (1981). A description of relational patterns of movement during "ritualized fighting" in wolves. *Anim. Behav.*, **29**, 1146–1165.

Neudorf, D. L. and S. G. Sealy (1992). Reactions of four passerine species to threats of predation and cowbird parasitism: enemy recognition or generalized responses? *Behaviour*, **123**, 84–105.

Neudorf, D. L. and S. G. Sealy (1994). Sunrise nest attentiveness in cowbird hosts. *Condor*, **96**, 162–169.

Norman, R. F. and R. J. Robertson (1975). Nest-searching behavior in the Brown-headed Cowbird. *Auk*, **92**, 610–611.

Onnebrink, H. and E. Curio (1991). Brood defence and age of young: a test of the vulnerability hypothesis. *Behav. Ecol. Sociobiol.*, **29**, 61–68.

Ortega, C. P. and A. Cruz (1988). Mechanisms of egg acceptance by marsh-dwelling blackbirds. *Condor*, **90**, 349–358.

Ortega, C. P. and A. Cruz (1991). A comparative study of cowbird parasitism in Yellow-headed Blackbirds and Red-winged Blackbirds. *Auk*, **108**, 16–24.

Patterson, T. L., L. Petrinovich, and D. K. James (1980). Reproductive value and appropriateness of response to predators by White-crowned Sparrows. *Behav. Ecol. Sociobiol.*, **7**, 227–231.

Payne, R. B. (1967). Interspecific communication signals in parasitic birds. *Amer. Natur.*, **101**, 363–376.

Payne, R. B. (1977). The ecology of brood parasitism in birds. *Annu. Rev. Ecol. Syst.*, **8**, 1–28.

Peer, B. D. (1993). An investigation of the host specificity of the Brown-headed Cowbird. M.S. Thesis. Eastern Illinois University, Charleston, Illinois.

Pettingill, O. S., Jr (1976). Observed acts of predation on birds in northern lower Michigan. *Living Bird*, **15**, 33–41.

Post, W. and J. W. Wiley (1977). Reproductive interactions of the Shiny Cowbird and the Yellow-shouldered Blackbird. *Condor*, **79**, 176–184.

Redondo, T. and J. Carranza (1989). Offspring reproductive value and nest defense in the Magpie (*Pica pica*). *Behav. Ecol. Sociobiol.*, **25**, 369–378.

Regelmann, K. and E. Curio (1983). Determinants of brood defense in the Great Tit *Parus major* L. *Behav. Ecol. Sociobiol.*, **13**, 131–145.

Robertson, R. J. and R. F. Norman (1976). Behavioral defenses to brood parasitism by potential hosts of the Brown-headed Cowbird. *Condor*, **78**, 166–173.

Robertson, R. J. and R. F. Norman (1977). The function and evolution of aggressive host behavior towards the Brown-headed Cowbird (*Molothrus ater*). *Can. J. Zool.*, **55**, 508–518.

Roëll, A. and I. Bossema (1982). A comparison of nest defense by jackdaws, rooks, magpies and crows. *Behav. Ecol. Sociobiol.*, **11**, 1–6.

Rohwer, S. and C. D. Spaw (1988). Evolutionary lag versus bill-size constraints: a comparative study of the acceptance of cowbird eggs by old hosts. *Evol. Ecol.*, **2**, 27–36.

Rohwer, S., C. D. Spaw, and E. Roskaft (1989). Costs to Northern Orioles of puncture-ejecting parasitic cowbird eggs from their nests. *Auk*, **106**, 734–738.

Røskaft, E., G. H. Orians, and L. D. Beletsky (1990). Why do Red-winged Blackbirds accept eggs of Brown-headed Cowbirds? *Evol. Ecol.*, **4**, 35–42.

Rothstein, S. I. (1975). An experimental and teleonomic investigation of avian brood parasitism. *Condor*, **77**, 250–271.

Rothstein, S. I. (1976). Experiments on defenses Cedar Waxwings use against cowbird parasitism. *Auk*, **93**, 657–691.

Rothstein, S. I. (1977). Cowbird parasitism and egg recognition of the Northern Oriole. *Wilson Bull.*, **89**, 21–32.

Rothstein, S. I. (1978). Mechanisms of avian egg-recognition: additional evidence for learned components. *Anim. Behav.*, **26**, 671–677.

Rothstein, S. I. (1982). Mechanisms of avian egg recognition: which parameters elicit responses by rejecter species? *Behav. Ecol. Sociobiol.*, **11**, 229–239.

Rothstein, S. I. (1990). A model system for coevolution: avian brood parasitism. *Annu. Rev. Ecol. Syst.*, **21**, 481–508.

Scott, D. M. (1977). Cowbird parasitism on the Gray Catbird at London, Ontario. *Auk*, **94**, 18–27.

Scott, D. M. (1991). The time of day of egg laying by the Brown-headed Cowbird and other icterines. *Can. J. Zool.*, **69**, 2093–2099.

Scott, D. M., P. J. Weatherhead, and C. D. Ankney (1992). Egg-eating by female Brown-headed Cowbirds. *Condor*, **94**, 579–584.

Scott, P. E. and B. R. McKinney (1994). Brown-headed Cowbird removes Blue-gray Gnatcatcher nestlings. *J. Field Ornith.*, **65**, 363–364.

Sealy, S. G. (1992). Removal of Yellow Warbler eggs in association with cowbird parasitism. *Condor*, **94**, 40–54.

Sealy, S. G. (1994). Observed acts of egg destruction, egg removal, and predation on nests of passerine birds at Delta Marsh, Manitoba. *Can. Field-Nat.*, **108**, 41–51.

Sealy, S. G. (1995). Burial of cowbird eggs but parasitized Yellow Warblers: an empirical and experimental study. *Anim. Behav.*, **49**, 877–889.

Sealy, S. G. (1996). Evolution of host defenses against brood parasitism: Implications of puncture-ejection by a small passerine. *Auk*, **113**, 346–355.

Sealy, S. G. and R. C. Bazin (1995). Low frequency of observed cowbird parasitism on Eastern Kingbirds: host rejection, effective nest defense, or parasite avoidance? *Behav. Ecol.*, **6**, 140–145.

Sealy, S. G. and D. L. Neudorf (1995). Male Northern Orioles eject cowbird eggs: implications for the evolution of rejection behavior. *Condor*, **97**, 369–375.

Sealy, S. G., D. L. Neudorf, and D. P. Hill (1994). Rapid laying by Brown-headed Cowbirds *Molothrus ater* and other parasitic birds. *Ibis*, **137**, 76–84.

Sealy, S. G., D. G. McMaster, S. A. Gill, and D. L. Neudorf. Yellow Warbler nest attentiveness before sunrise: anti-parasite strategy or onset of incubation? In: *Ecology and management of cowbirds* (eds T. Cook, S. R. Robinson, S. I. Rothstein, S. G. Sealy, and J. N. M. Smith), University of Texas Press, Austin, Texas (in press).

Searcy, W. C. (1979). Female choice of mates: a

general model for birds and its application to Red-winged Blackbirds. (*Agelaius phoeniceus*). *Amer. Natur.*, **114**, 77–100.

Seppä, J. (1969). The Cuckoo's ability to find a nest where it can lay an egg. *Ornis Fenn.*, **46**, 78–79.

Smith, J. N. M. (1981). Cowbird parasitism, host fitness, and age of the host female in an island Song Sparrow population. *Condor*, **83**, 152–161.

Smith, J. N. M., P. Arcese, and I. G. McLean (1984). Age, experience, and enemy recognition by Song Sparrows. *Behav. Ecol. Sociobiol.*, **14**, 101–103.

Smith, N. G. (1968). The advantage of being parasitized. *Nature*, **219**, 690–694.

Sordahl, T. A. (1986). Evolutionary aspects of avian distraction display: variation in American Avocet and Black-necked Stilt antipredator behavior. In: *Deception: perspectives on human and nonhuman deceit* (eds R. W. Mitchell and N. S. Thompson), pp. 87–712. State University of New York Press, Albany, New York.

Thompson, J. N. (1982). *Interaction and coevolution.* Wiley, New York.

Thompson, J. N. (1992). *The coevolutionary process.* University of Chicago Press, Chicago, Illinois.

Weatherhead, P. J. (1989). Sex ratios, host-specific reproductive success, and impact of Brown-headed Cowbirds. *Auk*, **106**, 358–366.

Webster, M. S. (1994). Interspecific brood parasitism of Montezuma Oropedolas by Giant Cowbirds: parasitism or mutualism? *Condor*, **96**, 794–798.

11

Impact of Brood Parasitism

Why Do House Wrens Accept Shiny Cowbird Eggs?

GUSTAVO H. KATTAN

Because brood parasitism usually depresses reproductive success of the host, hosts are expected to evolve defenses against parasitism (Payne 1977; Rothstein 1990). One such defense is rejection of parasitic eggs. Hosts of parasitic cowbirds (*Molothrus* spp.) can be classified into two discrete groups, "accepters" and "rejecters," according to their responses to natural and experimental parasitism (Rothstein 1975*a*; Mason 1986). Parasitic eggs may be rejected by ejecting the egg, deserting the nest, or constructing a new nest floor over the parasitized clutch.

Why do some hosts accept parasitic eggs? Rothstein (1975*a*, 1982, 1990) suggested that lack of rejection in some cowbird hosts is best explained by absence of genetic variants, within populations, that cause a bird to recognize and reject a foreign egg (evolutionary lag hypothesis). An alternative hypothesis suggests that for some cowbird hosts, rejection incurs costs that exceed any benefits, thereby making acceptance the most adaptive option (cost of rejection or evolutionary equilibrium hypothesis). For example, egg ejection may be costly for hosts with small bills because they are unable to grasp a cowbird egg between the mandibles. The only way to eject the egg would be to puncture it and hold it by the perforation. Cowbird eggshells, however, are unusually thick and therefore very strong and some of the host's own eggs could be damaged in the puncturing process (Rohwer and Spaw 1988; Rohwer et al. 1989; Røskaft et al. 1990; Røskaft and Moksnes, this vol.). This hypothesis may explain why most small-billed hosts accept eggs of the brown-headed cowbird (*Molothrus ater*) (Rothstein 1975*a*, 1982; Rohwer and Spaw 1988). In addition, hosts might mistakenly reject their own eggs. Such mistaken rejections might result in rejection behavior being less adaptive than acceptance, at least in certain circumstances (Lotem et al. 1992, 1995; Lotem and Nakamura, this vol).

Rejection of parasitic eggs by nest desertion also incurs costs that may exceed the costs of accepting parasitism. Nest desertion involves a time cost, which may be significant if duration of the breeding season is restricted. Besides, good nest sites may be in short supply and the replacement nest may also be parasitized (Røskaft et al. 1990; Petit 1991). Thus, to determine the best strategy for a host to follow, the costs of parasitism must be contrasted with the costs of rejection.

The cost of rejection hypothesis assumes that hosts have the ability to recognize foreign eggs, but tolerate parasitism because the costs of rejecting exceed the costs of acceptance. Here I test this hypothesis by examining the impact of shiny cowbird (*Molothrus bonariensis*) parasitism and host responses in a cavity-nesting species, the house wren (*Troglodytes aedon*). It has been hypothesized that cavity-nesting species experience low rates of parasitism, and thus these species have not evolved egg discrimination behavior (Friedmann 1963; Friedmann et al. 1977). An alternative hypothesis is that wrens can

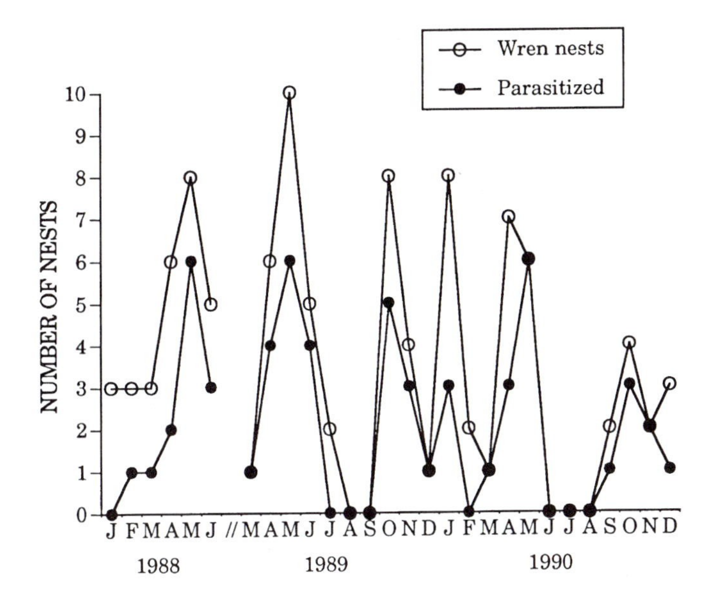

Figure 11.1 Reproductive seasonality of house wrens and shiny cowbirds in the Cauca Valley, southwestern Colombia. Open symbols indicate the number of wren nests with eggs; closed symbols indicate the number of nests parasitized.

discriminate foreign eggs, but are unable to grasp-eject cowbird eggs because they are large and heavy. I tested the following hypotheses and predictions:

1. Wrens can discriminate foreign eggs but can not lift cowbird eggs; thus they should eject artificial odd-looking eggs that are smaller and lighter than cowbird eggs, but similar in size and weight to wren eggs.
2. Wrens can grasp-eject cowbird eggs but do not recognize them as foreign; thus wrens should accept cowbird-sized artificial eggs but should eject odd-looking objects as large and heavy as cowbird eggs.

I also examined the potential costs of nest desertion and contrasted them with the costs of accepting parasitism, to test the hypothesis that wrens tolerate parasitism because the costs of deserting are higher than the costs of accepting.

STUDY AREA AND METHODS

The study was conducted between January and June 1988 and March 1989 and April 1991 at a dairy farm 15 km south of the city of Cali, in the Cauca Valley (1000 m elevation), Colombia. The annual rain regime in the area is bimodal, with peaks of rainfall in April and October, and two dry seasons, a mild one in January and a pronounced one in July–August. Local house wren and shiny cowbird populations breed nearly year-round, with peaks of reproductive activity in the rainy seasons and very low activity during the mid-year dry season (fig. 11.1; see also Alvarez et al. 1984; Kattan 1993; Kattan 1997). Wrens in this area nest in the hollow internodes of fence posts made of *Guadua angustifolia* (a thick bamboo), in farm buildings, and in a variety of natural places such as bromeliads and on the ground in dense clumps of tall grass. They also used wood nest boxes (10 × 10 × 15 cm) that I placed at random in the area. This study is based on 34 natural nests and 105 nests in nest boxes.

The study area encompassed 14 wren territories in well-shaded pastures and around farm buildings. At the beginning of the study I mapped wren territories and color-banded 39 wrens and 37 female cowbirds. I followed the nesting histories of all pairs of wrens in the area and obtained data on incidence (proportion of nests parasitized) and intensity (number of parasitic eggs per nest) of cowbird parasitism. I evaluated the effects of parasitism on nesting success of wrens by comparing clutch size, brood size, and number of wren fledglings in parasitized and unparasitized nests.

To test the discriminatory abilities of wrens, and whether they are capable of grasp-ejecting cowbird eggs, I placed real and artificial cowbird eggs in nests during the pre-laying (1–4 days before wrens laid their first egg) and egg-laying stages (table 11.1). I tested wrens during the pre-laying stage because this is the period when most cowbird eggs are laid (Kattan 1993). For artificial eggs I used commercially obtained plastic egg models. These models are hollow and very light. To make them heavier I filled them with

Table 11.1 Characteristics of wren and cowbird eggs and egg models placed in house wren nests in the Cauca Valley

Model	Dimensions (mm)	Mass (g)	Color of background/markings
Wren eggs	18.5 ± 0.5 × 13.8 ± 0.5	1.9 ± 0.2	Creamy white/reddish-brown
Cowbird eggs	24.1 ± 1.4 × 18.2 ± 0.8	4.2 ± 0.4	Creamy-white/brown
Artificial eggs			
Large	23.5 × 18.4		White/brown
Heavy		4.2	
Light		1.0	
Small	20.0 × 15.7		White/orange
Heavy		2.6	
Light		0.6	
Glass beads	16.0 × 16.0	5.2	Varied

water (table 11.1). Models were scored as accepted when they were incubated, and rejected when they disappeared from a nest that was not abandoned or lost to predation. Hypothesis 1 was tested by placing small artificial eggs in wren nests. These eggs were only slightly larger than a wren egg but were very different looking because of their orange-colored markings (table 11.1). Hypothesis 2 was tested by placing artificial cowbird-sized eggs in wren nests (large artificial eggs in table 11.1). I placed both light and heavy large eggs to test whether wrens could grasp-eject a large but light egg. To test whether both size and weight constrained the wren's ability to lift a cowbird egg, I placed glass beads in wren nests. The diameter of glass beads was similar to the breadth of a cowbird egg, but they were 23% heavier (table 11.1). I expected wrens to eject these odd-looking objects, and thus acceptance would indicate that wrens can not lift them.

RESULTS

Impact of parasitism

Parasitism was extremely common in this population of house wrens (Kattan 1997), and nest boxes were more frequently parasitized than natural nests. Rate of parasitism in natural cavities was 58.8% (20 or 34 nests), versus 94% (99 of 105 nests) in nest boxes ($G = 22.46$, d.f. = 1, $P < 0.001$; table 11.2). Nests in boxes were also more intensely parasitized than natural nests. Eleven of 20 (55%) parasitized nests in natural cavities received more than two cowbird eggs. In contrast, 84 of 99 parasitized nests (84.8%) in boxes received more than two cowbird eggs ($G = 7.80$, d.f. = 1, $P < 0.01$; table 11.2). The average number of cowbird eggs per nest was also significantly higher in boxes than in natural cavities (Mann–Whitney U-test, $P < 0.001$; table 11.2).

Parasitism had a strong negative effect on hatching success of wrens. Wren eggs hatched in only 55% of parasitized nests, compared with 92% of unparasitized nests (table 11.3). Most wren mortality occurred during incubation. Cowbirds did not remove host eggs, and wren clutch size at the beginning of incubation was not different between

Table 11.2 Incidence and intensity of shiny cowbird parasitism on house wrens

	Number of nests (%)	
No. of eggs/nest	Natural	Boxes
0	14 (41.2)	6 (5.7)
1–2	9 (26.5)	15 (14.3)
3–12	11 (32.3)	84 (80.0)
Mean ± S.D.*	1.97 ± 2.2	5.65 ± 3.2

*Mann–Whitney U-test, $P < 0.001$.

Table 11.3 Hatching success of wrens (excluding predation)

	No. of nests	Hatching success* (%)	G	P
Parasitized	20	11 (55.0)	5.9	<0.025
Unparasitized	13	12 (92.3)		

*Number of nests in which at least one wren hatched.

Table 11.4 Reproductive success of wrens in parasitized and unparasitized nests (excluding predation)

	Parasitized	Unparasitized	P*
Initial clutch size	2.97 ± 0.5 (20)	3.12 ± 0.5 (13)	0.3
Clutch size at hatching	0.99 ± 1.0 (15)	2.83 ± 0.9 (11)	<0.01
Fledglings/nest	0.54 ± 0.9 (11)	2.78 ± 0.7 (9)	<0.001

*Mann–Whitney *U*-tests between parasitized and unparasitized nests.

Note: Parasitized nests include only nests that received one cowbird egg. Numbers indicate mean, standard deviation and in parentheses, sample size.

parasitized (2.97 ± 0.5) and unparasitized nests (3.12 ± 0.5; table 11.4). Clutch size at hatching, however, was significantly reduced in parasitized nests (0.99 ± 1.0 vs. 2.83 ± 0.9 wren eggs; sample includes only nests parasitized with one cowbird egg; table 11.4). Clutch reduction was due to wren eggs disappearing during incubation, probably removed by wrens because they broke when jostled against the heavy, thick-shelled cowbird egg. During incubation I frequently found dented wren eggs in parasitized nests, and these eggs always disappeared in the next two or three days.

Fledging success of wrens was also significantly reduced by cowbird parasitism (Kattan 1996). Wrens fledged 2.78 ± 0.7 young per nest in unparasitized nests, while in nests parasitized with one cowbird egg they fledged 0.54 ± 0.9 young per nest (table 11.4). Most wren nestlings in parasitized nests died of starvation within three to four days of hatching. Only four of 11 parasitized nests (36.4%) that survived the entire nesting cycle produced wren fledglings, as opposed to eight of nine (88.9%) unparasitized nests ($G = 6.22$; d.f. = 1, $P < 0.02$).

Host responses to parasitism

Wrens abandoned 92 of 95 nests that received more than two cowbird eggs but showed no rejection at 24 nests that received one or two parasitic eggs ($G = 100.8$, d.f. = 1, $P < 0.001$). There were three cases in which wrens accepted and incubated large cowbird clutches (two with six and one with seven cowbird eggs), all in nest boxes. Two of these nests were abandoned halfway through incubation. Only one cowbird hatched in the third nest. Two other cowbird eggs in this nest had advanced embryos, but did not hatch.

Thirteen nests received one cowbird egg and were successfully incubated by wrens. In nests parasitized with two cowbird eggs, usually one failed to hatch. Only in two of 11 nests with two cowbird eggs did both cowbirds hatch. In one of these, the second cowbird hatched two days after the first one and died three days later. In the other nest, both cowbirds hatched the same day and 10 days later both were growing normally, but the nest was preyed upon. Probably both would have fledged because the cowbirds usually fledge at 12 days of age (Kattan 1993).

Wrens accepted and incubated all 14 real cowbird eggs that I placed in their nests, both before they initiated laying and during their laying period (table 11.5). During the pre-laying period, they also accepted both large and small artificial eggs that were heavy, but were significantly more likely to eject light eggs of both sizes ($P < 0.01$ for both 7 of 7 vs 1 of 10 and for 8 of 8 vs 0 of 6, Fisher exact tests). Wrens seemed more likely to accept large light eggs during laying than during pre-laying but the sample size for the former period was too small to show a conclusive result ($P > 0.1$ for 3 of 5 vs. 1 of 10; table 11.5). Overall, these egg addition experiments show that wrens can grasp objects as large as a cowbird egg with their bills (the artificial plastic eggs were too strong to be pierced by the wrens) and that they discriminate eggs on the basis of weight. Wrens also ejected glass beads (table 11.5), which were heavier than real cowbird eggs, thereby showing that failure to reject real cowbird eggs and small and large heavy artificial eggs was not due to an inability to grasp and lift such eggs. These experiments indicate that wrens have the ability to grasp-eject parasitic eggs, but do not discriminate among eggs based on color or size.

Potential costs of nest desertion

Several factors could limit the benefits of nest desertion. First, wrens could be limited by a short breeding season. At my study site, however, wrens

Table 11.5 Responses of house wrens to experimental eggs placed in their nests, during the prelaying and laying stages of the nesting cycle

	Prelaying			Laying		
	Accept	Reject	*P**	Accept	Reject	*P**
Cowbird eggs	6	0	<0.02	8	0	<0.01
Artificial eggs						
Large						
Heavy	7	0	<0.01	5	0	<0.03
Light	1	9	<0.01	3	2	0.5
Small						
Heavy	8	0	<0.01	—	—	
Light	0	6	<0.02	—	—	
Glass beads	0	4	<0.05	—	—	

*Chi-square test.

had an extended breeding season, nesting almost year-round (fig. 11.1). A similar pattern was reported by Alvarez et al. (1984) for a site about 10 km north of my study site. Birds that deserted nests because of multiple parasitism always started building a new nest within two to three weeks (except when desertion occurred close to the onset of the mid-year dry season). During 1989 and 1990, ten banded females made 3.5 ± 1.35 nesting attempts per year, and some females made up to six attempts per year. Therefore, length of the breeding season is not likely to be a factor that limits renesting possibilities.

A second potential factor limiting renesting possibilities is availability of nest sites. Because my study site was saturated with nest boxes and all wrens had at least six boxes within their territories, I can not tell whether desertion of parasitized nests was contingent upon availability of nest sites, as Petit (1991) did for prothonotary warblers (*Protonotaria citrea*). Most of the boxes remained in the field for 2.5 years and during this time only one new territory was formed, when in early 1990 a pair of wrens nested in a box placed between two old territories. This suggests that nest site availability was not limiting the establishment of territories. Territories, however, probably varied in the number and quality of nesting cavities. Most wrens used both natural cavities and nest boxes during the study, but in some territories wrens never used nest boxes (suggesting an abundance of nesting sites), while in others they always used nest boxes (suggesting a scarcity of nesting sites). Several instances of wrens destroying eggs and nestlings of spectacled parrotlets (*Forpus conspicillatus*) and taking over the nest box, suggest competition for cavities (N. Gómez and G. Kattan, unpubl. data). In relation to brood parasitism, a good-quality cavity is probably one with a small opening. In natural nests, diameter of the entrance of parasitized nests (57.0 ± 9.7 mm, $n = 12$) was significantly larger than the entrance of unparasitized nests (23.0 ± 5.8 mm, $n = 15$; $t = 8.96$, $P < 0.001$). All my nest boxes had openings larger than 35 mm, as cowbirds could not enter boxes with smaller openings. Thus, at least in some territories, wrens may be limited by availability of good-quality cavities.

A third factor that would limit the benefits of desertion is probability of parasitism of a replacement nest. Because nest boxes were more frequently and intensely parasitized than natural nests, I separated the sample of birds renesting after deserting a parasitized nest into birds renesting in boxes and in natural sites. Of 17 cases of pairs renesting in boxes, all were parasitized, while one of six pairs that renested in natural cavities were parasitized ($G = 13.6$, d.f. $= 1$, $P < 0.01$). Parasitism rate of replacement nests in natural cavities (17%) was significantly lower than the 59% overall parasitism rate of natural nests ($G = 3.88$, d.f. $= 1$, $P < 0.05$). After deserting a multiply parasitized nest, birds became wary and difficult to observe, and took a long time to renest (interval between desertion and initiation of a new clutch, $\bar{x} = 31.1 \pm 14.8$ days, $n = 23$; sample does not include nests abandoned just prior to the beginning of the dry season). In contrast, birds that abandoned for other reasons (human interference, nest flooding, predation) renested in 13.1 days (S.D. $= 7.2$, $n = 18$; Mann–Whitney U-test, $P < 0.01$).

DISCUSSION

Shiny cowbird parasitism substantially depressed the nesting success of house wrens. Most of the cost of parasitism was paid during incubation, probably because wren eggs broke when jostled against the thick-shelled cowbird eggs. Disappearance of wren eggs during incubation resulted in a hatching success of only 55% in parasitized nests, as opposed to 92% in unparasitized nests (table 11.3). Cowbird parasitism also reduced the fledging success of wrens. Wrens that hatched soon died of starvation, outcompeted by the more aggressively begging cowbird nestling (Gochfeld 1979, pers. obs.). Wrens fledged in only 36% of parasitized nests, while 90% of unparasitized nests produced fledglings.

The frequency of cowbird parasitism reported in this study is very high for a cavity nester. Petit (1991), for example, reported an incidence of parasitism of 21% for the prothonotary warbler in Tennessee, USA, in both nest boxes and natural cavities. Shiny cowbirds are very abundant in the Cauca Valley, a region with extensive plantations of rice and sorghum, two crops on which they feed. In contrast to natural nests, the nest boxes I used in this study were conspicuous and revealed the high potential for cowbird parasitism. Probably a large proportion of wren nests in the Cauca Valley are found by cowbirds, and the only defense wrens have is to nest in cavities with small openings. Woodward and Woodward (1979) also reported high levels of parasitism on a cavity nester, eastern bluebirds (*Sialia sialis*) in Virginia, USA, and they also attributed it to the conspicuousness of nest boxes, large size of openings, and high density of cowbirds.

Although cowbird parasitism may have little impact in some cases (Smith 1981; Weatherhead 1989), it is usually detrimental to the reproductive success of the host. Detailed studies of both the brown-headed cowbird and the shiny cowbird show that parasitized birds produce less young than unparasitized birds (e.g., Klaas 1975; Fraga 1985; Wiley 1985). The magnitude of the impact is variable and losses may occur at the incubation stage or at the nestling stage. Brown-headed cowbirds are reported habitually to remove a host egg when they lay (Sealy 1992), a habit that has also been reported for some populations of shiny cowbirds (e.g., Fraga 1985). At my study site I found no evidence of cowbirds removing wren eggs, and very rarely did they puncture wren or other cowbird eggs. Instead, losses during incubation probably occurred because host eggs broke when jostled against cowbird eggs, as has also been reported in other studies (Sick 1958; Blankespoor et al. 1982). Host eggs may also fail to hatch because of improper incubation (e.g., Klaas 1975; Petit 1991). Parasitism also reduces fledging success of the host because nestlings are unable to compete with cowbirds. This occurs primarily with small hosts (e.g., Fraga 1978, 1983; Marvil and Cruz 1989). Large hosts are better able to compete, and thus costs of parasitism at this stage are less severe for them (Fraga 1985; Weatherhead 1989; Røskaft et al. 1990).

Given that cowbird parasitism has a negative impact, wrens are expected to exhibit defenses. Wrens rejected parasitism by abandoning multiply parasitized nests but accepted when parasitized with only one or two cowbird eggs. This response could be explained if the cost of parasitism was an increasing function of the number of cowbird eggs in the nest. It could be argued that perhaps when parasitized with one or two cowbird eggs, the cost of rejecting is higher than the cost of accepting. This was not the case, however, because losses were total for wrens parasitized with two or more cowbird eggs. Even when parasitized with only one cowbird egg, the cost of accepting was probably higher than the cost of rejecting.

Wrens parasitized with one cowbird egg have the option of rejecting parasitism, either by ejecting the egg or abandoning the nest. Both of these two alternative modes of rejection have potential costs, which must be balanced against the costs of accepting a parasitic egg. Acceptance of the parasitic egg would be adaptive only if losses due to parasitism are not total and if the reproductive success of individuals that accept is higher than the success of individuals that reject. This may occur only for certain hosts for which nest success is not severely depressed by parasitism. For hosts that suffer very high losses, such as house wrens (see also Klaas 1975; Rothstein 1975*a*; Fraga 1978, 1983), almost any option should be better than accepting parasitism. Besides, wrens should be able to grasp-eject cowbird eggs. Light-weight plastic eggs were ejected, probably because they were perceived as empty shells, and this shows that wrens have the ability to discriminate eggs by weight, and grasp with the bill an object as large as a cowbird egg. Ejection of glass beads indicated that wrens can eject objects as large and heavy as cowbird eggs. Despite this ability, wrens never ejected real cowbird eggs or artificial eggs of the same size and mass. They also accepted small eggs despite their contrasting colors. These experiments indicate that wrens do not recognize cowbird eggs as foreign. Ortega and Cruz (1988) also showed experimentally that the red-winged blackbird (*Agelaius phoenicius*), an accepter species, had the ability to reject objects as large as a cowbird egg. They argued that redwings accept because the frequency and consequences of cowbird parasitism were too slight to be a significant selection pressure. They further suggested that if there are costs associated with ejection that they were unable to detect, selection might even favor acceptance over rejection (see also Røskaft et al. 1990). This is not likely to be true for wrens because the cost of parasitism was very high.

Nest abandonment is another option for wrens. As cavity nesters, wrens may be limited by the availability of nest sites. Acceptance of cowbird eggs by prothonotary warblers was found to be dependent on the opportunity to renest (Petit 1991). In this case, nest site limitation and a short breeding season made it more adaptive for females to accept the relatively low cost of parasitism (Petit 1991). At my study site, however, it is unlikely that these factors limit the benefits of nest desertion as opposed to the cost of accepting parasitism. Although wrens in some territories may be limited by the availability of good nesting sites, the long breeding season should make it more adaptive for a female to abandon and make a renesting attempt later. This idea is supported by the fact that wrens readily abandoned multiply parasitized nests, and replacement natural nests were less likely to be parasitized. Furthermore, the cost of accepting cowbird eggs was very high, i.e., 16 of 20 nests that were parasitized before the initiation of incubation failed to fledge any host young because of parasitism. When a host experiences complete reproductive failure, as occurs with all hosts of the common cuckoo (*Cuculus canorus*) (Davies and Brooke 1989*a*) any form of rejection is likely to be favored by selection. In such circumstances, the only factor that can make acceptance more adaptive than rejection is the occurrence of costs when a host nest is not parasitized, as would happen if hosts make recognition errors and eject their own eggs (Rothstein 1990). While this may explain acceptance by cuckoo hosts, which typically experience low rates of parasitism (Davies and Brooke 1989*a,b*), it is unlikely to do so in the cowbird–wren system I studied as most nests (58.8%) in natural cavities were parasitized.

Given this evidence, it is intriguing to ask why house wrens do not reject cowbird eggs. The selective advantage of rejection depends on the frequency and cost of parasitism (Rothstein 1975*b*; Kelly 1987; Davies and Brooke 1989*b*). If either one or both factors are low, individuals exhibiting the rejection response may have little or no advantage. However, if both factors are high, as is the case for house wrens, the rejection response would be expected rapidly to become common in the population (Rothstein 1975*b*). One factor that may explain lack of rejection in house wrens is that egg discrimination may be difficult inside a dark cavity. However, Mason and Rothstein (1986) found that rufous horneros (*Furnarius rufus*), which place their eggs in a dark enclosed dome made of mud, used tactile perception to discriminate cowbird eggs based on size. Also, in my study wrens rejected when parasitized with three or more cowbird eggs. The stimulus for desertion when three or more cowbird eggs are laid may be the total clutch volume (or a high frequency of cowbird intrusions). Because there is no overlap in egg sizes of cowbirds and wrens, and my experiments indicate that wrens can discriminate eggs by weight, rejection on the basis of size or weight alone should incur little or no cost. Thus, evolutionary lag, that is, the absence of genetic variants with the ability to recognize foreign eggs (Rothstein 1982, 1990), seems to be the best explanation for acceptance of parasitic eggs in house wrens.

ACKNOWLEDGMENTS I thank Sr. Alfonso Madriñán for permission to work on his property, and Dr. Humberto Alvarez for many delightful hours spent discussing shiny cowbirds and house wrens over countless cups of coffee. Natalia Gómez provided very efficient field assistance and many good ideas. Special thanks go to Carolina Murcia for support and encouragement. This work is part of a doctoral dissertation submitted to the University of Florida. I thank my doctoral committee, L. J. Guillette, D. J. Levey, M. L. Crump, R. A. Kiltie, S. K. Robinson, and J. W. Fitzpatrick, for their advice. The very detailed comments of S. Rothstein and S. Sealy greatly improved the manuscript. Financial support was provided by the Fundación para la Promoción de la Investigación y la Tecnología (Banco de la República, Bogotá), and the Fondo de Investigaciones Científicas "Francisco José de Caldas" (COLCIENCIAS, Bogotá, Colombia). I thank the staff of the Instituto Vallecaucano de Investigaciones Científicas, especially Mrs. B. Narváez, for logistical support.

REFERENCES

Alvarez, H., M. D. Heredia, and M. C. Hernández (1984). Reproducción del Cucarachero Común (*Troglodytes aedon*) en el Valle del Cauca. *Caldasia*, **14**, 86–123.

Blankespoor, G. W., J. Oolman, and C. Uthe (1982). Eggshell strength and cowbird parasitism of Red-winged Blackbirds. *Auk*, **99**, 363–365.

Davies, N. B. and M. L. Brooke (1989*a*). An experimental study of coevolution between the Cuckoo, *Cuculus canorus*, and its hosts: I. Host egg discrimination. *J. Anim. Ecol.*, **58**, 207–224.

Davies, N. B. and M. L. Brooke (1989*b*). An experimental study of coevolution between the Cuckoo, *Cuculus canorus*, and its hosts: II. Host egg markings, chick discrimination and general discussion. *J. Anim. Ecol.*, **58**, 225–236.

Fraga, R. M. (1978). The Rufous-collared Sparrow as a host of the Shiny Cowbird. *Wilson Bull.*, **90**, 271–284.

Fraga, R. M. (1983). Parasitismo de cría del Renegrido, *Molothrus bonariensis*, sobre el Chingolo, *Zonotrichia capensis*: nuevas observaciones y conclusiones. *Hornero* (número extraordinario) **1983**, 245–255.

Fraga, R. M. (1985). Host–parasite interactions between Chalk-browed Mockingbirds and Shiny Cowbirds. *Neotropical Ornithology* (eds P. A. Buckley et al.), pp. 829–844. Ornithological Monograph 36, American Ornithologists' Union, Washington, D.C.

Friedmann, H. (1963). Host relations of the parasitic cowbirds. *U.S. Nat. Mus. Bull.*, **233**, 273 pp.

Friedmann, H., L. F. Kiff, and S. I. Rothstein (1977). A further contribution to knowledge of host relations of the cowbirds. *Smithsonian Contributions to Zoology*, **235**, 75 pp.

Gochfeld, M. (1979). Begging by nestling Shiny Cowbirds: adaptive or maladaptive? *Living Bird*, **17**, 40–50.

Kattan, G. H. (1993). Reproductive strategy of a generalist brood parasite, the Shiny Cowbird, in the Cauca Valley, Colombia. Ph.D. dissertation, University of Florida, Gainesville.

Kattan, G. H. (1996). Growth and provisioning of shiny cowbird and house wren host nestlings. *J. Field Ornith.*, **67**, 434–444.

Kattan, G. H. (1997). Shiny cowbirds follow the 'shotgun' strategy of brood parasitism. *Anim. Behav.*, **53**, 647–654.

Kelly, C. (1987). A model to explore the rate of spread of mimicry and rejection in hypothetical populations of cuckoos and their hosts. *J. Theor. Biol.*, **125**, 283–299.

Klaas, E. E. (1975). Cowbird parasitism and nesting success in the Eastern Phoebe. *Occasional Papers Museum of Natural History University of Kansas*, **41**, 1–18.

Lotem, A. and H. Nakamura (this vol.). Evolutionary equilibria in avian brood parasitism: an alternative to the "arms race–evolutionary lag" concept.

Lotem, A., H. Nakamura, and A. Zahavi (1992). Rejection of cuckoo eggs in relation to host age: A possible evolutionary equilibrium. *Behav. Ecol.*, **3**, 128–132.

Lotem, A., H. Nakamura, and A. Zahavi (1995). Constraints on egg discrimination and cuckoo–host co-evolution. *Anim. Behav.*, **49**, 1185–1209.

Marvil, R. E. and A. Cruz (1989). Impact of Brown-headed Cowbird parasitism on the reproductive success of the Solitary Vireo. *Auk*, **106**, 476–480.

Mason, P. (1986). Brood parasitism in a host generalist, the Shiny Cowbird. I. The quality of different species as a host. *Auk*, **103**, 52–60.

Mason, P. and S. I. Rothstein (1986). Coevolution and avian brood parasitism: Cowbird eggs show evolutionary response to host discrimination. *Evolution*, **40**, 1207–1214.

Ortega, C. P. and A. Cruz (1988). Mechanisms of egg acceptance by marsh-dwelling blackbirds. *Condor*, **90**, 349–358.

Payne, R. B. (1977). The ecology of brood parasitism in birds. *Annu. Rev. Ecol. Syst.*, **8**, 1–28.

Petit, L. J. (1991). Adaptive tolerance of cowbird parasitism by Prothonotary Warblers: a consequence of nest site limitation? *Anim. Behav.*, **41**, 425–432.

Rohwer, S. and C. D. Spaw (1988). Evolutionary lag versus bill-size constraints: a comparative study of the acceptance of cowbird eggs by old hosts. *Evol. Ecol.*, **2**, 27–36.

Rohwer, S., C. D. Spaw, and E. Røskaft (1989). Costs to Northern Orioles of puncture-ejecting parasitic cowbird eggs from their nests. *Auk*, **106**, 734–738.

Røskaft, E. and A. Moksnes (this vol.). Coevolution between brood parasites and their hosts: an optimality theory approach.

Røskaft, E., G. H. Orians, and L. D. Beletsky (1990). Why do Red-winged Blackbirds accept eggs of Brown-headed Cowbirds? *Evol. Ecol.*, **4**, 35–42.

Rothstein, S. I. (1975*a*). An experimental and teleonomic investigation of avian brood parasitism. *Condor*, **77**, 250–271.

Rothstein, S. I. (1975*b*). Evolutionary rates and host defenses against avian brood parasitism. *Amer. Natur.*, **109**, 161–176.

Rothstein, S. I. (1982). Successes and failures in avian

egg recognition with comments on the utility of optimality reasoning. *Amer. Zool.*, **22**, 547–560.

Rothstein, S. I. (1990). A model system for coevolution: avian brood parasitism. *Ann. Rev. Ecol. Syst.*, **21**, 481–508.

Sealy, S. G. (1992). Removal of Yellow Warbler eggs in association with cowbird parasitism. *Condor*, **94**, 40–54.

Sick, H. (1958). Notas biológicas sôbre o Gaudério *Molothrus bonariensis* (Gmelin) (Icteridae, Aves). *Rev. Bras. Biol.*, **18**, 417–431.

Smith, J. N. M. (1981). Cowbird parasitism, host fitness and age of the host female in an island Song Sparrow population. *Condor*, **83**, 152–161.

Weatherhead, P. J. (1989). Sex-ratios, host-specific reproductive success and impact of Brown-headed Cowbird parasitism. *Auk*, **106**, 358–366.

Wiley, J. W. (1985). Shiny Cowbird parasitism in two avian communities in Puerto Rico. *Condor*, **87**, 167–176.

Woodward, P. W. and J. C. Woodward (1979). Brown-headed Cowbird parasitism on Eastern Bluebirds. *Wilson Bull.*, **91**, 321–322.

IV

MODELS OF HOST–PARASITE COEVOLUTION: EQUILIBRIUM VERSUS LAG

This part presents general models of host–parasite evolution. All three chapters address the question of whether accepting parasitism is adaptive (making the best of a bad situation) in some circumstances (the evolutionary equilibrium hypothesis), or is the result of a lag in the evolution of anti-parasite adaptations or of cognitive limitations on the part of the host. This issue is also addressed in many of the chapters in parts II and III. For a comprehensive overview of this issue and some alternative interpretations, see pages 23–31 in chapter 1.

12

Evolutionary Equilibria in Avian Brood Parasitism

An Alternative to the "Arms Race–Evolutionary Lag" Concept

ARNON LOTEM, HIROSHI NAKAMURA

The contributions of research on avian brood parasitism to the study of coevolution have been recently demonstrated (Rothstein 1990; Davies and Brooke 1991). Numerous experimental studies have shown the existence of coevolved adaptations in parasites and hosts (reviewed in Rothstein 1990). However, whereas most of these studies conclusively demonstrate the occurrence of coevolution, the details of the mechanisms on which these coevolved systems are based are still a matter of controversy. A major query is: are such systems continuously coevolving as in an arms race or, rather, are they in an evolutionary equilibrium? Under the "arms race" view, the acceptance of parasitic eggs or nestlings, which reduces host fitness, is a result of an evolutionary lag in the development of counteradaptations by the host (Rothstein 1975*a*, 1982*a*; Dawkins and Krebs 1979; Davies and Brooke 1988, 1989*b*; Moksnes et al. 1990). The lag, in this case, might be due to the absence of a new genetic variant (Rothstein 1975*b*) or to the time it takes for such a variant to spread in the host population (Kelly 1987; Davies and Brooke 1989*b*). On the other hand, under the equilibrium view, acceptance is an inevitable result of an equilibrium among various selective pressures (Zahavi 1979; Rohwer and Spaw 1988; Brooker and Brooker 1990; Petit 1991; Lotem et al. 1992). The two alternate views differ in the predictions they present for the future. The evolutionary lag model predicts further evolutionary change in the system (i.e., evolution of new counteradaptations or improvements in such adaptations), whereas the evolutionary equilibrium model predicts that no evolutionary change will occur as long as the system's conditions do not change (temporal fluctuations may occur in a dynamic type of equilibrium, but not a consistent evolutionary change in one direction).

Since the 1970s, the "arms race–evolutionary lag" concept has dominated the field of avian brood parasitism to the extent that the possibility of an evolutionary equilibrium has been almost completely ruled out in the major discussions (Rothstein 1975*a*, 1982*a*, 1990; Davies and Brooke 1988, 1989*b*; Harvey and Partridge 1988). However, excluding a few cases in which the interaction between the parasite and the host is known to have begun within recent decades (Cruz et al. 1985; Nakamura 1990; Soler 1990; but see Zuniga and Redondo 1992), there is no conclusive evidence for an evolutionary lag other than the absence of more convincing explanations (Rothstein 1982*a*, 1990; Davies and Brooke 1989*b*). Thus, the main question in this debate is whether all the possibilities for the existence of an evolutionary equilibrium have been considered and properly tested. In this chapter we argue that the role of equilibrium in the coevolution of parasitic birds and their hosts has been underestimated.

We limit most of our discussion to egg rejection, which seems to be of primary importance. Although it is not the only defense mechanism used by hosts, it is the most common, the best studied and probably the most effective one (reviewed in Rothstein 1990). Egg rejection protects the host young from competition with the parasite nestling or from being ejected by them (in the case of some cuckoo species). As it occurs early in the breeding cycle, egg rejection is also more likely to save time required for renesting if nest desertion or egg burial is involved. Because of its effectiveness, egg rejection might have a higher selective advantage and is expected to evolve faster than other defenses such as nestling rejection (Dawkins and Krebs 1979; Davies and Brook 1988). Hence, as long as it is believed that the absence of egg rejection is due to a lag in evolution, the lack of nestling rejection or any other defense mechanisms simply results from the same lag. On the other hand, if the absence of egg rejection is found to represent an evolutionary equilibrium, further investigation of the absence of other defenses is especially relevant (see McLean and Maloney, this vol.).

We will now review the possible costs and benefits of egg rejection, present them in a combined model, and suggest some possible ways in which an equilbrium might be expressed in a host population. We then discuss available data on avian brood parasitism in relation to the equilibrium hypothesis.

COSTS AND BENEFITS OF EGG REJECTION: A COMBINED MODEL

Because parasitism has been shown to reduce host fitness (Payne 1977; May and Robinson 1985), rejecting the parasitic egg is expected to be advantageous. Egg rejection, however, might also incur several possible costs (Rothstein 1976; Zahavi 1979; Davies and Brooke 1988) that reduce the selective advantage of rejection and result in a slower evolutionary rate and a longer evolutionary lag (Davies and Brooke 1989*b*). If the sum of all costs of egg rejection exceeds the benefit, then accepting the parasitic egg would be adaptive and could be explained as an evolutionary equilibrium. Davies and Brooke (1989*b*) presented a model for acceptance as an evolutionary equilibrium based on a single rejection cost (the cost of recognition errors). Other theoretical models (May and Robinson 1985; Takasu et al. 1993) assumed the existence of a general cost factor, but did not investigate its biologic components and the interaction between them. To understand better the combined quantitative effect of all the costs and benefits of egg rejection, we describe the rejection costs and benefits suggested in the literature, and propose a combined cost–benefit model.

The model

The model, illustrated in fig. 12.1, compares two pure strategies — an accepter and a rejecter. Each arrow in the scheme represents the probability of a certain event (i.e., the probability of being parasitized, of correctly identifying the parasitic egg, etc.). The payoff at the end of each course of events represents the reproductive success of the host under the specific circumstances. The overall payoff of each strategy is the sum of all the payoffs, each multiplied by its probability of occurrence.

The payoff to an accepter

The payoff of an accepter when not parasitized is the average reproductive success of an unparasitized nest and is designated as X. When an accepter is parasitized, it pays the parasitism cost "PC" and thus its payoff is X − PC. The value of PC usually varies between zero and X (Payne 1977), but could also be negative, if being parasitized is advantageous (Smith 1968), or greater than X, if parasitism reduces also the host's future reproductive success.

The payoff to a rejecter — costs when parasitized

Several costs can reduce the benefit of rejection in parasitized nests:

1. Parasite laying damage (PLD): If a parasitized rejecter ejects the parasitic egg without any rejection cost, its payoff is still reduced by PLD, the damage already caused by the parasite during the laying period (i.e., by removal, pecking, or accidental breakage of host eggs; Payne 1977; Wyllie 1981; Soler 1990). PLD is actually part of PC (parasitism cost).
2. Ejection cost (EC): The ejection of the parasitic egg itself might be accompanied by an accidental breakage of host eggs, which reduces the payoff (Rothstein 1976, 1977; Davies and Brook 1988; Rohwer and Spaw 1988; Rohwer et al. 1989; Moksnes et al. 1991).
3. Desertion cost (DC): Deserting a parasitized nest, which is a different strategy than egg ejection, may incur a desertion cost

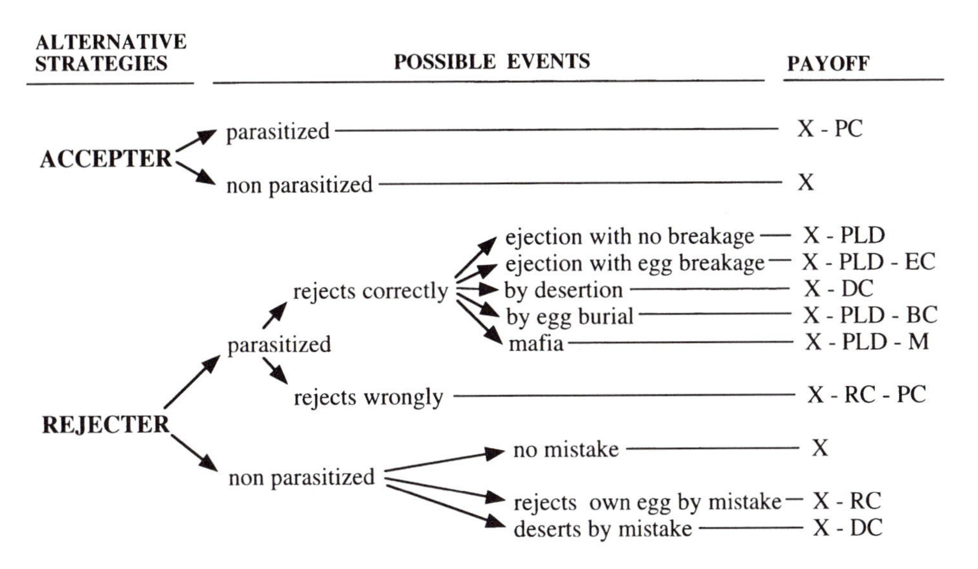

Figure 12.1 A combined cost–benefit model of egg rejection. The model compares two alternative strategies, an ACCEPTER and a REJECTER. Each arrow in the scheme represents the probability of a certain event (i.e., the probability of being parasitized, of correctly identifying the parasitic egg, etc.). The payoff at the end of each course of events represents the reproductive success of the host under the specific circumstances (see detailed explanations in the text). The overall payoff of each strategy is the sum of all the payoffs, each multiplied by its probability of occurring. REJECTS CORRECTLY: rejects the parasite egg. REJECTS WRONGLY: rejects own egg instead of the parasite egg. X: the average reproductive success of an unparasitized host assuming no rejection cost. PC: parasitism cost, the reduction in host reproductive success due to parasitism. PLD: parasite laying damage, the damage already caused by the parasite during the laying period by removal, pecking or breakage of host eggs (PLD is part of PC). EC: ejection cost, egg loss due to accidental breakage of host eggs when ejecting the parasite eggs. DC: desertion cost, the reduction in host reproductive success due to loss of time and energy, smaller clutch size and lower fledgling survival in the renest, or even inability to renest. BC: burial cost. The reduction in host reproductive success due to the loss of time, energy and of a preferred breeding date, when burying a parasitic egg with nest materials. MAFIA: the probability that the parasite revisits the parasitized nest and preys upon the host eggs or nestlings if the parasite egg has been removed. M: the reduction in brood size due to such "mafia" predation by the parasite. RC: recognition cost, egg loss due to mistaken ejections of host's own eggs.

resulting from loss of time and energy, smaller clutch size and lower fledgling survival in the renest, or even inability to renest (Zahavi 1979; Rower and Spaw 1988; Davies and Brook 1989*b*; Petit 1991; Moksnes et al. 1993).

4. Burial cost (BC): As in desertion, burial of a parasitic egg with nest materials may also incur a cost due to the loss of time, energy, and of a preferred breeding date (Clark and Robertson 1981; Davies and Brooke 1988).
5. There is a probability, which we term the "mafia" effect, that the parasite revisits the parasitized nest and preys upon the host eggs or nestlings if the parasitic egg has been removed (Zahavi 1979; Soler et al. 1995). The payoff in this case is reduced by M, the number of host young lost due to predation by the parasite. In a more general form, this arrow in the scheme can be modified to account for any kind of effect (including positive) on the success of a parasitized nest resulting from the parasitism event. For example, parasitized nests from which the parasitic egg has been ejected could have a higher success rate than nonparasitized nests because the parasite does not prey upon nests it has parasitized but only upon other host nests (Davies and Brooke 1988; Arcese et al. 1996).

6. Recognition cost (RC): When egg mimicry by a parasite is well developed, a parasitized host may eject or bury one of its own eggs instead of the parasitic egg (Davies and Brooke 1988). In such a case ("rejects wrongly" in fig. 12.1) it loses the egg and also has to pay the cost of parasitism (PC). In cases where X = PC, the loss of an egg (RC) is not really a cost because a parasitized host that fails to reject the parasitic egg raises no young (Rothstein 1990). However, although in these cases the loss of the egg is meaningless, the probability of making such recognition errors can have a great effect on the cost–benefit balance of egg rejection. This is because every time that such an error is being made, the benefit of rejection is lost.

The payoff of a rejecter — costs when not parasitized

It has been suggested that recognition errors can also be made by rejecters when not parasitized (Zahavi 1979; Rothstein 1982*b*; May and Robinson 1985; Davies and Brooke 1988; Marchetti 1992). This may lead to rejection (by ejection or by egg burial) of one or more of the host's own eggs and thus to a recognition cost (RC), or to desertion of an unparasitized nest which incurs the cost of desertion (DC: see above).

Comments and conclusions related to the model

The distinction between costs when parasitized and costs when not parasitized

The relative importance of costs when parasitized and costs when not parasitized is influenced by the cost of parasitism (PC). In systems where hosts lose all their young if parasitized (X = PC), it is unlikely that rejection costs when parasitized alone could outweigh the benefit of rejection (Rothstein 1990). However, if there are costs when not parasitized, an equilibrium might occur even when parasitized hosts lose all their young (Zahavi 1979; May and Robinson 1985; Rothstein 1990; Lotem et al. 1995).

The distinction between costs when parasitized and costs when not parasitized is especially important when the effect of parasitism frequency is discussed. If costs when not parasitized are involved, the parasitism frequency has an important effect on whether rejection is eventually better than acceptance or vice versa (May and Robinson 1985; Davies and Brooke 1989*b*; Lotem et al. 1992). On the other hand, if costs when parasitized are the only costs involved, the frequency of parasitism can affect only the magnitude of the existing difference between a rejecter's and an accepter's payoffs, but cannot change the direction of selection (i.e., make a rejecter better than an accepter or vice versa). This argument might be better understood by looking at fig. 12.1: in the absence of costs when not parasitized, the "nonparasitized" segments of rejecter and accepter strategies have identical payoffs and therefore can be removed from an equation that compares the two strategies (assuming equal probability of being parasitized as an accepter or as a rejecter). In this case, parasitism frequency becomes a coefficient of the two sides of the equation and, as such, it affects only the magnitude of the absolute difference between the two sides but not the direction of the difference.

The combined effect and its implications

The rejection payoff is a consequence of the combined effect of several factors. Therefore, rejection rates may not be clearly correlated with a single factor. This may explain why some predictions of the equilibrium hypothesis, tested with a single factor, have been rejected. For example, Davies and Brooke (1989*b*) expected that rejection rates should be lower among common cuckoo hosts with small bills. Such birds suffer a higher ejection cost (egg breakage) and/or rely on nest desertion which may be a more expensive rejection method. The overall rejection cost, however, might not be higher among such birds. The cost of desertion could be compensated for by a lower probability of suffering other costs. For example, small hosts may lay small eggs relative to the cuckoo eggs (see Davies and Brooke 1989*a*, fig. 1) and therefore might be less likely to make recognition errors. In addition, desertion cost might be relatively low when parasitism cost is especially high. For small hosts, rearing a cuckoo may be especially expensive and is more likely to result in a loss of future reproductive potential. According to this reasoning, rejection rates should not necessarily be lower among small hosts.

The cost of misimprinting — an additional possible cost

If egg recognition is attained by an imprinting-like process (Rothstein 1974, 1978; Lotem et al. 1992,

1995), an additional possible cost, which was not included in the model, should be considered. A naive breeder that is parasitized during the learning period may learn to recognize the parasitic egg as one of its own (Rothstein 1974). This cost, which we termed the "misimprinting cost," may increase the probability of accepting such parasitic eggs for the host's entire lifetime, and thus reduce the potential benefit of the rejecter strategy. The cost of misimprinting can be reduced if hosts have an innate preference to be imprinted on egg types that are similar to their own (see Lotem et al. 1995). On the other hand, the cost of misimprinting can be very high if the host learns to recognize only parasitic eggs and consequently rejects all of its own eggs. This situation is unlikely to occur in natural conditions because usually both egg types are present in a parasitized nest. Such a misimprinting cost, however, may be a major constraint on nestling recognition, where the probability of having only the parasitic nestling in the nest is relatively high (Lotem 1993). The cost of misimprinting can greatly complicate the model illustrated in fig. 12.1. It may prevent first-time breeders from rejecting, and imposes a severe long-term cost on rejecters that were parasitized during their first breeding. In such circumstances learned recognition may become maladaptive (see Lotem 1993 for a specific model treating this case; also McLean and Maloney, this vol.).

POSSIBLE TYPES OF EQUILIBRIUM

An evolutionary equilibrium resulting from the cost–benefit balance described above may be expressed in a host population in several ways.

An accepter host species

A "rejecter mutant" would not spread in the host population when the overall payoff of an accepter is, on average, greater than that of a rejecter, resulting in an accepter host species.

Intermediate rejection rate

In many parasite–host systems, both rejections and acceptances are exhibited within the host population (Smith 1968; Rothstein 1976; Clark and Robertson 1981; Cruz et al. 1985; Mason and Rothstein 1986; Davies and Brooke 1989*a*; Higuchi 1989; Soler and Møller 1990; Moksnes et al. 1990; Petit 1991; Lotem et al. 1992; Fraga, this vol.). Although most workers agree that some of the cases (e.g., Smith 1968; Rothstein 1976; Clark and Robertson 1981) are likely to represent an evolutionary equilibrium, evolutionary lag is still a common explanation for most other cases. Under the arms race–evolutionary lag concept, the coexistence of rejections and acceptances represents a dimorphic population in which rejection has not yet reached fixation (Kelly 1987; Davies and Brooke 1989*b*), a monomorphic population in which rejection is not yet well developed (Davies and Brooke 1989*b*), or a combination of the two. Alternatively, such intermediate rejection rates may represent the following types of evolutionary equilibria:

A genotypic model

Two genotypes, rejecter and accepter, may coexist in the host population if they are equally adaptive. Coexistence can be stable as a result of a frequency-dependent mechanism in which rejecters have an advantage only when parasitism levels increase, but a high frequency of rejecters, in turn, reduces parasitism levels (May and Robinson 1985; Brooker and Brooker 1990). This mechanism can operate only in systems where there are costs when not parasitized and thus fluctuations in parasitism rate can change the direction of the selection (see above). It is also possible that such changes in selection could be maintained by annual fluctuations in other ecologic factors, such as food abundance or nest site availability, which can affect the cost and benefit balance of the two genotypes (Petit 1991). For example, in years when the host population is dense, nest sites may be limited and thus the cost of desertion might be higher. Food abundance, on the other hand, may affect the intensity of the competition between the parasite and the host nestlings and can thus affect the cost of parasitism.

Mixed strategy — a conditional response

In systems where being a rejecter is better on average than being an accepter, a conditional response may evolve if, under some circumstances rejection is more costly than acceptance. In such cases a rejecter mutant, which suppresses the rejection behavior when it results in a net cost, should spread in the host population. For example, in a case where hosts reject by nest desertion, and late in the season renesting success becomes lower than that of a parasitized nest, the mixed strategy "reject early in the season and accept at the end of the season" should be better than rejection. While such conditional (phenotypic) behavior would be the best strategy for the host, it might also give the parasite some opportunities to be successful. More examples of conditional response will be discussed later in this chapter.

Genotypic variability in the adjustment of the conditional response

The adjustment of the conditional behavior (or, in other words, the reaction norm to parasitism) is likely to be based on some genetic rules that were favored by selection. When the system's conditions (i.e., parasitism frequency, rejection cost, etc.) fluctuate, different adjustments (reaction norms) will be favored under different conditions. For instance, if we take the example discussed above, in some years, the best rule can be "reject during May and June but accept in July," whereas in other years it woud be "reject during May but accept during June and July." If an optimal mechanism, which monitors all possible changes in the system and modifies the adjustment of the conditional rules accordingly, is impossible or too costly to develop, the suboptimal result might be a genotypic variability in the adjustment of the conditional behavior. Such genetic variability in reaction norms can be stabilized in the population as a result of changes in the direction of selection as suggested for the genotypic model (see above).

EVIDENCE FOR EQUILIBRIUM IN AVIAN BROOD PARASITISM

Measuring the costs and benefits

Measuring the costs and benefits of egg rejection, and then entering the measured values in the theoretical model (fig. 12.1), provides a critical test of the equilibrium hypothesis. There is one case in which such direct measurements clearly suggest that the acceptance of a parasitic egg under some circumstances represents an evolutionary equilibrium (Smith 1968). However, this case is an exception, because it is the only one reported in which under certain circumstances being parasitized is advantageous ($X - PC > X$), and then, obviously, there is no value in rejection.

Among systems in which parasitism is deleterious and yet accepted by the host, there is no single study in which all costs and benefits of egg rejection have been measured and can be used for a critical test of the model. Several studies provide measurements of ejection cost due to egg breakage (Rothstein 1976*a*, 1977; Davies and Brooke 1988; Rohwer et al. 1989), of the frequenty of recognition errors in parasitized nest (Molnar 1944; Davies and Brooke 1988), and of the success rate of renests after deserting or burying a parasitized clutch (Clark and Robertson 1981; Burgham and Picman 1989). In a recent study of a cuckoo host (Lotem et al. 1995), we attempted to provide data on all rejection costs and benefits, including the cost of desertion, the "mafia" probability, and the probability of making recognition errors in parasitized and nonparasitized nests. However, as we explain below, some direct measurements might be misleading if host response is conditionally determined.

The problem of conditional behavior

In a conditional response, the host suppresses the rejection behavior when rejection is more costly than acceptance. However, the cost selecting for the conditional behavior cannot be measured because the host becomes an accepter, and avoids the cost of rejection. Ignoring this problem could lead to underestimating the relative cost of egg rejection because what will be measured is only the costs and benefits under the circumstances in which rejection is adaptive. For example, the response of great reed warbles *Acrocephalus arundinaceus* toward cuckoo eggs was found to be conditionally determined according to the host age (Lotem et al. 1992, 1995). Most rejecters were likely to be experienced breeders. In such a case, direct measurements of costs and benefits might be irrelevant; measuring the frequency of recognition errors made by adult rejecters would not indicate the cost that naive breeders might have to pay if they tried to discriminate between egg types. This problem does not exist if rejecters and accepters are two different genotypes. In this case direct measurements of rejection costs and benefits will reliably reflect the fitness of each genotype. Hence, knowing whether the two strategies in a population represent different genotypes or a conditional response is an essential prerequisite for conducting direct measurements of rejection costs and benefits.

In some cases, the problem of conditional response can be solved by experimental simulation of egg rejection. By removing the parasitic egg from a parasitized nest, one can measure the potential benefit of rejection and the "mafia" probability (see Lotem et al. 1995). For measuring the cost of desertion, parasitized nests can be destroyed to force renesting. However, it is not possible to measure directly the cost of egg ejection (due to egg breakage), or the probability of recognition errors, when the host itself does not perform the rejection. Unfortunately, without these values the model cannot provide a conclusive test of the equilibrium hypothesis.

Can the costs outweigh the benefits?

Although measuring the costs and benefits of egg rejection has not yielded complete quantitative results, the available data can help to assess whether equilibria seem reasonable. By entering the known values into the proposed model (fig. 12.1), one can calculate the unknown values required to create an equilibrium. For example, using the model with data collected in our study of great reed warblers in central Japan, we provided a range of conditions in which the cost of recognition errors can outweigh the benefit of rejecting a cuckoo egg (Lotem et al. 1995). Because we found no evidence for a mafia effect, and no major differences between the payoffs of rejection by ejection, by nest desertion or by egg burial, we simplified the original model (fig. 12.1) as illustrated by table 12.1. The equations derived from table 12.1, and describing the overall payoffs of each strategy are: $(1-p)X$ for accepter, and: $p(1-e)(X-1)+(1-p)(1-e)X+(1-p)e(X-1)$ for rejecter, where "p" is the probability of being parasitized, "X" is the clutch size (usually four or five eggs), and "e" is the frequency of recognition errors in parasitized and unparasitized nests. Based on these equations, the minimal frequency of errors required to justify acceptance (i.e., to make accepter and rejecter payoffs equal) is given by: $e = (p - pX)/(2p - pX - 1)$, and illustrated by

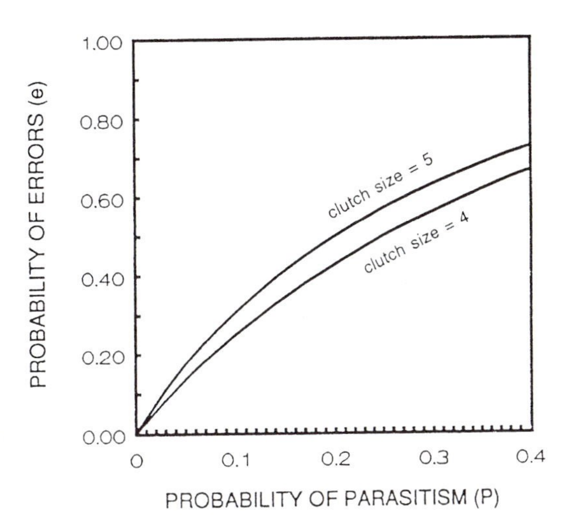

Figure 12.2 The minimal frequency of errors (e) required to justify acceptance for a given rate of parasitism (p), assuming clutch size of four and five eggs. (Based on Fig. 8 in Lotem et al. 1995.)

fig. 12.2 (see Lotem et al. 1995 for more details). The results of the model suggest that under parasitism frequency of 20%, the frequency of errors required to justify acceptance should exceed 43% (when $X = 4$) or 50% (when $X = 5$). On the other hand, if parasitism rate is around 1–5%, even an error frequency of 3–14% (when $X = 4$) or 4–18% (when $X = 5$) can make acceptance by a better strategy than rejection. Among cuckoo hosts in Europe, Britain, and Japan, local parasitism frequency can be as high as 20% or more (Wyllie 1981; Davies and Brooke 1988; Nakamura 1990; Rothstein 1990). Yet, it is likely that the regional parasitism frequency among many cuckoo hosts, and also in great reed warblers in Japan, is only around 1–6% (Moksnes

Table 12.1 The possible events and payoffs following acceptance and rejection of a common cuckoo egg by great reed warblers in Nagano, Japan

Strategy	Event	Payoff
Accepter	Parasitized	Zero
	Nonparasitized	Full clutch
Rejecter	Parasitized and rejected correctly	Full clutch − 1 egg
	Parasitized and rejected its own egg	Zero
	Nonparasitized, no error	Full clutch
	Nonparasitized and rejected its own egg	Full clutch − 1 egg

The overall payoff of each strategy is the sum of all the payoffs, each multiplied by its probability of occurrence. (See text and Lotem et al. 1995, for more details.)

and Røskaft 1987; Davies and Brooke 1989*b*; Lotem et al. 1995). If this is the case, the frequency of errors required to justify acceptance should not be very high (see fig. 12.2). In Europe, where cuckoo egg mimicry is well developed, the frequency of recognition errors in parasitized nests was measured as 15% (2/13) in reed warblers *Acrocephalus scirpaceus* (Davies and Brooke 1988), and 28% (13/46) in the great reed warbler *A. arundinaceus* (Molnar 1944). Although these data may not reflect the risk of recognition errors in nonparasitized nests, they suggest that an equilibrium in such systems is possible.

Similar calculations can be made for hosts of the brown-headed cowbird *Molothrus ater.* These may suggest that an equilibrium due to ejection or desertion costs is a reasonable hypothesis for hosts that do not lose most of their young when parasitized [e.g., song sparrow *Melospiza melodia* (Smith 1981), red-winged blackbird *Agelaius phoeniceus* (Røskaft et al. 1990), and prothonotary warbler *Protonotaria citrea* (Petit 1991)], but not for hosts that usually lose all of their young when parasitized (Rothstein 1990).

Another case in which the rejection costs might outweigh the benefits is that of the redhead duck *Aythya americana*, which parasitizes the canvasback *A. valisineria* (Sorenson, this vol.). In this case, the cost of parasitism cannot be recovered by egg ejection because host eggs are displaced from the nest during the parasite's egg laying (PC = PLD = 22.9%). Desertion of parasitized nests occurs only rarely, typically when there is a sudden reduction in the number of eggs in the nest or when the female is parasitized many times in a short period. In most cases, the cost of desertion may exceed the cost of parasitism because nest success of canvasback ducks may decline rapidly and substantially during the breeding season (Sorenson 1991).

In conclusion, the available measurements of rejection costs and benefits allow the equilibrium hypothesis to be supported or rejected only in a few cases. However, an equilibrium is as reasonable an interpretation as evolutionary lag for many of the cases in which hosts do not reject parasitic eggs.

Adaptive rules in the occurrence of rejections and acceptances within a host population

Strong support for the equilibrium hypothesis comes from several lines of evidence related to adaptive conditional rules evident from host responses. The conditional response model makes a unique prediction. Contrary to the genotypic model (either in equilibrium or during evolutionary lag), the conditional response model predicts that the occurrence of rejections and acceptances will not be random, but follows some adaptive rules (i.e., accept when rejection is too costly and reject when it is not). Evidence for a strong conditional effect would mean that the genotypic model cannot be the sole explanation for acceptance behavior. Such evidence is also likely to point out the factors selecting for the observed conditional behavior. The following are several examples in which host response was found to be conditionally determined and therefore a compromise among selective pressures.

Conditional response in relation to the time of parasitism

Yellow warblers *Dendroica petechia* are more likely to reject a cowbird egg (either by nest desertion or by egg burial) if it is laid early in the breeding season, or early during the host laying stage (Clark and Robertson 1981; Burgham and Picman 1989). These two conditional rules appear adaptive considering the changes in the cost-and-benefit balance under the different circumstances. The chances for a yellow warbler to renest successfully decrease toward the end of the breeding season (Clark and Robertson 1981), which implies a higher cost of desertion or egg burial that would make acceptance adaptive. Accepting a cowbird egg laid late in the host laying stage could be adaptive because the hatching probability of such an egg, like the cost of parasitism if it does eventually hatch, is relatively low (Rothstein 1976; Clark and Robertson 1981). When desertion or egg burial produces more costs than benefits, egg ejection might be an adaptive alternative. The lack of ejection in yellow warblers, however, might be due to high ejection costs incurred when small hosts try to use their small beak for ejecting cowbird eggs (Clark and Robertson 1981; Rohwer and Spaw 1988).

High acceptance rate during late incubation has been found recently also for a cuckoo host species. Moksnes et al. (1993) have shown that meadow pipits *Anthus pratensis* are more likely to accept cuckoo egg models introduced into their nests during late incubation. Moksnes et al. (1993) suggested that cuckoo eggs laid during late incubation are less likely to hatch, and that accepting them may be better than deserting a nest in a habitat where the breeding season is short and renesting is almost impossible. The inability of meadow pipits in this study to reject by egg ejection is yet to be explained.

Conditional response in relation to nest site availability

The tendency of prothonotary warblers *Protonotaria citrea* to desert nests parasitized by brown-headed cowbirds was higher when more than three nest sites were available in the territory (Petit 1991). Petit suggested that this relationship between desertion and the opportunity to renest following desertion, supports the view that under certain circumstances acceptance of cowbird parasitism may be adaptive because the cost of desertion is greater than the cost of parasitism. As with yellow warblers, egg ejection might not have evolved in prothonotary warblers because of the costs incurred when a small host tries to eject a cowbird egg from a nest cavity (Petit 1991).

Conditional response in the bay-winged cowbird

The bay-winged cowbird *Molothrus badius* does not discriminate between its own eggs and the eggs of its parasite, the screaming cowbird *M. rufoaxillaris*, based on differences in their appearance (Fraga, this vol.). Fraga suggested that discrimination has not evolved mostly because of the high risk of recognition errors in the poor light conditions of the host nests. Fraga (this vol.) found that bay-winged cowbirds may reject an entire clutch or decide to incubate it, according to the clutch size. Clutch size apparently is a good indicator of the proportion of parasitic eggs in the clutch, and thus of the expected parasitism cost. Following this decision rule therefore seems adaptive, and may be the best attainable solution in the absence of egg recognition.

Age, experience, and the cost of recognition errors

Acceptance of cuckoo eggs by great reed warblers in central Japan occurs mainly among the younger breeders (Lotem et al. 1992). These findings led us to suggest that the cost of mistakenly rejecting an odd egg from unparasitized nests selects for greater tolerance toward divergent eggs in young breeders. This cost also favors a prolonged learning mechanism in which a host can learn to recognize the range of variation of its own eggs. Further experimental work (Lotem et al. 1995) indicated that a learning mechanism is indeed involved, and that intra-clutch variability in egg coloration is higher among young females (suggesting that the risk of recognition errors is especially high for them). It therefore seems that acceptance of cuckoo eggs by young breeders can be explained as a compromise between the cost of parasitism and the cost of recognition errors.

Stimulus summation and the cost of recognition errors

Whereas American robins, *Turdus migratorius*, usually accept experimental eggs that differ from their own eggs in any one of the three parameters that differ between robin and cowbird eggs, they reject eggs that differ in any two of these parameters. Rothstein (1982*b*) suggested that such stimulus summation prevents recognition errors due to unusually sized or colored robin eggs, ensuring that only a highly divergent egg such as that of the cowbird will be rejected. Among cuckoo hosts, where egg mimicry is well developed, a stimulus that increases rejection probability is the sight of a cuckoo near the nest (Davies and Brooke 1988; Moksnes and Røskaft 1989; Moksnes et al. 1993). Interestingly, although the sight of the cuckoo cannot be helpful in deciding which egg to eject, it usually stimulates the host to eject the right one (Davies and Brooke 1988). This implies that in some of the cases, the host has the ability to identify the cuckoo egg, but suppresses its rejection behavior in the absence of the second stimulus. The evolution of such a mechanism cannot be explained without considering a counterselection pressure that could justify acceptance. The fact that the second required stimulus is the sight of the cuckoo near the nest, suggests that this selection factor is the cost of recognition errors (i.e., ejecting own eggs) when not parasitized.

To test this hypothesis experimentally, Davies and Brooke (1988) presented reed warblers with a stuffed cuckoo (the second stimulus) but with no cuckoo egg model at the nest. If reed warblers could be fooled by a stuffed cuckoo and reject some of their own eggs, it would indicate that the risk of recognition errors is real, and suggest that the stimulus summation mechanism is an adaptation to minimize this risk. Davies and Brooke's results showed that only two out of 16 hosts mistakenly ejected or deserted their own eggs in this experiment. Yet, this value (2/16) is relatively conservative because even when in another experiment, the two stimuli were provided (a mimetic cuckoo egg model and a stuffed cuckoo near the nest) only 47% (17/36) of the hosts exhibited rejection (Davies and Brooke 1988). Because only hosts that tend to reject mimetic eggs are faced with the risk of making mistakes, it seems more appropriate to evaluate the two cases of recognition errors in relation to these 47% of the tested birds. Accordingly, the rate of recognition errors prevented by this stimulus summation mechanism might be as high as 25% (two out of the 47% rejecters among the 16 tested birds). In any event, it seems that if hosts did not suppress their intolerance toward slightly divergent eggs, they would be likely to make recognition errors in about 10–25% of the nonparasitized nests. As was discussed in the previous section, such a rate of recognition errors may be sufficient to maintain an equilibrium. In other words, a host that detects a suspicious egg in the nest but fails to detect the cuckoo near the nest may do better by accepting the suspicious egg than by rejecting it. More such experiments with larger sample sizes could further support this view.

Comparative evidence for equilibrium

A comparative approach might be helpful in testing whether or not host responses toward parasitic eggs are adaptive. Some of the following comparative evidence is consistent with the evolutionary equilibrium view.

Rejection rates: cowbird hosts vs. cuckoo hosts

The hypothesis that the cost of recognition errors selects for some level of acceptance among common cuckoo hosts (see above) is supported by a comparison between cowbird and cuckoo systems. In contrast to cuckoo hosts, most cowbird hosts that exhibit egg rejection ("rejecter species") attain rejection rates of 90–100% (Rothstein 1975*a*, 1990, fig. 1). There are indications that some cowbird hosts learn to recognize their eggs from the first egg laid, and therefore even a yearling can reject (Rothstein, 1974, 1978). This may be possible because cowbird eggs usually differ greatly from the eggs of their host (Rothstein 1975*a*), and parasitism rates among cowbird hosts are usually higher than among cuckoo hosts (Rothstein, 1975*b*; Davies and Brooke 1989*b*). Discrimination is therefore easy and the risk of error is relatively low. Under these conditions, a prolonged learning mechanism is not necessary, and a stimulus summation mechanism should not suppress rejection of a parasitic egg.

Differences between populations

Soler and Møller (1990) showed that magpie *Pica pica* discrimination against mimetic cuckoo eggs was most pronounced in an area of ancient sympatry, whereas nonmimetic eggs were discriminated against both in areas of recent and ancient sympatry. They suggested that this might be explained by: (i) differences in the duration of sympatry (because rejection of mimetic eggs evolves in a later stage of the evolutionary arms race); and (ii) gene flow from the area of ancient sympatry and selection against rejecting mimetic eggs in the absence of the parasite (which used to be the situation in the area of recent sympatry). The second explanation, if correct, suggests that a recognition cost selected against sensitive discrimination ability when parasitism was rare or absent. This may explain why also in the area of ancient sympatry, magpies still accept some of the mimetic cuckoo eggs (Soler 1990).

Zuniga and Redondo (1992) provided additional data on these two particular populations and argued that there is no evidence for differences in the duration of sympatry between the two. Alternatively, they suggested that the higher rejection frequency in one of the populations is associated with a higher parasitism rate and a higher cuckoo density. This new interpretation of the system is also consistent with an equilibrium scenario: rejection of mimetic eggs is more likely to entail recognition errors and is therefore justified only where parasitism rates are sufficiently high.

Egg rejection by favorite and rarely used hosts

Davies and Brooke (1989*a*) found that among suitable common cuckoo hosts, rarely used hosts exhibit higher rejection rates of nonmimetic eggs than current favorite hosts. They raised the possibility that suitable species that are now rarely used are former cuckoo hosts that evolved strong rejection ability, whereas the current favorite hosts are still in an earlier stage of their arms race with the cuckoo (Davies and Brooke 1989*b*). Similar results have been presented by Moksnes et al. (1990) who argued that this trend is actually predicted by the arms race model. However, although these findings can be explained by the arms race model, they are also very much consistent with the idea of an evolutionary equilibrium. Hosts that are faced with relatively high rejection costs, as a result of their life history and ecologic circumstances (e.g., small bills, high desertion cost, or constraints on reducing egg variability within a clutch), are likely to accept cuckoo eggs more frequently and should therefore be favored by cuckoos. On the other hand, hosts that are able to attain higher rejection rates should be used by the cuckoo only rarely in the absence of better alternatives. An additional way to explain these findings as an evolutinary equilibrium is that, at present, rarely used hosts are secondary hosts for the cuckoo, and as such they are likely to be parasitized with nonmimetic cuckoo eggs. Under these conditions their best strategy might be to behave like some of the cowbird hosts that learn to recognize their eggs quickly from the first egg laid (see above). This strategy may allow them to attain high rejection rates of nonmimetic eggs (e.g., 80–100%) but very low rejection rates of highly mimetic eggs (e.g., 0–10%). The current favorite hosts, on the other hand, may adopt a prolonged learning strategy that causes them to exhibit lower rejection rates of nonmimetic eggs (e.g., 50–80%, because naive breeders tend to accept them) but higher rejection rates of mimetic eggs (e.g., 10–30%). This idea can be tested by comparing the response of all these host species to mimetic eggs.

CONCLUSIONS

When considering the available data on avian brood parasitism, the evolutionary equilibrium model seems to be both a reasonable option, and a concept supported by several lines of evidence. At least three species of cowbirds, two species of cuckoos, and one parasitic duck appear to have at least one host species that accepts their eggs as a result of an evolutionary equilibrium. Whereas in one of these examples rejection yields no benefit (Smith 1968), in the others acceptance may be favored by selection because of rejection costs. We believe that these cases are not exceptions, but, rather, represent a general phenomenon. The fact that rejection costs seem to favor some level of acceptance among yellow warblers and prothonotary warblers, suggests that such costs could favor complete acceptance (i.e., being an accepter species) among some other cowbird hosts. The risk of recognition errors may justify acceptance in many systems in which egg or nestling mimicry is involved. This might be true for a broad range of cuckoo hosts and for estrildid finches (Nicolai 1974). Yet, direct evidence for recognition errors in unparasitized nests is rare, and the importance of such errors is suggested mainly by indirect evidence (i.e., host conditional behavior and comparative evidence). Additional data on recognition errors are therefore needed to support further the equilibrium view.

It could be expected that a parasite might benefit from an evolutionary lag in host responses upon interacting with a new host population or with a new host species. In such a case the parasite is likely to be more successful, to increase in numbers, and to impose relatively high selection pressures on the host (Nakamura 1990; Soler and Møller 1990). An evolutinary lag is very likely to be the case in many passerine hosts recently exposed to the expanding cowbird population in North America and to the increasing intensity of parasitism (Rothstein and Robinson 1994). It is also possible that host response would not always be optimal because parasitism rates may fluctuate, ecological conditions may vary from year to year, and because young hosts may disperse into different environments. However, the fact that in several hosts some level of acceptance can be favored by selection, suggests that these hosts provide a stable niche for the parasite, and thus its existence no longer depends upon evolutionary lag.

In conclusion, we believe that the view that avian brood parasitism reflects a continuing evolutionary arms race, and that most variations between systems represent different stages on an evolutionary time scale, is oversimplified. The real picture seems to be composed of a mosaic of evolutionary stable and unstable systems. Possibly, the number of systems that have reached evolutionary equilibrium is far greater than was previously thought.

ACKNOWLEDGMENTS A. Zahavi inspired, encouraged, and assisted us in investigating avian brood parasitism from an evolutionary equilibrium perspective, and thus contributed greatly to this work. We are very grateful to S. I. Rothstein and N. B. Davies for many helpful comments and discussions, to Y. Yom-Tov, J. Wright, T. Redondo, H. Gefen, and M. Norman-Lotem who advised on an earlier draft of this paper, and to S. K. Robinson and S. I. Rothstein for greatly improving the current version. Our research on cuckoos and their hosts was supported by the United

States–Israel Binational Science Foundation (BSF), by a Grant in Aid for Scientific Research (C) from the Ministry of Education, Science and Culture, Japan, and by a doctoral dissertation fellowship from Tel Aviv University. We wish to thank the Tateiwa family and the Nagano International Friendship Club for hospitality in Japan.

REFERENCES

Arcese, P., J. N. M. Smith, and M. I. Hatch (1996). Nest predation by cowbirds and its consequences for passerine demography. *Proc. Natl. Acad Sci., USA*, **93**, 4608–4611.

Brooker, L. C. and M. G. Brooker (1990). Why are cuckoos host specific? *Oikos*, **57**, 301–309.

Burgham, M. C. J. and J. Picman (1989). Effect of brown-headed cowbirds on the evolution of yellow warbler anti-parasite strategies. *Anim. Behav.*, **38**, 298–308.

Clark, K. L. and R. J. Robertson (1981). Cowbird parasitism and evolution of anti-parasite strategies in the yellow warbler. *Wilson Bull.*, **92**, 244–258.

Cruz, A., T. Manolis and J. W. Wiley (1985). The shiny cowbird: a brood parasite expanding its range in the Caribbean region. In: *Neotropical Ornithology* (eds P. A. Buckley, M. S. Foster, E. S. Morton, R. S. Ridgely and F. G. Buckley), Ornithol. Monogr. No. 36, pp. 607–620. Am. Ornithol. Union, Washington, D.C.

Davies, N. B. and M. de L. Brooke (1988). Cuckoos versus reed warblers: adaptation and counteradaptions. *Anim. Behav.*, **36**, 262–284.

Davies, N. B. and M. de L. Brooke (1989*a*). An experimental study of co-evolution between the cuckoo, *Cuculus canorus*, and its hosts. I. Host egg discrimination. *J. Anim. Ecol.*, **58**, 207–224.

Davies, N. B. and M. de L. Brooke (1989*b*). An experimental study of co-evolution between the cuckoo, *Cuculus canorus*, and its hosts. II. Host egg markings, chick discrimination and general discussion. *J. Anim. Ecol.*, **58**, 225–236.

Davies, N. B. and M. de L. Brooke (1991). Coevolution of the cuckoo and its hosts. *Sci. Amer.*, **264**, 66–73.

Dawkins, R. and J. R. Krebs (1979). Arms races between and within species. *Proc. R. Soc. Lond. Ser. B.*, **205**, 489–511.

Fraga, R. M. The natural history and host–parasite interactions of screaming cowbirds (*Molothrus rufoaxillaris*) (this vol.).

Harvey, P. H. and L. Partridge (1988). Of cuckoo clocks and cowbirds. *Nature*, **335**, 586–587.

Higuchi, H. (1989). Responses of the bush warbler *Cettia diphone* to artificial eggs of *Cuculus* cuckoos in Japan. *Ibis*, **131**, 94–98.

Kelly, C. (1987). A model to explore the rate of spread of mimicry and rejection in hypothetical populations of cuckoos and their hosts. *J. Theor. Biol.*, **125**, 283–299.

Lotem, A. (1993). Learning to recognize nestlings is maladaptive for cuckoo *Cuculus canorus* hosts. *Nature*, **362**, 743–745.

Lotem, A., H. Nakamura, and A. Zahavi (1992). Rejection of cuckoo eggs in relation to host age: a possible evolutionary equilibrium. *Behav. Ecol.*, **3**, 128–132.

Lotem, A., H. Nakamura, and A. Zahavi (1995). Constraints on egg discrimination and cuckoo–host co-evolution. *Anim. Behav.*, **49**, 1185–1209.

Marchetti, K. (1992). Costs to host defence and the persistence of parasitic cuckoos. *Proc. R. Soc. Lond. Ser. B*, **248**, 41–45.

Mason, P. and S. I. Rothstein (1986). Coevolution and avian brood parasitism: cowbird eggs show evolutionary response to host discrimination. *Evolution*, **40**, 1207–1214.

May, R. M. and S. K. Robinson (1985). Population dynamics of avian brood parasitism. *Amer. Natur.*, **126**, 475–494.

Moksnes, A. and E. Røskaft (1987). Cuckoo host interaction in Norwegian mountain areas. *Ornis. Scand.*, **18**, 168–172.

Moksnes, A. and E. Røskaft (1989). Adaptations of meadow pipits to parasitism by the common cuckoo. *Behav. Ecol. Sociobiol.*, **24**, 25–30.

Moksnes, A., E. Røskaft, A. T. Braa, L. Korsnes, H. M. Lampe, and H. C. Pedersen (1990). Behavioural responses of potential hosts towards artificial cuckoo eggs and dummies. *Behaviour*, **116**, 64–89.

Moksnes, A., E. Røskaft, and A. T. Braa (1991). Rejection behavior by common cuckoo hosts towards artificial brood parasite eggs. *Auk*, **108**, 348–354.

Moksnes, A., E. Røskaft, and L. Korsnes (1993). Rejection of cuckoo (*Cuculus canorus*) eggs by meadow pipits (*Anthus pratensis*). *Behav. Ecol.*, **4**, 120–127.

Molnar, B. (1944). The cuckoo in the Hungarian plain. *Aquila*, **51**, 100–112.

Nakamura, H. (1990). Brood parasitism of the Cuckoo *Cuculus canorus* in Japan and the start of new parasitism on the Azure-winged Magpie *Cyanopica cyana*. *Jap. J. Ornith.*, **39**, 1–18.

Nicolai, J. (1974). Mimicry in parasitic birds. *Sci. Amer.*, **231**, 92–98.

Payne, R. B. (1977). The ecology of brood parasitism in birds. *Annu. Rev. Ecol. Syst.*, **8**, 1–28.

Petit, L. J. (1991). Adaptive tolerance of cowbird parasitism by prothonotary warblers: a consequence of nest-site limitation? *Anim. Behav.*, **41**, 425–432.

Rohwer, S. and C. D. Spaw (1988). Evolutionary lag versus bill-size constraints: a comparative study of the acceptance of cowbird eggs by old hosts. *Evol. Ecol.*, **2**, 27–36.

Rohwer, S., C. D. Spaw, and E. Røskaft (1989). Costs to northern orioles of puncture-ejecting parasitic cowbird eggs from their nests. *Auk*, **106**, 734–738.

Røskaft, E., G. H. Orians, and L. D. Beletsky (1990). Why do red-winged black-birds accept eggs of brown-headed cowbirds. *Evol. Ecol.*, **4**, 35–42.

Rothstein, S. I. (1974). Mechanisms of avian egg recognition: possible learned and innate factors. *Auk*, **91**, 796–807.

Rothstein, S. I. (1975*a*). An experimental and teleonomic investigation of avian brood parasitism. *Condor*, **77**, 250–271.

Rothstein, S. I. (1975*b*). Evolutionary rates and host defenses against avian brood parasitism. *Amer. Natur.*, **109**, 161–176.

Rothstein, S. I. (1976). Experiments on defences cedar waxwing use against cowbird parasites. *Auk*, **93**, 675–691.

Rothstein, S. I. (1977). Cowbird parasitism and egg recognition of the northern oriole. *Wilson Bull.*, **89**, 21–32.

Rothstein, S. I. (1978). Mechanisms of avian egg recognition: additional evidence for learned components. *Anim. Behav.*, **26**, 671–677.

Rothstein, S. I. (1982*a*). Success and failure in avian egg and nestling recognition with comments on the utility of optimality reasoning. *Amer. Zool.*, **22**, 547–560.

Rothstein, S. I. (1982*b*). Mechanisms of avian egg recognition: which egg parameters elicit host responses by rejecter species? *Behav. Ecol. Sociobiol.*, **11**, 229–239.

Rothstein, S. I. (1990). A model system for coevolution: avian brood parasitism. *Annu. Rev. Ecol. Syst.*, **21**, 481–508.

Rothstein, S. I. and S. K. Robinson (1994). Conservation and coevolutionary implications of brood parasitism by cowbirds. *Trends Ecol. Evol.*, **9**, 162–164.

Smith, J. N. M. (1981). Cowbird parasitism, host fitness, and age of the host female in an island song sparrow population. *Condor*, **83**, 152–161.

Smith, N. G. (1968). The advantage of being parasitized. *Nature*, **219**, 690–694.

Soler, M. (1990). Relationships between the great spotted cuckoo *Clamator glandarius* and its corvid hosts in a recently colonized area. *Ornis. Scand.*, **21**, 212–223.

Soler, M. and A. P. Møller (1990). Duration of sympatry and coevolution between the great spotted cuckoo and its magpie host. *Nature*, **343**, 748–750.

Soler, M., J. J. Soler, J. G. Martinez, and A. P. Møller (1995). Host manipulation by Great Spotted Cuckoos: evidence for an avian mafia? *Evolution*, **49**, 770–775.

Sorenson, M. D. (1991). The functional significance of parasitic egg laying and typical nesting in redhead ducks: an analysis of individual behaviour. *Anim. Behav.*, **42**, 771–796.

Takasu, F., K. Kawasaki, H. Nakamura, J. E. Cohen, and N. Shigesada (1993). Modeling the population dynamics of a cuckoo–host association and the evolution of host defenses. *Amer. Natur.*, **142**, 819–839.

Wyllie, I. (1981). *The cuckoo.* Batsford, London.

Zahavi, A. (1979). Parasitism and nest predation in parasitic cuckoos. *Amer. Natur.*, **113**, 157–159.

Zuniga, J. M. and T. Redondo (1992). No evidence for variable duration of sympatry between the great spotted cuckoo and its magpie host. *Nature*, **359**, 410–411.

13

Coevolution between Brood Parasites and Their Hosts

An Optimality Theory Approach

EIVIN RØSKAFT, ARNE MOKSNES

Among the 9000 or so living species of birds there are at least 80 brood parasites that lay their eggs in the nests of other bird species (Hamilton and Orians 1965). Their eggs are incubated and their young are raised by members of another species. Interspecific brood parasitism has evolved independently at least seven times in birds (Lack 1968). Two well-known examples are the parasitic cuckoos (Cuculidae) of the Old World and the parasitic cowbirds (Icterinae) of the New World.

The brown-headed cowbird (*Molothrus ater*), which is the most common brood parasite in North America, has parasitized at least 220 species of North American birds (Friedmann 1963; Friedmann et al. 1977; Friedmann and Kiff 1985), and at least 70 of these species are parasitized frequently (from 5% to 80% of all nests). The common cuckoo (*Cuculus canorus*), on the other hand, is the most common brood parasite in Europe. Cuckoo eggs have been found in the nests of more than 100 species, but only 15–20 have been frequently parasitized (Chance 1940; Baker 1942; Makatsch 1955; Wyllie 1981; Moksnes et al. 1990; Moksnes and Røskaft 1995). The parasitism frequencies of common cuckoo hosts are generally low, normally below 5% (Glue and Murray 1984; Brooke and Davies 1987; Moksnes and Røskaft 1987). Cowbird hosts, on the other hand, may occasionally suffer from rates that are considerably higher (between 20% and 100%) (Friedmann et al. 1977; Rothstein 1990). In contrast to the common cuckoo, which lays eggs that mimic those of the host species (Baker 1942; Lack 1968; Wyllie 1981), the eggs of the brown-headed cowbird are usually easily distinguishable from those of the host species (Rothstein 1975*a*).

In many instances, brood parasitism is exceedingly harmful to the host. The nestlings of some cuckoos and honeyguides are known to kill all the host's own young (Friedmann 1948, 1955). In contrast, the chicks of some cuckoos (Mundy and Cook 1977; Soler 1990) and cowbirds tolerate the presence of both the eggs and the young of the host species and may therefore have little effect on reproductive success of the host (Payne 1977*b*; see also Smith 1968, 1979, 1980). Despite this tolerance, the presence of the young of a cuckoo or a cowbird often increases the mortality of the host's nestlings. Parasitism may thus have a significant negative effect on the breeding success of the host.

Investigations of coevolutionary adaptations and interactions between the parasite and the host are necessary for a proper understanding of the mechanisms underlying brood parasitism in birds. Extreme adaptations such as thicker eggshells, mimicry of the gape or call patterns of the young, or mimicry of the parasite's eggs with those of the appropriate host, are all to be encountered among brood parasites (Baker 1942; Smith 1968; Nicolai 1974; Payne 1982; Mundy 1973; McLean and Waas 1987; Redondo and Arias-de-Reyna 1988; Spaw and Rohwer 1987). Counteradaptations, such as rejection of the parasite's eggs and aggression by the host, are found among the hosts (Rothstein 1975*a*; Davies and Brooke 1988; 1989*a*; Moksnes et al. 1990).

In this chapter, we use optimality models to evaluate the reactions by different host species that can be expected when they are faced with an egg of a parasite in their nest. We are thus dealing mainly with only one facet of the coevolutionary process (or arms race). The reaction will serve only as a defense if it reduces the impact of parasitism and has been evolved in response to, or is currently maintained by, the selection pressure arising from the act of parasitism (Rothstein 1990). We then use published and unpublished data to test these models.

Before using the above data to test such models, some of the implicit assumptions need to be clarified. 1) The host species being tested must be known to have been involved in a struggle with the brood parasite. The expansion of some American cowbirds into new areas, for instance, may have resulted in frequencies of parasitism that have driven some host species close to extinction because some of these species have not yet evolved anti-parasite strategies (Mayfield 1961, 1962, 1973; Post and Wiley 1977; Wiley 1985). 2) The host must be capable of recognizing the egg of the parasite, or the adult parasite itself, as an enemy. Many cuckoo hosts are probably unable to recognize the cuckoo's egg because of successful egg mimicry. There is little evidence for egg mimicry by cowbirds, although the brown-headed cowbird lays an egg that resembles those of some of its hosts (Elliott 1977). The shiny cowbird (*Molothrus bonariensis*) also lays an egg of a type that, in many respects, can be regarded as exhibiting mimicry (Mason and Rothstein 1986). Rothstein (1975*a*, 1982, 1990) has further argued that many hosts do not recognize the parasite's egg because of an evolutionary lag, that is, rejection would be adaptive but has become neither common nor even detectable, because it takes time for the necessary genetic changes to occur and to spread within the host population.

OPTIMALITY MODELS

May and Robinson (1985), Kelly (1987), Brooker et al. (1990), Takasu et al. (1993), and Yamauchi (1995) have previously modeled parasite–host interactions. These models must not be confused with the following models that only predict the expected responses of a host towards brood parasitism when faced with a parasite's egg in the nest. The optimality models used in this paper are based on ESS (Evolutionary Stable Strategy) models (cf. Maynard Smith 1974; 1982; Maynard Smith and Price 1973) and predict an adaptive response towards brood parasitism. Should available data not support such models, then an alternative explanation for the host behavioral pattern might be the time-lag model proposed by Rothstein (1982, 1990).

As argued above, any bird species that is at risk of being parasitized by a brood parasite has two possible options. First, it can *guard* the nest whenever possible. So far, however, there is little evidence that the birds actively guard their nests (Briskie and Sealy 1987; Burgham and Picman 1989; but see also Slack 1976). Second, the alternative for most species is therefore to leave the nest unguarded, face the risk of being parasitized, and take action subsequently.

Given that a bird is able to recognize a parasitic egg in its nest, then there are three possible strategies that a host can adopt: acceptance (i.e., the bird accepts the egg and broods the eventual parasite's nestling); ejection (i.e., the host ejects the parasitic egg, either by grasp- or by puncture-ejection); or finally desertion (i.e., the primary nest is deserted, or the first clutch buried).

The fitnesses of each strategy are W(A), W(E) and W(D), respectively. The probability of being parasitized at time i is p_i, and at time j, p_j. Furthermore, let the reproductive outcome of an unparasitized nest be U_i or U_j, and that of a parasitized nest be P_i or P_j at times i and j, respectively. R_i is U_i minus the number of eggs removed by the parasite when laying. C_e is the cost of ejection, and D_i is the benefit of desertion at time i. The respective fitnesses of the different strategies will then be

$$W(A) = p_i P_i + (1 - p_i)U_i$$
$$W(E) = p_i(R_i - C_e) + (1 - p_i)U_i$$
$$W(D) = p_i D_i + (1 - p_i)U_i$$

Is ejection an optimal strategy?

We assume that a population consists of A strategists. Is it possible for this population to invoke an E strategy? A will represent an optimal strategy if W(A) ≥ W(E). Thus,

$$p_i P_i + (1 - p_i)U_i \geq p_i(R_i - C_e) + (1 - p_i)U_i$$

which means that

$$C_e \geq R_i - P_i \qquad (13.1)$$

If $C_e \geq 0$, E will invade for any value of $C_e < R_i - P_i$.

Is desertion an optimal strategy?

Let us still assume that a population consists of A strategists, and that the population can not invoke an E strategy (viz. $C_e \geq R_i - P_i$). Is it possible for this population to invoke a D strategy?

A is an optimal strategy if $W(A) \geq W(D)$. Thus, $p_i P_i + (1 - p_i)U_i \geq p_i D_i + (1 - p_i)U_i$, or $P_i \geq D_i$. Because desertion gives possibilities for renesting we have

$$\begin{aligned} D_i &= (1 - p_j)U_j + p_j(1 - p_k)U_k + p_j p_k(1 - p_l) \\ &\quad U_l + p_j p_k p_l(1 - p_m)U_m + p_j p_k p_l p_m(1 - p_n) \\ &\quad U_n + p_j p_k p_l p_m p_n D_n \\ &= U_j - p_j(U_j - U_k) - p_j p_k \\ &\quad (U_k - U_l) - p_j p_k p_l(U_l - U_m) - p_j p_k p_l p_m \\ &\quad (U_m - U_n) - \cdots, \end{aligned}$$

which means that unless

$$\begin{aligned} P_i \geq U_j - [&p_j(U_j - U_k) + p_j p_k(U_k - U_l) \\ &+ p_j p_k p_l(U_l - U_m) + \cdots], \end{aligned} \qquad (13.2)$$

a D strategy will be invoked. This means that a D strategy will invade if the benefit of desertion is greater than the reproductive success of a parasitized nest.

Consider the situation when the reproductive success (U) of unparasitized nests is constant for renesters during the breeding season

According to the optimality model (13.2), A will be an optimal strategy if $P_i \geq U_j$. Therefore, given that the reproductive success of unparasitized renesters is constant during the breeding season, D will be an optimal strategy if $P_i < U$.

Consider the situation when $U_i - U_j$ is constant (C) during the breeding season

If the chances of reproductive success of a renester are constantly decreasing throughout the breeding season, such that $U_i - U_j = C$, we obtain the following requirement for A to be an optimal strategy according to equation (13.2):

$$P_i \geq U_j - p_j C(1 + p_k + p_k p_l + p_k p_l p_m + \cdots).$$

Including recognition costs

In these models we have assumed that the host is capable of recognizing the parasitic egg in its nest. However, many hosts — especially those parasitized with mimetic cuckoo eggs — may not always be able to distinguish the parasitic egg. These hosts may therefore in some cases reject their own eggs by mistake instead of the parasitic egg. If we include such recognition costs, C_r (see, e.g., Davies and Brooke 1988; Marchetti 1992; Lotem et al. 1995), in the model for ejection we have:

$$W(E) = p_i(R_i - C_e - C_r) + (1 - p_i)(U_i - C_r).$$

A will represent an optimal strategy if $W(A) \geq W(E)$. Thus,

$$\begin{aligned} p_i P_i + (1 - p_i)U_i \geq\ & p_i(R_i - C_e - C_r) \\ & + (1 - p_i)(U_i - C_r), \end{aligned}$$

which means that

$$C_e \geq R_i - P_i - C_r/p_i.$$

If $C_e \geq 0$, E will invade for any value of $C_e < R_i - P_i - C_r/p_i$. Models including recognition costs will not be considered any further in this paper.

ESTIMATING THE PROBABILITY OF BEING PARASITIZED, p_i

Breeding season of the brown-headed cowbird and frequencies of parasitism

It has been argued that the peak of the cowbird breeding season falls at a time when most of its host species have reached their breeding peak (Payne 1977*a*; Wiley 1988). For the northern populations this peak is normally in late May and continues throughout June. However, for the southern populations, this peak is much longer, lasting up to two months, depending on the composition of the community of potential host species (Payne 1973). Scott and Ankney (1979, 1980, 1983), from examination of weekly samples of cowbird oviducts in Ontario, showed that the number of eggs laid reached its peak in late May and continued throughout June. A number of weekly collections of cowbird eggs from Kansas, Oklahoma, and Pennsylvania also reached a peak in late May/early June (Norris 1947; Wiens 1963; Hill 1976; Lowther 1977). Furthermore, data for

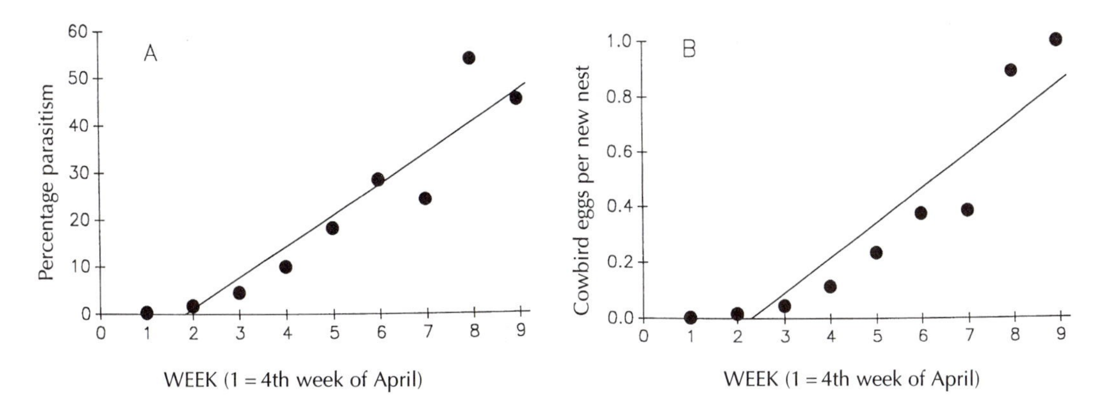

Figure 13.1 The probabilities of (A) red-winged blackbirds being parasitized by cowbirds during the breeding season in eastern Washington, USA, and (B) the number of cowbird eggs laid in each new red-winged blackbird nest during the breeding season. (Data from Orians et al. 1989.)

the red-winged blackbird (*Agelaius phoeniceus*) in Washington showed that most cowbird eggs were laid in late May and early June (Orians et al. 1989). The probability of a red-winged blackbird being parasitized also increased during late May and early June (fig. 13.1; Orians et al. 1989; Freeman et al. 1990). Data for Brewer's blackbird (*Euphagus cyanocephalus*; in Friedmann et al. 1977) in Washington showed a similar pattern, with the probability of being parasitized increasing during late May and early June. A number of other studies of individual host species have yielded a similar pattern [e.g., Abert's towhee, *Pipilo aberti*, in Arizona (Finch 1983); northern cardinal, *Cardinalis cardinalis*, in Ontario (Scott 1963); dickcissel, *Spiza americana*, in Kansas (Zimmermann 1983); prairie warbler, *Dendroica discolor*, in Indiana (Nolan 1978); solitary vireo, *Vireo solitarius*, in Colorado (Marvil and Cruz 1989)]. For song sparrows (*Melospiza melodia*), Smith and Arcese (1994) found that the number of cowbird eggs laid and the length of their laying period both increased with increasing host density. The frequency of parasitism varies from zero to 100% among cowbird hosts, depending on the state and the composition of the host community (Hann 1937; Nice 1937; Schrantz 1943; Twomey 1945; Norris 1947; Berger 1951; Hofslund 1957; Southern 1958; Mayfield 1960; Graber 1961; Terrill 1961; Overmire 1962; Wiens 1963; Young 1963; Newman 1970; McGeen 1972; Klaas 1975; Hill 1976; Middleton 1977; Elliott 1978; Nolan 1978; Finch 1983; Zimmermann 1983; Marvil and Cruz 1989; Weatherhead 1989; Petit 1991; Smith and Arcese 1994; Sealy 1995). High frequencies of parasitism of shiny cowbird hosts have also been reported from areas that have only recently had contact with this brood parasite (Post et al. 1990).

Common cuckoo breeding season and rates of parasitism

Except for some high frequencies of parasitism reported for particular localized host populations of different cuckoos in Europe, Japan, Africa, and Australia (Capek 1896; Owen 1933; Payne and Payne 1967; Jensen and Clinning 1974; Vernon 1984; Yamagishi and Fujioka 1986; Fry et al. 1988; Nakamura 1990; Soler and Møller 1990), the frequencies of parasitism by the common cuckoos seem to be generally low [0–4% in Britain (Lack 1963, 1968; Glue and Morgan 1972; Glue and Murray 1984; Brooke and Davies 1987; Davies and Brooke 1989*b*); 0–6% in Norway (Moksnes and Røskaft 1987)]. Studies from Australia and New Zealand have shown that the frequencies of parasitism of these host species by Australian cuckoos are also generally low (Gaston 1976; Gill 1983; McLean 1988; Brooker and Brooker 1989).

In a study of a sample comprising more than 12 000 eggs of cuckoos held in European museums (Moksnes and Røskaft 1995), we have found that the peak of egg laying for the common cuckoo in Europe corresponds quite well with that of the cowbird in the USA with the cuckoo egg-laying period mainly in June. The probability of being parasitized is therefore highest for those host species that breed in June. On the Danish island of Zealand, with a northern cuckoo population, the

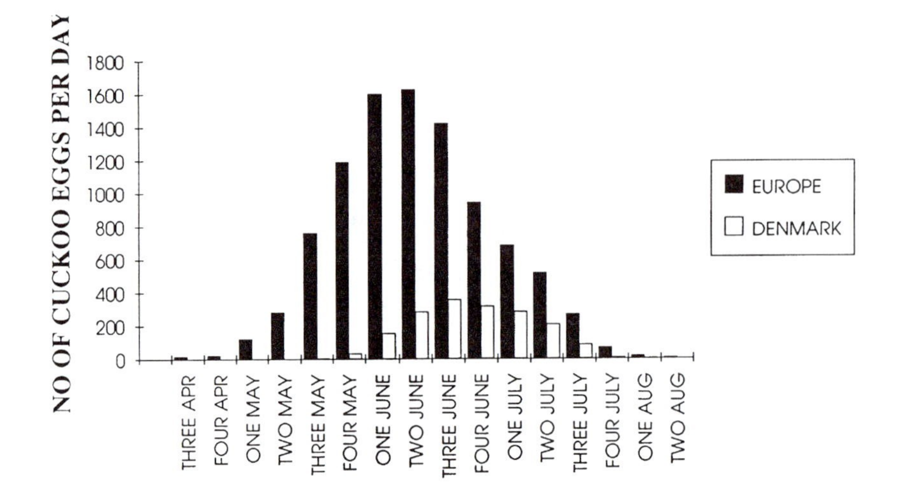

Figure 13.2 Number of cuckoo eggs collected in different weeks throughout the breeding season: data from collections in European museums. From all of Europe (filled bars; n = 9877) and from Zealand in Denmark (open bars, n = 1734). (Data from Moksnes and Røskaft, unpubl.)

peak of the breeding season falls later than that shown by the European material as a whole (fig. 13.2). In Japan, the frequency of cuckoo parasitism of the azure-winged magpie (*Cyanopica cyana*) was 11.1% in early May and as high as 68.8% in late May (Yamagishi and Fujioka 1986).

ESTIMATING THE COSTS OF EJECTION (C_e)

The costs of egg ejection by cowbird hosts have been reported by Rothstein (1976*b*, 1977) from the results of studies of the ejection behavior of northern orioles (*Icterus galbula*) and cedar waxwings (*Bombycilla cedrorum*), respectively. Davies and Brooke (1988) have studied the cost to reed warblers (*Acrocephalus scirpaceus*) of ejecting a cuckoo egg from their nests.

Spaw and Rohwer (1987) and Rohwer and Spaw (1988) suggested that the thick shell of the egg of the parasitic American cowbird species has evolved so as to resist puncture ejection by small host species. An implicit assumption of this hypothesis was that small-billed host species would be unable to eject the thick-shelled egg of the cowbird. They therefore measured the length and the width of the bills (the multiple of these two measurements they termed the *grasp index*) of a variety of brown-headed cowbird hosts which had previously been classified as either acceptors or rejectors. Rohwer and Spaw (1988), when considering the possibility of physical constraints on ejection by those host species parasitized by brown-headed cowbirds, made a distinction between two types of ejection. They compared the characteristic type of ejection (grasp or puncture) or lack of ejection response (acceptance) and the bill size for each of 40 parasitized passerine species. Their results suggest that a small bill prevents some species from properly grasping a cowbird's egg to eject it, and that the thickness of the cowbird's eggshell prevents successful puncture ejection by most of those species. They proposed that the costs associated with these constraints have selected for acceptance.

Rohwer et al. (1989) found that the northern oriole occasionally damaged some of their own eggs when puncture-ejecting the cowbird's egg from the nest. Grasp ejectors such as the American robin (*Turdus migratorius*) and the eastern kingbird (*Tyrannus tyrannus*) did not damage any of their own eggs when ejecting the cowbird's egg (table 13.1).

So far, few researchers have estimated such costs of puncture ejection. In a study of northern orioles, Røskaft et al. (1993) estimated such costs to be 0.26 host eggs per ejection. This estimate was calculated from the damage caused by the orioles to their own eggs when ejecting the cowbird egg. They thereafter estimated the hatching success of such eggs and used these combined data to arrive at a more accurate calculation of the costs of

Table 13.1 Cost of ejection (as number of host eggs lost per ejection) for different puncture (P) and grasp (G) ejector host species of a brown-headed cowbird (BHC) and a common cuckoo (CC) egg

Species	Type of ejection	Parasite species	Costs of ejection	Source
Northern oriole	P	BHC	0.26	Røskaft et al. (1993)
Eastern kingbird	G	BHC	0	Rohwer et al. (1989)
American robin	G	BHC	0	Rohwer et al. (1989)
Cedar waxwing	P/G	BHC	0.37*	Rothstein (1976*b*)
Gray catbird	G	BHC	0.02*	S. Rothstein (pers. commun.)
Reed warbler	P	CC	0.38*	Davies and Brooke (1988)

Note: *Estimated as number of host eggs broken per rejection; this estimate is therefore somewhat higher than for the northern orioles.

ejection. Although, as yet, no serious rearing experiments have been carried out on northern orioles, provisional results indicate that the cost of rearing a young cowbird exceeds the cost of ejecting the egg (Røskaft et al. 1993). Feeding these data into the optimality model (Eq. 13.1) gives a value of $0.26 < R_i - P_i$, which means that the observed ejection behavior by northern orioles is an optimal strategy. Table 13.1 indicates the estimated costs of ejection for some hosts.

Davies and Brooke (1988) have estimated the ejection costs for some hosts of the common cuckoo. Moksnes et al. (1991) found that many common cuckoo hosts damaged their own eggs when ejecting artificial cuckoo eggs, indicating that puncture ejection may be a costly process, even for cuckoo hosts. Some of these host species even damaged some of their own eggs when puncture ejecting conspecific eggs (Braa et al. 1992). Davies and Brooke (1988) have also pointed out that their reed warblers frequently ejected some of their own eggs when parasitized by the common cuckoo (table 13.1).

Brown-headed cowbird and common cuckoo eggs are similar in many respects (Moksnes et al. 1991), but because the rearing costs for cuckoo hosts are higher than those for cowbird hosts, we would expect to find that the cuckoo hosts will tolerate a higher cost of ejection than the North American cowbird hosts. This prediction was tested on a sample of 27 cowbird hosts and 17 similar-sized cuckoo hosts by Moksnes et al. (1991), who found that only two out of 27 cowbird hosts were puncture-ejectors, whereas seven out of the 17 cuckoo hosts were puncture-ejectors ($P < 0.01$). Davies and Brooke (1989*a*), likewise, found that the smaller-billed cuckoo hosts in their study suffered greater costs when rejecting model cuckoo eggs than those with larger bills. So far, however, no really accurate estimates of the costs of ejection by common cuckoo hosts have been made.

Davies and Brooke (1988) considered that one kind of ejection cost is that involved in the prior recognition of the cuckoo egg. This sometimes leads to mistaken rejection of the host's own eggs, in some few cases even when the nest has not been parasitized by the common cuckoo (e.g., see Marchetti 1992); however, this hypothesis is difficult to verify. On the other hand, if the pressure of parasitism causes hosts occasionally to reject their own eggs, those populations which are not subject to parasitism would be under selection pressure favouring acceptance, that is, the A-genotype would prevail and the host behavior become an optimal strategy. Recognition costs could explain why *Ploceus* weavers on Hispaniola, which have been free of parasites for over 200 years, have lower rejection rates than conspecific populations in Africa that are exposed to parasitic cuckoos (Cruz and Wiley 1989).

To conclude, few data exist regarding the cost of ejection by either cowbird or cuckoo hosts. However, significantly more cuckoo host species are ejectors than are cowbird hosts (Moksnes et al. 1991), which indicates that cuckoo hosts may be capable of tolerating a higher cost of ejection. This is also predicted by the optimality model. To evaluate the high rate of acceptance by cowbird hosts, some suitably designed experiments need to be conducted. It is difficult to estimate the ejection costs among many cowbird hosts because they are acceptors. However, calculation of these costs is possible if cowbird eggs would be introduced into

Table 13.2 The mean clutch size and mean number of young fledged by parasitized and unparasitized hosts in relation to the body size of the host

		Parasitized nests		Unparasitized nests			
Species	Bodyweight (g)	Egg	Fledged	Egg	Fledged	R_F	Source
Red-winged blackbird	52.7	2.8	1.8	3.7	2.7	0.88	Røskaft et al. (1990)
		3.1	2.8	4.2	3.2	1.19	Weatherhead (1989)
Dickcissel	29.6	2.4	1.8	4.0	3.7	0.81	Zimmermann (1983)
		2.4	2.0	4.0	3.2	1.04	
Eastern phoebe	19.8	3.4	1.7	4.3	3.8	0.57	Klaas (1975)
Red-eyed vireo	16.7	2.3	0.9	3.2	3.1	0.40	Southern (1958)
American goldfinch	12.9	4.4	2.3	5.2	2.9	0.94	Middleton (1977)
Yellow warbler	9.5	4.1	2.8	4.4	3.3	0.91	Goosen and Sealy (1982)
		3.7	1.9	4.6	3.6	0.66	Weatherhead (1989)
Solitary vireo	16.5	3.2	0.5	3.7	2.4	0.24	Marvil and Cruz (1989)
Prothonotary warbler	15.0	4.6	3.5	5.0	4.5	0.85	Petit (1991)
		3.9	2.9	4.6	3.9	0.88	

Note: Weights from Dunning (1984). R_F = % of eggs that resulted in fledged young in parasitized nests relative to % fledged in unparasitized nests.

nests of European ejector species of similar size as the cowbird hosts.

ESTIMATING THE BREEDING SUCCESS OF PARASITIZED P_i AND UNPARASITIZED U_i NESTS

When using optimality models to evaluate the significance of host adaptations towards brood parasitism in birds, one of the most important variables to be calculated is the breeding success of parasitized and of unparasitized nests, respectively. Ideally, this estimate should be based on the number of recruits to the population produced by the two categories (Gustafsson and Surtherland 1989) or be expressed in terms of reproductive costs (Røskaft 1985; Linden and Møller 1989). Such data may well be difficult to obtain for most species.

Many authors consider that a great cost is involved for cowbird hosts in rearing the young cowbird (e.g., Nice 1937, 1943; Mayfield 1961; Walkinshaw 1961; Scott 1963; Newman 1970; Klaas 1975; Hill 1976; Elliott 1978; Nolan 1978; Clark and Robertson 1981; Sedgewick and Knopf 1988), whereas others maintain that the cost is much less (Smith 1981; Zimmermann 1983; Smith and Arcese 1994). According to Røskaft et al. (1990), the cost of rearing a cowbird chick by a female red-winged blackbird is relatively small, a finding that has been supported by Weatherhead (1989). However, the cowbird often removes one host egg during the act of parasitism (see e.g., Rothstein 1990; Sealy 1992; Smith and Arcese 1994). The hosts of the shiny cowbird frequently show a lower rate of fledging success from parasitized than from unparasitized nests (Wiley 1985).

If nestling competition is the main reason for the lower rate of fledging success reported for parasitized nests, one would expect the chicks of the smaller-sized host species to compete less successfully with the cowbird chick. A reasonable prediction would therefore be that a negative correlation should exist between the size of the host and the cost involved in rearing a young cowbird. Rothstein (1975*b*) listed 14 different host species in regard to the harm caused by the young cowbird in the nest. By this ranking, a highly significant negative correlation was found between the mean bodyweight of the host species and the harm caused by the young cowbird ($r_{sp} = 0.84$, $n = 14$, $P < 0.001$, one-tailed). Differences in the breeding successes of parasitized and unparasitized nests of different cowbird hosts are shown in table 13.2. However, only weak support was found for the prediction that smaller-sized hosts suffered higher overall costs (see data presented in table 13.2; Kendall's $t = -0.214$, $n = 8$, $P = 0.229$, one-tailed; excluding the American goldfinch (*Carduelis tris-*

Table 13.3 The hatching success of eggs in unparasitized and parasitized nests in relation to the body size of the host

		Parasitized nests		Unparasitized nests			
Species	Bodyweight (g)	Egg	Hatch	Egg	Hatch	R_H	Source
Red-winged blackbird	52.7	2.7	1.8	3.7	3.2	0.77	Røskaft et al. (1990)
Abert's towhee	47.4	2.9	1.7	3.0	2.8	0.63	FInch (1983)
Northern oriole	33.8	3.7	2.5	4.9	4.6	0.78	E. Røskaft, S. Rohwer and C. D. Spaw unpubl. data
Eastern phoebe	19.8	3.1	2.4	4.2	4.0	0.82	Klaas (1975)
Red-eyed vireo[a]	16.7	2.3	1.4	3.2	3.1	0.63	Southern (1958)
American goldfinch	12.9	4.4	2.8	5.2	3.3	1.00	Middleton (1977)
Solitary vireo	16.5	2.8	1.7	3.4	2.8	0.74	Marvil and Cruz (1989)
Prothonotary warbler	15.0	4.4	3.3	4.8	4.3	0.84	Petit (1991)

Bodyweights from Dunning (1984); R_H = % hatching success in the parasitized nests relative to the % hatching success in the unparasitized nests.

[a]Given in the source as the number of small nestlings, and may therefore differ somewhat from the number of eggs that actually hatched.

tis) because it is an unsuitable (seed-eating) host; Kendall's $t = -0.429$, $n = 7$, $P = 0.088$, one-tailed). (Note that each host species was considered to have been tested only once, that is, if more than one data point existed for one host species, then the mean value was used in the test.) The reason for the lack of significance might be that too few data were used in this analysis.

An important factor stressed by Røskaft et al. (1990) is that because the cowbird itself is a puncture ejector, this brood parasite may also occasionally damage some of the host's eggs, in addition to those that it removes from the nest before laying its own (but see also Blankespoor et al. 1982; Dolan and Wright 1984). Carey (1986) has reported finding holes in the eggs of red-winged blackbirds, house finches (*Carpodacus mexicanus*) and eastern phoebes (*Sayornis phoebe*) in parasitized nests, probably caused by the cowbird. Similar observations have been made by Smith and Arcese (1994) for song sparrows. Both the shiny and the bronzed cowbird (*Molothrus aeneus*) also seem to pierce some of their hosts' eggs and lesser proportions of cowbird eggs laid by other females (Mason 1980; Carter 1986).

It could be argued that the cowbirds may be under a selection pressure to destroy some of the host's eggs in addition to those it removes from the nest, in order to reduce the subsequent nestling competition. This behavior may ensure that there will be enough eggs in the nest to prevent it from being deserted by the host, because, if the clutch size is reduced too much, the nest may be deserted (Rothstein 1982; Davies and Brooke 1988).

In their work on the northern oriole, Røskaft et al. (1993) also discovered that the cowbird occasionally damaged some additional oriole eggs when ejecting a host egg, damage that obviously reduces the hatching success of the oriole's eggs. It is thus not only the oriole that reduces its own hatching success in a parasitized nest by damage caused when ejecting the cowbird's egg, but the cowbird also may damage some additional host eggs. These costs affect the situation in the nest at the time the host is parasitized. Thus, when a host puncture ejects the parasite's egg, the ejection costs are added to the damage already caused by the cowbird. In fact, the hatching success of parasitized nests is normally lower than that of unparasitized nests (table 13.3), a pattern that also applies to hosts of the shiny cowbird (Wiley 1985). If the cowbird is under a selection pressure to damage some of the host's eggs so as to reduce nestling competition, one would expect that they should damage more of the eggs of a large-sized than of a small species. An almost significant support exists for this prediction because we found that the hatching success of the smaller hosts was relatively higher than that of the larger hosts (Kendall's $t = -0.473$, $n = 8$, $P = 0.053$, one-tailed; table 13.3).

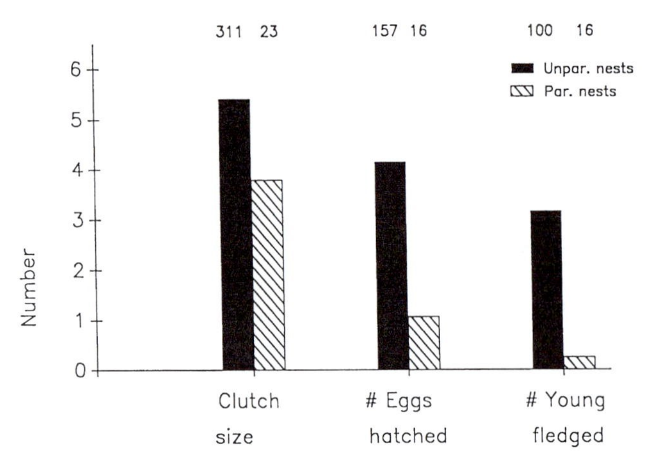

Figure 13.3 The clutch size, numbers of eggs hatched, and numbers of young fledged per nest in unparasitized and parasitized nests of meadow pipits in Norway. Numbers indicate sample size. All differences are statistically significant (*t*-tests, $P < 0.001$). (Data from A. Moksnes and E. Røskaft, unpubl.)

It is normally argued that cuckoos often cause a considerable reduction in the reproductive output of their hosts (Jensen and Vernon 1970). Once again, however, the magnitude of the effect is not well known because there are very few data to support this assumption. Data for meadow pipits indicate that although the common cuckoo has a dramatic effect on this species, the breeding success of a parasitized nest is still not zero, even though the sample size is small (fig. 13.3). Soler (1990) furthermore showed that the great spotted cuckoo (*Clamator glandarius*) has a significant effect on the breeding success of the magpie (*Pica pica*), even though the young cuckoo does not eject either the eggs or the young of the host. Data for reed warblers indicate that the energetic costs to the hosts of rearing a common cuckoo chick are similar to those involved in rearing a normal brood (Brooke and Davies 1989).

To sum up, cowbird hosts seem to suffer from parasitism, though at present no strong support exists for the postulate that smaller-sized hosts suffer more than larger-sized ones. Cowbirds also reduce the hatching success of hosts by piercing eggs in the nests in addition to those they eject, and there is almost significant evidence that they damage more eggs of the larger-sized hosts. We possess few data on the overall effects that the cuckoos have on their hosts. However, considerable harm seems to be done to the few hosts for which we have any data.

ESTIMATING THE DESERTION RATE

There are some reports of host species that occasionally desert their nest as a response to cowbird parasitism in nature (Rothstein 1975*b*; Nolan 1978; Clark and Robertson 1981; Wiley 1985; Graham 1988; Sedgwick and Knopf 1988; Weatherhead 1989; Petit 1991; Sealy 1995). It is unclear, however, why some of the host species should more frequently respond by nest desertion under natural conditions than under experimental parasitism. Desertion under experimental conditions is only documented by Rothstein (1976*a*) for cedar waxwings and by Sealy (1995) for a few individuals of yellow warbler (*Dendroica petechia*). Perhaps hosts that previously saw a live cowbird in or near their nest under natural conditions are more likely to inspect their nest more carefully (cf., Davies and Brooke 1988; Moksnes and Røskaft 1989). In one experimental study, however, only the timing of egg insertion and not the sight of a stuffed adult cowbird contributed to nest desertion (Sealy 1995). Further studies are needed before we can understand the reason for this important difference between natural and experimental parasitism.

To date, very few data exist with regard to the time in the breeding season at which nest desertion is used as an anti-parasite response. According to Clark and Robertson (1981) yellow warblers are more likely to reject (by desertion or egg burial) when parasitized early as opposed to late in their egg-laying season (Clark and Robertson 1981; Sealy 1995). Cuckoo hosts, on the other hand, frequently respond to both natural and experimental parasitism by nest desertion (Löhrl 1979; Davies and Brooke 1989*a*; Higuchi 1989; Moksnes et al. 1990).

Under natural conditions, birds that have their nests destroyed on the day after laying finishes or during egg laying normally lay about one egg fewer in the repeat clutch than on their first attempt (table 13.4). Furthermore, they have a

Table 13.4 Mean clutch sizes of two passerine species for their initial clutch and repeat clutches necessitated by loss of the initial clutch on the day after the first nest was completed

	Clutch-size		Number of days	
Species	First	Second	Between the two clutches	After the loss of the first clutch to the laying of the second clutch
Rook	4.4 (12)	3.7 (12)	15.3 (12)	10.4 (12)
Pied flycatcher	6.3 (7)	5.0 (7)	13.8 (9)	7.8 (9)

The mean interval between the loss of the first clutch and the laying of the second clutch, and the mean number of days elapsing between the starts of the two clutches are also indicated (own unpubl. data. ( ) = number of clutches).

between-clutch interval of about one week, which results in two weeks' delay in breeding (table 13.4; the data apply only to birds that remated with the same female; costs of an eventual remating are not included). The data presented in table 13.4 show about the same results for two markedly different passerine species, the rook (*Corvus frugilegus*) and the pied flycatcher (*Ficedula hypoleuca*). We therefore consider these results as representative also for host species of cowbirds and cuckoos.

The only cowbird host species for which we have data on the different desertion strategies was the prothonotary warbler (*Protonotaria citrea*), which tended to desert at a higher frequency when they had access to a new nest site than they did if there was no access to a new nest site (Petit 1991).

In conclusion, cuckoo hosts appear to desert the nest on many occasions when parasitized, but desertion by cowbird hosts is still not well documented. Cowbird hosts parasitized under natural conditions seem to desert to some degree, but seem more tolerant of cowbird eggs when experimentally parasitized.

TESTING THE MODELS

Results for a model species

We will start by considering a model host species of the brown-headed cowbird. We have assumed that the egg-laying period of this species is six weeks, which means that if the species is parasitized during the first few days of the breeding season it will have a maximum of two to three possible renesting attempts (according to the data in table 13.4). For most cowbird hosts, the rate of parasitism seems to increase during the early part of the nesting period, becoming stable in June. Parasitism frequencies of 30–40% also seem to be quite normal for many host populations. We therefore assumed that $p = 0.3$ during the first two weeks, and 0.4 during the final four weeks of the breeding season. Because most species lay up to one egg less in the clutch after each renesting attempt (table 13.4), we assumed that $U_i - U_j = 0.5$. According to the optimality equation (13.2), A will be an optimal strategy if

$$P_i \geq U_j - [p_j(U_j - U_k) + p_j p_k(U_k - U_l)];$$

which means that

$$P_i \geq U_j - [0.4 \times 0.5 + 0.4 \times 0.4 \times 0.5],$$

but since

$$U_i - U_j = 0.5,$$

we have

$$P_i \geq U_i - 0.78.$$

Thus, A is an optimal strategy if the difference in reproductive success between an unparasitized and a parasitized nest is less than 0.78 young fledged. According to table 13.2, this value seems to be close to some of the values found in nature. Comparable data for most host species of the American cowbird species are therefore needed.

Why does the red-winged blackbird accept eggs of the brown-headed cowbird?

In eastern Washington, the red-winged blackbird is frequently parasitized by the cowbird during May and June (Orians et al. 1989). The predation rate for this population is also very high (Orians et al. 1989). Renesting is therefore a frequent occur-

Table 13.5 Breeding success rates (fledged young/eggs laid) in unparasitized (U) and in parasitized (P) nests (note that all types of nests are included, also predated nests), and rate of parasitism (p) on red-winged blackbirds from Washington, USA

Week	U	P	p	W(D)	W(A)
1	0.86	0.42	0.004	0.8600	0.8582
2	0.86	0.42	0.02	0.8600	0.8512
3	0.86	0.42	0.08	0.8600	0.8248
4	0.86	0.42	0.13	0.8600	0.8028
5	0.86	0.42	0.20	0.8600	0.7720
6	0.86	0.42	0.27	0.8600	0.7412
7	0.86	0.42	0.33	0.8600	0.7148
8	0.86	0.42	0.40	0.0000	0.6840
9	0.86	0.42	0.46	0.0000	0.6576

Some unpublished data and others from Orians et al. (1989); Røskaft et al. (1990). W(D) and W(A) are estimates of U, P and p made from the data, assuming that a deserting host will lose about two weeks of the breeding season (as indicated in table 13.4).

rence. In spite of this there is no decline in clutch size during the breeding season (Røskaft et al. 1990), that is, renesters lay the same number of eggs as they do on the initial attempt. The data for this population can therefore also be used to test the optimality models.

The red-winged blackbird accepts the eggs of the brown-headed cowbird (Rothstein 1975*b*). The bill size and morphology of the female red-winged blackbird are not very different from those of the female northern oriole (Spaw and Rohwer 1987). The red-winged blackbird, therefore, probably incurs some small costs in ejecting the cowbird's egg comparable with those incurred by the northern oriole. Furthermore, the hatching success in parasitized nests of the red-winged blackbird is reduced (table 13.3). Assessing the fledging success from successfully hatched eggs is therefore the best way of determining the costs of acceptance. In fact, no difference in fledging success is noted between parasitized and unparasitized nests in consideration of the number of young fledged from the number of hatched eggs (Weatherhead 1989; Røskaft et al. 1990). Even a minor extra cost in rejecting the cowbird's egg would therefore have led to selection for acceptance by the red- winged blackbird. Nevertheless, could desertion be an adaptive strategy for the red-winged blackbird?

Using the model previously described we are able to estimate the values of U_i, P_i, p_i, W(A) and W(D) for red-winged blackbirds during the breeding season in Washington (table 13.5). No decline in breeding success occurred during the breeding season, either for successful unparasitized ($r = 0.004$, $n = 498$, N.S.), all unparasitized (i.e., when also predated nests are included, $r = 0.062$, $n = 1583$, $P < 0.05$), successful parasitized ($r = -0.244$, $n = 50$, N.S.) or for all parasitized nests together ($r = -0.061$, $n = 197$, N.S.). Using the data shown in table 13.5, W(D) > W(A) during most of the breeding season. According to the optimality model, the red-winged blackbird in Washington therefore should desert its nest when parasitized by the brown-headed cowbird. It should be stressed, however, that the difference between W(D) and W(A) is small, and even minor changes in the values could therefore lead to a different result.

The red-winged blackbird is a species with relatively low costs of brood parasitism (table 13.2). For smaller host species, therefore, it might be even more beneficial to desert in response to parasitism, and the rate of parasitism needs to be very high (close to 100%) before desertion would become unprofitable. Thus, at present it is difficult to understand why so few cowbird host species desert their nest when parasitized by the brown-headed cowbird. Although the available data would seem to support Rothstein's (1990) time-lag model, we would stress the fact that further data, which include all the necessary variables (table 13.6), should be collected.

The time-lag model, nevertheless, seems to provide the only reasonable explanation for the reaction to brood parasitism shown by some hosts

Table 13.6 Outline of variables necessary to test the optimality models for coevolution between brood parasites and their hosts. For all variables there is a need for data throughout the whole breeding season.

Laying dates for parasite and host
Frequency of parasitism
Host clutch size
Number of parasitic eggs laid per nest
Number of host eggs removed by the parasite
Host reproductive success in parasitized and unparasitized nests
Length of incubation and nestling periods
Frequency of desertion, ejection and acceptance
Costs of ejection (number of host eggs damaged or lost per ejection)
Time elapsing from desertion to laying in a new nest

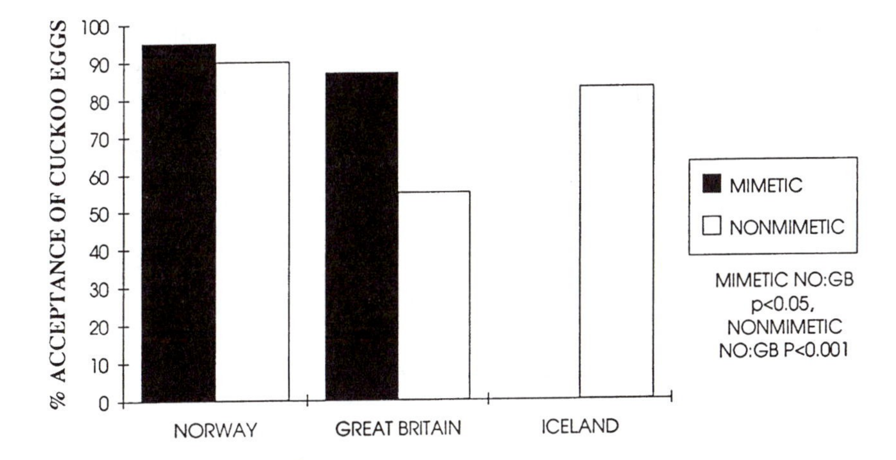

Figure 13.4 The rate of acceptance of mimetic and of nonmimetic cuckoo eggs by meadow pipits in Great Britain and Norway. The differences are statistically significant. ($P < 0.001$) for the nonmimetic cuckoo eggs ($c^2 = 11.7$, df = 1). The acceptance rate for meadow pipits in Iceland (22/27 nonmimetic eggs only) was statistically significantly higher than that for the British pipits ($c^2 = 6.87$, df = 1, $P < 0.01$), but not for that of the Norwegian pipits (22/27 vs. 22/24; Fisher's exact probabilities test; N.S. (Data from Davies and Brooke 1989*a*; Moksnes et al. 1993.)

of the common cuckoo such as the dunnock (*Prunella modularis*, Davies and Brooke 1989*a*; Moksnes et al. 1990). This species, which is commonly used as a host, does not seem to have evolved any counterdefenses against cuckoo parasitism. When experimentally parasitized with nonmimetic model cuckoo eggs, dunnocks show an acceptance rate of 100% and they show no aggression towards a cuckoo dummy at their nest (Davies and Brooke 1989*a*; Moksnes et al. 1990).

Reactions by meadow pipits (*Anthus pratensis*) to common cuckoo parasitism in Norway and Great Britain

The breeding success of nests parasitized by the common cuckoo is not necessarily zero. A successful meadow pipit nest produces on average 3.2 fledglings, whereas a parasitized nest still produces 0.3 meadow pipit fledglings, mainly because the cuckoo egg does not always hatch (fig. 13.3; Moksnes and Røskaft, unpubl.).

British meadow pipits are stronger rejectors than Norwegian ones (fig. 13.4; Davies and Brooke 1989*a*; Moksnes et al. 1993). The Norwegian meadow pipits rejected nonmimetic cuckoo eggs at the same rate as Icelandic meadow pipits, which have never experienced parasitism by the cuckoo (Davies and Brooke 1989*a*). Moksnes et al. (1993) have speculated about the difference in this frequency of rejection by meadow pipits in Norway and Great Britain. A possible answer may lie in the costs of desertion. During a 13-year study in the mountain areas of Norway, the egg-laying date of 88.7% of the clutches (n = 450) fell within a 15-day period around the median laying date for each year (Moksnes et al. 1993). Only five clutches (1.1%) were initiated before this period and 46 clutches (10.2%) after the period. These data indicate that the meadow pipit in this mountain area has a very concentrated breeding season, which probably makes renesting difficult. Any bird that deserts its nest, therefore, probably has limited possibilities of renesting ($U_j = 0$). The optimal solution for such a bird may be to take the risk and accept the cuckoo's egg because $P_i > 0$. The desertion rate under experimental parasitism with a nonmimetic cuckoo egg in this area increases to about 40% if the hosts at the same time are presented with a model cuckoo dummy (table 13.7). This indicates that a bird may only desert when it is certain that it has been parasitized during the egg-laying or early incubation periods.

RESPONSE TO PARASITISM IN RELATION TO THE STAGE IN THE HOST'S BREEDING CYCLE

According to the optimality models, a bird should accept the egg of a brood parasite when accept-

Table 13.7 Outcomes of experiments on meadow pipits (in central Norway) using dummy cuckoos and artificial cuckoo eggs in relation to the stage during the breeding cycle at which the experiments were made (Stage I = egg-laying and the first three days of incubation, Stage II = fourth to eighth day of incubation, Stage III = ninth day of incubation to hatching, R = number of nests where the cuckoo egg was rejected (mostly by desertion), T = total number of nests)

Experiment	Stage I R/T	Stage II R/T	Stage III R/T
Dummy cuckoo + nonmimetic egg	6/7	3/11	0/4[a]
Dummy cuckoo + mimetic egg	6/13	1/4	0/0[b]
Total	12/20	4/15	0/4[c]

[a]The rejection rate was statistically significantly higher in stage I than in either stage II or stage III (Fisher's Exact probabilities tests, $P < 0.05$).

[b]The differences between the different stages were not statistically significant (Fisher's Exact probabilities tests, N.S.).

[c]The rejection rate was statistically significantly higher in stage I than in stage II ($\chi^2 = 3.84$, $P = 0.05$).

ance will be more beneficial than rejection. If a host is parasitized during the egg-laying period, or a few days after the clutch is completed, there will be a high probability that the parasite's egg will hatch synchronously with the host's eggs, as most often occurs for both cowbird and cuckoo brood parasitism (Löhrl 1979; Mason 1980; Wiley 1985; Moksnes and Røskaft 1987; Brooker and Brooker 1989; Rothstein 1990). If the parasitic egg is laid late on during the incubation period, it will have a low probability of hatching. At that stage, therefore, a bird should normally accept the foreign egg. Rothstein (1976*a*) found support for this assumption in a study on the cedar waxwing, a species that normally rejects the cowbird's egg, although at some cost. Waxwings accepted at a higher frequency if the cowbird's egg was laid late during the incubation period. Many yellow warblers rejected cowbird eggs when parasitized before or early in their egg-laying period, but most parasitic eggs were accepted later during egg laying and incubation (Sealy 1995; see also Weatherhead 1989).

Little further support has been found, however, for the prediction that a host species should increase its rate of acceptance later on during the incubation period. Even among hosts of the common cuckoo, there is only weak support (table 13.8). Perhaps the strongest support for this prediction comes from a study of meadow pipits in Central Norway (Moksnes et al. 1993). In this study the meadow pipits accepted a high frequency (>90%) of both mimetic and nonmimetic model cuckoo eggs introduced into their nests. However, they rejected the cuckoo egg, normally by desertion, at a higher frequency (about 40%) if they were parasitized at the same time as being presented with a stuffed cuckoo dummy at their nests (table 13.7; Moksnes et al. 1993). In 60% of the cases (total $n = 58$), the meadow pipits showed aggressive behavior (mobbing or physical attack) towards the cuckoo dummy (Moksnes et al. 1990). However, this aggressive behavior was not correlated with the disposition to desert (Moksnes and Røskaft 1989). When the meadow pipits in central Norway were given a model cuckoo egg at the same time as they were presented with a cuckoo dummy, none of them deserted its nest when the experiments were made during the final stage of the incubation period before hatching (stage III; table 13.7). The rejection frequency for nonmimetic eggs was statistically significantly higher during egg laying and the first three days of incubation (stage I) than in either stage II (four to eight days of incubation) or stage III (Fisher's exact probabilities tests, $P < 0.05$). The same relationship between the stages of the breeding period, although not significant, was also found for mimetic model eggs (table 13.7; Fisher's exact probabilities test, N.S.). When the data for the nonmimetic and mimetic model eggs were pooled, there was a statistically significantly higher rejection rate during egg laying and early incubation (stage I) than during the middle of incubation (stage II) (table 13.7; $\chi^2 = 3.84$, d.f. = 1, $P = 0.05$).

A similar relationship between the rejection

Table 13.8 Frequency of rejection (%) of non-mimetic model eggs by cuckoo hosts in relation to the stage reached during the breeding season

Species	Laying	Incubation first week	Incubation second week	Source
Reed warbler	63 (43)	17 (6)	58 (12)	Davies and Brooke (1989*a*)
Meadow pipit	73 (11)	47 (17)	43 (47)	Davies and Brooke (1989*a*)
Meadow pipit[a]	86 (7)	27 (11)	0 (4)	Moksnes et al. (1993)
Pied wagtail	79 (14)	33 (9)	67 (24)	Davies and Brooke (1989*a*)
Blackbird	48 (21)	82 (11)	85 (13)	Davies and Brooke (1989*a*)
Song thrush	53 (30)	73 (11)	73 (11)	Davies and Brooke (1989*a*)
Bluethroat[a]	83 (6)	75 (4)	14 (7)	Moksnes et al. (1990) Moksnes and Røskaft (1992)

[a]Stages in the breeding cycle as in table 13.7.

( ) = number of experiments.

frequency and the stage in the breeding cycle was also found for some host species normally rejecting cuckoo eggs by ejection. The bluethroat (*Lucinia svecica*) rejected nonmimetic model cuckoo eggs at a statistically significantly higher rate during egg laying and the first three days of incubation than during the last five to six days before hatching (table 13.8; Fisher's exact probabilities test, $P < 0.05$; see also Moksnes et al. 1990). Such a relationship was also found for chaffinches (*Fringilla coelebs*) and bramblings (*F. montifringilla*) (combined) concerning their reaction towards foreign genuine and artificial conspecific eggs introduced into their nests (Braa et al. 1992).

It is therefore possible that modification of rejection behavior to minimize rejection costs is more common among cuckoo hosts than previously believed. The reason for the lack of further documentation of this behavior might be that too few experiments have been carried out late on during the incubation period.

CONCLUSIONS

We have argued that optimality models are a useful tool for studying the responses by a host to brood parasitism. They are applicable to any coevolutionary system between a brood parasite and its host, but cowbirds and cuckoos represent very suitable models because they exhibit two extremes of response. Unfortunately, few relevant data exist for testing the models. We have therefore given an outline of the kind of data needed and recommend that researchers collect such data in the future.

Adaptations against brood parasitism normally occur once the parasite's egg has been laid in the nest. Host defenses occurring at other stages of the nesting cycle seem to evolve with greater difficulty (Rothstein 1975*b*, 1990). Hamilton and Orians (1965) pointed out that the adaptive value of a positive response towards a nestling begging for food is apparently so great that host adaptations based on deliberate neglect of the parasitic nestling are largely precluded (but see also Davies and Brooke 1988).

Host species of the common cuckoo that we have tested seem to behave in accordance with an adaptive model when faced with a parasite's egg in their nest. Although few cowbird hosts have been tested in regard to optimal ejection behavior, some seem to respond adoptively as regards acceptance versus ejection. The results of the optimality models, however, would seem to indicate that the host birds should desert their nest more frequently than has been found in experimental studies.

According to the optimality models, a relatively large-sized host species, such as the red-winged blackbird in Washington, should desert its nest — a behavior pattern that is rarely found — even in naturally parasitized nests. Smaller-sized host species may incur even greater costs in accepting a parasite's egg. Only high rates (close to 100%) of parasitism should prevent these species from adopting the nest desertion solution. Although there are some indications that naturally parasitized nests are more frequently deserted than experimentally parasitized nests, the data so far seem to favor Rothstein's (1982, 1990) time-lag model. Before accepting such a maladaptive hypothesis, however, we need more carefully designed experimental studies, which we describe in this chapter.

ACKNOWLEDGMENTS We are indebted to Steinar Engen for help with constructions of the optimality models, to Gordon H. Orians and Les D. Beletsky for permission to use their unpublished data, and to Philip A. Tallantire for improving the English.

REFERENCES

Baker, E. C. S. (1942). *Cuckoo Problems*. Witherby Ltd., London.

Berger, A. J. (1951). The cowbird and certain host species in Michigan. *Wilson Bull.*, **63**, 26–34.

Blankespoor, G. W., J. Oolman, and C. Uthe (1982). Egg shell strength and cowbird *Molothrus ater* parasitism of red-winged blackbirds *Agelaius phoeniceus*. *Auk*, **99**, 363–365.

Braa, A. T., A. Moksnes, and E. Røskaft (1992). Adaptations of bramblings and chaffinches towards parasitism by the common cuckoo. *Anim. Behav.*, **43**, 67–78.

Briskie, J. V. and S. G. Sealy (1987). Responses of least flycatchers to experimental inter- and intraspecific brood parasitism. *Condor*, **89**, 899–901.

Brooke, M. De L. and N. B. Davies (1987). Recent changes in host usage by cuckoos *Cuculus canorus* in Britain. *J. Anim. Ecol.*, **56**, 873–883.

Brooke, M. De L. and N. B. Davies (1989). Provisioning of nestling cuckoos *Cuculus canorus* by reed warbler *Acrocephalus scirpaeus* hosts. *Ibis*, **131**, 250–256.

Brooker, L. C., M. G. Brooker, and A. M. H. Brooker (1990). An alternative population/genetics model for the evolution of egg mimesis and egg crypsis in cuckoos. *J. Theor. Biol.*, **146**, 123–143.

Brooker, M. G. and L. C. Brooker (1989). Cuckoo hosts in Australia. *Aust. Zool. Rev.*, **2**, 1–67.

Burgham, M. C. J. and J. Picman (1989). Effect of brown-headed cowbirds on the evolution of yellow warbler antiparasite strategies. *Anim. Behav.*, **38**., 298–308.

Capek, V. (1896). Beiträge zur Fortpflanzungsgeschichte des Kuckucks. *Orn. Jahrb.*, **7**, 41–181.

Carey, C. (1986). Possible manipulation of eggshell conductance of host eggs by brown-headed cowbirds. *Condor*, **88**, 388–390.

Carter, M. D. (1986). The parasitic behavior of the bronzed cowbird in south Texas. *Condor*, **88**, 11–25.

Chance, E. (1940). *The truth about the Cuckoo*. Country Life, London.

Clark, K. L. and R. J. Robertson (1981). Cowbird *Molothrus ater* parasitism and evolution of anti-parasite strategies in the yellow warbler *Dendroica petechia*. *Wilson Bull.*, **93**, 249–258.

Cruz, A. and J. W. Wiley (1989). The decline of an adaptation in the absence of a presumed selection pressure. *Evolution*, **43**, 55–62.

Davies, N. B. and M. De L. Brooke (1988). Cuckoos versus reed warblers, adaptations and counteradaptations. *Anim. Behav.*, **36**, 262–284.

Davies, N. B. and M. De L. Brooke (1989*a*). An experimental study of co-evolution between the cuckoo, *Cuculus canorus*, and its hosts. I. Host egg discrimination. *J. Anim. Ecol.*, **58**, 207–224.

Davies, N. B. and M. De L. Brooke (1989*b*). An experimental study of co-evolution between the cuckoo, *Cuculus canorus*, and its hosts. II. Host egg markings, chick discrimination and general discussion. *J. Anim. Ecol.*, **58**, 225–236.

Dolan, P. M. and P. C. Wright (1984). Damaged western flycatcher eggs in nests containing brown-headed cowbird chicks. *Condor*, **86**, 483–485.

Dunning, J. B., Jr. (1984). A Table of Body Weights for 687 species of North American Birds. *West. Birds Assoc. Monogr.*, **1**, 1–38.

Elliott, P. F. (1977). Adaptive significance of cowbird egg distribution. *Auk*, **94**, 590–593.

Elliott, P. F. (1978). Cowbird parasitism in the Kansas tallgrass prairie. *Auk*, **95**, 161–167.

Finch, D. (1983). Brood Parasitism of the Abert's towhee: timing, frequency, and effects. *Condor*, **85**, 355–359.

Freeman, S., D. F. Gori and S. Rohwer (1990). Red-winged blackbirds and brown-headed cowbirds: some aspects of a host parasite relationship. *Condor*, **92**, 336–340.

Friedmann, H. (1948). The parasitic cuckoos of Africa. *Wash. Acad. Science Monogr.*, **1**, 1–204.

Friedmann, H. (1955). The Honeyguides. *U.S. Nat. Mus. Bull.*, **208**.

Friedmann, H. (1963). Host relations of the parasitic cowbirds. *Smithson. Inst. U.S. Nat. Mus. Bull.*, **233**, 1–276.

Friedmann, H. and L. F. Kiff (1985). The parasitic cowbirds and their hosts. *Proc. West. Found. Vert. Zool.*, **2**, 225–302.

Friedmann, H., L. F. Kiff, and S. I. Rothstein (1977). A further contribution to knowledge of the host relations of the parasitic cowbirds. *Smithson. Contrib. Zool.*, **235**, 225–304.

Fry, C. H., S. Keith, and E. K. Urban (1988). *The Birds of Africa. Vol. III*. Academic Press, London.

Gaston, A. J. (1976). Brood parasitism by the pied

crested cuckoo *Clamator jacobinus*. *J. Anim. Ecol.*, **45**, 331–348.

Gill, B. J. (1983). Brood parasitism by the shining cuckoo *Chrysococcyx lucidus* at Kaikoura, New Zealand. *Ibis*, **125**, 40–55.

Glue, D. and R. Morgan (1972). Cuckoo hosts in British habitats. *Bird Study*, **19**, 187–192.

Glue, D. and E. Murray (1984). Cuckoo hosts in Britain. *Brit. Trust Ornith. News*, **5**, 134.

Goossen, J. P. and S. G. Sealy (1982). Production of young in a dense nesting population of yellow warblers *Dendroica petechia* in Manitoba. *Can. Field Natur.*, **96**, 189–199.

Graber, J. W. (1961). Distribution, habitat requirements and life history of the black-capped vireo (*Vireo atricapilla*). *Ecol. Monogr.*, **31**, 313–336.

Graham, D. S. (1988). Responses of five host species to cowbird parasitism. *Condor*, **90**, 588–591.

Gustafsson, L. and W. J. Sutherland (1989). The costs of reproduction in the collared flycatcher *Ficedula albicollis*. *Nature*, **335**, 813–815.

Hamilton, W. J., III and G. H. Orians (1965). Evolution of brood parasitism in altricial birds. *Condor*, **67**, 361–382.

Hann, H. W. (1937). Life history of the ovenbird in southern Michigan. *Wilson Bull.*, **49**, 145–237.

Higuchi, H. (1989). Responses of the bush warbler *Cettia diphone* to artificial eggs of *Cuculus* cuckoos in Japan. *Ibis*, **131**, 94–98.

Hill, R. A. (1976). Host–parasite relationships of the brown-headed cowbird in a prairie habitat of West-central Kansas. *Wilson Bull.*, **88**, 555–565.

Hofslund, P. B. (1957). Cowbird parasitism of the Northern yellow-throat. *Auk*, **74**, 42–48.

Jensen, R. A. and C. J. Vernon (1970). On the biology of the Didric cuckoo (*Crysococcyx caprius*) in Southern Africa. *Ostrich*, **41**, 237–246.

Jensen, R. A. C. and C. F. Clinning (1974). Breeding biology of two cuckoos and their hosts in Southwest Africa. *Living Bird*, **13**, 5–50.

Kelly, C. (1987). A model to explore the rate of spread of mimicry and rejection in hypothetical populations of cuckoos and their hosts. *J. Theor. Biol.*, **125**, 283–299.

Klaas, E. E. (1975). Cowbird parasitism and nesting success in the Eastern phoebe. *Occ. Pap. Mus. Nat. Hist. Univ. Kansas*, **41**, 1–17.

Lack, D. (1963). Cuckoo hosts in England. With an appendix on the cuckoo hosts in Japan, by T. Royama. *Bird Study*, **10**, 185–203.

Lack, D. (1968). *Ecological Adaptations in Breeding Birds*. Methuen, London.

Linden, M. and A. P. Møller (1989). Cost of reproduction and covariation of life history traits in birds. *Trends Ecol. Evol.*, **4**, 367–371.

Löhrl, H. (1979). Untersuchungen am Kuckuck, *Cuculus canorus*. (Biologie, Ethologie und Morphologie). *J. Ornith.*, **120**, 139–173.

Lotem, A., H. Nakamura and A. Zahavi (1995). Constraints on egg discrimination and cuckoo-host co-evolution. *Anim. Behav.*, **49**, 1185–1209.

Lowther, P. E. (1977). Old cowbird breeding records from the Great Plains Region. *Bird-Banding*, **48**, 358–369.

Makatsch, W. (1955). *Der Brutparasitismus in der Vogelwelt*. Neumann, Radebeul, Berlin.

Marchetti, K. (1992). Costs to host defence and the persistence of parasitic cuckoos. *Proc. R. Soc. Lond. B*, **248**, 41–45.

Marvil, R. E. and A. Cruz (1989). Impact of brown-headed cowbird parasitism on the reproductive success of the solitary vireo. *Auk*, **106**, 476–480.

Mason, P. (1980). Ecological and evolutionary aspects of host selection in cowbirds. Ph.D. Thesis. University of Texas, Austin.

Mason, P. and S. I. Rothstein (1986). Coevolution and avian brood parasitism, cowbird eggs show evolutionary response to host discrimination. *Evolution*, **40**, 1207–1214.

May, R. M. and S. K. Robinson (1985). Population dynamics of avian brood parasitism. *Amer. Natur.*, **126**, 475–494.

Mayfield, H. (1960). The Kirtland's warbler. *Cranbrook Inst. Sci. Bull.*, **40**, 242 pp.

Mayfield, H. (1961). Vestiges of a proprietary interest in nests by the brown-headed cowbird parasitizing the Kirtland's warbler. *Auk*, **78**, 162–167.

Mayfield, H. (1962). 1961 decennial census of the Kirtland's warbler. *Auk*, **79**, 173–182.

Mayfield, H. F. (1973). Census of Kirtland's warbler in 1972. *Auk*, **90**, 684–685.

Maynard Smith, J. (1974). The theory of games and the evolution of animal conflicts. *J. Theor. Biol.*, **47**, 209–221.

Maynard Smith, J. (1982). *Evolution and the theory of games*. Cambridge University Press, Cambridge.

Maynard Smith, J. and G. R. Price (1973). The logic of animal conflict. *Nature*, **246**, 15–18.

McGeen, D. S. (1972). Cowbird host relationships. *Auk*, **89**, 360–380.

McLean, I. G. (1988). Breeding behaviour of the long-tailed cuckoo on Little Barrier Island. *Notornis*, **35**, 89–98.

McLean, I. G. and J. R. Waas (1987). Do cuckoo chicks mimic the begging calls of their hosts? *Anim. Behav.*, **35**, 1896–1898.

Middleton, A. L. A. (1977). Effect of cowbird parasitism on American goldfinch nesting. *Auk*, **94**, 304–307.

Moksnes, A. and E. Røskaft (1987). Cuckoo host

interactions in Norwegian mountain areas. *Ornis Scand.*, **18**, 168–172.

Moksnes, A. and E. Røskaft (1989). Adaptations of meadow pipits to parasitism by the common cuckoo. *Behav. Ecol. Sociobiol.*, **24**, 25–30.

Moksnes, A. and E. Røskaft (1992). Responses of some rare cuckoo hosts to mimetic model cuckoo eggs and to foreign conspecific eggs. *Ornis Scand.*, **23**, 17–23.

Moksnes, A. and E. Røskaft (1995). Egg-morphs and host preference in the common cuckoo *Cuculus canorus*; an analysis of cuckoo and host eggs from European museum collections. *J. Zool.*, **236**, 625–648.

Moksnes, A., E. Røskaft, A. T. Braa, H. Lampe, and H. C. Pedersen (1990). Behavioural responses of potential hosts towards artificial cuckoo eggs and dummies. *Behaviour*, **116**, 64–89.

Moksnes, A., E. Røskaft, and A. T. Braa (1991). Rejection behavior by common cuckoo hosts towards artificial brood parasite eggs. *Auk*, **108**, 348–354.

Moksnes, A., E. Røskaft, and L. Korsnes (1993). Rejection of cuckoo (*Cuculus canorus*) eggs by meadow pipits (*Anthus pratensis*). *Behav. Ecol.*, **4**, 120–127.

Mundy, P. J. (1973). Vocal mimicry of their hosts by nestlings of the great spotted cuckoo and the striped crested cuckoo. *Ibis*, **115**, 602–604.

Mundy, P. J. and A. W. Cook (1977). Observations on the breeding of the pied crow and great spotted cuckoo in northern Nigeria. *Ostrich*, **48**, 72–84.

Nakamura, H. (1990). Brood parasitism by the cuckoo *Cuculus canorus* in Japan and the start of new parasitism on the azure-winged magpie *Cyanipica cyana*. *Japan J. Ornith.*, **39**, 1–18.

Newman, G. A. (1970). Cowbird parasitism and nesting success of lark sparrows in Southern Oklahoma. *Wilson Bull.*, **82**, 304–309.

Nice, M. M. (1937). Studies in the life history of the song sparrow, *Transactions of Linnean Society, New York*, **2**, 246 pp.

Nice, M. M. (1943). Studies in the life history of the song sparrow, *Transactions of Linnean Society, New York*, **6**, 328 pp.

Nicolai, J. (1974). Mimicry in parasitic birds. *Sci. Amer.*, **231**, 92–98.

Nolan, V., Jr. (1978). The ecology and behavior of the prairie warbler *Dendroica discolor*. *Ornith. Monographs*, **26**, 1–595.

Norris, R. T. (1947). The cowbirds of Preston Frith. *Wilson Bull.*, **59**, 83–103.

Orians, G. H., E. Røskaft, and L. D. Beletsky (1989). Do brown-headed cowbirds lay their eggs at random in the nests of red-winged blackbirds? *Wilson Bull.*, **101**, 599–605.

Overmire, T. G. (1962). Nesting in the dickcissel in Oklahoma. *Auk*, **79**, 115–116.

Owen, J. H. (1933). The cuckoo in the Felsted district. *Reports of Felsted School of Science Society*, **33**, 25–39.

Payne, R. B. (1973). The breeding season of a parasitic bird, the brown-headed cowbird in Central California. *Condor*, **75**, 80–99.

Payne, R. B. (1977*a*). The ecology of brood parasitism in birds. *Annu. Rev. Ecol. Syst.*, **8**, 1–28.

Payne, R. B. (1977*b*). Clutch size, egg size and the consequences of single vs multiple parasitism in parasitic finches. *Ecology*, **58**, 500–513.

Payne, R. B. (1982). Species limits in the indigobirds (Ploceidae, *Vidua*) of West Africa: mouth mimicry, song mimicry, and description of new species. *Misc. Publ. Univ. Michigan Museum Zool.*, **162**, 90 pp.

Payne, R. B. and K. Payne (1967). Cuckoo hosts in southern Africa. *Ostrich*, **38**, 135–143.

Petit, L. J. (1991). Adaptive tolerance of cowbird parasitism by prothonotary warblers: a consequence of nest-site limitation? *Anim. Behav.*, **41**, 425–432.

Post, W. and J. W. Wiley (1977). Reproductive interactions of the shiny cowbird and the yellow-shouldered blackbird. *Condor*, **79**, 176–184.

Post, W., T. K. Nakamura, and A. Cruz (1990). Patterns of shiny cowbird parasitism in St. Lucia and southwestern Puerto Rico. *Condor*, **92**, 461–469.

Redondo, T. and L. Arias-de-Reyna (1988). Vocal mimicry of hosts by great spotted cuckoo *Clamator glandarius*: further evidence. *Ibis*, **130**, 540–544.

Rohwer, S. and C. D. Spaw (1988). Evolutionary lag versus bill-size constraints: a comparative study of acceptance of cowbird eggs by old hosts. *Evol. Ecol.*, **2**, 27–36.

Rohwer, S., C. D. Spaw, and E. Røskaft (1989). Costs to northern orioles of puncture ejecting parasitic cowbird eggs from their nests. *Auk*, **106**, 734–738.

Røskaft, E. (1985). The effect of enlarged brood size on the future reproductive potential of the rook. *J. Anim. Ecol.*, **54**, 255–260.

Røskaft, E., G. H. Orians, and L. D. Beletsky (1990). Why do red-winged blackbirds accept eggs of brown-headed cowbirds. *Evol. Ecol.*, **4**, 35–42.

Røskaft, E., S. Rohwer, and C. D. Spaw (1993). Cost of puncture ejection compared with costs of rearing cowbird chicks for northern orioles. *Ornis Scand.*, **24**, 28–32.

Rothstein, S. I. (1975*a*). Evolutionary rates and host

defence against avian brood parasitism. *Amer. Natur.*, **109**, 161–176.

Rothstein, S. I. (1975*b*). An experimental and teleonomic investigation of avian brood parasitism. *Condor*, **77**, 250–271.

Rothstein, S. I. (1976*a*). Experiments on defenses cedar waxwings use against cowbird parasitism. *Auk*, **93**, 675–691.

Rothstein, S. I. (1976*b*). Cowbird parasitism of the cedar waxwing and its evolutionary implications. *Auk*, **93**, 498–509.

Rothstein, S. I. (1977). Cowbird parasitism and egg recognition of the northern oriole. *Wilson Bull.*, **89**, 21–32.

Rothstein, S. I. (1982). Success and failures in avian egg and nestling recognition with comments on the utility of optimal reasoning. *Amer. Zool.*, **22**, 547–560.

Rothstein, S. I. (1990). A model system for coevolution: avian brood parasitism. *Annu. Rev. Ecol. Syst.*, **21**, 481–508.

Schrantz, F. G. (1943). Nest life of the eastern yellow warbler. *Auk*, **60**, 367–387.

Scott, D. M. (1963). Changes in the reproductive activity of the brown-headed cowbird within the breeding season. *Wilson Bull.*, **75**, 123–129.

Scott, D. M. and C. D. Ankney (1979). Evaluation of a method for estimating the laying rate of brown-headed cowbirds. *Auk*, **96**, 483–488.

Scott, D. M. and C. Ankney (1980). Fecundity of the brown-headed cowbird in Southern Ontario. *Auk*, **97**, 677–683.

Scott, D. M. and C. D. Ankney (1983). The laying cycle of brown-headed cowbirds *Molothrus ater*. *Auk*, **100**, 583–592.

Sealy, S. G. (1992). Removal of yellow warbler eggs in association with cowbird parasitism. *Condor*, **94**, 40–54.

Sealy, S. G. (1995). Burial of cowbird eggs by parasitized yellow warblers: an empirical and experimental study. *Anim. Behav.*, **49**, 877–889.

Sedgwick, J. A. and R. L. Knopf (1988). A high incidence of brown-headed cowbird parasitism of willow flycatchers. *Condor*, **90**, 253–256.

Slack, R. D. (1976). Nest guarding behaviour by male gray catbirds. *Auk*, **93**, 292–300.

Smith, J. N. M. (1981). Cowbird *Molothrus ater*, parasitism, host fitness, and age of the host female in an island song sparrow *Melospiza melodia* population. *Condor*, **83**, 152–161.

Smith, J. N. M. and P. Arcese (1994). Brown-headed cowbirds and an island population of song sparrows: a 16-year study. *Condor*, **96**, 916–934.

Smith, N. G. (1968). The advantage of being parasitized. *Nature*, **219**, 690–694.

Smith, N. G. (1979). Alternate responses by hosts to parasites which may be helpful or harmful. In: *Host–Parasite Interfaces* (ed. B. B. Nickel), pp. 7–15. Academic Press, New York.

Smith, N. G. (1980). Some evolutionary, ecological, and behavioural correlates of communal nesting by birds with wasps or bees. In: *Acta XVII Congress International Ornithology* (ed. R. Nohring), vol. 2, pp. 1199–1205. Deutschen Ornithol.-Gessellschaft, Berlin.

Soler, M. (1990). Relationships between the great spotted cuckoo *Clamator glandarius* and its corvid hosts in a recently colonized area. *Ornis Scand.*, **21**, 212–223.

Soler, M. and A. P. Møller (1990). Duration of sympatry and coevolution between the great spotted cuckoo and its magpie host. *Nature*, **343**, 748–750.

Southern, W. E. (1958). Nesting of the red-eyed vireo in the Douglas Lake region, Michigan, pt. 2. *Jack-Pine Warbler*, **36**, 185–207.

Spaw, C. D. and S. Rohwer (1987). A comparative study of eggshell thickness in cowbirds and other passerines. *Condor*, **89**, 307–318.

Takasu, F., K. Kawasaki, H. Nakamura, J. E. Cohen, and N. Shigesada (1993). Modeling the population dynamics of a cuckoo–host association and the evolution of host defenses. *Amer. Natur.*, **142**, 819–839.

Terrill, L. M. (1961). Cowbird hosts in southern Quebec. *Can. Field Natur.*, **75**, 2–11.

Twomey, A. C. (1945). The bird population of an elm–maple forest. *Ecol. Monographs*, **15**.

Vernon, C. J. (1984). The breeding biology of the thickbilled cuckoo. *Proceedings of the 5th Pan-African Ornithology Congress*, pp. 825–840.

Walkinshaw, L. H. (1961). The effect of parasitism by the brown-headed cowbird on Empidonax flycatchers in Michigan. *Auk*, **78**, 266–268.

Weatherhead, P. J. (1989). Sex ratios, host-specific reproductive success, and impact of brown-headed cowbirds. *Auk*, **106**, 358–366.

Wiens, J. A. (1963). Aspects of cowbird parasitism in Southern Oklahoma. *Wilson Bull.*, **75**, 130–139.

Wiley, J. W. (1985). Shiny cowbird parasitism in two avian communities in Puerto Rico. *Condor*, **87**, 165–176.

Wiley, J. W. (1988). Host selection by the shiny cowbird. *Condor*, **90**, 289–303.

Wyllie, I. (1981). *The Cuckoo*. Batsford, London.

Yamagishi, S. and M. Fujioka (1986). Heavy brood parasitism by the common cuckoo *Cuculus canorus* on the azure-winged magpie *Cyanopica cyana*. *Tori*, **34**, 91–96.

Yamauchi, A. (1995). Theory of evolution of nest parasitism in birds. *Amer. Natur.*, **145**, 434–456.

Young, H. (1963). Breeding success of the cowbird. *Wilson Bull.*, **75**, 115–122.

Zimmermann, J. L. (1983). Cowbird *Molothrus ater* parasitism of dickcissels *Spiza americana* in different habitats and at different nest densities. *Wilson Bull.*, **95**, 7–22.

14

Brood Parasitism, Recognition, and Response

The Options

IAN G. McLEAN, RICHARD F. MALONEY

This chapter is about recognition. The context is one of encountering enemies; specifically, brood parasites. For a bird, recognition of an enemy involves detecting objects in the environment and determining whether that object represents a threat. Linked to recognition are the allied notions of coping, assessment and response. "Coping" is what animals do when dealing with enemies and involves two general possibilities. First, the bird may exhibit a variety of life history tactics designed to minimize the chances of encountering an enemy (e.g., behave cryptically, avoid cover that might conceal a predator). No recognition of the enemy is required although the bird presumably constantly assesses its immediate environment in terms of potential threat. Second, when actually confronted by an enemy the bird can exhibit a variety of behaviors (e.g., give alarm calls, fly away, abandon the nest). Here, recognition is required as well as some assessment of immediate threat.

Ornithologists studying enemy recognition are necessarily limited to measurements of response, from which they interpret both recognition and coping (a similar argument for mammals is made by Miller et al. 1990). With respect to brood parasites, it is the relationship between response patterns and the threat represented by the brood parasite that must be investigated.

McLean and Rhodes (1991) describe a simple cognitive model outlining the major steps in the process of moving from perception of a stimulus to some form of response. A stimulus is perceived by the sensory apparatus, where it is transformed into a representation that is stored. Next, the stored representation is compared with previously stored images, with recognition occurring if a sufficiently similar representation is already available. If recognition occurs, several steps must still be taken before a response is initiated. These steps, involving threat appraisal, selecting among possible responses, and assessing the risks associated with each possible response, provide the perceiver with response flexibility. A combination of limits to perceptual and recognition abilities and constraints on response options may allow brood parasites to outwit likely hosts. On the other hand, flexibility of response due to well-developed cognitive abilities allows the host to use numerous counter-tactics that may allow it to win in the end.

Notions such as "outwitting" and "winning" have been portrayed metaphorically in the brood parasite literature as a coevolutionary arms race (Rothstein 1990; Soler and Møller 1990; Lotem 1993). Clearly, there is likely to be ongoing evolutionary tension between brood parasites and their hosts. Unfortunately, we necessarily view the arms race between host and parasite as a snapshot in time. Normally we can only ascertain the evolutionary history that has led to the current status of the relationship by inference (e.g., Payne 1977), except in those rare occasions where changes in parasite or host distribution have been documented and the history of the relationship can be tracked through recent time (e.g., Brooke and Davies 1988; Cruz and Wiley 1989; Briskie et al. 1992; Soler et al. 1994; Nakamura et al. this vol.; Soler et al. this vol.) Here, we will focus on proximate issues that influence the outcome of host–parasite interactions. The notions of outwitting or winning are potentially applicable in a proximate sense, but this context has received little attention in the literature on brood parasites.

ANIMAL COGNITION

The notion that animals process information, and make decisions using that information, has emerged as a fundamental principle of animal behavior in recent years. The notion of the animal as a black box which responds to the environment in predictable and manipulable ways (behaviorism) has receded as a dominating paradigm (Roitblatt et al. 1983), and even animal experimental psychologists now consider a cognitive perspective when interpreting data (e.g., Pearce 1988).

Cognitive abilities include learning, memory, problem solving, rule and concept formation, perception, and recognition (Roitblatt 1987). These concepts are now frequently referred to as mental processes (Pearce 1988; Griffin 1992). The principle that animals may be viewed as having mental processes has gained considerable validity through the efforts of philosophers; for example, animals have been granted an "intentional stance" (Dennett 1983). With rare exceptions (e.g., Jamieson 1989), the shifting paradigmatic ground towards notions of mentality combined with the declining influence of behaviorism has led psychologists and biologists to reject or ignore the principles of animal learning (conditioning) that dominated behaviorist thinking in earlier decades. We consider that rejection to be unfortunate, because animal learning theory provides a rich source of information about the influence of the environment on animal behavior.

BIRDS AND BROOD PARASITES

It seems contradictory to suggest that birds have complex cognitive abilities for recognizing and dealing with enemies when the sight of a host bird feeding a huge and alien cuckoo evocatively argues that such abilities are not being used. Our objective is to search for those limits to the cognitive abilities of host birds that brood parasites currently exploit. The discussion is focused on interspecific brood parasitism because some of the issues we consider have markedly different options in the inter- and intraspecific frameworks (S. I. Rothstein, pers. commun.).

Any bird dealing with an enemy will use a mix of adaptive responses derived in evolutionary time, and proximate options based on learning and experience. In evolutionary terms, brood parasites may significantly decrease reproductive success of preferred host species, and selection is expected to act to improve the hosts' chances of avoiding or preventing parasitism (Davies and Brooke 1988; Rothstein 1990; Yamauchi 1995). This is the principle underlying the notion of an arms race (see above). Cryptic behavior near a nest, inspecting the contents of the nest before settling to incubate, locating the nest in inaccessible or hidden places, building a convoluted nest, and breeding early or late, are all options that may have been selected for by brood parasitism (and in some cases by other enemies in the local habitat). Active rejection behaviors such as building over a parasitized clutch, or ejecting a dissimilar egg, may also have been selected for (see Rothstein 1990 for a review). These responses do not require direct recognition of the brood parasite as an enemy or particular behavioral responses when that enemy is encountered, although both are possible. In the language of animal cognition, they do not require any comprehension of the threat posed by an enemy if it is physically encountered, even if it is encountered parasitizing the nest.

Evolutionarily programmed defense options may also be regarded as historic constraints, even when they are adaptive. For example, only certain types of alarm calls will be in the species' repertoire; or optimality theory requires that risk-taking should be higher when young are approaching fledging than when they are only recently laid (this prediction has been well supported by data; Montgomerie and Weatherhead 1988). Essentially, historical constraints provide the framework within which decisions about response options will be made by the bird when it is functioning as a cognitive organism. Because of their cognitive abilities, birds can learn from experience and they can assess and reassess the threat implied by the presence of an enemy. We hypothesize that the particular response(s) observed during the current interaction is likely to be influenced primarily by contextual and proximate factors, and that ultimate constraints provide a generalized response framework only. An example of this theme has been developed elsewhere (McLean and Rhodes 1991).

We will explore in some detail the proximate theme underlying host–brood parasite relationships: that of the options available to a bird actually faced with a brood parasite. The chosen response will reflect the particular individual's analysis of the current threat, and the response is likely to change with time. For example, the bird may respond inappropriately initially because it has not had time properly to assess the threat. It may adjust its response as more information is gathered about the enemy, but once a particular response has occurred (such as alarm calling), other responses that might have been more appropriate (such as hiding) may be excluded.

We begin with a test conducted on a passerine bird designed to investigate the contributions of evolutionary and developmental (experiential) factors to recognition of two enemies, a brood parasite and a predator.

THE TOMTIT AND THE CUCKOO

Our thoughts on the questions of recognition and response when faced with a brood parasite have been influenced by research on a species (the New Zealand tit *Petroica macrocephala*) that is occasionally parasitized by the long-tailed cuckoo *Eudynamys taitensis* (see McLean 1988). The genus *Petroica* consists of a group of small insectivorous, often ground-feeding birds in Australasia and southeast Asia. Historically, the genus must have had a long association with traditional enemies of small birds such as predatory birds and mammals, and more specifically with *Eudynamys* cuckoos that occasionally eat small birds (*Eudynamys* cuckoos occur throughout the range of *Petroica* species).

When faced with a long-tailed cuckoo, a New Zealand tit is dealing with an enemy that can be both a predator and a brood parasite (McLean 1987). Long-tailed cuckoos eat small birds, and presumably the contents of nests. If the tit has eggs in the nest, the cuckoo represents two different kinds of enemy, whereas when there are young in the nest the cuckoo can only be a predator. Thus, the role played by the cuckoo changes with time and the threat that it represents to the tit may similarly change.

The New Zealand tit is endemic to mainland New Zealand and several offshore islands (Bull et al. 1985). The yellow-breasted tit *P. m. macrocephala* is endemic to South Island where it routinely faces traditional enemies such as owls and both species of New Zealand cuckoos, and introduced predatory mammals. The black tit *P. m. dannefaerdi* is endemic to the isolated Snares Islands, 90 km south of the nearest land mass (Stewart Island). Except for short and infrequent visits by owls and long-tailed cuckoos, the black tit has no enemies on the Snares (McLean and Miskelly 1988). Although it is not known how long black tits have been isolated from mainland tit populations, it is unlikely that there has been genetic interchange since the Pleistocene glaciations (Fleming 1950; New Zealand *Petroica* species are nonmigratory, do not venture beyond forest margins, and are poor over-water dispersers). Individual black tits alive today are extremely unlikely to have interacted with an enemy of any sort. Nor do they have a recent heritage of enemy interactions, although they share the same genetic origins as mainland tits.

Because black tits must be naive about enemies, we tested the hypothesis that they would not recognize and respond to typical mainland enemies. Any evidence supporting recognition implies some genetic basis to enemy recognition because learning can be ruled out. We compared the responses of black tits and mainland tits to the same enemies in order to determine the influence of genetic recognition and experience on response patterns.

In November 1987, realistic stuffed dummies of a long-tailed cuckoo, a little owl *Athene noctua*, and a tinamou *Tinamus* sp. were presented at 24 black tit nests. The model species are all brown, of a similar size, and were unknown to the birds tested. In October and November 1989, the models (with the tinamou replaced by a song thrush *Turdus philomelos*) were similarly presented at nine yellow-breasted tit nests in the Waiho river valley, west coast, South Island. Here, the presented species were again brown and of a similar size, but all were resident in the study area and were presumably known to the experimental birds.

All nests contained eggs that were being incubated. Before any model presentations we observed the female return to the nest one to three times and term these "natural" returns. Thus four presentation conditions were analyzed: natural (no model), control (thrush or tinamou), owl, and cuckoo.

Human activity near the nests of these birds has little influence on their behavior and we were able to sit quietly about 5 m away and monitor the birds

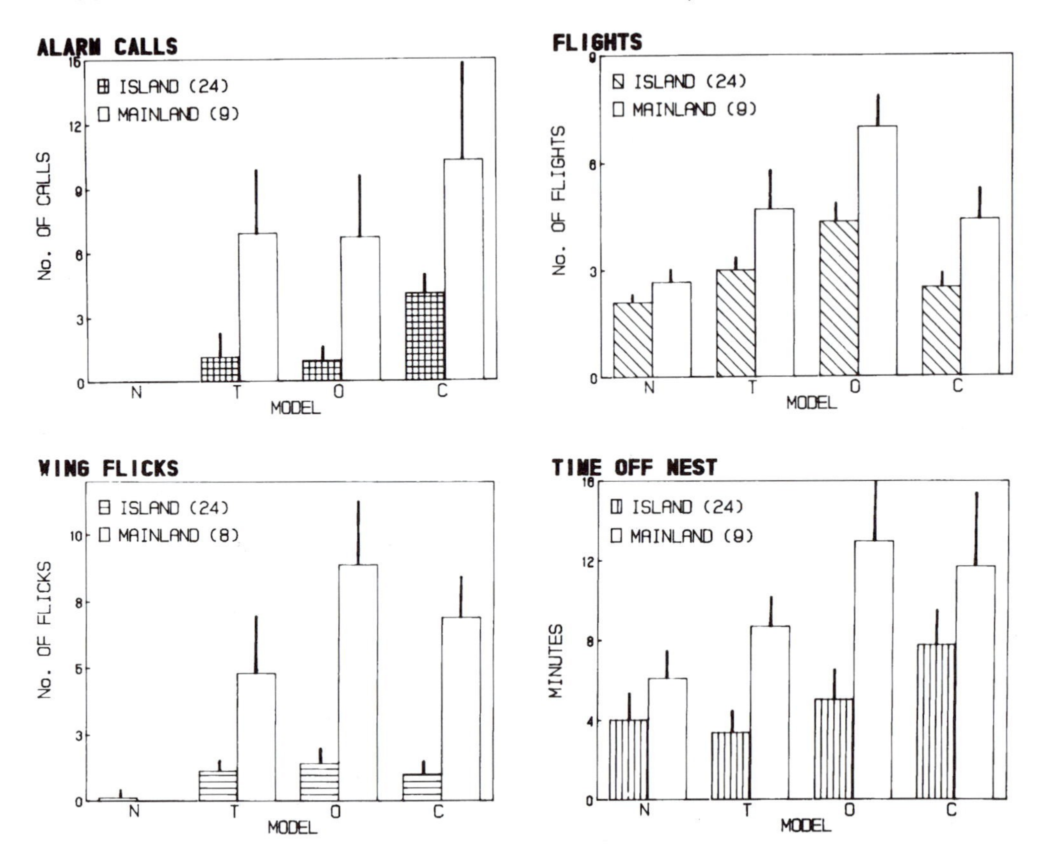

Figure 14.1 Behavioral responses of island (black tits) and mainland tits when returning to their nests naturally (N) and when three models were present: a control tinamou or thrush (T), a predator (owl, O), and a brood parasite (cuckoo, C). All between-study area differences were significantly different ($P < 0.05$). Within-study area differences were significant for alarm calls (both), flights (island only), and wing flicks (mainland only). See text for tests used. Sample size in parentheses indicates number of nests tested.

without effect. Black tits nest in somewhat more enclosed sites than mainland tits, but there is considerable overlap in the types of sites chosen (McLean and Miskelly 1988; R. Maloney and I. McLean, pers. obs.). Both species choose sites where nests are at least partially enclosed.

Model presentations began either immediately or the day after recording of natural returns. One model was presented per day with order of presentation randomized. Models were placed on a branch 1–2 m from the nest entrance (black tits), or were hauled to a position 1–2 m from the nest on previously placed strings (mainland tits; these nests were usually higher). The model faced towards the nest at about nest height. Female tits (who do all incubation) were not aware of model placement. In most tests they were away feeding (all tests for mainland tits); for some black tit tests the female was inside the nest chamber but could not see the model placement site.

An incubating female was usually called off the nest by the male who approached with food. She flew immediately towards the male (no female black tits saw models that were in position when they left the nest; no models were in position when mainland females left the nest). After being fed, the female either returned immediately to the nest, or flew off to feed for a short time. Females occasionally left the nest to feed before the male arrived. For all natural observations and tests, we recorded the total time off the nest from when the female left until when she resumed incubation ("time off nest" in fig. 14.1).

When the female approached the nest, the model was shaken for a few seconds using a string. Shaking stopped as soon as a bird clearly saw the model, the criteria being either an obvious change in behavior, or the bird stopping and looking directly at the model from a distance of no more than 2 m. If the bird did not fulfil these criteria, we waited for her to complete the next bout of incubation, leave the nest and return again (this only happened for black tits, as described below). Data were gathered for 1 minute from when the bird saw the model, after which the model was quickly removed. We continued to monitor the bird(s) until the female returned to the nest and continued incubating, if she had not already done so. The Snares Islands have special status as a nature reserve and the experimental period was limited to 1 minute to minimize the chance of a nest being abandoned.

All behavioral responses to the models during the 1-minute test were monitored. We counted number of vocalizations, flights, wing flicks, crest raises, and open wing displays. We interpret alarm calling as high-intensity mobbing behavior, and flights as low-intensity mobbing, with other behaviors indicating agitation. If both sexes responded we used a mean of the count for individual responses. Any of these behaviors was considered to indicate a response to the model as long as it was accompanied by clear physical orientation towards the model. The distance of both members of the pair from the model was monitored continuously.

RESULTS

Male black tits rarely reacted to the models even if they approached close enough to see them. Thus, all measured responses (i.e., any of the counted behaviors) were by females. Overall patterns of response were clear: seven females responded to both the cuckoo and the owl, nine responded to the cuckoo only, five responded to the owl only, and one responded to the cuckoo and the tinamou. Thus, of 24 females, 17 responded to the cuckoo, 12 responded to the owl, and one responded to the tinamou. Responses were variable with a few birds giving alarm calls or displays indicating high intensity alarm, and most birds appearing agitated but otherwise confused.

Significant variation in response by black tits was found for alarm calls (with response to the cuckoo being highest; fig. 14.1, $P < 0.05$, Friedman's ANOVA) and flights (with response to the owl being highest; $P < 0.05$, Friedman's ANOVA). No significant variation was found for wing flicks or time off the nest, and other displays were too infrequent to analyze. The models did not generally cause female black tits to delay resuming incubation and many females returned to the nest while models were in place (i.e., within 1 min).

These data suggest that black tits responded to the models in the order cuckoo > owl > tinamou, and they have an ability, which is presumably genetically based, to recognize traditional enemies. However, there is a more subtle aspect of this study that cannot be seen in the data. It took up to three returns to each nest before black tits even saw the models, despite the models being shaken as the female approached. Black tits were simply not alert to the possibility that an enemy might appear near their nests, and when an enemy did appear their response was unpredictable and erratic. While the tests were underway it became increasingly apparent that although the birds realized that the models represented something about which they should be concerned, *they did not know what to do.*

Mainland tits always saw the models immediately upon the first return to the nest, and both females and males responded. Significant variation was found for alarm calls and wing flicks ($P < 0.05$, Friedman's ANOVA, $n = 9$), with response to the cuckoo highest for alarm calls and response to the owl highest for wing flicks. In contrast to island tits, when mainland tits returned to the nest and found an enemy present, not only did they see the enemy immediately, *they knew what to do.*

Comparison between responses of island and mainland tits highlights the low level of response by island tits. We equalized sample sizes for this analysis in order to run a 2-factorial design. Thus, nine of the black tit nests were chosen randomly from the available 24 to give equal samples of island and mainland tests. Natural approach data were not used in the analysis. Study site (island/mainland) and model (cuckoo/owl/control) were factors in the ANOVA. The response of females only was used in the statistical analysis because male black tits exhibited no detectable response on the rare occasions in which they saw the models.

Response by mainland birds was significantly higher than response of island birds for all behaviors measured (calls: $F(1,53) = 5.33$, $P = 0.025$; flights: $F(1,53) = 6.19$, $P = 0.016$; wing flicks: $F(1,53) = 20.60$, $P < 0.001$; time off nest: $F(1,53) = 10.28$, $P = 0.002$). No interaction terms were significant. The pattern of response towards the models was therefore very similar in both study areas, in that the strongest responses were to the cuckoo and owl. However, the scale of response was very different, indicating that the ability to recognize enemies apparent in the naive island tit has been enhanced by experience in the mainland tit.

This study shows, first, that recognition of a cuckoo may be genetically based, and second, that the existence of enemies in their environment means that birds will be more alert to the possibility of that enemy appearing near the nest. The ability of birds to learn about enemies has been demonstrated experimentally in the laboratory (Vieth et al. 1980), in semi-natural conditions (Ellis et al. 1977) and in the field (Maloney and McLean 1995), and learning is the most likely explanation for response patterns in at least one host–brood parasite system (Smith et al. 1984). We consider it most likely that the difference between island and mainland New Zealand tits is a consequence of experience, although this study does not exclude the possibility that the island birds have secondarily lost genetically programmed responses in the absence of selection maintaining those responses in the population (cf. Cruz and Wiley 1989).

THE OPTIONS FOR BIRDS FACED WITH A BROOD PARASITE

The relevant literature for this section has been reviewed by Payne (1977), Rothstein (1990), and Sealy et al. (this vol.). Consequently, no detailed review is provided here and the reader is referred to these authors for background material. Rothstein (1990) particularly emphasized the more experimental approach that has been apparent since Payne (1977) was published.

Potential hosts must cope with brood parasites appearing in three different forms. First, *adult* brood parasites are a threat as a parasite if the potential host is building, laying, or early in incubation. They may be an additional threat as a predator at all stages of the breeding cycle, as is known for cuckoo species that may destroy nests even if they do not eat the contents (e.g., Payne 1977; Wyllie 1981). Second, as an *egg*, brood parasites appear in the nest without warning, and they can appear remarkably similar to the host's eggs. Third, as a *chick*, the brood parasite changes dramatically in appearance between hatching (when the young of all birds with altricial chicks are virtually indistinguishable, except for a few cues such as gape coloration; see Starck 1993) and the advanced nestling stage when it looks very similar to the adult brood parasite and usually looks very dissimilar to the host. Numerous (mostly anecdotal) reports of visual or auditory mimicry by brood parasite chicks are available (see Rothstein 1990), although the only available experimental test of apparent vocal mimicry (McLean and Waas 1987) suggested that it was ineffective (McLean and Rhodes 1991).

If the host cannot recognize the brood parasite, its egg or chick, then defensive responses are restricted to tactics that minimize the opportunity for the adult parasite to find or gain access to the nest. Once the egg has been laid, both it and the chick will be accepted and raised as the host's own. If, on the other hand, any of the forms of the parasite are recognized, then a wide range of additional response options become available.

RESPONSE TO THE ADULT BROOD PARASITE

If a potential host is not breeding, there is little to be gained by mobbing or driving off a brood parasite that appears in the vicinity. Unless it is a predator of small birds, or likely to settle in the area, the brood parasite does not represent a threat in the short term and will probably leave the area in the long term. The possibility of being parasitized sometime in the future may result in small birds mobbing brood parasites at any time of year, and may have resulted in the remarkably cryptic behavior of many cuckoo species (Wyllie 1981). However, nonbreeding birds sometimes accept brood parasites when they are encountered (e.g., Bell 1986, who noted Australian cuckoos joining mixed-species flocks of passerines), and mobbing outside the breeding season may have many different explanations anyway (e.g., the similarity in appearance between many cuckoos and predatory birds; Payne 1977). We suggest that responding to brood parasites by nonbreeding birds is only likely if it has little cost, as it is likely to carry little benefit.

Once breeding begins, a brood parasite represents a significant threat and should be responded to in any way that minimizes its ultimate impact on the breeding attempt. Likely responses are:

1. Mobbing, or other discouraging, distracting or aggressive behaviors. Although it might drive the enemy away, this response could also inform the brood parasite about the presence of a nest and perhaps even about the quality of the potential hosts (see Rothstein 1990). Even if driven away, the brood parasite can easily return later

and a mobbing response may signal that there is a nest in the vicinity.

2. Hide from the brood parasite (and presumably watch it). This response will allow the potential host to learn if the brood parasite knows if the nest is there, and what its immediate intentions are. Seeing the nest being parasitized allows the host to respond with any of the well-documented options of egg removal or destruction (if the egg can be distinguished) or nest abandonment. Ornithologists who work with nests quickly learn that most birds are particularly sensitive to any form of disturbance at nests during laying and/or early in incubation. From the birds' perspective, abandoning a nest that has been discovered is a sensible precaution when the future costs for that nest are still high, even if the nest contents are not disturbed. Increased rates of egg rejection if a brood parasite is seen nearby when a new egg appears in the nest has been documented for hosts of the common cuckoo *Cuculus canorus* (Davies and Brooke 1988; Moksnes and Røskaft 1989; Moksnes et al. 1991*a*), although this result is not always found (Braa et al. 1992).

3. Avoid the brood parasite, presumably by leaving the nest vicinity (ignoring the brood parasite and returning to the nest suggests that the parasite was not recognized as an enemy, although see option 5 below). This response will not draw attention to the nest, but if the brood parasite remains in the vicinity, then access to the nest is prevented and the host will gather no information about the brood parasite's intentions or knowledge of the nest.

4. Guard the nest. Several constraints suggest that this option will be rare. Most small birds need to feed continuously, and the immediate vicinity of the nest is unlikely to provide enough food for prolonged nest guarding. Females particularly are likely to need relatively large quantities of food because they are manufacturing eggs, and the male may guard the female or may even feed her during the courtship phase. If a brood parasite appears nearby, the nest guard will have to resort to options 1 or 2. Guarding may not prevent parasitism, but it is likely to increase the probability of detecting parasitism. Although nest guarding is occasionally hypothesized as a tactic used by birds to prevent brood parasitism (e.g., Slack 1976; Briskie and Sealy 1989), it is extremely difficult to demonstrate experimentally. Arcese and Smith (1988) explained a lowered rate of parasitism in nesting song sparrows *Melospiza melodia* that were provided with supplemental food by suggesting that the fed birds had more time to guard nests. Some male gray gerygones *Gerygone igata* guard second brood nests that are frequently parasitized but do not guard first brood nests that are rarely parasitized (Taylor 1991; Gill, this vol.).

5. Remain on the nest. Many birds spend at least some time sitting at the nest during laying, although they are not incubating. Also, incubation by both parents can result in incubation rates approaching 100%, and it seems likely that shared incubation has evolved in at least some species as a response to brood parasitism. Shared incubation will not prevent brood parasitism before laying is completed, but it may be an effective nest-guarding strategy once incubation has begun. This variant of option 4 has the same limitations, and may also require the development of incubation by males. However, the guarding bird has a dual role once incubation has begun and sitting on the nest may effectively prevent a brood parasite from gaining access even if the nest is discovered.

6. There are rare instances in which brood parasitism appears to benefit the host (Smith 1968) or the cost of brood parasitism is tolerated by the host because alternative breeding options are not available (Petit 1991). In the former instance host species are likely either to encourage brood parasitism, or at least those tactics designed to prevent brood parasitism will tend to disappear. We know of no situation where the adult brood parasite benefits the host directly, although the possible nest protection behavior described by Arias-de-Reyña (this vol.) may be an example of such a situation. This theoretical possibility is most likely where the brood parasite is resident during the breeding period, and its chick is raised alongside those of the host (costs to the host are low and the host may tolerate the presence of the brood parasite due to familiarity).

7. Parasitism insurance. Power et al. (1989) suggested that birds could "leave room" in the nest for a parasite egg by laying a reduced clutch. This scenario is only appropriate where parasitic eggs are additional to the hosts' own (many brood parasites remove a host egg when they lay) and where the brood parasite is raised alongside the hosts' own young. The insurance hypothesis is most relevant to intraspecific brood parasitism because recognition options are rarely available as host and parasitic eggs are typically identical (but see Jackson, this vol., for an important exception). However, it is conceivable that interspecific brood parasitism could fulfil the required conditions for the parasitism insurance hypothesis if egg mimicry is very highly developed.

As a final comment in this section we suggest

that birds will adjust their behavior depending on what species are encountered locally. Birds that routinely leave the nest vicinity to feed may remain nearby if a brood parasite is heard calling, or is encountered in the potential hosts' range. For example, female gray gerygones routinely feed out of view of the nest, but they remain close by to feed and stay off the nest for shorter periods if a cuckoo is seen near the nest (McLean and Rhodes 1991).

RESPONSE TO EGGS OF BROOD PARASITES

If a brood parasitic egg is a perfect mimic, then the cues available to the host for detecting brood parasitism are limited either to seeing the adult brood parasite lay, or to detecting that it has been at the nest, for example, because the nest has been disturbed. Many birds cover the eggs with feathers or other material when they leave the nest, or build the nest from materials that are easily ruffled or displaced, and the indirect detection of a visit by a stranger is theoretically possible. However, the potential for disturbance by weather or mate is high and the high cost of a mistake makes this tactic unlikely. It is therefore unsurprising that no instance of a species using the tactic of indirect detection has been documented.

If egg mimicry is less than perfect, then the host may be able to distinguish the foreign egg (apparently by recognizing own eggs; Rothstein 1978; see Rothstein and Robinson, this vol., for review). The problem of responding appropriately by rejecting the strange egg is not trivial because: (i) small birds may not be able to lift out or break the egg; (ii) if mimicry is good then the wrong egg may be removed by accident; (iii) breaking the foreign egg may result in cracks to the host's own eggs and foul the nest; and (iv) nest abandonment is costly. In most situations, the probability of the brood parasite egg not hatching is presumably too low for hosts to accept the cost of incubating a clutch containing a brood parasite egg on that chance. The question of egg recognition is discussed extensively elsewhere in this volume (Rothstein and Robinson, chapter 1), and little more will be said here, except to note that the question of whether birds evolve or develop (for example by short- or long-term dietary shifts) individually distinctive eggs as a response to brood parasitism has not yet been resolved (even though some parasitized bird species have individually distinctive eggs; Freeman 1988; Jackson, this vol.). Where brood parasitism is interspecific, there is no apparent difference in the distinctiveness of egg markings between parasitized and nonparasitized species (Davies and Brooke 1989*a*). However, where brood parasitism is intraspecific, parasitized species show greater between-individual variation in distinctiveness than for species in which parasitism does not occur (Møller and Petrie 1991). This difference, if real, remains unexplained.

Abandoning a nest because it has been parasitized is necessarily costly. Where nest architecture or some other factor prevents outright ejection, parasitized birds can potentially push the parasite egg to the edge of the nest (reported interspecifically for two duck species, Mallory and Weatherhead 1993; and intraspecifically for ostriches *Struthio camelus*, Bertram 1992). However, an alternative for birds that cannot eliminate parasitic eggs is to incubate the clutch, then discriminate against the parasitic chick after hatching (see below).

RESPONSE TO THE PARASITIC NESTLING

As noted by Davies and Brooke (1988), Røskaft et al. (1990), Rothstein (1990), Lotem (1993), and many others, it is extremely curious that there are no documented records of host birds physically ejecting the brood parasite chick (Nicolai 1974 described preferential feeding of own chicks in the Viduinae). Egg rejecters (species that remove or abandon parasitic eggs) invariably do so within a short time of the egg appearing in the nest. It seems likely that once the parasitic egg has been in the nest for several days, it becomes "one's own" (Rothstein's 1978 demonstration of egg imprinting notwithstanding) and will be nurtured as intensively as the rest of the nest contents. If birds can imprint on eggs, why not on chicks?

Lotem (1993) presented a model suggesting that it might be maladaptive for hosts of European cuckoos to recognize chicks. Fundamental to Lotem's model was an assumption: parents imprint on the first brood of chicks; and an inference: cuckoo nestlings look different from the host's young (imprinting will be maladaptive depending on the probability of the first brood containing a brood parasite). The model is convincing if one accepts the assumptions. However, we disagree with both the assumption and the inference for reasons that provide useful insights into the problem of chick recognition.

With respect to Lotem's (1993) assumption, the

Figure 14.2 A five- to-six-day-old brown-headed cowbird nestling (right) taken from a flycatcher *Sayornis phoebe* (phoebe) nest alongside a similarly aged phoebe chick from a different nest, and a surviving phoebe chick from the nest containing the cowbird. The cowbird is dramatically different in appearance to its nest-mate, primarily because it hatched several days earlier and has taken most of the food. However, even at this age it is not dramatically different in appearance to the well-fed phoebe nestling. Relative to the phoebe, the cowbird has darker feathers emerging from the sheaths (the phoebe is about one day more advanced), less down, a slightly wider gape, and a flesh-colored mouth lining, whereas the phoebe has a yellow mouth lining.

possibility that adult birds imprint on chicks has not yet been substantiated, and there is an urgent need for research on this question. However, a point ignored by Lotem is that adult birds are usually raised alongside siblings and would have ample opportunity as chicks to learn about the appearance of chicks of their own species. Presumably, evolution could exploit this experience if misimprinting is likely in early adulthood when the first brood is being raised. The option of imprinting as a chick is most likely to evolve in a system such as described by Lotem, where the cuckoo ejects all other nest contents.

Lotem's analysis focused on one brood parasite that is invariably larger than its hosts. However, at hatching, young of altricial birds look remarkably similar (see Starck 1993 for a detailed analysis of development in birds). An example of five- to six-day-old nestlings that are still quite similar looking is provided in fig. 14.2. Ignoring size differences (and it seems unlikely that imprinting would be on size alone), birds will need to use quite subtle cues if they are to imprint on newly hatched chicks. Clearly, such imprinting could generate recognition errors. However, chicks of different species usually diverge in appearance as time passes (see Fraga, this vol., for an exception). Thus, it would be safer to wait until a later stage of development before the imprinting event occurs. If so, the eventual decision about acceptance, whether based on imprinting or any other learning process, must be made against the backdrop of "yesterday's" decision to accept and in effect involves a reversal of that decision.

We suggest that decisions about chicks present a considerably more complex problem than decisions about eggs, and it seems likely that either evolution (selection) or cognition (a decision made in context) may not be able to generate error-free decisions in relation to this problem.

Of the many species known to reject dissimilar eggs (reviewed in Rothstein 1990), only three have been experimentally offered chicks of other species, and all three accepted the alien chick (Davies and Brooke 1989*b*). Of course, mimetic parasite eggs that are accepted still hatch into nonmimetic chicks, or more specifically into chicks that become nonmimetic as time passes, and these chicks are routinely accepted by hosts. Two fundamental differences between eggs and chicks are: (i) chicks change dramatically in appearance through time whereas eggs do not; and (ii) the external appearance of eggs is probably much more labile than that of newly hatched chicks. To solve the problem that hatchlings of altricial birds are virtually identical, hosts could focus on a relatively malleable cue such as the color of the lining of the mouth (as some apparently do, Nicolai 1974). Such cue-specific discrimination could be dangerous, first because it may require the host to ignore all other cues provided by the hatchling (such as general appearance), and second because it might only provide a short-term advantage anyway (because the brood parasite may quickly evolve the preferred mouth color).

The above comments aside, species that have not evolved egg rejection could still have evolved chick rejection, and there is no reason to assume that species known to accept eggs should also accept nestlings of other species (although they do: e.g., *Hirundo rustica*, *Passer domesticus*, Eastzer et al. 1980; *Prunella modularis*, Davies and Brooke 1989*b*; *Petroica macrocephala chathamensis*, and *Gerygone albofrontata*, Cemmick and Veitch 1985). Under some conditions, chick rejection may even be less costly than egg rejection, as egg ejection can result in damage to other eggs in the nest (Rohwer et al. 1989) and the cost of incubating one parasitic egg in a clutch of several will be low. For species where the parasite is raised alongside the hosts' chicks, discrimination should be a successful option and its absence begs explanation.

Should discrimination appear as a host tactic, the brood parasite will presumably be selected to evolve nestling mimicry, and it may be that known instances of nestling mimicry indicate an evolutionary response to discrimination by the parent against the parasite chick (see Rothstein 1990).

Our research group has found a possible example of incipient discrimination against cuckoo chicks by gray gerygones. We previously reported apparent vocal mimicry in the relationship between the shining cuckoo *Chalcites lucidus* and the gray gerygone (McLean and Waas 1987), and there may be partial visual mimicry of the gerygone chick by the cuckoo chick (Gill, this vol.). We observed individually marked gray gerygone pairs feeding 15 shining cuckoo chicks at Kaikoura, New Zealand. Each cuckoo chick was observed for a minimum of one hour, most were observed on several occasions, and observations were made on both nestlings and fledglings. In seven cases the female alone fed the chick (in all cases the female had a mate who could usually be found nearby). In seven other cases both the female and male fed the cuckoo chick.

The fifteenth case was the only record known of a cuckoo and gerygone chick being raised together. In this case the cuckoo egg was apparently laid about one week after incubation began, and the two gerygone chicks were partially feathered (about eight days old) when ejected from the nest by the cuckoo. One gerygone chick survived for a week on the ground below the nest before fledging (at which time the cuckoo was presumably abandoned, as it left the nest but was not found with its host parents despite several checks). At this nest the female fed the cuckoo chick, and the male fed the gerygone chick.

Our observations suggest that at least some male gray gerygones discriminate against cuckoo chicks, and that at about half the nests the cuckoo is raised entirely by the female. However, alternative explanations to discrimination by the male are possible: (i) one cuckoo chick makes demands on the parents equivalent to about 2.5 gerygone chicks (Gill 1982; this vol.), and females can fulfil this demand without assistance from the male, who has other important duties such as territory maintenance; or (ii) perhaps where demands are small (less than three gerygone chicks) the male never provides parental care. The first of these is clearly true, in that a female gerygone apparently can raise a cuckoo chick to fledging without help from the male. However, such an ability does not preclude male parental care. In relation to the second explanation, both female and male gerygones have fed the chicks at all one- and two-chick broods that we have observed (one-chick broods are rare in this species, but we observed many two-chick broods over a six-year period). Unfortunately, we did not investigate parental care in this species in enough detail to address this issue quantitatively. Although further work is required, particularly as Gill (this vol.; pers. commun.) found that male gray gerygones invariably fed cuckoo chicks, these observations provide preliminary evidence for discrimination by hosts against a cuckoo chick, and we suggest that other researchers search for evidence that some host individuals discriminate.

If brood parasite chicks are discriminated against, they might compensate by providing a supernormal stimulus (Dawkins and Krebs 1979), e.g., by begging loudly and having colorful mouths. Davies and Brooke (1988) provided reed warblers (*Acrocephalus scirpaceus*) with a choice of a cuckoo and their own young, and did not find that cuckoos were fed preferentially. Rather, the parents appeared to "simply feed the chicks who were begging the most" (p. 281) and the authors concluded that the cuckoo chick did not provide a supernormal stimulus. Alternative interpretations to the notion of supernormal stimulus are available: hungry chicks tend to be noisy (presumably because calling stimulates the parents to feed the chick more frequently; von Haartman 1957; Mondloch 1995) and large chicks tend to be loud; a large cuckoo chick may be continuously hungry and therefore continuously noisy.

Birds appear to raise their young by applying simple rules, including: (i) nurture the contents of your nest at all costs; (ii) feed the chick begging the most; and (iii) respond to stimuli that suggest hunger (for example by adjusting the feeding rate to the begging rate). Even if the nest is disturbed, moved, or replaced with an artificial nest that does not bear any resemblance to the original nest, birds will still continue to feed the chicks (De Hamel and McLean 1989). Such rules provide a simple stimulus–response framework that encompasses all behavioral aspects of the provision of parental care to the contents of a nest. No appeal to the cognitive abilities of birds allowing them to assess the cost represented by a brood parasite chick is required. The many examples of successful brood parasitism suggest that birds do not use such cognitive abilities, even if they are available.

RESPONSE TO THE PARASITIC FLEDGLING

Passerine birds are clearly devoted to their nest and its contents, and in the final reckoning it is perhaps unsurprising that they are prepared to care for the contents of that nest, whatever they look and sound like. However, once the chick fledges, the nest is no longer available as a stimulus and focus for the provision of food. Yet host birds continue to feed brood parasite fledglings, often for some weeks. Many hosts of brood parasites will have raised a brood of their own species either recently or in previous seasons, and they could potentially recognize that the newly fledged chick was alien.

Even more incongruous is that by the time they leave the nest, most fledgling brood parasites look much more similar to their own species than to the hosts' young (see Fraga, this vol., for an exception). And most hosts of brood parasites recognize and respond negatively to adult brood parasites (e.g., Edwards et al. 1949; Robertson and Norman 1976; Smith et al. 1984; McLean and Rhodes 1991, the tomtit study described here).

Experiments addressing the issue of recognition of alien chicks after fledging are rare, and inconclusive. Eastzer et al. (1980) found that barn swallows *Hirundo rustica* that had raised three alien species, a brown-headed cowbird *Molothrus ater*, a red-winged blackbird *Agelaius phoeniceus* and a gray catbird *Dumetella carolinensis* fed only the cowbird once the chicks left the nest, despite persistent calling by all three chicks. These results are anecdotal, it was not determined if the cowbird survived, and the greater success of the cowbird may have been due to its ability to take food on the wing rather than any form of discrimination by the parents or greater ability of the cowbird to provide a stimulus. However, black robin *Petroica australis* chicks raised by New Zealand tits are known to survive to independence (Cemmick and Veitch 1985). Clearly, more cross-fostering experiments at different chick ages would be of value.

Vocalizations are clearly an important component of the relationship between adults and fledglings, and are a likely mechanism for developing discrimination (by the host) or to ensure that feeding continues (for the chick). Vocal "mimicry" of host chick calls by brood parasite calls is occasionally reported, but the only experimental test of such mimicry indicated that the host birds were not "fooled" by the calls of the brood parasite (McLean and Waas 1987; McLean and Rhodes 1991). The question of whether birds with altricial young learn to recognize the calls given by individuals in the current brood has rarely been addressed, and has only been tested for (and found) in colonial-breeding species where there is a clear need for an ability to locate own young after fledging (Medvin and Beecher 1986; Lessells et al. 1991). Clearly, such an ability could be exploited by brood parasites and parasitism could select against the development of the ability. The question of whether host passerine species learn to recognize the calls given by their chicks (nestlings or fledglings) has not yet been addressed (Rothstein 1990), and urgently needs attention. If such learning does occur, then the stimulus–response framework may well explain why host parents feed alien young even after they have left the nest.

In a study of the structure of calls given by young birds in relation to habitat structure and predator pressure, McLean and Griffin (1991) found that the structure of loud begging calls (which are likely to attract predators) was not correlated with the habitat in which the species nested. Rather, the calls given by young birds show remarkable variation in structure. McLean and Griffin hypothesized that the needs of fledglings (primarily identification and locatability) provide the main selection pressures on the structure of calls given by young birds, whereas the needs of nestlings are largely irrelevant (any noise will do; cf. Fraga, this vol.). This scenario makes the apparently ineffective vocal mimicry described by McLean and Rhodes (1991) more reasonable. Calls given by host chicks are presumably well suited to their needs and those of their parents; they are therefore the best calls to be given by cuckoo chicks and the question of mimicry need not arise.

CONCLUSION

The issue of why host birds accept and raise a brood parasite chick remains a mystery. It is clear that selection has resulted in some instances of mimicry (of eggs and chicks), presumably due to the coevolution of host and parasite. We suggest that a straightforward answer to the question of why hosts accept brood parasite chicks can be found in the notion of parental care being generated by a simple stimulus–response system. Such a system can be viewed metaphorically as a small set of simple rules used by birds to guide their parental care behavior. Host parents may additionally learn to identify their young prior to fledging, although this is not required by the stimulus–response framework.

Clearly, even if the stimulus–response framework is the dominant proximate mechanism controlling the provision of parental care, evolution might still be able to modify the rules so as to minimize responses to inappropriate stimuli. Just as clearly, evolution has operated on behaviors such as egg ejection, abilities such as egg recognition, and egg morphology (e.g., color). Why can it not do the same for chicks?

Arguably, the benefits of versatility were a primary selective force in the evolution of cognition in animals (Griffin 1992). When faced with a complex world, there are benefits to be gained from assessing a current problem in context, and making decisions in that context. Birds are cognitive animals, but there are many respects in which they are also quite inflexible (e.g., the construction of nests is generally species specific, Collias and Collias 1984; the structure of alarm calls is generally invariant, Thorpe 1961). A cognitively functioning bird with a newly hatched brood parasite chick is faced with the combined problems that: (i) it has already made the decision to accept and nurture that chick in its former manifestation (as an egg); (ii) superficially, the chick will appear very similar to own-chicks (except possibly where specific cues such as mouth color are under selection); and (iii) rejection of the chick using subtle cues is likely to be costly (mistakes will be difficult to avoid). We suggest that evolutionary or cognitive assessment of newly hatched chicks is likely to result in mistakes, and should not occur.

Unfortunately, although it might be safer to defer the decision on acceptance to a later and safer stage of development, once the chick has been accepted any decision to reject is always being made against the backdrop of "yesterday's" decision to accept. With respect to eggs, there clearly is a specific moment (the arrival of the alien egg) when a decision about acceptance or rejection can be programmed (selected) or made (a cognitive decision made in context). It will be much more difficult for selection to find a point in the development of a brood parasite nestling where the previous decision to accept can be reliably reversed. All chicks change in appearance as they develop; the difference between own- and brood parasite chicks is that they change divergently. Frequently, the direct comparison is not even available if the parasite chick ejects or smothers other nest contents.

The stimulus–response framework is essentially hard wired and inflexible, in that any cognitive skills are bypassed and birds are programmed to respond to stimuli proffered by the nest and contents. It is therefore simple, conservative, and safe. But it is also easily exploited, perhaps explaining why so many bird species accept brood parasite chicks. However, no system will be so hard wired as to be impervious to invasion, and male gray gerygones and some Viduinae may provide incipient examples of rejection of a brood parasite chick. We suspect that careful field work will reveal other examples of individual variability in the apparent universal tendency to accept alien chicks uncritically.

We have attempted to outline the ways in

which a brood parasite represents problems for a potential host. Some birds clearly learn to recognize adult brood parasites as enemies (Smith et al. 1984), although they may also carry some genetic recognition capabilities (the tit example above). Egg discrimination abilities have apparently evolved in many different species (although not in all, e.g., dunnock *Prunella modularis*, Brooke and Davies 1988), and brood parasites have responded with many behavioral (e.g., rapid laying) and structural (e.g., egg mimicry) tactics. Virtually all species seem to accept alien nestlings and fledglings indiscriminately, although some individuals will discriminate against a brood parasite chick, for example by refusing to feed it. A particularly challenging area for research would be to test the possibility that host birds somehow understand that a brood parasite chick is inappropriate, but are unable to overcome the stimulus–response rules that drive them to feed the contents of their nest, and the products of that nest.

ACKNOWLEDGMENTS We thank the many people who have helped with field work on New Zealand tits and gray gerygones, especially H. Cameron, B. Lingard, C. Miskelly, and N. Wells. The manuscript has benefited from discussions with, or review by, D. Owen, H. Power, F. Proffitt, G. Rhodes, S. Robinson, and S. Rothstein. Funding for our research is provided by the New Zealand University Grants Committee, the New Zealand Lotteries Board, the Dept. of Conservation, Earthwatch, the Toyota Foundation, and the University of Canterbury.

REFERENCES

Arcese, P. and J. N. M. Smith (1988). Effects of population density and supplemental food on reproduction in Song Sparrows. *J. Anim. Ecol.*, **57**, 119–136.

Bell, H. L. (1986). The participation by cuckoos in mixed-species flocks of insectivorous birds in south-eastern Australia. *Emu*, **86**, 249–253.

Bertram, B. C. (1992). *The ostrich communal nesting system*. Princeton University Press, Princeton.

Braa, A. T., A. Moksnes, and E. Røskaft (1992). Adaptations of bramblings and chaffinches towards parasitism by the common cuckoo. *Anim. Behav.*, **43**, 67–78.

Briskie, J. V. and S. G. Sealy (1989). Changes in nest defense against a brood parasite over the breeding cycle. *Ethology*, **82**, 61–67.

Briskie, J. V., S. G. Sealy, and K. A. Hobson (1992). Behavioral defenses against avian brood parasitism in sympatric and allopatric host populations. *Evolution*, **46**, 334–340.

Brooke, M. de L. and N. B. Davies (1988). Egg mimicry by cuckoos *Cuculus canorus* in relation to discrimination by hosts. *Nature*, **335**, 630–632.

Bull, P. C., P. D. Gaze, and C. J. R. Robertson (1985). *The atlas of bird distribution in New Zealand*. Government Printer, Wellington.

Cemmick, D. and R. Veitch (1985). *Black Robin country*. Hodder and Stoughton, Auckland.

Collias, N. E. and E. C. Collias (1984). *Nest building and bird behavior*. Princeton University Press, Princeton.

Cruz, A. and J. W. Wiley (1989). The decline of an adaptation in the absence of a presumed selection pressure. *Evolution*, **43**, 55–62.

Davies, N. N. and M. de L. Brooke (1988). Cuckoos versus Reed Warblers: adaptations and counteradaptations. *Anim. Behav.*, **36**, 262–284.

Davies, N. N. and M. de L. Brooke (1989*a*). An experimental study of co-evolution between the Cuckoo, *Cuculus canorus*, and its hosts. I. Host egg discrimination. *J. Anim. Ecol.*, **58**, 207–224.

Davies, N. N. and M. de L. Brooke (1989*b*). An experimental study of co-evolution between the Cuckoo, *Cuculus canorus*, and its hosts. II. Host egg markings, chick discrimination, and general discussion. *J. Anim. Ecol.*, **58**, 225–236.

Dawkins, R. and J. R. Krebs (1979). Arms races between and within species. *Proc. R. Soc. Lond., Ser. B*, **205**, 489–511.

De Hamel, R. and I. G. McLean (1989). Caging as a technique for rearing wild passerine birds. *J. Wildlf. Manag.*, **53**, 852–856.

Dennett, D. C. (1983). Intentional systems in cognitive ethology: the "Panglossian paradigm" defended. *Behav. Brain Sci.*, **6**, 343–390.

Eastzer, D., P. R. Chu and A. P. King (1980). The young cowbird: average or optimal nestling. *Condor*, **82**, 417–425.

Edwards, G., E. Hosking, and S. Smith (1949). Reactions of some passerine birds to a stuffed cuckoo. *British Birds*, **42**, 13–19.

Ellis, D. H., S. J. Dobrott, and J. G. Goodwin (1977). Reintroduction techniques for masked bobwhites. In *Endangered birds. Management techniques for preserving threatened species* (ed. S. A. Temple), pp. 345–354. Croom Helm, London.

Fleming, C. J. (1950). New Zealand flycatchers of the genus *Petroica Swainson*. Part 1. *Trans. R. Soc. N.Z.*, **78**, 14–47.

Freeman, S. (1988). Egg variability and conspecific nest parasitism in *Ploceus* weaverbirds. *Ostrich*, **59**, 49–53.

Gill, B. J. (1982). Breeding of the Grey Warbler *Gerygone igata* at Kaikoura, New Zealand. *Ibis*, **124**, 123–147.

Griffin, D. R (1992). *Animal minds*. University of Chicago Press, Chicago.

Jamieson, I. G. (1989). Behavioral heterochrony and the evolution of birds' helping at the nest: an unselected consequence of communal breeding? *Amer. Natur.*, **133**, 394–406.

Lessells, C. M., N. D. Coulthard, P. J. Hodgson, and J. R. Krebs (1991). Chick recognition in European bee-eaters: acoustic playback experiments. *Anim. Behav.*, **42**, 1031–1033.

Lotem, A. (1993). Learning to recognize nestlings is maladaptive for cuckoo *Cuculus canorus* hosts. *Nature*, **362**, 743–745.

Mallory, M. L. and P. J. Weatherhead (1993). Responses of nestling mergansers to parasitic common goldeneye eggs. *Anim. Behav.*, **46**, 1226–1228.

Maloney, R. M. and I. G. McLean (1995). Historical and experimental learned predator recognition in a free-living bird. *Anim. Behav.*, **50**, 1193–1201.

McLean, I. G. (1987). Response to a dangerous enemy: should a brood parasite be mobbed? *Ethology*, **75**, 235–245.

McLean, I. G. (1988). Breeding biology and behaviour of the Long-tailed Cuckoo on Little Barrier Island. *Notornis*, **35**, 89–98.

McLean, I. G. and J. M. Griffin (1991). Structure, function, and mimicry in begging calls of passerines and cuckoos. Proc. 20th Int. Orn. Con., Wellington. New Zealand Ornithological Congress Trust Board, pp. 1273–1284.

McLean, I. G. and C. M. Miskelly (1988). Breeding biology and behaviour of the Black tit (*Petroica macrocephala dannefaerdi*) on the Snares Islands, New Zealand. *N.Z. Natur. Sci.*, **15**, 51–59.

McLean, I. G. and G. I. Rhodes (1991). Enemy recognition and response in birds. *Curr. Ornith.*, **8**, 173–211.

McLean, I. G. and J. R. Waas (1987). Do cuckoo chicks mimic the begging calls of their hosts? *Anim. Behav.*, **35**, 1896–1897.

Medvin, M. B. and M. D. Beecher (1986). Parent-offspring recognition in barn swallows (*Hirundo rustica*). *Anim. Behav.*, **34**, 1627–1639.

Miller, B., D. Biggens, C. Wemmer, R. Powell, L. Calvo, L. Hanebury, and T. Wharton (1990). Development of survival skills in captive-raised Siberian polecats (*Mustela eversmanni*) II: predator avoidance. *J. Ethol.*, **8**, 95–104.

Moksnes, A. and E. Røskaft (1989). Adaptations of meadow pipits to parasitism by the common cuckoo. *Behav. Ecol. Sociobiol.*, **24**, 25–30.

Moksnes, A., E. Røskaft, and A. T. Braa (1991*a*). Rejection behavior by common cuckoo hosts towards artificial brood parasite eggs. *Auk*, **108**, 348–354.

Moksnes, A., E. Røskaft, A. T. Braa, L. Korsnes, H. M. Lampe, and H. C. Pedersen (1991*b*). Behavioural responses of potential hosts towards artificial cuckoo eggs and dummies. *Behaviour*, **116**, 64–89.

Møller, A. P. and M. Petrie (1991). Evolution of intraspecific variability in birds' eggs: is intraspecific nest parasitism the selective agent? Proc. 20th Int. Orn. Con., Wellington: New Zealand Ornithological Congress Trust Board, pp. 1041–1048.

Mondloch, C. J. (1995). Chick hunger and begging affect parental allocation of feedings in pigeons. *Anim. Behav.*, **49**, 601–613.

Montgomerie, R. D. and P. J. Weatherhead (1988). Risks and rewards of nest defence by parent birds. *Q. Rev. Biol.*, **63**, 167–187.

Nicolai, J. (1974). Mimicry in parasitic birds. *Sci. Amer.*, **321**, 92–98.

Payne, R. B. (1977). The ecology of brood parasitism in birds. *Annu. Rev. Ecol. Syst.*, **8**, 1–28.

Pearce, J. M. (1988). *An introduction to animal cognition*. Lawrence Erlbaum Assoc., London.

Petit, L. J. (1991). Adaptive tolerance of cowbird parasitism by Prothonotary Warblers: a consequence of nest-site limitation? *Anim. Behav.*, **41**, 425–432.

Power, H. W., E. D. Kennedy, L. C. Romagnano, M. P. Lombardo, A. S. Hoffenberg, P. C. Stouffer, and T. R. McGuire (1989). The parasitism insurance hypothesis: why starlings leave space for parasitic eggs. *Condor*, **91**, 753–765.

Robertson, R. J. and R. F. Norman (1976). Behavioral defenses to brood parasitism by potential hosts of the Brown-headed Cowbird. *Condor*, **78**, 166–173.

Rohwer, S., C. D. Spaw, and E. Røskaft (1989). Costs to Northern Orioles of puncture-ejecting parasitic cowbird eggs from their nests. *Auk*, **106**, 734–738.

Roitblatt, H. L. (1987). *Introduction to animal cognition*. W. H. Freeman, New York.

Roitblatt, H. L., T. G. Bever, and H. S. Terrace (eds) (1983). *Animal Cognition*. Lawrence Erlbaum, Hillsdale N.J.

Røskaft, E., G. H. Orians, and L. D. Beletsky (1990). Why do Red-winged Blackbirds accept eggs of Brown-headed Cowbirds? *Evol. Ecol.*, **4**, 35–42.

Rothstein, S. I. (1978). Mechanisms of avian egg recognition: additional evidence for learned components. *Anim. Behav.*, **26**, 671–677.

Rothstein, S. I. (1990). A model system for coevolution: avian brood parasitism. *Annu. Rev. Ecol. Syst.*, **21**, 481–508.

Slack, R. D. (1976). Nest guarding behavior by male Gray Catbirds. *Auk*, **93**, 292–300.

Smith, J. N. M., P. Arcese, and I. G. McLean (1984). Age, experience, and enemy recognition by wild Song Sparrows. *Behav. Ecol. Sociobiol.*, **14**, 101–106.

Smith, N. G. (1968). The advantage of being parasitized. *Nature*, **219**, 690–694.

Soler, M. and A. P. Møller (1990). Duration of sympatry and coevolution between the great spotted cuckoo and its magpie host. *Nature*, **343**, 748–750.

Soler, M., J. J. Soler, J. G. Martinez, and A. P. Møller (1994). Micro-evolutionary change in host response to a brood parasite. *Behav. Ecol. Sociobiol.*, **35**, 295–301.

Starck, J. M. (1993). Evolution of avian ontogenies. *Curr. Ornith.*, **10**, 275–366.

Taylor, K. (1991). Enemy response patterns and parental care in the Rifleman and Grey Warbler. Unpubl. MSc. Thesis. University of Canterbury, Christchurch, New Zealand.

Thorpe, W. H. (1961). *Bird song: the biology of vocal communication and expression in birds*. Cambridge University Press, Cambridge.

Vieth, W., E. Curio, and E. Ulrich (1980). The adaptive significance of avian mobbing. III. Cultural transmission of enemy recognition in Blackbirds: cross species tutoring and properties of learning. *Anim. Behav.*, **28**, 1217–1229.

Von Haartman, L. (1957). Adaptations in hole-nesting birds. *Evolution*, **11**, 284–347.

Wyllie, I. (1981). *The Cuckoo*. Universe Books, N.Y.

Yamauchi, A. (1995). Theory of evolution of nest parasitism in birds. *Amer. Natur.*, **145**, 434–456.

V

EFFECTS OF PARASITISM ON HOST POPULATION DYNAMICS

This part addresses the possible effects of parasitism on host population dynamics. All three chapters assess the documented or potential impact of cowbirds on heavily parasitized populations and the factors that determine when hosts are vulnerable to extirpation by parasites. This topic has become a major conservation issue in the Americas where cowbirds are more abundant than many of their hosts, and three species are expanding their breeding ranges. For a comprehensive overview that includes potential impacts of other brood parasites, see pages 31–35 in chapter 1.

15

Consequences of Brown-Headed Cowbird Brood Parasitism for Host Population Dynamics

CHERYL L. TRINE, W. DOUGLAS ROBINSON, SCOTT K. ROBINSON

Generalist brood parasites such as the *Molothrus* cowbirds potentially decrease the reproductive success of some host species they parasitize to such an extent that they can threaten entire populations or even species (Mayfield 1977; Rothstein and Robinson 1994; Robinson et al. 1995*a*,*b*; Smith et al., in press). In contrast to cowbirds, the population dynamics of specialized parasites may be tightly coupled to those of the few hosts they parasitize so that a decline in a host species inevitably leads to a decline in the parasite and a recovery of the host (May and Robinson 1985; Takasu et al. 1993). Specialists therefore are unlikely to drive a host population to extinction. But the population dynamics of a generalist such as a cowbird can be uncoupled from those of some hosts (May and Robinson 1985). The result is that constant or increasing numbers of cowbirds can drive an uncommon host to extinction because cowbirds are dependent on more abundant hosts that are less affected by parasitism. The impact of cowbirds on hosts, however, depends partly on the lack of host specialization among individual females, which appears to be the case (Fleischer 1985). If female cowbirds did specialize on host species, then an evolutionary equilibrium such as that described in Davies and Brooke (this vol.) would be more likely.

Because brown-headed cowbird (for scientific names, see appendix) populations have increased dramatically within the past two centuries in North America (Mayfield 1965; Lowther 1993), some populations of host species are experiencing heavy parasitism and reduced reproductive success (reviewed in Robinson et al. 1993, 1995*a*). As an edge species, brown-headed cowbirds have responded positively to fragmentation of the eastern deciduous forest, the increase in crop and pasture lands, and, possibly, an increase in winter food availability (Brittingham and Temple 1983; Lowther 1993; Robinson et al. 1993). Cowbirds feed in pastures, crops, and other areas with short grass and commute up to 7 km between feeding and breeding areas where they search for host nests in the morning (Friedmann 1929; Rothstein et al. 1984; Thompson 1994; reviewed in Robinson et al. 1993, 1995*a*). As a result of the spread of cattle and agriculture, at least three species of cowbirds have expanded their geographic range and some host populations that were not previously sympatric with cowbirds have been exposed for the first time (Rothstein 1994; Cruz et al., in press) to such negative effects of brood parasitism as egg loss, reduced hatching and fledging success, and low fledgling weights (Rothstein 1975*a*; Payne 1977; Smith 1981; Trine, in press).

Cowbirds are more abundant and wide-ranging than most of their host species (Rich et al. 1994; Wiedenfeld, in press; Peterjohn et al., in press). Some host species with restricted ranges have already been pushed to the brink of extinction, at least in part due to cowbird parasitism. Black-capped vireos (Grzybowski et al. 1986, 1994; Hayden et al., in press), least Bell's vireos (Griffith and Griffith, in press) and Kirtland's warbler (Walkinshaw 1983; DeCapita, in press), for example, were nearing extinction and began to recover or stabilize only after cowbird control and habitat management programs were instituted (see below). Likewise, expanding populations of shiny cowbirds may have virtually eliminated yellow-shouldered blackbirds from most of Puerto Rico (Post and Wiley 1977; Post et al. 1990; Cruz et al., in press, this vol.). Nevertheless, population limitation is very complex and some workers question whether all of these host declines are due to cowbirds and whether cowbird control has produced tangible benefits (see Rothstein and Cook in press).

It has yet to be demonstrated, however, that parasitism pressure might cause population declines in species with large geographic ranges, although Böhning-Gaese et al. (1993) argued that parasitism was a likely cause of the declines of many North American species (see also Mayfield 1977; Brittingham and Temple 1983; but see Peterjohn et al., in press and Wiedenfeld, in press). Friedmann (1929, 1963), Friedmann et al. (1977), and Friedman and Kiff (1985) exhaustively documented parasitism levels of over 200 cowbird host species, many of which are neotropical migrants with large geographic ranges. Increasing cowbird populations may be a threat to declining populations of neotropical migrants, which are particularly vulnerable to parasitism because many species build open-cup nests, accept cowbird eggs, and raise only a single brood per year (Mayfield 1977; Whitcomb et al. 1981; Askins et al. 1990; Robinson et al. 1995*a*,*b*, in press). Many species of neotropical migrants that experience heavy parasitism levels in parts of their ranges, however, are not declining (Peterjohn et al. 1995; Wiedenfeld, in press), which raises the question of what sustains heavily parasitized populations (Brawn and Robinson 1996). In this chapter we examine the possible consequences of cowbird parasitism for host populations dynamics on a variety of scales, from local populations of endangered species to widespread species that vary geographically in their levels of brood parasitism. We relate cowbird parasitism to models of host population regulation, including single-population models, and discuss the need for metapopulation models (e.g., Pulliam 1988) that operate at the landscape level (e.g., Donovan et al. 1995*a*,*b*).

POPULATION MODELS

Single-population models

Cowbird parasitism usually causes a marked decrease in host reproductive output (Rothstein 1975*a*,*b*; Payne 1977). In many cases, the negative effects of brood parasitism can be even more detrimental than the effects of nest predation (May and Robinson 1985), in part because of the differing responses of the hosts. After successfully raising an offspring, whether it be a host or cowbird, many host species will not renest (e.g., single-brooded neotropical migrants; Nolan 1978). In contrast, if a host nest is depredated, most hosts will attempt to breed again. Thus, when hosts fail to discriminate between their own young and brood parasitic young and view the fledging of any young as successful, brood parasitism can be even more detrimental than nest predation, even if it affects a much smaller proportion of nests (Rothstein 1990). Furthermore, parasitized nests often fail to fledge any host young because cowbirds hatch earlier, beg more loudly to commandeer a disproportionate share of food, and, consequently, grow more quickly than the young of most hosts (Friedmann 1929; Rothstein 1975*a*). The situation would be even more complicated if cowbirds depredate unparasitized host nests (Smith and Arcese 1994; Arcese et al. 1996).

Empirical evidence demonstrating the negative effects of parasitism on host fledging success is abundant. Among a wide range of host species, fledging success is severely diminished in parasitized nests both when factoring in nest predation (table 15.1) and when only nests that escape predation are considered (table 15.2). The most severe reductions in fledging success are typically in small (<15 g) hosts (table 15.1: Acadian flycatcher, least Bell's vireo, Kirtland's warbler, prairie warbler, field sparrow; table 15.2: Acadian flycatcher, eastern wood-pewee, white-eyed vireo, hooded warbler). Larger hosts (>30 g), however, usually suffer reductions of less than 30%, despite often being multiply parasitized (see below). Multiple parasitism of small hosts such as Acadian flycatchers results in an almost complete failure to produce host young (table 15.2). The costs of multiple parasitism for a well-studied larger host species, wood thrush, are detailed in Trine (in press).

Given the substantial reductions in nesting success of some host species (tables 15.1 and 15.2), high frequencies of parasitism could prevent local populations from being self-sustaining. Popula-

Table 15.1 Some examples of reported fledging rates for unparasitized nests and nests parasitized by brown-headed cowbirds. All values include nests that escaped predation plus nests that failed totally because of predation

Host species (mass, g)	Mean number of host young fledged from nests: Parasitized	Unparasitized	% reduction	Proportion of nests parasitized	Reference
Acadian flycatcher (12.9)	0.38	1.68	77	0.24	Walkinshaw (1961)
Least Bell's vireo (9.0)	0.29	1.40	79	0.58	Goldwasser et al. (1980)
Solitary vireo (16.5)	0.50	1.52	67	0.49	Marvil and Cruz (1989)
Red-eyed vireo (16.8)	0.79	2.92	73	0.72	Southern (1958)
Yellow warbler (9.6)	0.88	2.28	61	0.30	Weatherhead (1989)
Kirtland's warbler (13.8)	0.28	1.28	78	0.55–0.71	Walkinshaw (1983)
Prairie warbler (7.8)	0.06	0.64	91	0.24	Nolan (1978)
Indigo bunting (14.5)	0.40	1.50	73	0.56	Berger (1951)
Lark sparrow (29.0)	0.60	1.78	66	0.45	Newman (1970)
Field sparrow (12.5)	0.60	2.38	75	0.18	Berger (1951)
Song sparrow (20.8)	0.81	1.91	58	0.44	Nice (1937)

tion-level consequences of reduced nesting success, however, are difficult to predict without information on predation rates, mortality rates of adults and juveniles, and number of renesting attempts per season (May and Robinson 1985; Clobert and Lebreton 1991; Pease and Grzybowski 1995; also see below).

May and Robinson (1985) developed a set of equations for predicting the level of parasitism a population could withstand before being driven to extinction. Their model relates the adult female host population in one year to that in the next by summing the number of surviving adult females and the number of female offspring that survive to reproductive maturity. In their terminology,

$$N_{t+1} = (1 - \mu)N_t + [1/2\gamma(1 - p) + 1/2\gamma' p]N_t \quad (15.1)$$

Table 15.2 Comparison of mean host productivity from parasitized and unparasitized nests that escape predation (fledge at least one host and/or cowbird young).

Species (mass, g)	Host young fledged from: Unparasitized nests	All parasitized nests	% reduction	Host young fledged from: Parasitized nests that fledge ≥Two cowbirds	Parasitized nests that fledge One cowbird
Acadian flycatcher (12.9)	2.52 (116)	0.42 (66)	83	0.20 (5)	0.19 (57)
Eastern wood-pewee (14.1)	2.80 (5)	0.25 (4)	91	–	–
White-eyed vireo (11.4)	2.50 (16)	0.20 (5)	92	–	0.0 (4)
Red-eyed vireo (16.7)	3.00 (2)	1.67 (3)	44	–	–
Worm-eating warbler (13.0)	4.23 (13)	1.43 (7)	66	1.33 (3)	–
Ovenbird (17.5)	3.67 (3)	2.50 (6)	32	1.00 (2)	2.0 (5)
Louisiana waterthrush (19.8)	3.94 (18)	0.50 (2)	87	0.50 (2)	–
Kentucky warbler (14.0)	3.99 (75)	1.34 (18)	66	0.94 (18)	1.25 (50)
Hooded warbler (7.7)	3.00 (3)	0.86 (7)	71	–	0.43 (6)
Summer tanager (28.2)	2.50 (4)	2.00 (4)	20	–	–
Northern cardinal (45.0)	2.43 (51)	1.74 (23)	28	1.67 (3)	1.67 (15)
Rufous-sided towhee (40.8)	2.90 (10)	2.10 (10)	28	1.00 (2)	1.50 (6)
Indigo bunting (14.5)	2.58 (31)	0.79 (14)	69	–	0.67 (12)

Data from the Shawnee National Forest area of southern Illinois, 1989–1992 (S. Robinson, unpubl. data). Numbers in parentheses refer to sample sizes.

where N_t is the adult female population in year t; N_{t+1} is the adult female population in the next year; μ is the adult mortality rate; γ is the number of offspring produced that survive to reproductive maturity from *unparasitized* females; γ' is the same as γ, but for *parasitized* females; and p is the probability of nests being parasitized. Assuming a 1:1 sex ratio in the offspring, the term 1/2 is present because only female offspring are considered.

A critical level of parasitism (p_c) above which the population begins to decline can be calculated:

$$p_c = (\gamma - 2\mu)/(\gamma - \gamma') \qquad (15.2)$$

It is difficult to estimate μ because birds disperse widely after fledging. May and Robinson (1985) added a new variable, λ, that represents the number of young fledged/female and related it to μ by

$$\gamma = \lambda(1 - \mu_o)$$

where μ_o is the mortality rate during the first year of life. Likewise,

$$\gamma = \chi'(1 - \mu_o)$$

for parasitized nests. Equation (15.2), therefore, can be rewritten as follows:

$$p_c = \{\lambda - [2\mu/(1 - \mu_o)]\}/\lambda - \lambda' \qquad (15.3)$$

When the observed parasitism rate exceeds p_c, equation (15.1) predicts that the host population will decline and may go extinct unless it is sustained by immigrants from another population.

Unfortunately, even λ and λ' are difficult to measure because most songbirds will renest several times during a breeding season, especially if nests are depredated. Few studies have followed color-marked females throughout all of their nesting attempts (for exceptions, see Nolan 1978; Roth and Johnson 1993). Most field studies present only a simple count of the number of young fledged from each nest both with (table 15.1) and without (table 15.2) the inclusion of depredated nests, that is, these studies focus on the overall proportion of nests, rather than females, that are parasitized. Because of multiple nesting attempts within a season, using data from single nests underestimates the true proportion of females parasitized (p) but overestimates the costs (λ') of parasitism because not all nesting attempts of a "parasitized" female will be parasitized. Sedgwick and Knopf (1988) provide a simple illustration of the need to keep these variables separated. For species with long nesting seasons and high predation and renesting rates, the calculations necessary to measure the impact of parasitism can be extremely complex (e.g., prairie warblers: Nolan 1978: 393–394). In response to these complexities, Pease and Grzybowski (1995) developed a model that incorporates "snapshot" daily parasitism rates during the laying period (not proportions), nesting season length, predation rates, inter-clutch intervals, and productivity from parasitized nests (among other variables). This model promises to provide an improved method for determining critical levels of parasitism (p_c) below which populations are no longer sustaining. In the meanwhile, however, we have to rely on the rare studies that directly measure reproductive output from color-marked pairs throughout entire breeding seasons (see below).

Metapopulation models

Most natural populations occupy a variety of qualitatively different habitats, some of which are incapable of supporting self-sustaining demes. Levins (1970) coined the term "metapopulation" to refer to a population of subpopulations undergoing local extinctions and recolonizations. Pulliam (1988) extended this model to situations in which the maintenance of populations in habitats where mortality usually exceeds productivity (population "sinks") depends upon immigration from populations where productivity exceeds mortality (population "sources") (see also Brown and Kodrick-Brown 1977; Hanski 1985; Howe et al. 1991). In Pulliam's model, patches are connected by density-dependent dispersal from source populations in good habitats to sink populations in inferior habitats. A source/sink model incorporating habitat selection processes (Pulliam and Danielson 1991) showed that reduction of high-quality habitat may cause extinction of the entire metapopulation if individuals only sample a few sites when choosing nesting territories. Habitat loss may increase the distance between potential breeding sites and limit an individual's ability to choose and to reach isolated sites (Urban and Shugart 1986; Askins and Philbrick 1987). Thus, increasing distance among patches and habitat degradation may limit opportunities for individuals to choose among habitats and therefore cause rapid metapopulation declines. Such metapopulation models have relevance to host population dynamics in fragmented landscapes because cowbird parasitism and nest predation increase as a result of habitat fragmentation (see below), which may create population sinks in small habitat patches that are only maintained by immigration from source populations in larger habitat patches (Porneluzi et al. 1993; Robinson et al. 1995*b*; Donovan et al., 1995*a*,*b*, 1997, in press).

POPULATION DYNAMICS AS A FUNCTION OF SCALE

Population dynamics on a local scale

Cowbird parasitism can severely affect local host populations and therefore may endanger species with restricted geographic ranges. The California population of least Bell's vireo, for example, declined within the past century to an estimated total of 200 pairs (Goldwasser et al. 1980; Rothstein 1994) concomitant with an increase in the brown-headed cowbird population. Given that parasitized nests fledged an average of 0.29 host young and unparasitized nests fledged an average of 1.4 young and a parasitism frequency of 0.55 (Goldwasser et al. 1980), May and Robinson (1985) estimated that the adult mortality rate would have to be 0.14 or less for the population to sustain itself without cowbird control. Franzreb (1989) estimated an annual adult mortality rate of 0.53, nearly four times higher than May and Robinson's calculation. Cowbirds may indeed be driving least Bell's vireo toward extinction. Habitat destruction and degradation, however, also have influenced population declines. Besides greatly reducing the number of individuals who are able to breed, habitat effects also have increased both nest predation (Franzreb 1989; Rothstein 1994) and parasitism levels (Gray and Greaves 1984) among those that do breed. At one California site where cowbird control measures were instituted, the parasitism level decreased from 47% to less than 10% and the vireo population increased from 27 in 1981 to 420 in 1993 (Salata 1983; Griffith and Griffith, in press). Cowbird control measures also appear to be locally successful in reducing parasitism of southwestern willow flycatchers (*Empidonax traillii extimus*) (Whitfield, in press), which nest in similar habitats in California. However, flycatcher breeding populations have so far (as of 1997) shown no marked increases, despite the improved productivity.

Cowbird control programs were begun in Michigan in 1972 after a 60% decline (to about 200 pairs) in the Kirtland's warbler population was noted between 1961 and 1971 (Probst 1986). Approximately 59% of all nests were parasitized, and parasitized nests fledged significantly fewer young than unparasitized nests (see Table 15.1 and also Kelly and DeCapita 1982; Walkinshaw 1983). Subsequently, thousands of cowbirds were trapped each year and the frequency of nest parasitism declined abruptly to 3% (Kelly and DeCapita 1982; DeCapita, in press). The Kirtland's warbler population stabilized, but did not increase as would be expected if parasitism were the primary factor depressing population growth. Availability of appropriate breeding habitat (Probst 1986) and wintering habitat (Ryel 1981) were suggested as limiting factors (see also Rothstein and Cook, in press). A recent tripling of the population appeared to result from a substantial increase in breeding habitat following succession after a major burn (DeCapita, in press).

Unlike least Bell's vireo and Kirtland's warbler, the black-capped vireo probably coevolved with cowbird parasitism; its range overlaps extensively with the original range of brown-headed cowbirds. Black-capped vireos are rapidly disappearing from many areas, especially at the periphery of their historical range (Grzybowski 1990; Grzybowski et al. 1994). They are highly susceptible to parasitism; 80–100% of their nests were parasitized in Oklahoma (Grzybowski 1990; Grzybowski et al. 1994; Hayden et al., in press). A parasitized population in the Wichita Mountains produced 0.41 host young per pair per breeding season, a value unlikely to compensate for adult mortality. After cowbird removal, the population at the same site produced three times as many young per pair (1.31; Grzybowski 1990). In uncontrolled situations, however, few nests escape parasitism and many black-capped vireos compensate by nesting more than once each season. Consequently, 60% of the pairs fledge at least one host young by the end of the breeding season (Graber 1961). Habitat deterioration and succession may also be contributing to the decline (Graber 1961; Grzybowski 1990; Grzybowski et al. 1994). As is the case for the least Bell's vireo and Kirtland's warbler, the combination of habitat limitation and cowbird parasitism may be driving the black-capped vireo towards extinction. Cowbird removal programs (reviewed in Robinson et al. 1993) may be essential for stabilizing populations (Hayden et al., in press).

Local populations of some widespread species have been studied in enough detail to assess the impact of cowbird parasitism. May and Robinson (1985) examined population demographics for Nolan's (1978) prairie warbler population, and song sparrow populations in Ohio (Nice 1937) and British Columbia (Smith 1981). Based on results from their single-population model, May and Robinson suggested that Nolan's prairie warbler population was barely replacing itself largely because individuals renested an average of five times per season in the face of heavy nest predation. Likewise, Nice's (1937) population of song sparrows was near equilibrium; females nested 2.8 times per breeding season, which is barely enough to compensate for the losses due to parasitism. In an island population in British Columbia, parasitized song sparrows made more nesting attempts than unparasitized song sparrows, partially offsetting the decreased fledging success per nest caused by cowbirds (Smith 1981; Smith and Arcese 1994; Smith and Myers-Smith, this vol.). In this system, however, cowbirds also may act as nest predators in order to increase the availability of host nests to parasitize (Arcese and Smith 1994; Arcese et al. 1996). Trail and Baptista (1993) also provide a demographic study of the effects of cowbirds on white-crowned sparrows in California.

Landscape-level metapopulations

Severely fragmented landscapes

The patchy distribution of forest-breeding birds in fragmented landscapes (Whitcomb et al. 1981; Robbins et al. 1989) makes these species particularly well suited for metapopulation models (Hanski and Gilpin 1991; Freemark 1992). In fragmented landscapes, habitat patches are spatially isolated and dominated by edge habitat where nest predation and brood parasitism levels are often higher than in interiors (e.g., Mayfield 1977; Gates and Gysel 1978; Chasko and Gates 1982; Brittingham and Temple 1983; Wilcove 1985; Wilcove et al. 1986; Ratti and Reese 1988; Small and Hunter 1988; Temple and Cary 1988; Yahner and Scott 1988; Johnson and Temple 1990; Wilcove and Robinson 1990; Winslow et al., in press). Consequently, habitat fragmentation may increase the ratio of population sinks to sources by increasing the proportion of edge habitat.

Temple and Cary (1988) modeled effects of habitat fragmentation on avian populations in which natality increases with the distance from forest edge. Compared with unfragmented forests, simulation results for populations in moderately fragmented landscapes showed greater annual variation and lower population densities. In highly fragmented landscapes, however, the whole population became extinct when no immigration occurred from outside populations.

Because Temple and Cary's model indicated such severe consequences of habitat fragmentation for avian population dynamics in highly fragmented landscapes, more studies are needed across a broad geographic range. In a preliminary study of severely fragmented forests in central Illinois (Robinson 1992), productivity for many forest-breeding species was negligible. Parasitism rates were among the highest ever recorded (50–100% depending on species), and most recently fledged families included at least one cowbird fledgling. Nest predation rates were also extremely high (*c*.80%). The low productivity was associated with a high annual turnover rate of individuals (mean annual return rate of marked birds was 20.0%) and considerable population fluctuations between years for many species (Brawn and Robinson 1996). Pulliam (1988) and Howe et al. (1991) predicted that sink populations are quite variable from year to year. These factors led Robinson (1992) to conclude that these woodlots are population sinks for most neotropical migrants and must be maintained by immigrants from other populations (see also Porneluzi et al. 1993; Donovan et al. 1995*a*,*b*; Brawn and Robinson 1996). Because of the small sizes of these woodlots (13–65 ha), they consist entirely of edge habitat. These results agree with the view that populations nesting in tracts without an undisturbed core will suffer reproductively (Temple and Cary 1988) and a number of recent studies have shown similar trends (Porneluzi et al. 1993; Hoover et al. 1995; Robinson et al. 1995; Donovan et al., in press; Dowell et al., in press; Petit and Petit, in press). Small central Illinois woodlots appear to be "ecological traps" (*sensu* Gates and Gysel 1978) where reproduction is negligible for long-range dispensers that are attracted to them, a result that has been replicated in other agricultural landscapes in the American Midwest (Robinson et al. 1995; Brawn and Robinson 1996; Donovan et al., in press; Thompson et al., in press).

Elliott's (1978) results for small prairie fragments in central Kansas suggest that cowbirds may create severe problems for grassland species in some areas (see also Johnson and Temple 1990; Davis and Sealy, in press; Koford et al., in press). He found that many species in a grazed grassland were parasitized at very high levels, including

Table 15.3 Characteristics of the four study areas in the Shawnee National Forest

Study area	Tract size (ha)[a]	Vegetative characteristics
Dutch Creek	1100	Mixed hardwoods along intermittent stream; oak/hickory on slopes and ridges. Many openings including eight clearcuts (8–12 ha), seven wildlife openings (0.2–1.5 ha) and a camp with mowed grass.
Pine Hills	2000	Mixed hardwoods along intermittent stream; oak/hickory slopes. Closed-canopy dirt road, and a primitive campground at southern edge. No clearcuts.
Ripple Hollow	1000	Mixed hardwoods along intermittent stream; oak/hickory slopes. No clearcuts.
Trail of Tears	2000	Mixed hardwoods along several intermittent streams; oak hickory on slopes and ridges. Selective logging of varying ages in four ravines; four other ravines uncut. Closed canopy road.

[a]Tract size is difficult to define precisely because all tracts are connected to other tracts. For the purpose of this chapter, we use roads and agricultural fields to define tract boundaries.

many with multiply parasitized nests. Robinson et al. (in press *b*), however, found that grassland birds in Illinois were rarely parasitized. Parasitism levels for some hosts of Shiny (Post et al. 1990) and bronzed cowbirds (Clotfelter and Brush 1995) are as high as those reported in Illinois.

Moderately fragmented landscapes

For the past six years, we have been studying the breeding ecology of birds in the moderately fragmented, 104 000-ha Shawnee National Forest (SNF) in extreme southern Illinois. Forested areas of SNF are relatively homogeneous and are dominated by oak/hickory ridgetops and steep mixed-hardwood ravines. Privately owned inholdings are common and are usually maintained as pasture or cropland. Approximately 50% of the region still remains forested, with fragment size ranging up to 2000 ha (Iverson 1989).

We concentrated our work in four medium-sized (1000–2000-ha) tracts (table 15.3), all of which are within 30 km of each other in the Illinois Ozarks section of forest. Our aims were to assess the impacts of cowbird parasitism and nest predation on host reproductive success. To do this, we established census routes through each tract to survey cowbird and host densities. We also found nests of as many hosts as possible and monitored their contents and fates (see Robinson et al. 1995 for methods). By color marking individual birds, we were able to follow the success or failure of some individuals and obtain measurements such as the number of nesting attempts made per season per female.

Unlike other localities in the Midwest where brown-headed cowbird abundance declines dramatically a few hundred meters into the forest interior (Brittingham and Temple 1983), cowbirds were just as common deep in the interior of the largest tracts of SNF as they were on the edges (table 15.4). Similarly, transects with extensive edge (e.g., Dutch Creek) had no more cowbirds than those consisting largely of contiguous forest (e.g., Pine Hills, Ripple Hollow) (table 15.4). Because cowbirds were uniformly abundant with respect to edge, few potential hosts escaped parasitism. In fact, parasitism rates in SNF were among the highest ever reported (table 15.5). Not only were most nests parasitized, but many species typically received more than one cowbird egg per nest (table 15.5).

The most severely affected species in the SNF appear to be scarlet and summer tanagers, red-eyed vireos, hooded warblers, and wood thrushes (table 15.5). Our sample sizes for the vireo and tanagers are small (table 15.5) because they place their nests in the canopy where monitoring is difficult. Wood thrush nests, however, are routinely placed less than 7 m high and are generally easy to find and monitor (Roth and Johnson 1993). In the SNF we have been able to measure most of the parameters necessary to predict whether wood thrush populations can sustain themselves. Unfortunately, as is the case with nearly every study of migratory passerines, we lack good estimates of adult and juvenile mortality rates. Given the mean number of fledglings per female (table 15.6), however, we can estimate the maximum mortality rate that wood

Table 15.4 Relative abundances of brown-headed cowbirds in relation to all edges, which are defined as openings in the cnopy of at least 0.2 ha. Each transect consists of at least 15 points separated by 150 m

Transect: Habitat	Distance from edge (m)	Independent cowbird detections/ 10 point counts
Dutch Creek:		
Ravine Bottom	<25	9.6
	25–100	10.9
	<100	12.4
Dutch Creek:		
High Ridge	<25	5.0
	25–100	5.8
	>100	5.2
Dutch Creek:		
Low Ridge	<100	14.4
	>100	14.4
Ripple Hollow:		
Ravine Bottom	<200	12.1
	200–400	16.5
	>400	14.4
Ripple Hollow:		
Ridge	<100	7.7
	>100	9.7
Pine Hills:		
Ravine Bottom	<300	18.1
	>300	15.0

Data include all cowbirds heard within a 70-m radius of each point (excluding birds flying overhead) during 6-min census intervals from 15 May to 5 July 1989. Groups of two or more cowbirds are counted as a single registration. None of the edge effects is significant at the 0.05 level (Mann–Whitney *U*-test).

thrushes could sustain while still maintaining their current population level.

To estimate the maximum annual adult mortality rate, we had to make some assumptions. First, we assumed that juvenile mortality (μ_o) is twice that of adults (μ; Nice 1937; Nolan 1978; May and Robinson 1985). Because wood thrushes renest after nest failures and are double-brooded, most females nest more than once each season. We therefore assigned to λ and λ' the mean number of fledglings per territory for unparasitized and parasitized females, respectively. Then, because so few nests were unparasitized in the SNF (and most of those were depredated), we had to estimate λ in order to use May and Robinson's model. Cowbirds routinely remove host eggs before laying their own; Trine (in press) estimated that unparasitized clutches averaged 3.3 eggs. We used data gathered from our own nests (Trine, in press) to arrive at estimates of the following parameters necessary to compute productivity of unparasitized nests: egg survival rate during the entire incubation period (0.99); hatching success for eggs in the few unparasitized nests for which we have data (0.87); and nestling survival over the entire nestling period in nests that escaped predation and had no cowbird nestlings (0.97). The mean number of fledglings per successful unparasitized nest is therefore the product of clutch size, egg survival rate, hatching success, and nestling survival, which equals 2.76. To calculate the number of young fledged/territory of unparasitized females, we multiplied 2.76 by the number of successful nests/territory. To calculate the μ necessary for population maintenance, we set $p = p_c$. We then compared the values of μ with values obtained from the literature (e.g., Karr et al. 1990).

Demographic parameters varied among our three main study sites in 1990 (table 15.6). At Dutch Creek, the parasitism rate was low relative to the other two sites. Fledging success, however, was also low because of high (>80%) levels of nest predation compared with the other two sites (55–65%). Given the 1990 fledging rates, we estimated the adult mortality necessary to sustain the population to be 15% or less per year, an extremely low value for a migratory passerine (Ricklefs 1973; Karr et al. 1990). Therefore, we conclude that, at least in 1990, productivity of the wood thrush population at Dutch Creek was insufficient to compensate for adult mortality. Pine Hills and Ripple Hollow had higher parasitism levels than Dutch Creek, but because of low levels of nest predation, territories at those two sites produced more young (table 15.6). Nevertheless, the estimated mortality rates of 27% for Pine Hills and 30% for Ripple Hollow are at the lower limits of estimated annual mortality rates for migratory songbirds (Ricklefs 1973; Greenberg 1980). We suggest, then, that the wood thrush populations at Pine Hills and Ripple Hollow probably did not reproduce at a rate that compensates for adult mortality during the study. Wood thrushes were producing less than a third the number of young per pair than black-throated blue warblers nesting in the unfragmented White Mountains of New Hampshire (Holmes et al. 1992). Trine (in prep.), however, determined that wood thrush populations in Ripple Hollow and Pine Hills would have been self-sustaining even with high parasitism levels, had the nest predation rate been 60% or less, a typical value for passerines (Ricklefs 1973).

Table 15.5 Parasitism levels from three Shawnee National Forest study areas, 1989–1993. All tracts were at least 900 ha and all nests were at least 250 m from the nearest agricultural opening

	Pine Hills		Ripple Hollow		Dutch Creek	
Species	% parasitized (*n*)	Cowbird eggs/ parasitized nest	% parasitized (*n*)	Cowbird eggs/ parasitized nest	% parasitized (*n*)	Cowbird eggs/ parasitized nest
Acadian flycatcher	28.5 (181)	1.2	47.4 (97)	1.3	39.0 (146)	1.2
Eastern wood pewee	27.3 (11)	1.0	—	—	—	—
Wood thrush	87.8 (221)	2.9	89.0 (91)	3.5	75.4 (57)	2.2
Red-eyed vireo	100 (2)	3.0	—	—	80.0 (5)	2.0
Yellow-throated vireo	—	—	—	—	75.0 (4)	2.0
White-eyed vireo	83.3 (6)	1.2	33.3 (3)	1.0	27.8 (72)	1.3
Ovenbird	—	—	—	—	80.0 (5)	1.0
Louisiana waterthrush	16.7 (6)	1.0	100 (2)	2.0	8.7 (23)	1.5
Kentucky warbler	48.3 (58)	1.9	63.0 (27)	1.9	21.9 (73)	1.7
Hooded warbler	—	—	100 (4)	1.5	72.2 (18)	1.2
Worm-eating warbler	16.7 (6)	3.0	66.7 (3)	1.0	37.0 (27)	1.7
Yellow-breasted chat	—	—	—	—	28.6 (14)	1.0
Scarlet tanager	100 (4)	3.0	—	—	100 (1)	2.0
Summer tanager	100 (3)	3.0	—	—	50 (8)	2.6
Northern cardinal	30.6 (36)	1.4	37.5 (40)	1.2	29.7 (111)	1.4
Rufous-sided towhee	63.6 (22)	1.7	66.7 (15)	1.6	20.0 (25)	1.0
Indigo bunting	50.0 (18)	1.4	20.0 (5)	1.0	29.3 (92)	1.3

Because of poor productivity on all sites, these wood thrush populations may have to be augmented by immigration from some other locale in order to persist. For the wood thrush, most of the SNF may represent a population "sink" even in large (2000-ha) tracts that are usually considered secure from edge effects (Wilcove 1985; Wilcove et al. 1986). The SNF is also a likely sink population for many of the other host species that are heavily parasitized (e.g., tanagers, vireos, and hooded warblers in table 15.5).

One of the major reasons why wood thrushes are so vulnerable to cowbird parasitism is that they are usually multiply parasitized (table 15.5; Robinson 1992). Trine (in press) found that overall host fledging success was reduced by about 18% per cowbird egg received, largely as a result of host egg removal by cowbirds (0.3–0.4 wood thrush eggs removed per egg laid). If it were not for this egg loss, wood thrushes would be relatively unaffected by parasitism because they are double-brooded and can raise mixed broods

Table 15.6 Demographic parameters for wood thrushes at three study sites in the Shawnee National Forest, 1990

Site	Young/ unparasitized female[a]	Young/ parasitized female	Proportion of nests parasitized[b]	Number successful nests/ season/female[c]	Estimated adult mortality[d]
Pine Hills	3.3	1.0	0.944	1.20	0.27
Ripple Hollow	3.4	1.4	0.933	1.40	0.30
Dutch Creek	1.0	0.2	0.737	0.36	0.15

[a]Estimated; see text for method of calculation.

[b]Proportion of nests parasitized over an entire season; every female in all three study sites was parasitized at least once per breeding season (C. Trine, unpubl. data).

[c]A successful nest is one that fledged at least one host *or* cowbird.

[d]Estimated maximum mortality required to maintain a stable population (see text for method of calculation).

Table 15.7 Annual variation in demographic productivity of Acadian flycatchers in the Shawnee National Forest area of southern Illinois

Site	Year	Propn nests parasitized[a]	Propn nests successful[b]	Propn nests depredated	Hosts fledged/pair/season: No. host fledged/nest built[c]	Potential productivity: with cowbirds[d]	Potential productivity: without cowbirds[e]	Estimated % cost of parasitism[f]
Dutch Creek	89	0.33	0.36	0.64	0.62	1.1	1.5	−27
	90	0.29	0.33	0.67	0.61	1.0	1.4	−29
	91	0.32	0.45	0.55	0.83	1.3	1.8	−28
	92	0.25	0.24	0.76	0.48	0.8	1.1	−27
	93	0.22	0.75	0.25	1.55	1.9	2.4	−21
Ripple Hollow	89	0.60	0.27	0.73	0.34	0.7	1.2	−42
	90	0.48	0.43	0.57	0.65	1.0	1.7	−41
	91	0.18	0.55	0.45	0.90	1.6	2.0	−20
	92	0	0.57	0.43	1.43	2.1	2.1	0
Pine Hills	89	0.53	0.71	0.29	0.99	1.3	2.3	−43
	90	0.35	0.43	0.57	0.77	1.2	1.7	−29
	91	0.18	0.64	0.36	1.38	1.9	2.2	−16
	92	0.26	0.54	0.46	1.06	1.6	2.0	−20
Trail of Tears	90	0.12	0.27	0.73	0.61	1.1	1.2	−8
	91	0.36	0.21	0.79	0.37	0.7	0.9	−22
	92	0.37	0.23	0.77	0.40	0.7	1.0	−30

[a]Includes all nests throughout the 15 May–5 August nesting season.

[b]Based on Mayfield (1975) method.

[c]Assuming 2.52 hosts fledged/successful unparasitized nest and 0.42 hosts/successful parasitized nest (table 15.2). Calculated by ([Propn unparasitized nests × 2.52 fledged/unparasitized nest] + [propn parasitized nests × 0.42 fledged/parasitized nest]) × Propn nests that fledge. This assumes no difference in predation rate between parasitized and unparasitized nests (S. Robinson, unpubl. data).

[d]Assuming that renesting rate increases with predation rate. We arbitrarily set the average number of nesting attempts per season at 2 minus the proportion of nests that fledged and multiplied it by the number of hosts fledged per nest built. These numbers are generally similar to those found in color-marked populations in nearby central Indiana (D. Whitehead, pers. commun.). Although some females nest more than twice, those that fledge young usually nest only once (R. Lu-Olendorf, unpubl. data).

[e]Calculated by assuming that all nests were unparasitized, but using survival rates generated by Mayfield estimates of survival.

[f]Calculated as the % reduction of potential hosts fledged/territory without cowbirds divided by the actual productivity.

of cowbird and host young (Roth and Johnson 1993). Cowbirds had very little effect on host productivity in a wood thrush population where most nests received only single cowbird eggs (Roth and Johnson 1993). Therefore, in the East, where parasitism levels are low (Peck and James 1987; Hoover and Brittingham 1993) and only weakly correlated with forest fragmentation (e.g., Hoover and Brittingham 1993; Dowell et al., in press; Petit and Petit, in press), cowbirds may have no significant effect on metapopulation dynamics.

Data from Acadian flycatchers in the same sites illustrate the potential effects of annual and site-by-site variation in both parasitism and predation levels on the source/sink status of populations (table 15.7). Parasitism levels varied little among years in some sites (e.g., Dutch Creek), but varied more than threefold in others (e.g., Ripple Hollow, Trail of Tears). Similarly, predation levels were relatively constant in some sites (e.g., Trail of Tears), but varied more than twofold in other sites (e.g., Dutch Creek). As a result of this variation, the number of hosts fledged per nest built varied from 0.34 to 1.55 among sites and years (table 15.7) and the estimated number of young produced per pair (see table 7 for assumptions) varied more than threefold. Assuming adult survival rates of 0.6–0.7 and juvenile survival rates of 0.3–0.35 ("typical" estimates; Martin 1992), Acadian flycatchers must produce 1.7–2.7 young/pair/season to break even. Productivity estimates/territory/year were generally well below this estimate; only three of 16 site–year combinations were above the 1.7 level and all were at the lower end of the 1.7–2.7 range. In the absence of cowbird parasitism, nine instead of three of 16 site–year combinations were in the 1.7–2.7 range, although none was above this range. Predation rates were so high in the Trail of Tears site (table 15.7) that all populations were well below the source-sink threshold. On average, cowbird parasitism reduced productivity by an estimated $25.2 \pm 11.2\%$ ($n = 16$ site–year combinations) (table 15.7).

Data in table 15.7 therefore suggest that cowbird parasitism can have profound consequences for metapopulation dynamics within landscapes. Several caveats, however, should be emphasized. First, Acadian flycatchers have small clutches (table 15.2), nest late, and are rarely (<20%) double-brooded in this area (R. Lu-Olendorf and L. Chapa, pers. commun.), which may make them unusually vulnerable to increased parasitism and predation levels. Second, we know very little about survival rates of any forest passerine; it is quite possible that Acadian flycatchers have a much higher survival rate than we can detect because of both juvenile and adult dispersal (Martin 1992; S. Robinson, unpubl. data). And third, the estimated numbers of nesting attempts/season are not based on observations of color-marked females as are those of wood thrushes in table 15.6. Changes in these assumptions could change the interpretation of the source/sink status of many of the populations listed in table 15.7. Nevertheless, the generally low productivity of Acadian flycatchers in all sites in all years suggests that they may be particularly vulnerable to the increases in predation and parasitism rates that characterize fragmented landscapes. The regional metapopulation may depend upon occasional years of low parasitism and predation (table 15.7) and upon immigration from areas where parasitism and predation rates are much lower (e.g., Robinson et al. 1995; Winslow et al., in press).

The continued existence of large, unfragmented forests within regions (Robinson et al. 1995*b*; Brawn and Robinson 1996) and continentally (Flather and Sauer 1996; Peterjohn et al., in press; Thompson et al., in press; Wiedenfeld, in press) may explain why many (if not most) neotropical migrants are not declining (James and McCulloch 1995; Peterjohn et al. 1995; Wiedenfeld, in press). Even migrants nesting in woodlots where their reproductive success is almost zero do not have declining populations within these woodlots (Brawn and Robinson 1996). In the absence of immigration most of the populations of forest birds in Illinois woodlots would become extinct within a few years or decades, if they had to depend upon young born in that woodlot (an unlikely assumption; Weatherhead and Forbes 1994). A population of 20 pairs of wood thrushes in Dutch Creek, for example, would decline by 27% annually and be extinct (less than one pair) within 10 years, assuming 70% adult and 35% juvenile survival annually and given the productivity estimates in table 15.6. Comparable estimates for Pine Hills and South Ripple Hollow are 23 years (12.5% annual decrease) and 53 years (5.5% annual decrease). The continued existence of regional metapopulations therefore may depend upon large tracts; even apparently stable populations may be extremely vulnerable to fragmentation of current and distant "source" habitat.

An additional issue that may complicate population analyses is the possibility that cowbirds may depredate unparasitized nests (Smith and Arcese 1994; Arcese et al. 1996). The high nest predation rates in the SNF, for example, may result from cowbird predation. However, a preliminary analysis of predation rates on parasitized and unparasitized nests in the SNF (S. Robinson, unpubl. data) showed no consistent difference in the predation rates. As Arcese et al. (1996) acknowledge, their result only may apply to situations in which the number of cowbirds and hosts are very limited such as often occurs on islands. A complete test of this hypothesis, however, will require experimental manipulations and tests of predation rates on singly versus multiply parasitized nests.

Populations in unfragmented landscapes

Nesting success and annual productivity of forest birds appear to be much higher in large, unfragmented forest tracts (Holmes et al. 1992; Robinson et al. 1995). At the Hubbard Brook Experimental Forest, an unfragmented temperate deciduous forest in New Hampshire, only two parasitized nests have been found out of thousands examined over more than 20 years including several species that also occur in Illinois (e.g., wood thrush, scarlet tanager, ovenbird, red-eyed vireo) (Sherry and Holmes 1992). Coker and Capen (in press) and Yamasaki et al. (in press) found that cowbirds were essentially absent from the interior of large forest tracts in nearby areas of New England. Cowbird populations are generally much lower in northern New England than they are in the Midwest (Wiedenfeld, in press; Peterjohn et al., in press) and are declining. The few cowbirds that do live in the region are restricted to the immediate vicinity of feeding areas (Coker and Capen, in press; Yamasaki et al., in press). Cowbird parasitism in the extensively forested areas of the Missouri Ozarks, southcentral Indiana and northern Wisconsin are also very low even though the bird communities are similar to more fragmented sections of the Midwest (Robinson et al. 1995; Thompson et al., in press). Wilcove (1988) reported an absence of cowbirds in the extensively forested Great Smoky Mountains National Park. On the western slope of the Sierra Nevada mountains of California, cowbirds were virtually absent from unlogged forest tracts but made use of meadows, clearcuts, and incompletely logged forests (Verner and Ritter 1983). These studies, as well as others that identify cowbird parasitism as an edge effect (Gates and Gysel 1978; Brittingham and Temple 1983; Temple and Cary 1988; Rich et al. 1994), indicate that unfragmented forests can serve as a refuge from cowbird parasitism even in regions such as the Midwest where cowbirds are abundant (Peterjohn et al., in press; Wiedenfeld, in press). Landscapes in which cowbird feeding opportunities are rare (e.g., primarily forested landscapes: Robinson et al. 1995; Coker and Capen, in press; Thompson et al., in press; Yamasaki et al., in press) may act as the source populations for colonization and maintenance of population sinks in fragmented habitats. Nest predation rates also decline with increasing forest cover and tract size in midwestern (U.S.) forests (Robinson et al. 1995*b*), which enhances the value of large tracts as population sources.

Ironically, large, relatively unfragmented or moderately fragmented forests also may be "source" habitat for cowbirds. Donovan et al. (in press) found that cowbird nesting success per egg was much lower in highly fragmented landscapes as a result of high nest predation rates and competition among cowbird nestlings in multiply parasitized nests (see also Trine, in press). Severely fragmented landscapes also may be ecological "traps" for cowbirds because nearby unlimited feeding habitat may attract high densities of cowbirds that fail to produce enough young to compensate for adult mortality. Cowbird fecundity, however, may be so high (Scott and Ankney 1980; Holford and Roby 1994) that their populations can sustain very heavy losses to predation.

HABITAT EFFECTS WITHIN LANDSCAPES

Parasitism levels also vary among habitats within landscapes (Robinson et al., in press; Ward and Smith, in press; Winslow et al., in press). As already mentioned, grassland birds are rarely parasitized anywhere in Illinois, even in areas where birds nesting in adjacent forests are heavily parasitized. Edge/shrubland species in the SNF are also much less heavily parasitized on average than the forest species in Illinois (table 15.8; see also table 15.5). Average parasitism percentages for edge/shrubland nesters were 15.1 ± 16.9 (S.D.)% (range = 0–50%, $n = 17$, table 15.8), compared with 57.3 ± 23.9 (S.D.)% (range = 27.3–100%, $n = 17$) for forest habitats (table 15.5; three sites averaged together to obtain a composite parasitism rate for each species) ($P < 0.001$, Wilcoxon Test). Robinson et al. (in press) found that parasitism levels of edge/shrubland birds were inversely related to the percentage forest cover within 10-km radii around the study sites, which suggests that cowbirds prefer to parasitize forest hosts and only visit edge/shrublands when the supply of forest hosts may be limiting. Indeed, forest birds appear to have no defenses against cowbird parasitism; they accept cowbird eggs in proportion to the size of the host (Robinson et al., in press *a*). In contrast, many of the edge/shrubland birds in table 15.8 have well-developed defenses against brood parasitism. Gray catbirds, American robins, and warbling vireos (S. G. Sealy, pers. commun.) all eject cowbird eggs (Rothstein 1975*b*; Rothstein 1990; Scott 1980; Sealy, pers. commun.). Field sparrows abandon nests that have been visited by cowbirds (Burhans, in press). Red-winged blackbirds mob cowbirds (Freeman et al. 1990; Carello and Snyder, in press). The low parasitism levels of shrubland/edge species therefore may reflect cowbird egg rejection by several hosts and avoidance by cowbirds of a habitat that may contain more resistant hosts than are found in forests (Hahn and Hatfield 1995; Robinson et al., in press *a*). Experimental manipulations of host community composition may be necessary to determine if community composition affects community wide levels of cowbird parasitism.

WITHIN-HABITAT VARIATION IN PARASITISM

Parasitism levels in the SNF appeared to vary little among vertical strata within the forest.

Table 15.8 Parasitism levels of edge/second-growth species in the Shawnee National Forest area

Species	% parasitized (n)	Cowbird eggs/ parasitized nest
Brown thrasher	16.6 (6)	1.0
Gray catbird	0 (5)	–
Carolina wren	12.0 (25)	1.3
Eastern phoebe	3 (32)	1.0
American robin	0 (18)	–
Warbling vireo	0 (5)	–
White-eyed vireo	40 (20)[a]	1.3
Yellow-breasted chat	27 (11)[a]	1.7
Common yellowthroat	0 (5)	–
Orchard oriole	42 (12)	1.0
Red-winged blackbird	0 (26)	–
Northern cardinal	22 (9)[a]	1.0
Rufous-sided towhee	50 (12)[a]	1.5
Indigo bunting	33 (36)[a]	1.6
Chipping sparrow	11.1 (9)	1.0
Field sparrow	0 (7)	–
Song sparrow	0 (5)	–

[a]These nests are from more open and shrubby study areas than those from the same species in table 15.5, all of which are from small forest openings.

Ground nesters (ovenbird, Kentucky warbler, Louisiana waterthrush, and worm-eating warbler) were somewhat less parasitized than canopy nesters (e.g., tanagers and vireos) (table 15.5). Shrub/sapling nesters included a mix of moderately (e.g., indigo bunting, white-eyed vireo) and heavily parasitized (e.g., hooded warbler and wood thrush) species (table 15.5). In contrast, Hahn and Hatfield (1995, in press) found that cowbirds showed a strong preference for ground nesters. Such preferences for microhabitats therefore may not be consistent among regions, a conclusion also reached by Robinson et al. (1995a). Local vegetation around nests may also affect conspicuousness to cowbirds (reviewed in Martin 1992; Howe and Knopf, in press; Uyehara and Whitfield, in press).

GEOGRAPHIC VARIATION

Cowbird parasitism rates vary greatly among different populations of some species (table 15.9; see also Friedmann 1963; Friedmann et al. 1977; Hoover and Brittingham 1993; Smith and Myers-Smith, this vol.). A variety of hypotheses, not mutually exclusive, can be invoked to explain these differences in parasitism rates (see Smith and Myers-Smith, this vol.).

Regional cowbird densities

Smith and Myers-Smith (this vol.) and Hoover and Brittingham (1993) showed that parasitism levels of song sparrows and wood thrushes were positively correlated with continent-wide patterns in cowbird abundance (see Peterjohn et al., in press for a detailed account of the cowbird's breeding distribution). Grassland bird parasitism levels also appear to be higher in the Great Plains where cowbird abundance is greatest (Davis and Sealy, in press; Koford et al., in press) than in Illinois (Robinson et al., in press, *a*). In locales where cowbirds have just arrived, they may still be in the process of increasing population size (Cruz et al., this vol.). Where preferred hosts are not saturated, less-preferred hosts may be virtually unparasitized (Friedmann 1963; Friedmann et al. 1977; Freeman et al. 1990).

Host nest density

Fretwell (1977) hypothesized an inverse relationship between host nest density and the intensity of cowbird parasitism. Both he and Zimmerman (1983) found that in habitats where dickcissels were less abundant, their incidence of cowbird parasitism was higher, a result also obtained by Carello and Snyder (in press). Neither study, however, considered the total density of all host species combined. In several other studies, cowbird parasitism levels and/or cowbird densities were positively correlated with total host densities (Gates and Gysel 1978; Verner and Ritter 1983; Rothstein et al. 1984; Airola 1986; Thompson et al., in press; but see Smith and Myers-Smith, this vol.). The influence of host densities on cowbird parasitism rates requires further research.

Constellation of host species available

Cowbirds apparently prefer certain species over others (Friedmann 1929). In an area where preferred hosts are few in number, parasitism on less-preferred hosts may be higher than in an area where preferred host densities are high (Wiens 1963; Southern and Southern 1980). If preferred host nests are already saturated, less-preferred hosts will be used more frequently (Freeman et al. 1990; Smith and Myers-Smith, this vol.).

Table 15.9 Cowbird nest parasitism rates from different studies

Species	Location	% parasitism (n)	Reference
Song sparrow	Pennsylvania	41 (27)	Norris (1947)
	Ohio	34 (398)	Hicks (1934)
	Ohio	44 (223)	Nice (1937)
	Ohio	29 (52)	Trautman (1940)
	Michigan	59 (63)	Berger (1951)
	Michigan	15 (20)	Sutton (1960)
	Michigan	7 (61)	Southern and Southern (1980)
	Quebec	13 (486)	Terrill (1961)
	Ontario	91 (22)	Scott (1977)
	British Columbia	22 (452)	Smith (1981)
Dickcissel	Kansas	50 (28)	Hill (1976)
	Kansas	95 (19)	Elliott (1978)
	Kansas	60 (395)	Zimmerman (1983)
	Kansas	85 (235)	Zimmerman (1983)
	Nebraska	53 (17)	Hergenrader (1962)
	Oklahoma	31 (61)	Overmire (1962*a*)
	Oklahoma	21 (29)	Wiens (1963)
Red-winged blackbird	Maryland	1 (367)	Meanley, in Friedmann (1963)
	Louisiana	2 (754)	Goertz (1977)
	Michigan	0 (44)	Southern and Southern (1980)
	Michigan	5 (99)	Berger (1951)
	Oklahoma	2 (110)	Wiens (1963)
	Kansas	22 (228)	Hill (1976)
	Kansas	31 (29)	Facemire (1980)
	Nebraska	54 (59)	Hergenrader (1962)
	North Dakota	76 (17)	Houston (1977)
	North Dakota	42 (258)	Linz and Bolin (1982)
	Colorado	23 (78)	Oretega and Cruz (1991)
	Washington	8 (709)	Røskaft et al. (1990)
	Washington	8 (1325)	Freeman et al. (1990)
	California	0 (>300)	K. Steele and S. I. Rothstein, pers. commun.
Eastern phoebe	Michigan	0 (97)	Southern and Southern (1980)
	Kansas	24 (391)	Klaas (1975)
	Kansas	10 (68)	Hill (1976)
Indigo bunting	Ohio	40 (43)	Hicks (1934)
	Ohio	43 (14)	Phillips (1951)
	Michigan	15 (26)	Sutton (1959)
	Michigan	43 (37)	Southern and SOuthern (1980)
	Illinois	36 (33)	Twomey (1945)
	Quebec	20 (30)	Terrill (1961)
Bell's vireo	Indiana	54 (13)	Mumford (1952)
	Texas	74 (23)	Webster (1971)
	Oklahoma	30 (61)	Overmire (1962*b*)
	Oklahoma	71 (31)	Wiens (1963)
	Kansas	69 (35)	Barlow (1962)
	California	50 (14)	Goldwasser et al. (1980)
Eastern meadowlark	Wisconsin	16 (38)	Lanyon (1957)
	Louisiana	1 (91)	Goertz (1977)
	Kansas	70 (40)	Elliott (1978)
	Nebraska	16 (31)	Hergenrader (1962)
	Quebec	2 (52)	Terrill (1961)
Red-eyed vireo	Pennsylvania	86 (14)	Norris (1947)
	Ohio	36 (231)	Hicks (1934)
	Michigan	69 (257)	Southern and Southern (1980); Southern (1958)
	Ontario	0 (30)	Lawrence (1953)
	Ontario	90 (10)	Rice (1974)
	Quebec	41 (63)	Terrill (1961)

Table 15.9 (*continued*)

Species	Location	% parasitism (*n*)	Reference
Wood thrush	Delaware	0 (142)	Longcore and Jones (1969)
	Maryland	8 (37)	Brackbill (1958)
	Pennsylvania	18 (11)	Norris (1947)
	Illinois	100 (19)	Robinson (1992)
	Wisconsin	80 (15)	Brittingham and Temple (1983)
	Louisiana	0 (20)	Goertz (1977)
Acadian flycatcher	Michigan	21 (121)	Walkinshaw (1966)
	Wisconsin	14 (37)	Brittingham and Temple (1983)
	Illinois	50 (12)	Graber et al. (1974)

Overlap of cowbird and host breeding seasons

Breeding seasons of cowbirds vary from location to location, as do those of potential host species (Berger 1951). When a host's breeding season overlaps only slightly with the cowbird's, parasitism levels are lower than when the breeding seasons of the two species coincide (Elliott 1978; Petrinovich and Patterson 1978).

Habitat structure

Wiens (1963) and Berger (1951) noticed that parasitized nests of field-nesting species were near thickets and woodlots whereas nests away from such vegetation were rarely parasitized (see also Johnson and Temple 1990). Gates and Gysel (1978) suggested that parasitism may be higher where appropriate perch sites are available from which cowbirds may observe host activities. This may be one reason for increased cowbird parasitism levels near forest edges and near woody vegetation in marshes (Freeman et al. 1990).

Proximity to cowbird feeding areas

In some locations, cowbird presence appears to depend upon human-based food supplies; the further from these food supplies, the lower the densities of cowbirds (Verner and Ritter 1983; Rothstein et al. 1984; Airola 1986; Smith and Myers-Smith, this vol.). Lack of an adequate food supply may explain the virtual absence of cowbirds in extensive forest tracts (Verner and Ritter 1983; Wilcove 1988; Robinson et al. 1995). Because cowbirds are known to travel up to 7 km from feeding areas to sites where they parasitize nests (Rothstein et al. 1984), the tract size required for a refuge from cowbird parasitism is probably very large.

Host nest defense

A species may vary across its geographic range with respect to nest defense against cowbirds, although the only clear evidence for this comes from Briskie et al. (1992). This difference may be a result of evolutionary differentiation among populations, whereas in other cases it may be a result of different population structures such as colonial versus solitary nesting. Colonial defense in dense colonies may be effective in reducing cowbird parasitism, whereas individual defense of widely scattered nests may allow higher overall parasitism rates (Friedmann 1963; Robertson and Norman 1977; Wiley and Wiley 1980; Freeman et al. 1990; Carello and Snyder, in press).

IMPLICATIONS OF REGIONAL VARIATION IN PARASITISM

As shown here, regional variation in host productivity is influenced by cowbird parasitism rates. Thus, metapopulations of neotropical migrants may operate on a regional or even a continental scale. For example, wood thrush populations are only lightly parasitized (0–8%) in the Mid-Atlantic Coast region even though other species (e.g. red-eyed vireo, *Vireo olivaceus*) are regularly parasitized (Brackbill 1958; Longcore and Jones 1969; Dowell et al., in press; Petit and Petit, in press). On the other hand, midwestern populations are much more heavily parasitized: 80% in Wisconsin (Brittingham and Temple 1983), 100% in central Illinois (Robinson 1992), and 89% in southern Illinois (table 15.4). Populations in neighboring Iowa are also severely parasitized (Friedmann 1929, 1963). Although more data are needed from other locations within the region to obtain a clearer picture, much of this region may be a population sink for the wood thrush and large parts of the Midwest may depend upon production in and emigration from populations elsewhere (Robinson et al. 1995b). Birds are relatively mobile compared with most terrestrial organisms and should be capable of long-distance dispersal. Even species having long-distance migration, however, do not necessarily have long-distance dispersal (Jones and Diamond 1976; Diamond 1984). Whitcomb et al. (1981) analyzed the small amount of Bird Banding Laboratory data from birds banded as nestlings and recovered in subsequent years during the breeding season. For both resident and migratory species, most recoveries were within 20–40 km of the nest from which birds fledged. Considering the extremely low productivity for some species in central Illinois, however, immigrants must be coming from at least 150 km for continued persistence in the woodlots (Robinson 1992; Brawn and Robinson, 1996). Certainly, more information is needed on dispersal distances in migratory birds to determine the spatial scale of metapopulation dynamics.

CONCLUSIONS

We have long known that local populations of host species such as the Kirtland's warbler, least Bell's vireo, and black-capped vireo can be threatened by intense cowbird parasitism. In the absence of cowbird control and habitat management, these taxa might become extinct. Recent studies from the Midwest, however, suggest that cowbird parasitism may also threaten regional metapopulations of more abundant, widespread species. Population declines of several neotropical migrants such as the wood thrush may be at least partly a result of severe and possibly increasing parasitism by cowbirds and the high levels of nest predation that usually occur in areas where cowbirds are also abundant (Robinson et al. 1995b). Habitat fragmentation carried to the extremes seen in Illinois can create regional population sinks that can only be maintained by distant populations in larger, unfragmented forests, such as those in Missouri and Indiana. These data provide strong arguments for the preservation of the largest remaining unfragmented forest and prairie tracts in cowbird-saturated parts of North America. Protecting songbirds from the combined effects of cowbird parasitism and nest predation may require preserving and restoring much larger tracts than we originally thought were necessary. Before we can understand the impact of cowbirds on widely distributed hosts, however, we must know more about the scale of metapopulation dynamics, geographic variation in cowbird parasitism, and survival rates.

ACKNOWLEDGMENTS We thank all those who have assisted in the field work in Illinois: Kris Bruner, Terry Chesser, Steve Bailey, Steve Amundson, Rob Olendorf, Rob Brumfield, Lonny and Caleb Morse, Rhetta Jack, Mike Ward, Tony Berto, Dori Rogg, John Knapstein, Dave Enstrom, and Todd Freeberg, and many others. The research was supported by the American Museum of Natural History (Chapman Memorial Fund); American Ornithologists Union (Wetmore Award); Sigma Xi (Grant-in-aid of Research); Champaign County Audubon Society (Kendeigh Memorial Fund); University of Illinois, Graduate College; University of Illinois, Department of Ecology, Ethology, and Evolution (all to C. L. Trine; National Science Foundation, BSR 9101211 DIS (to C. Trine); Illinois Department of Energy and Natural Resources; Illinois Department of Conservation; and U.S. Forest Service (to S. Robinson).

REFERENCES

Airola, D. A. (1986). Brown-headed Cowbird parasitism and habitat disturbance in the Sierra Nevada. *J. Wildlf. Manag.*, **50**, 571–575.

Arcese, P., J. N. M. Smith, and M. I. Hatch. (1996). Nest predation by cowbirds and its consequences for passerine demography. *Proc. Natl Acad. Sci. USA*, **93**, 4608–4611.

Askins, R. A. and M. J. Philbrick (1987). Effects of changes in regional forest abundance on the decline and recovery of a forest bird community. *Wilson Bull.*, **99**, 7–21.

Askins, R. A., J. F. Lynch and R. Greenberg (1990). Population declines in migratory birds in eastern North America. In: *Current Ornithology, volume 7* (ed. D. M. Power), pp. 1–57. Plenum Press, New York.

Barlow, J. C. (1962). Natural history of the Bell Vireo,

Vireo bellii Audubon. *University of Kansas Publications of the Museum of Natural History*, **12**, 241–296.

Berger, A. J. (1951). The cowbird and certain host species in Michigan. *Wilson Bull.*, **63**, 26–34.

Böhning-Gaese, K., M. L. Taper, and J. H. Brown (1993). Are declines in North American breeding birds due to causes on the breeding range? *Conserv. Biol.*, **7**, 76–86.

Brackbill, H. (1958). Nesting behavior of the Wood Thrush. *Wilson Bull.*, **70**, 70–89.

Brawn, J. D. and S. K. Robinson (1996). Source-sink population dynamics may complicate the interpretation of long-term census data. *Ecology*, **77**, 3–11.

Briskie, J. V., S. G. Sealy, and K. A. Hobson (1992). Behavioral defenses against avian brood parasitism in sympatric and allopatric host populations. *Evolution*, **46**, 334–340.

Brittingham, M. C. and S. A. Temple (1983). Have cowbirds caused forest songbirds to decline? *BioScience*, **33**, 31–35.

Brown, J. H. and A. Kodrick-Brown (1977). Turnover rates in insular biogeography: effect of immigration on extinction. *Ecology*, **58**, 445–449.

Burhans, D. E. Dawn nest arrivals in cowbird hosts: their role in aggression, cowbird recognition, and response to parasitism. In: *Ecology and management of cowbirds* (eds J. N. M. Smith, T. Cook, S. K. Robinson, S. I. Rothstein, and S. G. Sealy), University of Texas Press, Austin (in press).

Carello, C. A. and G. K. Snyder. The effects of host numbers on cowbird parasitism of Red-winged Blackbirds. In: *Ecology and management of cowbirds* (eds J. N. M. Smith, T. Cook, S. K. Robinson, S. I. Rothstein, and S. G. Sealy), University of Texas Press, Austin (in press).

Chasko, G. G. and J. E. Gates (1982). Avian habitat suitability along a transmission-line corridor in an oak-hickory forest region. *Wildlf. Monogr.*, **82**, 1–41.

Clobert, J. and J. D. Lebreton (1991). Estimation of demographic parameters in bird populations. In: *Bird population studies: relevance to conservation and management* (eds C. M. Perrins, J. D. Lebreton, and G. J. M. Hirons), pp. 74–104. Oxford Ornithology Series No. 1, Oxford University Press, Oxford.

Clotfelter, E. D. and T. Brush (1995). Unusual parasitism by the Bronzed Cowbird. *Condor*, **97**, 814–816.

Coker, D. R. and D. E. Capen. Distribution and habitat associations of Brown-headed Cowbirds in the Green Mountains of Vermont. In: *Ecology and management of cowbirds* (eds J. N. M. Smith, T. Cook, S. K. Robinson, S. I. Rothstein, and S. G. Sealy). University of Texas Press, Austin (in press).

Cruz, A., J. W. Prather, W. Post, and J. W. Wiley. The spread of Shiny and Brown-headed Cowbirds into the Florida region. In: *Ecology and management of cowbirds* (eds J. N. M. Smith, T. Cook, S. K. Robinson, S. I. Rothstein, and S. G. Sealy). University of Texas Press, Austin (in press).

Davis, S. K. and S. G. Sealy. Cowbird parasitism and predation in grassland fragments of southwestern Manitoba. In: *Ecology and management of cowbirds* (eds J. N. M. Smith, T. Cook, S. K. Robinson, S. I. Rothstein, and S. G. Sealy). University of Texas Press, Austin (in press).

DeCapita, M. E. Brown-headed Cowbird control on Kirtland's Warbler nesting areas. In: *Ecology and management of cowbirds* (eds J. N. M. Smith, T. Cook, S. K. Robinson, S. I. Rothstein, and S. G. Sealy). University of Texas Press, Austin (in press).

Diamond, J. (1984). Normal extinctions of isolated populations. In: *Extinctions* (ed. M. Nitecki). University of Chicago Press, Illinois.

Donovan, T. M., R. H. Lamberson, F. R. Thompson III, and J. Faaborg (1995*a*). Modeling the effects of habitat fragmentation on source and sink demography of neotropical migrant birds. *Conserv. Biol.*, **9**, 1396–1407.

Donovan, T. M., F. R. Thompson III, J. Faaborg, and J. R. Probst (1995*b*). Reproductive success of migratory birds in habitat sources and sinks. *Conserv. Biol.*, **9**, 1380–1395.

Donovan, T. M., P. W. Jones, E. M. Annand, and F. R. Thompson III (1997). Variation in local-scale edge effects: mechanisms and landscape context. *Ecology*, **78**, 2064–2075.

Donovan, T. M., F. R. Thompson III, and J. Faaborg. Ecological trade-offs and the influence of scale on Brown-headed cowbird distribution. In: *Ecology and management of cowbirds* (eds J. N. M. Smith, T. Cook, S. K. Robinson, S. I. Rothstein, and S. G. Sealy). University of Texas Press, Austin (in press).

Dowell, B. A., J. E. Fallon, C. S. Robbins, D. K. Dawson, and F. W. Fallon. Impacts of cowbird parasitism on Wood Thrushes and other neotropical migrants in suburban Maryland forests. In: *Ecology and management of cowbirds* (eds J. N. M. Smith, T. Cook, S. K. Robinson, S. I. Rothstein and S. G. Sealy). University of Texas Press, Austin (in press).

Elliott, P. F. (1978). Cowbird parasitism in the Kansas tallgrass prairie. *Auk*, **95**, 161–167.

Facemire, C. F. (1980). Cowbird parasitism of marsh-nesting Red-winged Blackbirds. *Condor*, **82**, 347–348.

Flather, C. H. and J. R. Sauer (1996). Using landscape ecology to test hypotheses about large-scale abundance patterns in migratory birds. *Ecology*, **77**, 28–35.

Fleischer, R. C. (1985). A new technique to identify and assess dispersion of eggs of individual brood parasites. *Behav. Ecol. Sociobiol.*, **17**, 91–99.

Franzreb, K. E. (1989). Ecology and conservation of the endangered Least Bell's Vireo. U.S. Fish and Wildlife Service Biological Report.

Freeman, S., D. F. Gori, and S. Rohwer (1990). Red-winged Blackbirds and Brown-Headed Cowbirds: some aspects of a host-parasite relationship. *Condor*, **92**, 336–340.

Freemark, K. E. (1992). Landscape ecology of temperate forest birds. In: *Ecology and conservation of neotropical migrant landbirds* (eds J. M. Hagan and D. W. Jonston). Smithsonian Institution Press, Washington, D.C.

Fretwell, S. D. (1977). Is the Dickcissel a threatened species? *American Birds*, **31**, 923–932.

Friedmann, H. (1929). *The cowbirds: a study in the biology of social parasitism*. Charles C. Thomas, Springfield, Illinois.

Friedmann, H. (1963). Host relations of the parasitic cowbirds. *U.S. National Museum Bulletin*, **233**, 1–276.

Friedmann, H. and L. F. Kiff (1985). Parasitic cowbirds and their hosts. *Proceedings Western Foundation for Zoology*, **2**, 226–304.

Friedmann, H., L. F. Kiff, and S. I. Rothstein (1977). A further contribution to knowledge of the host relations of the parasitic cowbirds. *Smithson. Contrib. Zool.*, **235**, 1–75.

Gates, J. E. and L. W. Gysel (1978). Avian nest dispersion and fledging success in field-forest ecotones. *Ecology*, **59**, 871–883.

Goertz, J. W. (1977). Additional records of Brown-headed Cowbird nest parasitism in Louisiana. *Auk*, **94**, 386–389.

Goldwasser, S., D. Gaines and S. R. Wilbur (1980). The Least Bell's Vireo in California: a *de facto* endangered race. *American Birds*, **34**, 742–745.

Graber, J. W. (1961). Distribution, habitat requirements, and life history of the Black-capped Vireo (*Vireo atricapilla*). *Ecol. Monogr.*, **341**, 313–336.

Graber, R. R., J. W. Graber, and E. L. Kirk (1974). Illinois Birds: *Tyrannidae" Illinois Natural History Survey Biological Notes*, **86**, 1–56.

Gray, M. V. and J. M. Greaves (1984). Riparian forest as habitat for the Least Bell's Vireo. In: *California Riparian systems* (eds R. E. Warner and K. M. Hendrix), pp. 605–611. University of California Press, Berkeley.

Greenberg, R. (1980). Demographic aspects of long-distance migration. In: *Migrant birds in the Neotropics: Ecology, behavior, distribution, and conservation* (eds A. Keast and E. S. Morton), pp. 493–504. Smithsonian Inst. Press, Washington, D.C.

Griffith, J. T. and J. C. Griffith. Cowbird parasitism and the endangered Least Bell's Vireo: a management success story. In: *Ecology and management of cowbirds* (eds J. N. M. Smith, T. Cook, S. K. Robinson, S. I. Rothstein, and S. G. Sealy). University of Texas Press, Austin (in press).

Grzybowski, J. A. (1990). Ecology and management of the Black-capped Vireo in the Wichita Mountains, Oklahoma: 1990. Report, Wichita Mountains National Wildlife Refuge, U.S. Fish and Wildlife Service, Indiahome, Oklahoma.

Grzybowski, J. A., R. B. Clapp, and J. T. Marshall, Jr. (1986). History and current population status of the Black-capped Vireo in Oklahoma. *American Birds*, **40**, 1151–1161.

Grzybowski, J. A., D. J. Tazik and G. D. Schnell (1994). Regional analysis of Black-capped Vireo breeding habitats. *Condor*, **96**, 512–544.

Hahn, D. C. and J. S. Hatfield (1995). Parasitism at the landscape scale: cowbirds prefer forests. *Conserv. Biol.*, **9**, 1415–1424.

Hahn, D. C. and J. S. Hatfield. Host selection in the forest interior: Cowbirds target ground-nesting species. In: *Ecology and management of cowbirds* (eds J. N. M. Smith, T. Cook, S. K. Robinson, S. I. Rothstein, and S. G. Sealy). University of Texas Press, Austin (in press).

Hanski, I. (1985). Single-species spatial dynamics may contribute to long-term rarity and commonness. *Ecology*, **66**, 335–343.

Hanski, I. and M. Gilpin (1991). Metapopulation dynamics: brief history and conceptual domain. *Biol. J. Linnaean Soc.*, **43**, 3–16.

Hayden, T. J., D. J. Tazik, R. H. Melton, and J. D. Cornelius. Cowbird control program on Fort Hood, Texas: lessons for mitigation of cowbird parasitism on a landscape scale. In: *Ecology and management of cowbirds* (eds J. N. M. Smith, T. Cook, S. K. Robinson, S. I. Rothstein, and S. G. Sealy). University of Texas Press, Austin (in press).

Hergenrader, G. L. (1962). The incidence of nest parasitism by the Brown-headed Cowbird (*Molothrus ater*) on roadside nesting birds in Nebraska. *Auk*, **79**, 85–88.

Hicks, L. E. (1934). A summary of cowbird host species in Ohio. *Auk*, **51**, 385–386.

Hill, R. A. (1976). Host–parasite relationships of the Brown-headed Cowbird in a prairie habitat of west-central Kansas. *Wilson Bull.*, **88**, 555–565.

Holford, K. C. and D. D. Roby (1993). Factors limiting fecundity of captive Brown-headed Cowbirds. *Condor*, **95**, 536–545.

Holmes, R. T., T. W. Sherry, P. P. Marra, and K. E. Petit (1992). Multiple brooding and productivity of a neotropical migrant, the Black-throated Blue Warbler (*Dendroica caerulescens*) in an unfragmented temperate forest. *Auk*, **109**, 321–333.

Hoover, J. P. and M. C. Brittingham (1993). Regional variation in brood parasitism of Wood Thrushes. *Wilson Bull.*, **105**, 228–238.

Hoover, J. P., M. C. Brittingham, and L. J. Goodrich, (1995). Effects of forest patch size on nesting success of Wood Thrushes. *Auk*, **112**, 146–155.

Houston, C. S. (1977). Northern Great Plains region. *American Birds*, **27**, 882–886.

Howe, R. W., G. J. Davis, and V. Mosca (1991). The demographic significance of 'sink' populations. *Biol. Conserv.*, **57**, 239–255.

Howe, W. H. and F. L. Knopf. The role of vegetation in cowbird parasitism of Yellow Warblers. In: *Ecology and management of cowbirds* (eds J. N. M. Smith, T. Cook, S. K. Robinson, S. I. Rothstein, and S. G. Sealy). University of Texas Press, Austin (in press).

Iverson, L. R. (1989). *Forest resources of Illinois: An atlas and and analysis of spatial and temporal trends.* Illinois Natural History Survey Special Publication 11.

James, F. C. and C. E. McCulloch (1995). The strength of inference about causes of population trends. In: *Ecology and management of neotropical migratory birds* (eds T. E. Martin and D. M. Finch), pp. 40–54. Oxford University Press, Oxford.

Johnson, R. G. and S. A. Temple (1990). Nest predation and brood parasitism of tallgrass prairie birds. *J. Wildlf. Manag.*, **54**, 106–111.

Jones, H. L. and J. M. Diamond (1976). Short-time-base studies of turnover in breeding bird populations on the California Channel Islands. *Condor*, **78**, 526–549.

Karr, J. R., J. D. Nichols, M. K. Kilmkiewicz, and J. D. Brawn (1990). Survival rates of birds of tropical and temperate forests: will the dogma survive? *Amer. Natur.*, **36**, 277–291.

Kelly, S. T. and M. E. Decapita (1982). Cowbird control and its effect on Kirtland's Warbler reproductive success. *Wilson Bull.*, **94**, 363–365.

Klaas, E. E. (1975). Cowbird parasitism and nesting success in the Eastern Phoebe. Occasional Papers of the Museum of Natural History, University of Kansas, Lawrence.

Koford, R. R., B. S. Bowen, J. T. Lokemoen, and A. D. Kruse. Cowbird parasitism in grassland and cropland in the northern Great Plains. In: *Ecology and management of cowbirds* (eds J. N. M. Smith, T. Cook, S. K. Robinson, S. I. Rothstein, and S. G. Sealy). University of Texas Press, Austin (in press).

Lanyon, W. E. (1957). *The comparative biology of the meadowlarks (Sturnella) in Wisconsin.* Cambridge, Publications of the Nuttall Ornithological Club, No. 1.

Lawrence, L. de K. (1953). Nesting life and behavior of the Red-eyed Vireo. *Can. Field-Natur.*, **67**, 47–87.

Levins, R. (1970). Extinction. In: *Some mathematical questions in biology. Lectures on mathematics in the life sciences, volume 2*, pp. 77–107. American Mathematical Society, Providence.

Linz, G. M. and S. B. Bolin (1982). Incidence of Brown-headed Cowbird parasitism on Red-winged Blackbirds. *Wilson Bull.*, **94**, 93–95.

Longcore, J. R. and R. E. Jones (1969). Reproductive success of the Wood Thrush in a Delaware woodlot. *Wilson Bull.*, **81**, 39–46.

Lowther, P. E. (1993). Brown-headed Cowbird. In: *Birds of North America, No. 47* (eds A. Poole and F. Gill), pp. 1–28. Academy of Natural Sciences, Philadelphia, PA.

Martin, T. E. (1992). Breeding productivity considerations: What are the appropriate habitat features for management? In: *Ecology and conservation of neotropical migrant landbirds* (eds J. M. Hagan III and D. W. Johnston). pp. 455–493. Smithsonian Institution Press, Washington, D.C.

Marvil, R. E. and A. Cruz (1989). Impact of Brown-headed Cowbird parasitism on the reproductive success of the Solitary Vireo. *Auk*, **106**, 476–480.

May, R. M. and S. K. Robinson (1985). Population dynamics of avian brood parasitism. *Amer. Natur.*, **126**, 475–494.

Mayfield, H. (1965). The Brown-headed Cowbird, with old and new hosts. *Living Bird*, **4**, 13–28.

Mayfield, H. (1975). Suggestions for calculating nest success. *Wilson Bull.*, **87**, 456–466.

Mayfield, H. (1977). Brown-headed Cowbirds: agent of extermination. *American Birds*, **31**, 107–113.

Mumford, R. E. (1952). Bell's Vireo in Indiana. *Wilson Bull.*, **64**, 224–233.

Newman, G. A. (1970). Cowbird parasitism and nesting success of Lark Sparrows in southern Oklahoma. *Wilson Bull.*, **82**, 304–309.

Nice, M. M. (1937). Studies in the life history of the Song Sparrow, Part I. *Trans. of the Linnaean Soc. of New York*, **4**, 1–247.

Nolan, V., Jr. (1978). *The ecology and behavior of*

the Prairie Warbler Dendroica discolor. Ornith. Monogr. 26. American Ornithologists' Union.

Norris, R. T. (1947). The cowbirds of Preston Frith. *Wilson Bull.*, **59**, 83–103.

Ortega, C. P. and A. Cruz (1991). A comparative study of cowbird parasitism in Yellow-headed Blackbirds and Red-winged Blackbirds. *Auk*, **108**, 16–24.

Overmire, T. G. (1962*a*). Nesting of the Dickcissel in Oklahoma. *Auk*, **9**, 115–116.

Overmire, T. G. (1962*b*). Nesting of the bell's Vireo in Oklahoma. *Condor*, **64**, 75.

Payne, R. B. (1977). The ecology of brood parasitism in birds. *Annu. Rev. Ecol. Syst.*, **8**, 1–28.

Pease, C. M. and J. A. Grzybowski (1995). Assessing the consequences of brood parasitism and nest predation on seasonal fecundity in passerine birds. *Auk*, **112**, 343–363.

Peck, G. K. and R. D. James (1987). *Breeding birds of Ontario. Nidiology and distribution, Vol. 2*: Passerines. Royal Ontario Museum, Toronto, Canada.

Peterjohn, B. G., J. R. Sauer, and C. S. Robbins (1995). Population trends from the North American Breeding Bird Survey. In: *Ecology and management of neotropical migratory birds* (eds T. E. Martin and D. M. Finch), pp. 3–39. Oxford University Press, Oxford.

Peterjohn, B. G., J. R. Sauer, and S. Orsillo. Temporal and geographic patterns in population trends of Brown-headed Cowbirds. In: *Ecology and management of cowbirds* (eds J. N. M. Smith, T. Cook, S. K. Robinson, S. I., Rothstein, and S. G. Sealy). University of Texas Press, Austin (in press).

Petit, L. J. and D. R. Petit. Brown-headed Cowbird parasitism of migratory birds: effects of forest area and surrounding landscape. In *Ecology and management of cowbirds* (eds J. N. M. Smith, T. Cook, S. K. Robinson, S. I. Rothstein, and S. G. Sealy). University of Texas Press, Austin (in press).

Petrinovich, L. and T. G. Patterson (1978). Cowbird Parasitism on the White-crowned for Sparrow. *Auk*, **95**, 415–417.

Phillips, R. S. (1951). Nest location, cowbird parasitism and nesting success of the Indigo Bunting. *Wilson Bull.*, **63**, 206–207.

Porneluzi, P., J. C. Bednarz, L. J. Goodrich, N. Zawada, and J. P. Hoover (1993). Reproductive performance of territorial Ovenbirds occupying forest fragments and a contiguous forest in Pennsylvania. *Conserv. Biol.*, **7**, 618–622.

Post, W. and J. W. Wiley (1977). Reproductive interactions of the Shiny Cowbird and the Yellow-shouldered Blackbird. *Condor*, **79**, 176–184.

Post, W., T. K. Nakamura, and A. Cruz (1990). Patterns of Shiny Cowbird parasitism in St. Lucia and southwestern Puerto Rico. *Condor*, **92**, 461–492.

Probst, J. R. (1986). A review of factors limiting the Kirtland's Warbler on its breeding grounds. *American Midl. Natur.*, **116**, 87–100.

Pulliam, H. R. (1988). Sources, sinks, and population regulation. *Amer. Natur.*, **132**, 652–661.

Pulliam, H. R. and B. J. Danielson (1991). Sources, sinks, and habitat selection: a landscape perspective on population dynamics. *Amer. Natur.*, **137**, S50–66.

Ratti, J. T. and K. P. Reese (1988). Preliminary test of the ecological trap hypothesis. *J. Wildlf. Manag.*, **52**, 484–491.

Rice, J. C. (1974). Social and competitive interactions between two species of vireos (Aves: Vireonidae). PhD Thesis, University of Toronto, Canada.

Rich, A. C., D. S. Dobkin, and L. J. Niles (1994). Defining forest fragmentation by corridor width: the influence of narrow forest-dividing corridors on forest-nesting birds in southern New Jersey. *Conserv. Biol.*, **8**, 1109–1121.

Ricklefs, R. E. (1973). Fecundity, mortality, and avian demography. In: *Breeding biology of birds* (ed. D. S. Farner), pp. 366–434. National Academy of Sciences, Washington D.C.

Robbins, C. S., D. K. Dawson, and B. A. Dowell, (1989). Habitat area requirements of breeding forest birds in the middle Atlantic states. *Wildlf. Monogr.*, **103**, 1–34.

Robertson, R. J. and R. F. Norman (1977). The function and evolution of aggressive host behavior towards the Brown-headed Cowbird (*Molothrus ater*). *Can. J. Zool.*, **55**, 508–518.

Robinson, S. K. (1992). Population dynamics of breeding birds in a fragmented Illinois landscape. In: *Ecology and conservation of neotropical migrant landbirds* (eds J. M. Hagan and D. W. Johnston), pp. 408–418. Smithsonian Institution Press, Washington, D.C.

Robinson, S. K. and D. S. Wilcove (1994). Forest fragmentation in the temperate zone and its effects on migratory songbirds. *Bird Conservation International*, **4**, 233–249.

Robinson, S. K., J. A. Grzybowski, S. I. Rothstein, M. C. Brittingham, L. J. Petit, and F. R. Thompson III (1993). Management implications of cowbird parasitism on neotropical migrant songbirds. In: *Status and management of neotropical migratory birds* (eds D. M. Finch and P. W. Stangel), pp. 93–102. General Tech. Report RM-229. U.S. Forest Service, Rocky Mountain Forest and Range Experiment Station, Fort Collins, Colorado.

Robinson, S. K., S. I. Rothstein, M. C. Brittingham, L. J. Petit, and J. A. Grzybowski (1995*a*). Ecology and behavior of cowbirds and their impacts on host populations. In: *Ecology and management of neotropical migratory birds* (eds T. E. Martin and D. M. Finch), pp. 428–460. Oxford University Press, Oxford.

Robinson, S. K., F. R. Thompson III, J. M. Donovan, D. R. Whitehead, and J. Faaborg (1995*b*). Regional forest fragmentation and the nesting success of migratory birds. *Science*, **267**, 1987–1990.

Robinson, S. K., J. P. Hoover, J. R. Herkert, and R. Jack. Cowbird parasitism in a fragmented landscape: effects of tract size, habitat, and abundance of cowbirds and hosts. In: *Ecology and management of cowbirds* (eds J. N. M. Smith, T. Cook, S. K. Robinson, S. I. Rothstein, and S. G. Sealy). University of Texas Press, Austin (in press).

Røskaft, E., G. H. Orians, and L. D. Beletsky (1990). Why do Red-winged Blackbirds accept eggs of Brown-headed Cowbirds? *Evol. Ecol.*, **4**, 35–42.

Roth, R. R. and R. K. Johnson (1993). Long-term dynamics of a Wood Thrush population breeding in a forest fragment. *Auk*, **110**, 37–48.

Rothstein, S. I. (1975*a*). Evolutionary rates and host defenses against avian brood parasitism. *Amer. Natur.*, **109**, 151–176.

Rothstein, S. I. (1975*b*). An experimental and teleonomic investigation of avian brood parasitism. *Condor*, **77**, 250–271.

Rothstein, S. I. (1990). A model system for coevolution: avian brood parasitism systems. *Annu. Rev. Ecol. Syst.*, **21**, 481–508.

Rothstein, S. I. (1994). The cowbird's invasion of the Far West: history, causes, and consequences experienced by host species. *Studies Avian Biol.*, **15**, 301–315.

Rothstein, S. I. and S. K. Robinson (1994). Conservation and coevolutionary implications of brood parasitism by cowbirds. *Trends Ecol. Evol.*, **9**, 162–164.

Rothstein, S. I., J. Verner, and E. Stevens (1984). Radio-tracking confirms a unique diurnal pattern of spatial occurrence in the parasitic Brown-headed Cowbird. *Ecology*, **65**, 77–88.

Rothstein, S. I. and T. L. Cook. Cowbird management, host population limitation and efforts to save endangered species. In: *Ecology and Management of Cowbirds* (eds J. N. M. Smith, T. Cook, S. K. Robinson, S. I. Rothstein, and S. G. Sealy). University of Texas Press, Austin (in press).

Ryel, L. A (1981). Population change in the Kirtland's Warbler. *Jack-Pine Warbler*, **59**, 77–90.

Salata, L. (1983). Status of the least bell's Vireo on Camp Pendleton, California: report on research done in 1983. Unpublished report, U.S. Fish and Wildlife Service, Laguna Niguel, California.

Scott, D. M. (1977). Cowbird parasitism on the Grey Catbird at London, Ontario. *Auk*, **94**, 18–27.

Scott, D. M. and C. D. Ankney (1980). Fecundity of the Brown-headed Cowbird in southern Ontario. *Auk*, **97**, 677–683.

Sedgwick, J. A. and F. L. Knopf (1988). A high incidence of Brown-headed Cowbird parasitism of Willow Flycatchers. *Condor*, **90**, 253–256.

Sherry, T. W. and R. T. Holmes (1992). Demography of a long-distance migrant: causes and consequences of variable yearling recruitment in the American Redstart. In: *Ecology and conservation of neotropical migrant landbirds* (eds J. M. Hagan and D. W. Johnston), pp. 431–442. Smithsonian Institution Press, Washington, D.C.

Small, M. F. and M. L. Hunter (1988). Forest fragmentation and avian nest predation in forested landscapes. *Oecologia*, **76**, 62–64.

Smith, J. N. M. (1981). Cowbird parasitism, host fitness, and age of the host female in an island Song Sparrow population. *Condor*, **83**, 152–161.

Smith, J. N. M. and P. Arcese (1994). Brown-headed Cowbirds and an island population of Song Sparrows: a 16-year study. *Condor*, **96**, 916–934.

Smith, J. N. M. and I. H. Myers-Smith. Spatial variation in parasitism of song sparrows by Brown-headed Cowbirds. In: *Parasitic birds and their hosts: studies in coevolution* (eds S. I. Rothstein and S. K. Robinson), Oxford University Press, Oxford (in press).

Smith J. N. M., T. Cook, S. K. Robinson, S. I. Rothstein, and S. G. Sealy (eds), *The ecology and management of cowbirds*. University of Texas Press, Austin (in press).

Southern, W. E. (1958). Nesting of the Red-eyed Vireo in the Douglas Lake region, Michigan, 2. *Jack-Pine Warbler*, **36**, 185–207.

Southern, W. E. and L. K. Southern (1980). A summary of the incidence of cowbird parasitism in northern Michigan. *Jack-Pine Warbler*, **58**, 77–84.

Sutton, G. M. (1959). The nestling fringillids of the Edwin S. George Reserve, southeastern Michigan, pt III. *Jack-Pine Warbler*, **37**, 77–101.

Sutton, G. M. (1960). The nestling fringillids of the Edwin S. George Reserve, southeastern Michigan, pt VII. *Jack-Pine Warbler*, **38**, 124–139.

Takasu, F., K. Kawasaki, H. Nakamura, J. E. Cohen, and N. Shigesada (1993). Modeling the population dynamics of a cuckoo-host association and the evolution of host defenses. *Amer. Natur.*, **142**, 819–839.

Temple, S. A. and J. R. Cary (1988). Modeling dynamics of habitat-interior bird populations in fragmented landscapes. *Conserv. Biol.*, **2**, 340–347.

Terrill, L. M. (1961). Cowbird hosts in southern Quebec. *Can. Field Natur.*, **75**, 2–11.

Thompson, F. R., III (1994). Temporal and spatial patterns of breeding in Brown-headed Cowbirds in the midwestern United States. *Auk*, **111**, 979–990.

Thompson, F. R. III, S. K. Robinson, T. M. Donovan, J. Faaborg, and D. R. Whitehead. Biogeographic landscape and local factors affecting cowbird abundance and host parasitism levels. In: *Ecology and management of cowbirds* (eds J. N. M. Smith, T. Cook, S. K. Robinson, S. I. Rothstein, and S. G. Sealy). University of Texas Press, Austin (in press).

Trail, P. W. and L. F. Baptista (1993). The impact of Brown-headed Cowbird parasitism on populations of the Nuttall's White-crowned Sparrow. *Conserv. Biol.*, **7**, 309–315.

Trautman, M. B. (1940). The birds of Buckeye Lake Ohio. Miscellaneous Publications of the Museum of Zoology, No. 44, University of Michigan.

Trine, C. L. Effects of multiple parasitism on cowbird and Wood Thrush nesting success. In: *Ecology and management of cowbirds* (eds J. N. M. Smith, T. Cook, S. K. Robinson, S. I. Rothstein, and S. G. Sealy). University of Texas Press, Austin (in press).

Twomey, A. C. (1945). The bird populations of an elm-maple forest: with special reference to aspection, territorialism, and coactions. *Ecol. Monogr.*, **15**, 173–205.

Urban, D. L. and H. H. Shugart, Jr (1986). Avian demography in mosaic landscapes: modeling paradigm and preliminary results. In: *Wildlife 2000: Modeling habitat relationships of terrestrial vertebrates* (eds J. Verener, M. L. Morrison, and C. J. Ralph), pp. 273–279. University of Wisconsin Press, Madison, Wisconsin.

Uyehara, J. C. and M. J. Whitfield. Vegetative cover and parasitism in territories of the Southwestern Willow Flycatcher (*Empidonax traillii extimus*). In: *Ecology and management of cowbirds* (eds J. N. M. Smith, T. Cook, S. K. Robinson, S. I. Rothstein, and S. G. Sealy). University of Texas Press, Austin (in press).

Verner, J. and L. V. Ritter (1983). Current status of the Brown-headed Cowbird in the Sierra National Forest. *Auk*, **100**, 355–368.

Walkinshaw, L. H. (1961). The effect of parasitism by the Brown-headed Cowbird on *Empidonax* flycatchers in Michigan. *Auk*, **78**, 266–268.

Walkinshaw, L. H. (1966). Studies of the Acadian Flycatcher in Michigan. *Bird-Banding*, **37**, 227–257.

Walkinshaw, L. H. (1983). *Kirtland's Warbler.* Cranbrook Institute of Science, Bloomfield Hills.

Ward, D. and J. N. M. Smith. Inter-habitat differences in parasitism frequencies by Brown-headed cowbirds in the Okanagan Valley, British Columbia. In: *Ecology and management of cowbirds* (eds J. N. M. Smith, T. Cook, S. K. Robinson, S. I. Rothstein, and S. G. Sealy). University of Texas Press, Austin (in press).

Weatherhead, P. J. (1989). Sex ratios, host-specific reproductive success, and impact of Brown-headed Cowbirds. *Auk*, **106**, 358–366.

Weatherhead, P. J. and M. R. L. Forbes (1994). Natal philopatry in passerine birds: genetic or ecological influences? *Behav. Ecol.*, **5**, 426–433.

Webster, F. S., Jr. (1971). South Texan region. *American Birds*, **25**, 875–878.

Whitcomb, R. F., C. S. Robbins, J. F. Lynch, B. L. Whitcomb, K. Klimkiewicz, and D. Bystrak (1981). Effects of forest fragmentation on avifauna of the eastern deciduous forest. In: *Forest island dynamics in man-dominated landscapes* (eds R. L. Burgess and D. M. Sharpe), pp. 125–205. Springer-Verlag, New York.

Whitfield, M. J. Results of a Brown-headed Cowbird control program for the Southwestern Willow Flycatcher. In: *Ecology and management of cowbirds* (eds J. N. M. Smith, T. Cook, S. K. Robinson, S. I. Rothstein, and S. G. Sealy). University of Texas Press, Austin (in press).

Wiedenfeld, D. A. Cowbird population changes and their relationship to changes in some host species. In: *Ecology and management of cowbirds* (eds J. N. M. Smith, T. Cook, S. K. Robinson, S. I. Rothstein, and S. G. Sealy). University of Texas Press, Austin (in press).

Wiens, J. A. (1963). Aspects of cowbird parasitism in southern Oklahoma. *Wilson Bull.*, **75**, 130–138.

Wilcove, D. S. (1985). Nest predation in forest tracts and the decline of migratory songbirds. *Ecology*, **66**, 1211–1214.

Wilcove, D. S. (1988). Changes in the avifauna of the Great Smoky Mountains: 1947–1983. *Wilson Bull.*, **100**, 256–271.

Wilcove, D. S. and S. K. Robinson (1990). Effects of habitat fragmentation on forest bird communities. In: *Biogeography and ecology of forest bird communities* (ed. A. Keast). SPB Academic, The Hague, The Netherlands.

Wilcove, D. S., C. H. McClellan, and A. P. Dobson (1986). Habitat fragmentation in the temperate zone. In: *Conservation biology: the science of scarcity and diversity* (ed. M. E. Soule), pp. 237–256. Sinauer Associates, Sunderland, Massachusetts.

Wiley, R. H. and M. S. Wiley (1980). Spacing and timing in the nesting ecology of a tropical blackbird: comparison of populations in different environments. *Ecol. Monogr.*, **50**, 153–178.

Winslow, D. E., D. R. Whitehead, C. Frazer Whyte, M. A. Koukal, G. M. Greenberg, and T. B. Ford. Within-landscape variation in patterns of cowbird parasitism in the forests of southcentral Indiana. In: *Ecology and management of cowbirds* (eds J. N. M. Smith, T. Cook, S. K. Robinson, S. I. Rothstein, and S. G. Sealy). University of Texas Press, Austin (in press).

Yahner, R. M. and D. P. Scott (1988). Effects of forest fragmentation on depredation of artificial nests. *J. Wildlf. Manag.*, **52**, 158–161.

Yamasaki, M., T. M. McLellan, R. M. DeGraaf, and C. A. Costello. Effects of land-use and management practices on the presence of Brown-headed Cowbirds in the White Mountains of New Hampshire and Maine. In: *Ecology and management of cowbirds* (eds J. N. M. Smith, T. Cook, S. K. Robinson, S. I. Rothstein, and S. G. Sealy). University of Texas Press, Austin (in press).

Zimmerman, J. L. (1983). Cowbird parasitism of Dickcissels in different habitats and at different nest densities. *Wilson Bull.*, **95**, 7–22.

Appendix 15.1 Common and scientific names of species mentioned in the text

Common name	Scientific name
Acadian flycatcher	*Empidonax virescens*
Southwestern willow flycatcher	*E. traillii extimus*
Eastern phoebe	*Sayornis phoebe*
Eastern wood-pewee	*Contopus virens*
American robin	*Turdus migratorius*
Wood thrush	*Hylocichla mustelina*
Brown thrasher	*Toxostoma rufum*
Gray catbird	*Dumetella carolinensis*
Carolina wren	*Thryothorus ludovicianus*
Black-capped vireo	*Vireo atricapillus*
Bell's vireo	*V. bellii*
Least Bell's vireo	*V.b. pusillus*
Warbling vireo	*V. gilvus*
Solitary vireo	*V. solitarius*
Yellow-throated vireo	*V. flavifrons*
Red-eyed vireo	*V. olivaceus*
White-eyed vireo	*V. griseus*
Yellow warbler	*Dendroica petechia*
Kirtland's warbler	*D. kirtlandii*
Prairie warbler	*D. discolor*
Worm-eating warbler	*Helmitheros vermivorus*
Ovenbird	*Seiurus aurocapillus*
Louisiana waterthrush	*S. motacilla*
Common yellowthroat	*Geothlypis trichas*
Yellow-breasted chat	*Icteria virens*
Kentucky warbler	*Oporornis formosus*
Hooded warbler	*Wilsonia citrina*
Summer tanager	*Piranga rubra*
Scarlet tanager	*P. olivacea*
Northern cardinal	*Cardinalis cardinalis*
Rufous-sided towhee	*Pipilo erythrophthalmus*
Indigo bunting	*Passerina cyanea*
White-crowned sparrow	*Zonotrichia leucophrys*
Lark sparrow	*Chondestes grammacus*
Field sparrow	*Spizella pusilla*
Chipping sparrow	*S. passerina*
Song sparrow	*Melospiza melodia*
Dickcissel	*Spiza americana*
Eastern meadowlark	*Sturnella magna*
Orchard oriole	*Icterus spurius*
Yellow-shouldered blackbird	*Agelaius xanthomus*
Red-winged blackbird	*A. phoeniceus*
Brown-headed cowbird	*Molothrus ater*
Shiny cowbird	*M. bonariensis*

16

Spatial Variation in Parasitism of Song Sparrows by Brown-Headed Cowbirds

JAMES N. M. SMITH, ISLA H. MYERS-SMITH

Most avian brood parasites use more than one host species (Friedmann 1948, 1960, 1963; Friedmann et al. 1977). Brown-headed and shiny cowbirds (*Molothrus ater* and *M. bonariensis*) are extreme generalists among brood parasites, and use a variety of hosts in each habitat and geographic area (Friedmann 1963; Friedmann et al. 1977; Mason 1986; Rothstein 1990). Host use by generalist brood parasites, however, can also vary greatly among areas within a species. Good examples of spatially variable parasitism by the brown-headed cowbird occur in the red-winged blackbird (*Agelaius phoeniceus*; Friedmann 1963; Robertson and Norman 1977; Facemire 1980; Linz and Bolin 1982; Ortega and Cruz 1988; Weatherhead 1989), and the wood thrush (*Hylocichla mustelina*; Hoover and Brittingham 1993).

In this chapter we document spatially variable parasitism in song sparrows (*Melospiza melodia*), and explore four general ideas to explain spatial variation in host use within a host species by generalist brood parasites.

1. The *Parasite Density Hypothesis*. According to this hypothesis, use of a particular host in an area is driven primarily by the density of brood parasites there (McGeen 1972; Fretwell 1977; Hoover and Brittingham 1993). The hypothesis assumes that parasites lack preferences for host species or habitats, and parasitism can be predicted by parasite density alone, that is, it is basically a null hypothesis. The hypothesis predicts strong associations between parasite density and the frequency and intensity of parasitism for a given host species, independent of variation in either the mix of habitats present, or of differences in the mix of local host species.

2. The *Parasite Habitat Preference Hypothesis*. Generalist brood parasites sometimes show habitat preferences (e.g., Lowther and Johnston 1977; Robinson et al., in press; Ward and Smith, in press). Accordingly, hosts that use a variety of habitats in a particular area may be differentially exposed to parasitism in different habitats. Friedmann (1963) proposed that the varying use of large marshes by eastern and mid-western red-winged blackbirds accounts for more frequent parasitism in midwestern individuals, which usually nest in the uplands preferred by cowbirds.

3. The *Host-Selection Hypothesis*. This idea can take several forms. Under *active host selection* (hypothesis 3a), different host species vary in their "value" to the brood parasite (i.e., they are differentially successful in rearing parasite young; e.g., Mason 1986; Briskie et al. 1990) and parasites actively choose the more valuable species. As an example of host selectivity, brown-headed cowbirds clearly avoid cuckoos and doves as hosts, because these species are very rarely parasitized, yet they accept cowbird eggs if these are placed in their nests (Rothstein 1976). This hypothesis might seem to be concerned only with differences in parasitism levels across species, but this is not the case. If host communities differ between areas, a host species can be used heavily in one area, but

rarely used in a second, because preferred host species are available only in the second area. This hypothesis predicts that use of a nonpreferred host will increase when other preferred hosts are absent or rare locally.

A second form of active host selection is *frequency-dependent host selection* (hypothesis 3b), where brood parasites have no set preferences for particular host species, but form a "search image" for the commonest suitable host(s) in each area, and thus under-use all rarer hosts locally. Thus, variation in host use of a particular species across areas could depend on the proportion it makes up of the total community of hosts. A third mechanism, *passive host selection* (hypothesis 3c), results in hosts being used differently across areas simply because of differing mixes of local habitats there.

4. The *Host Defense Hypothesis*. According to this hypothesis, hosts vary geographically in defenses against parasitism (e.g., Robertson and Norman 1977), and brood parasites respond adaptively to this variation, by avoiding species in areas where they defend themselves effectively by aggression or egg ejection (Briskie et al. 1992). This hypothesis is related to hypothesis 3a, as it also implies active host selection. Even if a host species ejects parasitic eggs, a low observed parasitism rate does not confirm this hypothesis, as parasite eggs may have been ejected before observers could detect parasitism (Rothstein 1976).

These ideas are not exclusive alternatives, and could interact in various ways. For example, parasite population density might interact with host preferences of the parasite. Thus, a parasite could be rare in one area, and might use only preferred hosts there. In a second area with a higher density of parasites, the parasite might use nonpreferred hosts, because they cannot find unparasitized nests of preferred hosts. Similarly, if complex functional responses to host density occur commonly in brood parasites (Smith and Arcese 1994, see below), host population density could influence host selection. A second type of interaction could take place if parasites selected for certain host species and habitats, but used others as predicted by hypothesis 1 (S. I. Rothstein 1976, pers. commun.).

In this chapter, we test the parasite density hypothesis at two spatial scales, and consider how well the other three hypotheses explain the frequency and intensity of parasitism of song sparrows by brown-headed cowbirds. The breeding range of the song sparrow includes most of the northern two-thirds of the North American continent, and also the southwest, where the species extends into Mexico. The cowbird had virtually completed its colonization of its large joint range with the song sparrow (Rothstein 1994) before the data presented here were gathered. Cowbirds are most abundant in agricultural landscapes, particularly where livestock are abundant, and are less common in heavily forested landscapes (Robinson et al. 1995). They can, however, parasitize hosts at very high frequencies within forest habitat in primarily agricultural landscapes (Robinson et al., in press; Thompson et al., in press; Trine, in press). They reach their maximum abundance in the Northern Great Plains states, especially North and South Dakota (Peterjohn et al., in press). Song sparrows are acceptors of cowbird eggs wherever they have been studied (Rothstein 1975), making it likely that nest record cards provide a reliable index of levels of parasitism for the species (Friedmann et al. 1977).

The song sparrow is regarded to be among the most commonly parasitized of all cowbird hosts (e.g., Nice; 1937, Friedmann et al. 1977). Friedmann (1963) noted that song sparrows have been recorded as cowbird hosts in all provinces of Canada and states of the U.S.A. where the two species co-occur. In a survey of all nest records in Ontario, however, Peck and James (1987) found the song sparrow to be only the ninth most commonly parasitized host species (23.2% of 1812 nests), among species with at least 50 nest records, and only sixteenth-ranked among species with at least 20 records. Because song sparrows are abundant, live near humans, and nest near the ground, they contribute absolutely more records of parasitism to nest-monitoring schemes than most other host species. They probably also contribute absolutely more recruits to cowbird populations than most other hosts.

We first explore variation in parasitism across the North American Continent using nest record cards from the Laboratory of Ornithology, Cornell University, Ithaca, U.S.A. Second, we explore patterns and causes of more local variation in detailed field studies at three areas in coastal British Columbia, Western Canada.

CONTINENTAL-SCALE VARIATION

Methods

Nest record cards from the Cornell Laboratory of Ornithology were examined for geographic and temporal variation in parasitism. Most records available were from the period from 1960 to 1989. Each record was assigned to a state and region, and records for this period were assigned to one of three 10-year time blocks (1960–69, 1970–79, 1980–89) to explore temporal variation in levels of parasitism. Seven regions within the U.S.A. were defined as follows: New England (Connecticut, Maine, Massachusetts, New Hampshire, Rhode Island, Vermont); Mid-Atlantic (Delaware, Maryland, New Jersey, New York, North Carolina, Pennsylvania, Virginia, West Virginia); Great Lakes (Michigan, Minnesota, Ohio, Illinois, Indiana, Wisconsin); Plains (North Dakota, South Dakota, Nebraska, Iowa); South Central (Kentucky, Missouri, Tennessee); Mountain (Colorado, Montana, Idaho, Wyoming, Utah); and Pacific (California, Oregon, Washington). States not included in these regions (save for Georgia with three records only) had no nest records. Some data were available for Canadian provinces, but coverage in Canada was poor relative to most U.S. states. Canadian data from other sources, however, are noted below.

When selecting records, we used only nests with at least two song sparrow eggs or young to estimate the frequency (per cent of nests parasitized) and intensity (mean number of cowbird eggs per parasitized nest) of parasitism. We did this because song sparrow clutches usually exceed three eggs, and nests with two or fewer eggs may represent breeding attempts that failed during the laying period. Nests containing only two sparrow young may have been parasitized, but the parasite egg or chick may have disappeared before the nest was found. Both these situations could yield biases in the incidence of parasitism. Parasitism was recorded only when cowbird eggs or young were actually noted by the observer, or when the recorder's description of a nest's contents made it clear that parasitism had occurred (e.g., "odd" eggs noted in a clutch and host eggs on the nest rim). Eggs on the nest rim or on the ground below the nest were considered as being part of the clutch. As New York had many more nest records than other states, we only examined approximately the first 100 New York nest cards in the card file for each of the two decades from 1960 to 1969, and 1970–1979. We felt that a subsample of 100 was sufficient to characterize patterns for the state in that decade. In all, 2282 nests were scored for parasitism status, 2210 of them from the U.S.A. A previous analysis of this data base for cowbird parasitism of song sparrows was reported by Friedmann et al. (1977). Some data for Ontario, Canada, were taken from Peck and James (1987).

Per cent parasitism was related to per cent landscape cover (forest, cropland, rangeland, pasture, urban) calculated by counting 1-mm squares, after laying a grid on an acetate sheet over a National Geographic Society Land Use Map prepared in 1993 (see appendix 15.1 for data). We also related frequencies and intensities of parasitism to data on cowbird abundance from the eastern U.S.A., gathered by the North American Breeding Bird Survey, and reported in Hoover and Brittingham (1993), and in slightly updated form by S. Droege (unpubl. data).

Results

Temporal stability of levels of parasitism

We first explored the temporal stability of brood parasitism for the five states with the most data (table 16.1). For three of the states, Ohio, Wisconsin, and Michigan, parasitism frequency remained fairly constant over the 30-year period, although fewer data were available for the

Table 16.1 Temporal variation in cowbird parasitism of song sparrows in five well-sampled U.S. States

	1960–1969			1970–1979			1980–1989		
State	Unpar.	Par.	(%)	Unpar.	Par.	(%)	Unpar.	Par.	(%)
Michigan	60	20	(25.0)	73	11	(13.1)	71	12	(14.4)
New York	89	10	(10.1)	102	12	(10.5)	218	12	(5.2)
Ohio	51	6	(10.5)	57	11	(16.2)	14	2	(14.3)
Pennsylvania	66	4	(5.7)	76	4	(5.0)	121	2	(1.6)
Wisconsin	63	18	(22.2)	71	16	(18.4)	1	1	(50.0)

Source: Cornell Laboratory of Ornithology Nest Record Scheme.

Unpar. = unparasitized, Par. = parasitized. (%) = percent of parasitized nests in sample.

Table 16.2 Temporal variation in cowbird parasitism of song sparrows in five U.S. regions. Definitions as in table 16.1. States shown in table 16.1 have been excluded from their respective regions

	1960–1969			1970–1979			1980–1989		
Region	Unpar.	Par.	(%)	Unpar.	Par.	(%)	Unpar.	Par.	(%)
New England	27	0	(0.0)	34	0	(0.0)	11	0	(0.0)
Mid-Atlantic	65	0	(0.0)	85	1	(1.2)	82	5	(5.7)
Great Lakes	41	8	(16.3)	23	3	(11.5)	13	1	(7.1)
Plains	6	0	(0.0)	11	6	(35.3)	14	2	(12.5)
South Central	25	2	(7.4)	17	1	(5.6)	14	2	(12.5)

third decade in Ohio and Wisconsin. There was a moderate decline after the 1960s in Michigan, perhaps in association with extensive cowbird control programs to protect the Kirtland's warbler, *Dendroica kirtlandii*, there (Walkinshaw 1983; Weinrich 1989; DeCapita, in press).

In the two remaining states, New York and Pennsylvania, parasitism rates were stable for the first two decades, but fell by 50% or more in the third decade. In both states, this drop was associated with a reduced number of observers reporting records, and an increase in the numbers of records from a few enthusiastic recorders in one or two counties, raising the possibility of biases in the records. In Allegany County, New York, in which two frequent reporters contributed most records for the state after 1979, the frequency of parasitism declined from 10.1% (n = 39 records) before 1980, to 4.1% (n = 278 records) from 1980 on, in parallel with the state-wide decline. Most data from Pennsylvania in 1980–1989 (78%) came from a county (Greene) that contributed only 19% of pre-1980 records. The frequency of parasitism was low in Greene County both before and after 1980. In summary, there is evidence that the decline in New York, at least, was real. It is of interest that cowbird numbers also declined in New York and in the eastern Great Lakes states after the 1960s (Peterjohn et al., in press).

Data on temporal stability by region are presented in table 16.2, with the five states in table 16.1 omitted from their respective regions. Parasitism in the New England region was consistently absent. Parasitism in the Mid-Atlantic region rose from zero in the 1960s to about 5% two decades later, the opposite of the trend seen in the two largest states of the region, Pennsylvania and New York (table 16.1). Parasitism in the Great Lakes region fell slightly, in agreement with the pattern seen for Michigan in table 16.1. There were few data from the Plains, where the pattern was less stable, and from the South Central region, where parasitism remained in the 5–10% range. There were no data from the Mountain region after 1980, and only two records before 1970; so temporal trends therefore could not be analyzed for this region.

In summary, strong temporal trends in the frequency of parasitism were generally lacking, although there were declines in the frequency of parasitism in the 1980s in three well-sampled states, New York, Michigan, and Pennsylvania. We therefore combined all data regardless of date when exploring spatial variation in frequency of parasitism. We also included records gathered before 1960 and after 1990 in the overall frequency for each state, as did Hoover and Brittingham (1993). In no case did these additional data diverge strongly from the patterns in data from 1960 to 1989, but they augmented sample sizes usefully for some poorly covered states and regions.

Geographic variation in parasitism

The frequencies of parasitism of song sparrows by cowbirds in the U.S.A. are shown by state in fig. 16.1, and listed in appendix 16.1. The mean frequency of parasitism over all records was 9.5%. Few records are available for the western half of the continent, save for the three Pacific states. States with the most frequent parasitism (15–23%) are grouped around, and just to the west of, the Great Lakes, and close to the Northern Great Plains states where the cowbird reaches its maximum abundance (Peterjohn et al., in press). All heavily forested states had low frequencies of parasitism, but the frequency of parasitism varied greatly among states with smaller amounts of

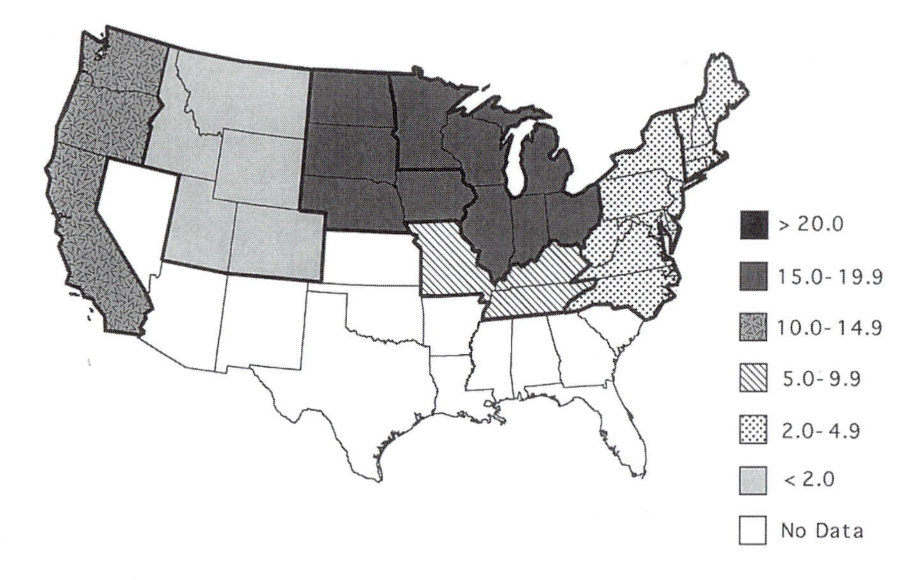

Figure 16.1 Frequencies of brown-headed cowbird parasitism in nests of song sparrows in the continental United States. States with fewer than 10 records are shown as no data.

forest cover. The only Canadian provinces with enough records to assess frequency of parasitism were Ontario and Quebec (nearly all records were from the south of each province). Data from Ontario agreed closely with data from neighboring U.S. Great Lake states (eight of 44 nests parasitized, 18.2%). Much additional data from Ontario have been summarized up to 1986 (Peck and James 1987), and these reveal a frequency of 23.2% (n = 1812 nests), equal to the highest value for any U.S. state. Quebec had a relatively high frequency in a small sample (three of 19 nests, 15.8%), compared with contiguous New England states. Additional nest record data from the Canadian prairie provinces (n = 152 nests) showed a frequency of 19.7% (Friedmann et al. 1977), similar to that of nearby Minnesota (23.3%).

Among regions, the most frequent parasitism occurred in the center of the cowbird's range in the Plains states (although sample sizes here are low) followed by slightly less frequent parasitism near the Great Lakes (fig. 16.2). Parasitism averaged about 10% in the Pacific region, and about 5% in both the South Central and Mid-Atlantic regions. There was little parasitism in either the New England or Mountain regions, though the sample size was very low in the latter case.

As simple correlations might not reflect the associations between parasitism and other variables accurately, we conducted a multiple regression analysis using states as sampling units, with the following independent variables: (i) cowbird density; (ii) linear distance from the center of each state to center of the area of maximum cowbird abundance (Mobridge, South Dakota); (iii) per cent forest cover; and (iv) per cent pasture and rangeland combined. We chose states as spatial units, rather than regions, to boost sample sizes. Values for the variables are given in appendix 16.1. We did not use per cent farmland per state as an independent variable, as it was strongly negatively correlated ($r = -0.847$, $n = 23$, $P < 0.001$) with per cent forest cover.

Multiple regression analysis with these four independent variables was a strong predictor of per cent parasitism (multiple $r^2 = 0.834$, $n = 23$, $P < 0.001$). Distance from the center of cowbird abundance was negatively correlated with per cent parasitism, and was the only significant variable in the regression (fig. 16.3, $P = 0.017$). Cowbird density was also positively associated with per cent parasitism in the regression, but this variable did not add a significant effect to distance ($P - 0.141$). These two variables were strongly negatively correlated ($r = -0.842$, $n = 23$, $P < 0.001$), whereas other variables were not closely inter-correlated ($r < 0.359$, $P > 0.05$ in all comparisons). When distance from the center of cowbird abundance was dropped from the set of independent variables, the regression remained highly significant (multiple $r^2 = 0.878$), with only

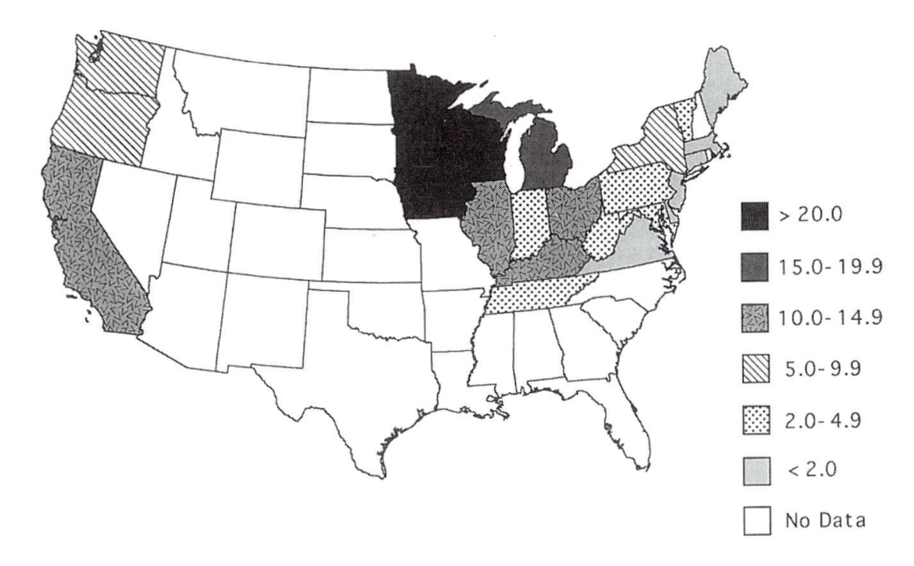

Figure 16.2 Regional frequency of parasitism by brown-headed cowbirds in nests of song sparrows within the U.S.A. For definitions of regions, see text.

cowbird density being a significant predictor of per cent parasitism (fig. 16.4, $P<0.001$). When only forest cover was used as a predictive variable, a smaller, but still significant proportion of the variation in per cent parasitism was explained (fig. 16.5, $r^2 = 0.32$, $P<0.01$). Inspection of the scatter plots of the variables did not reveal obvious nonlinearities, and the all assumptions of the analysis were met adequately.

We interpret this analysis to mean that cowbird abundance is a strong declining function of distance from the Northern Great Plains, and that average cowbird abundance is the main predictor of per cent parasitism, independent of the general habitat mix (per cent forest or grassland) in a state.

Hoover and Brittingham (1993) conducted a similar analysis for wood thrushes, so we compared the per cent parasitism per state in the two species for the 15 states where data were available for both species. We found a remarkably close correlation between the frequencies of parasitism in the two species per state ($r^2 = 0.79$, $P<0.001$, fig. 16.6), although wood thrushes were parasitized over twice as often as song sparrows on average.

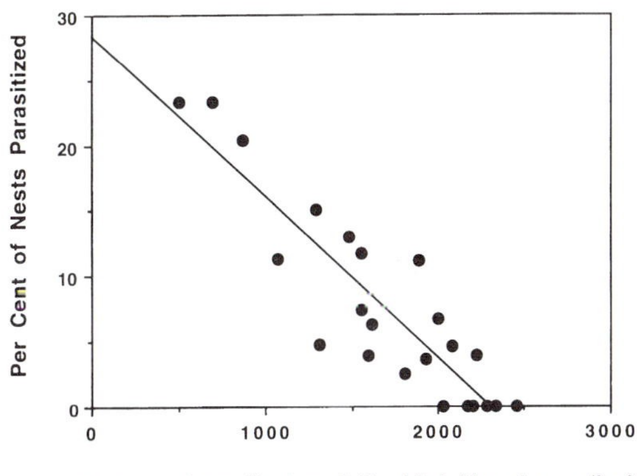

Figure 16.3 Frequency of parasitism of song sparrow nests in relation to distance from the center of abundance of the brown-headed cowbird.

Multiple parasitism

The amount of multiple parasitism can be viewed as an index of: (i) host preference; and (ii) the shortage of suitable alternative host nests for the parasite. Multiple parasitism occurred in 33.8% of 218 cases, and was more frequent in states with higher frequencies of parasitism (fig. 16.7, $r^2 = 0.34$, $n = 16$, $P = 0.01$). Among states with 20 or more nest records, the highest intensities of multiple parasitism (>1.75 eggs per parasitized nest) occurred in Wisconsin, Minnesota, Iowa, and Indiana. Multiple parasitism was about equally common in each of the three decades available for analysis. Spatial patterns in multiple parasitism thus were similar to those in the frequency of parasitism.

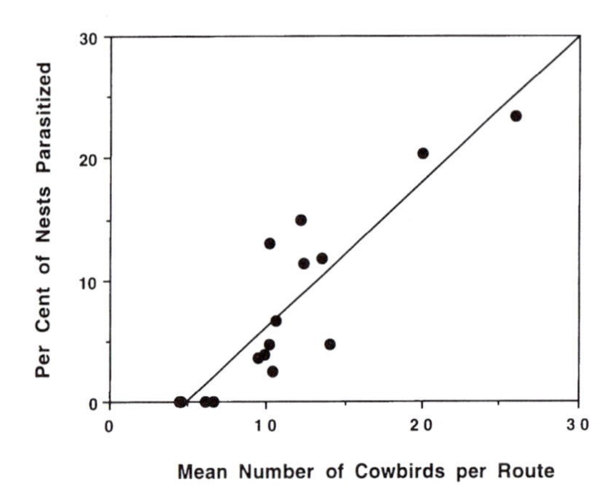

Figure 16.4 Frequency of cowbird parasitism of song sparrow nests per state in relation to the mean number of cowbirds detected on Breeding Bird Survey routes within the same state. Cowbird data from Hoover and Brittingham (1993).

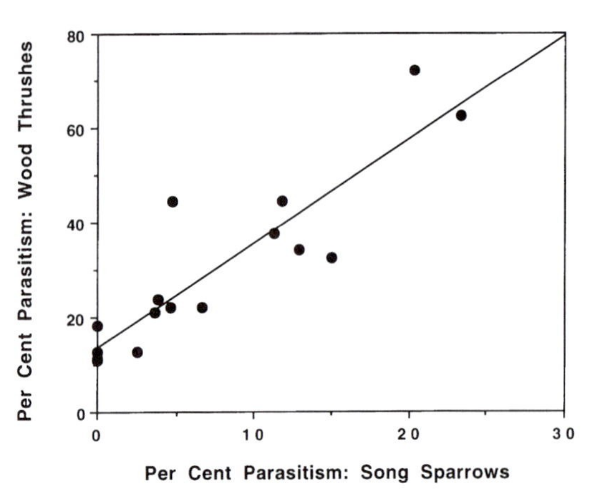

Figure 16.6 Frequencies of parasitism in nests of song sparrows and wood thrushes for the same states in the U.S.A. Thrush data from Hoover and Brittingham (1993).

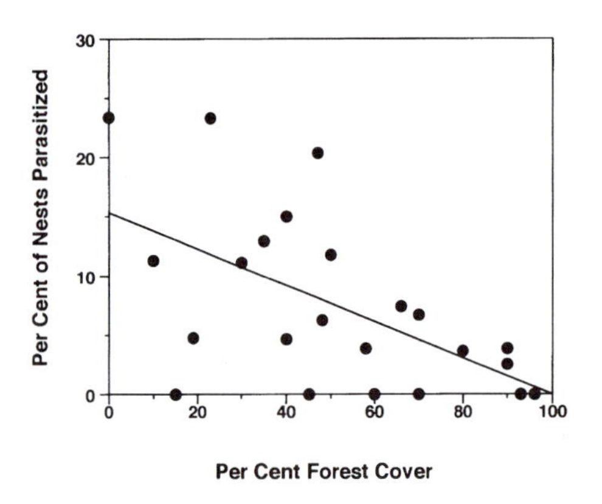

Figure 16.5 Percent parasitism by cowbirds in relation to percent forest cover per state.

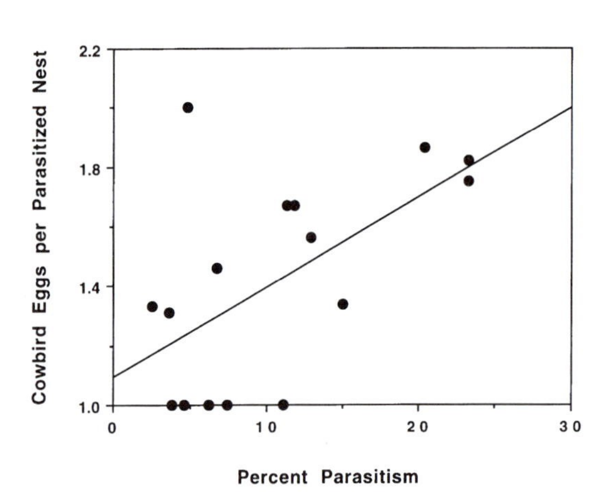

Figure 16.7 Relationship between frequency (percent parasitism) and intensity of cowbird parasitism of nests of song sparrows in U.S. states.

Discussion

Data from nest record schemes are not ideal for the investigation of patterns of brood parasitism (Friedmann et al. 1977). Their usefulness is restricted by uneven coverage, the possible unreliability of nest recorders, and by shifts in the identity and activity of recorders over time. A further limitation in our analysis was the use of arbitrary political units (states, and regions made up of sets of states) to make spatial comparisons; ecologically defined areas would have made more appropriate spatial units for analysis. Despite these shortcomings, the nest record data analyzed here allowed a large-scale spatial analysis, and revealed some clear patterns.

We found that song sparrows were parasitized most frequently in the north central part of the continent close to the center of abundance of the cowbird (Peterjohn et al., in press). A similar result was reported for wood thrushes by Hoover and Brittingham (1993), using the same data base and general methods. This close correlation is notable, because the two species occupy virtually nonoverlapping habitats, with wood thrushes mainly in hardwood forests, and song sparrows in scrub lands, wetlands, and forest-edge habitats. These results strongly suggest that common variables at the regional scale drive the frequency of parasitism in both these host species. The high correlation between cowbird abundance and the frequency of parasitism in both song sparrows and wood thrushes suggests that one such factor is statewide cowbird density. The greater frequency of multiple parasitism in wood thrushes (Trine, in press) and their more frequent use as hosts than song sparrows, however, also prove that cowbird density is not the only factor involved.

We now comment briefly on how the other three hypotheses might explain variable use of a particular host on a broad geographic scale. For the Parasite Habitat Preference Hypothesis to explain the patterns seen here, the states with higher frequencies of parasitism would have to have higher proportions of preferred breeding habitat for cowbirds. We have no data with which to test this suggestion. Cowbirds seem to prefer forested areas for breeding in mixed agricultural and forested landscapes (Thompson et al., in press; Ward and Smith, in press), but they require open areas for feeding (Rothstein et al. 1980). They are also most abundant in the Northern Plains states, where there are generally few trees.

For the Parasite Host Selection Hypothesis, the preferred use of wood thrushes over song sparrows, in states where both occur, suggests the operation of this hypothesis. Wood thrushes may be preferred as hosts because of their highly conspicuous nest (Trine, in press), or as a by-product of their habitat preferences for forest. Both the wood thrush (Trine, in press) and the song sparrow are superior hosts in terms of their ability to rear cowbird young (Smith and Arcese 1994; Trine, in press). In general, a strong analysis of host preferences in generalist brood parasites requires estimates of relative abundances of hosts in communities, as well as estimates of cowbird numbers and levels of parasitism (e.g., Robinson et al., in press). Such an analysis could be carried out by combining data from breeding bird surveys and nest record schemes, but it was beyond the scope of our study.

For the Host-Defense Hypothesis, song sparrows have a behavioral defense against cowbirds (Nice 1937; Robertson and Norman 1976); they recognize, mob, and occasionally attack cowbirds near their nests. This defense, however, does not vary greatly in intensity between eastern, central, and western Canada, nor is it effective at deterring cowbird parasitism (Robertson and Norman 1976, 1977; Smith et al. 1984). We therefore consider it unlikely that variable host defense drives variation in parasitism of song sparrows at the continental scale.

LOCAL VARIATION IN PARASITISM IN BRITISH COLUMBIA

Our goal in the second part of this chapter is to explore some determinants of the frequency and intensity of cowbird parasitism at a local scale, using detailed studies of the frequency and intensity of parasitism of song sparrows on near-shore islands in two areas of coastal British Columbia, Western Canada. Song sparrows are among the most abundant cowbird hosts in coastal British Columbia, and cowbirds are moderately abundant in the region (Peterjohn et al., in press). The results of a long-term study of cowbird–sparrow interactions at one study island, Mandarte, are reported in detail in Smith and Arcese (1994). We now compare findings there with patterns of parasitism elsewhere in the region.

Methods

Study areas

We studied cowbirds and song sparrows in three areas with distinctly different local habitats. The longest-running study was conducted from 1975 to 1992 on Mandarte Island, a 6-ha island in the San Juan/Gulf Islands Archipelago (Smith 1981; Smith and Arcese 1994). Mandarte is isolated from neighboring islands by 1.3 km of open water, and is 6.8 km from the nearest point on the very large Vancouver Island (32 137 km^2). The dominant vegetation on Mandarte is a mosaic of sea island scrub (mostly rose *Rosa nutkana*, and snowberry *Symphoricarpos albus*), and grassy meadows. Standing snags emerging from the scrub provide perches for cowbirds, but there are no live trees above 10 m in height. From four to 72 female song sparrows bred on Mandarte in spring each year (median = 46.5) during the study.

From 1988 to 1991 we also worked on a set of eight smaller islets (0.5 to 4 ha) within 7 km of Mandarte, and within 8.2 km of Vancouver Island. We refer to these collectively as the Small Islands, and to these islands together with Mandarte, as the Offshore Islands. Vancouver Island provides good feeding habitat for cowbirds (open pastures, hobby farms, a horse racetrack) near the points closest to the Small Islands. Rogers et al. (1991) presented a map of the locations of these eight islets relative to Mandarte and Vancouver Island. Six islets had similar vegetation to Mandarte, and all but one, Imrie, were within 0.5 km of larger forested islands. The largest of these islets, Ker and Rum, had much denser forest cover dominated by mature Douglas fir (*Pseudotsuga menziesii*) trees. In most analyses presented here, data for the small islands were pooled, because so few sparrows (0–10 females per year) nested on any one island.

Mandarte Island supports only two suitable alternative hosts for cowbirds, the fox sparrow (*Passerella iliaca*), and the red-winged blackbird. Fox sparrows were about one-tenth as abundant as song sparrows, and redwings bred only irregularly in very small numbers. The small shrub-dominated islets near Mandarte supported breeding song sparrows and very small numbers (one to three breeding females over all sites) of white-crowned sparrows, *Zonotrichia leucophrys*, and red-winged blackbirds. The two forested islands supported much more diverse breeding communities of alternative hosts, including fox sparrows, white-crowned sparrows, rufous-sided towhees (*Pipilo erythropthalmus*), Pacific-slope flycatchers (*Empidonax difficilis*), golden-crowned kinglets (*Regulus satrapa*), purple finches (*Carpodacus purpureus*), and orange-crowned warblers (*Vermivora celata*). Other breeding passerines included American robins, *Turdus migratorius*, and house wrens, *Troglodytes aedon*. The song sparrow, however, was the most abundant breeding songbird on all these islets.

The third site was at the Alaksen National Wildlife Area on Westham Island, a 12 km^2 island in the delta of the Fraser River, about 52 km north of Mandarte. Vegetation at this site consisted of shrub thickets (mainly *Rubus procerus*) on flood control dikes, pasture fields, alder (*Alnus rubra*) and Douglas fir woodland, and salt and freshwater marshes. Most of Westham Island consists of intensively farmed fields. Adjacent lands across the Fraser River are a mix of agricultural land, hobby farms, and suburban housing. Work on Westham Island took place from 1988 to 1991. Between 35 and 40 female song sparrows bred on two intensively studied plots (total area 12.6 ha) each year, and the song sparrow was the most abundant breeding bird at the two study areas, save perhaps for the tree swallow (*Tachycineta bicolor*), a species that we did not count. Other common potential host species at Westham included red-winged blackbirds, house finches (*Carpodacus mexicanus*), American goldfinches (*Carduelis tristis*), and common yellowthroats (*Geothlypis trichas*). Marsh wrens (*Cistothoris palustris*) and American robins were also common, but were presumably rarely parasitized by cowbirds. Rarer potential cowbird hosts at Westham included the savannah sparrow (*Passerculus sandwichensis*), orange-crowned warbler, Bewick's wren (*Thryomanes bewickii*), and yellow warbler (*Dendroica petechia*). Parasitism of hosts other than song sparrows was not monitored at Westham.

Field methods

In all three areas, most adult song sparrows were individually color banded. Most host nests were found early during incubation and were monitored every 2–7 days until all young left the nest, or the nest had failed. Daily nest failure rates were calculated for song sparrow nests at each site. Nests found after the eggs had hatched were not used in calculating the frequency or intensity of parasitism. We searched for nests for the entire host breeding season (March to late July) at all sites, save for a few very late (August) nests that we missed on Westham Island. Nests were checked carefully for presence of parasite eggs on each visit. In one year, 1991, we studied rates of turnover of eggs in nests at Westham Island by marking eggs with a varying number of inconspicuous dots using a fine felt-tip pen. The first egg laid was given one such dot, the second two dots, etc. The sparrow and cowbird lay similar-sized eggs in coastal British Columbia, but usually have a different spot pattern and shape (Smith and Arcese 1994). In east and central North America, size differences between eggs of the two species exist (cowbird eggs are larger), although they cannot always be easily distinguished (Nice 1937; Rothstein 1975). Adult female cowbirds were removed from Mandarte Island in 1977, and data from this year are excluded here. Also, no breeding data were collected on Mandarte in 1980.

A small fraction of nests on Westham Island were hard to see into or reach, and the identity of eggs in these nests could not be assigned with confidence. In cases of uncertainty, eggs were assumed to be host eggs. Several observers also monitored nests at Westham, and less experienced observers probably made more errors in egg identification (usually failure to recognize cowbird eggs). Cowbird nestlings were easier to distinguish from sparrows because of their lighter-colored

Table 16.3 Frequency and intensity of cowbird parasitism on song sparrows at three sites in coastal British Columbia

Site	Number of nests monitored	Number parasitized	Frequency of parasitism	Mean number of cowbird eggs/parasitized nest (S.E.)
Mandarte Island	1509	327	21.7	1.05 (0.01)
Small Islands	148	40	22.2	1.45 (0.11)
Westham Island	388	256	66.0	1.81 (0.06)

down and plumage, and their larger size for a given developmental stage.

At each site, we conducted 50-m radius point counts for 5 minutes two to three times each summer from 1989 to 1991 to obtain an index of the abundances of cowbirds and potential hosts. All counts were conducted in the morning (06:00–13:00 h), the time of day when cowbirds visited breeding areas to search for host nests (Rothstein et al. 1980; Smith and Arcese 1994). Most individual birds were detected by sound on the forested islands, but many were detected visually on the more open smaller islands. The number of points surveyed per site varied from two on the very smallest islets to 14 on Westham Island. All counts for the small islands were treated as independent samples to increase sample sizes. Although this procedure is not ideal for resident host species, it may be reasonable for estimating numbers of the mobile cowbird. In 1989, we also conducted morning counts in song sparrow habitat at three sites on Vancouver Island within 4 km of the nearest point to the small islands. Using data from these counts, we calculated cowbird-to-host ratios using the total numbers of hosts and cowbirds (of either sex) detected at each site on all counts. When estimating the total abundance of cowbird hosts, we included both known acceptor species of cowbird eggs, and rejecters such as the American robin that are sometimes parasitized (Rothstein 1975), but we excluded cavity-nesting species.

Results

Levels of parasitism across sites

We present data on frequency and intensity of parasitism across the three areas in table 16.3. The most detailed information was available for Mandarte Island. Frequencies of parasitism on Mandarte varied from zero to 45% per year. There was no parasitism in four of 17 study years. Multiple parasitism was rare on Mandarte Island (15 of 327 cases, 4.6%) in comparison with other sites, but Mandarte had a similar overall frequency of parasitism (just over 20%) to the nearby Small Islands. The number of cowbird eggs laid on Mandarte Island was greater, and the laying period of cowbirds was more extended, in years of higher host density. That is, the cowbird acted like a typical predator by adjusting its searching and laying effort in relation to the payoff provided by the song sparrow as a host (Smith and Arcese 1994).

We now examine variation in parasitism levels on the Offshore Islands in relation to distance from the large Vancouver Island. Two pairs of islets, Little Shell and Ker, and Dock 1 and Dock 2, were each combined for this analysis because they were so close to each other (<50 m apart at low tide) that we felt they should not be considered as independent observations. Song sparrows on islands closest to Vancouver Island experienced higher frequencies of parasitism than those on more distant islands (fig. 16.8, $r^2 = 0.82$, $n = 7$, $P < 0.01$). Multiple parasitism was also more common (nine of 15 parasitized nests, 60%) on the pair of islets closest to Vancouver Island than on more distant islets (three of 23 parasitized nests, 13%), excluding Mandarte Island.

Frequencies and intensities of parasitism on Westham Island close to the Continental mainland, were both much higher than on the Offshore Islands (table 16.3). The frequency distribution of cowbird eggs per nest also differed across sites (fig. 16.9). The data from Mandarte Island deviated significantly from a truncated Poisson distribution ($\chi^2 = 16.87$, d.f. = 2, $P < 0.001$); there were too many nests with only a single egg, and too few multiply parasitized nests. Data from the Small Islands and Westham did not differ from a random expectation in either case ($P > 0.05$, χ^2 tests, d.f. = 2, 3 respectively). These data show a gradient of parasitism with increasing distance from mainland areas, and suggest that cowbirds on

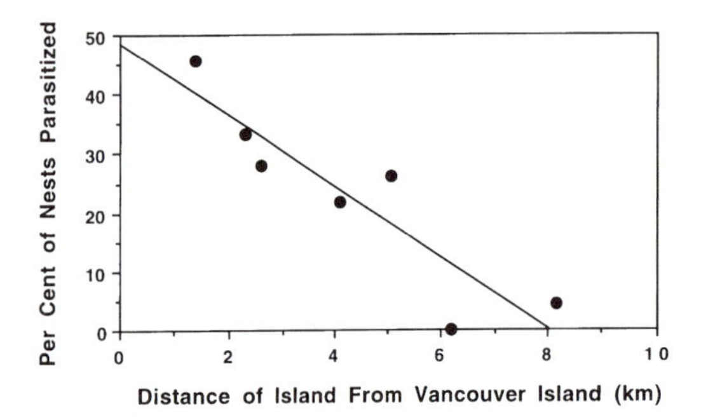

Figure 16.8 Frequency of cowbird parasitism of song sparrow nests on Small Islands in relation to distance from the large Vancouver Island.

Mandarte Island, but not at the other study areas, avoided laying more than once in a nest.

Egg removal and turnover

In 1991, we measured losses of marked host and parasite eggs at Westham. Eggs were marked in 59 song sparrow nests. Twenty-two of 158 marked sparrow eggs (13.9%) disappeared before hatching, excluding all eggs that disappeared at the time of nest failure. In contrast, only three of 73 cowbird eggs (4.1%) laid early in the host incubation period disappeared. These data suggest that female cowbirds preferentially removed sparrow eggs rather than cowbird eggs, but we do not present a formal statistical test here, as we did not check nests frequently enough to know how many eggs of each species were in the nest at the actual time that egg removals took place. As a result of such turnover of eggs in nests, the clutch sizes of hosts, and the numbers of cowbird eggs laid, were both underestimated to a moderate degree.

Host reproductive success and parasitism level

The success of song sparrows in rearing their own and cowbird young at each of the three sites varied with the frequency and intensity of parasitism (table 16.4). Data on fledging success for Mandarte came from 1981 to 1990, excluding 1982. No cowbird eggs were laid on Mandarte in 1982, 1991, or 1992, and our data on fledging success were less

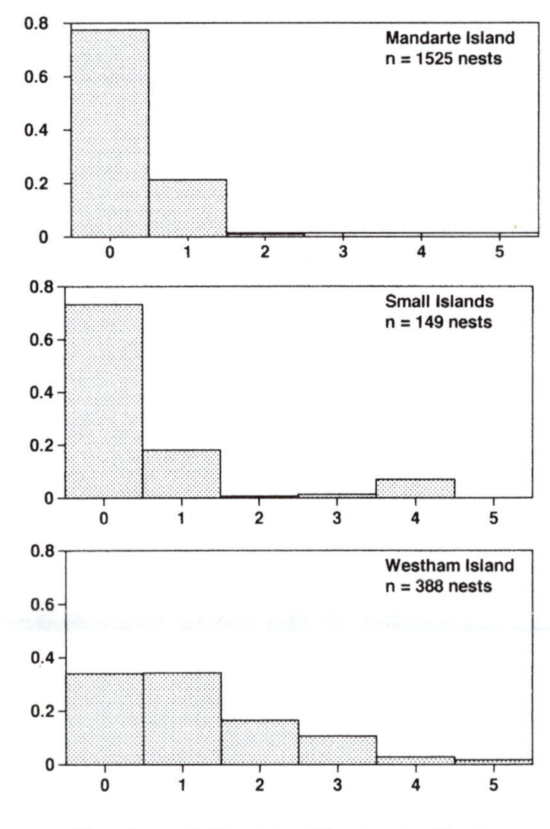

Figure 16.9 Frequency distributions of numbers of cowbird eggs per song sparrow clutch at the three study areas.

Table 16.4 Fledging success (mean number ± S.E. of individuals leaving the nest successfully) of song sparrows and brown-headed cowbirds in relation to parasitism at each of three study sites

	Unparasitized nests	Parasitized nests	
Study site	Sparrows (n)	Sparrows	Cowbirds
Mandarte Island	1.37 ± 0.05 (701)	0.74 ± 0.07 (214)	0.43 ± 0.04 (214)
Small Islands	1.78 ± 0.15 (103)	1.40 ± 0.21 (38)	0.61 ± 0.11 (38)
Westham Island	0.63 ± 0.09 (143)	0.21 ± 0.04 (246)	0.25 ± 0.04 (247)

complete before 1981. The average unparasitized sparrow nest produced about 25% fewer fledglings on Mandarte Island than on the Small Islands. Parasitized nests on the Offshore Islands produced about half as many cowbird fledglings as they did sparrows, and parasitized nests on Mandarte fledged only about half as many young as on the Small Islands. The sample size of parasitized nests on the Small Islands, however, was small and concentrated on the islets closest to shore.

On Westham Island, unparasitized song sparrow nests were less than half as successful as on the offshore islands, because of frequent nest failures at Westham. Slightly more cowbirds than song sparrows were fledged from parasitized nests at Westham. Multiple parasitism was particularly deleterious to host reproductive success at Westham. Only four of 52 clutches containing more than two cowbird eggs at Westham fledged any host young, although 13 of these fledged cowbird young.

Cowbird abundance and parasitism levels

In general, the number of cowbirds detected during counts corresponded well to the per cent parasitism detected at a study site. Cowbirds were detected frequently during counts on Vancouver Island and Westham Island, but were seen or heard on only two of the nine Offshore Islands (table 16.5). One of these was the close-to-shore Ker Island [0.13 ± 0.08 (S.E.) detections per count, $n = 16$], which had the highest frequency and intensity of parasitism of all Small Islands. The other was the distant but song sparrow-rich Mandarte Island, which supported only three host species, song sparrows, fox sparrows, and red-winged blackbirds. We have no records of parasitism for the latter two species in about 25 nests followed over a 15-year period, but detailed studies were not conducted, and we could have missed parasitism if it had taken place. In summary, cowbird density and parasitism levels were highest on Westham Island. On the Offshore Islands, parasitism was most common closest to the dense cowbird populations on Vancouver Island.

Discussion

Our results revealed a gradient of parasitism in relation to isolation from cowbird feeding sites. Cowbird foraging areas are abundant near Westham Island, and on eastern Vancouver Island. Cowbirds were counted frequently at Westham, and parasitism there was both frequent and in-

Table 16.5 Results of point counts conducted at each of the three study areas, and on Vancouver Island. Cowbird/host ratio is the total number of cowbirds detected at an area divided by the total number of host individuals detected at the same area. n = total number of counts conducted at all points

Study site	n	No. of species detected	Mean no. of cowbirds detected (S.E.) per point	Mean no. of sparrows detected (S.E.) per point	Cowbird/host ratio
Mandarte Island	36	7	0.08 (0.06)	1.78 (0.23)	0.04
Small Islands	112	25	0.04 (0.02)	1.74 (0.13)	0.005
Westham Island	96	23	0.60 (0.09)	2.04 (0.14)	0.11
Vancouver Island	31	25	0.55 (0.15)	0.68 (0.15)	0.11

tense. Nesting success at Westham Island was low because of very high levels of: (i) complete predation; and (ii) partial predation plus brood parasitism, and consequent nest desertion (C. M. Rogers et al., unpubl. results). Daily nest failure rates on Westham Island (0.074/day) averaged three times as high as on the Offshore Islands (0.025/day). Losses of marked host eggs at Westham were frequent in 1991, but marked cowbird eggs disappeared infrequently from nests, suggesting that cowbirds avoided removing their own eggs from nests, or avoided removing the eggs of other cowbirds. Cowbirds on Mandarte Island tended to avoid laying more than one egg in the same host nest, perhaps because only one or two individuals were laying on Mandarte each year (Smith and Arcese 1994), but no such pattern was evident on the other areas, perhaps because cowbirds there were short of host nests to lay in. Cowbird eggs are highly variable in coloration, and nests on the mainland frequently contained two or more similar-looking cowbird eggs, suggesting repeated use of the same host nest by individual female cowbirds.

In contrast, sparrows on the Offshore Islands experienced much lower parasitism. On Mandarte Island, numbers of song sparrows fluctuated strongly, and the frequency of parasitism varied from year to year. Nice (1937) also found annually variable parasitism levels during her long-term study in Ohio. She found that parasitism increased in association with a decline in song sparrow numbers, whereas cowbird numbers remained relatively stable.

Higher numbers of cowbird eggs were laid on Mandarte over a longer laying period, in years with a higher number of host females. That is, cowbirds on Mandarte showed a functional response to host density (Smith and Arcese 1994). Across the Offshore Islands, both the frequency and intensity of parasitism were higher closer to Vancouver Island. Mandarte Island was an outlier from this pattern. Presumably the dense song sparrow population on Mandarte in some years made the long commute worth while to the few cowbirds that visited the island (Smith and Arcese 1994).

There were some unexplained differences in the effects of parasitism within the Offshore Islands. For example, parasitized nests on Mandarte fledged fewer song sparrows than nests on the Small Islands. The opposite result might have been expected, given the higher frequency of multiple parasitism in parasitized nests on the Small Islands.

A second possible reason for the differences between nest failure rates on the Offshore Islands and the large Westham Island, was differences in predator communities. The identity of nest predators was studied at Westham using previously used song sparrow nests containing wax or plasticene model sparrow eggs (C. M. Rogers et al., unpubl. results). Deer mice (*Peromyscus maniculatus*), shrews, and blackbird-sized birds were the most common predators of dummy eggs at Westham (C. M. Rogers et al., unpubl. results). Deer mice are also abundant on Mandarte Island, but Mandarte lacks shrews. The open Small Islands had few terrestrial predators to our knowledge. The forested Small Islands supported coast garter snakes, *Thamnophis ordinoides*, but no detailed studies of the predator communities were done there. Northwestern crows (*Corvus caurinus*) were abundant predators on Mandarte Island, occurred commonly on the Small Islands, but were rare at Westham.

We now discuss these results in relation to the four hypotheses introduced earlier. The number of species of alternative hosts was similar on Westham Island and the forested Offshore Islands, but was much higher than on Mandarte Island and the small nonforested islets, where parasitism levels were low. Forest birds on Westham and the small forested islands thus did not draw off cowbirds from parasitizing song sparrows there. This result disagrees with a prediction of the Parasite Host Selection Hypothesis that parasitism of a nonpreferred host should decrease as the number of preferred alternative hosts increases. The song sparrow, however, is probably a preferred host of the cowbird, so that this prediction may not apply here. The Host-Selection Hypothesis, however, might play a role in variable use of less common, and perhaps smaller, host species of cowbirds (Mason 1986). However, some accounts of parasitism by brown-headed (e.g., Friedmann 1963) and shiny cowbirds (Mason 1986) suggest that host selection is the exception (e.g., Briskie et al. 1990), not the rule.

The Host Habitat Hypothesis also received little support. Among the Offshore Islands, song sparrows dominated host communities on scrub-dominated islands, but comprised a smaller fraction of hosts on the two forested islands, Ker and Rum. Song sparrow nests, however, were parasitized heavily on the forested Ker, and also on the more open Mandarte in some years. Isolation by distance seemed to be a more important influence on levels of parasitism than variation in local habitat, although further data are needed to confirm this result.

The Host Defense Hypothesis is unlikely to explain local differences in parasitism, as host defenses probably vary only on a broad geographic scale. Of the four hypotheses considered here, only the cowbird density hypothesis consistently predicted the use of the song sparrow as a host. Parasitism of song sparrows was more frequent and intense, and was associated with much lower host reproductive success, at sites where cowbirds were seen more often, regardless of the local habitat or of the abundances of alternative host species.

GENERAL DISCUSSION

In this study, we found marked spatial variation in the frequency and intensity of parasitism of song sparrows by cowbirds at both the continental and local scales. At both spatial scales, the Parasite Density Hypothesis was strongly supported. This hypothesis may apply particularly well to the song sparrow, because it is an edge-loving species, and thus is likely to be encountered by cowbirds in virtually every landscape in which it occurs. In addition, the song sparrow may be a preferred cowbird host, which may increase the applicability of this hypothesis (see earlier).

Studies of the use of cowbird hosts in states the upper Mississippi Valley (Lowther and Johnston 1977, Koford et al., in press; Robinson et al., in press; Thompson et al., in press), however, suggest that other explanations, particularly the Parasite Habitat Preference Hypothesis, may also play a role in variable use of particular hosts. We could not conduct strong tests of the Parasite Habitat Preference Hypothesis, or of the other two hypotheses listed above. These hypotheses, however, may help to explain the preferred use of wood thrushes over song sparrows (fig. 16.6).

A noteworthy finding of this and other studies of parasitism of song sparrows by cowbirds is that average frequencies of parasitism [$\bar{x} = 34.7 \pm 6.6$ (S.E.)] recorded in 10 detailed studies (Nice 1937; Johnston 1956; Friedmann 1963; Friedmann et al. 1977; this study) lie well above the maximum value (23%) recorded for any well-studied U.S. state (Iowa, Minnesota) in the Cornell data. There are two likely explanations for this result. First, variation within a local region can be large (see earlier), and statewide frequencies of parasitism are average values across varying habitats and densities of hosts and cowbirds. Detailed studies, in contrast, may have been conducted where high host densities generated "hot spots" for cowbirds.

A second explanation for the higher parasitism levels in detailed studies is that parasitism frequency may have been underestimated in nest record data, because not all recorders discriminated cowbird eggs from those of song sparrows. The song sparrow egg resembles that of the cowbird more closely than the egg of any other cowbird host (S. I. Rothstein, pers. commun.), and it seemed likely to us from information on some nest record cards (damaged eggs, up to three eggs on the nest rim or ground beside nest, comments on odd-colored eggs, abnormally large clutch sizes), that parasitism had probably occurred, yet it was not detected. It is noteworthy here that frequencies of parasitism were over twice as high in nest record data for wood thrushes than in data for song sparrows (fig. 16.6). It should have been much simpler for recorders to discriminate the unspotted and bright blue wood thrush eggs from the brown speckled and drab cowbird eggs.

Our finding that islands distant from a mainland source suffer reduced parasitism may have application to conservation of threatened hosts. Song sparrows breeding on Westham Island near the continental mainland, and in prime cowbird habitat, reproduced so poorly that Westham Island may be a "population sink" (Pulliam 1988) for song sparrows. Island birds reared nearly four times as many young as required to fill breeding vacancies on average, but mainland birds reared fewer young than needed, even before post fledging survival was taken into account. Thus, sparrow populations at Westham Island could only maintain their numbers there because of immigration from sites elsewhere (C. Rogers et al., unpublished results). Host populations in small areas of habitat far from cowbird feeding areas may thus be at less risk than hosts near good cowbird feeding habitat. While this result might not apply when land rather than water separates "islands" of host habitat, Trine (in press) and Robinson et al. (in press) noted similar results for other host species in islands of forested habitat in agricultural landscapes in the American mid-west.

Further studies are needed to test if brood parasite density is a major determinant of variation in frequency and intensity of parasitism for other cowbird hosts, and in other generalist brood parasites and their hosts. Our results provide some encouragement for the practice of protecting host populations by trapping cowbirds (e.g., DeCapita, in press; Whitfield, in press). If, however, frequency-dependent host selection occurs commonly (see above) in host communities, the benefits of removing cowbirds to protect rare host species may be limited. Cowbird counts, however, can serve as a useful first step in judging if the species is a conservation threat in an area (Miles and Buehler, in press; Rothstein et al., in press). If cowbirds are not common, they are unlikely to be a conservation problem.

A standardized methodology would be valuable in cowbird counts. We counted cowbirds using 5-min point counts with a 50-m fixed radius. We selected the 50-m radius so that hosts could be censused readily in high-density populations. A larger radius is probably more appropriate for detecting the noisy and mobile male cowbird, but the secretive female is easily overlooked even at close range (Rothstein et al., in press). Because of the male-biased sex ratios typical of brown-headed cowbird populations (Weatherhead 1989; Robinson et al. 1995), counts based on female cowbirds alone are preferable to the counts of both sexes that we used. Others have used a 6-min count (e.g., Robinson et al. 1995), or a 10-min count (Rothstein et al. 1980). Further studies of methods of counting cowbirds (e.g., Miles and Buehler, in press), and/or of alternative counting methods, would be timely.

ACKNOWLEDGMENTS We thank Jim Low of the Cornell University Laboratory of Ornithology for assistance. Many colleagues and assistants helped collect data on song sparrows. Special thanks are due to Peter Arcese, Christine Chesson, Lorne Gould, Gwen Jongejan, Durrell Kapan, Marlene Machmer, Chris Rogers, Mary Taitt, and David Westcott. Dolph Schluter helped to begin the studies on the Small Islands and Vancouver Island, and Judy Myers gave encouragement and helped with graphics. Steve Rothstein and Scott Robinson made many helpful comments on the manuscript. Grant Gilchrist helped with statistical analyses, and Don Kroodsma provided hospitality and intellectual stimulation. The Tsawout and Tseycum Indian Bands kindly allowed us to work on Mandarte Island. Our studies were supported by operating grants from the Natural Sciences and Engineering Research Council (Canada).

REFERENCES

Briskie, J. V., S. G. Sealy, and K. A. Hobson (1990). Differential parasitism of Least Flycatchers and Yellow Warblers by the Brown-headed Cowbird. *Behav. Ecol. Sociobiol.*, **27**, 403–410.

Briskie, J. V., S. G. Sealy, and K. A. Hobson (1992). Behavioral defenses against avian brood parasitism in sympatric and allopatric host populations. *Evolution*, **46**, 334–340.

DeCapita, M. F. Brown-headed Cowbird control on Kirtland's Warbler nesting areas in Michigan, 1972–1993. In: *Ecology and management of cowbirds* (eds J. N. M. Smith, T. Cook, S. K. Robinson, S. I. Rothstein, and S. G. Sealy). University of Texas Press, Austin (in press).

Facemire, C. F. (1980). Cowbird parasitism of marsh-dwelling Red-winged Blackbirds. *Condor*, **82**, 347–348.

Fretwell, S. D. (1997). Is the Dickcissel a threatened species? *American Birds*, **31**, 923–932.

Friedmann, H. (1948). *The parasitic cuckoos of Africa. Washington Academy of Sciences Monographs*, Washington, D.C. Vol. 1, 1–204.

Friedmann, H. (1960). The parasitic weaverbirds. *U.S. National Museum Bull.*, **223**, 1–196.

Friedmann, H. (1963). Host relations of the parasitic cowbirds. *U.S. National Museum Bull.*, **233**, 1–276.

Friedmann, H., L. F. Kiff, and S. I. Rothstein (1977). A further contribution to the knowledge of the host relations of the parasitic cowbirds. *Smithsonian Contributions to Zoology*, **235**, 1–75.

Hoover, J. P. and M. C. Brittingham (1993). Regional variation in cowbird parasitism of Wood Thrushes. *Wilson Bull.*, **105**, 228–238.

Johnston, R. F. (1956). Population structure of salt marsh Song Sparrows. Part I. Environment and annual cycle. *Condor*, **58**, 24–44.

Koford, R. R., B. S. Bowen, J. T. Lokemoen, and A. D. Kruse. Cowbird parasitism in grassland and cropland in the northern Great Plains. In: *Ecology and Management of Cowbirds* (eds J. N. M. Smith, T. Cook, S. K. Robinson, S. I. Rothstein, and S. G. Sealy). University of Texas Press, Austin (in press).

Linz, G. M. and S. B. Bolin (1982). Incidence of Brown-headed Cowbird parasitism on Red-winged Blackbirds. *Wilson Bull.*, **94**, 93–95.

Lowther, P. E. and R. F. Johnston (1997). Influences of habitat on cowbird nest selection. *Bull. Kansas Ornithol. Soc.*, **28**, 36–40.

Mason, P. (1986). Brood parasitism in a host generalist, the shiny cowbird. II. Host selection. *Auk*, **103**, 61–69.

McGeen, D. S. (1972). Cowbird–host relationships. *Auk*, **89**, 360–380.

Miles, R. K. and D. A. Buehler. An evaluation of point count and playback techniques for censusing brown-headed cowbirds. In: *Ecology and Management of Cowbirds* (eds J. N. M. Smith, T. Cook, S. K. Robinson, S. I. Rothstein, and S. G. Sealy). University of Texas Press, Austin (in press).

Nice, M. M. (1937). Studies in the life history of the Song Sparrow. Part I. *Trans. Linn. Soc. N.Y.*, **4**, 1–247.

Ortega, C. P. and A. Cruz (1988). Mechanisms of egg acceptance in marsh-nesting blackbirds. *Condor*, **90**, 349–358.

Peck, G. K. and R. D. James (1987). *Breeding Birds of Ontario. Nidiology and distribution. Volume 2: Passerines*. Royal Ontario Museum, Toronto.

Peterjohn, B. G., J. R. Sauer, and S. Orsillo. Temporal and Geographic Patterns in Population Trends of Brown-headed Cowbirds. In: *Ecology and Management of Cowbirds* (eds J. N. M. Smith, T. Cook, S. K. Robinson, S. I. Rothstein, and S. G. Sealy). University of Texas Press, Austin (in press).

Pulliam, H. R. (1988). Sources, sinks and population regulation. *Amer. Natur.*, **132**, 652–661.

Robertson, R. J. and R. F. Norman (1976). Behavioral defenses to brood parasitism by potential hosts of the Brown-headed Cowbird. *Condor*, **78**, 166–173.

Robertson, R. J. and R. F. Norman (1977). The function and evolution of aggressive host behavior towards the Brown-headed Cowbird (*Molothrus ater*). *Can. J. Zool.*, **55**, 508–518.

Robinson, S. K., J. P. Hoover, J. R. Herkert, and R. Jack. Cowbird parasitism in a fragmented landscape: effects of tract size, habitat and abundance of cowbirds and hosts. In: *Ecology and Management of Cowbirds* (eds J. N. M. Smith, T. Cook, S. K. Robinson, S. I. Rothstein, and S. G. Sealy). University of Texas Press, Austin (in press).

Robinson, S. K., S. I. Rothstein, M. C. Brittingham, L. J. Petit, and J. A. Grzybowski (1995). Brood parasitism in the Brown-headed Cowbird: effects on host populations. In: *Ecology and Management of Neotropical Migrant Birds* (eds T. E. Martin and D. M. Finch), pp. 428–460. Oxford University Press, Oxford.

Rogers, C. M., J. N. M. Smith, W. M. Hochachka, A. L. E. V. Cassidy, M. J. Taitt, and D. Schluter (1991). Spatial variation in winter survival of Song Sparrows. *Ornis Scand.*, **22**, 387–395.

Rothstein, S. I. (1975). An experimental and teleonomic investigation of avian brood parasitism. *Condor*, **77**, 250–271.

Rothstein, S. I. (1976). Cowbird parasitism of the Cedar Waxwing and its evolutionary implications. *Auk*, **93**, 498–509.

Rothstein, S. I. (1990). A model system for coevolution: avian brood parasitism. *Annu. Rev. Ecol. Syst.*, **21**, 481–508.

Rothstein, S. I. (1994). The cowbird's invasion of the far west: history, causes and consequences experienced by host species. *Studies in Avian Biology*, **15**, 301–315.

Rothstein, S. I., J. Verner, and E. Stevens (1980). Range expansion and diurnal change in dispersion of the Brown-headed Cowbird in the Sierra Nevada. *Auk*, **97**, 253–267.

Rothstein, S. I., C. Farmer, and J. Verner. Cowbird vocalizations: an overview and the use of playbacks to enhance cowbird detectability. In: *Ecology and Management of Cowbirds* (eds J. N. M. Smith, T. Cook, S. K. Robinson, S. I. Rothstein, and S. G. Sealy). University of Texas Press, Austin (in press).

Smith, J. N. M. (1981). Cowbird parasitism, host fitness and age of the host female in an island population of Song Sparrows. *Condor*, **83**, 152–161.

Smith, J. N. M. and P. Arcese (1994). Brown-headed Cowbirds and an island population of Song Sparrows: a 16-year study. *Condor*, **96**, 916–934.

Smith, J. N. M., P. Arcese, and I. G. McLean (1984). Age, experience, and enemy recognition by wild Song Sparrows. *Behav. Ecol. Sociobiol.*, **14**, 101–106.

Thompson, F. R., III, S. K. Robinson, T. M. Donovan, J. Faaborg, and D. R. Whitehead. Biogeographic, landscape, and local factors affecting cowbird abundance and host parasitism levels. In: *Ecology and Management of Cowbirds* (eds J. N. M. Smith, T. Cook, S. K. Robinson, S. I. Rothstein, and S. G. Sealy). University of Texas Press, Austin (in press).

Trine, C. L. Effects of multiple parasitism on cowbird and Wood Thrush nesting success. In: *Ecology and Management of Cowbirds* (eds J. N. M. Smith, T. Cook, S. K. Robinson, S. I. Rothstein, and S. G. Sealy). University of Texas Press, Austin (in press).

Walkinshaw, L. (1983). *The Kirtland's Warbler*. Cranbrook Inst. of Sci., Bloomfield Hills, Michigan.

Ward, P. and J. N. M. Smith, Inter-habitat differences in parasitism frequencies by Brown-headed Cowbirds in the Okanagon Valley, British Columbia. In: *Ecology and Management of Cowbirds* (eds J. N. M. Smith, T. Cook, S. K. Robinson, S. I. Rothstein, and S. G. Sealy). University of Texas Press, Austin (in press).

Weatherhead, P. J. (1989). Sex-ratios, host-specific

reproductive success, and impact of Brown-headed Cowbirds. *Auk*, **106**, 358–366.
Weinrich, J. A. (1989). Status of the Kirtland's Warbler, 1988. *Jack-Pine Warbler*, **67**, 69–72.
Whitfield, M. J. Results of a Brown-headed Cowbird Control Program for the Southwestern Willow Flycatcher. In: *Ecology and Management of Cowbirds* (eds J. N. M. Smith, T. Cook, S. K. Robinson, S. I. Rothstein, and S. G. Sealy). University of Texas Press, Austin (in press).

Appendix 16.1 Frequencies and intensities of parasitism of song sparrows' nests by brown-headed cowbirds in states of the U.S.A., and values used in the multiple regression analyses

State	No. of nests recorded	% Nests parasitized	Mean no. of cowbird eggs per parasitized nest	Cowbird density	Distance (km) from center of cowbird abundance	% Forest cover	% Grassland and pasture cover
California	45	11.1	1.00	5.04	1888	30.0	40.0
Connecticut	19	0.0	–	6.50	2288	60.0	0.0
Delaware	18	0.0	–	4.86	2176	15.0	0.0
Indiana	20	4.8	2.00	12.97	1312	19.0	5.0
Illinois	53	11.3	1.67	12.86	1072	10.0	1.0
Iowa	30	23.3	1.75	26.05	688	0.0	2.0
Kentucky	33	1.7	1.67	13.52	1552	50.0	32.0
Maine	15	0.0	–	5.60	2464	96.0	0.0
Maryland	43	4.7	1.00	9.69	2080	40.0	0.0
Massachusetts	17	0.0	–	6.08	2336	70.0	0.0
Michigan	373	15.0	1.34	11.25	1206	40.0	3.0
Minnesota	43	23.3	1.82	17.60	496	23.0	0.0
Nebraska	3	33.3	2.00	–	–	–	–
New Hampshire	1	0.0	–	–	–	–	–
New Jersey	35	0.0	–	5.57	2208	45.0	0.0
New York	583	6.7	1.46	9.52	2000	70.0	0.0
North Carolina	6	16.7	1.00	–	–	–	–
Ohio	147	12.9	1.56	10.05	1488	35.0	2.0
Oregon	16	6.3	1.00	7.88	1616	48.0	39.0
Pennsylvania	275	3.6	1.31	8.39	1936	80.0	0.0
Rhode Island	2	0.0	–	–	–	–	-
Tennessee	26	3.9	1.00	10.38	1600	58.0	13.0
Virginia	39	0.0	–	5.32	2032	90.0	6.0
Vermont	26	3.9	1.00	6.36	2224	90.0	0.0
Washington	26	7.4	1.00	10.20	1552	66.0	7.0
West Virginia	119	2.5	1.33	9.60	1808	90.0	3.0
Wisconsin	172	20.4	1.86	19.00	864	47.0	0.0

17

Potential Impacts of Cowbird Range Expansion in Florida

ALEXANDER CRUZ, WILLIAM POST, JAMES W. WILEY,
CATHERINE P. ORTEGA, TAMMIE K. NAKAMURA, JOHN W. PRATHER

Avian brood or social parasitism is a reproductive strategy in which parasites lay their eggs in the nests of other birds and depend on the hosts to incubate the eggs and care for the young. Host reproductive success is often reduced by parasitism because of: (i) egg removal or puncturing by the adult parasite; (ii) host desertion of the parasitized nest; (iii) breakage of host eggs when cowbird eggs are laid; (iv) short incubation period of cowbird eggs, 11 days compared with 12–14 days of most hosts, which give nestling cowbirds a head start; (v) reduced egg hatchability and fledgling success; and (vi) cowbird nestlings grow faster, beg more loudly and have larger gapes than host nestlings (Friedmann 1929; Payne 1977; Post and Wiley 1977a; Wiley 1985; Spaw and Rohwer 1987; Cruz et al. 1989; Weatherhead 1989; Rothstein 1990; Ortega and Cruz 1991; Sealy 1992; Robinson et al. 1993). In addition, Holford and Roby (1993) found that brown-headed cowbirds in captivity can lay up to 77 eggs per season. Annual fecundity of shiny cowbirds can be as high as 120 eggs/year (Kattan 1993; A. Cruz, pers. obs.). Relatively small numbers of cowbirds can therefore parasitize many nests (Robinson et al. 1993).

Recent changes in the range of two brood parasites, the shiny cowbird (*Molothrus bonariensis*) and the brown-headed cowbird (*M. ater*) have brought them into contact with avian communities that have never experienced brood parasitism. Originally confined to South America, Trinidad, and Tobago, the shiny cowbird has spread north into the West Indies and Florida during the past century (Post and Wiley 1977*b*; Cruz et al. 1985, 1989) (figs 17.1 and 17.2). From the opposite direction, the brown-headed cowbird has expanded its range in eastern North America. The brown-headed cowbird is presently colonizing Florida (Hoffman and Woolfenden 1986; fig. 17.2). A third species, the bronzed cowbird (*Molothrus aeneus*) has spread east from south Texas into Louisiana and in the future may pose a threat to Florida passerines (Robbins and Easterla 1981) (fig. 17.3). The presence of these cowbird species is expected to have important negative consequences for the breeding passerines of Florida, some of which are patchily distributed and breed in small populations that may be vulnerable to extirpation by cowbirds (Hoffman and Woolfenden 1986; Smith and Sprunt 1987; Stevenson and Anderson 1994; Prather and Cruz 1995).

The arrival of shiny and brown-headed cowbirds in Florida presents a unique opportunity to study the colonization of Florida by brood parasites at an early interfacing of host and parasite populations. The colonization process is made especially interesting because both cowbird species are obligate brood parasites, and are host generalists with more than 200 host species recorded for each cowbird species (Friedmann and Kiff 1985; Cruz et al. 1989; Post et al. 1993).

We began our studies of shiny cowbirds in the West Indian region in the 1970s, of brown-headed

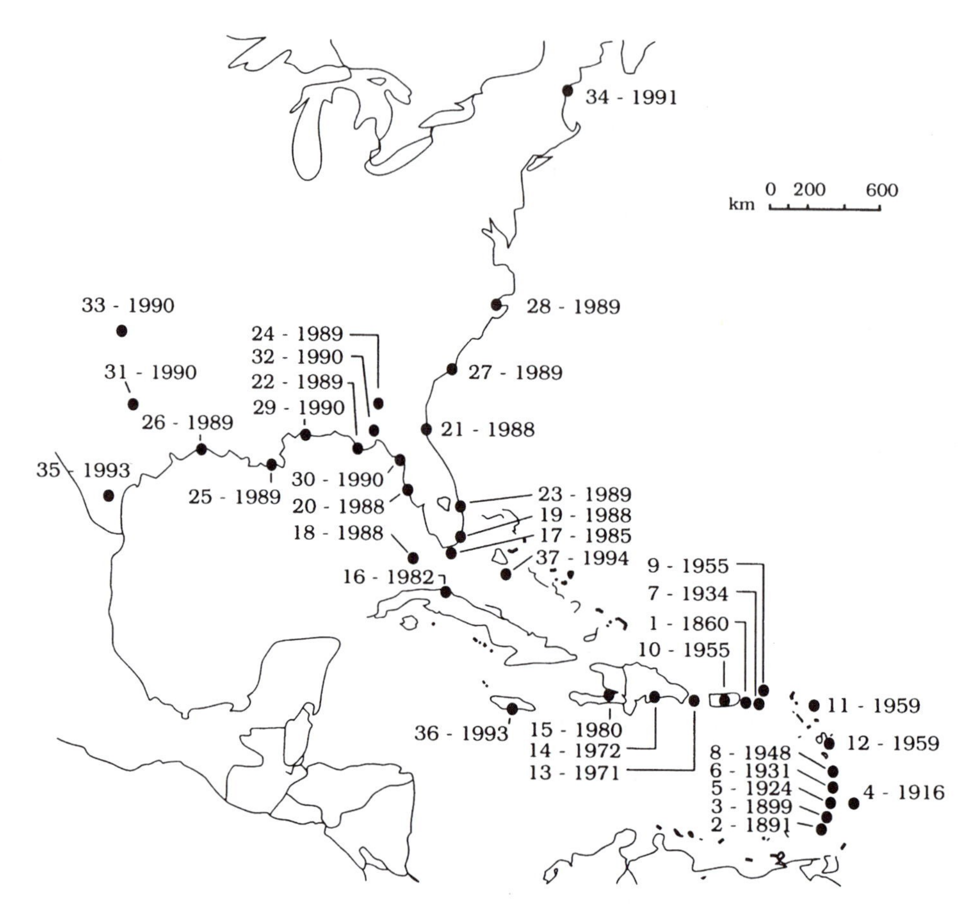

Figure 17.1 Range expansion of the shiny cowbird in the Greater Caribbean Basin and North America. Islands are numbered in chronological order by year of first sightings. In North America, except for Flamingo, Florida, the first mainland locality, only range extensions of 100 km or more are shown. 1. Vieques, 1860, 2. Grenada, 1891, 3. Grenadines, 1899, 4. Barbados, 1916, 5. St. Vincent, 1924, 6. St Lucia, 1931, 7. St. Croix, 1934, 8. Martinique, 1948, 9. St John, 1955, 10. Puerto Rico, 1955, 11. Antigua, 1959, 12. Maria-Galante, 1959, 13. Mona Island, 1971, 14. Dominican Republic, 1972, 15. Haiti, 1980, 16. Cuba, 1982, 17. Lower Matecumbe Key, Florida, 1985, 18. Dry Tortugas, Florida, 1988, 19. Homestead, Florida, 1988, 20. Fort Desoto, Florida, 1988, 21. Black Hammock Island, Florida, 1989, 22. Cape San Blas, Florida, 1989, 23. Delray Beach, Florida, 1989, 24. Warner Robbins, Georgia, 1989, 25. Port Fourchon, Louisiana, 1989, 26. Cameron, Louisiana, 1989, 27. Sullivan's Island, South Carolina, 1989, 28. Aurora, North Carolina, 1990, 29. Bon Secour, Alabama, 1990, 30. Levy County, Florida, 1990, 31. Fort Hood, Texas, 1990, 32. Tallahassee, Florida, 1990, 33. Winborn Spring, Oklahoma, 1991, 34. Monhegan Island, Maine, 1991, 35. Goliad, Texas, 1993, 36. Yallahs Pond, Jamaica, 1993, and 37. Staniard Creek, Andros Island, 1994. Data based on field observations, personal communications, Florida Breeding Bird Atlas Project, and literature review, e.g., *American Birds*, *Florida Naturalist*, *Florida Field Naturalist*, and *Tropical Audubon Bulletin*.

cowbirds in the western United States (Colorado) in the 1980s, and of cowbirds in Florida in 1989. In this paper, we focus on Florida, a region that has a unique faunal assemblage with affinities to both the West Indies and North America (Robertson 1955; Robertson and Kushlan 1974). We document the spread of cowbird species into the Florida region, discuss the implications of this colonization for passerines breeding in Florida, and suggest management strategies.

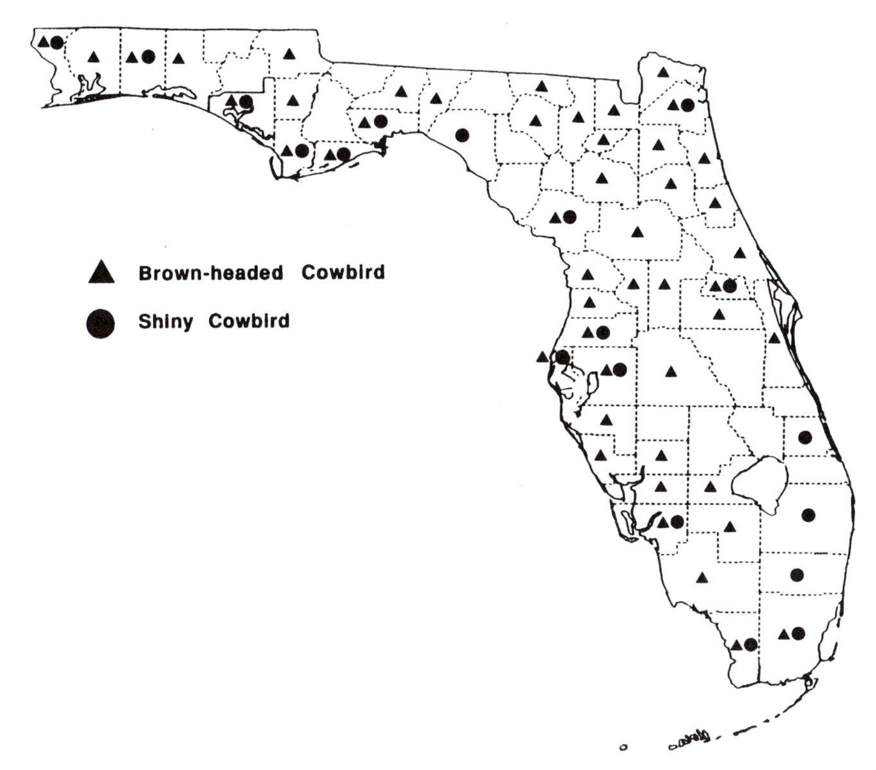

Figure 17.2 Map of Florida showing occurrence of cowbirds by counties. Brown-headed cowbird records represent confirmed breeding records, whereas shiny cowbird records represent counties where observed and include possible breeding records. Confirmed breeding records are based on the criteria used by the Florida Breeding Bird Atlas Project and include, e.g., female with egg in the oviduct, host nest with cowbird egg or nestling, recently fledged young, and host feeding cowbird young. Data based on field observations, personal communications, Florida Breeding Bird Atlas Project, and literature review, e.g., *American Birds*, *Florida Naturalist*, *Florida Field Naturalist*, and *Tropical Audubon Bulletin.*

STUDY AREAS

Field work in Florida is ongoing and is being conducted in mangroves, fresh-water marshes, upland hammocks, and human-modified areas, such as agricultural and disturbed areas, in Monroe, Dade, Collier, Lee, and Charlotte Counties (see Prather and Cruz 1995 for habitat descriptions). We conducted field work in the West Indies (Puerto Rico, Dominican Republic, and St. Lucia) from 1973 to 1986. This region is a composite of many habitats and vegetation types, most of which have been influenced by humans. Study areas were in mangrove forest, dry scrub, dry forest woodlands, and areas under cultivation and animal husbandry. Detailed information on these areas is provided by Post and Wiley (1977*a*), Post (1981), Wiley (1982, 1985), Cruz et al. (1985), and Post et al. (1990). In Colorado, field work was conducted in marshes, riparian areas, and upland forests from 1985 to 1994 (see Marvil and Cruz 1989; Ortega and Cruz 1988, 1991; Ortega 1991 for habitat descriptions).

METHODS

Documentation of the past and present distribution of cowbirds in Florida is based on extensive review of the literature, communications with individuals knowledgeable of the avifauna, and field sampling in Florida. We gathered information on the breeding biology of cowbirds and their potential and actual host species, the frequency of use of host species, and the effects of brood parasitism on host breeding success. We located nests by

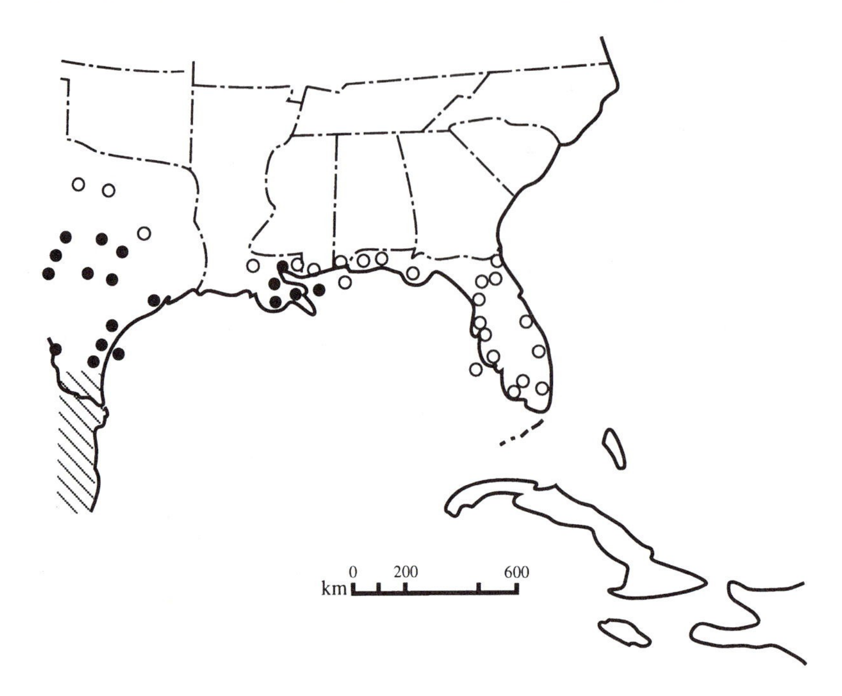

Figure 17.3 Range expansion of the bronzed cowbird into the southeastern United States. Diagonal lines represent breeding range in the early 1900s. Black circles represent extension of the breeding range as of 1994, whereas open circles represent nonbreeding records, spring through fall. Data based on personal communications and literature review, e.g., *Audubon Field Notes*, *American Birds*, *Florida Naturalist*, *Florida Field Naturalist*, and *Tropical Audubon Bulletin.*

regularly searching the study areas. At each visit to the study areas (usually at 2- to 4-day intervals), we inspected nests to determine the number of host eggs and, if present, parasite eggs and chicks. We defined a nest as active when the resident laid at least one egg or if the host incubated cowbird eggs, regardless of whether or not host eggs were present. We defined egg success as the proportion of the eggs hatched per egg laid, and nesting success as the proportion of nests fledging at least one young. If the nest failed, the area was searched for evidence of the cause of failure (animal tracks, punctured eggs, damage to the nest). Nest appearance and mode of disturbance were used to determine whether or not nests were disturbed by predators.

COWBIRDS IN THE FLORIDA REGION

Human-caused changes in landscape and land-use patterns have resulted in cowbirds expanding their ranges and using new host species. In particular, the fragmentation of forests and the spread of animal husbandry in North America and the West Indies over the past 150 years or so appears to have favored cowbirds, which are species of edge and open habitats (Mayfield 1965; May and Robinson 1985; Cruz et al. 1989). Cowbird populations have also increased within their range as a result of increasing winter food supply, primarily waste grain in agricultural fields (Brittingham and Temple 1983).

Shiny cowbird

The subspecies involved in the Caribbean basin expansion is *M. b. minimus*, which was originally confined to northern Brazil, the Guianas, eastern Venezuela, and Trinidad and Tobago. Its spread through most of the West Indian region (fig. 17.1) has been well documented (Bond 1966; Post and Wiley 1977*b*; Wiley 1982; Garrido 1984; Arendt and Vargas Mora 1984; Cruz et al. 1985, 1989).

In the western Caribbean, the shiny cowbird was first recorded from Cuba in 1983, where a bird was captured at Cardenas (Garrido 1984). In 1983, several more birds were captured at this locale.

More recent reports indicate that cowbirds are now widespread in Cuba, occurring in the majority of the province and the Isle of Pines (Guerra and Alayon 1987; O. Garrido, A. Kirkconnel, C. Wotzkow, pers. commun.) Other than a report of the black-cowled oriole (*Icterus dominicensis*) feeding a shiny cowbird young, however, very little is known about the host species use in the western Caribbean.

In Florida, the shiny cowbird was first sighted on Lower Matecumbe Key in 1985 (Smith and Sprunt 1987) (fig. 17.1). Cowbirds were again found in the Florida Keys in 1986 and 1987 (e.g., Islamorada; Paul 1986; Smith and Sprunt 1987). Subsequently, it has been recorded regularly in both the Upper and Lower Florida Keys (fig. 17.1). The first recorded occurrence for mainland Florida was from Flamingo, Everglades National Park, in 1987, where eight individuals, including adults and juveniles, were observed (fig. 17.1). During 1988, a minimum of 35 were seen in eight Florida localities, including two localities outside of south Florida, where males were observed as far north as Fort DeSoto, Pinellas County (Langridge 1988), and Jacksonville (Paul 1988). By 1991 cowbirds had been reported in other Florida localities (fig. 17.2), and as far north as the Carolinas and Maine and as far west as Texas and Oklahoma (fig. 17.1) (LeGrand 1989; Hutcheson and Post 1990; Lasley and Sexton 1991; Grzybowski and Fazio 1991; Perkins 1991; Surner 1992; Post et al. 1993). In the western Caribbean, the shiny cowbird continues its spread (fig. 17.1). In 1993, Fletcher (1993) observed four shiny cowbirds (one male and three females) in Yallahs Salt Pond, southwestern Jamaica, and P. P. Marra (pers. commun.) observed a pair of shiny cowbirds in southwestern Jamaica. In the Bahamas, M. Baltz (pers. commun.) observed a group of three males and two females on Staniard Creek, North Andros during July of 1994.

Several points indicate that shiny cowbirds arrived in Florida from Cuba. These include the fact that in the western Caribbean, the shiny cowbird was first recorded on the Bahamas in 1994, but has been found in Cuba since 1982. Secondly, the point of entry into the United States has been through the Keys, which are approximately 145 km north of Cuba (fig. 17.1).

Localities visited by us further confirm the range extension of shiny cowbirds into Florida, where cowbirds are now observed on a year-round basis. We found cowbirds in residential areas, agricultural areas, and modified habitats in the Keys and on the mainland. In Everglades National Park, cowbirds were most commonly observed on lawns and mangroves around the visitors center, motel, concession areas at Flamingo, and employee residence area where the maximum number seen was 13. Outside of the park, cowbirds were observed in agricultural areas near Florida City and Homestead. As in the West Indian region, where shiny cowbirds are more common in coastal areas (Cruz et al. 1985), the majority of Florida records have been from coastal sites (fig. 17.2).

On the Keys, we observed shiny cowbirds usually at feeders or foraging on lawns in residential areas of Plantation Key, Windley Key, and Key Largo. Additionally, National Audubon Personnel (T. Bancroft, R. Sawicki, A. Strong) have seen male and female shiny cowbirds in Florida Bay on the smaller Keys (e.g., Cowpens Keys), and flying to the larger Keys (e.g., Plantation Key). The Keys are used as breeding areas by several passerines, including black-whiskered vireos (*Vireo altiloquus*), prairie warblers (*Dendroica discolor*), yellow warblers (*D. petechia*), and red-winged blackbirds (*Agelaius phoeniceus*).

Observations of cowbirds flying to the mainland and cowbird feeding aggregations suggest that cowbird foraging and feeding sites may be separated in some instances, as in the case of the cowbirds in the Sierra Nevada (Rothstein et al. 1980, 1984), Illinois (Thompson 1994), and in Pennsylvania (Hoover and Brittingham 1993). Distances traveled between breeding and feeding sites by cowbirds are unknown. Clearly, some birds need not travel at all, because suitable breeding and feeding localities can be intermingled as in the Florida areas.

Shiny cowbirds are most common during April–July, suggesting a migratory population (Post et al. 1993). On Key West, for example, 44 cowbirds were seen on 16 May 1989; on the Dry Tortugas, 52 birds were observed on 24 May (Langridge 1989); and on Marathon Key, a flock of 20 individuals was observed on 17 June 1992.

As of 1994, we know of no substantiated report of parasitism by shiny cowbirds in Florida. We have, however, observed males displaying to females both on the mainland and on the Keys. In addition, juvenile cowbirds have been observed in Everglades National Park and nearby Homestead (P. Smith and A. Sprunt 1978, pers. obs.). In February 1992, National Audubon personnel in Collier County, southwestern Florida, collected four males and three females from a flock of at least 10 birds, suggesting that potentially breeding birds are present in late winter/early spring. On 24 March 1994, P. William Smith (pers. commun.) observed two female shiny cowbirds at their feeders. On 26 March, the females were joined by a singing shiny cowbird. Because fledgling brown-headed and shiny cowbirds are difficult to identify in the field, we believe that confirmation cannot be based on sight records, but will require collection of shiny cowbird eggs or fledglings (Post et al. 1993).

Brown-headed cowbirds

Mayfield (1965) documented the expanded distribution and increased abundance of brown-headed cowbirds in eastern North America over the past two centuries. By the 1800s, cowbirds were well established in many areas of the eastern states, primarily in cultivated areas (Wilson 1810; Bendire 1893; Friedmann 1929). Brown-headed cowbirds were not known to breed in Florida according to Howell (1932) and Sprunt (1954), but since the mid-1950s they have spread rapidly through Florida (Wetson 1965; Robertson and Kushlan 1974; Hoffman and Woolfenden 1986; Stevenson and Anderson 1994; Cruz et al. in press) and now occur as a breeding species in Florida (fig. 17.2).

The first record of the brown-headed cowbird breeding in Florida is a juvenile "caught in hand" in the northwestern region, near Pensacola, Escambia County, on 12 July 1956 (Newman 1957; Sprunt 1963). A specimen taken on 8 June 1957 contained eggs in advanced stage of development, the largest of which would have been laid the next day (Newman 1957). By 1965, cowbirds were reported breeding in the Tallahassee and Jacksonville regions (Ogden 1965), and by 1980 fledglings were reported from different areas in north and north central Florida (Edscorn 1980; Stevenson and Anderson 1994). On 2 June 1985, a fledgling brown-headed cowbird was found in central Florida (Hoffman and Woolfenden 1986).

Similar to shiny cowbirds, brown-headed cowbirds were seen at different sites in the Keys and the mainland, including residential areas, agricultural areas, and other modified habitats (fig. 17.3). In Everglades National Park, we observed brown-headed cowbirds in areas also frequented by shiny cowbirds. Brown-headed cowbirds were frequently seen foraging with shiny cowbirds, red-winged blackbirds, and European starling (*Sturnus vulgaris*). Foraging aggregations, sometimes dozens of birds, were seen feeding on grass seeds, phytophagous insects, and mosquitos. Outside the Park, cowbirds were observed in agricultural areas near Homestead and Florida City. On the Keys, brown-headed cowbirds were not observed as commonly as shiny cowbirds.

Although brown-headed cowbirds have been recorded on a year-round basis in Florida, numbers are augmented from mid-summer on with the arrival of post-breeding flocks from areas north of Florida (Robertson and Kushian 1974). In July–August 1988, at least 1000 cowbirds were observed in Dade County (Neville 1988*a*; Paul 1988). In August 1991, we observed a flock of approximately 170 brown-headed cowbirds in Fort Pierce, just north of Palm Beach County, and approximately 40 in Everglades City, Dade County. In 1992–1994, we located groups of 50–75 cowbirds near Homestead, Dade County. During mid April to late June, we observed groups of cowbirds in Flamingo, Everglades National Park, ranging from six males and 14 females in 1991 to six females and 10 males in 1992. On Sanibel Island, groups ranging from two males and three females to five males and six females were observed in 1994. In 1993 and 1994, we located red-winged blackbird nests that were parasitized by brown-headed cowbirds on Sanibel Island.

Bronzed cowbird (*Molothrus aeneus*)

Historically, the bronzed or red-eyed cowbird was found from extreme south Texas (Rio Grande Valley) through Mexico and Central America to western Panama (Friedmann 1929; Robbins and Easterla 1981). Since the 1900s, the species has expanded its range in the United States and now occurs from California to Louisiana as a breeding species (e.g., Phillips et al. 1964; Niles 1970; Easteria and Wauer 1972; Oberholser 1974; Stewart 1976; Robbins and Easterla 1981; Rosenberg et al. 1991). *M. a. aeneus*, the eastern subspecies, has spread eastward from the Lower Rio Grande Valley (fig. 17.3). Oberholser (1974) noted that its range increased significantly in Texas after large forest tracts along the Rio Grande Valley were converted into agricultural lands. Rappole and Blacklock (1985) note that the bronzed cowbird is a comparatively new species for the Texas Central Coast, and that egg collections taken before 1970 show no evidence of the presence of bronzed cowbirds along the Texas Central Coast.

In Texas, bronzed cowbirds have been recorded breeding from March to September, and have been reported to parasitize over 23 species of hosts (Goldman 1951; Friedmann 1963; Carter 1984, 1986; Friedman and Kiff 1985: Rappole and

Blacklock 1985). The data suggest that *M. aeneus* prefers larger species than *M. ater*. Common host species include northern mockingbirds (*Mimus polyglottos*), long-billed thrasher (*Toxistoma longirostre*), hooded (*I. cucullatus*), Audubon's (*I. graduacuada*) and orchard orioles (*I. spurius*), and northern cardinals (*C. cardinalis*). Brush (1993) noted that hooded, Audubon's, northern, and orchard orioles were once widespread breeders in the Lower Rio Grande Valley. Bronzed cowbirds probably contributed to the decline of Audubon's oriole, which now seems extirpated from portions of the lower Rio Grande Valley. Both bronzed and brown-headed cowbirds may also have contributed to declines in the other three species (Brush 1993).

In Louisiana, the bronzed cowbird was first recorded as a breeding species in the New Orleans area in 1976 (Stewart 1976). The number of breeding individuals has increased since that time, and it now occurs as a breeding species in southeastern Louisiana (Purrington 1991) (fig. 17.3), where it is known to parasitize the orchard oriole (Stewart 1976). Although not recorded as a breeding species, bronzed cowbirds have now been reported during the winter, spring, and summer in Mississippi (Imhof 1987, 1989; Jackson 1989; Purrington 1993). In Alabama, bronzed cowbirds have also been reported in April 1990 and 1992 from Dauphine Island and Fort Morgan (Jackson 1990; Muth 1992).

Prior to 1962, the species was not recorded in Florida (Sprunt 1963). The bronzed cowbird was first reported in 1962 in Sarasota (Truchot 1962), and first collected near Gainesville in 1968 (Matteson 1970). By 1995 there were more than 35 reports from throughout the state. The majority of reports have been from the winter, but some records also for spring through fall (fig. 17.3). The bronzed cowbird is now considered a rare and regular migrant and winter visitor throughout Florida (Stevenson and Anderson 1994). As the bronzed cowbird breeding population in Louisiana is only 20, 100, and 170 km, respectively, from the Mississippi, Alabama, and Florida boundaries, we feel that in the future bronzed cowbirds will become established as a breeding species in Mississippi, Alabama, and Florida.

Cowbird interactions

The association of both brown-headed cowbirds and shiny cowbirds, such as those at foraging aggregations in Everglades National Park, provides an opportunity to observe interactions between these two closely related species that only recently have come into contact. Although the possibility of hybridization exists, only intraspecific courtship behavior similar to that described for the shiny cowbird by Friedmann (1929) and by Laskey (1950), Darley (1978), and Teather and Robertson (1985) for the brown-headed cowbird was seen.

On several occasions, male shiny and brown-headed cowbirds were observed singing accompanied by display (song spread), involving the ruffling of feathers and spreading of wings. In shiny cowbirds, flight displays were also observed. Copulation was observed on two occasions in brown-headed cowbirds in Everglades National Park. A male mounted a female that was in the copulatory posture; that is, her back and neck arched, wings spread and quivering, and tail elevated. On both occasions mounting was preceded by vocalization and bowing.

Aggressive or intimidation displays involving "head down" displays with tails and wings spread and head-up (bill-tilt) displays, a standard icterid threat gesture (Lowther and Rothstein 1980; Orians 1989), and chasing, or combinations of those behaviors, were observed in both cowbird species. Although both species were occasionally observed together, the majority of aggressive encounters were intraspecific, involving males and, rarely, females. Most head-down and bill-up displays by adult brown-headed cowbirds and shiny cowbirds occurred when cowbirds approached one another while feeding in open situations. Among two groups of shiny cowbirds, one consisting of four males and two females and the other of three males, head-up displays and supplanting attacks, involving at least two males, were observed. P. W. Smith and S. A. Smith (pers. commun.) have observed Brown-headed cowbirds chase or displace shiny cowbirds from perches, food, and foraging areas.

POTENTIAL HOSTS IN THE FLORIDA REGION

The presence of these formerly allopatric cowbird species also provides a unique opportunity for understanding the mechanisms of host selection and the ecological requirements of each species. Until now, the data have demonstrated that brown-headed and shiny cowbirds show no preference for particular hosts, although in certain areas some species are more heavily parasitized than others (Post and Wiley 1977*a*).

These generalist parasites, however, do not occur sympatrically elsewhere, and until the present situation in Florida, there was little overlap in potential host species. If both shiny cowbird and brown-headed cowbird eggs appear in a nest, the two parasitic species are likely competitors. Alternatively, resource partitioning may occur, with each cowbird species showing particular preferences to certain hosts or temporal differences in nesting cycles. Differences in habitat preferences between shiny and brown-headed cowbirds may minimize host-use overlap. The shiny cowbird, for example, has been recorded from coastal areas in Florida, whereas brown-headed cowbirds have been found in both coastal and inland areas (fig. 17.2). The shiny cowbird breeding season in the West Indian region also extends from mid-March to September, whereas the brown-headed cowbird breeds from May to early July (Wiley 1985; Cruz et al. 1989; Marvil and Cruz 1989), suggesting that populations that breed in July and August will be parasitized by shiny cowbirds. Based on host use in South America and the West Indies, cavity-nesting species (e.g., *M. antillarum* and *T. ludovicianus*) and some mangrove-nesting species (e.g., *A. phoeniceus*) in south Florida, might be more heavily parasitized by shiny cowbirds than brown-headed cowbirds (Friedmann et al. 1977; Manolis 1982; Friedmann and Kiff 1985; Wiley 1985; Mason 1986*a*,*b*).

Many factors are important in the selection of hosts by cowbirds in a given community. A species may be unsuitable because it breeds during different seasons than the cowbirds, nests in habitats that are not used by the parasite, or has a diet that is physiologically incompatible with the needs of the parasite young (Eastzer et al. 1980; Wiley 1982; Cruz et al. 1989). Nesting cowbirds require animal protein in their diet, and those hosts that feed the young seeds are not suitable hosts (Eastzer et al. 1980; Mason 1986*a*,*b*). Of the species that are potentially capable of rearing parasite young, additional measures of suitability are important. These measures are generally partitioned into components of temporal and spatial availability of hosts: (i) accessibility of the egg-laying female cowbird to the nest (Rothstein 1975; Robertson and Norman 1977; Wiley 1982; Cruz et al. 1985; Ortega and Cruz 1991); (ii) cowbird egg acceptance (Rothstein 1975; Friedmann et al. 1977; Cruz et al. 1985); and (iii) high likelihood of cowbird fledgling success (Manolis 1982; Fraga 1985; Mason 1986*a*,*b*; Ortega and Cruz 1991).

Given the above criteria for host suitability, information on shiny cowbird and host interactions in the West Indies, comparable information on brown-headed and bronzed cowbirds and host interactions in North America, and the breeding biology of Florida birds, we found that a number of potential hosts are present in Florida (table 17.1). We excluded species determined to be unsuitable hosts based on literature review and our field surveys, including egg rejection experiments (Wiley 1985; Cruz et al. 1989; Post et al. 1990). Subspecific names are given for those North American species that have Florida races or for those West Indian species that also breed in Florida.

Eastern wood-pewee (*Contopus virens*)

The eastern wood-pewee is a fairly common to uncommon breeder in northern Florida (Kale and Maehr 1990; Stevenson and Anderson 1994). Friedmann (1929) considered the eastern wood-pewee as a frequent victim of brown-headed cowbird parasitism, and is also a likely cowbird host in Florida.

Acadian flycatcher (*Empidonax virescens*)

The Acadian Flycatcher is an uncommon to fairly common summer resident in the panhandle and south to north-central Florida (Kale and Maehr 1990; Robertson and Woolfenden 1992; Stevenson and Anderson 1994). Outside of Florida, the Acadian flycatcher is an infrequent cowbird host, although in Illinois, Graber et al. (1974) reported a 50% (6/12 nests) rate of parasitism, and S. Robinson (unpubl. data) reported a 40% (84/210) rate of parasitism. The Acadian flycatcher therefore is a likely cowbird host in Florida.

Great crested flycatcher (*Myiarchus crinitus*)

The great crested flycatcher occurs throughout Florida from late March until early fall. Some birds remain in extreme south Florida year-round (Kale and Maehr 1990). Outside of Florida, this cavity-nesting flycatcher is infrequently parasitized by brown-headed cowbirds. Only 12 incidences of parasitism have been reported, including three of 201 nest records from Ontario (Friedmann et al. 1977; Friedmann and Kiff 1985). Friedmann et al. (1977) felt that because brown-headed cowbirds infrequently parasitized cavity-nesting species, great crested flycatchers are low-suitability hosts for brown-headed cowbirds. Shiny cowbirds, on the other hand, frequently parasitize cavity-nesting species (Friedmann et al.

Table 17.1 Suitability of potential hosts in Florida to shiny cowbird and brown-headed cowbird parasitism

Species		Vulnerability[a]		Host[b] accessibility
		BHC	SC	
Eastern wood-pewee	*Contopus virens*	M	L	I-BHC
Acadian flycatcher	*Empidonax virescens*	M-H	L	I-BHC
Great crested flycatcher	*Myiarchus crinitus*	L	M	CN-SC
Carolina wren	*Thryothorus ludovicianus*	L	M	CN-SC
Blue-gray gnatcatcher	*Polioptila caerulea**	L-M	L-M	
Wood thrush	*Hylocichla mustelina*	M	L	I-BHC
White-eyed vireo	*Vireo griseus**	M-H	H	ENS-SC
Yellow-throated vireo	*Vireo flavifrons*	M	L	I-SC
Red-eyed vireo	*Vireo olivaceus**	M-H	M-H	
Black-whiskered vireo	*Vireo altiloquus**	M-H	H	C-SC
Northern parula	*Parula americana**	L	L	
Yellow warbler	*Dendroica petechia*	M	H	C-SC
Yellow-throated warbler	*Dendroica dominica**	L	VL	I-BHC
Pine warbler	*Dendroica pinus**	L	L	
Prairie warbler	*Dendroica discolor**	M	M-H	C-SC
American redstart	*Setophaga ruticilla*	M	L	I-BHC
Prothonotary warbler	*Protonotaria citrea*	L	M	CN-SC
Worm-eating warbler	*Helmitheros vermivorus*	M	L	I-BHC
Swainson's warbler	*Limnothlypis swainsonii*	L	VL	I-BHC
Louisiana waterthrush	*Seiurus motacilla*	L	VL	I-BHC, ENS-SC
Kentucky warbler	*Oporonis formosus*	M	VL	I-BHC
Common yellowthroat	*Geothlypis trichas**	M-H	M-H	
Hooded warbler	*Wilsonia citrina*	L	VC	I-BHC
Yellow-breasted chat	*Icteria virens*	L	VL	I-BHC, NDP
Summer tanager	*Piranga rubra**	M-H	L	I-BHC
Northern cardinal	*Cardinalis cardinalis**	H	H	
Blue grosbeak	*Guiraca caerulea*	L-M	VL	I-BHC
Indigo bunting	*Passerina cyanea*	H	L-M	I-BHC
Rufous-sided towhee	*Pipilo erythrophthalmus**	H	H	
Bachman's sparrow	*Aimophala aestivalis*	?	?	DN-SC
Chipping sparrow	*Spizella passerina*	H	L	I-BHC
Field sparrow	*Spizella pusilla*	L-M	VL	I-BHC
Grasshopper sparrow	*Ammodramus savannarum*	L-H	VL	I-BHC
Cape Sable seaside sparrow	*Ammodramus maritimus*		(see text)	C-SC
Eastern meadowlark	*Sturnella magna*	H	M	I-BHC
Common grackle	*Quiscalus quiscula*	VL	VL	ND
Red-winged blackbird	*Agelaius phoeniceus**	M-H	M-H	
Orchard oriole	*Icterus spurius*	H	M	
Spot-breasted oriole	*Icterus pectoralis*	L	L-M	C-SC

Host vulnerability is based on frequency of parasitism, including Florida, of that species or related species and host accessibility. Host accessibility includes factors that will make a host more accessible to one of the cowbird species. Those species that have been recorded as cowbird hosts in Florida are shown with an asterisk. Data based on field surveys, Florida Breeding Bird Atlas Project, and review of the literature (e.g., Sprunt 1954; Robertson 1955; Friedmann et al. 1977; Cruz et al. 1985; Friedmann and Kiff 1985; Wiley 1985; Cruz et al. 1989; Kale and Maehr 1990; Robertson and Woolfenden 1992; Stevenson and Anderson 1994; *Audubon Field Notes* and *American Birds*).

[a]VL = very low; L = low; M = moderate; H = high.

[b]For example, I-BHC = Inland (mainly)-brown-headed cowbirds—host more accessible to brown-headed cowbird parasitism because of preference for inland habitats in Florida; C-SC = Coastal (mainly)-shiny cowbird; CN-SC = Cavity nester-shiny cowbird; ENS-SC = Extended nesting season-shiny cowbird; NDP = Nest desertion of parasitized nests; ND = Nest defense; DN = domed nest.

1977; Cruz et al. 1985; Wiley 1985; Post et al. 1990), including the congeneric Puerto Rican flycatcher (*M. antillarum*), which has a parasitism rate of 34% (16/47) in Puerto Rico (Cruz et al. 1989). We thus feel that the great crested flycatcher will be a potential host in south Florida for the shiny cowbird, but possibly of low or marginal suitability for brown-headed cowbirds.

Carolina wren (*Thryothorus ludovicianus*)

Friedmann and Kiff (1985) state that nowhere is the Carolina wren a frequent host choice of the brown-headed cowbird, possibly because of its cavity-nesting habits, although in Illinois, S. Robinson (pers. commun.) found a 15% rate of parasitism. The shiny cowbird, however, is known to parasitize cavity-nesting species. In Trinidad, Tobago, and South America, the cavity-nesting house wren (*Troglodytes aedon*) is a frequent host of the shiny cowbird (Friedmann et al. 1977; Manolis 1982; Kattan 1993, this vol.) Accordingly, we feel that shiny cowbirds will have a greater impact on this species than brown-headed cowbirds. A record of the Carolina wren as a host for the bronzed cowbird in Texas also exists (Williams 1976).

Blue-gray gnatcatcher (*Polioptila caerulea*)

The blue-gray gnatcatcher breeds from the panhandle to south Florida, but it is absent from extreme south Florida (Stevenson and Anderson 1994). The gnatcatcher is a fairly well-documented host of the brown-headed cowbird (39 records in Friedmann 1963). Three records of cowbirds parasitizing gnatcatchers exist for Florida, each reporting gnatcatchers feeding fledgling cowbirds. A fledgling cowbird was observed being fed by a pair of gnatcatchers in Jacksonville (Duval County) in 1984 and in 1988. In 1990, a fledgling cowbird was observed being fed by gnatcatchers in Levy County (Paul 1984; H. Kale and W. Pranty, pers. commun.). We thus expect continued use of this species by brown-headed cowbirds in the Florida region.

Wood thrush (*Hylocichla mustelina*)

The wood thrush is confined primarily to northern Florida, but is nowhere abundant in Florida (Kale and Maehr 1990; Stevenson and Anderson 1994). Outside of Florida, levels of cowbird parasitism are not homogeneous over large geographic areas (Robinson et al. 1993). Rates of parasitism on wood thrushes differed significantly ($P < 0.001$) among the Midwest (42.1%), Mid-Atlantic (26.5%), and Northeast (14.7%) (Hoover and Brittingham 1993). We thus expect this species to be parasitized in Florida.

Mockingbird (*Mimus polyglottos*)

The northern mockingbird has been recorded infrequently as a host for brown-headed and shiny cowbirds. In egg rejection experiments, the species is known to be a rejecter (Cruz et al. 1985, unpubl. data). Brown-headed cowbirds and West Indian shiny cowbird eggs are strikingly different in coloration from mockingbird eggs. These cowbird species lay eggs that are off-white and speckled with reddish brown uniformly distributed over the egg (Ortega and Cruz 1988; Cruz and Wiley, 1989). Mockingbirds lay eggs from light blue green to dark blue green speckled with brown spots and blotches (pers. obs.). Carter (1984, 1986), however, found that the mockingbird in Texas accepts bronzed cowbird eggs and that it had a parasitism frequency of 43% (16/37 nests). The bluish green eggs of bronzed cowbirds resemble in background color those of the mockingbird host, possibly facilitating parasitism by bronzed cowbirds.

White-eyed vireo (*Vireo griseus*)

The white-eyed vireo is a common resident of Florida, where two subspecies occur, *V. g. griseus* on the mainland, and *V. g. maynardi* on the Keys. This species has been regarded as a fairly frequent host of brown-headed cowbirds in other parts of the United States (Friedmann 1963; Friedmann and Kiff 1985). In Louisiana, Goertz (1977) found that 40% (6/15) of vireo nests contained cowbird eggs, and in Illinois, S. Robinson (pers. commun.) found that 45% (14/32) of nests were parasitized. In Okaloosa County, north Florida, Stevenson (1963) reported a white-eyed vireo nest with one cowbird egg and three vireo eggs in June 1963, and another such instance was reported near Tallahassee (Stevenson and Anderson 1994). In Pasco County, central Florida, a fledgling cowbird was observed being fed by a white-eyed vireo (H. Kale and W. Pranty, pers. commun.). S. Robinson (pers. commun.) reported that white-eyed vireos in Illinois continue nesting after cowbirds are gone in late July and August, but in Florida they breed from April through July (Stevenson and Anderson 1994), overlapping the breeding season of both brown-headed and shiny cowbirds (Cruz et al. 1989).

In Puerto Rico, the closely related Puerto Rican vireo (*V. latimeri*) is heavily parasitized; Pérez-Rivera (1986) recorded a rate of parasitism of 36.3% (4/11), and we found an 87.2% (34/39) rate of parasitism. We thus feel that white-eyed vireos will be an important cowbird host in Florida for both shiny and brown-headed cowbirds.

Yellow-throated vireo (*Vireo flavifrons*)

This species is an uncommon to fairly common summer resident in the panhandle south to north-central Florida (Kale and Maehr 1990; Robertson and Woolfenden 1992; Stevenson and Anderson 1994). This vireo is known to be a brown-headed cowbird host with a reported rate of parasitism of 32.7% (17 of 52 nests) for Toronto (Friedmann et al. 1977). We thus feel that this species will serve as a cowbird host in Florida.

Red-eyed vireo (*Vireo olivaceus*)

In Florida, red-eyed vireos are common throughout northern and central Florida. They breed less frequently in south Florida and are absent from the Keys (Robertson and Kushlan 1974; Kale and Maehr 1990). Outside of Florida, red-eyed vireos are very commonly recorded as cowbird hosts. Friedmann (1963) and Friedmann et al. (1977) reported more than 875 nest records of cowbird parasitism on vireos. Norris (1947) reported 85.8% (12/14) of vireo nests parasitized in Pennsylvania. Southern (1958) reported 72.1% (75/104) of nests parasitized in Michigan, and S. Robinson (pers. commun.) found 75% (18/24) of nest, parasitized in Illinois. In July 1986, a parasitized vireo nest was reported near Gainesville, north-central Florida (Paul 1986). Thus, we feel that Florida red-eyed vireos will become a common cowbird host and should receive special monitoring in the years to come.

Black-whiskered vireo (*Vireo altiloquus altiloquus*)

The black-whiskered vireo, a predominantly West Indian species, occurs in North America only in Florida, where it is considered rare (Kale 1978). It breeds in the Florida Keys and along the coastline of Florida irregularly north to Tampa Bay on the west coast and New Smyrna Beach on the east coast. In Florida, it is confined to coastal mangroves, although south Florida populations also occur in subtropical hammocks (Robertson 1955; Robertson and Kushlan 1974; pers. obs.). In southern Florida, we found this species to be more common in subtropical hammocks than in mangrove habitats.

In our West Indian study sites, we found black-whiskered vireos to be common hosts for shiny cowbirds, with rates of parasitism of 88%, 73%, and 35% for St. Lucia, Puerto Rico, and Hispaniola, respectively (Wiley 1985; Cruz et al. 1989; Post et al. 1990). Clutch sizes of vireos were significantly smaller ($P < 0.001$, Mann–Whitney two-sample test; one-tailed) in parasitized (2.22 ± 0.83) than in nonparasitized nests (3.00 ± 0.0), as was host fledgling success (0.33 ± 0.05 vs. 2.00 ± 0.0; $P = 0.025$, Mann–Whitney two-sample test; one-tailed; Cruz et al. 1985).

In 1989, the first reported incidence of parasitism in black-whiskered vireo for North America was recorded (Kale 1989). A vireo was observed feeding a cowbird fledgling, identified as a brown-headed cowbird, in mangroves at Flamingo, Everglades National Park. The black-whiskered vireo proved very scarce in the Tampa Bay area, 1988–89, suggesting to local observers the possibility that they were victimized by the invading brown-headed cowbirds, but no direct relationship has been discovered to date (Stevenson and Anderson 1994). We predict that in the future cowbird parasitism will become an important threat to the survival of black-whiskered vireos in southern Florida.

Northern parula (*Parula americana*)

The northern parula is a breeding summer resident throughout all but the southernmost part of Florida (Robertson and Kushlan 1974; Anderson and Stevenson 1994). Although northern parulas have been recorded infrequently as cowbird hosts, possibly because the nest is well concealed and the semi-cavity pouch nest is very hard to enter (Friedmann 1929; S. Robinson pers. commun.), there are several records for Florida. In northern Florida, a parula was observed feeding a fledgling brown-headed cowbird in Pensacola in 1983, and a parula feeding a fledgling brown-headed cowbird was also reported from Hamilton County in 1988 (Stewart 1963; H. Kale and W. Pranty, pers. commun.). In central Florida, three instances of fledgling cowbirds being fed by parulas were reported, one from Volusia County in 1987 and two from Alachua County in 1988 (H. Kale and W. Pranty, pers. commun.). It appears from the data that the northern parula will continue to serve as a cowbird host in Florida.

Cuban yellow warbler (*Dendroica petechia gundlachii*)

Cuban yellow warblers are found in extreme south Florida, Cuba, Isle of Pines, and throughout the Bahamas. The Florida population has great biological interest because it represents an authentic case of recent, successful colonization of the Florida Keys by a West Indian bird (Robertson 1978; Prather and Cruz 1995). Its present distribution appears to be rather patchy, the birds occurring more commonly in mangrove habitat in the low outlying Keys than on mainline Keys or along the mainland shore (Robertson 1978; pers. obs.). The habitat of the Cuban yellow warbler in Florida is relatively secure as most of its habitat is found within Everglades National Park, Biscayne National Park, or the National Wildlife Refuge of the lower Florida Keys. Its limited range in Florida, however, makes it vulnerable to natural catastrophes such as severe hurricanes. The destruction of mangrove outside of protected areas has also reduced potential habitat in Florida.

The very high rate of parasitism reported for yellow warblers outside of Florida suggests that parasitism by cowbirds will probably be the most significant threat to these populations. In the West Indies, the yellow warbler is an important shiny cowbird host, with rates of parasitism of 55.2%, 63%, and 39% for St. Lucia, Puerto Rico, and Hispaniola, respectively (Cruz et al. 1989). Clutch sizes of warblers were significantly smaller ($P < 0.001$) in parasitized (2.0 ± 0.83) than in nonparasitized nests (2.65 ± 0.0) (Cruz et al. 1985; Wiley 1985). In North America, other subspecies are also heavily parasitized by brown-headed cowbirds (Friedmann et al. 1977; pers. obs.), and we anticipate that yellow warblers will be used as hosts in Florida (Friedmann et al. 1977; pers. obs.). In Illinois the available data suggest that the incidence of parasitism of yellow warbler nests has increased from about 6% before 1900 to 40.5% after 1900 (Graber et al. 1983). Cornell University nesting records show that 22.4% (115 of 313) of yellow warbler nests reported were parasitized (Friedmann et al. 1977), and Clark and Robertson (1979*a*) found 45% (41 of 109) parasitized in Ontario. The yellow warbler was a relatively uncommon breeding species in mangrove habitats in southern Florida. It was found only on small islets in Florida Bay and on a few of the lower Florida Keys. Thus far, we found no incidence of parasitism in the 19 active yellow warbler nests that we examined in Florida Bay and on the Keys. It is possible that once yellow warbler populations are parasitized in Florida they may evolve nest attentiveness, nest desertion, and nest burial as a strategy against brood parasitism (Clark and Robertson 1981; Burgham and Picman 1989; Hobson and Sealy 1989; Nakamura 1995).

Yellow-throated warbler (*Dendroica dominica*)

A rare to fairly common breeder in the panhandle south to central Florida (Kale and Maehr 1990; Stevenson and Anderson 1994), the yellow-throated warbler is an infrequent cowbird host (Friedmann and Kiff 1985). In Gainesville, northern Florida, a yellow-throated warbler was observed feeding a juvenile cowbird in June 1976. We thus expect that this species will be an infrequently used host in Florida.

Pine warbler (*Dendroica pinus florida*)

Pine warblers are year-round residents of the Florida mainland (Kale and Maehr 1990). Throughout its North American range, the pine warbler is considered as an infrequent cowbird host, with only 15 records (Friedmann and Kiff 1985). However, in 1988, a pine warbler was observed feeding a fledgling cowbird in Duval County, north Florida (H. Kale and W. Pranty, pers. commun.). Accordingly, we feel that it will probably be an infrequent host in south Florida.

Prairie warbler (*Dendroica discolor*)

The prairie warbler is a rare to fairly common breeding species in two disjunct sections of the state. The Florida prairie warbler (*D. d. paludicola*) breeds on the Florida Keys and northward along the coast to Volusia County on the Atlantic and Levy County on the Gulf. This subspecies is primarily restricted to mangrove habitats and is considered a species of special concern on the basis of its dependence on mangroves (Kale 1978). In recent decades the breeding range of the northern race, *D. d. discolor*, has pushed into northern Florida, where it has been recorded breeding from the panhandle to the Atlantic coast (Robertson and Woolfenden 1993; Stevenson and Anderson 1994).

Nolan's (1978) study of the nominate race of the prairie warbler in Indiana reveals that 27.4% (92/336) of nests were parasitized by brown-headed cowbirds. It should be noted that Nolan's studies were conducted in what has been generally considered the "original" habitat of the cowbird, where cowbirds and prairie warblers have coexisted for many years. Nolan also noted that parasitized nests were often abandoned.

In 1987 and in 1988, Atherton and Atherton (1988) found Florida prairie warblers feeding fledgling cowbirds in Pinellas and Sarasota Counties, confirming parasitism of this species in Florida. In 1989, a cowbird fledgling was observed being fed by a prairie warbler in Everglades National Park (Kale 1989), and in 1990, a prairie warbler was observed feeding a brown-headed cowbird young on Captiva Island, southwest Florida (V. C. McGrath, pers. commun.). In the Florida Keys, we have found no evidence of parasitism in the nests examined (n = 42) (Prather and Cruz 1995). However, this species should be carefully monitored in the years to come.

American redstart (*Setophaga ruticilla*)

The American redstart nests in the western panhandle, where it is an uncommon summer resident (Kale and Maehr 1990; Robertson and Woolfenden 1992; Stevenson and Anderson 1994). Graber et al. (1983) recorded a parasitism rate after 1900 of 37% (15/41). We thus expect that the redstart will be a frequent cowbird host.

Prothonotary warbler (*Protonotaria citrea*)

The prothonotary warbler is a rare to fairly common summer resident throughout all but the southernmost part of the Florida mainland (Robertson and Kushlan 1974; Toops and Dilley 1986; Kale and Maehr 1990). Unlike the majority of other hole-nesters in North America that are infrequently parasitized, prothonotary warblers suffer a relatively high incidence of parasitism by brown-headed cowbirds (Friedmann and Kiff 1985; Petit 1989, 1991). Petit (1989) observed a 20.9% rate of parasitism of 191 nests. We feel that this species is a potential cowbird host, especially for the shiny cowbird that parasitizes cavity nesters to a greater frequency than the brown-headed cowbird (Friedmann and Kiff 1985).

Worm-eating warbler (*Helmitheros vermivorus*)

The worm-eating warbler is known to breed in Florida only in northern Okaloosa County in the western panhandle (Stevenson and Anderson 1994). Robinson found that the worm-eating warbler was parasitized 41% of the time in Illinois (n = 27). We thus expect that this species will be a cowbird host in Florida.

Swainson's warbler (*Limnothlypis swainsonii*)

The Swainson's warbler is thought to be limited to the Florida panhandle as a breeding species (Robertson and Woolfenden 1992; Stevenson and Anderson 1994). Regarding nest parasitism of Swainson's warbler by brown-headed cowbirds, Friedmann (1963) considered it unusual and remarked that the warbler's breeding range was largely outside that of the cowbird. Since that time, the brown-headed cowbird's range has extended into the southeastern United States, including the Florida panhandle. Records of parasitism now exist for Louisiana, Mississippi, North Carolina, South Carolina, Oklahoma, and West Virginia (Friedmann et al. 1977; Friedmann and Kiff 1985). Meanley (1982) notes that in Virginia and North Carolina "it is a late nester and is apt to escape cowbird (brown-headed) parasitism." In Florida, Swainson's warbler nests were found in May and June and singing males were observed from late June to August (Stevenson and Anderson 1994). We thus expect that this species will be recorded as a cowbird host in Florida.

Louisiana waterthrush (*Seiurus motacilla*)

The Louisiana waterthrush breeds in the panhandle, and perhaps irregularly, south to north-central Florida (Stevenson and Anderson 1994). Friedmann (1929) found this species to be frequently parasitized by the brown-headed cowbird. S. Robinson (unpubl. data) found a frequency of parasitism of 20% in Illinois (3/15). We thus expect that the waterthrush will be a cowbird host in Florida.

Kentucky warbler (*Oporornis formosus*)

The Kentucky warbler breeds in the panhandle and extreme northern peninsula (Stevenson and Anderson 1994). The Kentucky warbler was considered by Friedmann (1929) to be a frequent victim of cowbird nest parasitism. S. Robinson (unpubl. data) found that the Kentucky warbler had a 40% (40/100) rate of parasitism in Illinois. Thus, we feel that this species will be a cowbird host in Florida.

Common yellowthroat (*Geothlypis trichas*)

The common yellowthroat is a common nesting species throughout mainland Florida. In Illinois, frequency of nest parasitism on this species ranged from 16.7% (5/30) to 75% (12/16) (Graber et al. 1983; S. Robinson, pers. commun.), in Ohio 46.3% (19/41), and in Michigan 38.8% (35/90) (Friedmann 1963). In 1987, a fledgling brown-headed cowbird being fed by a common yellowthroat was observed in Pasco County, central Florida (H. Kale and W. Pranty, pers. commun.). Thus, we feel that this species will be a common cowbird host.

Hooded warbler (*Wilsonia citrina*)

The hooded warbler is a fairly common breeder in the panhandle south to central Florida (Kale and Maehr 1990; Stevenson and Anderson 1994). There is some uncertainty concerning the role of brown-headed cowbirds as a parasite of hooded warbler nests. Friedmann (1929) considered it an uncommon cowbird victim, but A. B. Williams (in Bent 1953) asserted, "the cowbird is a frequent visitor to hooded warblers' nests, slightly over 50% of nests found with eggs or young, containing from one to three cowbird's eggs or young." We thus expect that the hooded warbler will be a potential cowbird host.

Yellow-breasted chat (*Icteria virens*)

The chat is an uncommon to fairly common summer resident in north Florida (Stevenson and Anderson 1994). Friedmann (1963) notes that the chat is imposed upon frequently by the brown-headed cowbird, but the degree to which it is affected appears to vary locally. In some areas it is said to be one of the main victims; in other areas it is an occasional host. Friedmann (1929) and S. Robinson (pers. commun.) note that the chat is known to desert its nest if parasitized. This species has also been recorded as a bronzed cowbird host, Friedmann et al. (1977) list five records from south Texas and Arizona. The two Arizona nests each contained eggs of the chat, bronzed cowbird, and brown-headed cowbird. Thus, we feel that this species will be a cowbird host in Florida.

Summer tanager (*Piranga rubra*)

The summer tanager is a rare to fairly common breeding species in the panhandle and peninsula south to south-central Florida (Robertson and Woolfenden 1992). Friedmann et al. (1977) concluded that the summer tanager was an infrequently reported host. However, Mengel (1965) considered it the most often chosen cowbird host in Kentucky, seven of 17 nests parasitized (41.25%). In Illinois, S. Robinson (unpubl. data) found a 71% (5/7) rate of parasitism in Illinois. In Florida, a parasitized tanager nest was found near Gainesville (Paul 1986). Accordingly, we feel that it will be a cowbird host in Florida.

Northern cardinal (*Cardinalis cardinalis*)

The cardinal is a fairly common to common resident throughout Florida, including the Keys, where two subspecies occur (Robertson and Woolfenden 1992; Stevenson and Anderson 1994). In other parts of its range, this species appears to be a common host of brown-headed cowbirds. Scott (1977) found that 84% (85/105) of nests examined in Ontario were parasitized. Mason (in Friedmann and Kiff 1985) found that of 52 nests examined in Texas, 42 (81%) were parasitized. In Illinois, S. Robinson (unpubl. data) found a 40% (53/133) rate of parasitism. S. Robinson (pers. commun.) suggests that the long breeding season of this species allows some individuals to escape parasitism in Illinois. In Florida, cardinals breed from March through September (Stevenson and Anderson 1994), overlapping the breeding season of both brown-headed and shiny cowbirds (Cruz et al. 1989). In northwestern Florida, a cardinal feeding a brown-headed cowbird was observed near Pensacola (Escambia County) in 1959, and a cardinal feeding a cowbird was also observed in Washington County in 1987 (Stevenson 1959; H. Kale and W. Pranty, pers. commun.). In south Florida, a fledgling cowbird being fed by a cardinal was reported from Corkscrew (Collier County) in June 1989 (Paul 1989). The northern cardinal has also been recorded as a bronzed cowbird host in Texas (Friedmann 1963), and Carter (1986) reported that two of five parasitized nests also contained brown-headed cowbird eggs. We feel that this common nesting species needs careful monitoring.

Blue grosbeak (*Guiraca caerulea*)

The blue grosbeak is a rare to locally common breeding bird in the panhandle and south to central Florida (Robertson and Woolfenden 1992; Stevenson and Anderson 1994). Friedmann (1963) found the species to be a rather frequent host to brown-headed and bronzed cowbirds. In Florida, one cowbird egg was found in the abandoned nest of a grosbeak with one host egg in Seminole county, central Florida (Stevenson and Anderson 1994). We thus feel that this species will be a cowbird host in Florida.

Indigo bunting (*Passerina cyanea*)

The indigo bunting is a common breeding bird throughout the panhandle and, more sparingly, in the peninsula south to at least north-central Florida (Stevenson and Anderson 1994). Friedmann (1963) listed the species as a very frequent victim of cowbird parasitism. In Illinois, Robinson found a 43% (16/37) rate of parasitism. We thus feel that this species will be an important cowbird host.

Rufous-sided towhee (*Pipilo erythrophthalmus*)

The rufous-sided towhee is a fairly common to common resident of mainland Florida, but it is not known to breed in the Keys (Robertson and Woolfenden 1992). Three subspecies of towhees breed in Florida, and one, *P. e. alleni*, is primarily confined to Florida (Howell 1932). The rufous-sided towhee is frequently reported as a brown-headed cowbird host. Friedmann (1971) listed 328 cases of parasitism of towhees and categorized them as "frequently" rearing cowbirds. Of the 29 host species recorded in the Carolinas, the one most frequently reported is the rufous-sided towhee (25% of nests) (Potter and Whitehurst 1981). In Illinois, S. Robinson (pers. commun.) found that 50% (33/66) of nests were parasitized, and that those individuals that nest late in the season escape parasitism. In Florida, towhees breed from April to September (Stevenson and Anderson 1994), suggesting that they will be potential hosts for both brown-headed and shiny cowbirds. In Gainesville, north-central Florida, an immature brown-headed cowbird was observed being fed by a female towhee in July 1980 (Edscorn 1980), and in Hernando County, central Florida, a towhee nest with a cowbird egg was found in 1990 (H. Kale and W. Pranty, pers. commun.). Thus, we feel that towhees will be an important cowbird host in Florida.

Bachman's sparrow (*Aimophala aestivalis*)

Bachman's sparrow is an uncommon resident of mainland habitats of Florida south to the Lake Okeechobee region (Kale and Maehr 1990). Friedmann (1929) found this species to be an uncommon cowbird host. However, the breeding biology of this species has not been well studied in any part of its range. Thus, predictions as to how this species will be affected by cowbird parasitism cannot be made at this time. However, the domed nest may make this species vulnerable to shiny cowbird parasitism (S. Robinson, pers. commun.).

Chipping sparrow (*Spizella passerina*)

The chipping sparrow is a rare, irregular breeding bird in northwestern Florida (Stevenson and Anderson 1994). Outside of Florida, the frequency of brood parasitism by the brown-headed cowbird is high. In Ohio, 60 of 115 nests (50%) were parasitized, and in Michigan 36.2% (30/83) were parasitized (Friedmann et al. 1977). We thus feel that this species will be parasitized in Florida.

Field sparrow (*Spizella pusilla*)

The field sparrow is a rare to uncommon breeding resident locally in the panhandle and north Florida (Kale and Maehr 1990; Robertson and Woolfenden 1992; Stevenson and Anderson 1994). Outside of Florida, the field sparrow is a frequent victim of the cowbird, with over 125 records of parasitism (Friedmann 1963; Friedmann et al. 1977). Accordingly, we feel that it will be a cowbird host in Florida.

Florida grasshopper sparrow (*Ammodramus savannarum floridanus*)

Although the species as a whole breeds over most of the United States and some parts of Canada, the Florida race is restricted to a small portion of the south-central peninsula (Delany et al. 1985). Hill (1976) found 22% (4/18) of nests parasitized in Kansas; Elliott (1978) found 50% (9/18) of nests parasitized in Kansas; and in Minnesota, Johnson and Temple (1990) found 71% (30/42) of nests parasitized. The high frequency of cowbird parasitism on this species suggests that Florida grasshopper sparrow populations ought to be carefully monitored.

Cape Sable seaside sparrow (*Ammodramus maritima*)

The seaside sparrow is a locally distributed breeding resident of salt and brackish marshes of Florida, where seven subspecies are known to occur (Kale and Maehr 1990; Robertson and Woolfenden 1992; Stevenson and Anderson 1994). The endangered Cape Sable seaside sparrow (*A. m. mirabilis*) is known from only three population centers in south Florida. The Taylor Slough population in Everglades National Park represents virtually all of the approximately 1000–2000 sparrows believed to exist (Werner 1976). Among the southern Florida cowbird sightings was that of an adult male brown-headed cowbird in the midst of the Cape Sable sparrow colony near Mahogany Hammock in June 1988 (Neville 1988*b*). Although no information is available on seaside sparrows as cowbird hosts, the related grasshopper sparrow (*A. savannarum*) has been recorded as a host. Other sparrow species (e.g., dickcissel, *Spiza americana*; Savanna sparrow, *Passerculus sandwichensis*; vesper sparrow, *Pooecetes gramineus*; lark sparrow, *Chondestes grammacus*; chipping sparrow, *Spizella passerina*; clay-colored sparrow, *S. pallida*; and song sparrow, *Melospiza melodia*) are also parasitized by brown-headed cowbirds (Friedmann 1929, 1963; Hergenrader 1962; Wiens 1963; Elliott 1978; Zimmerman 1983; Johnson and Temple 1990). Thus, the Cape Sable seaside sparrow, as well as other subspecies, warrant special attention in the years to come.

Eastern meadowlark (*Sturnella magna*)

The eastern meadowlark is resident throughout mainland Florida. While the nest, which is a natural or scraped depression lined with grasses and with a domed-shaped roof, is difficult to find, species in the genus *Sturnella* are parasitized by shiny cowbirds in South America, and eastern meadowlarks are parasitized by brown-headed cowbirds in North America (Friedmann 1929, 1963; Friedmann et al. 1977; Elliott 1978; Gochfeld 1979). Bowen (1976) and Elliott (1978) found a high rate of parasitism for this species; 93% (14/15) of nests and 70.0% (28/40) of nests were parasitized, respectively. Accordingly, we expect the eastern meadowlark to be parasitized by cowbirds in Florida.

Common grackle (*Quiscalus quiscula*)

The common grackle has been rarely reported as a brown-headed cowbird host (Friedmann et al. 1977; Friedmann and Kiff 1985). Egg rejection experiments on this species by Rothstein (1975), however, have shown this species to be an accepter species. Therefore, the small number of records of parasitism cannot result from rejection behavior by the host, but must reflect an actual absence of parasitism by brown-headed cowbirds (Friedmann et al. 1977). The related greater Antillean grackle (*Q. niger*) and Carib grackle (*Q. lugubris*) have been recorded as uncommon shiny cowbird hosts in the West Indies. In Puerto Rico, parasitism of the Antillean grackle was only 9% (25/281), and in St. Lucia, parasitism of the Carib grackle was 1.7% (2/115) (Cruz et al. 1985; Wiley 1985; Post et al. 1990). In Venezuela, 30% (6/20) of Carib grackle nests were parasitized (Ramo and Busto 1980); however, four Carib grackle nests found parasitized by Cruz and Andrews (1989) had been abandoned. In contrast to common grackles, rejection of cowbird eggs by Antillean grackles was noted in 32 of 36 trials (88.9%) in Puerto Rico (Cruz et al. 1989), and by Carib grackles in 19 of 30 trials (63.3%) in St. Lucia (Post et al. 1990). The low frequency of parasitism by cowbirds on grackles is probably related to vigilant nest defense by grackles combined with colonial breeding (Cruz et al. 1985; Carello, 1993; Nakamura 1995; S. Robinson, pers. commun.) We thus expect that the common grackle will serve as an infrequent cowbird host in Florida.

Red-winged blackbird (*Agelaius phoeniceus*)

The red-winged blackbird is a permanent resident over the entire state except the Dry Tortugas and some other small islets (Stevenson and Anderson 1994). Four subspecies of red-winged blackbirds breed in Florida, including two subspecies confined to Florida, *A. p. floridanus* and *A. p. mearnsi*. Despite vigilant nest defense (Clark and Robertson 1979*b*; Ortega 1991; Ortega and Cruz 1991; S. Robinson, pers. commun.), red-winged blackbirds are parasitized by cowbirds. The incidence of brown-headed cowbird parasitism on redwings has been reported to be as low as 1.6% (Brown and Goertz 1978) and as high as 54% (Hergenrader 1962). The rate varies geographically, locally, annually, and within habitat (Ortega and Cruz 1988; Freeman et al. 1990). In our study sites in Colorado, overall parasitism rates in active red-wing nests were 4.9% (13/182) in 1984, 12.8% (41/274) in 1985, and 9.8% (39/346) in 1986 (Ortega 1991; Ortega and Cruz 1991). Clutch size was significantly smaller in parasitized nests than in nonparasitized nests in 1985 and 1986 (1985: $P = 0.001$, $U = 1887.5$, $Z = 3.171$; 1986: $P = 0.034$, $U = 2929.5$, $Z = 2.123$, Mann–Whitney U-test, two-tailed, corrected for ties). In 1984, clutch size tended to be smaller in parasitized nests than in nonparasitized nests ($P = 0.093$, $U = 483.0$, $Z = 1.681$). The red-winged blackbird is also parasitized by the bronzed cowbird. Friedmann (1963, 1971) reported six instances of parasitism from southeastern Texas. One nest contained one blackbird nestling, one brown-headed cowbird egg, and six eggs of bronzed cowbirds. Carter (1984, 1986) found that the red-winged blackbird was an infrequent bronzed cowbird host in Texas.

In Seminole County, central Florida, a red-winged blackbird was recorded feeding an immature cowbird in 1989, and near Florida City, southern Florida, a red-winged blackbird was observed feeding a fledgling cowbird in 1991, identified as a shiny cowbird by its vocalization (H. Kale and W. Pranty, pers. commun.).

Southern Florida redwings breed primarily in freshwater marshes and mangroves. The latter habitat is also used by the yellow-shouldered blackbird of Puerto Rico and the yellow-hooded blackbird (*A. icterocephalus*) of Trinidad and South America, both of which are heavily parasitized by shiny cowbirds (Cruz et al. 1989, 1990). Shiny cowbird parasitism is considered to be the main cause of endangerment of the yellow-shouldered blackbird (Post and Wiley 1977*a*; Post 1981; Cruz et al. 1989; Wiley et al. 1991).

Of particular significance is the discovery in 1993 of a parasitized red-winged blackbird nest on Sanibel Island, southwestern Florida. The nest contained one cowbird egg and two blackbird eggs. In 1994, we found a red-winged blackbird nest containing one cowbird young and two host nestlings on Sanibel Island. Because both cowbird species have been reported in the area, and their eggs are very similar, we are not sure of the identity. However, brown-headed cowbirds were observed more frequently in the study areas than shiny cowbirds. In 1991, near Florida City, south Florida, a red-winged blackbird was recorded feeding an immature cowbird, identified as a shiny cowbird by its vocalization (H. Kale and X. Pranty, pers. commun.). Despite their abundance, red- winged blackbirds should be carefully monitored in the future.

Orchard oriole (*Icterus spurius*)

The orchard oriole is a common summer resident in northern Florida (Stevenson and Anderson 1994). In 1990, an orchard oriole was observed feeding a brown-headed cowbird in Levy County, central Florida (H. Kale and W. Pranty, pers. obs.). According to naturalists in Leon County, the orchard oriole is heavily parasitized in northern Florida by the brown-headed cowbird (Stevenson and Armstrong 1994). Hill (1976) reported that eight of 15 nests found in Kansas were parasitized, an incidence of 53.3%. Goertz (1977) noted that 20 of 71 nests in Louisiana were similarly affected, an incidence of 28.2%. S. Robinson (pers. commun.) found that 90% (45/50) of nests in Illinois were parasitized. The orchard oriole is also parasitized by bronzed cowbirds, with records from southern Texas, eastern Mexico, and southern Louisiana (Stewart 1976; Friedmann and Kiff 1977). In Kingsville, Texas, a nest containing eggs of both the brown-headed and bronzed cowbirds with those of the oriole was found (Friedmann et al. 1977). We thus feel that the orchard oriole should be carefully monitored in Florida for evidence of cowbird parasitism.

Spot-breasted oriole (*Icterus pectoralis*)

The spot-breasted oriole, native to Middle America, became established in south Florida in the 1950s (Kale and Maehr 1990). Friedmann (1963, 1966) recorded this species as a bronzed cowbird host in Mexico and El Salvador. In addition, both brown-headed and shiny cowbirds are known to parasitize other oriole species. Orchard orioles and hooded orioles (*I. cucullatus*) have been recorded as hosts for brown-headed cowbirds (Friedmann et al. 1975; Hill 1976; Friedmann and Kiff 1985), and in the West Indies, the black-cowled oriole (*I. dominicensis*), St. Lucia oriole (*I. bonana*), and troupial (*I. icterus*) have been recorded as shiny cowbird hosts (Wiley 1985; Pérez-Rivera 1986; Cruz et al 1989; Post et al. 1990). We thus feel that this species will be a potential cowbird host.

OPPORTUNITIES FOR RESEARCH AND MANAGEMENT OF COWBIRDS

The arrival of cowbirds in Florida provides unique opportunities for research and management. The cowbirds' invasion of southern Florida means that their host populations have no recent history of exposure to nest parasitism as is the case in South America and in western North America. Consequently, these populations are unlikely to have evolved defensive strategies and thus may be particularly vulnerable to reproductive failure as a result of cowbird parasitism. We recommend studies on: (i) the distribution and status of the expanding cowbird populations; (ii) age and sex ratios of the parasite populations; (iii) host use by cowbirds; (iv) reproductive biology of potential cowbird hosts; (v) timing of breeding in shiny cowbirds; and (vi) defense strategies of hosts. Results from these studies can be used as baseline data with which to gauge future effects of cowbird parasitism on avian populations. Indeed, we have already begun monitoring cowbird populations and have begun studies on the breeding biology of potential host species in south Florida.

Research implications

Cowbirds are particularly well suited for ecological and evolutionary studies, from both practical and theoretical standpoints. Brood parasitism constitutes a selection pressure favoring the evolution of antiparasite defenses by host species (e.g., Rothstein 1975; Cruz and Wiley 1989). Because nearly all the major adaptations related to parasite–host interactions are focused in or near the nest, avian brood parasitism is an excellent system for studies of coevolution (Cruz and Wiley 1989; Rothstein 1990). One of the problems of measuring effects in such a coevolved system is that current interactions give us insufficient clues to past events. The current colonization of the Florida region by cowbirds allows us an opportunity to observe coevolution as an ongoing and dynamic process. More specifically, it provides a unique opportunity to examine adaptations against brood parasitism in the actual and potential cowbird hosts in areas where cowbirds are recent arrivals (Florida) and where cowbirds are native members of the avifauna or have been established for longer periods of time (e.g., Venezuela, Trinidad, St. Lucia, Puerto Rico, Hispaniola, and western North America). Additionally, some of the events crucial to parasite–host interactions can be replicated experimentally [e.g., avoiding versus facilitating parasitism, host egg rejection, and adaptations during the nesting stage (Wiley 1982; Harvey and Partridge 1988; Cruz and Wiley 1989; Rothstein 1990)].

Management strategies and conservation implications

Monitoring the spread and abundance of cowbirds in Florida, as well as undertaking studies on the levels of parasitism of various host species, will provide data valuable in attaining a better understanding of the evolution and ecology of brood parasitism. These studies may also provide information that may be used to manage declining host populations.

In our studies of the yellow-shouldered blackbird in southeastern Puerto Rico, we have found that trapping of cowbirds has proved successful in efforts to manage shiny cowbirds in yellow-shouldered blackbird nesting areas (Wiley et al. 1991). This technique is modeled after that employed to control brown-headed cowbirds in Kirtland's warbler (*Dendroica kirtlandii*) and least Bell's vireo (*Vireo bellii pusillus*) nesting grounds (Shake and Mattsson 1975; Mayfield 1977; Kelly and DeCapita 1982).

The possibility that cowbirds can be eliminated from a large geographic region, such as Florida, is unlikely, because of continued colonization by brown-headed cowbirds and shiny cowbirds from areas outside the region. In Florida, localized trapping may be the best available strategy for reducing the impact of cowbird parasitism from the nesting areas of targeted species, such as breeding areas of yellow warblers in the Keys. The shooting of cowbirds near the breeding grounds of host species can also be used to remove a substantial number of cowbirds, and may be useful and more cost-effective than trapping in some areas with small or scattered groups of sensitive species (Laymon, 1987; Robinson et al. 1993).

Many natural ecosystems in Florida have been rapidly fragmented because of the pressure of economic development. Land required by the growing human population of greater Miami and the Florida Keys, for example, has been obtained largely by clearing pinelands, hardwood hammocks, and wetlands, thereby threatening or endangering the existence of a large number of species (Myers and Ewel 1990).

Humans have accelerated the rate of successful cowbird colonization of Florida by the destruction and fragmentation of habitats, which has created the open and edge habitat favored by cowbirds. Fragmentation creates small patches of forest surrounded by open habitat and increases the area of forest edge habitat available for cowbird nest searching. In studies of brown-headed cowbirds, brood parasitism was more intense along the edges of wooded tracts than in forest interiors, where nests are presumably closer to the feeding areas of cowbirds and are therefore more likely to be found (Gates and Gysel 1978; Chasko and Gates 1982; Brittingham and Temple 1983). Nest predation and parasitism by cowbirds increased with forest fragmentation in nine midwestern landscapes that varied from 6% to 95% forest cover (Robinson et al. 1995).

Perhaps the best and most permanent way to reduce the impact of cowbirds on host species is through landscape-level management, which can be effective at a much larger scale than trapping. Because cowbirds are frequently associated with agriculture, human settlements, and internal and external edge, the best management strategy is to maintain large areas of contiguous habitats (Robinson et al. 1993, 1995). Managers must also keep in mind the landscape surrounding the area being managed. Landscapes with few feeding opportunities for cowbirds (e.g., feedlots, feeders) may not have problems with cowbird parasitism even along edges and small openings. Landscapes with abundant cowbird feeding habitat may have cowbird populations that saturate breeding habitat regardless of proximity to edge (Robinson et al. 1993, 1995).

The data indicate that some small passerines, especially members of the Vireonidae, Parulinae, and Fringillidae, are heavily parasitized outside of Florida and thus will be preferred hosts for cowbird parasitism. The data presented also suggest that at least some species ought to receive special monitoring in the years to come. Clearly, the black-whiskered vireo, white-eyed vireo, prairie warbler, yellow warbler, and perhaps the northern cardinal, rufous-sided towhee, grasshopper sparrows, Cape Sable seaside sparrow, and red-winged blackbird deserve investigation in this regard.

In addition to Florida, shiny cowbirds have been recorded in other areas in North America, particularly the southeast (fig. 17.1). Post et al. (1993) observed several males displaying to female shiny cowbirds on Sullivan's Island, South Carolina, in 1991, and, in Alabama, at least 43 cowbirds were reported from four localities in the state (Jackson 1990). Along the Gulf Coast the situation will be further compounded by the movement of bronzed cowbirds eastward from Texas, where it has already been recorded breeding in Louisiana (Stewart 1976; Purrington 1988) (fig. 17.3).

As most records of shiny cowbirds for North America have been from coastal areas (fig. 17.1), we believe that this species will continue to expand its range, especially along the Atlantic and Gulf Coasts. We predict that northern-dispersing populations will return to southern Florida, and possibly the West Indies, during the winter (Post et al. 1993), as the subspecies involved in the expansion, *M. b. minimus*, is primarily a tropical species, originating in northern South America.

The dramatic expansion of the Florida breeding range of the brown-headed cowbird in the past 35 years suggests that this species is likely to continue its spread into south Florida, including the Keys. The possibility exists that the spread of the brown-headed cowbird will not stop at the Keys, but will continue southward into Cuba.

ACKNOWLEDGMENTS The work was supported by an NSF Grant PRTM-8112194 to the University of Colorado, A. Cruz, principal investigator; the U.S. Fish and Wildlife Service; the National Geographic Society; the New York Zoological Society; the U.S. Forest Service; and the University of Miami. We wish to thank the Field Research Department of the National Audubon Society for all the logistical help, in particular Alexander Sprunt, IV, G. Thomas Bancroft, Wayne Hoffman, Richard J. Sawicki, and Allen M. Strong. Herb W. Kale, II and William Pranty of the Florida Audubon Society generously provided cowbird records for the Florida Breeding Bird Atlas project. We are grateful to the National Park Service and the United States Fish and Wildlife Service for allowing us to work within lands under their jurisdiction. We wish to thank W. Jackson and S. K. Robinson for their helpful comments on the manuscript.

REFERENCES

Arendt, W. J. and T. A. Vargas Mora (1984). Range expansion of the Shiny Cowbird in the Dominican Republic. *J. Field Ornith.*, **55**, 104–107.

Atherton, L. S. and B. H. Atherton (1988). Florida region. *American Birds*, **42**, 60–63.

Bendire, C. E. (1893). The Cowbirds. *Rep. U.S.N.M.*, pp. 587–624.

Bent, A. C. (1953). Life histories of North American wood warblers. *U.S. Natl Mus. Bull.*, **203**.

Bond, J. (1966). Eleventh Supplement to the Check-list of the Birds of the West Indies (1956). *Academy of Natural Sciences of Philadelphia.*

Bowen, D. E. (1976). *Community and density effects on habitat distribution and regulation of breeding Upland Sandpipers* (Bartramia longicauda). Ph.D. Thesis, Kansas State University, Manhattan, Kansas.

Brittingham, M. C. and S. A. Temple (1983). Have cowbirds caused forest birds to decline? *Bioscience*, **33**, 31–35.

Brown, B. T. and J. W. Goertz (1978). Reproduction and nest site selection by Red-winged Blackbirds in north Louisiana. *Wilson Bull.*, **90**, 396–403.

Brush, T. (1993). Declines in orioles in the Lower Rio Grande Valley, Texas – possible effects of cowbirds. P. 63 (abstract) North American Research Workshop on the Ecology and Management of Cowbirds. Austin, Texas.

Burgham, M. C. J. and J. Picman (1989). Effect of

brown-headed cowbirds on the evolution of yellow warbler anti-parasite strategies. *Anim. Behav.*, **38**, 298–308.

Carello, C. (1993). Comparison of rates of Brown-headed Cowbird brood parasitism between two Red-Winged Blackbird Nesting sites. In: *North American Research Workshop on the Ecology and Management of Cowbirds*. Abstract, p. 64.

Carter, M. D. (1984). The social organization and parasitic behavior of the Bronzed Cowbird in south Texas. Ph.D. Thesis. Univ. of Minnesota, Minneapolis, Minnesota.

Carter, M. D. (1986). The parasitic behavior of the Bronzed Cowbird in south Texas. *Condor*, **85**, 11–25.

Chasko, G. G. and J. E. Gates (1982). Avian habitat suitability along a transmission-line corridor in an oak-hickory forest region. *Wildl. Monogr.*, **82**, 1- 41.

Clark, K. L. and R. J. Robertson (1979*a*). Responses of Yellow warblers to the threat of cowbird parasitism. *Anim. Behav.*, **38**, 510–519.

Clark, K. L. and R. J. Robertson (1979*b*). Spatial and temporal multi-species nesting aggregations in birds as anti-parasite and anti-predator defenses. *Behav. Ecol. Socio-biol.*, **5**, 359–371.

Clark, K. L. and R. J. Robertson (1981). Cowbird parasitism and evolution of anti-parasite strategies in the Yellow Warbler. *Wilson Bull.*, **93**, 249–258.

Cruz, A. and R. W. Andrews (1989). Observations on the breeding biology of passerines in a seasonally flooded savanna in Venezuela. *Wilson Bull.*, **101**, 62–76.

Cruz, A. and J. W. Wiley (1989). The decline of an adaptation in the absence of a presumed selection pressure. *Evolution*, **43**, 55–62.

Cruz, A., T. D. Manolis, and R. W. Andrews (1990). Reproductive interactions of the Shiny Cowbird *Molothrus bonariensis* and the Yellow-hooded Blackbird *Agelaius icterocephalus* in Trinidad. *Ibis*, **132**, 436–444.

Cruz, A., T. D. Manolis, and J. W. Wiley (1985). The Shiny Cowbird: a brood parasite expanding its range in the Caribbean region. In: *Neotropical Ornithology* (eds P. A. Buckley, M. S. Foster, E. S. Morton, R. S. Ridgely, and F. G. Buckley), pp. 607–619. American Ornithologists' Union, Ornithol., Monogr. No. 36.

Cruz, A., J. W. Wiley, T. K. Nakamura, and W. Post (1989). The Shiny Cowbird in the Caribbean region – biogeographical and ecological implications. In: *Biogeography of the West Indies–past, present and future*, pp. 519–540. Sandhill Crane Press, Gainesville, Florida.

Cruz, A., J. W. Prather, W. Post, and J. W. Wiley. The spread of the shiny and Brown-headed Cowbirds into the Florida region. In: *Ecology and Management of Cowbirds* (eds J. N. M. Smith, T. Cook, S. K. Robinson, S. I. Rothstein, and S. G. Sealy). Univ. Texas Press, Austin (in press).

Darley, J. A. (1978). Pairing in captive Brown-headed Cowbirds (*Molothrus ater*). *Can. J. Zool.*, **56**, 2249–2252.

Delany, M. F., H. M. Stevenson, and R. McCracken (1985). Distribution, abundance, and habitat of the Florida Grasshopper sparrow. *J. Wildlf. Manag.*, **49**, 626–631.

Easterla, D. A. and R. Wauer (1972). Bronzed Cowbirds in west Texas and two bill abnormalities. *Southwest Nat.*, **17**, 293–312.

Eastzer, D., P. R. Chu and A. P. King (1980). The young cowbird: average or optimal nestling? *Condor*, **82**, 417–425.

Edscorn, J. B. (1980). The nesting season: Florida region. *American Birds*, **34**, 887–889.

Elliott, P. F. (1978). Cowbird parasitism in the Kansas tallgrass prairie. *Auk*, **95**, 161–167.

Fletcher, J. (1993). Is the Shiny Cowbird in Jamaica? *Gosse Bird Club Broadsheet*, **61**, 5–7.

Fraga, R. M. (1985). Host–parasite interactions between Chalk-browed Mockingbirds and Shiny Cowbird, pp. 829–944. In: *Neotropical Ornithology* (eds P. A. Buckley, M. S. Foster, E. S. Morton, R. S. Ridgely, and F. G. Buckley). American Ornithologists' Union, Ornithol. Monogr. No. 36.

Freeman, S. D., F. Gori, and S. I. Rohwer (1990). Red- winged Blackbird and Brown-headed Cowbirds: some aspects of a host-parasite relationship. *Condor*, **92**, 336–340.

Friedmann, H. (1929). *The cowbirds: a study in the biology of social parasitism*. C. C. Thomas, Springfield, IL.

Friedmann, H. (1963). Host relations of the parasitic cowbirds. *U.S. Natl Mus. Bull.*, **233**, 273 pp.

Friedmann, H. (1966). Additional data on the host relations of the parasitic cowbirds. *Smith. Misc. Coll.*, **149**, 1–12.

Friedmann, H. (1971). Further information on the host relations of the parasitic cowbirds. *Auk*, **88**, 239–255.

Friedmann, H. and L. F. Kiff (1985). The parasitic cowbirds and their hosts. *Proceedings of the Western Foundation of Vertebrate Zoology*, **2**, 225–302.

Friedmann, H., L. F. Kiff, and S. I. Rothstein (1977). A further contribution to knowledge of the host relations of the parasitic cowbirds. *Smithsonian Contributions to Zoology*, **235**, 1–75.

Garrido, O. H. (1984). *Molothrus bonariensis* (Aves: Icteridae). Nuevo record para Cuba. *Misc. Zoologica, Academia de Ceincias de Cuba*, Num. 19:3.

Gates, J. E. and L. W. Gysel (1978). Avian nest dispersion and fledgling success in field-forest ecotones. *Ecology*, **59**, 871–883.

Gochfeld, M. (1979). Brood parasite and host coevolution: interactions between Shiny Cowbirds and two species of meadowlarks. *Amer. Natur.*, **113**, 855–870.

Goertz, J. W. (1977). Additional records of Brown-headed Cowbird nest parasitism in Louisiana. *Auk*, **94**, 386–389.

Goldman, L. (1951). South Texas region. *Audubon Field Notes*, 5.

Graber, J. W., R. R. Graber, and E. L. Kirk (1974). Illinois birds: Tyrannidae. *Illinois Natural History Survey, Biological Notes*, **86**, 1–56.

Graber, J. W., R. R. Graber, and E. L. Kirk (1983). Illinois birds: wood warblers. *Illinois Natural History Survey, Biological Notes*, **118**, 1–144.

Grzybowski, J. A. and V. W. Fazio (1991). Shiny Cowbird reaches Oklahoma. *American Birds*, **45**, 50–52.

Guerra, F. and G. Alayon (1987). Cowboy bird invades Cuba – Cuban ornithologist Orlando Garrido discusses what to do about this threat. *Gramma*, **8** (22 February 1987).

Harvey, P. H. and L. Partridge (1988). Of cuckoo clocks and cowbirds. *Nature*, **335**, 586–587.

Hergenrader, G. L. (1962). The incidence of parasitism by Brown-headed Cowbirds on roadside nesting birds. *Auk*, **79**, 85–88.

Hill, R. A. (1976). Host–parasite relationships of the Brown-headed Cowbird in a prairie habitat of west-central Kansas. *Wilson Bull.*, **85**, 555–565.

Hobson, K. A. and S. G. Sealy (1989). Mate guarding in the Yellow Warbler *Dendroica petechia*. *Ornis Scand.*, **20**, 241–249.

Hoffman, W. and G. E. Woolfenden (1986). A fledgling Brown-headed Cowbird from Pinellas County. *Florida Field Naturalist*, **14**, 18–20.

Holford, K. C. and D. D. Roby (1993). What limits the fecundity of Brown-headed Cowbirds? In: *North American Research Workshop on the Ecology and Management of Cowbirds*. Abstract, p. 21.

Hoover, J. P. and M. C. Brittingham (1993). Regional variation in cowbird parasitism of wood thrushes. *Wilson Bull.*, **105**, 228–238.

Howell, J. C. (1932). *Florida Bird Life*. Florida Dept. Game and Fresh Water Fish, Tallahassee, Florida.

Hutcheson, W. M. and W. Post (1990). Shiny Cowbird collected in South Carolina: first North American specimen. *Wilson Bull.*, **102**, 561.

Imhof, T. A. (1987). Central Southern Region. *American Birds*, **41**, 442–448.

Imhof, T. A. (1989). Central Southern Region. *American Birds*, **43**, 491–495.

Jackson, G. D. (1990). Central Southern Region. *American Birds*, **44**, 439–444.

Jackson, J. A. (1989). Central Southern Region. *American Birds*, **43**, 1299–1330.

Johnson, R. G. and S. A. Temple (1990). Nest predation and brood parasitism of tallgrass prairie birds. *J. Wildl. Manag.*, **54**, 106–111.

Kale, H. W., II (1978). *Rare and Endangered Plants and Animals*, Vol. 2, Univ. of Florida Press.

Kale, H. W., II (1989). Florida birds. *Florida Naturalist*, **62**, 14.

Kale, H. W., II and D. S. Maehr (1990). *Florida's birds, a handbook reference*. Pineapple Press, Inc., Sarasota, Florida.

Kattan, G. H. (1993). Extraordinary annual fecundity of Shiny Cowbirds at a tropical locality and its energetic trade-off. In: *North American Research Workshop on the Ecology and Management of Cowbirds*. Abstract, p. 20.

Kelley, S. T. and M. E. DeCapita (1982). Cowbird control and its effect on Kirtland's Warbler reproductive success. *Wilson Bull.*, **94**, 363–365.

Langridge, H. P. (1988). The Florida region. *American Birds*, **42**, 424–426.

Langridge, H. P. (1989). The Florida region. *American Birds*, **43**, 467–470.

Laskey, A. R. (1950). Cowbird behavior. *Wilson Bull.*, **62**, 157–174.

Lasley, G. W. and C. Sexton (1991). Texas region. *American Birds*, **45**, 1135–1139.

Laymon, S. A. (1987). Brown-headed Cowbirds in California: historical perspectives and management opportunities in riparian habitats. *Western Birds*, **18**, 63–70.

LeGrand, H. E., Jr (1989). Southern Atlantic Coast Region. *American Birds*, **43**, 1303–1306.

Lowther, P. E. and S. I. Rothstein (1980). Head-down or "preening invitation" displays involving juvenile Brown-headed Cowbirds. *Condor*, **82**, 459–460.

Manolis, T. D. (1982). Host relationships and reproductive strategies of the Shiny Cowbird in Trinidad and Tobago. Unpublished Ph.D. Thesis. University of Colorado, Boulder, Colorado.

Marvil, R. E. and A. Cruz (1989). Host-parasite interactions between Solitary Vireos (*Vireo solitarius*) and Brown-headed Cowbirds (*Molothrus ater*). *Auk*, **106**, 476–480.

Mason, P. (1986*a*). The ecology of brood parasitism

in a host generalist, the Shiny Cowbird. I. The quality of different host species. *Auk*, **103**, 52–60.

Mason, P. (1986*b*). The ecology of brood parasitism in a host generalist, the Shiny Cowbird. II. Host selection. *Auk*, **103**, 61–69.

Matteson, R. E. (1970). Bronzed Cowbird taken in Florida. *Auk*, **87**, 588.

Maurer, B. A. (1993). Are cowbirds increasing in abundance and expanding their geographic range? Evidence from the breeding bird survey. In: *North American Research Workshop on the Ecology and Management of Cowbirds*. Abstract, p. 10.

May, P. M. and S. K. Robinson (1985). Population dynamics of avian brood parasitism. *Amer. Natur.*, **126**, 475–494.

Mayfield, H. (1965). The Brown-headed Cowbird with old and new hosts. *Living Bird*, **4**, 13–28.

Mayfield, H. (1977). Brown-headed Cowbirds: agent of extermination? *American Birds*, **31**, 107–113.

Meanley, B. (1982). Swainson's warbler and the cowbird in the Dismal Swamp. *Raven*, **53**, 47–49.

Mengel, R. M. (1965). *The birds of Kentucky*. American Ornithologists' Union, Ornithol. Monogr. No. 3.

Muth, D. P. (1992). Central-Southern region. *American Birds*, **46**, 434–437.

Myers, R. L. and J. J. Ewel (1990). *Ecosystems of Florida*. University of Central Florida Press, Orlando, Florida.

Nakamura, T. K. (1995). Nest defense and nest attentiveness as an adaptation against brood parasitism. M. A. Thesis. University of Colorado, Boulder, Colorado.

Neville, B. (1988*a*). *The monthly bird report*. Tropical Audubon Bulletin, newsletter of the Tropical Audubon Society. October 1988: p. 4.

Neville, B. (1988*b*). *Florida Ornithological Society News*. Florida Audubon Society – Florida Ornithological Society Newsletter 14: 5–6.

Newman, G. (1957). The Florida region. *Audubon Field Notes*, **11**, 412–413.

Niles, D. M. (1970). A record of clutch size and breeding in New Mexico for the Bronzed Cowbird. *Condor*, **72**, 500.

Nolan, V., Jr. (1978). Ecology and behavior of the Prairie Warbler, *Dendroica discolor*. American Ornithologists' Union, Ornith. Monogr. No. 26.

Norris, R. T. (1947). The cowbirds of Preston Frith. *Wilson Bull.*, **59**, 83–103.

Oberholser, H. C. (1974). *The Birdlife of Texas*. University of Texas Press, Austin, Texas.

Ogden, J. C. (1965). Florida region. *Audubon Field Notes*, **19**, 634–637.

Orians, G. (1989). *Blackbirds of the Americas*. University of Washington Press, Seattle, Washington.

Ortega, C. P. (1991). Ecology of Blackbird/cowbird interactions in Boulder County, Colorado. Ph.D. Thesis University of Colorado, Boulder, Colorado.

Ortega, C. P. and A. Cruz (1988). Mechanisms of egg acceptance behavior by marsh dwelling blackbirds. *Condor*, **90**, 349–358.

Ortega, C. P. and A. Cruz (1991). A comparative study of cowbird parasitism in Yellow-headed Blackbirds and Red-winged Blackbirds. *Auk*, **108**, 16–24.

Paul, R. T. (1984). Florida region. *American Birds*, **38**, 1011–1013.

Paul, R. T. (1986). The Florida region. *American Birds*, **40**, 1193–1197.

Paul, R. T. (1988). Florida region. *American Birds*, **42**, 1278–1281.

Paul, R. T. (1989). Florida region. *American Birds*, **43**, 1307–1310.

Payne, R. E. (1977). The ecology of brood parasitism in birds. *Annu. Rev. Ecol. Syst.*, **8**, 1–28.

Peréz-Rivera, R. A. (1986). Parasitism by the Shiny Cowbird in the interior parts of Puerto Rico. *J. Field Ornith.*, **57**, 99–104.

Perkins, S. A. (1991). New England region. *American Birds*, **45**, 415–419.

Petit, L. J. (1989). Breeding biology of Prothonotary Warblers (*Protonotaria citrea*) in riverine habitat in Tennessee. *Wilson Bull.*, **101**, 51–61.

Petit, L. J. (1991). Adaptive tolerance of cowbird parasitism by Prothonotary Warblers: a consequence of nest-site limitation? *Anim. Behav.*, **41**, 425–432.

Phillips, A., J. Marshall, and G. Monson (1964). *The Birds of Arizona*. University of Arizona Press, Tucson, Arizona.

Post, W. (1981). Biology of the Yellow-shouldered Blackbird – *Agelaius* on a tropical island. *Bull. Florida State Mus., Bio., Sci.*, **25**, 125–202.

Post, W. and J. W. Wiley (1977*a*). The Shiny Cowbird in the West Indies. *Condor*, **79**, 119–121.

Post, W. and J. W. Wiley (1977*b*). Reproductive interactions of the Shiny Cowbird and the Yellow-shouldered Blackbird. *Condor*, **79**, 176–184.

Post, W., T. K. Nakamura, and A. Cruz (1990). Patterns of Shiny Cowbird parasitism in St. Lucia and southwestern Puerto Rico. *Condor*, **92**, 461–469.

Post, W., A. Cruz, and D. B. McNair (1993). The North American invasion pattern of the Shiny Cowbird. *J. Field Ornith.*, **64**, 32–41.

Potter, E. F. and G. T. Whitehurst (1981). Cowbirds in the Carolinas. *Chat*, **45**, 57–68.

Prather, J. W. and A. Cruz (1995). Breeding biology of the Florida Prairie Warbler and the Cuban Yellow Warbler. *Wilson Bull.*, **107**, 475–484.

Purrington, R. D. (1988). Central southern region. *American Birds*, **42**, 85–52.

Purrington, R. D. (1991). Central-Southern Region. *American Birds*, **45**, 1124–1128.

Purrington, R. D. (1993). Central-Southern Region. *American Birds*, **47**, 1115–1118.

Ramo, C. and B. Busto (1980). Biologia reproductiva de la viudita (*Fluvicola pica*) en el Llano Venezolano. *Natura*, **86**, 22–25.

Rappole, J. H. and G. W. Blacklock (1985). *Birds of the Texas Coast Bend – abundance and distribution*. Texas A & M University Press, College Station, Texas.

Robbins, M. B. and D. A. Easterla (1981). Range expansion of the Bronzed Cowbird with the first Missouri records. *Condor*, **83**, 270–272.

Robertson, R. J. and R. F. Norman (1977). The function and evolution of aggressive host behaviour towards the brown-headed cowbird (*Molothrus ater*). *Can J. Zool.*, **55**, 508–518.

Robertson, W. B., Jr (1955). An analysis of the breeding bird populations of tropical Florida in relation to the vegetation. Ph.D. Thesis. University of Illinois, Urbana, Illinois.

Robertson, W. B., Jr (1978). Cuban Yellow Warbler. In: *Rare and Endangered Biota of Florida, Vol. 2 Birds* (ed. H. W. Kale, II), pp. 61–62. University Presses of Florida.

Robertson, W. B., Jr and J. A. Kushlan (1974). The southern Florida avifauna. In: *Environments of south Florida: present and past*, (ed. P. J. Gleason, II) pp. 414–452. Miami Geological Soc. Memoir 2.

Robertson, W. B., Jr and G. E. Woolfendent (1992). *Florida bird species – an annotated list*. Florida Ornithological Society, Special Publication No. 6.

Robinson, S. K., J. A. Grzybowski, S. I. Rothstein, M. C. Brittingham, L. J. Petit, and F. R. Thompson (1993). Management implications of cowbird parasitism on neotropical migrant songbirds. In: *Status and Management of Neotropical Migratory Birds* (ed. D. M. Finch and P. W. Stangel), U.S. Dept. of Agriculture, Forest Service, Rocky Mountain Forest and Range Experiment Station.

Robinson, S. K., F. R. Thompson, III, T. M. Donovan, D. R. Whitehead, and J. Faaborg (1995). Regional forest fragmentation and the nesting success of migratory birds. *Science*, **267**, 1987–1990.

Rosenberg, K. V., R. D. Ohmart, W. C. Hunter, and B. W. Anderson (1991). *Birds of the Lower Colorado River Valley*. University of Arizona Press, Tucson, Arizona.

Rothstein, S. I. (1975). An experimental and teleonomic investigation of avian brood parasitism. *Condor*, **77**, 250–271.

Rothstein, S. I. (1990). A model system for coevolution: avian brood parasitism. *Annu. Rev. Ecol. Syst.*, **21**, 481–508.

Rothstein, S. I., J. Verner, and E. Stevens (1980). Range expansion and diurnal changes in dispersion of the Brown-headed Cowbird in the Sierra Nevada. *Auk*, **97**, 253–267.

Rothstein, S. I., J. Verner, and E. Stevens (1984). Radio-tracking confirms a unique diurnal pattern of spatial occurrence in the parasitic Brown-headed Cowbird. *Ecology*, **65**, 777–788.

Scott, D. M. (1977). Cowbird parasitism on the Gray Catbird at London, Ontario. *Auk*, **94**, 18–27.

Sealy, S. G. (1992). Removal of Yellow Warbler eggs in association with cowbird parasitism. *Condor*, **94**, 40–54.

Shake, W. F. and J. P. Mattsson (1975). Three years of cowbird control: an effort to save the Kirtland's Warbler. *Jack-Pine Warbler*, **53**, 48–53.

Smith, P. W. and A. Sprunt, IV (1987). The Shiny Cowbird reaches the United States. *American Birds*, **41**, 370–371.

Southern, W. E. (1958). Nesting of the red-eyed vireo in the Douglas Lake region, Michigan, part 2. *Jack-Pine Warbler*, **36**, 185–207.

Spaw, C. D. and S. Rohwer (1987). A comparative study of eggshell thickness in cowbirds and other passerines. *Condor*, **89**, 307–318.

Sprunt, A., Jr (1954). *Florida bird life*. Coward-McCaan, Inc., New York.

Sprunt, A., Jr (1963). *Addendum to Florida birdlife, 1954*. Coward-McCann, Inc., New York.

Stevenson, H. M. (1959). Florida region. *Audubon Field Notes*, **13**, 426–429.

Stevenson, H. M. (1963). Florida region. *Audubon Field Notes*, **20**, 564–565.

Stevenson, H. M. and B. H. Anderson (1994). *Florida Birdlife*. University of Florida Press, Gainesville, Florida.

Stewart, J. R., Jr (1963). Central-Southern region. *Audubon Field Notes*, **17**, 465–467.

Stewart, J. R., Jr. (1976). Central-Southern region. *American Birds*, **30**, 965–969.

Surner, S. (1992). First record of the Shiny Cowbird (*Molothrus bonariensis*) in Maine. *Maine Bird Notes*, **5**, 1–2.

Teather, J. L. and R. J. Robertson (1985). Female spacing patterns in Brown-headed Cowbirds. *Can. J. Zool.*, **63**, 218–222.

Thompson, F. R., III (1994). Temporal and spatial

patterns of breeding Brown-headed Cowbirds in the midwestern United States. *Auk*, **111**, 979–990.

Toops, C. and W. E. Dilley (1986). *Birds of south Florida*. River Road Press, Conway, Arkansas.

Truchot, E. (1962). Bronzed or red-eyed cowbird (*Tangavius aeneus*). *Fla. Natur.*, **35**, 135.

Weatherhead, P. J. (1989). Sex ratios, host-specific reproductive success, and impact of Brown-headed Cowbirds. *Auk*, **106**, 358–366.

Werner, H. W. (1976). Distribution, habitat, and origin of the Cape Sable Seaside Sparrow. Master's Thesis. Dept. of Biology, University of South Florida, Tampa.

Weston, F. M. (1965). A survey of the birdlife of northwestern Florida. *Bull. Tall Timbers Research Stat.*, **5**, 1–147.

Wiens, J. A. (1963). Aspects of cowbird parasitism in southern Oklahoma. *Auk*, **75**, 130–139.

Wiley, J. W. (1982). Ecology of avian brood parasitism at an early interfacing of host and parasite populations. Ph.D. Thesis. University of Miami, Cral Gables, Florida.

Wiley, J. W. (1985). Shiny Cowbird parasitism in two avian communities in Puerto Rico. *Condor*, **87**, 165–176.

Wiley, J. W., W. Post, and A. Cruz (1991). Conservation of the Yellow-shouldered Blackbird – an endangered West Indian species. *Biol. Conserv.*, **55**, 119–138.

Williams, F. (1976). Southern Great Plains. *American Birds*, **30**, 972–975.

Wilson, A. (1810). *American Ornithology*, **2**, 145–150, 158.

Zimmerman, J. L. (1983). Cowbird parasitism of dickcissels in different habitats and at different nest densities. *Wilson Bull.*, **95**, 7–22.

VI

CONSEQUENCES OF PARASITISM FOR THE MATING SYSTEMS AND LIFE HISTORIES OF BROOD PARASITES

This part includes only a single chapter that emphasizes the consequences of brood parasitism for aspects of the parasite's mating system. Freed from parental care, the mating systems, spatial systems, and diets of parasitic birds are much less constrained than those of most birds. In spite of this lack of constraints, many avian brood parasites appear to be monogamous. Since they are raised by other species, some aspects of the behavior of parasites must be genetically programmed. Because this section focuses on mating systems, a more inclusive overview of the consequences of parasitism is found in pages 35–38 in chapter 1.

18

Variability in the Mating Systems of Parasitic Birds

PHOEBE BARNARD

Brood parasitic birds are perplexing creatures in many ways. Evolutionarily, their strategy of reproductive freeloading has been highly successful. Many species, such as some of the cowbirds (Icterinae), have combined a strategy of generalist parasitism with a behavioral plasticity that has allowed them to expand their range greatly in historic times. Parasites and their hosts also make intriguing studies in coevolution (Rothstein, 1990), as explored by other contributors to this book.

Obligate brood parasitism, like other types of parasitism, has profound consequences for the evolutionary ecology of parasites. Reproduction often dominates the lives of nonparasitic birds, and the establishment of pair bonds, building of nests, and raising of young to independence are demanding and energetically costly tasks. Parasitic birds, of course, are freed from many of these constraints. Therefore, we should expect that parasitism is a pivotal strategy that influences many other aspects of a bird's behavior and ecology. Because brood parasites avoid the demands and constraints imposed by parental behavior, study of their lifestyles can help clarify related areas of evolutionary biology by testing the generality of current life history theories (Rothstein et al. 1986; Yokel 1986).

Because there are so many other interesting questions to study in brood parasites, we tend to overlook one of the potentially most intriguing. Parasites have a puzzling variety of mating systems. At first this may seem logical, given that the strategy of interspecific brood parasitism has arisen at least five or six times, in five or six rather different families (Hamilton and Orians 1965; Payne 1977; Rothstein 1990). Yet, the minimal parental investment of brood parasites should itself minimize the need for prolonged association between male and female (Teather and Robertson 1986; Yokel 1986). Indeed, monogamy, which is still the prevalent social mating system in birds despite the covert sexual habits of individual partners (Gowaty and Mock 1985; Westneat et al. 1990), theoretically should not occur among brood parasites at all (Yokel 1986). And yet it does. Why?

Mating systems theory, as applied to birds, has been developed largely from studies of polygynous, parental species (e.g., Orians 1961, 1969; Crook 1965; Verner and Willson 1966; Emlen and Oring 1977; Oring 1982; Rubenstein and Wrangham 1986). The mating systems of parental birds are usually strongly influenced by variation in the availability and quality of specific ecological resources. These resources are often straightforward, such as safe nest sites, high-quality food, the help of a mate in feeding young, and so on. But what breeding resources are important to parasitic birds? In the absence of male parental care or territorial resources, are female parasites influenced by male phenotype or genetic quality? To answer these questions, we must identify both the constraints acting on parasites and the benefits that accrue from their lifestyle. In particular, we need to investigate the selective pressures fostering prolonged association between male and female parasites. To date, very little has been done to clarify such issues.

This chapter addresses the variability in mating systems of obligate avian brood parasites, and attempts to find common threads that help explain this diversity. The discussion is confined to obligate, interspecific brood parasites, because the evolutionary leap from facultative intraspecific parasitism to obligate parasitism is a major one, made by relatively few species. Whereas individuals of many other species practice intraspecific parasitism or "egg dumping" (Andersson 1984; MacWhirter 1989; other chapters in this book), its frequency is often low and may have little effect on the overall biology of these species (Payne 1977), including the mating system. I first review the available information on mating systems and sexual dimorphism in brood parasites, which is patchy for many groups. I then focus on some constraints facing brood parasites that may influence their mating and pairing behavior. In closing, I emphasize the many profitable areas of research that remain open. While this analysis may be a bit premature, given the patchiness of our knowledge of parasites' ecology and social behavior, I hope it will stimulate readers to help fill in the gaps.

TERMINOLOGY

For lack of sophisticated maternity and paternity data with which to infer "genetic" mating systems, this paper follows Wittenberger's (1979, 1981) "first-tier" behavioral classification of mating systems. With few exceptions, there is not yet enough information to categorize brood parasites by detailed behavioral criteria such as dominance interactions or resource/mate monopolization (Emlen and Oring 1977; Oring 1982). Even fewer data are available to judge the extent of extrapair copulations (EPCs) and other cryptic forms of variation in reproductive success (Westneat et al. 1990). The discussion is therefore by necessity restricted to a traditional behavioral classification of mating systems. Thus, monogamy is a "prolonged and (apparently) essentially exclusive mating relationship between one male and one female at a time"; polygyny is the same with one male and two or more females (simultaneous or sequential mate acquisition), and polyandry is the same with one female and two or more males. By contrast, promiscuity involves no prolonged association between the sexes and multiple matings (with different mates) by members of at least one sex (Wittenberger 1981). This classification, emphasizing the presence or absence of prolonged associations or pair bonds (>20–25% of the breeding season; Wittenberger 1981), is most appropriate here, because male and female parasites theoretically need not associate with each other past copulation (Ankney and Scott 1982; Yokel 1986). The fact that some do in intensively observed populations needs explanation.

Reviews of mating systems, however, have been made difficult by semantic bias and ambiguity. As Payne and Payne (1977) noted, it is better for fieldworkers to describe than to classify mating systems. The term polygamy, for example, has often been used in different, sometimes contradictory ways (Wittenberger 1981). Authors have also sometimes used mutually exclusive, species-level terms to classify the habits of males and females (e.g., male polygyny and female promiscuity). Of course, some individuals in a population may be promiscuous whereas others have a form of pair bond (Ankney and Scott 1982), and individuals may combine both strategies with different mates (Westneat et al. 1990), but intensive study of marked birds is needed to establish this.

For systematic terminology, I have followed Campbell and Lack (1985) as a general source, and Wyllie (1981), Beehler et al. (1986), King et al. (1986) and Maclean (1993) for generic classification. Where views conflict, I have mentioned both in the text to avoid confusion.

THE MATING SYSTEMS OF OBLIGATE BROOD PARASITES

Our understanding of the social and sex lives of brood parasites is extremely patchy. Descriptions of mating systems based on intensive observations of color-marked birds have, to my knowledge, been published only for six parasitic species (references below and in tables 18.1–18.4): brown-headed cowbird *Molothrus ater* (Icterinae), European cuckoo *Cuculus canorus* (Cuculidae), orange-rumped honeyguide *Indicator xanthonotus* (Indicatoridae), village indigobird *Vidua chalybeata*, pin-tailed whydah *V. macroura*, and shaft-tailed whydah *V. regia* (all Viduidae, often regarded as Ploceidae). Mating systems of the brown-headed cowbird and the viduines are relatively well known. Yet even in these species, more intensive study is needed to establish the extent of variation in mating systems within and between populations.

For most obligate brood parasites, however, the mating system is unknown. In some cases, authors have made a largely subjective guess on the basis of opportunistic observation of unmarked individuals. Many parasitic birds are secretive, especially the females, and following individuals is difficult in woodland or forest habitats. Yet with large, color-marked populations and intensive observations, it should be possible to fill many of the gaps mentioned below.

Parasitic cuckoos

Cuckoos are the archetypal brood parasites, after which other animals of similar lifestyles are rather quaintly named (Becking and Snow 1985). The cuckoo family, Cuculidae, is large and diverse, with about 130 species in 34 genera (Payne 1985*a*). At least 50 of these species are parasitic (Wyllie 1981), occurring in two subfamilies, the Cuculinae (47 species) and Neomorphinae (three species) (table 18.1). Sexual dimorphism is generally subtle in this family. In most genera, males and females are not easily distinguishable in the field, whereas in the shiny or bronze cuckoos [*Chrysococcyx* (*Chalcites*) spp.] males are often more bright and iridescent. Most species of cuckoo parasitize a small range of hosts; some have well-developed egg mimicry and "gentes" which specialize on one host (reviewed by Payne 1985a).

In Wyllie's (1981) six-year study of the European cuckoo *Cuculus canorus*, males occupied regular call-sites in host nesting areas during the breeding season, and more than one male sometimes sang from the same site. Males and females had overlapping home ranges, and appeared to have dominance hierarchies rather than true territories. Females were secretive: Wyllie saw only three copulations despite intensive study. Wyllie tentatively concluded, based mainly on home range information, that European cuckoos are promiscuous. Vigorous courtship flights by competing males may, he suggested, inform females about male fitness. Brown-headed cowbirds also engage in vigorous courtship flights during which several male cowbirds chase a female for 5 minutes or more while vocalizing frequently (Friedmann 1929) and these too may inform females about male quality (S. Rothstein, pers. commun.). This may also be true of *C. micropterus* in Russia and Java (Becking and Snow 1985). More recently, Dröscher (1988) and Nakamura and Miyazawa (1990) radio-tracked *C. canorus* in Germany and Japan, respectively, and found exclusive male territories overlapping the laying areas of multiple females. In Japan, subordinate males often intruded into these territories in the absence of the owners.

By contrast, Riddiford's (1986) study of unmarked *C. canorus* suggested the formation of pair bonds: a male may cooperate with a female to defend her host-rich breeding territory, although he may then attempt to mate with unpaired females that have been chased out of the territory. This is, of course, typical of many pair-bonded birds (Westneat et al. 1990); brood parasites of both sexes may similarly practice a mixed strategy of short-term monogamy and EPCs. Prolonged association of male and female in the genus *Cuculus* is also suggested by Vernon's statement (in Ginn et al. 1989) that male red-chested cuckoos *C. solitarius* establish a territory, find the host nests within, call to attract females and then lead them to the nests.

Less still is known of the mating systems of other cuckoos. Males and females of the large *Clamator* cuckoos (four species) cooperate in finding and gaining access to host nests (Liversidge 1971; Steyn 1973; von Frisch 1973; Gaston 1976; Rowan 1983; Soler 1990). Noisy contact calling between the sexes is mentioned for this genus (Rowan 1983), but males do not call after an initial courtship period (Gaston 1976). Short- to medium-term pair bonds (more than about one-fourth of the breeding season), with the capacity for diverse cooperative behaviors (Rowan 1983; Soler 1990; Arias-de-Reyña, this vol.) therefore seem to occur in this genus. Liversidge (1971), however, suggested that although a male *Clamator jacobinus* may associate with a female during and immediately after laying, the presumed 48-hour laying cycle allows him to consort with other females on the intervening day. Female *C. jacobinus* appear to have overlapping home ranges (Rowan 1983). Liversidge therefore regarded this species as "promiscuous," at least in high-density populations; work with color-marked birds would clarify the nature and extent of consortships or pair bonds.

In at least some of the iridescent *Chrysococcyx* cuckoos, males call from song posts throughout the breeding season and seem to defend exclusive "laying territories" or home ranges (Ford 1963; Jensen and Vernon 1970; Jensen and Clinning 1974; Gill 1983). No studies of the copulation patterns of marked birds have been published, but

Table 18.1 Mating systems and sexual dimorphism of the parasitic cuckoos

Taxon	Color dimorphic?	Size dimorphic?	Mating system	References
Clamator[a,b]	No	No	Promiscuity?	Liversidge (1971)
			Monogamy?	Gaston (1976); Rowan (1983); Arias-de-Reyña (this vol.)
Pachycoccyx	No	Minor/no	Promiscuity?	Rowan (1983); Vernon (1984)
Cuculus[a,b,c]	Minor/no	Minor/no	Promiscuity?	Jensen and Clinning (1974); Wyllie (1981),[4] Rowan (1983)
			Monogamy?	Riddiford (1986)
Cercococcyx	No	No	Unknown	Rowan (1983); Ginn et al. (1989); Maclean (1993)
Cacomantis[a]	No	No	Unknown	Beehler et al. (1986)
Rhamphomantis	Yes	Minor/no?	Unknown	Beehler et al. (1986)
Misocalius	No	No	Unknown	Beehler et al. (1986)
Chrysococcyx[a,b,c]	Yes	Minor/yes	Polygyny?	Rowan (1983); Ginn et al. (1989)
[Chalcites]			Monogamy?	Friedmann (1948); Gill (1983)
			Promiscuity?	Maclean (1990)
Caliechthrus	No	No	Unknown	Beehler et al. (1986)
Surniculus	No	No	Unknown	King et al. (1986)
Microdynamis	Yes	?	Unknown	Beehler et al. (1986)
Eudynamis[a]	Yes	?	Polygyny?	Wyllie (1981)
Urodynamis	No?	?	Unknown	Beehler et al. (1986)
Scythrops	No?	?	Unknown	Beehler et al. (1986)
Tapera	No	No	Unknown	Meyer de Schauensee and Phelps (1978)
Dromococcyx	No	No	Unknown	Meyer de Schauensee and Phelps (1978)

[a]Apparent cooperation of sexes in distracting host at nest recorded in this group.
[b]Apparent supplementary (courtship) feeding recorded in this group.
[c]Male use of habitual call-sites recorded in this group.
[d]Mating system categorization based at least partly on observation of copulations by color-marked birds.

monogamy (Friedmann 1948; Gill 1983), short-term polygyny (Jensen and Vernon 1970; Rowan 1983) and promiscuity (Maclean 1990) have all been implied. Whereas the use of prominent call sites and the extended courtship season suggests promiscuity, several authors (Zim, in Friedmann 1968; Jensen and Jensen 1969; Vernon, in Ginn et al. 1989) mention cooperation by males and females of *Chrysococcyx* in luring hosts from the nest. Male *C. caprius* and *C. klaas* also supply hairy caterpillars to females after, or rarely before, copulation (Friedmann 1948; Rowan 1983; Vernon, in Ginn et al. 1989). Rowan (1983) cites evidence, from unmarked birds, that pair bonds form for up to ten days in *C. klaas*. However, R. B. Payne (pers. commun.) has observed males consecutively feeding several ovulating females in this way at a regular site, apparently in a typical example of call-site promiscuity.

Information about other cuckoo genera is mainly limited to systematics, host records, and distribution. Some species will undoubtedly remain little-known for decades, as a result of their rarity, inaccessibility, secretive habits, and lack of sexual dimorphism which tends to confuse behavioral observation. However, with only a little effort to color-mark cuckoo populations, observers can make a good contribution to our understanding of their sexual behavior and mating strategies. This is as true of the genera mentioned above as of those that are completely unknown.

Parasitic finches

There are two types of parasitic finches, both African. They are the viduine finches (Viduidae, often regarded as Ploceidae: Viduinae) and the rather aberrant cuckoo-finch, *Anomalospiza imberbis* (Ploceidae). There are 14 or more viduines (Payne 1985*b*), of three general forms: the glossy black, short-tailed indigobirds (sometimes called subgenus *Hypochera*), the long- and slender-tailed whydahs (subgenus *Tetraenura* or *Vidua*), and the long- and broad-tailed paradise whydahs (subgenus *Steganura*) (Payne 1985*b*). All parasitize grassfinches of the family Estrildidae. Nearly all are host-specific, and mimic the nestling gape patterns, begging calls, behavior, and sexual song of their host (Payne 1973*a*, 1985*b*,*c*).

The cuckoo finch is a generalist parasite on

Table 18.2 Mating systems and sexual dimorphism of the parasitic finches

Taxon	Color dimorphic?	Size dimorphic?	Mating system	References
Anomalospiza[a]	Minor	No	Unknown	Friedmann (1960); Vernon (1964); Ginn et al. (1989)
Vidua[a]	Yes	Yes	Promiscuity	Payne (1973a); Payne and Payne (1977);[b] Barnard (1989);[b] Barnard and Markus (1989)[b]
			Monogamy	Friedmann (1960)

[a]Male use of habitual call-sites recorded in this group.
[b]Mating system categorization based at least partly on observation of copulations by color-marked birds.

grass warblers, *Prinia* and *Cisticola* (Friedmann 1960; Vernon 1964). It is not highly sexually dimorphic (Maclean 1993), and does not mimic its hosts (Payne 1985*c*; Ginn et al. 1989). As its generic name implies, its systematic affiliations are uncertain and its parasitism probably evolved independently of the viduines (Friedmann 1955, 1960). Its mating system is unknown (table 18.2). Unlike some parasites, the cuckoo finch lives in an open habitat and study of its habits should be fairly easy.

Mating systems have been studied primarily in three viduine finches, which are all sexually dimorphic in size and color (table 18.2). Payne and Payne (1977) analyzed the mating system of village indigobirds (*Vidua chalybeata*) in a Zambian woodland, and concluded that it represented features of both polygyny and promiscuity. Males formed "dispersed leks" with well-spaced, traditional song perches (call-sites). Females visit these call-sites and mate with only a few of the resident males (three of 14 males in one population had 86% of the copulations). Few individual females were seen to copulate with multiple males. Female choice in their study seemed to be based mainly on song output, and secondarily on the presence of neighboring males and defended resources. Song seemed to be used primarily in male–male contexts, with females basing their mating decisions on the outcome of male competition. Other papers (Payne 1973*a*, 1982, 1983, 1985*c*; Payne and Groschupf 1984) suggest that essentially the same mating systems are found in the other indigobirds.

Mating systems have also been studied in two long-tailed viduines, the pin-tailed whydah *V. macroura* and shaft-tailed whydah *V. regia* (Shaw 1984; Barnard 1989*a*,*b*, 1990; Barnard and Markus 1989). The pin-tailed whydah is a nonspecialist parasite of African grasslands (Macdonald 1980; Payne 1985*b*). The shaft-tailed whydah is a specialist parasite of *Acacia* savanna, and mimics the gape and song of its host, the violet-eared waxbill *Uraeginthus granatinus*.

The mating systems of both whydahs, just as in the village indigobird, are based on male defense of a traditional call-site. Males and females are promiscuous, but males are more extensively so; some males mate with 12 or more females in a breeding season, but females apparently mate with only one or a few males (Barnard 1989*a*,*b*; Barnard and Markus 1989). Females compete aggressively for copulations in certain contexts (Barnard and Markus 1989). As in the indigobirds (Payne 1979), copulation rates of male whydahs are highly skewed, and almost certainly much more variable than those of individual females. Defense of a call-site appears necessary for successful copulation by males. Nonterritorial, satellite males repeatedly challenge call-site holders; holders of poor-quality sites also attempt to take over better sites (Barnard 1989*a*,*b*). If males leave their call-sites temporarily to drink or inspect neighboring sites, satellites often arrive to sing from the song post. However, they rarely if ever obtain copulations. Many satellites acquire breeding plumage much later than territorial males, perhaps reflecting poor condition or reduced fertility (Barnard 1991, 1995). It is not known whether they are always younger than territory-holders.

The seasonal mating success of male whydahs is linked to the defense of resources such as perennial water, and to the centrality of the site (Barnard 1989*a*,*b*). Whydahs at isolated call-sites usually have low copulation rates and few visits by receptive females or challenging males (see Bradbury et al. 1986). If these isolated territorial males are experimentally removed, their sites stand vacant longer than central sites. Therefore, attractive sites seem to be defined by the presence of water and by their location. Neither body size nor tail length is correlated with mating success, although male shaft-tailed whydahs with longer

Table 18.3 Mating systems and sexual dimorphism of the honeyguides

Taxon	Color dimorphic?	Size dimorphic?	Mating system	References
Indicator[a,b]	Minor/no	Minor/yes	Polygyny?	Short and Horne (1985)
			Promiscuity?	Friedmann (1955); Ranger (1955); Cronin and Sherman (1977)[c]
Prodotiscus	No	Minor/no	Unknown	Friedmann (1955); Short and Horne (1985)
Melignomon	No	Unknown	Unknown	Friedmann (1955); Short and Horne (1985)
Melichneutes	No	No	Unknown	Friedmann (1955); Short and Horne (1985)

[a]Male use of habitual call-sites recorded in this group.
[b]Apparent cooperation of sexes in distracting host at nest recorded in this group.
[c]Mating system categorization based at least partly on observation of copulations by color-marked birds.

tails were preferred by females in choice experiments (Barnard 1989*a*,*b*, 1990). Mating systems in the viduines are probably therefore best described as call-site-based promiscuity.

Where males controlling call-sites also defend a patchy resource (such as water), receptive females may mate on a resources-for-sex basis (Barnard 1989*a*,*b*), although such males are also likely to be dominant (Payne and Payne 1977). Grass seed, which is more evenly distributed, cannot be exclusively controlled by males, but females may feed, drink, and mate at a male's call-site through economic convenience. There is no evidence that females form postcopulatory pair bonds with any male, although Van Someren (in Friedmann 1960: 100) saw a male pin-tailed whydah accompany a laying female to a host nest. Indeed, although most females habitually visit the call-sites of one or two males, the behavior of both sexes is notably opportunistic. The term "polygyny" (Payne and Payne 1977; Shaw 1984; Savalli 1990) thus seems a misnomer. Payne's term "dispersed lek" (Payne and Payne 1977; Payne 1984) has much merit in describing how males form song populations, are visited by receptive females, and have highly skewed mating success. However, the term "lek" implies that males defend no resources of significance to females. This is not strictly true of the whydahs (Barnard 1989*a*,*b*; Savalli 1990). The term may also, by implication, exaggerate the extent to which females "sample" males before copulating. There is a need for focused study of female movements and behavior in this genus.

The honeyguides

The parasitic honeyguides (Indicatoridae) are about 17 species of small piciform birds, which get their name from the cooperative association between humans and *Indicator indicator* in finding bees' nests (Friedmann 1955; Short and Horne 1985). Most species are Afrotropical and two are Asian. Most parasitize hole-nesting birds, especially other piciforms (woodpeckers and barbets), although the smallest genus, *Protodiscus*, parasitizes white-eyes, swallows, swifts, and warblers. The honeyguides are drably colored, usually sexually monomorphic birds whose forest and woodland habitats, large home ranges, and secretive behavior make them very difficult to study (Short and Horne 1985). Many species, however, feed at bees' nests or other predictable sites, some of which are defended (Short and Horne 1985), so the potential exists to study mating at defended sites.

Cronin and Sherman's (1977) early study of mating systems in the orange-rumped honeyguide *Indicator xanthonotus* is unfortunately still the only published analysis of honeyguide mating systems and resource defense using color-marked birds (table 18.3). In their study in eastern Nepal, males aggressively defend territories, centered around *Apis dorsata* hives, from other males throughout the year. Females that copulate with a territorial male, and juveniles accompanying them, are allowed access to the male's wax resources in a kind of "food-for-sex" arrangement. Nonterritorial males at the edge of established territories try to intercept females on their way to hives, but appear unsuccessful at insemination because females struggle during these mountings.

Cronin and Sherman (1977) classified this system as "resource-based non-harem polygyny," as their focal male copulated with at least 18 (and perhaps more than 28) females, and some females also mated with more than one male. Yet, at least superficially, this system has very strong similarities to the promiscuous system of the viduines.

Table 18.4 Mating systems and sexual dimorphism of the parasitic cowbirds and black-headed duck

Taxon	Color dimorphic?	Size dimorphic?	Mating system	References
Molothrus	Mostly	Yes	Variable/ monogamy	Friedmann (1929); Ankney and Scott (1982); Darley (1982);[a] Dufty (1982);[a] Teather and Robertson (1986);[a] Yokel (1986, 1989);[a] Rothstein et al. (1986); Mason (1987); Robinson (1988)
Scaphidura	Yes	Yes	Polygyny/ promiscuity?	Meyer de Schauensee and Phelps (1978); Robinson (1986)
Heteronetta[b,c]	Yes	Minor/yes	Monogamy/ polyandry?	Rees and Hillgarth (1984) Madge and Burn (1988); Todd (1979)

[a]Mating system categorization based at least partly on observation of copulations and/or consortships (Yokel 1989) by color-marked birds.

[b]Apparent defense of females and trios recorded in this group in captivity.

[c]Apparent cooperation of sexes in distracting host at nest recorded in this group.

The main differences seem to be that: (i) bee hives are a more patchy, rare resource than the grass seed and water important to female whydahs; and (ii) male honeyguides do not appear to advertise their presence by calling (but see Ranger 1955; Short and Horne 1985). However, as in the whydahs and indigobirds, control of a territorial post appears necessary for successful mating: upon experimental removal of territorial males at a hive, Cronin and Sherman (1977) found that previously peripheral males were able to copulate successfully with receptive females.

Short and Horne (1985) found that Kenyan honeyguides show extended pair bonds and complex cooperative behaviors. Males assist females to enter host nests (see section "Cooperation to gain access to host nests"), and may actively monitor and defend duetting host pairs from other males, even in the nonbreeding season. Two adult lesser honeyguides, *Indicator minor*, also monitored the parasitized nest of a barbet until the young honeyguide left the nest (Short and Horne 1985).

Honeyguide males may hold territories and show intraspecific aggression year-round, although many species remain to be studied intensively (Short and Horne 1985). Copulation at the habitual call-sites of *I. indicator* and *I. minor* males has been repeatedly noted (Friedmann 1955; Ranger 1955; pers. commun. to R. B. Payne; Payne, 1992), and females visit these sites during their fertilization window (Payne 1992). By contrast, Short and Horne (1985) found no evidence for mating at Kenyan call-sites. Males sing for up to eight months per year from these call-sites. Intraspecific competition in honeyguides seems high in many areas, perhaps reflecting the scarcity of host nest sites.

Short and Horne (1985) also observed dominance interactions in scaly-throated honeyguides *I. variegatus* and their congeners by providing honey, wax, and hive debris at a site in Kenya. Their studies, and those of Cronin and Sherman (1977), show that honey and related products are important, predictable resources to which honeyguides are attracted. Although such studies require hard work, resource-based mating systems are intriguing and warrant further study. There is a particular need to disentangle competition for food resources and hosts.

Parasitic cowbirds

Five cowbirds in the New World subfamily Icterinae are obligate parasites. These include four of five species in the genus *Molothrus* (screaming *M. rufoaxillaris*, bronzed *M. aeneus*, shiny *M. bonariensis* and brown-headed cowbirds *M. ater*), plus the giant cowbird, *Scaphidura orizivora* (Friedmann 1929; Payne 1977; Gochfeld 1979; Robinson 1988). As in many nonparasitic icterines, the parasitic cowbirds (except *M. rufoaxillaris*) are sexually dimorphic (table 18.4). They parasitize a wide range of mainly insectivorous hosts (Payne 1977); the brown headed cowbird parasitizes nearly all sympatric passerines in its expanding range, with host records from over 200 species (Friedmann and Kiff 1985). Detailed studies of mating systems have been published only for the brown-headed cowbird. However, Robinson's (1988; pers. commun.) work on the social behavior of the very dimorphic, gregarious giant cowbirds provides indirect evidence of male–female cooperation, in the apparent absence of pair-bonds. Mason (1987) has also provided some evidence for monogamy in the screaming cowbird (see Fraga, this vol.).

The brown-headed cowbird's behavior has been the subject of much debate (e.g., Ankney and Scott 1982; Teather and Robertson 1986). This bird represents a warning to those of us working on mating systems not to become complacent. In different studies, it has been termed promiscuous, polygynous, monogamous, and polyandrous (table 18.4). Recent reviews of this variability have analyzed whether it is due to semantic or observational bias (Ankney and Scott 1982) or to actual geographic diversity (Teather and Robertson 1986; Yokel 1986). It is striking that although many general sources reported this cowbird's mating system as promiscuous or polygamous, nearly all of the intensive studies with marked birds reported that social monogamy was predominant (Darley 1982; Dufty 1982; Teather and Robertson 1986; Yokel 1986, 1989; but cf. Elliott 1980).

In these color-marked populations, there appears to be a complex blend of conflicting strategies in response to different environments and resource dispersion patterns (Rothstein et al. 1984, 1986, 1987; Yokel 1986, 1989; see also Ankney and Scott 1982). Whereas in some cases virtually all birds in a population have proven to be monogamous (Dufty 1982) or promiscuous (Elliott 1980), other intensive studies have shown that there is general monogamy, with some individuals either polygynous or promiscuous (e.g., Teather and Robertson 1986; Yokel 1986, 1989). Even in a predominantly monogamous situation, pair bonds in some individuals are short term and dissolve despite the continued survival and breeding of both partners (Yokel 1986). Different populations may also have different rates of extrapair copulation (Rothstein et al. 1986). For example, Yokel (1986) found that females had ample opportunity to engage in EPCs, but did not, and mate guarding by males was weak. By contrast, males in Darley's (1982) and Dufty's (1982) populations guarded their mates closely.

Regional variation in cowbird population density, sex ratio, home range size, and the spatial distribution of host nests creates a varying potential for monogamy (Elliott 1980; Dufty 1982; Rothstein et al. 1986; Yokel 1986, 1989). In the breeding season, brown-headed cowbirds of both sexes have a "morning-breeding/afternoon feeding" daily distribution, spending most of the morning in host-rich areas, and most of the afternoon at feeding sites (Rothstein et al. 1984, 1987). This pattern — the reverse of that found in nonparasitic icterines — involves dispersal from clumped feeding and roosting sites, where cowbirds are highly gregarious, to scattered host breeding areas where they are asocial and intent on nest finding (Rothstein et al. 1984, 1986). Females thus have opportunities to mate with other males on a daily basis, but do not do so, in part because virtually all copulations occur when a single male and female are alone, which is usually in the morning and on breeding areas (Rothstein et al. 1988). Males may remain paired because the high reproductive output of their mates assures them of continuous mating opportunities (Robinson 1986). Nevertheless, paired males will court females other than their mate if they are alone with such females; this courtship, however, rarely results in matings (Yokel and Rothstein 1991). Females are aggressive to other females within their range, but do not defend exclusive territories (Payne 1977:14; Teather and Robertson 1985; Yokel 1986; but cf. Dufty 1982).

Male cowbirds, like most other parasites, do not defend breeding resources of importance to females (Yokel and Rothstein 1991). Females choosing copulation partners may therefore pay closer attention to male phenotypic traits, which perhaps reflect genetic quality. Ankney and Scott (1982) found that apparently paired males were heavier than seemingly unpaired males, and often had better fat and protein reserves. This relationship could reflect male dominance interactions rather than female choice, and indeed Yokel and Rothstein (1991) concluded that dominance interactions were used by female cowbirds as proximate cues for male genetic quality. Arguments in favor of female choice of genetic quality have generated much controversy (e.g., Jones 1987), but do seem difficult to dispute in these cowbirds (see below). No direct data on the attractiveness of male phenotypes are available, although females appear to "cue into" variation in the two song types given by male cowbirds (O'Loghlen and Rothstein 1995), which may reflect dominance (West et al. 1981; Rothstein and Fleischer 1987; Rothstein et al. 1988).

Some authors have proposed that the chief benefit of monogamy to female cowbirds is protection from harassment by other males (see subsection "Assurance of paternity and reduction of female harassment"; Ankney and Scott 1982; Darley 1982). However, such a benefit was not clearly apparent in experimental conditions in California, where females were often alone with other courting males but virtually never copulated with them (Yokel and Rothstein 1991). Paired male cowbirds do not appear to help females find food, or alert them to the presence of predators. Nor do male cowbirds assist females in finding or gaining access to host nests (Yokel and Rothstein 1991). At least in the California population studied by Yokel, Rothstein and coworkers, therefore, there seem to be no direct, tangible benefits to females in pairing. It is possible that pairing behavior is adaptive mainly for males, through some paternity assurance by mate guarding (e.g., Darley 1982; Dufty 1982), and has beneficial side effects for females (see section "Assurance of paternity and reduction of female harassment"). Alternatively, the large home ranges occupied by cowbirds as a result of their requirements for feeding and breeding may make monopolization of more than one mate impractical for males (Yokel and Rothstein 1991; see subsection "Economic monopolization of mates").

As Yokel (1986) has noted, a large measure of the confusion surrounding cowbird mating systems may arise from workers' tendencies to cubbyhole behavior into the discrete terms of mating systems theory. This point deserves emphasis. Until we recognize that natural variation in animal behavior often lies along a complex continuum, our efforts to understand it will only cause confusion (see also Wickler and Seibt 1983). The mating and spacing systems of cowbirds appear unusually plastic (Teather and Robertson 1986; Yokel 1989), but this may only be because they are much better studied than other parasites.

The black-headed duck

The black-headed duck *Heteronetta atricapilla* is the only obligately parasitic duck, unique in a family that otherwise shows great flexibility in social, mating, and laying strategies (Todd 1979; Madge and Burn 1988; Sorenson, this vol.). Interestingly, it is also the only precocial bird among the 80 or so species of obligate brood parasites (Rothstein 1990). Because the duckling can feed and care for itself a day or two after hatching (Weller 1968), its demands on the host parent are few and there should be less stringent counter-selection on host defenses. Indeed, in Weller's view, the duck "borders on commensalism rather than parasitism" and is "the most perfect of avian parasites." This suggests that obligate parasitism should be a more common strategy of precocial birds than it is, although facultatively parasitic ducks may have a negative impact on host fitness (Payne 1977). The black-headed duck parasitizes mainly coots and other ducks, but 18 species of host have been recorded, including gulls and raptors (Madge and Burn 1988).

There is little information on the sexual behavior of this slightly dimorphic duck (table 18.4). In a captive study, Rees and Hillgarth (1984) observed male defense of females, mutual preening, joint patrolling, and consistent pair association, although paired males pursued other females for EPCs. Weller (as quoted by Rees and Hillgarth 1984) also noted seasonal pair-bonds. The presence of a one-female/two-male "trio" in this population, and possible forced copulations, show that the mating system deserves much more detailed, intensive study such as that being conducted by B. Lyon (unpubl. data).

SELECTIVE PRESSURES FAVORING PAIR BONDS IN PARASITIC BIRDS

What factors influence the mating systems of brood parasites? This review suggests that there are a variety of factors and constraints that have shaped the strategies of different brood parasites. Because of the phylogenetic and ecological diversity of parasites, it may be unrealistic to look for universal causation in the evolution of their mating systems.

Brood parasitism is a risky business, despite the freedom from parental duties that parasites have. Host nests are not always present or accessible at the right time; parasites may be strongly mobbed when they approach nests; successfully laid eggs may be rejected or abandoned by the host; and host nest predation may be substantial, especially in the tropics (Rowan 1983; Barnard and Markus 1990). In this context, assistance from males may often increase the reproductive success of both sexes. Distinct and largely exclusive pair bonds have evolved in some groups, presumably to increase the success of male and female. In this section, I summarize what appear to be the main factors promoting pair bonds in parasites.

Cooperation to gain access to host nests

Female parasites need not only tune their breeding physiology to that of suitable hosts in their home range; they must also gain access to nests within a short period, timing their approach to encounter as little resistance as possible. Laying opportunities may represent severe constraints for individual females (Barnard and Markus 1989).

Cuckoos normally lay eggs at 48-hour intervals and can probably retain their eggs in the oviduct for a maximum of a day (Payne 1973*b*). If no suitable nests are accessible during this time, the egg has to be dumped. In some cases, it may help to have the cooperation of a male in luring the hosts away from their nest. This may apply especially to cases where the hosts are larger than the parasite, where they are communal or gregarious, where they have vigorous mobbing behavior, or where they readily abandon nests in response to disturbance by a parasite. Possibly, sexual monomorphism in parasites may aid in confusing the hosts, although it may also be a nonadaptive consequence of phylogeny.

In the *Clamator* cuckoos, cooperation of males and females in nest access is indeed associated with both monomorphism and "difficult," often large, hosts (Arias-de-Reyña, this vol.). These cuckoos parasitize communally breeding, noisy babblers (*Turdoides*), large crows (*Corvus*), laughing-thrushes (*Garrulax*), and bulbuls (*Pycnonotus*) (Wyllie 1981). Even bulbuls, which are small to medium-sized passerines with rather weak flight, are noisy and conspicuous when in groups, alarm-calling incessantly and attracting other birds (Maclean 1993; pers. obs.). Some *Cacomantis*, *Cuculus*, and *Surniculus* cuckoos also parasitize communally breeding Asian babblers (Wyllie 1981). These genera show little or no sexual dimorphism, although it is not yet known whether the species concerned have male–female cooperation. *Surniculus* and *Cuculus gularis* also parasitize drongos (*Dicrurus*), which are particularly bold, pugnacious birds (Rowan 1983; Tarboton 1986); monomorphic *Pachycoccyx audeberti* parasitizes a pugnacious communal breeder, the red-billed helmet-shrike *Prionops retzii* (Rowan 1983). One cuckoo for which cooperation has been reported, the koel *Eudynamis scolopacea*, parasitizes large crows but is sexually dimorphic in color, the black male being thought to mimic the host (Wyllie 1981).

In the monomorphic or subtly dimorphic honeyguides, cooperation between male and female has occasionally been noted in attempts to lure barbets (Capitonidae) and woodpeckers (Picidae) from their well-defended hole nests (Short and Horne 1985). In these cases, a lone female may be unable to withstand attack by hosts, and a surreptitious approach is called for. Experimentation would help to establish the generality of the hypothesis that parasites have developed pair bonds to facilitate access to the host nest.

Assurance of paternity and reduction of female harassment

The benefits of paternity assurance presumably accrue mainly to males, whereas both sexes may benefit from reduced harassment of females if harassment would otherwise reduce female efficiency at finding and parasitizing nests (see also Rothstein et al. 1986; Yokel and Rothstein 1991). Therefore, although mate-guarding may be a male reproductive strategy, females may accept it if it simultaneously reduces interference from other males and if this benefit outweighs any from engaging in EPCs. This may apply both to the socially feeding cowbirds and to the black-headed duck.

Mate-guarding by male brown-headed cowbirds has been emphasized by Darley (1982), Dufty (1982) and Teather and Robertson (1986). As most cowbird populations studied so far have male-biased sex ratios (see subsection "Economic monopolization of mates"), paired females are often subject to consortship and EPC attempts by other males. Teather and Robertson (1986) removed a trigamous male from their population and showed sharply increased courtship of his females thereafter, implying that male consorts are effective at guarding. However, overall these authors found that females were courted by males other than their mate in 30% of all courtship events, apparently giving ample opportunity for extrapair copulations. Yokel (1986) found a low rate of mate guarding in cowbirds; females had frequent opportunities for EPCs but virtually never engaged in them. Yokel and Rothstein (1991) also suggested that males were ineffective at guarding and protecting females from harassment, as females mated to bigamous males had similar consort behavior to those mated to monogamous males.

Although it seems possible in theory that female cowbirds could accept pairing in terms of these direct "service" benefits (Yokel and Rothstein 1991), the evidence is weak that females actually do benefit. In any case, this hypothesis probably does not apply to parasites that are not socially gregarious. A focused study of the black-headed duck, in which forced copulations may occur (Rees and Hillgarth 1984), might shed light on the subject.

Economic monopolization of mates

At the proximate level, the ability of males to monopolize more than one female should vary with local population density and sex ratio (Yokel

1989). Most well-studied cowbird populations have a male-biased sex ratio (up to 1.6:1; reviewed by Teather and Robertson 1986), which may influence both the costs and benefits of mate guarding, and the ease with which two or more females can be closely attended (Yokel 1989). Teather and Robertson (1986) noted that at equal sex ratios, although EPCs do occur, less pressure is put on paired males by satellites, and thus there is less need to guard females.

The population density of parasites, relative to those of their hosts, may influence the monopolizability of females through its influence on female home range size (Elliott 1980; Rothstein et al. 1986; Teather and Robertson 1986; Yokel 1989), but the direction of effect is not clear. If parasite density is high and host density low, female monopolization may be uneconomic, both because of constant pressure from competing males and large female home range sizes. An alternative view is that at high parasite densities, constant mate guarding is the only viable strategy for a male (Robinson 1986). Where parasite density is low and host density high, males may be able to manage exclusive access to one or two females (Dufty 1982; Teather and Robertson 1986; Yokel 1989).

What is the empirical support for an economic-monopolization scenario? Liversidge's (1971) proposal that Jacobin cuckoos should be promiscuous at high densities and monogamous or polygynous at low densities has been only very weakly supported in other species. Comparing two cowbird populations of similar sex ratio, Yokel (1989) found that although monogamy was predominant in both, female–female aggression was higher and male mate guarding weaker in the high-density area. In normally promiscuous parasites such as the whydahs, female sexual competition for copulations is apparent, and both sexes are more extensively promiscuous at high densities (Barnard and Markus 1989; P. Barnard, unpubl. data). At low whydah densities, females have a very limited range of economically accessible mates, although this does not lead to the formation of a pair bond. There is a great deal of room for rewarding and profitable research in this area, especially with African and Australasian parasites.

CONCLUSIONS

At present, our patchy information on the mating systems of brood parasites allows only tentative conclusions, not only about how best to explain these systems, but even how best to classify them. Among the cuckoos, whose mating systems have seldom been studied in marked populations, there seem to be several different systems. The *Clamator* cuckoos, and probably several other groups (*Pachycoccyx*, *Eudynamis*, and other Australasian and tropical genera), seem to have extended and primarily monogamous associations between the sexes, including cooperation in gaining access to host nests. Other genera, such as *Chrysococcyx*, may have brief associations. An enormous amount remains to be learned about the habits, population ecology and mating strategies of cuckoos.

For the honeyguides, there is a dichotomy between the reports of call-site-based, resource-defense promiscuity (Cronin and Sherman 1977; H. A. Isack, as quoted by R. B. Payne, pers. commun.), and that of Short and Horne (1985) which implies some form of pair bond (polygamous or monogamous). Because many species are attracted to patchy resources such as beeswax and honey, the honeyguides can be productively studied despite their otherwise secretive habits. There is a particular need to disentangle selection pressures based on the distribution of food and hosts. The viduine finches have a call-site-based promiscuity: sites with a central location and abundant critical resources nearby are more active than those without, although it is unclear how important male-controlled resources are in the economy of individual females. In the precocial black-headed duck, apparent defense of females by males and prolonged association suggest the formation of pair bonds, but may better be regarded as paternity assurance because males pursue other females and sometimes force them to copulate (Rees and Hillgarth 1984). An intensive study of sexual behavior with marked birds still needs to be done.

Of the cowbirds, the mating systems of only the brown-headed cowbird have been explicitly studied with marked birds. These studies form a valuable model for those of us working on less well-known parasites. The cowbird's mating system is geographically variable, apparently in response to changing sex ratios, population densities, and host availability. The intriguingly flexible responses of this species to changing environmental conditions may not be unique to the cowbirds. Students of other brood parasites, particularly those in tropical or Australasian habitats where resource use and availability may differ radically from that in northern climates, should be alert to potentially dynamic patterns of intraspecific variability in their subjects.

In general terms, how can the patchiness of our knowledge of brood parasite mating systems be remedied? There is no single substitute for intensive observation of color-marked populations, preferably with simultaneous study of laying periodicity (e.g., Payne, 1992). However, our knowledge and understanding can probably be far advanced by the use of powerful supplementary techniques such as molecular parentage markers and radiotelemetry (Gibbs et al. 1990; Nakamura and Miyazawa 1990; Westneat et al. 1990). It can probably be said that the study of reproduction in brood parasites has always been more difficult than that of reproduction in nesting birds, because the eggs were never thought to be "all in one basket." However, while biologists working on birds with parental care have recently experienced an unsettling revolution in their assumptions and methods, those of us who study parasites should already be hardened to the view of taking nothing for granted, either about parentage or mating systems.

Although it seems difficult to understand any form of pair bonding in brood parasites, such bonds do exist, possibly for easier host nest access, assurance of paternity, protection from harassment, or other, unexplored reasons. Many of these factors probably relate to the availability of resources which, under our present paradigms, we cannot immediately identify. If we want brood parasite studies to make a good contribution to mating systems theory, we must ask much more explicit and probing "why" questions about the costs and benefits to individual parasites of different mating tactics.

ACKNOWLEDGMENTS I thank the other faculty members of the Zoology Department, University of Namibia, for shouldering the teaching load while I wrote this review. My research on whydahs was partly supported by the universities of Uppsala, Namibia, and the Witwatersrand. Robert Payne, Stephen Rothstein, and Scott Robinson provided uncommonly helpful, detailed, and stimulating reviews, editorial input and unpublished information, for which I am extremely grateful. Thanks also go to Guy Cowlishaw, Arne Lundberg, Paul Sherman, Rob Simmons, Staffan Ulfstrand, and Fredrik Widemo for discussing earlier drafts.

REFERENCES

Andersson, M. (1984). Brood parasitism with species. In: *Producers and scroungers: strategies of exploitation and parasitism* (ed. C. J. Barnard), pp. 195–228. Croom Helm, London.

Ankney, C. D. and D. M. Scott (1982) On the mating system of brown-headed cowbirds. *Wilson Bull.*, **94**, 260– 268.

Barnard, P. (1989*a*). Territoriality and the determinants of male mating success in the southern African whydahs (*Vidua*). *Ostrich*, **60**, 103–117.

Barnard, P. (1989*b*). Comparative mating systems and reproductive ecology of the African whydahs (*Vidua*). M.Sc. Thesis. University of the Witwatersrand, Johannesburg.

Barnard, P. (1990). Male tail length, sexual display intensity and female sexual response in a parasitic African finch. *Anim. Behav.*, **39**, 652–656.

Barnard, P. (1991). Ornament and body size variation and their measurement in natural populations. *Biol. J. Linn. Soc.*, **42**, 379–388.

Barnard, P. (1995). Timing of ornament growth, phenotypic variation, and size dimorphism in two promiscuous African whydahs (Ploceidae: *Vidua*). *Biol. J. Linn. Soc.*, **55**, 129–141 .

Barnard, P. and M. B. Markus (1989). Male copulation frequency and female competition for fertilizations in a promiscuous brood parasite, the pin-tailed whydah *Vidua macroura*. *Ibis*, **131**, 421–425.

Barnard, P. and M. B. Markus (1990). Reproductive failure and nest site selection of two estrildid finches in *Acacia* woodland. *Ostrich*, **61**, 117–124.

Becking, J. H. and D. W. Snow (1985). Brood parasitism. In: *A dictionary of birds* (eds B. Campbell and E. Lack), pp. 67–70. T. and A. D. Poyser Ltd, Calton, UK.

Beehler, B. M., T. K. Pratt, and D. A. Zimmerman (1986). *Birds of New Guinea*. Princeton University Press, Princeton, New Jersey.

Bradbury, J. W., R. Gibson, and I. M. Tsai (1986). Hotspots and the dispersion of leks. *Anim. Behav.*, **34**, 1694–1709.

Campbell, B. and E. Lack (eds) (1985). *A dictionary of birds*. T. and A. D. Poyser Ltd, Calton, UK.

Cronin, E. W., Jr and P. W. Sherman (1977). A resource-based mating system: the orange-rumped honeyguide. *Living Bird*, **15**, 5–37.

Crook, J. H. (1965). The adaptive significance of avian social organisations. *Symp. Zool. Soc. Lond.*, **14**, 181–218.

Darley, J. A. (1982) Territoriality and mating behavior of the male brown-headed cowbird. *Condor*, **84**, 15–21.

Dröscher, L. (1988). A study on radio-tracking of the European cuckoo (*Cuculus canorus canorus*). *Proc. Int. 100 Deutsche Ornithologische Gesellschaft meeting. Current Topics in Avian Biology*, Bonn 1988, pp. 187–193.

Dufty, A. M., Jr (1982). Movements and activities of radio-tracked brown-headed cowbirds. *Auk*, **99**, 316–327.

Elliott, P. F. (1980). Evolution of promiscuity in the brown-headed cowbird. *Condor*, **82**, 138–141.

Emlen, S. T. and L. W. Oring (1977). Ecology, sexual selection, and the evolution of mating systems. *Science*, **197**, 215–223.

Ford, J. (1963). Breeding behaviour of the yellow-tailed thornbill in south-western Australia. *Emu*, **63**, 185–200.

Friedmann, H. (1929). *The cowbirds, a study in the biology of social parasitism.* C. C. Thomas, Springfield, Illinois.

Friedmann, H. (1948). *The parasitic cuckoos of Africa.* Washington Academy of Sciences, No. 1, Washington, DC.

Friedmann, H. (1955). *The honey-guides.* United States National Museum Bulletin No. 208.

Friedmann, H. (1960). *The parasitic weaverbirds.* Smithsonian Institution Bulletin No. 223.

Friedmann, H. (1968). *The evolutionary history of the avian genus* Chrysococcyx. United States National Museum Bulletin No. 265. Washington, D.C.

Friedmann, H. and L. F. Kiff (1985). The parasitic cowbirds and their hosts. *Proc. West. Found. Vert.*, **2001.2**, 225–304.

Gaston, A. J. (1976). Brood parasitism by the pied crested cuckoo *Clamator jacobinus. J. Anim. Ecol.*, **45**, 331–348.

Gibbs, H. L., P. J. Weatherhead, P. T. Boag, B. N. White, L. M. Tabak, and D. J. Hoysak (1990). Realized reproductive success of polygynous red-winged blackbirds revealed by DNA markers. *Science*, **250**, 1394–1397.

Gill, B. J. (1983). Brood-parasitism by the shining cuckoo *Chrysococcyx lucidus* at Kaikoura, New Zealand. *Ibis*, **125**, 40–55.

Ginn, P. J., W. G. McIlleron and P. le S. Milstein (eds) (1989). *The complete book of southern African birds.* Struik Winchester, Cape Town.

Gochfeld, M. (1979). Brood parasite and host coevolution: interactions between shiny cowbirds and two species of meadowlarks. *Amer. Natur.*, **113**, 855–870.

Gowaty, P. A. and D. W. Mock (eds) (1985). *Avian monogamy.* Ornithological Monographs No. 37.

Hamilton, W. J. and G. H. Orians (1965). Evolution of brood parasitism in altricial birds. *Condor*, **67**, 361–382.

Jensen, R. A. C. and C. F. Clinning (1974). Breeding biology of two cuckoos and their hosts in South West Africa. *Living Bird*, **13**, 5–50.

Jensen, R. A. C. and M. K. Jensen (1969). On the breeding biology of southern African cuckoos. *Ostrich*, **40**, 163–181.

Jensen, R. A. C. and C. J. Vernon (1970). On the biology of the didric cuckoo in southern Africa. *Ostrich*, **41**, 237–246.

Jones, J. S. (1987). The heritability of fitness: bad news for "good genes"? *Trends Ecol. Evol.*, **2**, 35–38.

King, B., M. Woodcock and E. C. Dickinson (1986). *A field guide to the birds of south-east Asia.* Collins, London.

Liversidge, R. (1971). The biology of the Jacobin cuckoo *Clamator jacobinus. Ostrich* (Suppl. 8), 117–137.

Macdonald, M. A. (1980). Observations on Wilson's widowfinch and the pintailed whydah in southern Ghana, with notes on their hosts. *Ostrich*, **51**, 21–24.

Maclean, G. L. (1990). *Ornithology for Africa: a text for users on the African continent.* University of Natal Press, Pietermaritzburg, South Africa.

Maclean, G. L. (1993). *Roberts' birds of southern Africa*, 6th edn. John Voelcker Bird Book Fund, Cape Town.

MacWhirter, R. B. (1989). On the rarity of intraspecific brood parasitism. *Condor*, **91**, 485–492.

Madge, S. and H. Burn (1988). *Wildfowl.* Christopher Helm, Bromley, UK.

Mason, P. (1987). Pair formation in cowbirds: evidence found for screaming but not shiny cowbirds. *Condor*, **89**, 349–356.

Meyer de Schauensee, R. and W. H. Phelps, Jr (1978). *A guide to the birds of Venezuela.* Princeton University Press, Princeton, New Jersey.

Nakamura, H. and Y. Miyazawa (1990). Social system among cuckoo males *Cuculus canorus* at Kayanodaira Heights, Central Japan. *Bull. Inst. Nature Educ., Shiga Heights, Shinsu Univ.*, **27**, 17–27.

O'Loghlen, A. L. and S. I. Rothstein (1995). Culturally correct song dialects are correlated with male age and female song preferences in wild populations of Brown-headed cowbirds. *Behav. Ecol. Sociobiol.*, **36**, 251–259.

Orians, G. H. (1961). The ecology of blackbird (*Agelaius*) social systems. *Ecol. Monogr.*, **31**, 285–312.

Orians, G. H. (1969). On the evolution of mating systems in birds and mammals. *Amer. Natur.*, **103**, 589–603.

Oring, L. W. (1982). Avian mating systems. In: *Avian biology*, Vol. 6 (eds D. Farner, J. King and K. Parkes), pp. 1–92. Academic Press, New York.

Payne, R. B. (1973*a*). Behavior, mimetic songs and song dialects, and relationships of the parasitic indigobirds (*Vidua*) of Africa. *Ornith. Monogr.*, **11**, 1–333.

Payne, R. B. (1973*b*). Individual laying histories and the clutch size and numbers of eggs of parasitic cuckoos. *Condor*, **75**, 414–438.

Payne, R. B. (1977). The ecology of brood parasitism in birds. *Annu. Rev. Ecol. Syst.*, **8**, 1–28.

Payne, R. B. (1979). Sexual selection and intersexual differences in variation of mating success. *Amer. Natur.*, **114**, 447–452.

Payne, R. B. (1982). Species limits in the indigobirds (Ploceidae, *Vidua*) of West Africa: mouth mimicry, song mimicry, and description of new species. Miscellaneous Publications of the Museum of Zoology, University of Michigan, 162.

Payne, R. B. (1983). Bird songs, sexual selection, and female mating strategies. In: *Social behavior of female vertebrates* (ed. S. K. Wasser), pp. 55–90. Academic Press, New York.

Payne, R. B. (1984). Sexual selection, lek and arena behavior, and sexual size dimorphism in birds. *Ornith. Monogr.*, **33**, 1–52.

Payne, R. B. (1985*a*). Cuckoo. In: *A dictionary of birds* (eds B. Campbell and E. Lack), pp. 123–126. T. and A. D. Poyser Ltd, Calton, UK.

Payne, R. B. (1985*b*). Whydah. In: *A dictionary of birds* (eds B. Campbell and E. Lack), pp. 652–654. T. and A. D. Poyser Ltd, Calton, UK.

Payne, R. B. (1985*c*). The species of parasitic finches in West Africa. *Malimbus*, **7**, 103–113.

Payne, R. B. (1992). Clutch size, laying periodicity and behaviour in the honeyguides *Indicator indicator* and *I. minor*. *Proceedings of the 7th Pan-African Ornithological Congress*, 537–547.

Payne, R. B. and K. D. Groschupf (1984). Sexual selection and interspecific competition: a field experiment on territorial behavior of nonparental finches (*Vidua* spp.). *Auk*, **101**, 140–145.

Payne, R. B. and K. Payne (1977). Social organization and mating success in local song populations of village indigobirds, *Vidua chalybeata*. *Zeitschrift für Tierpsychologie*, **45**, 113–173.

Ranger, G. A. (1955). On three species of honey-guide: the greater (*Indicator indicator*), the lesser (*Indicator minor*) and the scaly-throated (*Indicator variegatus*). *Ostrich*, **36**, 70–87.

Rees, E. C. and N. Hillgarth (1984). The breeding biology of captive black-headed ducks and the behavior of their young. *Condor*, **86**, 242–250.

Riddiford, N. (1986). Why do cuckoos *Cuculus canorus* use so many species of hosts? *Bird Study*, **33**, 1–5.

Robinson, S. K. (1986). The evolution of social behavior and mating systems in the blackbirds (Icterinae). In: *Ecological aspects of social evolution. Birds and mammals* (eds D. I. Rubenstein and R. W. Wrangham), pp. 175–200. Princeton University Press, Princeton, New Jersey.

Robinson, S. K. (1988). Foraging ecology and host relationships of giant cowbirds in southeastern Peru. *Wilson Bull.*, **100**, 224–235.

Rothstein, S. I. (1990). A model system for coevolution: avian brood parasitism. *Annu. Rev. Ecol. Syst.*, **21**, 481–508.

Rothstein, S. I. and R. C. Fleischer (1987). Vocal dialects and their possible relation to honest status signalling in the brown-headed cowbird. *Condor*, **89**, 1–23.

Rothstein, S. I., J. Verner, and E. Stevens (1984). Radio-tracking confirms a unique diurnal pattern of spatial occurrence in the parasitic brown-headed cowbird. *Ecology*, **65**, 77–88.

Rothstein, S. I., D. A. Yokel and R. C. Fleischer (1986). Social dominance, mating and spacing systems, female fecundity, and vocal dialects in captive and free-ranging brown-headed cowbirds. In: *Current Ornithology, Vol. 3* (ed. R. F. Johnston), pp. 127–185. Plenum Press, New York.

Rothstein, S. L., J. Verner, E. Stevens, and L. V. Ritter (1987). Behavioral differences among sex and age classes of the brown-headed cowbird and their relation to the efficacy of a control program. *Wilson Bull.*, **99**, 322–337.

Rothstein, S. I., D. A. Yokel and R. C. Fleischer (1988). The agonistic and sexual functions of male brown-headed cowbirds, *Molothrus ater*. *Anim. Behav.*, **36**, 73–86.

Rowan, M. K. (1983). *The doves, parrots, louries and cuckoos of southern Africa*. David Philip, Cape Town.

Rubenstein, D. I. and R. W. Wrangham (eds) (1986). *Ecological aspects of social evolution. Birds and mammals*. Princeton University Press, Princeton, New Jersey.

Savalli, U. M. (1990). Interspecific aggression for food by a granivorous bird. *Condor*, **92**, 1082–1084.

Shaw, P. (1984). The social behaviour of the pin-tailed whydah in northern Ghana. *Ibis*, **126**, 463–473.

Short, L. L. and J. F. M. Horne (1985). Behavioral notes on the nest-parasitic Afrotropical honeyguides (Aves: Indicatoridae). *American Museum Novitates*, **2825**, 1–46.

Soler, M. (1990). Relationships between the great spotted cuckoo, *Clamator glandarius*, and its corvid hosts in a recently colonized area. *Ornis Scand.*, **21**, 212–223.

Steyn, P. (1973). Some notes on the breeding biology of the striped cuckoo. *Ostrich*, **44**, 163–169.

Tarboton, W. (1986). African cuckoo: the agony and ecstasy of being a parasite. *Bokmakierie*, **38**, 109–111.

Teather, K. L. and R. J. Robertson (1985). Female spacing patterns in brown-headed cowbirds. *Can. J. Zool.*, **63**, 218–222.

Teather, K. L. and R. J. Robertson (1986). Pair bonds and factors influencing the diversity of mating systems in brown-headed cowbirds. *Condor*, **88**, 63–69.

Todd, F. S. (1979). *Waterfowl. Ducks, geese and swans of the world.* Seaworld Press, San Diego.

Verner, J. and M. F. Willson (1966). The influence of habitats on mating systems of North American passerine birds. *Ecology*, **47**, 143–147.

Vernon, C. J. (1964). The breeding of the cuckoo-weaver (*Anomalospiza imberbis* (Cabanis)) in Southern Rhodesia. *Ostrich*, **35**, 260–263.

Vernon, C. J. (1984). The breeding biology of the thick-billed cuckoo. *Proceedings of the Pan-African Ornithological Congress*, **5**, 825–840.

von Frisch, O. (1973). Ablenkungsmanöver bie der Eiablage des Häherkuckucks (*Clamator glandarius*). *Journal für Ornithologie, Leipzig*, **114**, 129–131.

Weller, M. W. (1968). The breeding biology of the parasitic black-headed duck. *Living Bird*, **7**, 169–207.

West, M. J., A. P. King and D. H. Eastzer (1981). Validating the female bioassay of cowbird song: relating differences in song potency to mating success. *Anim. Behav.*, **29**, 490–501.

Westneat, D. F., P. Sherman and M. L. Morton (1990). The ecology and evolution of extra-pair copulations in birds. In: *Current Ornithology, Vol. VII* (ed. D. M. Power), pp. 331–369. Plenum Press, New York.

Wickler, W. and U. Seibt (1983). Monogamy: an ambiguous concept. In: *Mate choice* (ed. P. Bateson), pp. 33–50. Cambridge University Press, Cambridge.

Wittenberger, J. F. (1979). The evolution of mating systems in birds and mammals. In: *Handbook of behavioral neurobiology, Vol. 3. Social behavior and communication* (eds P. Marler and J. Vandenbergh), pp. 271–349. Plenum Press, New York.

Wittenberger, J. F. (1981). *Animal social behavior.* Duxbury Press, Boston.

Wyllie, I. (1981). *The cuckoo.* Batsford, London.

Yokel, D. A. (1986). Monogamy and brood parasitism: an unlikely pair. *Anim. Behav.*, **34**, 1348–1358.

Yokel, D. A. (1989). Intrasexual aggression and the mating behavior of brown-headed cowbirds: their relation to population densities and sex ratios. *Condor*, **91**, 43–51.

Yokel, D. A. and S. I. Rothstein (1991). The basis for female choice in an avian brood parasite. *Behav. Ecol. Sociobiol.*, **9**, 39–45.

VII

CONSPECIFIC BROOD PARASITISM

This penultimate group of chapters concerns intraspecific brood parasitism, which shares many features with interspecific brood parasitism. Both hosts and parasites have evolved adaptations and counteradaptations to intraspecific brood parasitism, which is much more widespread than originally thought. Intraspecific parasitism may affect population dynamics and clutch sizes of many species and may be a necessary precursor to the evolution of interspecific brood parasitism. For a more comprehensive overview of this topic, see pages 38–41 in chapter 1.

19

Patterns of Parasitic Egg Laying and Typical Nesting in Redhead and Canvasback Ducks

MICHAEL D. SORENSON

Most studies of obligate brood parasitism have focused on the dynamics of host–parasite interactions. Of particular interest has been the coevolution of adaptations in the parasite that increase the effectiveness of parasitism with counter-adaptations in the host that minimize its negative effects (e.g., Davies and Brook 1988, 1989*a*,*b*; Rothstein 1990). Although similar issues can be addressed in cases of facultative brood parasitism (e.g., Lombardo et al. 1989; Power et al. 1989), there are perhaps two more basic questions about parasitic egg laying in species that both parasitize and nest. These are: (1) What is the functional significance of parasitic egg laying in the reproductive strategy of females? and (2) How are two qualitatively different reproductive tactics, parasitic egg laying and typical nesting, maintained within a species?

Facultative brood parasitism has been documented more often and is probably more common in the waterfowl (Anatidae) than in any other group of birds (Rohwer and Freeman 1989; Sorenson 1992). Although only a rare event in many species, parasitic egg laying is a prominent feature in the biology of several cavity-nesters, including wood ducks, *Aix sponsa* (Clawson et al. 1979), black-bellied whistling ducks, *Dendrocygna autumnalis* (Chronister 1985), goldeneyes, *Bucephala clangula* (Ericksson and Andersson 1982) and *B. islandica* (Eadie 1991), and common shelducks, *Tadorna tadorna* (Pienkowski and Evans 1982); and in several over-water-nesting species, including redheads, *Aythya americana* (Weller 1959), red-crested pochards, *Netta refina* (Amat 1985), and ruddy ducks, *Oxyura jamaicensis* (Joyner 1983). In other species, such as snow geese, *Chen caerulescens* (Lank et al. 1989*a*,*b*), and canvasbacks, *Aythya valisineria* (Sorenson 1993), parasitism occurs regularly but at lower frequency. The waterfowl also include one obligate brood parasite, the black-headed duck, *Heteronetta atricapilla* (Weller 1968).

Although the nesting biology of waterfowl has been studied extensively, and the occurrence of facultative parasitism has been documented in many species (Weller 1959; Rohwer and Freeman 1989), few field studies have focused on the evolutionary aspects of parasitic egg laying. Several alternative hypotheses for the function and maintenance of parasitic behavior in waterfowl have been proposed (Yom-Tov 1980; Andersson 1984; Eadie et al. 1988; Sayler 1992), but only recently have a few studies succeeded in systematically documenting the parasitic and typical nesting behavior of individual females, allowing these hypotheses to be tested (Eadie 1989; Sorenson 1991, 1993; Weigmann and Lamprecht 1991; see also Clawson et al. 1979; Heusmann et al. 1980). Extensive population-level data on the frequency and success of parasitism in relation to annual variation in environmental conditions have also been used to address functional hypotheses for parasitic egg laying in waterfowl (Lank et al. 1989*b*, 1990).

In this chapter, I provide an overview of the

reproductive tactics of female redheads and canvasbacks during a three-year study in Manitoba. I emphasize differences between the two species in individual and population-level patterns of parasitic and typical nesting behavior, in the relationship between age and reproductive tactics and in the effect of drought conditions on reproductive tactics. The implications of these data for understanding the functional significance of parasitic egg laying in redheads and canvasbacks are discussed.

HYPOTHESES FOR PARASITIC EGG LAYING IN WATERFOWL

Assuming that parasitic egg laying is an adaptive behavior, hypotheses for the function of facultative parasitism must explain the context(s) in which laying parasitic eggs enables an individual female to achieve greater lifetime reproductive success than would be possible through typical nesting alone. These hypotheses can be divided into two categories (table 19.1). "Best-of-a-bad-situation" (Gross 1984) hypotheses suggest that parasitic egg laying is employed when typical nesting is impossible or unprofitable. In this category are a variety of constraint hypotheses (table 19.1, Hypotheses I.A), which suggest that females lay parasitic eggs when their ability to nest in the typical manner is limited by some proximate factor, such as nest predation, energetic/nutrient limitations on breeding, or nest site competition (Yom Tov 1980; Eadie et al. 1988). Nest site limitation is most likely in cavity-nesting species but is unlikely in emergent-nesting species such as redheads and canvasbacks, in which females construct their own nests (Sayler 1992). A non-adaptive hypothesis, that "parasitism" is an inadvertent consequence of two females selecting the same nest site (Erskine 1990), is also applicable primarily to cavity nesters (but see Eadie 1991).

An alternative "best-of-a-bad-situation" hypothesis that has received less discussion is that parasitic egg laying epresents an adaptive reduction of reproductive effort (restraint) in response to poor prospects for successful nesting (Sorenson 1991). In contrast to the constraint hypotheses, which focus on external factors that may limit the ability of females to nest, the restraint hypothesis (table 19.1, I.B) suggests that females change reproductive tactics in relation to a basic trade-off between current and future reproduction. By laying parasitic eggs, females avoid the costs and risks of nest-building, incubation, and brood care,

Table 19.1 Hypotheses for the functional significance of facultative parasitic egg laying

I. The "Best-of-a-Bad-Situation" Hypotheses
   A. The constraint hypotheses—parasitic egg laying in response to:
      1. nest loss due to predation, weather or other factors: parasitic eggs laid after failed nesting attempts
      2. nutrient/energetic limitations on breeding: parasitic eggs laid by females in poor physiological condition and/or facing limited food resources
      3. nest site limitation or competition: parasitic eggs laid when there are more females than nest sites or when females compete for high-quality nest sites
   B. The restraint hypothesis—when conditions for nesting are unfavorable, laying only parasitic eggs results in higher annual survival and therefore greater lifetime reproductive success: nesting, however, is the better strategy when conditions are good

II. The Enhancement Hypotheses
   A. Parasitic egg laying instead of nesting allows:
      1. greater annual fecundity: females lay a large number of parasitic eggs and/or
      2. higher annual survival: same basic advantages as restraint hypothesis but parasitic laying is also the best strategy when conditions are good
   B. Parasitic egg laying in addition to nesting: annual reproductive success is increased by exploiting opportunities to lay additional eggs in others' nests
      1. dual ("renesting") strategy: one parasitic and one nonparasitic clutch
      2. mixed strategy: one clutch of both parasitic and nonparasitic eggs

Sources: Yom Tov (1980); Andersson (1984); Eadie et al. (1988); Sorenson (1991); Sayler (1992).

yet realize greater reproductive success than if they deferred from breeding completely until the next season. A simple model illustrating the restraint hypothesis is presented in figure 19.1.

"Enhancement" (after Kendra et al. 1988) hypotheses suggest that parasitic egg laying allows an improvement in annual or lifetime fecundity even when conditions for nesting are ideal. By laying parasitic eggs instead of nesting (table 19.1, II.A), females could increase their fecundity in one or both of two ways: 1) time and energy usually invested in incubation and brood-care could be reallocated to the production of parasitic eggs, thereby increasing annual fecundity, and 2)

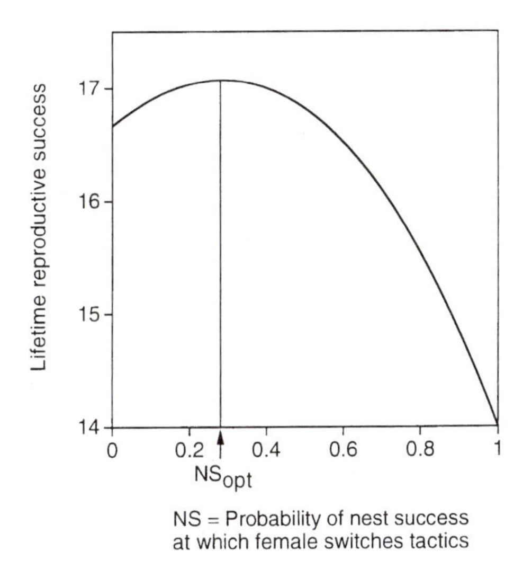

Figure 19.1 Assume 1) that a female duck can either lay 10 eggs in a nest of her own or lay 10 parasitic eggs each season, 2) that the success of parasitic eggs is only 70% that of nonparasitic eggs, 3) that the only cost of typical nesting is the risk that a female is killed by a predator during incubation and that there is no cost to laying parasitic eggs, 4) that the probability of successful nesting varies annually from 0 to 100% and that females can respond to cues that predict this probability prior to breeding. Although a female would achieve greater reproductive success in any given year by nesting, there is a level of nest success (NS_{opt}) at which females should switch between parasitism and nesting to maximize lifetime reproductive success (LRS). Expected LRS (in number of eggs hatched) of a female already surviving to breeding age is $(1 + S_{NS}/(1 - S_{NS})) \times (NS \times 10 \times 0.7 \times NS/2 + (1 - NS) \times 10 \times (NS + 1)/2)$, where S_{NS} is average annual survival, which varies with NS, the level of nest success at which a female switches from parasitism to nesting. If annual survival is 75% for parasitic females, but only 70% for nesting females because of the risks of incubation, then females should lay parasitic eggs instead of nesting when the probability of successful nesting is 28% or less.

avoidance of the risks and energetic costs associated with incubation and brood-care could increase survival of the adult female to the next breeding season, thereby increasing lifetime reproductive success. Alternatively, if females have the capacity to lay more eggs during a breeding season than they can effectively care for themselves, annual fecundity could be enhanced by laying parasitic eggs in addition to nesting (table 19.1, II.B; e.g., Brown 1984; Møller 1987).

"Best-of-a-bad-situation" hypotheses imply that parasitic and nesting behavior are maintained by condition-dependent selection (see Dawkins 1980; Dominey 1984). In other words, the best tactic for a given female depends on her particular phenotype (e.g., age, physiological condition) and/or the environmental conditions (including ecological interactions) she faces. Enhancement hypotheses, in contrast, are often coupled with the suggestion that parasitism and nesting are maintained as alternative reproductive strategies by frequency-dependent selection (e.g., Andersson 1984; Eadie et al. 1988). The clear expectation that parasitism should become less successful as it becomes frequent (because the number of parasitic eggs per host nest increases) raises the possibility of an equilibrium in which nesting and parasitic strategies have equal success. A variety of mechanisms might be involved in such an equilibrium (see Davies 1982; Dominey 1984), perhaps the most intriguing of which is a genetic polymorphism determining alternative strategies. It should be noted, however, that enhancement hypotheses are also compatible with condition dependence: for example, the ability of a female to capitalize on opportunities for parasitism may be dependent on her physiological condition. Although the questions of function and maintenance are intimately related, I focus on questions of function in this paper and generally assume that parasitic behavior in redheads and canvasbacks represents a conditional strategy. The role of frequency-dependent selection in the maintenance of alternative nesting tactics in waterfowl has been examined elsewhere (Lank et al. 1990; Sorenson 1990; Eadie and Fryxell 1992).

NATURAL HISTORY

Weller's (1959) monograph on parasitic egg laying in redhead ducks laid the groundwork for the study of a fascinating species. Parasitic egg laying represents a greater proportion of total reproductive effort in redheads than perhaps in any other nesting bird (see Discussion). Nonetheless, at least some redhead females build nests, incubate eggs, and care for their own young in typical waterfowl fashion. Canvasbacks have long been recognized as the primary host of redhead parasitism and generally suffer a high frequency of parasitism wherever the two species are sympatric (e.g., Erickson 1948; Weller 1959; Sugden 1980; Stoudt 1982; Bouffard 1983). Unexpectedly, however, intensive documentation of egg laying at canvasback host nests revealed that canvasbacks also frequently lay parasitic eggs (Sorenson 1993).

Redheads and canvasbacks are seasonally monogamous: new pairs form each year during late winter and spring migration, and males follow females back to their natal areas (Weller 1965, 1967; Rohwer and Anderson 1988). Nesting females of both species build nests in residual emergent vegetation and usually lay 7–10 eggs on consecutive days. Incubation and post-hatch parental care are provided by the female only. Parasitizing redhead and canvasback females generally lay parasitic eggs when the host female is present at the nest (Sayler 1985; M. Sorenson, unpubl. data). Parasitic females gain access to nests by tunneling under the host female and pushing her out of the nest bowl. Host females usually respond by aggressively pecking the parasite in the head (McKinney 1954; M. Sorenson, unpubl. data). Both host eggs and previously laid parasitic eggs are frequently displaced from nests during parasitic intrusions as host and parasite struggle on top of the clutch (M. Sorenson, in preparation). Although parasitism is costly to hosts because of egg displacement, there is no evidence that canvasback hosts discriminate against parasitic eggs or ducklings. Host and parasitic eggs are equally likely to be displaced from nests (Sorenson 1997), and there are no discernible differences in the behavior of brood-rearing females toward redhead and canvasback ducklings in mixed broods (Mattson 1973). Nesting females leave nests with their broods within 48 hours of hatching, abandoning any unhatched parasitic eggs that were laid during the incubation stage. Ducklings feed themselves but females lead and protect their broods for up to 60 days after hatch.

STUDY AREA AND METHODS

I studied redheads and canvasbacks in the prairie pothole region of Manitoba from 1986 to 1988. The study area, just south of Minnedosa, Manitoba, is characterized by a particularly high density and diversity of small wetlands called potholes, making it ideal habitat for breeding waterfowl (see Kiel et al. 1972; Stoudt 1982). My 10.4-km^2 study area included over 350 separate wetlands. The Minnedosa area hosts a relatively high-density population of canvasbacks (Bellrose 1980) which are locally about three times as numerous as redheads (about 5–8 pairs of canvasbacks/km^2 compared with about 2–3 pairs of redheads/km^2, Sorenson 1990; densities estimated by complete pair counts, Sugden and Butler 1980).

Rates of nest success for canvasbacks and redheads vary considerably among years depending on habitat conditions (Stoudt 1982). During this study, nest success was relatively high in 1986 and 1987 (47% and 52%, respectively, for redheads and canvasbacks combined) when high water levels provided extensive areas of flooded emergent vegetation for nesting. In 1988, however, wetland conditions deteriorated rapidly during June because of unusually high temperatures and low precipitation, resulting in significantly lower water levels ($F = 10.7$, $P < 0.001$) and narrower bands of emergent vegetation ($F = 20.0$, $P < 0.0001$) on 1 July than in the previous two years (for both comparisons: d.f. = 2, 22; $n = 12$ semi-permanent wetlands on an adjacent study area; M. G. Anderson, unpublished data). As result, nest success was much lower in 1988 (23%) than in the previous two years ($G = 19.0$, $P < 0.001$, d.f. = 2), and the breeding season ended early and abruptly (Sorenson 1990, 1991). Nonetheless, conditions in 1988 were still moderate in comparison with other years of more severe drought (Anderson et al. 1997).

The objective of this study was to obtain as much information as possible about parasitic egg laying and typical nesting by individual females. To this end, I trapped and color-marked 150 female canvasbacks and 74 female redheads during the three years of the study, of which 132 canvasbacks and 60 redheads were considered to be "resident" on the study area (see Sorenson 1991). Yearling and adult (≥ 2 years old) females of both species were distinguished by plumage characteristics (Dane and Johnson 1975; Serie et al. 1982). All wetlands on the study area were searched every 8–10 days during May and June to find canvasback and redhead nests as soon as possible after they were initiated and before they were parasitized. Nests in the first few days of laying were occasionally missed, reducing data available on parasitic egg laying by individual females. Because nests of both species become relatively large and conspicuous by the end of laying and are restricted to a well-defined habitat, however, essentially all nests on the study area were eventually found, including ones that were already abandoned or destroyed.

Parasitic egg laying by marked females was documented with time-lapse photography. I simultaneously monitored up to 25 potential host nests with Super-8 movie cameras set to expose 1 frame per minute. Films were viewed frame by frame to find cases of parasitic females intruding at nest sites. Film events, sequences of frames including a female redhead or canvasback other than the host female, were classified as nest visits (nonlaying intrusions) or egg-laying events (see Sorenson 1991 for criteria used to classify film events). Each egg-laying event included at least several frames during which the intruding female was on the nest, usually allowing intruding females with nasal markers to be identified.

I filmed a total of 1499 nest-days at 204 canvasback and redhead nests during the study. These films included 218 parasitic egg-laying events involving intruding redheads and 96 egg-laying events involving intruding canvasbacks, representing about 40% of all parasitic egg laying on the study area. Because the number of marked females increased and coverage of nests with cameras improved, the proportions of parasitic events recorded on film and of intruding females that could be individually identified were higher in 1987 and 1988 than in 1986. For 1986, 1987, and 1988, respectively, 7, 23, and 28% of all parasitic egg laying by redheads and 15, 25, and 21% of all parasitic egg laying by canvasbacks could be assigned to individual females. At least for 1987 and 1988, film data are of comparable quality for redheads and canvasbacks and should be a fairly representative sample of parasitism on the study area without any systematic seasonal bias.

Film records of parasitic egg laying and records of typical nesting for individual females were combined to produce individual egg-laying histories. Because only a portion of parasitic egg laying on the study area was assigned to individual females, egg-laying histories are incomplete, and some females that laid parasitic eggs may not have been recorded on film at all. Nonetheless, each resident redhead and canvasback female (see Sorenson 1991, 1993 for criteria) could be categorized according to the reproductive tactics each was known to have employed, allowing differences in reproductive tactics among age classes and years to be tested with log-linear models (Sokal and Rohlf 1981).

Nests were visited every other day during egg laying and the first week of incubation to document the addition (and loss) of host and parasitic eggs and to change film in cameras. Interspecific parasitic eggs could be identified unambiguously because canvasback and redhead eggs are easily distinguished by color and shell texture (Bellrose 1980). Intraspecific parasitism was documented directly with time-lapse photography or, at nests where cameras were not in operation, by irregularities in the laying sequence (i.e., more than one egg added per day or eggs laid during the incubation stage of the host). At a few nests not found until the incubation stage, obvious differences in the size and color of eggs, differences in the developmental stages of eggs within the clutch, and large clutch size were taken as indications of "probable" parasitism (see Sorenson 1993). Results on intraspecific parasitism reported in this paper include both definite and "probable" cases of parasitism.

RESULTS

Frequency of parasitism

Both intra- and interspecific parasitic egg laying were prominent features in the nesting biology of redheads and canvasbacks at Minnedosa. Although every possible host–parasite relationship was observed, redhead parasitism of canvasbacks was the most common phenomenon, whereas canvasback parasitism of redheads was quite rare (table 19.2). Intraspecific parasitism was relatively common in both species: nesting redheads were frequently parasitized by other redhead females, and although not often reported in previous studies, the combination of time-lapse photography and frequent nest checks employed in this study revealed a relatively high frequency of intraspecific parasitism among canvasbacks as well (see Sorenson 1993).

From a population-level perspective, parasitic egg laying represented a much smaller proportion of reproductive effort in canvasbacks than in redheads (table 19.3). At least 60% of all redhead eggs were laid parasitically during the three years of the study but only 10–13% of canvasback eggs were laid parasitically. Both species laid most of their parasitic eggs in the nests of canvasbacks.

Reproductive tactics of individual females

Although both redheads and canvasbacks laid parasitic eggs, histories of parasitic and typical nesting behavior for individual females differed greatly between the two species. When environmental conditions were favorable for nesting in 1986 and 1987, many redhead females laid

Table 19.2 Frequency of parasitism in redhead and canvasback host nests with completed clutches. Data for 1986–1988 combined

		Redhead parasites		Canvasback parasites	
Host	n	% Nests parasitized	Parasitic eggs/ parasitized nest	% Nests parasitized	Parasitic eggs/ parasitized nest
Redhead	33	61[a]	3.5[a]	6	2.0
Canvasback	179	65	3.8	41[a]	3.1[a]

[a]Estimates include "probable" cases of intraspecific parasitism.

parasitic eggs prior to nesting (Sorenson 1991). Details of one particularly well-documented egg-laying history are presented in figure 19.2. Redhead female #679 was recorded on film 11 times at five different canvasback nests over a 10-day period in 1986. This female laid at least seven parasitic eggs at four of these canvasback nests from 14 to 23 May and then was not recorded on film again. She was found with a brood of her own on 16 July just off the main study area. Given the age of the ducklings, female #679 must have initiated her own nest on or after 8 June, 17 days after she was last recorded laying a parasitic egg.

Although other egg-laying histories were less complete, eight individually marked redhead females were known to lay parasitic eggs prior to nesting, and indirect evidence (see Sorenson 1991) suggested that six additional females followed a similar pattern of behavior, accounting for 38% of female-years in 1986 and 1987 ($n = 37$). In all but one case, an interval of at least seven days separated the last parasitic egg detected for a given female and the initiation of her own nest. Other adult females were known only to lay parasitic eggs ($n = 9$, 24% of female-years) or only to nest ($n = 12$, 32% of female-years) in 1986 and 1987. It is likely, however, that I failed to detect either parasitic egg laying or nesting by many redhead females (see Sorenson 1991), and population-level data suggest that a much larger proportion of adult females both parasitized and nested in 1986 and 1987 (see below).

As expected, given the lower frequency of parasitic egg laying in canvasbacks, a much smaller proportion of canvasbacks were recorded on film laying parasitic eggs (14%, 22 of 162 resident-female-years) than redheads (61%, 49 of 80

Table 19.3 Population level frequency of parasitic egg laying by canvasbacks and redheads and species distribution of parasitic eggs

			Number of parasitic eggs laid in . . .		
Species/ year	Total number of eggs laid[a]	% Eggs parasitic[b]	Canvasback nests[b]	Redhead nests[b]	Other nests[c]
Canvasbacks					
1986	545	8.6	44	2	1
1987	667	13.8	90	2	0
1988	580	17.2	99	1	0
Redheads					
1986	458	62.7	231	47	9
1987	211	64.5	123	12	1
1988	192	73.4	119	13	9

[a]Total number of eggs laid by each species in all nests on the 10.4-km^2 main study area.
[b]Estimates include "probable" cases of intraspecific parasitism.
[c]Includes mallard, *Anas platyrhynchos*, and ring-necked duck, *Aythya collaris*, nests.

Source: Canvasback data from Sorenson (1993).

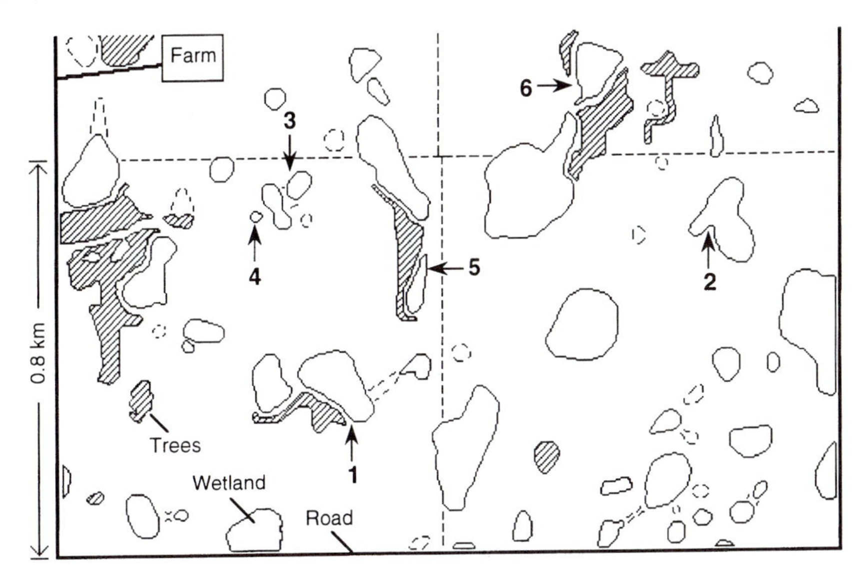

Figure 19.2 History of parasitic egg laying and typical nesting for female redhead #679 in 1986. (Arrows #1–5 show the locations of five canvasback nests.) 14 May: parasitic egg laid at 09:55 (nest #1). 15 May: parasitic egg laid at 09:45 (#1). 17 May: recorded in nest visit at 12:25 (#2). 18 May: parasitic egg laid at 10:39 (#3) and recorded in nest visits at 06:41 & 21:12 (#4). 19 May: parasitic egg laid at 11:01 (#5). 20 May: parasitic egg laid at 11:20 (#4). 22 May: parasitic egg laid at 14:30 (#5) and recorded in nest visit at 13:58 (#4). 23 May: parasitic egg laid at 14:58 (#4). 16 July: female first sighted with brood of seven ducklings (#6). Backdating for the age of the brood, incubation, and egg laying suggests a nest initiation date of 8–15 June.

resident-female-years). As with redheads, many canvasback females that did lay parasitically also initiated nests of their own (fig. 19.3). In comparison with redheads, however, parasitizing canvasbacks were recorded in many fewer egg-laying events prior to nesting. In 13 of 15 cases, females were known to lay only one or two parasitic eggs (Sorenson 1993). Also in contrast to the egg-laying histories of redheads, canvasback females initiated their own nests almost immediately after laying parasitically. Female canvasbacks were known to initiate their own nest the day after laying a parasitic egg in five of 15 cases and may have initiated a nest within two days of laying a parasitic egg in 12 of 15 cases (Sorenson 1993).

During the three years of the study, only seven canvasback females were known to lay only parasitic eggs in a given season, and five of these cases occurred in 1988 (see below). Canvasback females known only to lay parasitically were recorded in more egg-laying events on film than females laying parasitically prior to nesting, suggesting that many of these females did, in fact, only lay parasitic eggs (Sorenson 1993).

Individual reproductive tactics in relation to female age and environmental conditions

Reproductive effort in both redheads and canvasbacks was strongly related to female age. More three-year-old and older females were known to initiate nests than younger females in both species (fig. 19.4; for Redheads, $G = 11.0$, $P = 0.027$, d.f. = 4; for canvasbacks, $G = 82.8$, $P < 0.001$, d.f. = 4). Reproductive effort was also related to environmental conditions and was much reduced in 1988 when drought conditions reduced the probability of successful nesting: fewer canvasback females initiated nests in 1988 than in 1987 ($G = 16.1$, $P = 0.001$, d.f. = 3).

Interestingly, age-related variation in reproductive effort in redheads was reflected in qualitative differences in reproductive tactics. In 1987, seven older redhead females were known to lay parasitic eggs prior to initiating a typical nest. Many younger females also nested but were less likely to employ a dual strategy (fig. 19.4, $G = 3.64$, $P = 0.056$, d.f. = 1). As in canvasbacks,

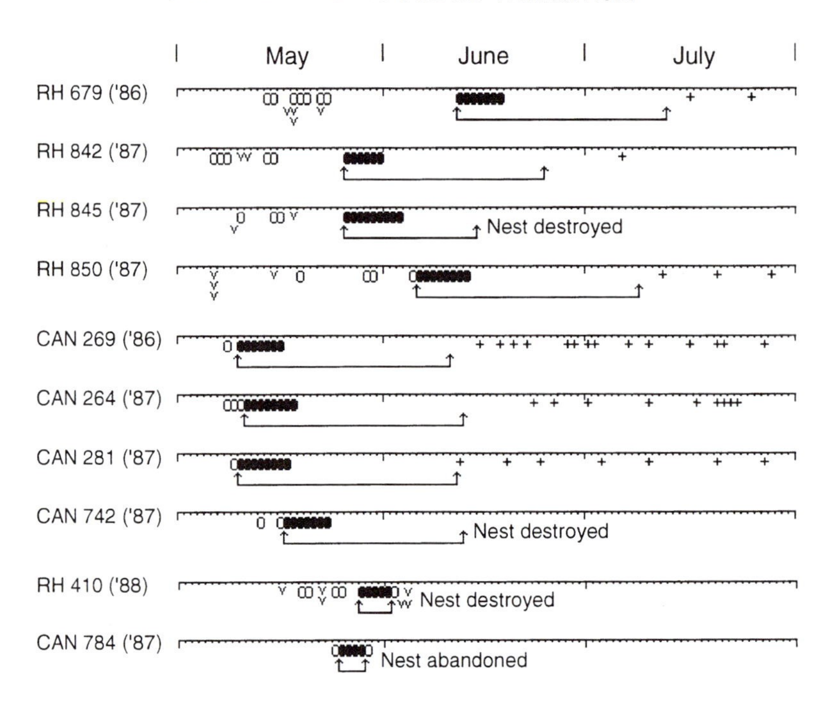

Figure 19.3 Egg-laying histories for individual redhead (RH) and canvasback (CAN) females known to lay parasitic eggs prior to nesting. Dates of laying for known parasitic eggs (○) and eggs laid in the female's own nest (●), and dates of nest visits (v), initiation and termination of the female's own nest (arrows), and sightings of the female with her brood (+) are indicated below each dateline. Note that redhead female #410 and canvasback female #784 also laid at least one parasitic egg after their own nests were terminated in the laying stage. (Redhead data from Sorenson 1991.)

drought conditions in 1988 resulted in a general reduction of reproductive effort in redheads and age-dependent changes in reproductive tactics. Although females of all ages were less likely to nest in 1988 than in 1987 ($G = 23.3$, $P < 0.001$, d.f. = 3), the effect of year on the proportion of redhead females parasitizing was dependent on age (significant three-way interaction AGE * PARASITIZE * YEAR, $G = 8.60$, $P = 0.014$, d.f. = 2): older females (at least three years old) were less likely to lay parasitic eggs in 1988 but younger females were more likely to do so. The six redheads known to nest in 1988 were all at least three years old, and only one was known to lay parasitic eggs prior to nesting. In contrast, almost all younger females were known only to lay parasitically in 1988.

Drought conditions had a greater effect on the frequency of nesting in redheads than in canvasbacks (significant three-way interaction SPECIES * NEST * YEAR, $G = 4.93$, $P = 0.026$, d.f. = 1), as many redhead females apparently switched from typical nesting in 1987 to parasitic egg laying in 1988. Although canvasback females also reduced reproductive effort in 1988, there was no clear change in reproductive tactics. Although most older canvasbacks still made at least one nesting attempt in 1988, the proportion of younger females that initiated nests was much lower, and only two of 21 yearling females showing any evidence of breeding at all. Only five younger canvasback females were known to lay only parasitic eggs, suggesting that few canvasbacks responded to drought conditions by switching to parasitic egg laying.

One case history further illustrates the changes in reproductive tactics shown by redheads in 1988, apparently in response to drought conditions. Female #850 was color-marked in 1986 while incubating her own nest relatively late in the season. She returned to the study area in both 1987 and 1988 and was sighted frequently in both years. Prior to nesting in 1987, she was recorded on film eight times at six different canvasback nests and appeared to lay parasitic eggs in five of these film events. As in 1986, she then initiated her own nest

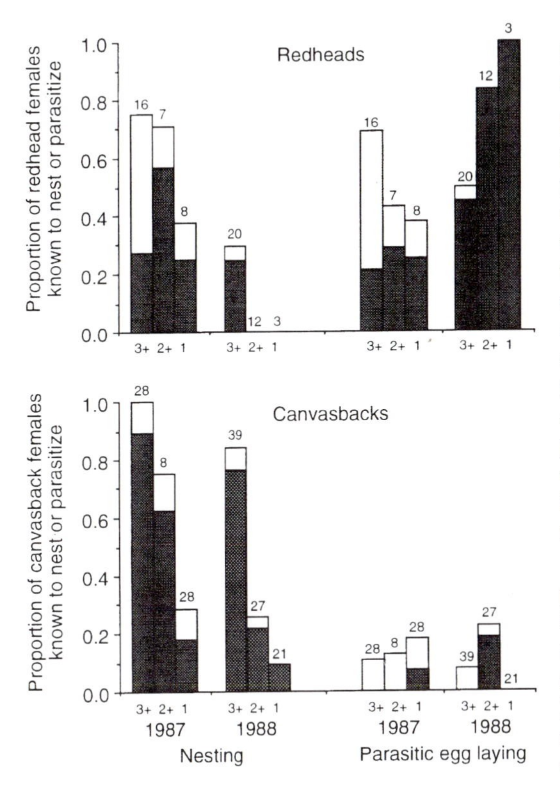

Figure 19.4 Reproductive tactics of decoy-trapped and returning redhead and canvasback females in 1987 and 1988. (Nest-trapped females are not included because they were not marked until late in the incubation stage of their own nests and therefore could not have been detected laying parasitic eggs.) For each species and year, females are divided into three age categories: 1) newly marked yearlings, 2+) returning females captured as yearlings the year before, and newly marked, unknown-age adults, and 3+) returning females previously marked as unknown-age adults and 3-year-olds in 1988 marked as yearlings in 1986. The lightly shaded area at the top of each bar represents the proportion of females in each category that employed both reproductive tactics in the same season. (Based on data from Sorenson 1991, 1993.)

relatively late in the season. In 1988, however, this female nested 27 days earlier than in 1987 (relative to mean nest initiation dates for canvasbacks, see Sorenson 1991) and was not recorded on film prior to nesting.

Seasonal chronologies of parasitism and nesting

Because females may range over a wide area and parasitize a number of different host nests, complete egg-laying histories for individual females were extremely difficult to obtain. Most parasitic egg laying and also a few typical nest initiations that occurred on the study area could not be assigned to individual females. Population-level data collected during the study, however, strongly suggest that the patterns of behavior observed in individually marked females were representative of the reproductive tactics of most females in the population. Population-level data were consistent with differences in reproductive tactics observed between the two species, between females of different age groups, and among years.

The seasonal chronology of typical nesting and parasitic egg laying by redheads and canvasbacks in 1986 perhaps represents close to an "average" season (see Arnold and Sorenson 1988) with favorable conditions for nesting in Manitoba. Most canvasback nests were initiated during the first two weeks of May (fig. 19.5). Then between the 10th and 25th of May a very large number of parasitic redhead eggs were laid, primarily in canvasback nests but also in the few redhead nests that were initiated early in the season. Then starting about the 25th of May and continuing through June, most of the redhead nests were initiated. The seasonal progression from parasitic egg laying to typical nesting by redheads is clearly consistent with observations of individual females laying parasitic eggs prior to nesting. In addition, high per-capita rates of both parasitic egg laying and typical nesting in 1986 (table 19.4), accounting for 10 parasitic eggs plus a typical nest per redhead female, suggest that almost all redhead females laid parasitic eggs prior to nesting in 1986.

Differences among years and age classes in reproductive tactics were also reflected in seasonal chronologies of parasitic egg laying and typical nesting. The seasonal timing of parasitism and nesting by redheads was similar in 1987 but with less separation in the dates of parasitism and nesting (fig. 19.5) and lower per-capita production of parasitic eggs (table 19.4). Both of these differences were probably attributable to a greater proportion of yearling females in the redhead population in 1987 (see Sorenson 1991). Like most other birds, yearling canvasbacks initiated nests much later than older females. In contrast, yearling redheads nested relatively early in 1987 (fig. 19.5). This unusual relationship between age and

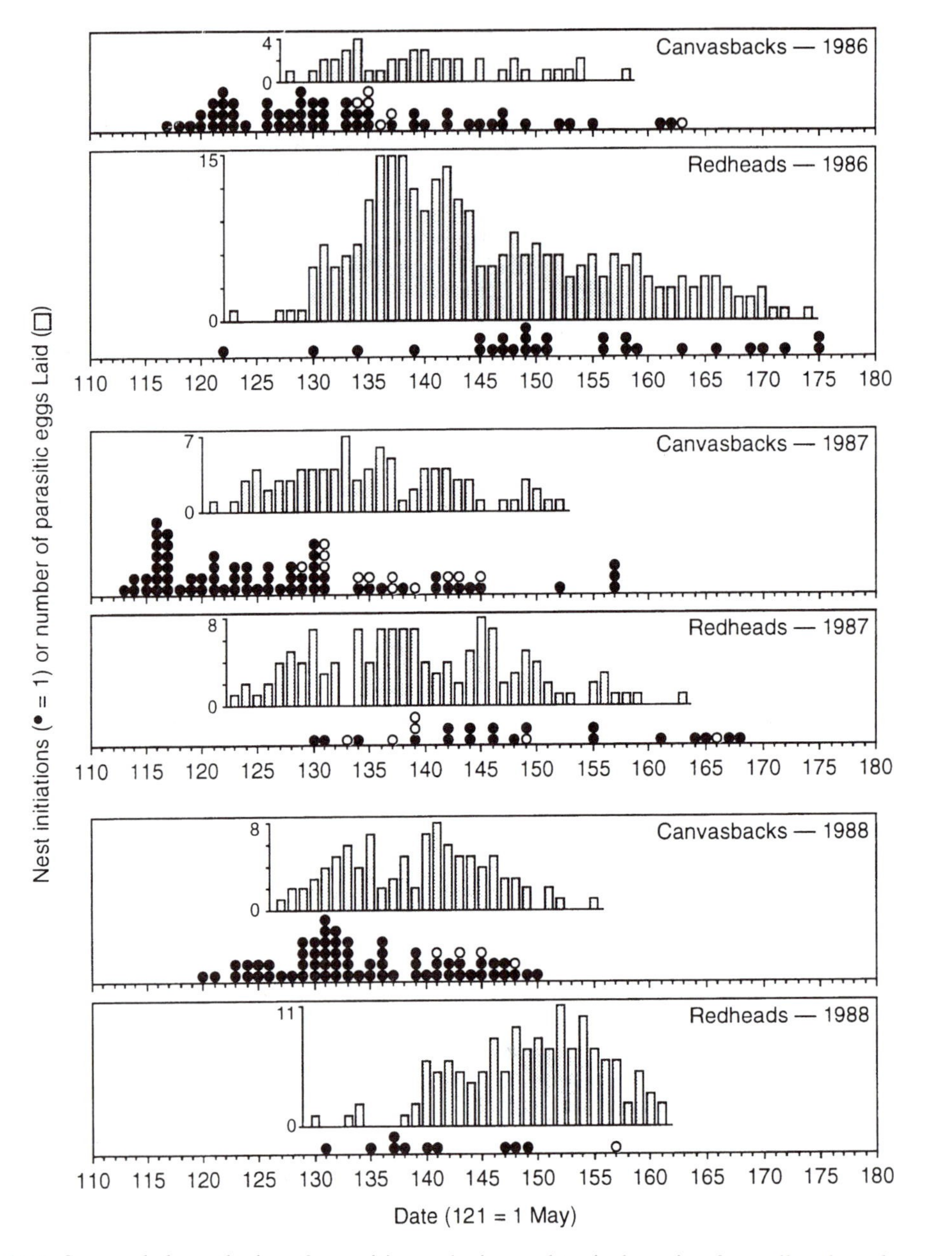

Figure 19.5 Seasonal chronologies of parasitic egg laying and typical nesting for redheads and canvasbacks at Minnedosa. Laying dates for every parasitic egg laid by each species on the 10.4-km^2 study area (histograms) and dates of nest initiations (●,○) for each species are shown (redhead nests initiations in a 6.3-km^2 area surrounding the main study area are also included). Nests initiated by females known to be yearlings are indicated by open circles (○). Both definite and probable cases of intraspecific parasitism (as well as parasitic eggs laid interspecifically) are included in the figure. Note that a series of nonparasitic eggs would be laid on 6–9 consecutive days after each nest initiation.

Table 19.4 Per-capita production of parasitic eggs and typical nests by canvasbacks and redheads at Minnedosa

Year	Females[a]	Parasitic eggs	Parasitic eggs/female	Nests[b]	Nests/female
Canvasbacks					
1986	~50	47	~0.9	75	~1.5
1987	~70	92	~1.3	85	~1.2
1988	~80	100	~1.3	80	~1.0
Redheads					
1986	23–29	287	9.9–12.5	24	0.8–1.0
1987	18–24	136	5.7–7.6	12[c]	0.5–0.7
1988	21–29	141	4.9–6.7	8	0.3–0.4

[a]Rough estimates obtained from pair counts of the density of canvasbacks and redheads on the 10.4-km^2 study area.

[b]Total number of typical nests initiated by each species on the study area.

[c]The number of redhead nests found on the study area in 1987 probably underestimates the true level of redhead nesting activity (Sorenson 1991). Sixteen additional nests were found in a 6.3-km^2 area surrounding the main study area in 1987: only seven nests were found in this area in 1988.

Sources: Sorenson (1991, 1993).

nest initiation date can be explained by an age difference in reproductive tactics: most older redheads probably laid parasitic eggs prior to nesting, whereas many younger females initiated nests earlier because they laid eggs only in their own nests (Sorenson 1991).

In 1988, when drought conditions resulted in very low rates of nest success, the seasonal chronology of parasitism and nesting by redheads was quite different. A few redhead nests were initiated relatively early in this late season, whereas the majority of parasitic eggs were laid somewhat later (fig. 19.5). This pattern is consistent with the observation that individual females either laid parasitic eggs or nested, but did not do both, and also with an age difference in reproductive tactics: only older females nested in 1988 whereas younger females laid parasitic eggs (fig. 19.4). Relatively early initiation dates for females that did nest in 1988 are a clear indication of a conditional switch in reproductive tactics: instead of beginning the season by laying parasitic eggs, these females started directly with their own nests (see Sorenson 1991). The 1988 breeding season also ended relatively early and abruptly, as water levels dropped rapidly. Only one redhead nest was initiated on the study area after May 30th, compared with 14 and 8 nests in 1986 and 1987, respectively. It is very unlikely that females initiated nests in other areas late in 1988; marked females remained on the study area well into June and most were sighted without mates, male redheads having already departed to northern molting areas (see Bergman 1973).

In contrast to the data for redheads, seasonal chronologies of nesting and parasitism by canvasbacks varied relatively little during the three years of the study (fig. 19.5). In all three years, most parasitic canvasback eggs were laid well after the peak of canvasback nest initiations. A greater proportion of yearling females in the canvasback population in 1987 and 1988 (Sorenson 1993) and drought conditions in 1988, however, may have contributed to a higher per-capita frequency of parasitic egg laying and a lower frequency of nesting in 1987 and 1988 (table 19.4). In 1988, the number of renests initiated late in the season was also reduced by an early end to the breeding season.

DISCUSSION

Parasitic egg laying in redheads

Histories of parasitic egg laying and typical nesting suggest that individual redhead females lay an entire "clutch" of parasitic eggs in several different canvasback nests and then, after a short "renest interval," lay a second clutch in a nest of their own. The behaviors associated with the laying of parasitic and nonparasitic eggs are dramatically different. Parasitic eggs are laid during short visits to a number of different host nests, but the female's own clutch is laid in a single location on consecutive days and involves extensive time spent in nest building and a gradual increase in nest attentiveness through the laying stage (unpublished data). There was also no evidence that females interrupted their own clutches to lay parasitic eggs. Population-level data, including seasonal chronologies and per-capita rates of parasitic egg laying and typical nesting, also support the conclusion that a large proportion of redhead females lay parasitically prior to nesting (see also Weller 1959). This pattern of behavior is consistent with the hypothesis that redhead females increase their annual fecundity by laying parasitic eggs in addition to those in their own nests (table 19.1, II.B.1).

Given that the hatching success of parasitic redhead eggs is much lower than that of nonparasitic eggs (Weller 1959; Bouffard 1983; Joyner 1983; Sorenson 1991), however, it would seem that a female would increase her reproductive success to a greater degree by simply laying additional eggs in her own nest. Obviously, this raises the issue of clutch size limitation: employing both parasitic and typical nesting tactics should be advantageous only if the number of eggs or young that a female can care for herself is limited (Lyon 1993). In species in which brood size is tightly constrained by the ability of parents to feed young, laying just one additional egg parasitically could increase reproductive success (e.g., Jackson 1993). Waterfowl, however, have extremely precocial, self-feeding young and both experimental and natural (i.e., due to parasitism) manipulations of clutch size and brood size suggest that females can adequately incubate more eggs and care for more young than in the usual clutch (reviewed by Rohwer 1992).

Although the nature of clutch-size limitation in waterfowl is a controversial and unresolved issue (Ankney et al. 1991; Arnold and Rohwer 1991), prairie-nesting ducks are limited to about 10 eggs in a given clutch by factors operating during the laying stage (Rohwer 1992). Nonetheless, females have the capacity to produce more than one clutch in a season, readily renesting after first nests are destroyed (e.g., Sowls 1955; Alliston 1979*a*; Doty et al. 1984). Because the time required for successful incubation and brood rearing eliminates the possibility of raising two broods in one season (Rohwer 1992), however, the maximum annual reproductive success of these species is limited to the number of eggs in one clutch. In contrast, the dual strategy of redheads allows females to obtain care for more than one clutch in a given season, thereby avoiding the normal limits to clutch size and increasing fecundity above that possible through typical nesting alone (Sorenson 1991).

Because redheads may actually delay initiation of their own nests, however, this conclusion also requires that the benefits of laying parasitically outweigh the potential costs of delayed nesting. Rough estimates of reproductive success incorporating seasonal declines in clutch size and nest success suggest that the benefits of a dual strategy would have outweighed these costs in 1986 and 1987 but not in 1988, when nest success declined dramatically at the end of the season (Sorenson 1991). Too few data are available to evaluate the magnitude of other potential costs of delayed nesting such as reduced adult or juvenile survival.

Parasitic egg laying in redheads also appears to function as a low-cost alternative to typical nesting when environmental conditions are unfavorable. Consistent with either the constraint or restraint hypotheses, most redhead females were known only to lay parasitic eggs in 1988 when drought conditions resulted in rapidly deteriorating conditions. Although laying histories were incomplete, no female was known to lay more than five parasitic eggs in 1988 and film records of parasitic laying for a given female almost always fell within an interval of 10 days or less (Sorenson 1990), suggesting that parasitic females did not lay a large number of eggs (i.e., more than a nesting female). Given the low success of parasitic eggs (46% vs. 91% for nonparasitic eggs, Sorenson 1991), parasitic females had lower reproductive success than nesting females. They also greatly reduced their reproductive effort, however, by avoiding the energetic costs and predation risks of incubation and brood rearing. It could be argued that most redhead females did not really switch tactics in 1988, but began the season by laying parasitic eggs as usual, and then "decided" not to nest only after conditions deteriorated later in the season. Real, qualitative changes in reproductive tactics, however, are suggested by the tendency for yearling females to nest early in the season in 1987 but to parasitize only in 1988 and by older females that nested early in 1988 instead of laying parasitically prior to nesting.

A variety of proximate factors associated with drought may have influenced the reproductive tactics of females in 1988. It seems likely that females restrained reproductive effort in response to a reduction in the extent of flooded nesting cover which reduced the probability of successful nesting (table 19.1, I.B), although declining food resources may have also constrained the ability of females to nest (table 19.1, I.A.2). Distinguishing between constraint and restraint as the explanation for a reduction in reproductive performance is difficult (Rohwer 1992), and in reality both may play a role in determining reproductive tactics in redheads (see Sorenson 1991).

Overall, there is no single functional explanation for the parasitic behavior of redheads. Instead, parasitic egg laying appears to be part of a very flexible conditional reproductive strategy that includes four different options of increasing reproductive effort: 1) nonbreeding; 2) laying parasitic eggs only; 3) nesting only and 4) laying parasitic eggs prior to nesting. This conditional strategy results in age-related and annual variation in reproductive tactics as females adjust their reproductive effort in relation to the prospects for successful reproduction. Goldeneyes (Eadie 1989) and coots, *Fulica americana* (Lyon 1993), also appear to use parasitism in different contexts consistent with both enhancement and best-of-a-bad-situation hypotheses. Nonetheless, several specific hypotheses for parasitic behavior in redheads can be rejected. Parasitism was not primarily a response to nest loss (table 19.1, I.A.1) because parasitic laying generally preceded redhead nest initiations (fig. 19.5; see also Weller 1959). Parasitism is also not a high fecundity alternative to nesting (table 19.1, II.A.1) because the number of parasitic eggs laid and their success is too low. That individual females switched tactics among years also allows the hypothesis that nesting and parasitism are determined by a genetic polymorphism to be rejected (see Sorenson 1991).

Parasitic egg laying in canvasbacks

Some canvasback females also laid parasitic eggs prior to nesting but the pattern of behavior was quite different from that in redheads. Canvasbacks did not lay two separate clutches but instead laid only one or a few parasitic eggs just before initiating their own nests. This pattern of behavior is difficult to reconcile with either of the competing hypotheses for clutch size limitation in waterfowl (see Ankney et al. 1991) and therefore is hard to interpret as a tactic for increasing fecundity. Because laying parasitic eggs immediately prior to nesting does not reduce or spread out the demands of egg laying, it cannot be a strategy to avoid nutritional limitations on clutch size (e.g., Ankney and MacInnes 1978; Drobney and Fredrickson 1985; Barzen and Serie 1990). Given that only one or a few parasitic eggs are laid, this pattern of laying is also inconsistent with the egg viability hypothesis for clutch size limitation (Arnold et al. 1987): the incremental costs associated with increasing clutch size by one or two eggs (measured in terms of lower egg viability and increased exposure of the nest to predators for eggs already laid) would be insignificant in comparison to the large difference in success of parasitic and nonparasitic canvasback eggs (29% vs. 79%, Sorenson 1993). Finally, in six of 13 cases, canvasbacks laying parasitic eggs prior to nesting had unusually small clutches (3–6 eggs) in their own nests (Sorenson 1990).

Given these considerations, it is not clear why a female would not lay an additional egg(s) in her own nest where it would be more likely to hatch, especially if parasitic laying immediately precedes nesting. Sorenson (1993) suggested that a female canvasback losing a nest at an early stage might lay one or a few parasitic eggs prior to initiating a renest because a new nest structure would not be immediately ready to receive eggs. Consistent with this variation of the nest loss hypothesis (table 19.1, I.A.1), parasitic egg laying and nest initiations by parasitic females both followed the peak of typical nest initiations in the canvasback population (fig. 19.5; Sorenson 1993). Unfortunately, I obtained little information about the hypothesized initial nesting attempt of parasitic females, perhaps because females laid parasitic eggs only after nests were lost in the very early stages. Experiments in which nests of individually marked birds are destroyed and subsequent egg laying is monitored are needed to resolve this issue (e.g., Haramis et al. 1983; Feare 1991; Stouffer and Power 1991).

In canvasbacks, as in redheads, parasitic egg laying may also function as a low-cost alternative to typical nesting when drought conditions reduce the probability of successful nesting. The proportion of canvasback females that laid only parasitic eggs in 1988, however, was much lower than in redheads; almost all older canvasback females made at least one typical nesting attempt even when conditions were poor, whereas most younger females responded to drought conditions by not breeding at all (fig. 19.4). Although both species showed strong effects of age and environmental conditions on reproductive effort, these effects were manifest in qualitative changes in reproductive tactics primarily in redheads. Variation in reproductive effort in canvasbacks was reflected primarily in the proportion of females breeding and in the frequency of renesting.

Although intraspecific parasitism among canvasbacks was fairly frequent, parasitic egg laying does not play the central role in canvasback reproductive biology that it does in redheads. Most parasitic egg laying in canvasbacks appears to be associated with disruptions in the typical nesting cycle resulting from nest predation or other forms of disturbance (Sorenson 1993). Perhaps in response to other constraints on typical nesting but also consistent with the restraint hypothesis, some younger females may employ parasitic egg laying as a low-cost alternative to nesting. It seems likely that parasitic egg laying will be found to function in similar contexts in the wide variety of waterfowl species in which parasitism occurs regularly but much less frequently than in redheads (e.g., Lank et al. 1989*b*; Eadie 1991; Weigmann and Lamprecht 1991).

Life history strategies of redheads and canvasbacks

Basic differences in life-history strategy may explain the much lower frequency of parasitic egg laying and the lack of a dual strategy in canvasbacks as compared to redheads. Relatively large body size ($\bar{x} \pm$ S.E. = 1289 ± 17 g, $n = 39$ decoy-trapped, adult females), specialization on pondweed tubers (*Potamogeton* spp.) prior to breeding (Noyes and Jarvis 1985), and reliance on endogenous nutrient reserves for egg production (Barzen and Serie 1990) are probably adaptations that allow canvasbacks to initiate breeding shortly after arrival on the breeding grounds (Anderson 1985). At the time an older female canvasback would normally begin egg laying, there would be few, if any, laying-stage host nests available to her. In order to lay parasitic eggs prior to nesting, she would have to delay egg laying until many other canvasback females were in the laying stage, resulting in a further delay before the initiation of her own nest.

In contrast, the dual strategy of redheads is made possible by the opportunity to parasitize the earlier-nesting canvasback. Substantially smaller ($\bar{x} \pm$ S.E. = 1060 ± 12 g, $n = 41$ decoy-trapped, adult females), opportunistic in diet (Noyes and Jarvis 1985), and presumably adapted to breeding slightly later than canvasbacks, a redhead female can begin laying parasitic eggs as soon as she is physiologically ready because a large proportion of canvasback females have already initiated nests and are available as hosts. The delay in nest initiation due to parasitic laying is therefore relatively short in redheads compared with what it would be in canvasbacks if they attempted a similar strategy, and therefore the potential costs of delayed nesting may be less severe in redheads. The very low frequency of canvasback parasitism in redhead nests is probably also a consequence of the difference in nesting chronology: few redhead nests are available when most canvasback females are laying eggs. (Smaller population densities of redheads at Minnedosa and a lower per-capita rate of nest initiation in redheads, however, also reduce the potential availability of redhead nests for canvasback parasites.)

Whether and to what extent selection has modified the timing of breeding in redheads to facilitate parasitism of canvasbacks (or in canvasbacks to avoid parasitism) are intriguing but unanswered questions. Renesting has not been documented in prairie populations of redheads perhaps because, under favorable conditions in which other species would renest, redheads employ a dual strategy and have less time available for renesting. Dates of nest initiation appear to be earlier in two redhead populations that do not have interspecific hosts available (Alliston 1979*b*; Jobes 1980), and at least 86% of females renested after experimental clutch removals in one of these populations (Alliston 1979*a*), suggesting that earlier nesting is a conditional response to the lack of interspecific hosts. A truly valid comparison of nest initiation dates among redhead populations that controls for geographic and annual variation in climatological conditions, however, will require much more additional information.

Why the redhead?

Population-level frequencies of parasitic egg laying in redheads are among the highest recorded in any avian species. Although most studies report the proportion of nests parasitized (see Rohwer and Freeman 1989 for a review), better measures of the importance of parasitic egg laying as a reproductive tactic are the proportion of all eggs laid or the proportion of young hatched that are attributable to parasitism. In this study, over 60% of redhead eggs laid (table 19.3) and over 30% of redheads hatched (unpublished data) were parasitic. Similarly high proportions of redheads produced through parasitism (ranging from 20% to 50%) have been recorded in several in other studies (Erickson 1948; Olson 1964; Mattson 1973; Michot 1976; Johnson 1978; Sugden and Butler

1980; Bouffard 1983; Joyner 1983). Much lower rates of parasitic laying are indicated for other waterfowl for which similar data are available (proportion of eggs laid: canvasback, 13%, table 19.3; eider, *Somateria mollissima*, 9%, Robertson et al. 1992; goldeneyes, 17%, Eadie 1989; proportion of young hatched: snow goose, 5.3%, Lank et al. 1989*b*; ruddy duck, 7%, Joyner 1983). Although very high rates of intraspecific parasitism have been recorded in wood ducks (e.g., Clawson et al. 1979; Semel and Sherman 1986) and black-bellied whistling ducks (McCamant and Bolen 1979), these studies involved high-density populations nesting in conspicuous nest boxes, a situation shown to increase greatly the frequency of parasitism (Semel et al. 1988). The highest frequencies of facultative parasitism recorded in nonwaterfowl are around 10% of eggs (Jackson 1992; Lyon 1993) or young (Birkhead et al. 1990).

Clearly, what distinguishes redheads from other facultative brood parasites and what I suggest has allowed for the exceptional development of parasitic behavior in redheads is their extensive use of interspecific hosts. In addition to extending the seasonal duration of host availability, interspecific parasitism increases the absolute number of hosts available to a female in a given area and may increase parasitic egg success by spreading the negative effects of parasitism over a greater number of nests. Although Lyon and Eadie (1991) list 31 species of waterfowl as facultative interspecific parasites, many are classified as such on the basis of anecdotal records and are probably very rarely parasitic, whereas others are primarily intraspecific parasites (see Sayler 1992). The occurrence of parasitism in waterfowl is strongly related to nest site location and presumably to the ease with which females can find each other's nests (Rohwer and Freeman 1989; Eadie 1991). That canvasbacks build large, conspicuous nests in a restricted habitat (emergent vegetation) is one factor allowing a high frequency of interspecific parasitism in redheads. In addition. canvasbacks nest earlier than redheads (see above), are tenaciously faithful to their nests despite repeated intrusions by parasitic females, and are a congener that provides appropriate post-hatch care for redhead ducklings (see Amat 1987). Such a favorable combination of factors is apparently unusual among birds. Nonetheless, the redhead is not necessarily in the process of evolving into an obligate brood parasite: because parasitic eggs are relatively unsuccessful, redheads presumably do best by caring for their second clutch themselves (see also Lyon and Eadie 1991).

A host–parasite relationship similar to that of canvasbacks and redheads is possible in another pair of diving ducks, the pochard, *Aythya ferina*, and red-crested pochard, but results from one area suggest some major differences. First, rates of interspecific parasitism were relatively high for both species: red-crested pochards parasitized 31% of pochard nests (Amat 1985) and 42% of mallard nests (Amat 1991), whereas pochards parasitized 18% of red-crested pochard nests (Amat 1993) and 5% of mallard nests (Amat 1991). Second, the earlier nesting species of the pair, the red-crested pochard (Amat 1982), also exhibited the higher frequency of parasitic egg laying. Thus, in contrast to the biology of redheads and canvasbacks, the two species cannot be easily characterized as host and parasite. Given that there is no information on individual patterns of parasitic and nesting behavior for pochards or red-crested pochards, it is difficult to make any conclusions about the functional significance of parasitic laying in these species.

Although redheads parasitize canvasbacks over much of their breeding range, conclusions made here about the reproductive tactics of female redheads may not be applicable to all areas. Where interspecific hosts are not available or are greatly outnumbered by redheads, rates of intraspecific parasitism are higher and the success of parasitic eggs appears to be much lower (e.g., Weller 1959; Olson 1964; Lokemoen 1966; Jobes 1980; Sayler 1985) than in situations where canvasback hosts are relatively numerous (e.g., Erickson 1948; Olson 1964; Stoudt 1982; Bouffard 1983). The rate of intraspecific parasitism in one isolated redhead population, however, is very low and typical nesting is highly successful (Alliston 1979*b*). Females in this population nest in extensive sedge meadows rather than emergent vegetation, such that even the nests of other redheads are difficult to find and therefore unavailable to would-be parasites.

In some areas, however, redheads exploit interspecific hosts other than the canvasback and may successfully employ a dual strategy. Weller (1959) and Joyner (1975) presented seasonal chronologies of redhead parasitism and nesting very similar to those presented here (with most parasitic egg laying occurring before the peak of redhead nest initiations), except that mallards and cinnamon teal, *Anas cyanoptera*, served as hosts. Redheads parasitize mallards and cinnamon teal where these species nest in emergent vegetation (Weller 1959;

Michot 1976; Talent et al. 1981; Joyner 1983) as well as other dabbling ducks and lesser scaup, *Aythya affinis*, nesting at high density on natural or man-made islands (e.g., Giroux 1981; Lokemoen 1991). Geographic variation in the reproductive tactics of redhead females associated with differences in host species and population density presents an interesting problem for further research. Such differences are likely to be conditional rather than genetic, however, given that pair formation away from the breeding grounds and the extensive gene flow that results (Anderson et al. 1992) should limit genetic adaptation to local conditions (Rockwell and Cooke 1977; Lank et al. 1990).

The availability of many different host species may have been a primary factor in the evolution of the only obligate parasite among the Anatidae, the black-headed duck. Although obligate brood parasitism, by definition, must be interspecific, the black-headed duck's ability to parasitize successfully a wide variety of host species sets it apart from facultative parasites among the waterfowl, including the redhead. Black-headed ducks frequently parasitize rosybill pochards, *Netta peposaca*, coots, *Fulica* spp., and white-face ibis, *Plegadis falcinellus*, and have been recorded to parasitize at least 14 marsh-nesting species in five orders (Weller 1968). The evolution of obligate brood parasitism in black-headed ducks has also included the development of extreme precociality which undoubtedly facilitates successful parasitism of non-anatids, although it is not clear if this was a necessary condition for obligate parasitism or a later refinement.

ACKNOWLEDGMENTS I thank Scott K. Robinson, Stephen I. Rothstein, and an anonymous reviewer for comments on the manuscript. Funding for field work was provided by the North American Wildlife Foundation through the Delta Waterfowl and Wetlands Research Station and by the James Ford Bell Delta Waterfowl Fellowship administered by the Bell Museum of Natural History. This research could not have been completed without the help of many dedicated field assistants. I also gratefully acknowledge the farmers in the Minnedosa area for their hospitality and for permitting me to work on their land.

REFERENCES

Alliston, W. G. (1979*a*). Renesting by the redhead duck. *Wildfowl*, **30**, 40–44.

Alliston, W. G. (1979*b*). The population ecology of an isolated nesting population of redheads. Ph.D. Thesis, Cornell University, Ithaca, N.Y.

Amat, J. A. (1982). The nesting biology of the ducks in the Marismas of the Guadalquivir, south-western Spain. *Wildfowl*, **33**, 94–104.

Amat, J. A. (1985). Nest parasitism of pochard *Aythya ferina* by red-crested pochard *Netta rufina*. *Ibis*, **127**, 255–262.

Amat, J. A. (1987). Is nest parasitism among ducks advantageous to the host? *Amer. Natur.*, **130**, 454–457.

Amat, J. A. (1991). Effects of red-crested pochard nest parasitism on mallards. *Wilson Bull.*, **103**, 501–503.

Amat, J. A. (1993). Laying of common pochards in red-crested pochard nests. *Ornis Scand.*, **24**, 65–70.

Anderson, M. G. (1985). Social behavior of breeding canvasbacks (*Aythya valisineria*): male and female strategies of reproduction. Ph.D. Thesis. University of Minnesota, Minneapolis.

Anderson, M. G., J. M. Rhymer, and F. C. Rohwer (1992). Philopatry, dispersal, and the genetic structure of waterfowl populations. In: *Ecology and management of breeding waterfowl* (ed. B. D. J. Batt), pp. 365–395. Univ. of Minnesota, Minneapolis.

Anderson, M. G., R. B. Emery, and T. W. Arnold (1997). Reproductive success and female survival affect local population density of canvasbacks. *J. Wildlf. Manag.* **61**, 1174–1191.

Andersson, M. (1984). Brood parasitism within species. In: *Producers and Scroungers* (ed. C. J. Barnard), pp. 195–228. Croom Helm, London.

Ankney, C. D. and C. D. MacInnes (1978). Nutrient reserves and reproductive performance of female lesser snow geese. *Auk*, **95**, 459–471.

Ankney, C. D., A. D. Afton, and R. T. Alisauskas (1991). The role of nutrient reserves in limiting waterfowl reproduction. *Condor*, **93**, 1029–1032.

Arnold, T. W. and F. C. Rohwer (1991). Do egg formation costs limit clutch size in waterfowl? A skeptical view. *Condor*, **93**, 1032–1038.

Arnold, T. W. and M. D. Sorenson (1988). A record early season for marsh-breeding birds in southwestern Manitoba. *Blue Jay*, **46**, 133–135.

Arnold, T. W., F. C. Rohwer, and T. Armstrong (1987). Egg viability, nest predation, and the adaptive significance of clutch size in prairie ducks. *Amer. Natur.*, **130**, 643–653.

Barzen, J. A. and J. R. Serie (1990). Nutrient reserve dynamics of breeding canvasbacks. *Auk*, **107**, 75–85.

Bellrose, F. C. (1980). *Ducks, Geese and Swans of*

North America. Stackpole Books, Harrisburg, Pennsylvania.

Bergman, R. D. (1973). Use of southern boreal lakes by postbreeding canvasbacks and redheads. *J. Wildlf. Manag.*, **37**, 160–170.

Birkhead, T. R., T. Burke, R. Zann, F. M. Hunter and A. P. Krupa (1990). Extra-pair paternity and intraspecific brood parasitism in wild zebra finches *Taeniopygia guttata*, revealed by DNA fingerprinting. *Behav. Ecol. Sociobiol.*, **27**, 315–324.

Bouffard, S. H. (1983). Redhead egg parasitism of canvasback nests. *J. Wildlf. Manag.*, **47**, 213–216.

Brown, C. R. (1984). Laying eggs in a neighbor's nest: benefit and cost of colonial nesting in swallows. *Science*, **224**, 518–519.

Chronister, D. (1985). Egg-laying and incubation behavior of black-bellied whistling ducks. M.S. Thesis. University of Minnesota, Minneapolis.

Clawson, R. L., G. W. Hartman and L. H. Fredrickson (1979). Dump nesting in a Missouri wood duck population. *J. Wildlf. Manag.*, **43**, 347–355.

Dane, C. W. and D. H. Johnson (1975). Age determination of female redhead ducks. *J. Wildlf. Manag.*, **39**, 256–263.

Davies, N. B. (1982). Behavior and competition for scarce resources. In: *Current Problems in Sociobiology* (ed. King's College Sociobiology Group), pp. 363–380. Cambridge University Press, Cambridge.

Davies, N. B. and M. de L. Brook (1988). Cuckoos versus reed warblers: adaptations and counteradaptations. *Anim. Behav.*, **36**, 262–284.

Davies, N. B. and M. de L. Brook (1989*a*). An experimental study of co-evolution between the cuckoo, *Cuculus canorus*, and its hosts. I. Host egg discrimination. *J. Anim. Ecol.*, **58**, 207– 224.

Davies, N. B. and M. de L. Brook (1989*b*). An experimental study of co-evolution between the cuckoo, *Cuculus canorus*, and its hosts. II. Host egg markings, chick discrimination and general discussion. *J. Anim.* Ecol., **58**, 225–236.

Dawkins, R. (1980). Good strategy or evolutionary stable strategy? In: *Sociobiology: Beyond Nature/Nurture?* (eds G. W. Barlow and J. Silverberg), pp. 331–367. Westview Press, Boulder, Colorado.

Dominey, W. J. (1984). Alternative mating tactics and evolutionarily stable strategies. *Amer. Zool.*, **24**, 385–396.

Doty, H. A., D. L. Trauger, and J. R. Serie (1984). Renesting by canvasbacks in southwestern Manitoba. *J. Wildlf. Manag.*, **48**, 581–584.

Drobney, R. D. and L. H. Fredrickson (1985). Protein acquisition: a possible proximate factor limiting clutch size in Wood Ducks. *Wildfowl*, **36**, 122–128.

Eadie, J. M. (1989). Alternative reproductive tactics in a precocial bird: the ecology and evolution of brood parasitism in goldeneyes. Ph.D. Thesis. University of British Columbia, Vancouver.

Eadie, J. M. (1991). Constraint and opportunity in the evolution of brood parasitism in waterfowl. Proceedings of the 20th International Ornithologisal Congress, pp. 1031–1040, Vol. 2. New Zealand Ornitl. Cong. Trust Board: Wellington, New Zealand.

Eadie, J. M. and J. M. Fryxell (1992). Density-dependence, frequency-dependence and alternative nesting strategies in goldeneyes. *Amer. Natur.*, **140**, 621–641.

Eadie, J. M., F. P. Kehoe, and T. D. Nudds (1988). Pre-hatch and post-hatch brood amalgamation in North American Anatidae: a review of hypotheses. *Can. J. Zool.*, **66**, 1701–1721.

Erickson, R. C. (1948). Life history and ecology of the canvas-back, *Nyroca valisineria* (Wilson), in south-eastern Oregon. Ph.D. Thesis. Iowa State College, Ames.

Eriksson, M. O. G. and M. Andersson (1982). Nest parasitism and hatching success in a population of Goldeneyes *Bucephala clangula*. *Bird Study*, **29**, 49–54.

Erskine, A. J. (1990). Joint laying in *Bucephala* ducks – 'parasitism' or nest-site competition? *Ornis Scand.*, **21**, 52–56.

Feare, C. J. (1991). Intraspecific nest parasitism in starlings *Sturnus vulgaris*: effects of disturbance on laying females. *Ibis*, **133**, 75–79.

Giroux, J.-F. (1981). Interspecific nest parasitism by redheads on islands in southeastern Alberta. *Can. J. Zool.*, **59**, 2053–2057.

Gross, M. R. (1984). Sunfish, salmon, and the evolution of alternative reproductive strategies and tactics in fishes. In: *Fish Reproduction: Strategies and Tactics* (eds G. W. Potts and R. J. Wootton), pp. 55–75. Academic Press, London.

Haramis, G. M., W. G. Alliston, and M. E. Richmond (1983). Dump nesting in the wood duck traced by tetracycline. *Auk*, **100**, 729–730.

Heusmann, H. W., R. Bellville, and R. G. Burrell (1980). Further observations on dump nesting by wood ducks. *J. Wildlf. Manag.*, **44**, 908–915.

Jackson, W. M. (1992). Estimating conspecific nest parasitism in the northern masked weaver based on within-female variability in egg appearance. *Auk*, **109**, 435–443.

Jackson, W. M. (1993). Causes of conspecific nest parasitism in the northern masked weaver. *Behav. Ecol. Sociobiol.*, **32**, 119–126.

Jobes, C. R. (1980). Aspects of the breeding biology of the Redhead duck. M.S. Thesis. Texas A&M University, College Station.

Johnson, D. J. (1978). Age related breeding biology of the Redhead duck in southwestern Manitoba. M.S. Thesis. Texas A&M University, College Station.

Joyner, D. E. (1975). Nest parasitism and brood-related behavior of the ruddy duck (*Oxyura jamaicensis rubida*). Ph.D. Thesis. University of Nebraska, Lincoln.

Joyner, D. E. (1983). Parasitic egg laying in redheads and ruddy ducks: incidence and success. *Auk*, **100**, 717–725.

Kendra, P. E., R. R. Roth, and D. W. Tallamy (1988). Conspecific brood parasitism in the house sparrow. *Wilson Bull.*, **100**, 80–90.

Kiel, W. H., A. S. Hawkins, and N. G. Perret (1972). Waterfowl habitat trends in the aspen parkland of Manitoba. *Canadian Wildlife Service Report Series*, **18**, 1–63.

Lank, D. B., R. F. Mineau, R. F. Rockwell, and F. Cooke (1989*a*). Intra-specific nest parasitism and extra-pair copulation in lesser snow geese. *Anim. Behav.*, **37**, 74–89.

Lank, D. B., E. G. Cooch, R. F. Rockwell, and F. Cooke (1989*b*). Environmental and demographic correlates of intraspecific nest parasitism in lesser snow geese. *J. Anim. Ecol.*, **58**, 29–45.

Lank, D. B., R. F. Rockwell, and F. Cooke (1990). Frequency-dependent fitness consequences of intraspecific nest parasitism in snow geese. *Evolution*, **44**, 1436–1453.

Lokemoen, J. T. (1966). Breeding ecology of the redhead duck in western Manitoba. *J Wildlf. Manag.*, **30**, 668–681.

Lokemoen, J. T. (1991). Brood parasitism among waterfowl nesting on islands and peninsulas in North Dakota. *Condor*, **93**, 340–345.

Lombardo, M. P., H. W. Power, P. C. Stouffer, L. C. Romagnano, and A. S. Hoffenberg (1989). Egg removal and intraspecific brood parasitism in the Europea, starling (*Stunus vulgaris*). *Behav. Ecol. Sociobiol.*, **24**, 217–223.

Lyon, B. E. (1993). Conspecific brood parasitism as a flexible female reproductive tactic in American coots. *Anim. Behav.*, **46**, 911–928.

Lyon, B. E. and J. M. Eadie (1991). Mode of development and interspecific avian brood parasitism. *Behav. Ecol.*, **2**, 309–318.

Mattson, M. E. (1973). Host–parasite relations of canvasback and redhead ducklings. M.S. Thesis. University of Manitoba, Winnipeg.

McCamant, R. E. and E. G. Bolen (1979). A 12-year study of nest box utilization by black-bellied whistling ducks. *J. Wildlf. Manag.*, **43**, 936–943.

McKinney, F. (1954). An observation on Redhead parasitism. *Wilson Bull.*, **66**, 146–148.

Michot, T. C. (1976). Nesting ecology of the redhead duck on Knudson Marsh, Utah. M.S. Thesis. Utah State University, Logan.

Møller, A. P. (1987). Intraspecific nest parasitism and anti-parasite behavior in swallows, *Hirundo rustica. Anim. Behav.*, **35**, 247–254.

Noyes, J. H. and R. L. Jarvis (1985). Diet and nutrition of breeding female redhead and canvasback ducks in Nevada. *J. Wildlf. Manag.*, **49**, 203–211.

Olson, D. P. (1964). A study of canvasback and redhead breeding populations, nesting habits, and productivity. Ph.D. Thesis. University of Minnesota, Minneapolis.

Pienkowski, M. W. and P. R. Evans (1982). Clutch parasitism and nesting interference between Shelducks at Aberlady Bay. *Wildfowl*, **33**, 159–163.

Power, H. W., E. D. Kennedy, L. C. Romagnano, M. P. Lombardo, A. Hoffenberg, P. C. Stouffer, and T. McGuire (1989). The parasitism insurance hypothesis: why starlings leave space for parasitic eggs. *Condor*, **91**, 753–765.

Robertson, G. J., M. D. Watson, and F. Cooke (1992). Frequency, timing and costs of intraspecific nest parasitism in the common eider. *Condor*, **94**, 871–879.

Rockwell, R. F. and F. Cooke (1977). Gene flow and local adaptation in a colonially nesting dimorphic bird: the lesser snow goose (*Anser caerulescens caerulescens*). *Amer. Natur.*, **111**, 91–97.

Rohwer, F. C. (1992). The evolution of reproductive patterns in waterfowl. In: *Ecology and Management of Breeding Waterfowl* (ed. B. D. J. Batt). University of Minnesota Press, Minneapolis.

Rohwer, F. C. and M. G. Anderson (1988). Female-biased philopatry, monogamy, and the timing of pair formation in migratory waterfowl. *Curr. Ornith.*, **5**, 187–221.

Rohwer, F. C. and S. Freeman (1989). The distribution of conspecific nest parasitism in birds. *Can. J. Zool.*, **67**, 239–253.

Rothstein, S. I. (1990). A model system for coevolution: avian brood parasitism. *Annu. Rev. Ecol. Syst.*, **21**, 481–508.

Sayler, R. D. (1985). Brood parasitism and reproduction of canvasbacks and redheads on the Delta Marsh. Ph.D. Thesis, University of North Dakota, Grand Forks.

Sayler, R. D. (1992). Brood parasitism in waterfowl. In: *Ecology and Management of Breeding Waterfowl* (ed. B. D. J. Batt). University of Minnesota Press, Minneapolis.

Semel, B. and P. W. Sherman (1986). Dynamics of

nest parasitism in wood ducks. *Auk*, **103**, 813–816.

Semel, B., P. W. Sherman, and S. M. Byers (1988). Effects of brood parasitism and nest-box placement on wood duck breeding ecology. *Condor*, **90**, 920–930.

Serie, J. R., D. L. Trauger, H. A. Doty, and D. E. Sharp (1982). Age-class determination of canvasbacks. *J. Wildlf. Manag.*, **46**, 894–904.

Sokal, R. R. and F. J. Rohlf (1981). *Biometry*. Freeman, San Francisco.

Sorenson, M. D. (1990). Parasitic egg laying in redhead and canvasback ducks. Ph.D. Thesis, University of Minnesota, Minneapolis.

Sorenson, M. D. (1991). The functional significance of parasitic egg laying and typical nesting in redhead ducks: an analysis of individual behaviour. *Anim. Behav.*, **42**, 771–796.

Sorenson, M. D. (1992). Comment: Why is conspecific nest parasitism more frequent in waterfowl than in other birds? *Can. J. Zool.*, **70**, 1856–1858.

Sorenson, M. D. (1993). Parasitic egg laying in canvasbacks: frequency, success and individual behavior. *Auk*, **110**, 57–69.

Sorenson, M. D. (1997). Effects of intra- and interspecific brood parasitism on a precocial host, the canvasback, *Aythya valisineria*. *Behav. Ecol.*, **8**, 153–161.

Sowls, L. K. (1955). *Prairie Ducks*. Wildlife Management Institute, Washington, D.C.

Stoudt, J. H. (1982). Habitat use and productivity of canvasbacks in southwestern Manitoba, 1961–72. U.S. Fish and Wildlife Service Special Scientific Report. *Wildlife*, **248**, 1–31.

Stouffer, P. C. and H. W. Power (1991). Brood parasitism by starlings experimentally forced to desert their nests. *Anim. Behav.*, **41**, 537–539.

Sugden, L. G. (1980). Parasitism of canvasback nests by redheads. *J. Field Ornith.*, **51**, 361–364.

Sugden, L. G. and G. Butler (1980). Estimating densities of breeding canvasbacks and redheads. *J. Wildlf. Manag.*, **44**, 814–821.

Talent, L. G., G. L. Krapu, and R. L. Jarvis (1981). Effects of redhead nest parasitism on mallards. *Wilson Bull.*, **93**, 562–563.

Weigmann, C. and J. Lamprecht (1991). Intraspecific nest parasitism in bar-headed geese, *Anser indicus*. *Anim. Behav.*, **41**, 677–688.

Weller, M. W. (1959). Parasitic egg-laying in the redhead (*Aythya americana*) and other North American Anatidae. *Ecological Monographs*, **29**, 333–365.

Weller, M. W. (1965). Chronology of pair formation in some nearctic *Aythya* (Anatidae). *Auk*, **82**, 227–235.

Weller, M. W. (1967). Courtship of the redhead (*Aythya americana*). *Auk*, **84**, 544–559.

Weller, M. W. (1968). The breeding biology of the parasitic black-headed duck. *Living Bird*, **7**, 169–207.

Yom-Tov, Y. (1980). Intraspecific nest parasitism in birds. *Biol. Rev.*, **55**, 93–108.

20

Quality Control and the Important Questions in Avian Conspecific Brood Parasitism

HARRY W. POWER

Recent years have seen a plethora of papers on conspecific brood parasitism (CBP) which have been uneven in quality and variable in the questions they address. This chapter attempts to establish minimal standards of quality control in biochemically based studies, and provide a uniform list of important questions to be addressed in such studies. It is hoped that the chapter will improve the quality of future studies of CBP, as Hickey's (1943) *A Guide to Bird Watching* (Appendix C) improved life history studies. With both life history and CBP studies, the fundamental problem was and is the same: publication of a large body of work of such varying degree of quality and thoroughness that it presents an almost anarchic picture of natural phenomena, and in some cases confuses as much as it informs.

QUALITY CONTROL

Romagnano et al. (1989) stated the necessary ingredients of quality control in biochemical studies of avian parentage. I build on their recommendations and assess how published biochemical studies meet these standards and can thus be used to identify the true frequencies of CBP and its parallel form of conspecific genetic parasitism (CGP), cuckoldry (Power 1990*b*), as well as the perpetrators and products of CGP. Studies using other biochemical methods can be useful for a number of purposes (e.g., determining the simple presence or absence of CBP); however, such studies are more limited than biochemical studies because their methods have less resolving power.

Table 20.1 summarizes all biochemical studies of CGP available at the time of writing, and criticizes them according to a series of variables. All CGP studies are listed rather than just CBP studies because biochemical studies have the capacity to reveal both cuckoldry and CBP simultaneously, and cuckoldry and CBP are analogous parasitic tactics for the two sexes and one may not be fully understandable without knowledge of the other (Power 1990*b*). Species are listed phylogenetically following Sibley et al. (1988) for families, subfamilies, and tribes, and the Sixth AOU Checklist (1983) for species within tribes. No scientific names are given in the text for species listed in table 20.1 in order to avoid redundancy.

The variables

The variables and their states are described here rather than in footnotes to table 20.1 because of their complexity.

1. *Analytical method*. The biochemical technique.
   a. *DNA methods*:
      (i) DNAf = DNA fingerprinting (also called DNA profiling).

(ii) RFLP = Restriction fragment length polymorphism.

b. *Protein electrophoresis methods*:

(i) CAE = Cellulose acetate electrophoresis.

(ii) PAGE = Polyacrylamide gel electrophoresis.

(iii) SGE = Starch gel electrophoresis.

2. *Field design*. The distribution of nest sites used for tissue biopsies.

a. A = Artificial distribution, i.e., use of nestboxes or other human constructs.

b. N = Natural distribution.

c. S = Satisfactory control for layout of artificial nest sites, i.e., not too crowded and nest sites not less than 1 territory diameter apart (Romagnano et al. 1989).

d. U = Unsatisfactory control for nest site layout, i.e., nest density unnaturally high and/or nest sites less than 1 territory diameter apart.

e. ? = Layout of artificial nest sites insufficiently described to allow judgment as to whether or not it was satisfactory.

3. *Temporal laying pattern*. The time of day eggs were laid. This knowledge allows determination of whether hosts and parasite lay at the same or different times (Romagnano et al. 1990).

a. R = Time reported

b. U = Time unreported and/or unknown by the investigator.

4. *Egg of chick origin*. Which chick came from which egg. Important in determining where in the laying sequence CGP happened. If CBP is involved, this information will indicate whether it occurred before, during or after host laying; this will help establish the window of opportunity for the parasite and the window of vulnerability for the host. If cuckoldry is involved, this information may indicate when mate guarding is most or least effective.

a. NA = Not applicable, e.g., only eggs were sampled.

b. R = Reported.

c. U = Unreported.

5. *Sample size*

a. # *Broods* (or clutches). This is the more important because it is the unit of selection in clutch size determination, parental investment allocation and sex ratio (Fisher 1930). Moreover, the success of the parent generally depends on the success of the brood whether it is large or small. Nowhere is this more evident than in the quality/quantity trade-off of relatively small broods for relatively high-quality offspring in birds (Darwin 1871; Lack 1968; Trivers 1972).

b. # *Individuals*. Although less important than the number of broods, this is still important because each individual sampled is potentially a source (adults) or product (young) of CGP, giving the investigator an opportunity to discover CGP and its dynamics. However, focusing on the number of individuals can be misleading when the number of broods is small and the number of individuals per brood is large because the number of affected care-givers is then small, covering up the fact that the number of *host* decisions subject to negative selection by parasitism is also small. By contrast, when the number of broods is large irrespective of brood size, the number of parental decisions is large, and the true consequences of CGP unfold.

6. *Adult capture control*. The care that the investigator used to capture the adults actually providing the parental investment or actually sequestering the mate and territory, i.e., the putative parents. Capturing the wrong adult automatically generates a false mismatch between putative parent and offspring, inflating the apparent frequency and distorting the apparent pattern of CGP.

a. NA = Not applicable because only eggs examined.

b. S = Satisfactory.

c. U = Unsatisfactory. This is most evident when an obviously wrong adult has been captured, but it is also a potential problem when the criteria for establishing putative parents are not given.

7. *System resolution*. An ordinal judgment as to how capable a particular biochemical system is in revealing CGP.

a. H = High. Reserved for DNA studies.

b. L = Low. Electrophoresis with only one or two variable loci.

c. M = Moderate. Electrophoresis with more than two variable loci, each with two or more alleles in relatively high frequency, e.g., Wrege and Emlen (1987) had four loci with two to five alleles each and the most common allele at each locus

Table 20.1 Biochemical studies in avian conspecific genetic parasitism

	Field methods							Laboratory methods		
					Sample size					
Species	Analytical method	Field design	Temporal laying pattern	Eggs of chick origin	#Broods	#Individuals	Adult capture control	System resolution	Developmental effects	Tissue artifacts
Lesser snow goose (*Chen caerulescens caerulescens*)	RFLP	N	U	U	4	25	S	H	C	N
Barnacle goose (*Branta leucopsis*)	DNAf	N	U	U	9	35	S	H	C	N
Mallard (*Anas platyrhynchos*)	SGE	N	U	U	46	344	U	M	C	?
Acorn woodpecker (*Melanerpes formicivorus*)	SGE	N	U	U	3	16	S	?	N	N
	SGE	N	U	U	62	186+	S	L	C	N
Red-cockaded woodpecker (*Picoides borealis*)	DNAf	N	U	U	10	36	S	H	C	N
	DNAf	N	U	U	44	224	S	H	C	N
White-fronted bee-eater (*Merops bullockoides*)	SGE	N	U	U	65	255	S	M	C	N
European bee-eater (*Merops apiaster*)	DNAf	N	U	U	19	146	U	H	C	N
Yellow-billed cuckoo (*Coccyzus americanus*)	SGE	N	U	NA	2	10	N/A	L	NA	N
Shag (*Phalacrocorax aristotelis*)	DNAf	N	U	U	15	58	S	H	C	N
Black vulture (*Coragyps atratus*)	DNAf	N	U	U	16	68	S	H	C	N
Northern fulmar (*Fulmarus glacialis*)	DNAf	N	U	R	9	18+	S	H	C	N
Eastern kingbird (*Tyrannus tyrannus*)	SGE	N	U	U	19	98	S	M	C	N
Splendid fairy-wren (*Malurus splendens*)	CAE	N	U	U	?	152	S	M	N	N
Bull-headed shrike (*Lanius bucephalus*)	DNAf	N	U	U	24	158	S	H	C	N
Eastern bluebird (*Sialia sialis*)	SGE	AU	U	U	57	257	U	L	N	N
	SGE	A?	U	U	10	48	U	L	N	N
	SGE	AS	U	U	79	730	S	M	N	N
	DNAf	AU	U	U	21	83+	S	H	C	N
American robin (*Turdus migratorius*)	SGE	N	U	U	14	61	S	L	N	N
Pied flycatcher (*Ficedula hypoleuca*)	DNAf	A?	U	U	27	135+	S	H	C	N
	SGE	AS	U	U	22	174	S	L	C	N
	DNAf	AS	U	U	7	52	S	H	C	N

Conclusions				
	Overall quality control	CGP frequency estimate	Comments	References
CBP: 2/4 broods (50%); 3/17 chick (17.6%) CBP/CUCK: 2/4 broods (50%); 3/17 chicks (17.6%)	S	U	Sample too small but study showed value of RFLP. Lank et al. (1989) did a genetic marker study of the same population.	Quinn et al. (1987)
CBP: 1/8 broods (12.5%); 1/18 chicks (5.6%)	S	S	The CBP case may have involved quasiparasitism.	Choudhury et al. (1993)
CUCK: 8/46 broods (17.4%); 9/298 chicks (3.0%)	?	C	Adult males not sampled. Field methods insufficiently described.	Evarts and Williams (1987)
"CUCK": 1/3 broods (33.3%); unspecified number of chicks	S	U	Sample too small; not all group nesting males caught; allelic frequencies not given.	Joste et al. (1985)
"CUCK": 2/62 broods (3.2%); 4/186 chicks (2.2%)	S	C	# adults at 62 broods where paternity analysis done not clearly reported.	Mumme et al. (1985)
None	S	S	Study shows that combining DNAf with a known pedigree can help quantify population structure.	Haig et al. (1993)
CUCK: 1/44 broods (2.3%); 1/80 chicks (1.3%)	S	A	A careful study showing nearly true monogamy. The one case of extra-pair fertilization (EPF) was true CUCK because the cuckolder was external to the nesting group.	Haig et al. (1994)
CUCK: 2/65 broods (3.1%); 2/97 chicks (2.1%) CBP: 2/65 broods (3.1%); 2/97 chicks (2.1%) CBP/CUCK: 2/65 broods (3.1%); 3/97 chicks (3.1%)	S	C		Wrege and Emlen (1987)
CUCK: 1/19 broods (5.3%); 1/100 chicks (1.0%)	S	S	Field methods poorly described, e.g., adult capture control criteria not stated. CUCK occurred in a helped nest.	Jones et al. (1991)
CBP: 1/2 broods (50%); 2/10 eggs (20%)	S	U	Only egg proteins considered; thus it was impossible to detect CUCK. Sample too small.	Fleischer et al. (1985)
CUCK: 3/15 broods (20%); 4/28 chicks (14.3%) CBP: 1/15 broods (6.7%); 1/28 chicks (3.6%)	S	C	Estimate C because of loss of some chicks and samples. Authors deny that one case of mismatch between both parents and chick was a case of CBP.	Graves et al. (1992)
None	S	S	Species appears truly monogamous.	Decker et al. (1993)
None	S	A	No CUCK despite observation of EPCs at all sampled broods.	Hunter et al. (1992)
CBP/CUCK: 9/19 broods (47.4%); 18/60 chicks (30%)	S	C	Author argues the bulk of cases are due to quasiparasitism but does not support this conclusion with sufficient evidence.	McKittrick (1990)
CUCK: 59/91 chicks (64.8%); (No. broods not reported)	S	C	Results contradict earlier conclusion that inbreeding common in this species (Rowley et al. 1986).	Brooker et al. (1990)
CUCK: 4/24 broods (16.7%); 10/99 chicks (10.1%)	S	A	Source of EPFs sought among neighboring resident males but not found there.	Yamagishi et al. (1992)
CUCK: 1/29 broods (3.5%); 1/210 chicks (0.5%) CBP: 3/41 broods (7.3%); 3/210 chicks (1.4%)	U	I	Field design probably increased the frequency of CGP (Romagnano et al. 1989). I excluded family 35 from the authors' tabla 1 because the female mismatched all three chicks, implying the wrong female was caught. I calculated frequencies on the basis of broods, not the authors' "families," in order to make the results of this study comparable with others. This radically reduces the importance of the authors' claimed rate of CGP of 25% of "families."	Gowaty and Karlin (1984)
CUCK: frequency uncalculable because # broods where males sampled along with chicks unreported	U	U	Only one family complete. Closeness of boxes not reported. Sample size ambiguous.	Karlin et al. (1990)
CBP/CUCK: frequencies not clearly reported	S	I	Experimentally showed that packing nestboxes drives up the CGP rate as stated by Romagnano et al. (1989). Layout of results confusing. CUCK rate inflated by criteria for recognizing CGP that are not upheld in nature.	Gowaty and Bridges (1991)
CUCK: 5/21 broods (23.8%); 7/83 chicks (8.4%) CBP: 2/21 broods (9.5%); 2/83 chicks (2.4%)	S	S	# sampled adults not reported. Conclusion that mate guarding was unimportant exceeds the power of the experimental design.	Meek et al. (1994)
CBP:2/14 broods (14.3%); 4/46 chicks (8.7%)	U	U	Allelic frequencies not clearly given. Loci on which exclusions based not defined. Ambiguous presentation.	Gowaty and Davies (1986)
CUCK: 4/27 broods (14.8%); 6/135 chicks (4.4%)	S	S	Four cuckolds and one cuckolder were identified. All were older males	Lifjeld et al. (1991)
CUCK: 2/22 broods (9.1%); 2/131 chicks (1.5%)	S	C	Compare SGE results with DNAf results from same study (see next).	Gelter and Tegelström (1992)
CUCK: 3/7 broods (42.9%); 9/38 chicks (23.7%) CBP: 1/7 broods (14.3%); 1/38 chicks (2.6%)	S	S	The # sampled broods is smaller but the # detected cases is much larger than with SGE.	Gelter and Tegelström (1992)

(*continued*)

Table 20.1 (*continued*)

	Field methods							Laboratory methods		
					Sample size					
Species	Analytical method	Field design	Temporal laying pattern	Eggs of chick origin	#Broods	#Individuals	Adult capture control	System resolution	Developmental effects	Tissue artifacts
European starling (*Sturnus vulgaris*)	SGE	A?	U	U	260	?	S	L	C	N
	PAGE	AS	R	R	118	550	S	L	C	N
	DNAf	AS	U	U	14	62+	S	H	C	N
	DNAf	AS	U	U	22	92+	S	H	C	N
House wren (*Troglodytes aedon*)	SGE	AU	U	U	18	116	S	L	N	N
Stripe-backed wren (*Campylorhynchus nuchalis*)	DNAf	N	U	U	22(?)	260	S	H	C	?
Blue tit (*Parus caeruleus*)	DNAf	AS	U	U	36	314+	S	H	C	N
Black-capped chickadee (*Parus atricapillus*)	DNAf	?	U	U	8	53+	S	H	C	N
Purple martin (*Progne subis*)	DNAf	AS	U	U	30	?	S	H	C	N
Tree swallow (*Tachycineta bicolor*)	DNAf	AS	U	R	10	50+	S	H	C	N
	DNAf	AS	U	U	?	?	S	H	C	N
	DNAf	AS	U	U	17	86+	S	H	C	N
	DNAf	AS	U	U	18	95+	S	H	C	N
Cliff swallow (*Hirundo pyrrhonota*)	SGE	N	U	U	105	559	S	L	N	N
	SGE/PAGE	N	U	NA	54	?	NA	M	NA	N
Barn swallow (*Hirundo rustica*)	DNAf	N	U	U	11	112	S	H	C	N
Willow warbler (*Phylloscopus trochilus*)	DNAf	N	U	U	19	158	?	H	C	N
Wood warbler (*Phylloscopus sibilatrix*)	DNAf	N	U	U	13	82	?	H	C	N

	Conclusions			
	Overall quality control	CGP frequency estimate	Comments	References
CBP: 54/260 broods (20.8%); # chicks unspecified	S	U	Closeness of nestboxes not reported. Arbitrarily assigned all cases to CBP but CUCK might have accounted for some (see Power 1990). # individuals biopsied not reported.	Evans (1980, 1988)
CUCK: 2/118 broods (1.7%); 4/365 chicks (1.1%) CBP: 5/118 broods (4.2%); 6/365 chicks (1.6%) CBP/CUCK: 6/118 broods (5.1%); 6/365 chicks (1.6%)	S	C	Census and POF studies showed a much higher rate of CBP.	Hoffenberg et al. (1988); Kennedy et al. (1989); Power (1990b); Romagnano et al. (1990)
EPP: 4/14 broods (28.6%); 6/62 chicks (9.7%) CBP: 1/14 broods (7.1%); 1/62 chicks (1.6%)	S	S	Three of six cases of EPP involving two of four broods due to rapid mate replacement after predation of original male. Other three cases at other two broods due to CUCK. # adults unspecified.	Pinxten et al. (1993)
CUCK: 7/22 broods (31.8%); 8/92 chicks (8.7%)	S	A	Although CBP not found in DNA sample, it was found by observational methods.	Smith and von Schantz (1993)
?	U	I	Nestboxes tightly packed (up to 43/ha) possibly causing great increases in CGP. Estimated frequency of CBP (84.8% of broods) is inconsistent with the results of a larger study in NJ (0%) that followed marked eggs (Kennedy 1989). No justification given for assigning all ambiguous cases to CBP.	Price et al. (1989)
"CUCK": 4/22 "social groups" (18.2%); 6/69 chicks (8.7%)	S	S	Ambiguous sample size because a "social group" can produce more than one brood. Not CUCK because EPFs perpetrated by auxiliary males that helped at the nest.	Rabenold et al. (1990)
CUCK: 11/36 broods (30.6%); 33/314 chicks (10.5%)	S	S-A	# adults biopsied not clearly reported. Although only 10.5% of *all* chicks came from CUCK, 33 of 90 chicks (36.7%) in nests where CUCK occurred came from EPFs by males physically superior to the cuckold.	Kempenaers et al (1992)
CUCK: 3/8 broods (37.5%); 9/19 chicks (17%)	S	S	Unclear whether biopsied birds lived in nestboxes or natural cavities. # adults not reported. The cuckolder was always dominant to the cuckold at winter feeders. Confirms Smith (1988).	Otter et al. (1994)
CUCK: 7/14 broods (50%); 18/52 chicks (34.6%) CBP: 5/14 broods (35.7%); 10/52 chicks (19.2%)	S	S	DNAf used to show the special vulnerability of young males to CUCK.	Morton et al (1990)
CUCK: 5/10 broods (50%); 16/44 chicks (36.4%)	S	S	5/10 broods (50%) had CUCK involving all chicks before the original male was removed in an experiment. The other five females remained loyal to the original male even after his removal, i.e., subsequent eggs were still sired by his sperm. # individual biopsied not clearly reported.	Lifjeld and Robertson (1992)
CUCK: no clear statement of numerical result.	S	?	Sample sizes ambiguous: 12 broods of polygnous males listed in table 1 but only 7–10 referred to in text; no clear statement of # adults. Monogamous males compared with polygynous males from Lifjeld et al. (1993) (see next).	Dunn and Robertson (1993)
CUCK: 9/17 broods (52.9%); 33/86 chicks (384%)	S	S	# adults not reported. CUCK usually involved several chicks per brood. More than one cuckolder was indicated in 3/9 broods (33.3%) with CUCK.	Lefjeld et al. (1993)
CUCK: 13/18 broods (72.2%); 59/95 chicks (62.1%)	S	S	Report based on 57 broods but 39 of these originally reported in Lifjeld et al. 1993. # adults not clearly reported.	Dunn et al. (1994)
CBP/CUCK: 22/105 broods (20.9%); 35/349 chicks (10%)	S	C	32/35 chicks (91.4%) could have been either CUCK or CBP. Authors believed CBP to be more important.	Brown and Brown (1988)
CBP: 2/54 clutches (3.7%); unspecified # eggs	S	C	SGE used in first year of study, PAGE in second. Egg albumin study. CUCK undetectable by this method. # eggs biopsied not clearly reported.	Smyth et al. (1993)
CUCK: 5/11 broods (45.5%); 10/90 chicks (10.1%)	S	S	Males with artificially elongated tails cuckolded more often than males with artificially shortened tails.	Smith et al. (1991)
None	S	S	Some ambiguity in reported sample sizes, i.e., an additional 18 males were biopsied between the two species but the contribution of each species was not stated, and it is implied but not stated that both putative parents were biopsied for each family. Criteria for determining putative parents and capture method not described.	Gyllensten et al. (1990)
None	S	S		Gyllensten et al. (1990)

(*continued*)

Table 20.1 (*Continued*)

	Field methods							Laboratory methods		
					Sample size					
Species	Analytical method	Field design	Temporal laying pattern	Eggs of chick origin	#Broods	#Individuals	Adult capture control	System resolution	Developmental effects	Tissue artifacts
Dunnock (*Prunella modularis*)	DNAf	N	U	U	45	272	S	H	C	N
	DNAf	N	U	U	61	296	S	H	C	N
House sparrow (*Passer domesticus*)	DNAf	?	U	U	4	13	S	H	C	N
	DNAf	A?	U	U	6	43	S	H	C	N
	PAGE	A?	R	NA	17	?	NA	L	NA	N
	DNAf	AS	U	U	183	699	S	H	C	N
Zebra finch (*Taeniopygia guttata*)	DNAf	N	U	U	25	124	S	H	C	N
Chaffinch (*Fringilla coelebs*)	DNAf	N	R	U	13	67	S	H	C	N
Corn bunting (*Miliaria calandra*)	DNAf	N	U	U	18	56	S	H	C	N
Indigo bunting (*Passerina cyanea*)	SGE	N	U	U	98	449	?	L	N	N
	DNAf	N	U	U	25	108	S	H	C	N
Field sparrow (*Spizella pusilla*)	CAE	N	U	R	19	89	S	L	C	N
Mountain white-crowned sparrow (*Zonotrichia leucophrys oriantha*)	SGE	N	U	U	35	180	S	L	C	N
Hooded warbler (*Wilsonia citrina*)	DNAf	N	U	U	25	112	S	H	C	N
Northern cardinal (*Cardinalis cardinalis*)	DNAf	N	U	U	19	37+	S	H	C	N
Bobolink (*Dolichonyx oryzivorus*)	SGE	N	U	U	12	?	S	L	N	N
	SGE	N	U	U	191	1031	S	L	N	N
Red-winged blackbird (*Agelaius phoeniceus*)	RFLP/ DNAf	N	U	U	36	163	U	H	C	N
	DNAf	N	U	U	68	317	S	H	C	N
	DNAf	N	U	U	78	262+	S	H	C	N

	Conclusions			
	Overall quality control	CGP frequency estimate	Comments	References
CUCK: 1/45 broods (2.2%); 1/133 chicks (0.8%)	S	A	Only one case of true CUCK but paternity was shared between alpha and beta males in many families. Sharing was not CUCK because both males fed young and did so in proportion to paternity. # adults biopsied not clearly reported.	Burke et al. (1989)
CUCK: 13/27 broods (48%) and 28/74 chicks (37.8%) in the male-removed group, but only 2/34 (5.9%) and 2/100 chicks (2%) in the control group	S	A	Male parental investment in proportion to mating access in multi-male groups but monogamous males subject to possible CUCK provided as much parental investment as monogamous males not so subject.	Davies et al. (1992)
CUCK: 1/4 broods (25%); 1/11 chicks (9.1%)	S	U	Where birds nested unreported so cannot determine if field design was natural or artifactual. Method of parental assignment unreported.	Burke and Bruford (1987)
CUCK: 1/6 broods (16.7%); 1/25 chicks (4%)	S	U	Closeness of boxes not reported. Sample too small.	Wetton et al. (1987)
CBP: 5/42 broods (11.9%); unspecified number of eggs	S	U	Closeness of boxes not reported. Only eggs examined. # eggs not clearly reported. Cuckoldry could not have been detected if it had occurred. Sample too small.	Kendra et al. (1988)
CUCK: 49/183 broods (26.8%); 73/536 chicks (13.6%)	S	A	CUCK was most common in broods with some infertility.	Wetton and Parkin (1991)
CUCK: 2/25 broods (8.0%); 2/92 chicks (2.2%) CBP: 9/25 broods (36.0%); 10/92 chicks (10.9%)	S	A	Supplements other studies on same species led by senior author.	Birkhead et al. (1990)
CUCK: 3/13 broods (23.1%); 8/45 chicks (17.8%)	S	S	All cases assigned to CUCK but reasons for excluding CBP not sufficient to discard its possibility.	Sheldon and Burke (1994)
"CUCK": 1/15 broods (6.7%); 2/44 chicks (4.5%)	S	S	The one brood of male-chick mismatch involved both chicks and might have been due to misassignment of male and nest rather than misparentage. No male parental investment at this nest so no CUCK possible.	Hartley et al. (1993)
'CUCK": 24/98 broods (24.5%); 37/257 chicks (14.4%)	S	C	Criteria for determining putative parents and capture methods not described. Males provide little parental investment thus no real CUCK.	Westneat (1987)
"CUCK": 12/25 broods (48%); 22/63 chicks (34.9%)	S	A	Extends earlier study using SGE and shows greater resolving power of DNAf.	Westneat (1990)
CBP/CUCK: 5/17 broods (29.4%); 10/52 chicks (19.2%)	S	U	Only 12 broods complete. Sample too small. Assignment of misparentage to specific kinds of CGP ambiguous.	Petter et al. (1990)
CUCK: 9/35 broods (25.7%); 15/110 chicks (13.6%)	S	C		Sherman and Morton (1988)
CUCK: 8/25 broods (32.0%); 23/78 chicks (29.5%)	S	S	CUCK was strongly bimodally distributed such that either none occurred or a large fraction of a brood resulted from CUCK.	Stutchbury et al. (1994)
CUCK: 3/19 broods (15.8%); 5/37 chicks (13.5%)	S	S	# Adults biopsied not specified.	Ritchison et al. (1994)
CUCK: 2/12 broods (16.7%); unspecified # chicks	S	U	Sample too small considering seven broods came from one population and five from another. EPFs seem to involve true CUCK in that broods were apparently those of primary females. # individuals biopsied not specified.	Gavin and Bollinger (1985)
CUCK: 72/191 broods (37.7%); 36/840 chicks (4.3%)	S	C	A very careful large-scale study partially undercut by near fixation of the most common allele at each of four loci, generating a very conservative frequency estimate.	Bollinger and Gavin (1991)
"CUCK": 17/36 broods (47.2%); 31/111 chicks (27.9%)	S	S	Criteria for adult capture control vague leading to possibility of misassignment of adult males. Not true CUCK in all cases because males provide no care to young of secondary females.	Gibbs et al. (1990)
"CUCK": 26/68 broods (41.2%); 58/235 chicks (24.7%)	S	A	Some cases of EPFs were true CUCK, others were not because chicks were in the nests of both primary and secondary females.	Westneat (1993)
"CUCK": 34/78 broods (43.6%); 64/262 chicks (24.4%)	S	A	Paternity had no effect on male feedings of chicks. # adults biopsied not reported.	Westneat (1995)

occurred at a frequency of only 42–70%.
   d. ? = Judgment not possible because essential information not provided, e.g., allelic frequencies.
8. *Developmental effects.* The degree to which the investigator tried to determine whether young and adults have the *capacity* to have the same electrophoretic phenotypes. Romagnano et al. (1989) showed that young may have electrophoretic phenotypes inconsistent with their parents only because they have *not yet* developed adult phenotypes but will do so in a few days.
   a. C = Controlled for. The paper mentioned the possibility of developmental effects (e.g., Petter et al. 1990), indicating some effort had been made to see whether they occurred, or young were biopsied close to the expected time of fledging when developmental effects could be expected to be past (Hoffenberg et al. 1988), or the study used DNA techniques; DNA is unaffected by development.
   b. N = Not controlled for. Developmental effects were not mentioned in the paper and therefore may not have been considered as a potential source of error, or young were biopsied early in the nestling period when developmental effects could be expected to be important, especially in pectoral muscle (Romagnano et al. 1989), or the age of young at the time of biopsy was not given. It may be that developmental effects did not occur or were controlled for, but that can't be known if the paper does not provide the relevant information.
   c. NA = Not applicable because only eggs were examined.
9. *Tissue artifacts.* The degree to which different methods of tissue treatment generate false evidence of misparentage, e.g., if parental blood is frozen before separation into cellular and plasma components but offspring blood is not, then lysing of RBCs in the parental sample could produce false mismatches (Romagnano et al. 1989). However, if all tissue is treated the same, such artifactual misparentage does not occur.
   a. N = No. Artifacts not likely.
   b. Y = Yes. Artifacts likely because of techniques.
   c. ? = Insufficient information upon which to decide.
10. *Result.* The kind of CGP discovered and its frequency as a proportion of both # broods and # offspring.
   a. CBP = only CBP found.
   b. CUCK = only cuckoldry found.
   c. CBP/CUCK = could be *either* CBP or CUCK.
   d. None = neither cuckoldry nor CBP found.
   e. EPP = extra-pair parentage that includes both CUCK and rapid mate replacement.
   f. " " (quotation marks) = no true CGP because either: (i) the affected putative parent provided little or no parental investment even when all its young were legitimate, e.g. red-winged blackbird males never provide direct parental care to the young of secondary females (Orians 1980); or (ii) the perpetrator of CGP provided care along with the affected putative parent, e.g., beta male dunnocks helped alpha males feed young (Burke et al. 1989). Therefore these cases involve loss of parentage but not loss of parental investment, the more important currency of parasitism (Trivers 1972).
11. *Overall quality control.* The extent to which the study met the previously listed criteria, which provides an index of quality.
   a. S = Satisfactory.
   b. U = Unsatisfactory, especially because of uncontrolled artifactual design.
   c. ? = Insufficient information given to make a judgment.
12. *CGP frequency estimate.* My judgment as to the accuracy/reliability of the published frequency estimate.
   a. A = Accurate because the system has high resolving power and the sample size is intuitively reasonable.
   b. C = Conservative. The true frequency is probably higher than the published frequency because of the low resolving power of the system (L or M).
   c. I = Inflated. The field design has probably driven the CGP rate well above natural levels. This may be compounded by poor adult capture control.
   d. S = Satisfactory. The system had high resolution but the sample was too small to project an accurate frequency estimate.

e. U = Unsatisfactory. The sample size was too small relative to the resolving power of the system and/or the field design was actually or potentially too artifactual to reflect what was happening in nature.
f. ? = Can not make a determination because of reasons listed under "Comments" in table 20.1.

Studies with an S for overall quality control but a U for CGP frequency estimate are still considered good studies for purposes other than CGP frequency estimate — for example, for simply determining the presence or absence of CGP and its two forms (e.g., Petter et al. 1990), or for demonstrating the value of a technique (e.g., Burke and Bruford 1987; Quinn et al. 1987), or for showing the widespread existence of CGP (e.g., Evans 1980, 1988). Studies with an A are obviously best but those with a C are also credible and worthy of emulation in planning future studies.

Choice of system

An A was given to only some studies and all of these were DNA fingerprinting studies. But this does not mean that only DNA studies should be pursued in the future. DNA studies are potentially the most reliable, but they are also the most expensive and have the greatest intrinsic potential for laboratory errors because they require a greater number of procedural steps. However, they are also the only ones with the potential to *include* parentage rather than merely exclude it (Hill 1987).

Electrophoretic studies are inherently limited compared with DNA studies, but they are much cheaper in both time and money. This makes them still the method of choice if the questions are limited enough (e.g., does CBP ever occur in a population?), and the system variability is great enough (e.g., more than two loci each with two or more alleles of approximately equal frequency).

Clearly the choice of system must depend upon the goals of the project and its potential financing. Fortunately, many of the important questions of CGP can be addressed in a cost- and time-effective manner, some of them without the use of biochemical techniques, e.g., Smith and von Schantz (1993) were unable to find CBP in European starlings using DNA fingerprinting but found it using observational methods.

THE IMPORTANT QUESTIONS

In this section I list systematically the questions that should be considered in any study attempting to understand the whole phenomenon of CGP. In practice, few studies will be able to consider all of them because of logistical constraints in technology, manpower, and financing. But this should not prevent the attempt to answer as many as possible. My overall assessment of table 20.1 is that far more questions could have been answered than were answered by most studies, but that the attempt was never made, possibly because many of the questions never occurred to the investigators. It is my hope that the following list will help investigators avoid such oversights in future studies.

1. *Frequency questions.*
   a. Does CGP occur in the study population?
   b. If CGP occurs, what forms does it take (cuckoldry, CBP, quasiparasitism [Wrege and Emlen 1987], all)?
   c. At what frequency does each form occur?
      (i) The fraction of broods affected.
      (ii) The fraction of individuals affected.
   d. Does the frequency of each form vary?
      (i) Within seasons.
      (ii) Between years.
      (iii) Among populations of the same species.
      (iv) As a function of population density.
      (v) As a function of operational sex ratio (Emlen 1976).
      (vi) As a function of limiting resources, e.g., nest availability.
2. *Who are the victims of CGP?*
   a. Their age.
   b. Their nesting experience.
   c. Their place in the social mating system, e.g., polygynous males or secondary females.
   d. Their size or secondary sexual characters.
3. *Who are the perpetrators of CGP?*
   a. *Floaters*, i.e., birds without nests of their own engaging in an *alternative* reproductive strategy.
   b. *Residents*, i.e., birds with nests:
      (i) Near or distant residents (as measured by number of territories or nests between nest of residence and nest of parasitism).
      (ii) Special characteristics, e.g., secon-

dary sexual characters, size, age, experience, mating status.
   - (iii) Recent victims of nest loss but still with territories, and the ability and time to renest. Resident categories (i) and (ii) engage in a *supplementary* reproductive strategy because they have active nests of their own, but category (iii) engages in an *alternative* reproductive strategy since they have lost their eggs or nestlings.
4. *What are the costs of CGP to the victims?* Potential costs:
   - a. *Death of host young*
     - (i) Are eggs removed by brood parasites?
     - (ii) Effects of over-crowding: partial or complete
       - a) Hatching failure.
       - b) Fledging failure.
   - b. *Parasitism of host parental investment*
     - (i) Cuckoldry: *substitution* of parasitic young for host young results in reduced share of clutch for male host.
     - (ii) CBP: *addition* of parasitic young increases the burden of parental investment on the host female in all cases and on the host male in all cases excepting quasi-parasitism (Wrege and Emlen 1987).
   - c. *Phenotypic costs of hosts*
     - (i) Death or injury because excess workload causes host to take otherwise unnecessary risks or to suffer exhaustion on the current clutch.
     - (ii) Reduced ability to produce future clutches because excess workload permanently damages the host but does not kill it.
5. *What are the costs to the perpetrators?*
   - a. *Physical injury* at time of attempted CGP by affected resident(s).
   - b. *Poor choice of host*, i.e., the opportunistic nature of CGP may inevitably result in a high error rate in host choice, leading to a lower reproductive success than would have occurred had the parasite made a better choice. This is probably more important with CBP than cuckoldry because the cost of eggs exceeds the cost of sperm in birds.
   - c. *Loss of mate*. This may be especially important when mate betrayal is involved, e.g., loss of mate by the female if she is caught in cuckoldry by her mate, and by the mate if he is caught in quasi-parasitism.
6. *What are the defenses of the potential and actual victims and hosts?*
   - a. *Prevention of parasitism*
     - (i) Cuckoldry and quasi-parasitism: mate guarding.
     - (ii) CBP: nest guarding.
   - b. *Amelioration of the cost* of CGP after it occurs.
     - (i) Cuckoldry and quasi-parasitism:
       - (a) Abandonment of mate.
       - (b) Reduction of parental investment to compensate for reduced parentage.
     - (ii) CBP
       - (a) Egg recognition and rejection of parasitic eggs.
       - (b) Abandonment of nest.
       - (c) Adding a nest layer over the parasitic egg(s).
       - (d) Recognition of own young and rejection of parasitic nestlings.
       - (e) Parasitism insurance (Power et al. 1989).
       - (f) Addition indeterminacy (Kennedy and Power 1990).
7. *What are the offensive strategies and tactics of the parasites?*
   - a. *Cuckoldry and quasi-parasitism*
     - (i) Searching for birds with nests with whom to have extra-pair copulations (EPCs).
     - (ii) Advertising for EPCs.
   - b. *CPB*
     - (i) Searching for available nests.
     - (ii) Laying at the appropriate time as determined by
       - (a) *Time of season*, e.g., early clutches rather than late clutches in species where early nesting is more successful.
       - (b) *Time of day*, e.g., when the host is least likely to be around the nest.
       - (c) *Time in host laying cycle*, e.g., during the host's own laying period.

EUROPEAN STARLINGS: A CASE STUDY

In this section I use the case of European starlings (table 20.1) to illustrate some answers to the important questions. These answers are based on more than a decade of research on starlings on the Livingston (formerly Kilmer) Campus of Rutgers University in Piscataway, N.J., by my graduate students and myself.

We conducted our study with strict safeguards for quality control (Romagnano et al. 1989). Nestboxes were more than a territory diameter apart. Only adults that fed nestlings were considered to be putative parents. Developmental effects were controlled for by biopsying nestlings one day before the expected date of fledging. Tissue artifacts were controlled for by treating samples from all individuals in the same manner.

Here are the answers to those previously listed questions for which we have answers.

1. Frequency questions

CGP does occur in our population and takes the form of both CBP and cuckoldry (Hoffenberg et al. 1988). Cuckoldry affects at least 1.9–7.5% of broods and 1.6–3.3% of fledglings. These values are extremely conservative because of the low resolution of polyacrylamide gel electrophoresis in starlings (table 20.1).

CBP affects 12.7–23% of all broods (27.8–36.4% of early broods — those initiated in April), and 5.5% of all fledglings (Kennedy 1989; Kennedy et al. 1989; Power et al. 1989; Romagnano et al. 1990). CBP estimates are based only on parasitic eggs laid during the host laying period (Power 1990*a*). If parasitic eggs laid at other times are counted, the CBP frequency rises to 46% (Power et al. 1981).

The frequency of CGP varies in our population with several parameters:

a) *Seasonally*. Some 92% of all CBP occurred in early broods (Power et al. 1989). There were too few unambiguous cases of cuckoldry to detect seasonal variation (Hoffenberg et al. 1988).

b) *Between years*. No evidence for cuckoldry, but CBP increased every year of our study (Romagnano et al. 1990) (see below).

c) *Among-study populations*. Evans (1988) found, using starch gel electrophoresis, that 24.2% of "first nests" had CBP in England and Scotland. Using polyacrylamide gel electrophoresis (Romagnano et al. 1990), we found that 27.8% of early broods had CBP. These electrophoretic comparisons provide the most straightforward basis for comparison although postovulatory follicle analysis is more accurate and provides a higher estimate of CBP, 36.4% (Kennedy et al. 1989). There is no basis for comparing the frequencies of cuckoldry in Britain and New Jersey because Evans (1980, 1988) assigned all cases of CGP to CBP. We believe this was inappropriate because it was an apparently arbitrary decision (Power 1990*b*), and EPCs have been seen in Belgium by Eens and Pinxten (1990).

Results on the European mainland were qualitatively, as well as quantitatively, different from both our New Jersey population and its ancestor, the English population studied by Evans (1988). In Belgium, Pinxten et al. (1993) used DNA fingerprinting to find that 28.6% of broods and 9.7% of nestlings stemmed from EPP, but only half of these cases were ascribable to cuckoldry, the other half stemming from rapid mate replacement (table 20.1). Pinxten et al. (1993) also found that CBP occurred in 7.1% of broods and gave rise to 1.6% of nestlings.

In Sweden, Smith and von Schantz (1993) used DNA fingerprinting and found cuckoldry in 31.8% of broods involving 8.7% of nestlings (table 20.1). They found no evidence of CBP from DNA fingerprinting but observational methods showed that four of 117 eggs (3.4%) stemmed from CBP.

That a higher frequency of EPP and cuckoldry was found on the European mainland than in England or New Jersey is not surprising given the much higher resolving power of DNA fingerprinting compared with electrophoresis, and thus is not a matter of concern. But the much lower frequency of CBP on the European mainland is a cause for concern because it can not be accounted for by differences in the detection power of different biochemical techniques for if such differences were critical, then the frequency of CBP would have been higher in the populations where DNA fingerprinting was used, not lower as reported. Thus, it appears that the direction of difference in CBP rates is biological rather than artifactual.

The most obvious biological explanation of varying rates of CBP is nest site availability. Empty nestboxes (and hence available nest sites) occur in the Belgian and Swedish populations (pers. obs.), but few empty nestboxes occurred in either our population or Evans' (1988). The New Jersey population is much denser than the European mainland population (pers. obs.) and appears to increase when nest boxes are erected. So also does the English population (J. Wright, pers. commun.). But the Belgian and Swedish populations appear to be declining because of excessive mortality on their wintering grounds in North Africa caused by agricultural interests (H.

Källander, pers. commun.). The British population is not subject to this mortality because British starlings winter in England, not North Africa (Feare 1984).
d) *Population density*. Both Evans (1988) and Romagnano et al. (1990) found increases in CBP in every year of our separate studies, suggesting that nestbox trails attract more birds than can nest, that the number of floaters builds every year, and that such birds use CBP as an alternative reproductive strategy to nesting themselves. If this is true, then limited nest site availability drives up CBP. Population density and nest site limitation apparently interact.

2. The victims of CGP

We found too few unambiguous cases of cuckoldry to detect a pattern of victimization (Hoffenberg et al. 1988). Our evidence for CBP suggests that no group of resident female is more or less vulnerable than any other (Romagnano et al. 1990).

Neither Pinxten et al. (1993) nor Smith and von Schantz (1993) could find a meaningful pattern of victimization in their study populations. They compared yearling vs. older males and monogamous vs. polygynous males for susceptibility to cuckoldry, and primary vs. secondary females for susceptibility to CBP but could find no statistically significant patterns.

3. The perpetrators of CGP

We were unable to identify the perpetrators of cuckoldry because we had too few cases and our electrophoretic system had too low resolving power (Hoffenberg et al. 1988). But neither could Pinxten et al. (1993) nor Smith and von Schantz (1993) identify the perpetrators of cuckoldry in their populations despite their larger number of cases and use of DNA fingerprinting. However, Eens and Pinxten (1990) found that EPCs are perpetrated by neighboring territorial males in starlings in Belgium.

CBP is apparently largely perpetrated by females without nests (Power 1990*b*; Romagnano et al. 1990). However, some brood parasites may be females that lost their nests during the laying period. This was shown by Stouffer and Power (1991*a*) using tetracycline to follow the movements of laying females [females injected with tetracycline lay eggs that fluoresce under ultraviolet light (Haramis et al. 1983)], and by Feare (1991) using idiosyncratic egg features as markers of the laying histories of females. In both cases, females had their nesting disrupted during the laying cycle, and they abandoned their nests and laid in the nests of other birds. However, the infrequent failure of nests during the laying period in our population suggests that this kind of parasitism accounts for only a fraction of CBP (H. Power, unpubl. data).

Female starlings without nests of their own that commit CBP engage in an alternative reproductive strategy. Those that have nests of their own and lay in the nests of other females engage in a supplementary reproductive strategy.

4. The costs of CGP

There were too few cases of cuckoldry to know the pattern of cost, but it is interesting that the two unambiguous cases involved two nestlings each (Hoffenberg et al. 1988). Because there were five nestlings at each of the affected nests, each cuckolded male had his potential reproductive success reduced by 40%. I use the term "potential" here because the fate of the brood after fledging was unknown, leaving obscure the relative fates of the legitimate and illegitimate young.

In the cases of cuckoldry detected by Pinxten et al. (1993) and Smith and von Schantz (1993), all but one involved only one chick per brood. The exceptional case involved both of two chicks in the brood (Smith and von Schantz 1993).

CBP produced several costs for hosts:
a) *Overcrowding*, i.e., excess mortality among host young because there were more young to care for than the host had "intended." This was shown in the electrophoresis subsample where significantly fewer host young fledged from parasitized than nonparasitized nests (Romagnano et al. 1990). However, for the complete sample of clutches, there was no significant difference in the number of host young fledged from parasitized and nonparasitized nests (Power et al. 1989). The electrophoresis subsample was more likely to show the effects of parasitism than the complete sample because it included only those broods that survived to biopsy age (the 20th day after hatching). These were broods where competitive interactions among nestlings were more likely to be evident simply because there was more time for those competitive interactions to take effect. By contrast, broods that failed before biopsy age tended to show a diluted competitive effect because mortality was caused by noncompetitive factors such as weather.
b) *Egg removal*. This occurred at 34% of the

parasitized early nests (Lombardo et al. 1989*a*). Early nests are the ones that count most because 92% of all CBP occurred in them, and they produced most of the fledglings (77% of early broods produced at least one fledgling whereas only 45% of later broods did so) (Power et al. 1989). Altogether, 4.2% of all eggs were removed from early nests during the host laying period. In 92% of the cases, one egg was taken; in the other 8%, two eggs were taken. Lombardo et al. (1989*a*) showed that egg removal in the laying period was largely the work of brood parasites.

c) *Future costs*. There was no evidence of long-term phenotypic effects on host females (Power et al. 1989). Females with larger clutches were as successful over a lifetime as those with smaller clutches. This suggests that CBP at the level at which it occurs in our population does not so overburden hosts that they have fewer clutches in a lifetime or are less successful with subsequent clutches than are females never parasitized.

5. The costs to perpetrators

a) *Physical injury*. We have observed no physical injuries in the context of cuckoldry but we have frequently seen males chasing each other near fertile females. Some of these encounters were very aggressive, especially those in which a male following his foraging female encountered a displaying male on the ground advancing toward his mate (Power et al. 1981). The intensity of the mate-guarding male's attack suggested that injury might have occurred had the displaying male not fled.

We have seen physical injury in the contexts of apparently attempted CBP and nest usurpation. During the egg-laying and incubation periods, dyads of females have been captured in nestboxes holding one another with their legs and screaming at each other. These birds frequently had deep facial wounds apparently inflicted by beak stabs. Although none of these birds died and their wounds appeared to heal in a few days, discovery of injured dead birds in nestboxes during the nonbreeding season shows that fights within boxes can be lethal (Lombardo et al. 1989*b*). Captured females often included one marked bird (the resident) and one unmarked bird (the intruder), suggesting that the intruder was attempting either CBP or nest usurpation, and that the resident was resisting.

b) *Poor choice of host and loss of mate*. We found no evidence of these costs in the context of either form of CGP. About 50% of parasitic eggs, however, were laid at the wrong time, i.e., outside the host laying period (Romagnano et al. 1990). Parasitic eggs laid before the onset of host laying were tossed out by the host (Stouffer et al. 1987; Pinxten et al. 1991) while those laid after the cessation of host laying were at a competitive disadvantage with earlier-laid eggs and never fledged (Romagnano et al. 1990). But these mistakes in laying time by parasites appear to result from a combination of nest guarding by hosts and physiological constraints on laying by parasites (eggs cannot be held indefinitely or deposited at will) rather than to poor choice of host by the parasite.

6. The defenses of the victims and hosts

a) *Prevention*. Mate-guarding is well developed in starlings as a preventative against cuckoldry (Power et al. 1981; Pinxten et al. 1987). The low frequency of detected cuckoldry in our population (Hoffenberg et al. 1988) is consistent with both the hypothesis that mate-guarding is highly effective and the hypothesis that electrophoresis is very low in resolving power for parentage studies (Power 1990*b*). Pinxten et al. (1993) and Smith and von Schantz (1993) also believe that mate guarding is effective and causes cuckoldry in starlings to be less frequent than in some other species.

Nest guarding is well developed in starlings as a preventive against CBP and nest usurpation (Power et al. 1981; Romagnano et al. 1990). In practical terms, only the female effectively nest guards as the male escorts the female on her foraging and nest-building trips away from the nest in order to guard her from insemination by other males. This leaves the nest open to CBP during the pair's absence (Power 1990*b*).

b) *Amelioration*. There are no documented strategies to reduce the cost of cuckoldry after it occurs in starlings, but in dunnocks (Burke et al. 1989), pied flycatchers (Bjorklund and Westman 1983), and barn swallows (Møller 1988), males reduce the amount of food they provide to broods in proportion to their probability of paternity. This may also occur in starlings.

We have explored a number of possible and actual *strategies of amelioration* of the cost of CBP in starlings.

An effective strategy would be *egg recognition and ejection of parasitic eggs*. However, we found experimentally that females are able to do this only on a phenological basis, i.e., they reject any egg in their nests found before they themselves have begun to lay but accept any egg found after they have laid their own first egg (Stouffer et al. 1987). Our result was confirmed by Pinxten et al. (1991) who also found that males sometimes eject eggs on the same phenological basis.

A parallel strategy would be *recognition and rejection of parasitic nestlings*. However, an experimental study of the effects of asynchronous hatching that involved movement of nestlings between nests showed that parent starlings will accept any small nestling proffered (Stouffer and Power 1991*b*). Crossner (1977) also showed acceptance of cross-fostered nestlings in our population. Van Elsacker et al. (1988) showed starlings to be similar to many other birds in not recognizing their own nestlings until shortly before those nestlings are scheduled to fledge.

Another strategy to ameliorate the cost of CBP would be for females to *reduce their parental investment in parasitized broods* in a manner analogous to that found in males of some other species when challenged with reduced paternity (Bjorklund and Westman 1983; Møller 1988; Burke et al. 1989). The obvious problem with this response would be that females would risk starving their own young unless their mates automatically compensated for females' reduced parental investment by increasing their own. Compensation for reduced parental investment by mates when those mates were experimentally handicapped physically has been shown in starlings by Wright and Cuthill (1989, 1990). However, compensating for an obviously physically impaired mate may be different from compensating for an apparently lazy mate. This is because an apparently lazy bird may be attempting to exploit its mate, and that mate may be favored for resisting exploitation. By contrast, compensating for an impaired mate may be favored insofar as it salvages a bad situation. Clearly the whole question of intersexual exploitation needs experimental analysis. Power (1980) provided the formal criteria necessary for determining whether intersexual exploitation occurs and applied them in one species.

Much of our effort has focused on the possibility that female starlings ameliorate the cost of CBP after it happens by "leaving space" for parasitic eggs before they are laid (Power et al. 1989). We call this possibility the *parasitism insurance hypothesis* (PIH). This hypothesis assumes that CBP is a likely but undetectable event. As a result the host cannot tell whether or not it has been parasitized. Thus, it responds not to the fact of CBP but to the *possibility* of CBP by protecting itself against the negative consequences of CBP whether they occur or not. It does this by laying a clutch sufficiently small that if a parasitic egg is added there is little risk that the sum of the host's and parasite's eggs (what we call the "gross clutch size") will produce an overcrowded clutch, i.e., one where there is excess mortality because the number of dependent young exceeds the ability of the parents to care for them properly.

The PIH is a complex hypothesis of many requirements before it can even be considered as well as several testable predictions (Power et al. 1989). We applied it to our study population of starlings and found evidence both consistent with it, and inconsistent with alternative explanations of clutch size. Consequently I will treat it as tested and refer the reader to the original paper rather than reiterate its detailed content in the remaining space. I end my consideration of the PIH by relating the original paper to two important papers bearing on it that were published since it appeared. The first of these is Stouffer and Power (1991*b*).

An *apparent* exception to the conclusion of Power et al. (1989) that C/6 is the most productive clutch size is implicit in Stouffer and Power (1991*b*). The latter found that experimental asynchronous broods of size 6 (b/6) produced fledglings of significantly lighter weight than both experimental synchronous and experimental asynchronous b/5. However, experimental synchronous b/6 produced fledglings not significantly lighter than experimental synchronous b/5. Moreover, experimental b/6 produced more fledglings than experimental b/5, including a larger modal number of fledglings per clutch. Experimental b/7 was inferior to both b/5 and b/6 in both fledging success and fledgling weight.

In comparing the results between Power et al. (1989) and Stouffer and Power (1991*b*) it is important to distinguish between *clutch* and *brood*. Power et al. (1989) considered clutches whereas Stouffer and Power (1991b) considered broods. As shown in Fig. 1 of Power et al. (1989), no c/6 resulted in b/6 in unmanipulated nests. The b/6 and larger nests of Stouffer and Power (1991*b*) were all experimentally constructed by cross-fostering between nests. Thus, the somewhat equivocal relationship between b/5 and b/6 in Stouffer and Power (1991*b*) is not evidence that c/6 is not superior to c/5. To the contrary, the b/5 of Stouffer and Power (1991*b*) is analogous to the c/6 of Power et al. (1989) because the pattern of hatching in unmanipulated nests usually produced broods one nestling smaller than the clutches they came from.

Clutches were used as the unit in Power et al. (1989) because the PIH pertains to clutch size decisions, whereas broods were used as the unit in Stouffer and Power (1991*b*) because it tested Lack's (1954, 1968) brood reduction hypothesis.

The second paper of importance was Rothstein's (1990) review of the PIH. Rothstein (1990) found value in the idea, saying that it "could provide a partial solution to the fact that the most common clutch size is smaller than the most productive size in many birds," and that CBP may be "an important ultimate determinant of clutch size." However, he also voiced a number of caveats. The first is the calculation of net CBP.

Rothstein (1990) noted that parasitic eggs laid after the host's own laying period are only a small threat to the host because they produce nestlings that are not competitive with the host's young on account of their hatching after the host's young. He suggested that inclusion of such eggs in our estimate of net CBP would have inflated it and, by implication, possibly invalidated the comparison between the actual frequency of net CBP and the threshold frequency required by the third prediction.

As I stated in my reply (Power 1990*a*), we were acutely aware of the difference in impact of parasitic eggs laid *during* versus those laid *after* host laying, and that we based our measure of net CBP solely on parasitic eggs laid *during* the host's own laying period. This was clearly stated in early drafts of our paper but was omitted in later versions as we struggled to meet the demands of various reviewers while simultaneously trying to keep the paper from growing too large. This was an unfortunate omission for which I accept sole responsibility. I sincerely hope that this matter is now over, and that everyone will understand that the frequency of net CBP *must* be calculated only on the basis of parasitic eggs with the potential to damage the host's reproductive success.

Rothstein (1990) suggested that starlings could effectively protect themselves from the deleterious consequences of CBP "by beginning to incubate before egg laying is over. (This) would decrease the competitiveness of the last nestling to hatch, and thereby kill it off quickly in an overcrowded nest before it seriously depleted the food for other nestlings. Because starlings begin incubating with the penultimate egg (Feare 1984), an even earlier onset of incubation should be possible."

In fact, as we reported (Romagnano et al. 1990) not long after Rothstein's (1990) news article about the PIH, starlings do begin to incubate mid-way through the laying cycle, terminating the parasite's opportunity for effective CBP before the end of host laying by virtue of the fact that a nest with an incubating bird is not vulnerable to egg dumping. As noted above, we (Stouffer et al. 1987) had previously shown that starling hosts eject eggs added to their nests prior to the beginning of their own laying. The effect of such ejection coupled with early onset of incubation is to reduce the window of opportunity for the parasite (which is also the window of vulnerability for the host) to little more than the daylight hours of host laying days 1–3.

We have also shown that while nestling mortality is focused on last-hatched young ("runts") (Stouffer and Power 1991*b*; Litovich and Power 1992), this mortality is not sufficient to protect the host fully from the effects of overcrowding. The food limitation hypothesis of Lack (1954, 1968) is implicit in Rothstein's (1990) remark about the advantage of asynchronous hatching (i.e., early death of the runt), but we disproved this hypothesis in relation to our population of starlings (Stouffer and Power 1991*b*). We did this by experimentally showing that enlarged asynchronous broods fared no better than equally large synchronous broods in terms of both nestling survival and fledgling weight. Thus, asynchronous hatching is not sufficient protection against overcrowding from CBP, and there remains an advantage to leaving space for parasitic eggs.

7. The offensive strategies and tactics of the parasites

a) *Cuckoldry*. We (Hoffenberg et al. 1988) have never seen an EPC in starlings although we have seen males displaying toward females other than their own mates (Power et al. 1981). Consequently, we know neither the identity of cuckolders nor their strategies and tactics within our population. However, Eens and Pinxten (1990) saw EPCs in starlings in Belgium and found that they always involved neighboring, resident males rather than floaters, and that these males openly displayed at neighboring females while the latter's mates were briefly absent.

b) *CBP*. We have seen "sneaky" birds entering and leaving nests. Some of these birds were implicated in infanticide during the nestling period (Romagnano et al. 1986). During the laying period we believe such birds are scouting for suitable nests. Lombardo et al. (1989*a*) hypothesized that scouting females usurp nests if they are capable and parasitize them if they are not. In either case, a nest must be unattended at the time of the perpetrator's entry, and it may throw out some or all of the host's eggs. Finding an unattended nest would seem to involve some kind of systematic searching as random entry attempts could easily result in dangerous encounters of the kind described in section 5 above.

We have observed several tactics within a CBP strategy:

Time of season. Power et al. (1989) and Romagnano et al. (1990) showed that CBP is heavily concentrated within early clutches, the time when fledging success is highest and therefore CBP is most adaptive. The relative paucity of CBP later in the season probably reflects a poor pay-off, making the potential risks (e.g., physical injury or even death if a parasite is cornered by the resident female) too high.

Timing of laying. Romagnano et al. (1990) showed that CBP most often occurs on the second and third days of the host's laying cycle. This corresponds to the window of opportunity/vulnerability between the appearance of the first host egg and the onset of incubation. Perhaps parasites specifically search for nests in this stage, or perhaps the window of vulnerability itself creates this pattern of parasitic egg appearance in *willy-nilly fashion.*

Time of day. Romagnano et al. (1990) found that 56% of parasitic eggs laid during the host laying cycle were laid at a different time of day than that at which the host laid. This suggests that parasites may have staked out nests and waited for the hosts to be gone. Obviously this would be most likely at a time of day other than the time of host oviposition. Alternatively, parasites may tend to lay at those times of day when hosts are least likely to lay, and simply dump their parasitic eggs in the first nests they find available.

REFERENCES

Birkhead, T. R., T. Burke, R. Zann, F. M. Hunter, and A. P. Krupa (1990). Extra-pair paternity and intraspecific brood parasitism in wild Zebra Finches *Taeniopygia guttata*, revealed by DNA fingerprinting. *Behav. Ecol. Sociobiol.*, **27**, 315–324.

Bjorklund, M. and B. Westman (1983). Extra-pair copulations in the Pied Flycatcher (*Ficedula hypoleuca*): a removal experiment. *Behav. Ecol. Sociobiol.*, **13**, 271–275.

Bollinger, E. K. and T. A. Gavin (1991). Patterns of extra-pair fertilizations in Bobolinks. *Behav. Ecol. Sociobiol.*, **29**, 1–7.

Brooker, M. G., I. Rowley, M. Adams, and P. R. Baverstock (1990). Promiscuity: an inbreeding avoidance mechanism in a socially monogamous species? *Behav. Ecol. Sociobiol.*, **26**, 191–199.

Brown, C. R. and M. B. Brown (1988). Genetic evidence of multiple parentage in broods of Cliff Swallows. *Behav. Ecol. Sociobiol.*, **23**, 379–387.

Burke, T. and M. W. Bruford (1987). DNA fingerprinting in birds. *Nature*, **327**, 149–152.

Burke, T., N. B. Davies, M. W. Bruford, and B. J. Hatchwell (1989). Parental care and mating behavior of polyandrous Dunnocks *Prunella modularis* related to paternity by DNA fingerprinting. *Nature*, **338**, 249–251.

Choudhury, S., C. S. Jones, J. M. Black, and J. Prop (1993). Adoption of young and intraspecific nest parasitism in Barnacle Geese. *Condor*, **95**, 860–868.

Crossner, K. A. (1977). Natural selection and clutch size in the European Starling. *Ecology*, **58**, 885–892.

Darwin, C. R. (1871). *The Descent of Man and Selection in Relation to Sex*. John Murray, London.

Davies, N. B., B. J. Hatchwell, T. Robson, and T. Burke (1992). Paternity and parental effort in Dunnocks *Prunella modularis*: how good are male chick-feeding rules? *Anim. Behav.*, **43**, 729–745.

Decker, M. D., P. G. Parker, D. J. Minchella, and K. N. Rabenold (1993). Monogamy in Black Vultures: genetic evidence from DNA fingerprinting. *Behav. Ecol.*, **4**, 29–35.

Dunn, P. O. and R. J. Robertson (1993). Extra-pair paternity in polygynous Tree Swallows. *Anim. Behav.*, **45**, 231–239.

Dunn, P. O., L. A. Whittingham, J. T. Lifjeld, R. J. Robertson, and P. T. Boag (1994). Effects of breeding density, synchrony, and experience on extrapair paternity in Tree Swallows. *Behav. Ecol.*, **5**, 123–129.

Eens, M. and R. Pinxten (1990). Extra-pair courtship in the starling *Sturnus vulgaris*. *Ibis*, **132**, 618–619.

Emlen, S. T. (1976). Lek organization and mating strategies in the bullfrog. *Behav. Ecol. Sociobiol.*, **1**, 283–313.

Evans, P. G. H. (1980). Population genetics of the European starling. Ph.D. Thesis. University of Oxford, Oxford.

Evans, P. G. H. (1988). Intraspecific nest parasitism in the European Starling *Sturnus vulgaris*. *Anim. Behav.*, **36**, 1282–1294.

Evarts, S. and C. J. Williams (1987). Multiple paternity in a wild population of Mallards. *Auk*, **104**, 597–602.

Feare, C. J. (1984). *The Starling*. Oxford University Press, New York.

Feare, C. J. (1991). Intraspecific nest parasitism in starling *Sturnus vulgaris*: effects of disturbance on laying females. *Ibis*, **133**, 75–79.

Fisher, R. A. (1930). *The Genetical Theory of Nature Selection*. Oxford University Press, Oxford.

Fleischer, R. C., M. T. Murphy, and L. Hunt (1985). Clutch size increase and intraspecific brood parasitism in the Yellow-billed Cuckoo. *Wilson Bull.*, **97**, 125–127.

Gavin, T. A. and E. K. Bollinger (1985). Multiple paternity in a territorial passerine: the Bobolink. *Auk*, **102**, 550–555.

Gelter, H. P. and H. Tegelström (1992). High frequency of extra-pair paternity in Swedish Pied Flycatchers revealed by allozyme electrophoresis and DNA fingerprinting. *Behav. Ecol. Sociobiol.*, **31**, 1–7.

Gibbs, H. L., P. J. Weatherhead, P. J. Boag, B. N. White, L. M. Tabak, and D. J. Hoysak (1990). Realized reproductive success of polygynous Red-winged Blackbirds revealed by DNA markers. *Science*, **250**, 1394–1397.

Gowaty, P. A. and W. C. Bridges (1991). Nestbox availability affects extra-pair fertilizations and conspecific nest parasitism in Eastern Bluebirds, *Sialia sialis*. *Anim. Behav.*, **41**, 661–675.

Gowaty, P. A. and J. C. Davies (1986). Uncertain maternity in American Robins. In: *Behavioral ecology and population biology* (ed. L. C. Drickamer), pp. 65–70. Privat, I.E.C., Toulouse.

Gowaty, P. A. and A. A. Karlin (1984). Multiple maternity and paternity in single broods of apparently monogamous Eastern Bluebirds (*Sialia sialis*). *Behav. Ecol. Sociobiol.*, **15**, 91–95.

Graves, J., R. T. Hay, M. Scallan, and S. Rowe (1992). Extra-pair paternity in the Shag, *Phalacrocorax aristotelis* as determined by DNA fingerprinting. *J. Zool., Lond.*, **226**, 399–408.

Gyllensten, U. B., S. Jakobsson, and H. Temrin (1990). No evidence for illegitimate young in monogamous and polygynous warblers. *Nature*, **343**, 168–170.

Haig, S. M., J. R. Belthoff, and D. H. Allen (1993). Examination of population structure in Red-cockaded Woodpeckers using DNA profiles. *Evolution*, **47**, 185–194.

Haig, S. M., J. R. Walters, and J. H. Plissner (1994). Genetic evidence for monogamy in the cooperatively breeding Red-cockaded Woodpecker. *Behav. Ecol. Sociobiol.*, **34**, 295–303.

Haramis, G. M., W. G. Alliston, and M. E. Richmond (1983). Dump nesting in the Wood Duck traced by tetracycline. *Auk*, **100**, 729–730.

Hartley, I. R., M. Shephard, T. Robson, and T. Burke (1993). Reproductive success of polygynous male Corn Buntings (*Miliaria calandra*) as confirmed by DNA fingerprinting. *Behav. Ecol.*, **4**, 310–317.

Hickey, J. J. (1943). *A Guide to Bird Watching*. Oxford University Press, New York.

Hill, W. G. (1987). Genetic probes: DNA fingerprints applied to animal and bird populations. *Nature*, **327**, 98–99.

Hoffenberg, A. S., H. W. Power, L. C. Romagnano, M. P. Lombardo, and T. R. McGuire (1988). The frequency of cuckoldry in the European Starlings (*Sturnus vulgaris*). *Wilson Bull.*, **100**, 60–69.

Hunter, F. M., T. Burke, and S. E. Watts (1992). Frequent copulation as a method of paternity assurance in the Northern Fulmar. *Anim. Behav.*, **44**, 149–156.

Jones, C. S., C. M. Lessels, and J. R. Krebs (1991). Helpers-at-the-nest in European Bee-eaters (*Merops apiaster*): a genetic analysis. In: *DNA fingerprinting: approaches and aqpplications* (eds T. Burke, G. Dolf, A. J. Jeffreys, and R. Wolff), pp. 169–192. Birkhauser Verlag, Basel.

Joste, N., J. D. Ligon, and P. B. Stacey (1985). Shared paternity in the Acorn Woodpecker (*Melanerpes formicivorus*). *Behav. Ecol. Sociobiol.*, **17**, 39–41.

Karlin, A. A., K. G. Smith, M. C. Stephens, and R. A. Barnhill (1990). Additional evidence of multiple parentage in Eastern Bluebirds. *Condor*, **92**, 520–521.

Kempenaers, B., G. R. Verheyen, M. V. den Broeck, T. Burke, C. V. Broeckhoven, and A. A. Dhondt (1992). Extra-pair paternity results from female preference for high-quality males in the Blue Tit. *Nature*, **357**, 494–496.

Kendra, P. E., R. R. Roth, and D. W. Tallamy (1988). Conspecific brood parasitism in the House Sparrow. *Wilson Bull.*, **100**, 80–90.

Kennedy, E. D. (1989). Clutch size and reproductive success in House Wrens (*Troglodytes aedon*) and European Starlings (*Sturnus vulgaris*). Ph.D. Thesis. Rutgers University, New Brunswick, N.J.

Kennedy, E. D. and H. W. Power (1990). Experiments on indeterminate laying in House Wrens and European Starlings. *Condor*, **92**, 861–865.

Kennedy E. D., P. C. Stouffer, and H. W. Power (1989). Postovulatory follicles as a measure of clutch size and brood parasitism in European Starlings. *Condor*, **91**, 471–473.

Lack, D. (1954). *The Natural Regulation of Animal Numbers*. Clarendon Press, Oxford.

Lack, D. (1968). *Ecological Adaptations for Breeding in Birds*. Chapman & Hall, London.

Lank, D. B., P. Mineau, R. F. Rockwell, and F. Cooke (1989). Intraspecific nest parasitism and extra-pair copulation in Lesser Snow Geese. *Anim. Behav.*, **37**, 74–89.

Lifjeld, J. T. and R. J. Robertson (1991). Female control of extra-pair fertilization in Tree Swallows. *Behav. Ecol. Sociobiol.*, **31**, 89–96.

Lifjeld, J. T., T. Slagsvold, and H. M. Lampe (1991). Low frequency of extra-pair paternity in Pied Flycatchers revealed by DNA fingerprinting. *Behav. Ecol. Sociobiol.*, **29**, 95–101.

Lifjeld, J. T., P. O. Dunn, R. J. Robertson, and P. T. Boag (1993). Extra-pair paternity in monogamous Tree Swallows. *Anim. Behav.*, **45**, 213–229.

Litovich, E. and H. W. Power (1992). Parent–offspring conflict and its resolution in the European Starling. *Ornithol. Monogr.*, **47**.

Lombardo, M. P., L. C. Romagnano, A. S. Hoffenberg, P. C. Stouffer, and H. W. Power (1989*a*). Egg removal and intraspecific brood parasitism in the European Starling (*Sturnus vulgaris*). *Behav. Ecol. Sociobiol.*, **24**, 217–223.

Lombardo, M. P., L. C. Romagnano, P. C. Stouffer, A. S. Hoffenberg, and H. W. Power (1989*b*). The use of nest boxes as night roosts during the nonbreeding season in European Starlings in New Jersey. *Condor*, **91**, 744–747.

McKittrick, M. C. (1990). Genetic evidence for multiple parentage in Eastern Kingbirds. (*Tyrannus tyrannus*). *Behav. Ecol. Sociobiol.*, **26**, 149–155.

Meek, S. B., R. H. Robertson, and P. T. Boag (1994). Extrapair paternity and intraspecific brood parasitism in Eastern Bluebirds revealed by DNA fingerprinting. *Auk*, **111**, 739–744.

Møller, A. P. (1988). Paternity and paternal care in the swallow, *Hirundo rustica*. *Anim. Behav.*, **36**, 996–1005.

Morton, E. S., L. Forman, and M. Braun (1990). Extra pair fertilizations and the evolution of colonial breeding in Purple Martins. *Auk*, **107**, 275–283.

Mumme, R. L., W. D. Koenig, R. M. Zink, and J. A. Marten (1985). Genetic variation and parentage in a California population of Acorn Woodpeckers. *Auk*, **102**, 305–312.

Orians, G. H. (1980). *Some Adaptations of Marsh-nesting Blackbirds*. Princeton University Press, Princeton.

Otter, K., L. Ratcliffe, and P. T. Boag (1994). Extra-pair paternity in the Black-capped Chickadee. *Condor*, **96**, 218–222.

Petter, S. C., D. B. Miles, and M. W. White (1990). Genetic evidence of mixed reproductive strategy in a monogamous bird. *Condor*, **92**, 702–708.

Pinxten, R., L. van Elsacker, and R. F. Verheyen (1987). Duration and temporal pattern of mate guarding in the starling. *Ardea*, **75**, 263–269.

Pinxten, R., M. Eens, and R. F. Verheyen (1991). Responses of male starlings to experimental intraspecific brood parasitism. *Anim. Behav.*, **42**, 1028–1030.

Pinxten, R., O. Hanotte, M. Eens, R. F. Verheyen, A. A. Dhondt, and T. Burke (1993). Extra-pair paternity and intraspecific brood parasitism in the European Starling, *Sturnus vulgaris*: evidence from DNA fingerprinting. *Anim. Behav.*, **45**, 795–809.

Power, H. W. (1980). The foraging behavior of Mountain Bluebirds. *Ornithol. Monogr.*, **28**.

Power, H. W. (1990*a*). The calculation of net brood parasitism. *Trends Ecol. Evol.*, **5**, 200.

Power, H. W. (1990*b*). Genetic parasitism in the European Starling. In: *Population Biology of Passerine Birds: an Integrated Approach* (eds J. Blondel, A. Gosler, J. D. Lebreton, and R. McCleery), pp. 223–233. Springer-Verlag, Heidelberg.

Power, H. W., E. Litovich, and M. P. Lombardo (1981). Male starlings delay incubation to avoid being cuckolded. *Auk*, **98**, 386–389.

Power, H. W., E. D. Kennedy, L. C. Romagnano, M. P. Lombardo, A. S. Hoffenberg, P. C. Stouffer, and T. R. McGuire (1989). The parasitism insurance hypothesis: why starlings leave space for parasitic eggs. *Condor*, **91**, 753–765.

Price, D. K., G. E. Collier, and C. F. Thompson (1989). Multiple parentage in broods of House Wrens: genetic evidence. *J. Hered.*, **80**, 1–5.

Quinn, T. W., J. S. Quinn, F. Cooke, and B. N. White (1987). DNA marker analysis detects multiple maternity and paternity in single broods of the Lesser Snow Goose. *Nature*, **326**, 392–394.

Rabenold, P. P., K. N. Rabenold, W. H. Piper, J. Haydock, and S. W. Zack (1990). Shared paternity revealed by genetic analysis in cooperatively breeding tropical wrens. *Nature*, **348**, 538–540.

Ritchison, G., P. H. Klatt, and D. F. Westneat (1994). Mate guarding and extra-pair paternity in Northern Cardinals. *Condor*, **96**, 1055–1063.

Romagnano, L. C., M. P. Lombardo, P. C. Stouffer, and H. P. Power (1986). Suspected infanticide in the starling. *Condor*, **88**, 530–531.

Romagnano, L. C., T. R. McGuire, and H. W. Power (1990). Pitfalls and improved techniques in avian parentage studies. *Auk*, **106**, 129–136.

Romagnano, L. C., A. S. Hoffenberg, and H. W. Power (1990). Intraspecific brood parasitism in the European Starling. *Wilson Bull.*, **102**, 279–291.

Rothstein, S. I. (1990). Brood parasitism and clutch-size determination in birds. *Trends Ecol. Evol.*, **5**, 101–102.

Rowley, I., E. M. Russell, and M. G. Brooker (1986). Inbreeding: benefits may outweigh costs. *Anim. Behav.*, **34**, 939–941.

Sheldon, B. C. and T. Burke (1994). Copulation behavior and paternity in the Chaffinch. *Behav. Ecol. Sociobiol.*, **34**, 149–156.

Sherman, P. W. and M. L. Morton (1988). Extra-pair fertilization in Mountain White-crowned Sparrows. *Behav. Ecol. Sociobiol.*, **22**, 413–420.

Sibley, C. G., J. E. Ahlquist, and B. L. Monroe (1988). A classification of the living birds of the world based on DNA-DNA hybridization studies. *Auk*, **105**, 409–423.

Smith, H. G. and T. von Schantz (1993). Extra-pair paternity in the European Starling: the effect of polygyny. *Condor*, **95**, 1006–1015.

Smith, H. G., R. Montgomerie, T. Poldmaa, B. N. White, and P. T. Boag (1991). DNA fingerprinting reveals relation between tail ornaments and cuckoldry in Barn Swallows, *Hirundo rustica*. *Behav. Ecol.*, **2**, 90–98.

Smith, S. M. (1988). Extra-pair copulations in Black-capped Chickadees: the role of female. *Behaviour*, **107**, 15–23.

Smyth, A. P., B. K. Orr, and R. C. Fleischer (1993). Electrophoretic variants of egg white transferrin indicate a low rate of intraspecific brood parasitism in colonial Cliff Swallows in the Sierra Nevada, California. *Behav. Ecol. Sociobiol.*, **32**, 79–84.

Stouffer, P. C. and H. W. Power (1991*a*). Brood parasitism by starlings experimentally forced to desert their nests. *Anim. Behav.*, **41**, 537–539.

Stouffer, P. C. and H. W. Power (1991*b*). An experimental test of the brood reduction hypothesis in European Starlings. *Auk*, **108**, 519–531.

Stouffer, P. C., E. D. Kennedy, and H. W. Power (1987). Recognition and removal of intraspecific parasite eggs by starlings. *Anim. Behav.*, **35**, 1583–1584.

Stutchbury, B. J., J. M. Rhymer, and E. S. Morton (1994). Extrapair paternity in Hooded Warblers. *Behav. Ecol.*, **5**, 384–392.

Trivers, R. L. (1972). Parental investment and sexual selection. In: *Sexual Selection and the Descent of Man* (ed. B. Campbell), pp. 136–179. Aldine, Chicago.

Van Elsacker, L., R. Pinxten, and R. F. Verheyen (1988). Timing of offspring recognition in adult starlings. *Behavior*, **107**, 122–130.

Westneat, D. F. (1987). Extra-pair fertilizations in a predominantly monogamous bird: genetic evidence. *Anim. Behav.*, **35**, 877–886.

Westneat, D. F. (1990). Genetic parentage in the Indigo Bunting: a study using DNA fingerprinting. *Behav. Ecol. Sociobiol.*, **27**, 67–76.

Westneat, D. F. (1993). Polygyny and extrapair fertilizations in eastern Red-winged Blackbirds (*Agelaius phoeniceus*). *Behav. Ecol.*, **4**, 49–60.

Westneat, D. F. (1995). Paternity and paternal behaviour in the Red-winged Blackbird, *Agelaius phoeniceus*. *Anim. Behav.*, **49**, 21–35.

Wetton, J. H. and D. T. Parkin (1991). An association between fertility and cuckoldry in the House Sparrow, *Passer domesticus*. *Proc. R. Soc. Lond. B.*, **245**, 227–233.

Wetton, J. H., E. C. Royston, P. T. Parkin, and D. Walters (1987). Demographic study of a wild House Sparrow population by DNA fingerprinting. *Nature*, **327**, 147–149.

Wrege, P. H. and S. T. Emlen (1987). Biochemical determination of parental uncertainty in White-fronted Bee-Eaters. *Behav. Ecol. Sociobiol.*, **20**, 153–160.

Wright, J. and I. Cuthill (1989). Manipulation of sex differences in parental care. *Behav. Ecol. Sociobiol.*, **25**, 171–181.

Wright, J. and I. Cuthill (1990). Manipulation of sex differences in parental care: the effect of brood size. *Anim. Behav.*, **40**, 462–471.

Yamagishi, S., I. Nishiumi, and C. Shimoda (1992). Extrapair fertilization in monogamous Bull-headed Shrikes revealed by DNA fingerprinting. *Auk*, **109**, 711–721.

21

Density-Dependent Intraspecific Nest Parasitism and Anti-Parasite Behavior in the Barn Swallow *Hirundo rustica*

ANDERS P. MØLLER

One of the most fascinating recent discoveries of behavioral ecology is that the behavior of conspecifics is variable and can consist of several distinct entities, or strategies. Intraspecific nest parasitism, where individual females lay eggs in the nests of other females without caring for the offspring, is one such strategy. The alternative strategy is ordinary reproduction during which individuals only lay eggs in their own nest and care for their own offspring. Individuals may either adopt a fixed strategy or play different strategies at different times. Nest parasitism of conspecifics has been found to be relatively common in a large number of different bird species, and individuals may adopt a parasitic as well as a parental strategy (reviews in Yom-Tov 1980; Andersson 1984; Rohwer and Freeman 1989; Petrie and Møller 1991).

These so-called alternative strategies have been analyzed by means of evolutionary game theory (Maynard Smith 1982; Parker 1984) because the fitness of one individual depends on what other conspecifics are doing. Intraspecific nest parasitism has sometimes been considered a conditional strategy because individuals "make the best of a bad-job" (e.g., Yom-Tov 1980; Lank et al. 1989). In other words, females may lay eggs in other nests because they have lost their own nest or because they are unable to acquire a suitable nest site. Alternatively, intraspecific nest parasitism has been considered a mixed reproductive strategy with equal fitness payoffs for parasitic and self-reproducing females (e.g., Andersson and Eriksson 1982; Andersson 1984).

The fitness payoff of each strategy must be estimated to determine whether reproductive strategies are conditional or mixed. Reproductive success and thus fitness, however, often vary with population density because resources necessary for successful reproduction may become limiting at high densities. But, game theory models usually determine frequency-dependent fitness under the point assumptions that population density is stable and the population has a stable age distribution (Maynard Smith 1982). Theoretically, population density may have profound effects on the evolutionary stability of behavioral strategies (Sibly 1984; Parker 1985; Brown and Vincent 1987*a*,*b*), but field information on density-dependent variation in the frequency of alternative reproductive strategies is often unavailable.

The reproductive success of intraspecific parasites in species providing costly parental care has sometimes been found to exceed that of individuals reproducing entirely on their own (Møller 1987*a*; Brown and Brown 1989), and parasites appear to be able to locate host nests that are particularly likely to produce offspring successfully (Brown and Brown 1991). This suggests that parasitism is not always a conditional inferior alternative to the provisioning of own offspring. The average fitness of these populations, however, may become lowered as a consequence of parasitism, simply because it is costly for parental adults to raise parasitic offspring (May et al. 1991; Nee and May 1993). Overall population density will decrease as a consequence of the overall reduction in average fitness. Selection should favor the evolution of a reduced clutch size in hosts in order to reduce the costs of reproduction or prevent parasitism from producing an overcrowded clutch (Power et al. 1989). The frequency of parasitism may subsequently be reduced by evolution of anti-parasite strategies such as vigilant behavior (nest guarding and aggression directed towards potential parasites), reproductive synchrony, egg ejection, and nest desertion (reviewed in Petrie and Møller 1991). Anti-parasite behavior is also likely to be costly in terms of time and/or energy use, and vigilance should thus be adjusted to the potential levels of parasitism. Vigilant behavior could reduce the level of parasitism and therefore increase the average fitness of individuals in the population, and overall population density should consequently increase (May et al. 1991; Nee and May 1993). Intraspecific parasitism and anti-parasite behavior could therefore potentially cause fluctuations in population density and affect population dynamics. Again, there are virtually no data available on density-dependent variation in anti-parasite behavior.

Here I describe density-dependent intraspecific nest parasitism in the barn swallow *Hirundo rustica*, which is a small (c. 20 g), socially monogamous, semi-colonial passerine bird that feeds on airborne insect prey. Barn swallows most often breed inside buildings. The males arrive in my study area in Denmark during late April and May, whereas females, on average, arrive one week later (Møller 1994). Barn swallow males defend small territories of a few square meters without any resources other than nest sites. After pair formation the male and the female build a nest, and the female subsequently lays a clutch of 3–8 eggs, one per day (Glutz von Blotzheim and Bauer 1985). Nest parasitism by conspecifics is relatively common in large colonies and during years with high population density (von Vietinghoff Riesch 1955; Møller 1987*a*,*b*), as recorded from the laying of more than one egg per day in a nest and as evidenced by DNA fingerprinting (A. P. Møller and H. Tegelström, unpubl. data). Colonial barn swallows defend their nests against potential conspecific parasites by: (i) removing any eggs laid in their nest before they start laying their own eggs; (ii) behaving aggressively towards intruders; (iii) staying close to their nests from the start of egg laying; and (iv) laying slightly out of synchrony with the nearest neighbor (Møller 1987*a*, 1989). Most pairs have two clutches per season. Females incubate the eggs for two weeks, whereas nestlings are fed by both parents for three weeks.

The purpose of this chapter is two-fold. First, I describe density-dependent variation in the frequency of parasitism. Second, I describe density-dependent variation in anti-parasite behavior.

MATERIALS AND METHODS

I studied barn swallows throughout the breeding seasons 1982–1988 at Kraghede (57°12′ N, 10°00′ E), Denmark, which is an open farmland site with scattered plantations, ponds, and hedgerows. The swallows usually breed on farms either solitarily or in colonies of up to 50 pairs. The number of breeding pairs in the study area [and elsewhere in Western Europe (Glutz von Blotzheim and Bauer 1985)] has decreased considerably during the past 20 years, and I have therefore increased the size of the study area by a factor of three from 1987 onwards. A detailed description of the study population is given by Møller (1994). I visited all nest sites daily during the laying period and almost daily throughout the rest of the breeding seasons 1982–1988. Nests were checked for contents, and nest visits by color-ringed females and the intensity of nest guarding by nest owners were recorded during daily 2-hour observation periods during the egg-laying period (see below). [See Møller (1987*a*, 1994) for further details of the methods.]

Intraspecific nest parasitism was determined from the appearance of two or more eggs in a single nest during one day since the female nest owner can lay a maximum of one egg per day. The frequency of parasitism as estimated by this method was compared with the frequency of parasitism inferred from the occurrence of eggs with deviating spotting patterns, because the patterns of eggs and their size and shape are specific for individual females (see Møller 1987*a*). There was a high consistency in these two methods of estimating the frequency of nest parasitism. Recent DNA fingerprinting results have confirmed that multiple eggs laid during a single day are the result of nest parasitism (A. P. Møller and H. Tegelström, unpubl. data).

The intensity of nest guarding was estimated from the percentage of scans, made every second minute during daily 2-hour morning (04:00–12:00) observation periods, when at least one color-ringed nest owner was present within a distance of 5 m from its nest. Nest guarding intensity for each pair was estimated as mean guarding intensity for each day of the egg-laying period, which resulted in an average of 10 hours of observations for the modal clutch size of five eggs. Nest guarding during prelaying was not considered because parasite eggs laid during the prelaying period were ejected by nest owners (Møller 1987*a*). Nest guarding during incubation was not considered because potential parasites rarely visited nests at that stage of the nesting cycle (Møller 1987*a*: see fig. 21.2). The morning estimate of nest guarding was representative for the entire day, because pairs demonstrated a high degree of consistency in their intensity of nest guarding. One-hour estimates of nest guarding by 20 pairs made in the morning (04:00–10:00), around noon (12:00–16:00) and in the evening (18:00–21:00) during the 1988 breeding season were significantly highly positively correlated (Kendall's coefficient of concordance, $W = 0.84$, $\chi^2 = 59.24$, d.f. = 24, $P < 0.001$).

Population density was simply defined as the total number of breeding pairs in the original study area as determined from the capture of adult barn swallows and the number of nests recorded. Annually, more than 98% of all adult barn swallows were captured as determined from the frequency of birds with color rings. Colony size was simply defined as the total number of breeding pairs recorded on a single farm in a single year.

Variables were $\log_{10}$ or square root arcsine-transformed when necessary to acquire distributions not deviating significantly from normality. Individual pairs contributed only one observation to each data set, with the exception of a few cases in which paired tests were used. Values given throughout are means ± S.E.

RESULTS

Parasitism and density

The frequency of parasitism could potentially be affected by population density at a number of different levels: (i) overall population density; (ii) local population density (or colony size); and (iii) nearest-neighbor distance (these measures of density are not independent; see Møller 1991). The proportion of nests with parasite eggs was predicted to increase with each of the three measures of population density, simply because there would be more opportunities for parasitism (Brown 1984; Møller 1987*a*). This could be the case if more host nests were available, or if nearest-neighbor distances decreased at higher population densities. Alternatively, the frequency of naive birds less able to defend their nests could increase with population density, for example, if young, inexperienced birds comprised a larger fraction of the population at high densities.

The relative frequency of parasitism increased with overall population density in the study area when estimates for individual years were used as statistically independent data points (fig. 21.1). This was the case during both first and second clutches (first clutch: parasitism (%) = −16.29 + 0.72 (0.07) Density, $F = 101.57$, d.f. = 1,5, $P = 0.0002$; second clutch: parasitism (%) = −23.98 + 1.07 (0.22) Density, $F = 23.42$, d.f. = 1,5, $P = 0.0047$). These relationships indicate that an increase in population density is accompanied by a similarly sized increase in the frequency of parasitism.

The relative frequency of parasitism also increased with colony size when individual colonies were used as statistically independent data (fig. 21.2). This was also the case in both first and second clutches (first clutch: arcsine(Parasitism) = −(0.01 + 0.22 (0.03) log (Colony size), $F = 40.63$, d.f. = 1,55, $P < 0.0001$; second clutch: arcsine(Parasitism) = 0.37 + 0.26 (0.06) log(Colony size), $F = 15.63$, d.f. = 1,48, $P = 0.0003$). A similar relationship has previously been reported by Brown (1984). A closer inspection of fig. 21.2 reveals that the relationship between the frequency of parasitism and colony size could be due to the absence of parasitism among solitarily breeding pairs. Brown (1984, see also Brown and Brown 1989) found that the frequency of nest parasitism was unrelated to colony size among large colonies. However, the relationship was still positive following exclusion of solitary pairs of barn swallows (first clutch: arcsine(Parasitism) = −0.06 + 0.26 (0.11) log(Colony size), $F = 5.62$, d.f. = 1,30, $P = 0.024$; second clutch: arcsine(Parasitism) = −0.10 + 0.19 (0.08) log(Colony size), $F = 5.63$, d.f. = 1,30, $P = 0.024$). This suggests that there was an effect of colony size on the likelihood of parasitism. There was a positive and statistically significant effect of colony size on the frequency of parasitism in six out of seven years, which differs significantly from the expectation of equally many years with positive and negative relationships (binomial test, $P = 0.031$). Colony size therefore affects the frequency of parasitism independently of overall population density.

There is also an independent effect of nearest-neighbor distance on the probability of parasitism.

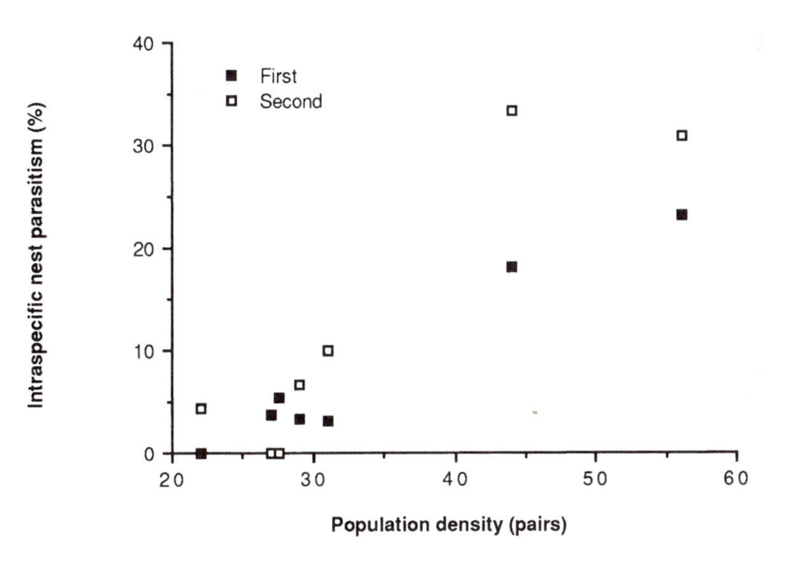

Figure 21.1 Relative frequency of intraspecific nest parasitism in barn swallows during first and second clutches in relation to the population size in the Kraghede study area during 1982–1988. Each data point is the estimate for a single year.

Barn swallow nests situated close to neighboring nests where owners were laying eggs at the same time were more likely to become parasitized than more distant nests: 65.1% of 63 nests less than 5 m from a focal nest were parasitized, whereas only 1.4% of 296 more distant nests were parasitized $G^2 = 147.1$, d.f. = 1, $P < 0.001$; Møller 1989). A similar relationship has been found in the cliff swallow *Hirundo pyrrhonota* (Brown and Brown 1989).

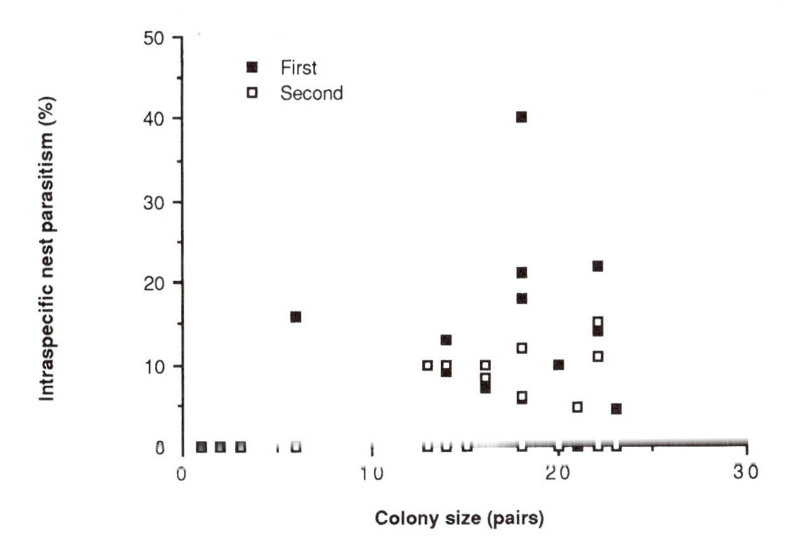

Figure 21.2 The relative frequency of intraspecific nest parasitism during first and second clutches in relation to colony size. Estimates for colonies are used as data points. The number of colonies with one pair was 25 and 18, with two pairs 10 and 7, and with three pairs 9 and 6 for first and second clutch, respectively.

In summary, the proportion of nests containing eggs of a parasite increased with three different measures of population density; overall density, colony size and nearest-neighbor distance.

Anti-parasite behavior

Barn swallows adopt a number of different strategies to avoid being parasitized. These include removal of parasite eggs laid in their nest before the start of own egg-laying, aggressive behavior, nest guarding, and slightly asynchronous egg laying relative to the nearest neighbor.

Egg removal and aggression

Egg removal occurs nearly ubiquitously during the prelaying period of the nest owner, and is not related to colony size (see Table IV in Møller 1987*a*; unpubl. results). This aspect of anti-parasite behavior is therefore not considered further here.

Nest owners behaved aggressively to all intruders, both males and females. Aggression consisted of threat displays, vigorous singing, chases during flight, or direct attacks – which often involved pecks at the opponent or birds fighting in the air while locked together by their claws. The function of aggressive behavior was apparently not defense of nest sites from takeover attempts, because I have never seen fights resulting in nest takeovers. Theft of nest material has never been observed in the study area or in other studies of the species (Glutz von Blotzheim and Bauer 1985), and theft thus can not account for aggressive behavior or nest guarding behavior. Both males and females were involved in aggressive interactions with intruding females (Møller 1987*a*, 1989), and it is therefore unlikely that aggressive behavior played a role in mate guarding. Aggressive responses varied in relation to the time of the nesting cycle of the nest owner. Levels of aggression were relatively high during the prelaying period; nearly half the intrusions by females approaching closer than 5 m to the nest site resulted in aggression (Møller 1989). This figure increased to nearly 90% during egg laying and decreased during incubation to nearly 30% (Møller 1989). There was also a clear relationship between the likelihood of aggressive responses and the approach distance by the potential parasite. Close approaches invariably elicited aggressive responses more often than more distant responses (Møller 1989). Data are not available to test whether these responses vary with population density.

Nest guarding

Efficiency of nest guarding If nest owners spend a considerable amount of time close to their nest, they should be able to interfere with any potential parasites approaching the nest (Emlen and Wrege 1986; Møller 1987*a*, 1989; Gowaty and Wagner 1988; Gowaty et al. 1989). There is some evidence to suggest that nest guarding is an efficient way of reducing the risk of parasitism. All barn swallows of both sexes spent time near their nests, although there was considerable intraspecific variation in the time spent near the nest. Females spent a larger proportion of their time within a distance of 5 m from their lest than their mates during the egg-laying period (males, 31.2 ± S.E.2.8%; females, 52.5 ± 3.5%; $n = 229$; Wilcoxon matched-pairs signed-ranks test, $P < 0.001$), and this may partly be due to time spent laying. Nest owners that were subsequently parasitized spent less time near their nests than nest owners having non-parasitized nests (parasitized nests, 55.3 ± 3.7% of time spent within 5 m of the nest, $n = 4$; non-parasitized nests, 74.9 ± 2.3%, $n = 34$, Mann–Whitney U-test, $P < 0.01$; Møller 1987*a*). This suggests that one function of time spent near the nest by nest owners was anti-parasite nest guarding. Parasitized nest owners, however, may differ from nonparasitized barn swallows in a number of other respects such as age or body condition, and it is thus difficult to know whether nest guarding alone accounts for the differences in parasitism.

Alternatively, the effect of nest guarding could be determined by comparing the rate of nest parasitism of nests which were guarded by barn swallows with the rate of parasitism of unguarded nests. I determined whether a total absence of nest guarding increased the probability of parasitism by putting up old swallow nests close to guarded, active nests either before or during egg laying (Møller 1989). The rate of parasitism of these nests was much higher for unguarded as compared with guarded nests, and this was the case independently of whether eggs were present in the nest (fig. 21.3). This result suggests that the presence of nest guarding does reduce the risk of parasitism.

Nestguarding and levels of parasitism The intensity of nest guarding during the egg-laying period showed a considerable amount of variation among individual barn swallows sampled in colonies of various sizes in different years (fig. 21.4). Nest guarding is costly because nest owners are unable to forage while guarding. Barn swallows should therefore optimize their nest guarding activities in

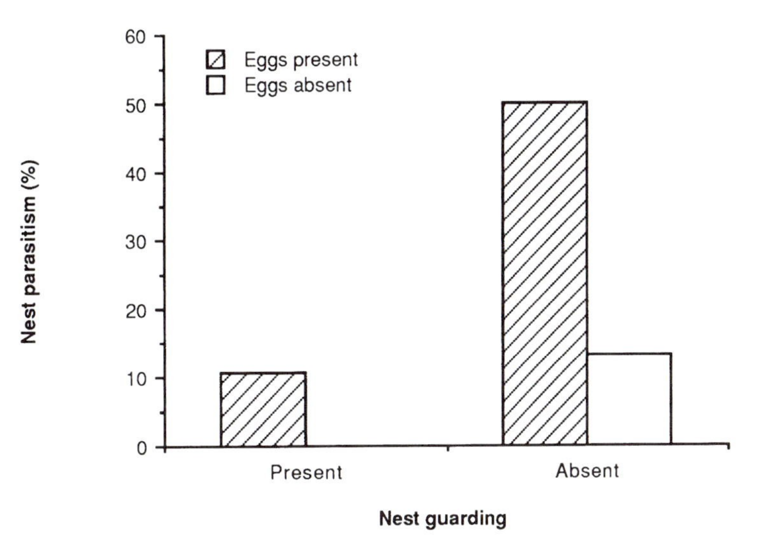

Figure 21.3 Frequency of intraspecific nest parasitism in experimental (no nest guarding) and active nests (nest guarding) (after seven days of exposure), in relation to the presence of swallow eggs. (Adapted from Møller 1989.)

relation to the risk of nest parasitism. The risk of parasitism increases with population density (see above), and it could therefore be predicted that the intensity of nest guarding should also vary with population density.

The average intensity of nest guarding increased with the relative frequency of nest parasitism in different years (fig. 21.5). This was the case during both first and second clutches (first clutch: arcsine(Nest guarding) = 0.80 + 0.78 (0.06) arcsine(Parasitism), $F = 162.35$, d.f. = 1,5, $P < 0.0001$; second clutch: arcsine(Nest guarding) = 0.79 + 0.57 (0.09) arcsine(Parasitism), $F = 42.61$, d.f. = 1,5, $P = 0.0012$). We would

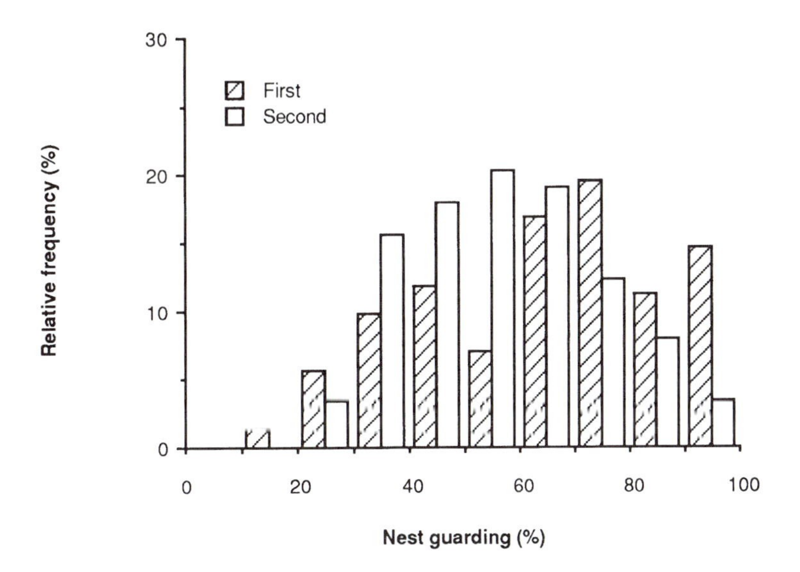

Figure 21.4 Frequency distribution of nest guarding (% time during the egg-laying period spent within 5 m of the nest) by barn swallow pairs of first (n = 140) and second clutch (n = 89) nests. Data from all observed nests during 1982–1988.

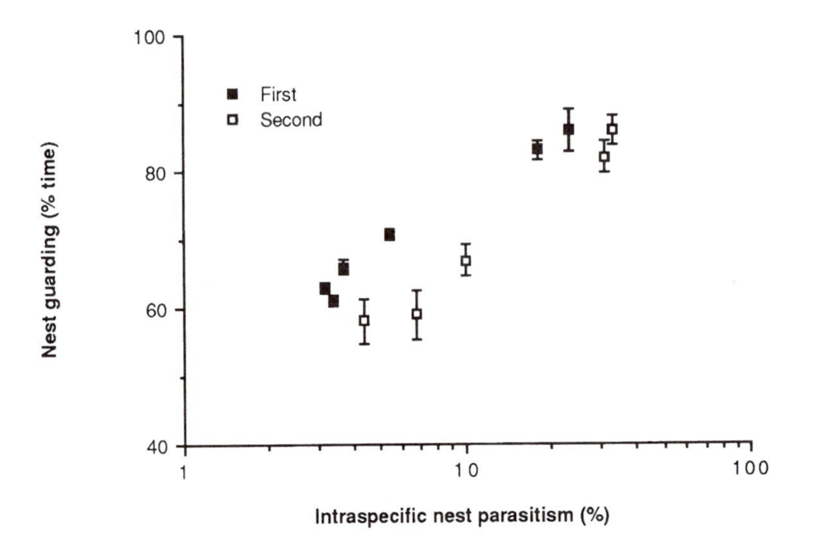

Figure 21.5 The intensity of nest guarding (% time during the egg-laying period spent within 5 m of the nest) during the first and second clutch in relation to the relative frequency of nest parasitism during 1982–1988. Values are means ±S.E. Each data point is the estimate for a single year.

therefore also expect the average intensity of nest guarding to increase with population density, which was the case (first clutch: arcsine(Nest guarding) = −0.37 + 0.90 (0.16) log(Density), $F = 32.30$, d.f. = 1,5, $P = 0.0024$; second clutch: arcsine(Nest guarding) = −0.49 + 0.95 (0.22) log(Density), $F = 19.02$, d.f. = 1,5, $P = 0.0073$). Nest guarding intensity increased with the risk of parasitism and thus with population density, and individuals therefore appeared to optimize their level of nest guarding at a higher intensity when the risk of nest parasitism was high. The positive relationship between nest-guarding intensity and the risk of nest parasitism may simply arise because individual barn swallows adjust their guarding intensity to the perceived risk of parasitism as inferred, for example, from the frequency of intrusions.

Nest guarding and colony size The probability of nest parasitism increased with colony size (see above) and barn swallows should therefore invest relatively more in nest guarding when the risk of nest parasitism was large. This prediction was supported because average nest guarding intensity per colony increased with the percentage of nests being parasitized (first clutch: arcsine(Nest-guarding) = 0.60 + 0.96 (0.17) arcsine(Parasitism), $F = 30.57$, d.f. = 1,21, $P < 0.0001$; second clutch: arcsine(Nest guarding) = 0.82 + 0.24 (0.12) arcsine(Parasitism), $F = 4.01$, d.f. = 1,10, $P = 0.076$). Therefore, barn swallows apparently adjusted their investment in nest guarding in relation to the local risk of parasitism as determined by colony size. The rate of parasitism at high densities, however, might have been even greater if there was no increase in nest guarding intensity with colony size, because nestguarding is effective in reducing parasitism.

Barn swallows breeding in larger colonies should invest relatively more in guarding activities than conspecifics breeding in small colonies or solitarily. The average intensity of nest guarding per colony increased asymptotically with colony size in both first and second clutches (fig. 21.6; first clutch: arcsine(Nest guarding) = 0.54 + 1.09 (0.34) log(Colony size) − 0.59 (0.27) (log(Colony size))^2, $F = 30.24$, d.f. = 2,20, $P < 0.0001$; second clutch: arcsine(Nest guarding) = 0.42 + 1.29 (0.23) log(Colony size) − 0.45 (0.22) (log(Colony size))^2, $F = 29.87$, d.f. = 2,20, $P < 0.0001$). The significant quadratic terms suggest that the pattern of nest guarding did not follow the pattern of parasitism, which increased directly with colony size (see subsection "Parasitism and density"). The intensity of nest guarding therefore may have reached a maximum level in medium-sized and large colonies. Alternatively, individual barn swallows unable to guard their nests may have aggregated in the large colonies. The average intensity of nest guarding also increased significantly with colony size in

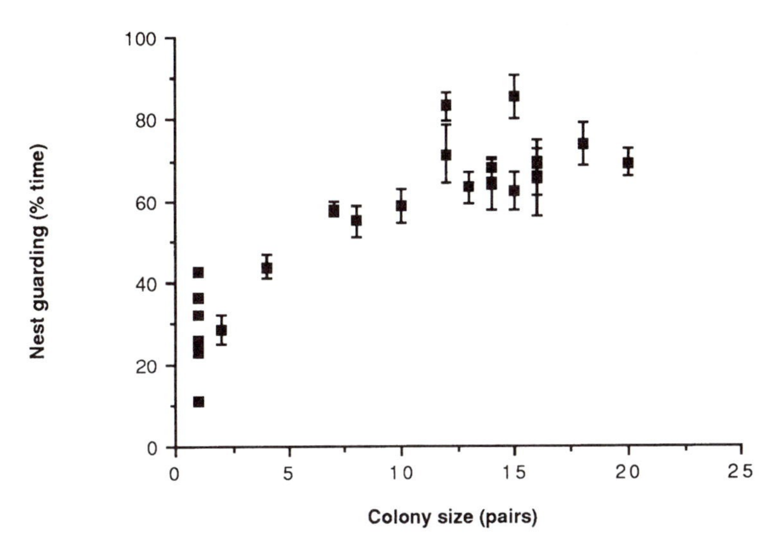

Figure 21.6 Nest guarding (% time during the egg-laying period spent within a distance of 5 m from the nest) in relation in colony size. Values are means ± S.E. Each data point is the estimate for a single colony during the first clutch.

seven out of seven years, which is statistically significant (binomial test, $P = 0.016$). Nest guarding intensity therefore increased independently with local population density as measured by colony size.

Exploitation of conspecific nest defense Finally, barn swallow pairs may be able to exploit the nest guarding and aggressive behavior of neighboring pairs by settling close to pairs that have laid eggs and thus do not pose a threat of intraspecific nest parasitism (Møller 1989). Barn swallows attack conspecifics that approach their nest within a distance of 5 m during the incubation period less intensively than those making more distant approaches during the egg-laying period (see Table 4 in Møller 1989). Any pair of barn swallows that is able to nest close to, but slightly out of synchrony with a focal pair therefore will be able to exploit the aggressive behavior of the focal pair. Nearest-neighbor distances between barn swallow nests with overlapping egg-laying periods were in fact greater than distances between nests initiated during the five-day period before start of egg laying (see Fig. 1 in Møller 1989). In other words, barn swallows were apparently able to increase the efficiency of their nest guarding by exploiting the guarding and aggressive behavior of the nearest, asynchronously breeding neighbor, without simultaneously running the risk of being parasitized by these neighbors which had already laid their eggs.

Individual variation in nestguarding There was some consistency in nest guarding intensity among pairs between clutches. Pairs that guarded their first clutch intensively also tended to guard their second clutch nest intensively ($r = 0.25$, $t = 2.15$, d.f. = 70, $P = 0.035$). There was a tendency for early-nesting pairs to guard their nests more intensively than later-nesting pairs even when controlling for year and colony size effects (ANCOVA: $F = 8.85$, d.f. = 7,135, $P < 0.0001$; date of start of egg laying: $F = 5.03$, $P < 0.05$; colony size: $F = 12.08$, $P < 0.001$; year: $F = 5.85$, $P < 0.001$). This effect was independent of age ($F = 0.85$, $P > 0.20$). This suggests that early-breeding individuals guarded their nests more intensively, despite the fact that fewer potential parasites were present. Barn swallows that laid many eggs also guarded their nests more intensively than pairs with small clutch sizes (ANCOVA: $F = 9.50$, d.f. = 7,135, $P < 0.0001$; clutch size: $F = 8.68$, $P < 0.001$; colony size: $F = 15.60$, $P < 0.001$; year: $F = 5.75$, $P < 0.001$). This again suggests that differences in nest guarding intensity may be associated with phenotypic quality of nest owners.

Finally, barn swallows performed activities other than nest guarding during the egg-laying period. For example, males spent a considerable amount of time guarding their mates during the egg-laying period in order to prevent risks of cuckoldry (Møller 1985). If females were away from the nest, males should follow them to guard them, but this would simultaneously expose the nest to potential parasites. I would therefore predict that mate guarding and nest guarding would be inversely correlated. There was in fact such a negative relationship between the intensity of nest guarding and the intensity of mate guarding (ANCOVA: $F = 13.87$, d.f. = 7,135, $P < 0.0001$; mate guarding: $F = 12.14$, $P < 0.001$; colony size: $F = 11.90$, $P < 0.001$; year: $F = 3.87$, $P < 0.001$). Not all males guarded their mates equally intensively, but males with short tails, which had mates that were often engaged in extra-pair copulations, did not guard their females more intensively than others (Møller 1994).

In conclusion, nest guarding was an efficient anti-parasite behavior. Investment in nest guarding was positively related to two measures of population density and thus to the probability of parasitism. The intensity of nest guarding was consistent between reproductive events and some of this consistency could apparently be explained in terms of phenotypic quality differences between pairs. Alternatively, different individuals may adopt alternative tactics of nest guarding behavior. There was a trade-off between nest guarding and mate guarding because intense mate guarding by the male was associated with a reduction in the intensity of nest guarding.

GENERAL DISCUSSION

Intraspecific nest parasitism may affect the population dynamics of hosts in a way similar to that of interspecific parasitism (May and Robinson 1985; Soler and Møller 1990). May et al. (1991) and Nee and May (1993) have suggested that intraspecific nest parasitism would reduce the average fitness of individuals provided that parental care is costly; an untested but likely assumption. A reduction in population density could be a consequence of this reduced average fitness. High levels of parasitism associated with a high population density would render costly anti-parasite behavior highly advantageous, which would result in an increase in the frequency of highly vigilant birds. The frequency of parasitism might subsequently become reduced as an effect of vigilant behavior, and costly anti-parasite behavior would tend to disappear. The positive relationships between nest guarding intensity and the frequency of intraspecific nest parasitism (figs 21.5 and 21.6) do not suggest that nest guarding is an inefficient tactic to prevent parasitism. These positive relationships suggest that individual pairs optimize nest guarding at a higher level when the risk of parasitism is high, such as in large colonies and in years with high population density. The consequence of a reduction in the rate of parasitism as caused by vigilant behavior might be a new increase in the frequency of parasites (May et al. 1991). This scenario rests on a number of untested assumptions.

First, the frequency of parasitism is assumed to increase with population density. The frequency of intraspecific nest parasitism in the barn swallow varied directly in relation to three measures of population density. Parasitism was common when barn swallows nested close to each other in large colonies during years with high population density. A positive relationship between the rate of parasitism and colony size has also been reported for the cliff swallow (Brown 1984; Brown and Brown 1989). Density-dependent nest parasitism could be due to either more opportunities for parasitism or the presence of a higher frequency of parasites under high-density conditions. For example, it is possible that birds are in better condition and therefore lay more eggs in years with high population densities. This latter explanation is less likely as the colony size effect remained in all years. However, barn swallows regularly reproduce during more than a single season, and a phenotypically fixed parasite strategy should result in relatively similar levels of parasitism in subsequent years. This did not appear to be the case (fig. 21.1), although confounding variables such as population density may have changed opportunities for parasitism. Some females appeared to be successful in repeatedly parasitizing conspecifics in subsequent clutches (Møller 1987a). Whereas this suggests that some females might be more inclined to become successful parasites, and thus represent a phenotypically fixed parasite strategy, there was no indication that particular females suffered from repeated cases of parasitism (Møller 1987a).

A second assumption is that high levels of parasitism reduce the average fitness of all individuals in the population. Parasitized barn swallow females lay smaller clutches than non-parasitized and parasite females (Møller 1987*a*) and this also appears to be the case in the cliff swallow (Brown and Brown 1989). This could be: (i) a direct response to parasitism, because parasitism should result in a reduction in the clutch size of the host (Andersson and Eriksson 1982; Andersson 1984); (ii) a result of the parasitized female parasitizing another nest herself; or (iii) these differences in clutch size could be explained by differences in phenotypic quality. Similarly, the nesting success of parasitized barn swallow females was lower than nonparasitized and, especially, parasitic females (Møller 1987*a*) as is the case in the cliff swallow (Brown and Brown 1989). If the small clutch size and the low breeding success of parasitized females are direct responses to parasitism, this would translate directly into reduced seasonal reproductive success during high population density.

A third assumption is that the frequency of vigilant behavior increases with the frequency of parasitism. Anti- parasite responses could either be genetically determined or, more likely, phenotypically plastic. The primary anti-parasite behavior of barn swallows was nest guarding, which appeared to be an efficient way of avoiding parasitism. Nest guarding intensity was positively related to population and colony levels of nest parasitism and thus to various measures of population density, which suggests that the barn swallows were able to adjust their investment in nest guarding to current threats of parasitism.

A final assumption of the models by May and Nee (May et al. 1991; Nee and May 1993) is that anti-parasite behavior is costly, which results in a reduction of its frequency or intensity during low levels of parasitism. Nest guarding could be costly in terms of time or energy. Barn swallows were unable to forage while guarding their nest because the small territory contains no or very little food. A long time spent guarding will thus invariably lead to a reduction in the amount of time spent foraging. I was unable to estimate directly the magnitude of this cost.

A second cost of nest guarding to male nest owners is the conflict between mate guarding and nest guarding. Male barn swallows guard their mates intensively during the prelaying and egg-laying periods, and a male usually follows its mate if she leaves the nest site. The nest is therefore left exposed to potential parasites. There was a strong negative relationship between the intensity of mate guarding and the intensity of nest guarding, and this indicates that nest guarding is costly in terms of the ability to guard a mate. If mate guarding is a more efficient paternity guard in some males than in others, differences in nest guarding intensity among individuals may then reflect alternative tactics adopted by different kinds of individuals.

I was able to test some of the assumptions of the model suggesting that intraspecific nest parasitism may affect the population dynamics of a species and cause cycles in population density (May et al. 1991; Nee and May 1993). However, much more work is needed before we will be able to determine the ways in which nest parasitism and other alternative reproductive strategies affect population dynamics.

ACKNOWLEDGMENTS I would like to thank M. Petrie, S. K. Robinson, and S. I. Rothstein for comments on previous versions of this chapter. I am grateful for the support provided by the Swedish Natural Science Research Council.

REFERENCES

Andersson, M. (1984). Brood parasitism within species. In *Producers and Scroungers: Strategies of Exploitation and Parasitism* (ed. C. J. Barnard), pp. 195–228. Croom Helm, London.

Andersson, M. and M. O. G. Eriksson (1982). Nest parasitism in goldeneye *Bucephala clangula*: some evolutionary aspects. *Amer. Natur.*, **120**, 1–16.

Brown, C. R. (1984). Laying eggs in a neighbor's nest: benefit and cost of colonial living in swallows. *Science*, **224**, 518–519.

Brown, C. R. and M. B. Brown (1988). A new form of reproductive parasitism in cliff swallows. *Nature*, **331**, 66–68.

Brown, C. R. and M. B. Brown (1989). Behavioural dynamics of intraspecific brood parasitism in colonial cliff swallows. *Anim. Behav.*, **37**, 777–796.

Brown, C. R. and M. B. Brown (1991). Selection of high-quality host nests by parasitic cliff swallows. *Anim. Behav.*, **41**, 457–465.

Brown, J. S. and T. L. Vincent (1987*a*). Coevolution as an evolutionary game. *Evolution*, **41**, 66–79.

Brown, J. S. and T. L. Vincent (1987*b*). A theory for the evolutionary game. *Theor. Pop. Biol.*, **31**, 140–166.

Emlen, S. T. and P. H. Wrege (1986). Forced copulations and intraspecific parasitism: two costs of social living in white-throated bee-eaters. *Ethology*, **71**, 2–29.

Glutz von Blotzheim, U. N. and K. M. Bauer (1985). *Handbuch der Vögel Mitteleuropas*. Vol. 10. AULA-Verlag, Wiesbaden.

Gowaty, P. A., J. H. Plissner, and T. G. Williams (1989). Behavioural correlates of uncertain parentage: mate guarding and nest guarding by eastern bluebirds, *Sialia sialis*. *Anim. Behav.*, **38**, 272–284.

Gowaty, P. A. and S. J. Wagner (1988). Breeding season aggression of female and male eastern bluebirds (*Sialia sialis*) to models of potential conspecific and interspecific egg dumpers. *Ethology*, **78**, 238–250.

Lank, D. B., E. G. Cooch, R. F. Rockwell, and F. Cooke (1989). Environmental and demographic correlates of intraspecific nest parasitism in lesser snow geese *Chen caerulescens caerulescens*. *J. Anim. Ecol.*, **58**, 29–44.

May, R. M. and S. K. Robinson (1985). Population dynamics of avian brood parasitism. *Amer. Natur.*, **126**, 475–494.

May, R. M., S. Nee, and C. Watts (1991). Could intraspecific brood parasitism cause population cycles? Proceedings of the XXth International Ornithological Conference, pp. 1012–1022.

Maynard Smith, J. (1982). *Evolution and the Theory of Games*. Cambridge University Press, Cambridge.

Møller, A. P. (1985). Mixed reproductive strategy and mate guarding in a semi-colonial passerine, the swallow *Hirundo rustica*. *Behav. Ecol. Sociobiol.*, **17**, 401–408.

Møller, A. P. (1987*a*). Intraspecific nest parasitism and anti-parasite behaviour in swallows, *Hirundo rustica*. *Anim. Behav.*, **35**, 247–254.

Møller, A. P. (1987*b*). Advantages and disadvantages of coloniality in the swallow *Hirundo rustica*. *Anim. Behav.*, **35**, 819–832.

Møller, A. P. (1989). Intraspecific nest parasitism in the swallow *Hirundo rustica*: the importance of neighbors. *Behav. Ecol. Sociobiol.*, **25**, 33–38.

Møller, A. P. (1991). Density-dependent extra-pair copulations in the swallow *Hirundo rustica*. *Ethology*, **87**, 316–329.

Møller, A. P. (1994). *The Barn Swallow and Sexual Selection*. Oxford University Press, Oxford.

Nee, S. and R. M. May (1993). Population-level consequences of conspecific brood parasitism in birds and insects. *J. Theor. Biol.*, **161**, 95–109.

Parker, G. A. (1984). Evolutionarily stable strategies. In: *Behavioural ecology: an evolutionary approach* (eds J. R. Krebs and N. B. Davies), pp. 30–61. Sinauer, Sunderland.

Parker, G. A. (1985). Population consequences of evolutionarily stable strategies. In: *Behavioural ecology: ecological consequences of adaptive behaviour* (eds R. M. Sibly and R. H. Smith), pp. 33–58. Blackwell Scientific Publications, Oxford.

Petrie, M. and A. P. Møller (1991). Laying eggs in others' nests: Intraspecific brood parasitism in birds. *Trends Ecol. Evol.*, **6**, 315–320.

Power, H. W., E. D. Kennedy, L. C. Romagnano, M. P. Lombardo, A. S. Hoffenberg, P. C. Stouffer, and T. R. McGuire (1989). The parasitism insurance hypothesis: why starlings leave space for parasitic eggs. *Condor*, **91**, 753–765.

Rohwer, F. C. and S. Freeman (1989). The distribution of conspecific nest parasitism in birds. *Can. J. Zool.*, **67**, 239–253.

Sibly, R. (1984). Models of producer/scrounger relationships between and within species. In: *Producers and Scroungers: Strategies of Exploitation and Parasitism* (ed. C. J. Barnard), pp. 267–287. Croom Helm, London.

Soler, M. and A. P. Møller (1990). Duration of sympatry and coevolution between the great spotted cuckoo and its magpie host. *Nature*, **343**, 748–750.

von Vietinghoff-Riesch, A. (1955). *Die Rauchschwalbe*. Duncker and Humblot, Berlin.

Yom-Tov, Y. (1980). Intraspecific nest parasitism in birds. *Biol. Rev.*, **55**, 93–108.

22

Egg Discrimination and Egg-Color Variability in the Northern Masked Weaver

The Importance of Conspecific versus Interspecific Parasitism

WENDY M. JACKSON

Many species of the *Ploceus* weaverbirds of Africa and Asia exhibit a dramatic degree of intraspecific egg-color variability (Mackworth-Praed and Grant 1955; Moreau 1960; Freeman 1988; W. Jackson, pers. obs.). Within a species, eggs can range from white to blue to brown and can be immaculate or have various types of speckling. Freeman (1988) reported variable eggs in 16 of the 34 *Ploceus* species whose eggs he examined in collections. These species differ from most species of birds, in which one egg looks very much like any other egg of that species and which typically show variability only in the speckling pattern. Even those birds that have been noted for their variable eggs, such as the common murre (*Uria aalge*), arguably do not compare with the weaverbirds in this regard.

Although eggs of weaverbirds may vary substantially within a species, eggs within a particular nest typically all look alike. This observation suggests that individual females lay only one egg-color type. Collias (1984) has shown for a captive colony of village weaverbirds (*Ploceus cucullatus*) that eggs produced over a lifetime by individual females show little, if any, variability in color. Likewise, in a field study of individually marked northern masked weavers (*Ploceus taeniopterus*; Jackson 1992*a*), I determined that within-female variability in egg color is significantly and substantially lower than between-female variability. Thus, within species of weaverbirds, eggs vary in color because different females lay differently colored eggs.

However, there are exceptions to the rule that all the eggs within a nest resemble one another. Freeman (1988) found that in those *Ploceus* species with variable eggs, anywhere from 2% to 29% of nests in collections contained an oddly colored egg. Because eggs produced by individual females do not vary substantially, the odd eggs noted by Freeman must have been laid by brood parasites. Freeman ruled out *Chrysococcyx* cuckoos, brood parasites of many tropical passerines, including weaverbirds, as being responsible for these odd eggs by showing that weaverbird species known to be frequent hosts of cuckoos did not have a higher incidence of odd-egg clutches than did species not known to be hosts. Rather, he concluded that conspecifics had laid these oddly colored eggs. In support of his conclusion, parasitism by cuckoos was rare in my study of northern masked weavers, occurring in fewer than 1% of the nests, while conspecific nest parasitism (CNP) was a common reproductive strategy, occurring in 23–35% of the nests (Jackson 1992*a*,*b*).

Northern masked weavers that parasitize conspecifics appear to be nesting females that lay an additional egg in another female's nest to enhance their reproductive success (Jackson 1993). Brood size in this species is essentially constrained to three: starvation in naturally occurring and artificially created four-chick broods is very common. The constraint on brood size is probably imposed by the extreme hatching asynchrony characteristic of this species: the last egg in a four-egg clutch hatches 72 hours after the first egg, and when a chick starves, it is almost always the youngest one. Late-hatching chicks also grow more slowly than do early-hatching chicks. If a female were to lay a fourth egg in her own nest, there is little chance that it would succeed. However, if she places it in another female's nest before that host lays her third egg, then it may have a greater chance of success. Conspecific nest parasitism is therefore potentially costly to a host's reproductive success. If a host lays a three-egg clutch, one of her own chicks may die if her nest is parasitized, and regardless of clutch size, her chicks may grow more slowly.

Egg-color variability in weaverbirds, then, may have evolved because producing eggs that appear different from those of conspecifics enables a female to avoid the cost of parasitism by distinguishing her eggs from those laid by conspecific nest parasites. To determine whether weavers discriminate between their own eggs and those of conspecifics, I performed egg-recognition experiments in a field study of the northern masked weaver and present my results here. Because CNP is the likely force behind egg recognition as well as the color variability, the experiments were designed to examine how effective egg-color variability is in deterring nest parasitism by conspecifics. Thus I examined females' abilities to discriminate between their eggs and those of conspecifics over a wide range of color differences. I found that difference in color was a strong, but imperfect, predictor of whether a female would reject a parasitic egg. In addition to examining the effects of background color and speckling pattern on host response, I also attempted to determine whether females distinguish among eggs on the basis of size: the importance of size in egg recognition has been documented in cowbird hosts (Rothstein 1982; Mason and Rothstein 1986) and cuckoo hosts (Davies and Brooke 1988). However, Victoria (1972) found size to be unimportant in egg-recognition experiments on the village weaver. Because eggs of the Didric cuckoo (*Chrysococcyx caprius*), the only likely heterospecific parasite at my study site, are larger than northern masked weaver eggs (Jackson 1992*b*), weavers should use egg size to detect parasitism if egg-recognition ability evolved in response to cuckoo parasitism.

A growing number of studies have documented the ability of birds to distinguish their own eggs from those of conspecifics. Among nonpasserines, recognition and rejection of experimentally introduced or naturally occurring parasitic eggs has been documented in several species, including the American coot (*Fulica americana*; Lyon 1993) and the sora (*Porzana carolina*; Sorenson 1995). Shugart (1987) showed that Caspian terns (*Sterna caspia*) could recognize their own clutches based on both maculation pattern and background color.

Among passerines, both bramblings (*Fringilla montifringilla*) and chaffinches (*F. coelebs*) recognize eggs of conspecifics, although they rejected such eggs in only 53% of the experimental trials of Braa et al. (1992). Moksnes (1992) demonstrated that the degree of contrast between host and foreign eggs was a good predictor of whether foreign eggs were rejected. Eastern kingbirds (*Tyrannus tyrannus*) can distinguish between their own eggs and those of both conspecifics and heterospecifics; however, the probability of rejection of conspecific eggs was lower than for heterospecific eggs (Bischoff and Murphy 1993). Both female and male European starlings (*Sturnus vulgaris*) could recognize conspecific eggs that were experimentally added to their nests, with females rejecting these eggs 98% of the time and males 61% of the time (Pinxten et al. 1991). Victoria (1972) performed egg-recognition experiments on captive village weavers and found that females discriminate their own eggs from those of conspecifics: the probability of ejection depended on the degree of difference in background color between the hosts' eggs and those added to their nests. In addition, hosts laying immaculate eggs were more likely to eject speckled eggs than other immaculate eggs, and vice versa. Despite this body of evidence for discrimination among conspecific eggs, no prior study has assessed the recognition behavior of *Ploceus* weavers in the field. This is an important omission in light of the unsurpassed level of intraspecific egg-color variation that occurs in this genus.

METHODS

Study organism and site

Egg-recognition experiments were performed in 1986–1988 on a population of northern masked weavers breeding at Lake Baringo, Rift Valley Province, Kenya (0°35′ N, 36°5′ E). Nests were built primarily in marsh grass (*Echinochloa haploclada*) and reeds (*Typha* sp.), and occasionally in small acacia trees (*Acacia* sp.), so they were accessible to me. In addition, this species does not build entrance tubes on its nests.

These tubes would have hindered my attempts to remove eggs for inspection. Although males build the nests in this species, they essentially never enter a nest after they have finished weaving it: all incubation and care of the young is performed by females (W. Jackson, pers. obs.). Thus, only females were tested for their ability to recognize and reject "parasitic" eggs based on the degree of difference between their own eggs and the parasitic eggs.

Assessing egg appearance

Three different aspects of egg appearance were tested: ground color, speckling pattern, and egg volume (a measure of egg size). In 1986, ground color was assigned to one of 24 subjective categories, ranging from blue-green to dark brown. In 1987–1988, color was quantified by matching eggs to the closest "color chip" from the matte finish version of the *Munsell Book of Color* (1976). When the color of an egg fell between two color chips, the values for the two chips were averaged. Because each color chip represents a point in three-dimensional space, three values are assigned to each chip (hue, value, and chroma). So that the experiments in 1986 could be combined with the others, I converted categories to Munsell values as follows: I plotted frequency distributions for the categories in 1986, and for the Munsell values in 1987–1988, and matched the distributions by inspection. Although the frequency distribution for the Munsell values had a greater number of divisions than did the distribution for the categories, so that within a category there was some variation in color, the technique does not result in any systematic bias in the differences between eggs. Thus I should neither underestimate nor overestimate the ability of females to distinguish between colors by combining all trials.

Although the Munsell variable for hue is circular, the eggs of the northern masked weaver span only 135° of the circle (i.e., less than half of the circle), from 5 Yellow-Red (Hue #15) to 2.5 Blue-Green (Hue #52.5), and do not straddle the arbitrary beginning point of the circle. Therefore I treated hue as a linear variable without overestimating differences between two eggs (which would have occurred if two eggs fell on opposite sides of the arbitrary beginning point yet in reality were quite similar in hue).

Hue, value, and chroma are highly intercorrelated for eggs of this species: bivariate correlations calculated from 1790 eggs examined during the three years of this study ranged from 0.81 to 0.88. Therefore I combined these three variables with a principal component analysis to derive a single variable I term "color." The first factor extracted in the analysis of all 1790 eggs accounted for 90.4% of the variance in these three variables, and the scores calculated using this factor were entered as the values for color.

Speckling pattern was determined for each egg by noting the amount of speckling and the distribution of the speckling over the egg. Amount of speckling was categorized as none (no speckling), sparse (some but no more than 25% of the egg speckled), medium (26–50% of the egg speckled) and heavy ($\geq 51\%$ of the egg speckled). If an egg was speckled, the distribution of the speckling was categorized as even, slightly graded (speckling slightly more concentrated at the blunt end), and graded (speckling concentrated at blunt end). Because amount of speckling and distribution of speckling are correlated ($r = 0.77$), I performed a principal component analysis to combine these two variables into one variable termed "speckling pattern." The first factor extracted in this analysis accounted for 88.7% of the variance in these two variables, and the factor scores calculated in this analysis were entered as speckling pattern.

Egg volume was calculated using the formula given in Spaw and Rohwer (1987): volume $= 0.498(\text{length})(\text{breadth})^2$. Length and breadth were measured to the nearest 0.1 mm in 1986 and usually to the nearest 0.05 mm in 1987–1988.

Egg-recognition experiments

Eggs used in the experiments were real northern masked weaver eggs, and all host and parasitic eggs were marked with indelible ink to allow for individual identification. Eggs were inserted in hosts' nests at various times throughout the day, but usually either between 10:00 and 11:00 hours or between 14:00 and 16:00 hours. I have no evidence that females of this species remove a host egg when laying parasitically (Jackson 1992*a*). Nevertheless, because some brood parasites remove host eggs when they lay whereas others do not (Friedmann 1963; Wyllie 1981), I performed two types of manipulation: in some cases parasitic eggs simply were added to a female's nest; and in other cases host eggs were replaced with parasitic eggs. This procedure allowed me to determine if hosts count eggs to decide if their nests have been parasitized, and if they use the discordancy rule when ejecting eggs (that is, if they eject the odd egg; see Rothstein 1975*a*).

In most cases, the stage of the host's nesting cycle was known. When nest stage was known, I inserted eggs in nests either during the laying period of the host female (including the day following clutch completion), or after clutch completion. This procedure allowed me to examine whether rejection was more likely to occur in the former nests than in the latter. Such a result might be expected if parasitic eggs laid after the host's clutch completion pose a lesser threat than those laid earlier because "late" parasitic eggs may receive insufficient incubation to hatch.

I added only one parasitic egg to the majority of nests, but some nests received two eggs. In seven nests where two eggs were inserted, the original host clutch of two eggs was completely replaced by parasitic eggs (average clutch size in this species is 2.57 ± 0.63; Jackson 1992*a*). I treated each of these nests as one trial, rather than two, because the two parasitic eggs in each of the seven nests were identical in egg color and speckling pattern, and the host response was always either to reject both eggs or accept both eggs. I did not remove any host eggs from the remaining 20 nests where I inserted two eggs. I treated each parasitic egg in these 20 cases as a separate trial because some females rejected one parasitic egg while accepting the second (the two eggs in these cases often differed in color and speckling pattern).

Eggs were added to nests randomly with regard to color, whether the eggs were added during or after laying, and whether parasitic eggs replaced or were added to host eggs. Trials lasted for varying lengths of time, but only responses that occurred within two days of the manipulation were included in the analyses. This cut-off was established because daily rates of nest loss and egg loss were very high. Thus, eggs or nests disappearing more than two days after a manipulation may have disappeared due to other reasons (e.g., predation). Eggs or nests disappeared after this cut-off time in only four trials, so the results are unlikely to change if these nests are included. In two trials, the experiment lasted just one day, but these are included in the analyses because, again, they are unlikely to alter the general results.

I divided host responses into five categories: (i) acceptance of parasitic egg; (ii) ejection of parasitic egg on Day 1 (i.e., within 24 hours); (iii) ejection of parasitic egg on Day 2 (i.e., within 48 hours); (iv) abandonment of the nest; and (v) ejection of a host egg (which may occur in addition to ejection of a parasitic egg). A nest was considered abandoned if it was torn down by the male after having contained eggs at my last previous visit (males tear down unused nests). If all eggs were missing (except for the seven trials in which the host's eggs were replaced entirely), I considered the nest to have been preyed upon and excluded it from my analyses.

I quantified the difference in egg color and speckling pattern between the host and parasitic eggs by taking the difference between the corresponding factor one scores. These values range from 0.000 to 2.877 for color, and from 0.000 to 3.022 for speckling pattern. In most cases there was little variability in either color or speckling pattern among the eggs in a host's nest. When eggs did vary, the average color and speckling pattern were computed and the differences between these values and the scores for the parasitic eggs were used in the analyses. Difference in egg volume between host eggs and parasitic eggs was the difference between the mean host egg volume and the parasitic egg volume. In the seven cases in which I treated a nest as one trial even though two parasitic eggs completely replaced the host's clutch, the average parasitic egg volume was used to calculate the volume difference.

Statistical analyses

The differences in color, speckling pattern, and volume between host and parasitic eggs are continuous variables whereas host female response to parasitic eggs is a categorical variable. Therefore, I used a logistic regression analysis to examine the effects of these independent, continuous variables on host response. I performed the analyses with the LOGIT module of SYSTAT for the Macintosh (SYSTAT 1985). The statistic computed in this analysis is a likelihood chi-square statistic (-2 times log likelihood ratio, or G) for the null hypothesis that the regression coefficient, β, for the independent variable is zero. In addition, this program computes the derivative of β, which is interpreted in this situation as the increased probability of rejection for the smallest possible finite increase in the difference of the independent variable (SYSTAT 1985). The odds ratio, e^{β}, can also be computed, which yields the relative probability of rejection for each unit increase in the independent variable.

RESULTS

Rejection responses were noted in 31 of 109 trials (table 22.1). Included as rejections are two cases in which only a host egg was ejected. Host response

Table 22.1 Outcomes of egg-recognition experiments

Accepted	78
Rejected	
Parasitic egg ejected on Day 1	14
Parasitic egg ejected on Day 2	7
Nest abandoned	4
Host egg ejected	
in addition to parasitic egg	3
instead of parasitic egg	2
before abandoning nest	1

to parasitic eggs did not depend on nest stage: 33% (7 of 21) of the parasitic eggs added after clutch completion were rejected, whereas 21% (16 of 76) of the eggs added during laying (including the day following clutch completion) were rejected ($G = 1.299$, 1 d.f., $P > 0.05$).

Likewise, host response to parasitic eggs did not depend on whether host eggs were replaced with parasitic eggs or parasitic eggs were added to host eggs: 38% (11 of 29) of the former trials resulted in rejection, while 25% (20 of 80) of the latter trials resulted in rejection ($G = 1.690$, 1 d.f., $P > 0.05$).

Difference in egg color was a strong predictor of whether a host rejected a parasitic egg (fig. 22.1). The goodness-of-fit test of the logistic regression model using the difference in egg color between the host and parasitic eggs as the independent variable is highly significant (table 22.2). The derivative of the regression coefficient for the difference in color is 0.337, which means that for the smallest finite increase in this difference the probability of rejection increases by 0.337. The odds ratio of the regression coefficient reveals that when the difference in color increases by 1 unit, the probability of rejection goes up nearly sixfold (the maximum difference was 2.877 units). Females with brownish eggs rejected only 16% of 76 brownish parasitic eggs, but all eight bluish-green eggs. Females with bluish-green eggs rejected only 24% of 17 bluish-green eggs, but seven of eight brownish eggs.

Difference in speckling pattern was not a strong predictor of whether a host would reject a parasitic egg, and the goodness-of-fit test of this model was not significant (table 21.2); however,

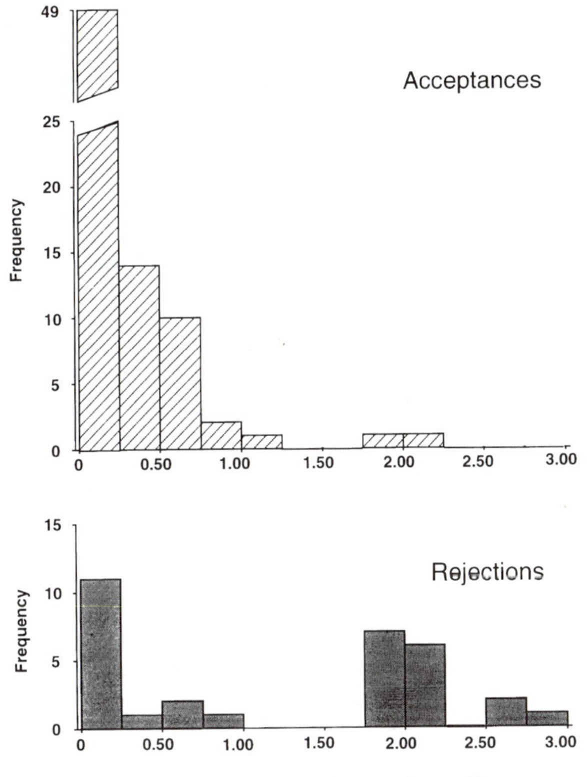

Figure 22.1 Frequency distributions for difference in color between host and parasitic eggs in trials that resulted in acceptances (upper graph) and trials that resulted in rejections (lower graph).

Table 22.2 Goodness-of-fit tests of logistic regression models examining the effects of difference in color, speckling pattern, or volume on host response to parasitic eggs

Independent variable	*G*	df	*P*	*n*	Regression coeff. (β)	Odds ratio (e^{β})
Color	33.949	1	<0.001	109	1.770	5.871
Speckling pattern	3.259	1	<0.10	109	0.496	1.642
Volume	5.864	1	<0.05	87	−0.004	0.996

there was a tendency for the probability of rejection to increase as the difference in speckling pattern increased ($P < 0.10$). The odds ratio of the regression coefficient for difference in speckling pattern reveals that for an increase in this difference of 1 unit, the probability of rejection increases approximately 64% (the maximum difference was 3.022 units).

The goodness-of-fit test of the model using difference in egg volume as the independent variable is statistically significant (table 22.2). However, inspection of the regression coefficient for difference in egg volume reveals that the smaller the difference, the more likely hosts were to reject the parasitic egg. This is the opposite result of what was expected. Although the model is significant at the $P = 0.05$ level, note that the odds ratio of the regression coefficient for egg volume is nearly 1, and the regression coefficient is nearly 0. That is, the probability of rejection goes down only very slightly with a decrease in the difference in egg volume. I conclude from this that difference in egg volume has little, if any, biologically meaningful effect on host response to parasitic eggs.

DISCUSSION

Egg recognition in weavers and other species

The only information that female northern masked weavers appear to use to determine if their nests have been parasitized is difference in egg color: none of the other variables measured in this study was correlated with females' responses in the egg-recognition experiments.

Whether females would reject parasitic eggs did not depend on when in their nesting cycles parasitic eggs were inserted. Thus, the probability of rejection was not greater when the potential cost of parasitism is thought to be higher. Davies and Brooke (1988) obtained similar results in their experimental study of cuckoo parasitism of reed warblers (*Acrocephalus scirpaceus*). However, the assumption that parasitic eggs inserted during the host's laying cycle have a higher success rate than eggs inserted after clutch completion may be incorrect. Brown and Brown (1988) documented that physical transfer of eggs from one nest to another is a frequently employed parasitic tactic in cliff swallows (*Hirundo pyrrhonota*), a species with a rate of CNP as high as 43%. In such instances, parasitic eggs transferred at any point in the host's cycle may succeed if they have received sufficient incubation, so the potential cost of CNP to the host may remain high throughout incubation. If CNP via egg transfer occurs in the northern masked weaver, host response to parasitism would not be expected to vary with nest stage. In fact, I detected three cases in which eggs were physically transferred between nests (described in Jackson 1992*a*).

In contrast to my results, Bischoff and Murphy (1993) found that eastern kingbirds are more likely to reject conspecific eggs prior to laying (50% were rejected) than during laying (17%), and that none was rejected post-laying. These results differed from their results for trials involving heterospecific eggs: 67%, 100%, and 43% were rejected when added during the prelaying, laying, and postlaying phases, respectively. They concluded that a parasitic kingbird egg is so similar to the host's eggs that there is a high chance of a bird mistakenly rejecting its own eggs if it attempts to discriminate amongst kingbird eggs. This cost of mistaken rejections results in hosts' accepting parasitic kingbird eggs once they have begun laying their own eggs. Likewise, Pinxten et al. (1991) found that the response of European starlings to foreign conspecific eggs depended on when in the laying cycle the eggs were added: almost 100% of eggs added prior to clutch initiation resulted in rejection of the foreign egg, while none of the eggs added after clutch initiation was rejected.

Host response in the current study did not depend on whether host eggs were removed when the parasitic eggs were inserted. This result rejects the hypothesis that females count eggs to determine if their nests have been parasitized. If they did use this method, there should have been a higher proportion of rejections in the trials where parasitic eggs were added to hosts' eggs than when they replaced hosts' eggs. Similarly, Rothstein (1975*b*) found that removal versus nonremoval of host eggs had no effect on responses of cowbird hosts. Davies and Brooke (1988) also detected no difference in the responses of reed warblers depending on whether their own eggs were removed in the experiment.

In addition, my results support the hypothesis that females can recognize their own eggs and do not simply reject odd eggs. If they did use the odd-egg rule (see Rothstein 1975*a*), then none of the seven cases where host eggs were completely replaced by parasitic eggs should have resulted in rejection, yet three cases did. In addition, in 19 nests, two eggs were added on the day the host laid her first egg, so that the host egg was the odd egg (at least until the host laid her second egg). However, in none of these nests did the host eject one of her own eggs. These results concur with those found for village weavers (Victoria 1972) and for chaffinches and bramblings (Moksnes 1992). Rothstein (1975*a*), in a study of egg-recognition ability in a variety of birds, also found no evidence that birds use the discordancy rule to eject parasitic eggs.

Host response did not depend greatly on the degree of difference in speckling pattern between host and parasitic eggs, although there was a tendency for females to reject eggs that had a different speckling pattern from their own. In contrast, in village weavers, a female's response to introduced eggs could be predicted based on the presence or absence of speckling. Why do female northern masked weavers not use (or use to a greater extent) speckling pattern as information about parasitism? Speckling pattern in this species is correlated with egg color ($r = 0.55$), primarily because eggs with blue-green or green backgrounds are always speckled (eggs of all other colors may be either speckled or immaculate). Therefore, it is possible that speckling pattern has evolved for some reason other than egg-recognition ability pertaining to CNP. For example, Oniki (1985) has shown that egg color and presence or absence of speckling is highly correlated across species with nest structure and/or nesting habitat. Lack (1958) obtained similar results for the Turdinae. Immaculate blue-green eggs may be unsuitable in the northern masked weaver for reasons relating to predation (although this seems unlikely given the enclosed architecture of weaverbird nests) or thermoregulation. These suggestions remain to be tested.

Response to parasitic eggs did not depend on differences in egg size in either the northern masked weaver or the village weaver (Victoria 1972; but see Cruz and Wiley 1989). In contrast, egg size has been shown to be important in other studies of egg-recognition ability in both cowbird (Rothstein 1982) and cuckoo hosts (Davies and Brooke 1988). Female weavers may not use information about egg volume to detect CNP because the variation is not great enough for females to detect a difference. In the studies where egg size has been shown to affect a female's response to a parasitic egg, the differences in size between host and parasitic eggs have been substantial. For example, American robins (*Turdus migratorius*) respond to the small size of cowbird eggs but these eggs have less than half the volume of robin eggs, so robins do not need to make subtle distinctions between eggs on the basis of size (Rothstein 1982).

As noted earlier, difference in egg color had a very strong effect on the probability of rejection. However, as can be seen in fig. 22.1, females often made mistakes and accepted parasitic eggs. Even when the difference in color was moderate (e.g., when the difference was 0.40, which corresponds to the difference between a greenish-tan egg and a medium brown egg), females often accepted parasitic eggs. Why are the birds often unresponsive to egg-color differences? As noted earlier, weaverbird nests are enclosed structures with bottom entrance holes. Thus, the amount of light that reaches the eggs may be relatively low, hindering a female's ability to distinguish between eggs. In support of this idea, Freeman (1988) showed that *Ploceus* species with elongated nest tubes, which presumably reduce the amount of light entering the nest to an even greater extent, are more likely to have plain white eggs than are species whose nests lack tubes. Oniki (1985) found similar results in a study of Amazonian birds: species with enclosed nests have plain white eggs more often than do species with exposed nests. Among the Turdinae, Lack (1958) discovered that species nesting in deep holes tend to have immaculate white eggs.

Possible costs of egg-recognition behavior

Another factor that may influence whether a female rejects a parasitic egg is the cost of ejecting one of her own eggs by mistake. As noted in table 22.1, a female presumably ejected one of her own eggs in six trials, either in addition to (three cases) or instead of (two cases) the parasitic egg, or prior to abandoning the nest (one case). The differences in color between the parasitic egg and the host eggs in these six cases were 0.040, 0.367, 0.771, 2.181, 2.221, and 2.606 and were similar to the range of differences for all combinations (see fig. 22.1). Thus, this type of mistake did not appear to be more likely to occur when the difference in color was small, as might have been expected. In their egg-recognition experiments on *Cuculus canorus* hosts, Davies and Brooke (1988) found that host eggs were just as likely to be rejected along with cuckoo eggs in trials where parasitic eggs were nonmimetic as in trials where they were mimetic. However, they found that host eggs were more likely to be rejected instead of cuckoo eggs in trials where parasitic eggs were mimetic than in trials where they were nonmimetic.

A note of caution in assessing the cost of ejection should be made. As noted earlier, the day-to-day rate of nest loss in this species is very high. It is possible that some of these six host eggs that I interpreted as having been mistakenly ejected by the host may have disappeared due to predation or some other factor. For example, Rothstein (1986) noted that single host eggs sometimes disappear from eastern phoebe (*Sayornis phoebe*) nests, although experiments show a lack of rejection behavior in this cowbird host. This same note of caution applies to the results in fig. 22.1, in which 11 of 60 eggs disappeared when the difference in color between the host and parasitic eggs was smallest, i.e., 0.00–0.24. Five of these 11 eggs were "ejected" on Day 2, rather than Day 1, whereas only two of 20 eggs were ejected on Day 2 when color differences were > 0.24. Thus, some of the eggs in the former sample may have disappeared for reasons unrelated to the experiments.

If there really is a cost of ejection, some of the females who accepted parasitic eggs when the color difference was small may have detected that their nest had been parasitized, but accepted the parasitic egg rather than risk accidentally ejecting one of their own. On the other hand, the fact that five nests were abandoned by hosts suggests that females may be unwilling to pay the cost of acceptance in some cases. Perhaps in these cases females determined that their nests were parasitized, but were uncertain about which egg was the parasitic one and opted to scrap the breeding attempt entirely rather than raise a parasite's chick or eject one of their own eggs by mistake. If these interpretations are correct, then the results of my experiments should be interpreted as reflecting females' abilities to identify parasitic eggs, and not simply their ability to tell whether they have been parasitized.

Egg recognition in response to interspecific versus conspecific parasitism

Egg appearance is used by individuals in many species to identify heterospecific eggs that appear in their nests. Many hosts of the European cuckoo (*Cuculus canorus*) detect cuckoo eggs in their nests and reject them (Davies and Brooke, this vol.), and in egg-recognition experiments, the greater the difference between cuckoo eggs and host eggs, the greater the probability of rejection (Davies and Brooke 1988, 1989*a*). This ability by hosts to recognize cuckoo eggs has led to selection favoring cuckoos that lay mimetic eggs. Among European cuckoos different groups of females (referred to as gentes) specialize on different host species, producing eggs that mimic those of their particular host (Chance 1940; Baker 1942; Wyllie 1981; Brooke and Davies 1988). Rothstein (1974, 1975*a*,*b*, 1978, 1982) has shown that egg color is an important variable in allowing several North American passerines to detect eggs of the brown-headed cowbird (*Molothrus ater*), a common brood parasite of these species.

In agreement with these cuckoo and cowbird studies—but contrary to the conclusions of my study—Victoria (1972) and Cruz and Wiley (1989) argued that egg recognition in village weavers has probably evolved in response to parasitism by *Chrysococcyx* cuckoos. However, there is reason to doubt that cuckoos have been the main selective force in the evolution of egg-recognition ability in weavers, or that cuckoo parasitism has led to the extreme intraspecific egg-color variability in weavers. First, the frequency of weaver nests parasitized by cuckoos (often less than 3%; see Freeman 1988 and references therein; Jackson 1992*a*,*b*) appears to be much lower than the frequency of nests parasitized by conspecifics. Davies and Brooke (1989*b*) developed a mathematical model to describe the spread of a gene for rejection of parasitic eggs among hosts of the European cuckoo. They estimated that at a parasitism rate of 3% or less, it would take thousands of generations for the gene to spread. This is so because the vast majority of hosts are never exposed to parasitic eggs. In contrast, at a parasitism rate of 20%, it would take only 200 generations or so for the rejection gene to spread. Kelly (1987) came to a similar conclusion, and noted that a gene for mimicry by cuckoos would spread faster than a gene for rejection of cuckoo eggs. Assuming that these general models apply to both conspecific nest parasitism among weaverbirds and parasitism by cuckoos of weavers, egg-recognition ability by weavers is more likely to have evolved in response to CNP than interspecific parasitism.

It should be noted that frequency of parasitism is not the only indicator of the relative strength of selection favoring host defenses against CNP versus parasitism by cuckoos. Because a host loses its entire brood when parasitized by a cuckoo, this form of parasitism is especially costly (Rothstein 1975*c*). At a given level of parasitism, the strength of selection for host defenses should be relatively greater when the host produces no young than when the host produces at least some young. Weavers parasitized by conspecifics may still produce young but, as indicated above, there are significant costs to a host imposed by CNP.

The second reason to doubt that cuckoos have been the main selective force on weaverbird egg color and egg-recognition ability is because Freeman (1988) found that the degree of intraspecific egg-color variability seen among *Ploceus* spp. did not correlate with the frequency of cuckoo parasitism reported in the literature. Likewise, Davies and Brooke (1989*b*) found no correlation between the degree of interclutch egg variability within *Cuculus canorus* host species in Europe and the frequency of parasitism these species experience. However, Øien et al. (1995) conducted a larger survey of European hosts and found evidence that *C. canorus* parasitism has selected for increased interclutch variation. Species that are suitable hosts had more variation than passerine species that are unsuitable as hosts and there was a significant correlation between interclutch variation and a host's rejection rate of nonmimetic eggs (although this difference disappeared when hole-nesting species were excluded from the analysis). Nevertheless, none of these European species has egg variation that rivals that seen in some species of weavers, so it appears that the most extreme cases of intraspecific egg color variability are due more to parasitism by conspecifics than by heterospecifics.

ACKNOWLEDGMENTS I thank the Office of the President of the Republic of Kenya and the Department of Zoology at the University of Nairobi for permission to conduct this research. Financial support for the research was provided by the National Geographic Society, Sigma Xi, the Wilson Ornithological Society, an NSF Grant to Sievert Rohwer, a National Geographic Society Grant to John Wingfield, the Graduate School Research Fund and the Department of Zoology at the University of Washington, the Burke Museum Endowment for Natural History Fieldwork, and my parents. Elena Thomas and Tamara Gordy helped with data collection. Sievert Rohwer, Gordon Orians, Steve Rothstein, and John Wingfield provided valuable comments on the manuscript.

REFERENCES

Baker, E. C. S. (1942). *Cuckoo Problems*. Witherby, London.

Bischoff, C. M. and M. T. Murphy (1993). The detection of and responses to experimental intraspecific brood parasitism in eastern kingbirds. *Anim. Behav*., **45**, 631–638.

Braa, A. T., A. Moksnes, and E. Røskaft (1992). Adaptations of bramblings and chaffinches towards parasitism by the common cuckoo. *Anim. Behav*., **43**, 67–78.

Brooke, M. de L. and N. B. Davies (1988). Egg mimicry by cuckoos *Cuculus canorus* in relation to discrimination by hosts. *Nature*, **335**, 630–632.

Brown, C. R. and M. B. Brown (1988). A new form of reproductive parasitism in cliff swallows. *Nature*, **331**, 66–68.

Chance, E. P. (1940). *The Truth about the Cuckoo*. Witherby, London.

Collias, E. C. (1984). Egg measurements and color throughout life in the Village Weaverbird, *Ploceus cucullatus*. Proceedings of the Vth Pan-African Ornithological Congress, pp. 461–475.

Cruz, A. and J. W. Wiley (1989). The decline of an adaptation in the absence of a presumed selection pressure. *Evolution*, **43**, 55–62.

Davies, N. B. and M. de L. Brooke (1988). Cuckoos versus reed warblers: adaptations and counteradaptations. *Anim. Behav*., **36**, 262–284.

Davies, N. B. and M. de L. Brooke (1989*a*). An experimental study of co-evolution between the cuckoo, *Cuculus canorus*, and its hosts. I. Host egg discrimination. *J. Anim. Ecol*., **58**, 207–224.

Davies, N. B. and M. de L. Brooke (1989*b*). An experimental study of co-evolution between the cuckoo, *Cuculus canorus*, and its hosts. II. Host egg markings, chick discrimination and general discussion. *J. Anim. Ecol*., **58**, 225–236.

Freeman, S. (1988). Egg variability and conspecific nest parasitism in the *Ploceus* weaverbirds. *Ostrich*, **59**, 49–53.

Friedmann, H. (1963). Host relations of the parasitic cowbirds. *Smithsonian Institution Bulletin*, 233.

Jackson, W. M. (1992*a*). Estimating conspecific nest parasitism in the Northern Masked Weaver based on within-female variability in egg appearance. *Auk*, **109**, 435–443.

Jackson, W. M. (1992*b*). Relative importance of parasitism by *Chrysococcyx* cuckoos versus conspecific nest parasitism in the Northern Masked Weaver *Ploceus taeniopterus*. *Ornis Scand.*, **23**, 203–206.

Jackson, W. M. (1993). Causes of conspecific nest parasitism in the Northern Masked Weaver. *Behav. Ecol. Sociobiol.*, **32**, 119–126.

Kelly, C. (1987). A model to explore the rate of spread of mimicry and rejection in hypothetical populations of cuckoos and their hosts. *J. Theor. Biol.*, **125**, 283–299.

Lack, D. (1958). The significance of the colour of Turdine eggs. *Ibis*, **100**, 145–166.

Lyon, B. E. (1993). Tactics of parasitic American coots: host choice and the pattern of egg dispersion among host nests. *Behav. Ecol. Sociobiol.*, **33**, 87–100.

Mackworth-Praed, C. W. and C. H. B. Grant, (1955). *African handbook of birds Series I, Volume II: Birds of eastern and north eastern Africa*. Longmans, London.

Mason, P. and S. I. Rothstein (1986). Coevolution and avian brood parasitism: cowbird eggs show evolutionary response to host discrimination. *Evolution*, **40**, 1207–1214.

Moksnes, A. (1992). Egg recognition in chaffinches and bramblings. *Anim. Behav.*, **44**, 993–995.

Moreau, R. E. (1960). Conspectus and classification of the Ploceine weaver-birds. *Ibis*, **102**, 298–321, 443–471.

Munsell Color, Inc. (1976). *Munsell Book of Color*. Kollmorgen, Baltimore.

Øien, I. J., A. Moksnes, and E. Røskaft (1995*a*). Evolution of variation in egg color and marking pattern in European passerines: adaptations in a coevolutionary arms race with the cuckoo, *Cuculus canorus*. *Behav. Ecol.*, **6**, 166–174.

Oniki, Y. (1985). Why robin eggs are blue and birds build nests: statistical tests for Amazonian birds. *Ornithological Monogr.*, **36**, 536–545.

Pinxten, R., M. Eens, and R. F. Verheyen (1991). Responses of male starlings to experimental intraspecific brood parasitism. *Anim. Behav.*, **42**, 1028–1030.

Rothstein, S. I. (1974). Mechanisms of avian egg recognition: possible learned and innate factors. *Auk*, **91**, 796–807.

Rothstein, S. I. (1975*a*). Mechanisms of avian egg-recognition: do birds know their own eggs? *Anim. Behav.*, **23**, 268–278.

Rothstein, S. I. (1975*b*). An experimental and teleonomic investigation of avian brood parasitism. *Condor*, **77**, 250–271.

Rothstein, S. I. (1975*c*). Evolutionary rates and host defenses against avian brood parasitism. *Amer. Natur.*, **109**, 161–176.

Rothstein, S. I. (1978). Mechanisms of avian egg-recognition: additional evidence for learned components. *Anim. Behav.*, **26**, 671–677.

Rothstein, S. I. (1982). Mechanisms of avian egg recognition: which egg parameters elicit responses by rejecter species? *Behav. Ecol. Sociobiol.*, **11**, 229–239.

Rothstein, S. I. (1986). A test of optimality: egg recognition in the Eastern Phoebe. *Anim. Behav.*, **34**, 1109–1119.

Shugart, G. W. (1987). Individual clutch recognition by Caspian terns, *Sterna caspia*. *Anim. Behav.*, **35**, 1563–1565.

Sorenson, M. D. (1995). Evidence of conspecific nest parasitism and egg discrimination in the sora. *Condor*, **97**, 819–821.

Spaw, C. D. and S. Rohwer (1987). A comparative study of eggshell thickness in cowbirds and other passerines. *Condor*, **89**, 307–318.

SYSTAT (1985). *Logit Analysis*. SYSTAT, Inc., Evanston, IL.

Victoria, J. K. (1972). Clutch characteristics and egg discriminative ability of the African Village Weaverbird *Ploceus cucullatus*. *Ibis*, **114**, 367–376.

Wyllie, I. (1981). *The Cuckoo*. Batsford, London.

VIII

MAJOR UNRESOLVED QUESTIONS

23

Major Unanswered Questions in the Study of Avian Brood Parasitism

STEPHEN I. ROTHSTEIN, SCOTT K. ROBINSON

We emphasized in chapter 1 that the heuristic nature of brood parasitism is one of the features that make it such an attractive phenomenon for research. Therefore, it is fitting to end this book with a brief section on major questions. Some of these questions have already been noted, at least implicitly, by various authors or by us in chapter 1, but restating them here in explicit terms will make them more identifiable and, we hope, better help to direct future research on parasitic birds. The 16 questions listed below are not the only "big" questions as regards brood parasitism but they are the important questions whose answers are the most uncertain. Fortunately, recent work is starting to answer other important questions such as whether facultative parasitism represents a best-of-a-bad-situation strategy or a fitness-enhancing strategy pursued by individuals able to rear their own young [it appears to be both (Sorenson, this vol.)] or whether parasitic birds, especially the viduines, have cospeciated with their hosts [the answer appears to be no (Klein et al. 1993)].

1. How did brood parasitism evolve? This is perhaps the biggest of all questions related to brood parasitism, although it has not figured prominently in this book because our emphasis has been on coevolution between parasites and their hosts. Also, because it deals with past events, this sort of question is less tractable than questions dealing with parasite–host interactions, which are ongoing. Nevertheless, there still is much that can be done to investigate the origin of brood parasitism. For example, Hamilton and Orians' (1965) suggestion, that obligate interspecific brood parasitism might arise from facultative conspecific brood parasitism (CBP), has been assessed with field experiments (Rothstein 1993) and modeling (Yamauchi 1993, 1995; Cichon 1996). The widespread CBP demonstrated in recent years shows that the condition necessary for the Hamilton–Orians model is not rare but the relatively small number of species and groups of obligately parasitic birds suggests that CBP alone does not readily lead to obligate parasitism.

2. Do hosts with egg rejection behavior commit recognition errors when they are not parasitized, and are the costs of the lost host eggs sufficient to make acceptance more adaptive than rejection? In our view, this is the most significant question dealing with parasite–host coevolution because it addresses the enigmatic acceptance of parasitic eggs by hosts with easily distinguishable eggs, that is, the issue of evolutionary lag versus equilibrium. Much recent research has focused on the costs of defenses when not parasitized because costs incurred when parasitized are rarely sufficient to make acceptance more adaptive than rejection. Although some authors have argued that occasional lost eggs from the unparasitized nests of rejecters is evidence of recognition errors, such egg losses occur in species with no rejection behavior (chapter 1). Therefore, other types of evidence are needed. The hypothesis that recognition errors make acceptance more adaptive than rejection may be falsified if experiments show that conspecific eggs are never rejected (e.g., Moksnes and Røskaft 1992). Unfortunately, rejection of such eggs is not sufficient to confirm this hypothesis because a bird's own eggs are not likely to have the range of variation shown by conspecifics. Furthermore, rejection can still be more adaptive than acceptance at certain levels of recognition errors and other parameters (Davies et al. 1996).

3. Do hosts lose rejection behavior in the absence of parasitism? Hosts that possess rejection may cease to be parasitized if they or a parasite undergo changes in geographic ranges or if a parasite shifts to another host. Whether rejecter hosts lose their rejection when they are no longer parasitized is vital to the long-range dynamics of parasite–host coevolution. If old hosts with good defenses lose their defenses when no longer parasitized, they will eventually become suitable hosts again and parasites will be able to shift back to using them. But if defenses are kept for long periods, then eventually all potential hosts will possess effective defenses, such as egg recognition, and a parasite will run out of easily exploited hosts. Parasites will then have to become highly specialized since they can develop counteradaptations, such as egg mimicry, to only one or a few species because these counteradaptations need to be tailored to the features of individual host species. Whether a host loses a defense it no longer needs will be determined by costs of the defense in the absence of parasitism, so this question is related to question 2. This problem can be addressed by a phylogenetic or comparative approach in which egg recognition is compared between closely related populations (conspecific or congeneric) that are sympatric versus allopatric with parasites. In some cases this approach has found reduced levels of egg recognition in allopatry (Cruz and Wiley 1989; Davies and Brooke 1989*a,b*, this vol.; Brown et al. 1990; Briskie et al. 1992). But no study has found a complete absence of rejection in allopatry and some have found very high levels of it (Cruz et al. 1985; Post et al. 1990; Rothstein, in prep.) so more tests are needed. Although Soler and Møller (1990) reported that magpies (*P. pica*) in Sweden that are allopatric with great spotted cuckoos (*Clamator glandarius*) show no rejection whatsoever, more recent work has shown rejection rates of at least 50% in each of six other allopatric populations (M. Soler, pers. commun.).

4. Why is it so difficult for hosts to evolve nestling recognition and rejection of parasitic nestlings? We will not elaborate on this question here other than to point out no host is known actively to reject (i.e., remove) nonmimetic nestlings from its nests. Although there are a number of examples of partial to well-developed nestling mimicry in some cuckoos, the screaming cowbird (*Molothrus rufoaxillaris*) and viduines, very little work has been done on the host discrimination of nestlings (but see Fraga, this vol.), if any, that occurs in these cases. The problematic lack of host rejection of nonmimetic nestlings is discussed in this volume by Davies and Brooke, and McLean and Maloney, and elsewhere by Lotem (1993) and Redondo (1993). In our view, none of these discussions, including our own (chapter 1), completely resolves the enigma.

5. How widespread is the "mafia" effect? Soler et al.'s (1995) evidence that great spotted cuckoos (*Clamator glandarius*) destroy nests from which their eggs have been removed is likely to stimulate additional work. Unlike most cuckoos, this species has nestlings that do not kill off all of the host young and is therefore a potential candidate for the mafia-like process envisaged by Zahavi (1979). However, other features of the great spotted cuckoo system indicate that the mafia effect is unlikely, as discussed in chapter 1. In addition to further studies on this system, mafia effects should be sought in other parasite–host interactions. Perhaps the best candidates for this effect are the brown-headed, shiny, and bronzed cowbirds (*Molothrus ater*, *M. bonariensis*, and *M. aeneus*) as they parasitize many hosts that accept parasitic eggs and that are able to raise some of their own young if they also raise a cowbird. However, a number of researchers have routinely removed cowbird eggs from parasitized nests and have not reported high failure rates in such nests (Rothstein 1982), although they did not carry out the manipulations to test for the mafia effect.

6. Do cowbirds regularly depredate unparasitized nests to force hosts to renest and make nests available for parasitism? Arcese et al. (1996) presented strong but circumstantial evidence that cowbirds destroy unparasitized song sparrow (*Melospiza melodia*) nests on Mandarte Island in order to force hosts to renest and thereby generate more nests at stages suitable for parasitism. Cuckoos are known to engage in such behavior and although Arcese et al.'s arguments are persuasive, more evidence is needed to confirm selective cowbird predation of unparasitized nests. Previous studies cited by Rothstein (1975) showed no consistent differences between predation rates of parasitized and unparasitized nests. Preliminary analyses of a very large sample of nests from the Midwest also show no consistent differences (S. Robinson, unpubl. data). These contrary data may mean that cowbirds practice selective predation only under the unusual circumstances, such as those on Mandarte Island, that is, the presence of only one or two female cowbirds and of only one or two likely hosts. So even if selective predation is confirmed on Mandarte, it may not occur in more typical complex, multi-host communities. Nevertheless, if selective predation is widespread, it could be highly significant to two other questions we have highlighted. As with the mafia effect, selective predation could decrease the adaptive value of host egg rejection if cowbirds identify unparasitized nests via the presence versus absence of a cowbird egg (but not if they remember which nests they have parasitized and selectively destroy other nests). Secondly, widespread selective predation would mean that cowbirds reduce host productivity by both parasitism and predation and would therefore greatly increase the likelihood that cowbirds affect the population dynamics of numerous host species (see question 11).

7. Is selective egg removal by cuckoos a significant selection pressure favoring mimicry of host eggs? If a second cuckoo parasitizes an already parasitized nest, it would be adaptive for it to remove the parasitic egg that is already there because this latter egg is a much greater threat to its fitness than are any of the host eggs. Such selective egg removal would favor mimicry of the host's eggs even if the host accepts nonmimetic eggs, an interpretation prompted by the discovery that certain Australian cuckoo hosts that are parasitized with mimetic eggs accept nonmimetic eggs (Brooker and Brooker 1989, Brooker et al. 1990). However, the egg types in question are possessed by a number of other host species whose egg recognition has not been tested and only one of these species need be a rejecter for host egg recognition to be the primary selective pressure for cuckoo egg mimicry. Furthermore, there is no doubt that host discrimination is currently the primary source of selection for egg mimicry in some cuckoo systems (Davies and Brooke 1988, 1989*a*,*b*) because hosts reject nonmimetic eggs at frequencies that are far higher than the frequency of multiple parasitism. Nevertheless, cuckoo egg removal is a reasonable hypothesis for egg mimicry in some systems and it can be falsified if its basic requirement is not confirmed, that is, second cuckoos must selectively remove eggs laid by other cuckoos, as opposed to host eggs. Although Davies and Brooke (1988) found a weak but insignificant trend towards selective removal, there are no conclusive data here. Perhaps the best data on the selective removal question will come from Japan, where cuckoos and therefore multiple parasitism are relatively common (Nakamura et al., this vol.).

8. What are the relative roles of interspecific and conspecific brood parasitism (CBP) in the evolution of egg recognition and interclutch variability, especially in ploceid weavers? Jackson (this vol.) argues that the extreme interclutch variation in egg coloration in ploceids is of greater benefit in combating CBP than cuckoo parasitism. This is a persuasive argument for the population she studied because CBP was far more common than cuckoo parasitism in that population. Jackson's hypothesis contrasts with the traditional view, which has attributed the egg variation and recognition to cuckoo parasitism. The key to determining whether Jackson's result is a general one is to determine the relative frequencies of CBP and interspecific parasitism at a number of localities in a species' range because parasitism can vary spatially.

9. What factors do brood parasites use to determine which host species will be parasitized? This question deals with one of the basic issues in ecology, namely resource use, and can be addressed by approaches similar to optimal foraging theory. Unless hosts are in limited supply, a parasite should first use all available nests of the highest-quality host species. Host quality depends on several parameters, including rates of nest predation (Mason 1986*a*). Although it is not clear that they choose the very best hosts, nearly all specialized parasites appear to avoid low-quality hosts that are likely to reject their eggs, that are the wrong size, or that feed their nestlings the wrong food. Although host imprinting could provide an easy mechanism facilitating the preferential use of good hosts, the only direct test indicates that imprinting does not occur (Davies and Brooke, this vol.) so the mechanism for recognizing suitable hosts is unclear. The situation in generalist cowbirds, the shiny and brown-headed, is even less certain because it is not even clear that they preferentially choose high-quality hosts. Some studies have shown high rates of parasitism on rejecter species or hosts that are unsuitable because of the food they feed their nestlings (chapter 1; see also Rothstein 1976; Mason 1986*b*). Some of the noise that seems to exist in cowbird host use may be due to high cowbird densities and competition for suitable hosts. Thus, comparisons between host selection in areas used by a single female cowbird and areas that several females share might be informative. Cowbird removal experiments in areas with high cowbird densities would also allow tests of the effects of reduced competition on cowbird host selection.

10. What local factors (edges, forest openings, tract size, vegetation structure) affect levels of parasitism within landscapes? Some brood parasites, especially shiny and brown-headed cowbirds and the common cuckoo, parasitize the same host species at greatly different rates in different regions. With cowbirds, there is no clear evidence that these hosts show spatial variation in their resistance to parasitism. Cowbirds may even use entire host communities to different extents in different regions. For example, grassland species are heavily parasitized in the Great Plains but rarely so in the Midwest (Robinson et al. 1995*a*). Among woodland hosts, ground nesters in Illinois are parasitized less than species that nest higher up but the reverse is true in New York (Martin 1992; Robinson et al. 1993, 1995*a,b*; Trine et al., this vol.). These variable patterns may occur in cowbirds because of variation in landscapes (Trine et al., this vol.), and not because of variation in the cowbirds themselves. These questions have great relevance to management strategies designed to increase avian nesting success. As more experimental and correlative studies become available (many are underway, e.g., Smith et al., in press), we may be able to detect consistent patterns.

11. Do brood parasites limit and even cause declines of widespread host species? This question applies especially to cowbirds (see overview for part V) and necessitates the collection of better demographic data for hosts. Currently, there is a great deal of uncertainty about the relative contributions of nest predation and parasitism (Martin 1992) and the importance of population limitation outside of the breeding season (Sherry and Holmes 1995). These issues must be addressed with large-scale, spatially explicit population models that assess population sources and sinks. Although more demographic data are needed, the plethora of recent studies of wood thrushes (*Hylocichla mustelina*) makes this species an ideal candidate for such models (e.g., Winker et al. 1990; Roth and Johnson 1993; Donovan et al. 1995*a,b*; Hoover et al. 1995; Trine, in press, this vol.; several papers in Smith et al., in press). Developing and testing such models will require a great deal of cooperation among scientists.

12. What course of events, as regards interactions among cowbird species and potential host declines, will ensue in the southeastern U.S. as shiny and brown-headed cowbirds increase their range overlap? Cruz et al. (this vol.) have outlined the remarkable situation now underway in the southeastern United States where two sister species, the two most generalized parasites in the world, have recently come into secondary contact. There seems little doubt that the shiny cowbird will expand its range in North America and the brown-headed cowbird may expand southwards into the Caribbean. This is an extremely fertile field for research involving possible competition for hosts, potential interbreeding between the two cowbirds, and possible threats to the avifaunas. The latter potential alone merits the action of a large team of governmental monitors and investigators.

13. Are population levels of parasitic birds limited by recruitment and host availability or by other factors such as winter mortality? Limitation by recruitment could be due to insufficient habitat, as with any species, or to an insufficient number of hosts in habitats a parasite occupies. It is clear that the distribution of some parasites is limited by habitat suitability as some regions with few or no cowbirds have numerous hosts but almost no feeding habitat (Robinson et al. 1995*a,b*; Trine et al., this vol.). Other regions have such high rates of parasitism (Robinson et al. 1995*a,b*) that it seems clear that there are insufficient numbers of hosts for the local cowbirds. Despite these findings, there may be so much dispersal that regional patterns may not reveal the determinants of the population size of an entire parasitic species. Thus, parasites limited by winter mortality could settle in local areas according to the availability of feeding habitat and/or hosts even if there is no cause-and-effect relation between the local production of parasites and numbers of adult parasites. The latter situation seems to be the case where cowbirds are removed to protect endangered species because the number of cowbirds removed each year has not declined despite the nearly complete absence of local cowbird recruitment over large contiguous areas (DeCapita, in press; Griffith and Griffith, in press). The continuing increase of cowbirds in the central part of their range, in spite of no obvious recent changes in habitat or host availability (Peterjohn, in press; Wiedenfeld et al., in press) also argues that local conditions are not closely related to local numbers of cowbirds. It is possible that the population sizes of more specialized parasites are more closely limited by local recruitment and host availability. Determining the mechanism of population limitation for parasitic birds is essential to determining the applicability of models that assume a close correspondence between host and parasite numbers (e.g., May and Robinson 1985; Kelly 1987; Takasu et al. 1993).

14. Is monogamy the prevalent mating system among parasitic birds and, if so, why? As discussed by Barnard (this vol.), most parasitic birds with known mating systems seem to be primarily monogamous, although there are some clear examples of promiscuity such as the viduines and one species of honeyguide (Cronin and Sherman 1977). Although consistent male–female associations and observed copulations indicate social monogamy, extra-pair copulations (EPCs) might be common, as in numerous nonparasitic birds that are socially monogamous. The first DNA fingerprinting on cowbirds (Alderson, Gibb and Sealy, in prep.) indicates that EPCs are less common than in most nonparasitic species.

15. Does CBP have population-level consequences? May et al.'s (1991) and Nee and May's (1993) models indicating population cycles due to CBP require that intraspecific brood parasitism is costly and that it lowers the average individual fitness within a population. But available data are insufficient to determine whether costs such as those documented by Møller (this vol.) are great enough to lower the average fitness. In terms of demographics, it is possible that losses incurred by some individuals are negated by gains made by conspecifics that benefit from CBP.

16. What are the hosts and other natural history features of numerous little-known brood parasites? The wealth of available information on some parasites, such as most of the cowbirds and several species of cuckoos, makes it relatively easy to identify the remaining critical gaps in our understanding, which in turn makes it possible to formulate explicit questions, such as those in this chapter. Yet, our knowledge of some parasites is extremely limited. For example, so little is known of the breeding mode of several species of honeyguides, including both species in Asia (*Indicator xanthonotus*, *I. archipelagicus*), that there is no proof that they are in fact parasitic. Similarly, it is assumed that all species in the subfamily Cuculinae are parasites but there are no definite hosts known for a number of species, such as *Callechthrus leucolophus* and several *Cercococcyx* spp. Even some congeners of the best-known cuckoo species, *Cuculus canorus*, have unknown hosts. Finally, although one or more hosts can be listed for most species of brood parasites, very few parasites other than the ones highlighted in chapters in this book have been the subjects of focused field studies that can reveal such basic features as the frequency of parasitism. We suspect that many more questions — and we hope some answers too — will emerge as natural history information accumulates on these little-known parasites.

REFERENCES

Arcese, P., J. N. M. Smith, and M. I. Hatch (1996). Nest predation by cowbirds and its consequences for passerine demography. *Proc. Natl Acad. Sci. USA*, **93**, 4608–4611.

Briskie, J. V., S. G. Sealy, and K. A. Hobson (1992). Behavioral defenses against avian brood parasitism in sympatric and allopatric host populations. *Evolution*, **46**, 334–340.

Brooker, L. C., M. G. Brooker, and A. M. H. Brooker (1990). An alternative population/genetics model for the evolution of egg mimesis and egg crypsis in cuckoos. *J. Theor. Biol.*, **146**, 123–143.

Brooker, M. G. and L. C. Brooker (1989). The comparative breeding behaviour of two sympatric species of cuckoos, Horsfield's bronze-cuckoo *Chrysococcyx basalis* and the shining bronze-cuckoo *C. lucidus* in Western Australia: a new model for the evolution of egg morphology and host specificity in avian brood parasites. *Ibis*, **131**, 528–547.

Brown, R. J., M. N. Brown, M. de L. Brooke, and N. B. Davies (1990). Reactions of parasitized and unparasitized populations of *Acrocephalus* warblers to model cuckoo eggs. *Ibis*, **132**, 109–111.

Cichon, M. (1996). The evolution of brood parasitism: the role of facultative parasitism. *Behav. Ecol.*, **7**, 137–139.

Cronin, E. W. and P. W. Sherman (1977). A resource-based mating system: the orange-rumped honeyguide, *Indicator xanthonotus*. *Living Bird*, **15**, 5–37.

Cruz, A. and J. W. Wiley (1989). The decline of an adaptation in the absence of a presumed selection pressure. *Evolution*, **43**, 55–62.

Cruz, A., T. Manolis, and J. W. Wiley (1985). The shiny cowbird: a brood parasite expanding its range in the Caribbean region. *Ornithol. Monogr.*, No. **36**, 607–620.

Davies, N. B. and M. de L. Brooke (1988). Cuckoos versus reed warblers: adaptations and counteradaptations. *Anim. Behav.*, **36**, 262–284.

Davies, N. B. and M. de L. Brooke (1989*a*). An experimental study of co-evolution between the cuckoo, *Cuculus canorus*, and its hosts. I. Host egg discrimination. *J. Anim. Ecol.*, **58**, 207–224.

Davies, N. B. and M. de L. Brooke (1989*b*). An experimental study of co-evolution between the cuckoo, *Cuculus canorus*, and its hosts. II. Host egg markings, chick discrimination and general discussion. *J. Anim. Ecol.*, **58**, 225–236.

Davies, N. B., M. de L. Brooke, and A. Kacelnik (1996). Recognition errors and probability of parasitism determine whether reed warblers should accept or reject mimetic cuckoo eggs. *Proc. R. Soc. Lond. B.*, **263**, 925–931.

DeCapita, M. E. Brown-headed cowbird control on Kirtland's warbler nesting areas in Michigan, 1972–1994. In: *Ecology and management of cowbirds* (eds J. N. M. Smith, T. Cook, S. K. Robinson, S. I. Rothstein, and S. G. Sealy). University of Texas Press, Austin, Texas (in press).

Donovan, T. M., F. R. Thompson, III, J. Faaborg, and J. R. Probst (1995*a*). Reproductive success of migratory birds in habitat sources and sinks. *Conserv. Biol.*, **9**, 1380–1395.

Donovan, T. M., R. H. Lamberson, A. Kimber, F. R. Thompson, III, and J. Faaborg (1995*b*). Modeling the effects of habitat fragmentation on source and sink demography of neotropical migrant birds. *Conserv. Biol.*, **9**, 1396–1407.

Griffith, J. T. and J. C. Griffith. Cowbird control and the endangered least Bell's vireo: a management success story. In: *Ecology and management of cowbirds* (eds J. N. M. Smith, T. Cook, S. K. Robinson, S. I. Rothstein, and S. G. Sealy). University of Texas Press, Austin, Texas (in press).

Hamilton, W. J. and G. H. Orians (1965). Evolution of brood parasitism in altricial birds. *Condor*, **67**, 361–382.

Hoover, J. P., M. C. Brittingham, and L. J. Goodrich (1995). Effects of forest patch size on nesting success of wood thrushes. *Auk*, **112**, 146–155.

Kelly, C. (1987). A model to explore the rate of spread of mimicry and rejection in hypothetical populations of cuckoos and their hosts. *J. Theor. Biol.*, **125**, 283–299.

Klein, N. K., R. B. Payne, and M. E. D. Nhlane (1993). A molecular genetic perspective on speciation in the brood parasitic *Vidua* finches. Proc. VIII Pan-African Ornithological Congr., pp. 29–39.

Lotem, A. (1993). Learning to recognize nestlings is maladaptive for cuckoo *Cuculus canorus* hosts. *Nature*, **362**, 743–745.

Martin, T. E. (1992). Breeding productivity considerations: what are the appropriate habitat features for management? In: *Ecology and conservation of neotropical migrant landbirds* (eds J. M. Hagan and D. W. Johnston), pp. 455–493. Smithsonian Inst. Press, Washington, D.C.

Mason, P. (1986*a*). Brood parasitism in a host generalist, the shiny cowbird: I. The quality of different species as hosts. *Auk*, **103**, 52–60.

Mason, P. (1986*b*). Brood parasitism in a host generalist, the shiny cowbird: II. Host selection. *Auk*, **103**, 61–69.

May, R. M. and S. K. Robinson (1985). Population dynamics of avian brood parasitism. *Amer. Natur.*, **126**, 475–494.

May, R. M., S. Nee, and C. Watts (1991). Could intraspecific brood parasitism cause population cycles? Proc. XX International Ornithological Congress, pp. 1012–1022.

Moksnes, A. and E. Røskaft (1992). Responses of some rare cuckoo hosts to mimetic model cuckoo eggs and to foreign conspecific eggs. *Ornis Scand.*, **23**, 17–23.

Nee, S. and R. M. May (1993). Population-level consequences of conspecific brood parasitism in birds and insects. *J. Theor. Biol.*, **161**, 95–109.

Peterjohn, B. and J. R. Sauer. Temporal and geographic patterns in population trends of Brown-headed Cowbirds. In: *Ecology and management of cowbirds* (eds J. N. M. Smith, T. L. Cook, S. K. Robinson, S. I. Rothstein, and S. G. Sealy). University of Texas Press, Austin, Texas (in press).

Post, W., T. K. Nakamura, and A. Cruz (1990). Patterns of Shiny Cowbird parasitism in St Lucia and southwestern Puerto Rico. *Condor*, **92**, 461–492.

Redondo, T. (1993). Exploitation of host mechanisms for parental care by avian brood parasites. *Etología*, **3**, 235–297.

Robinson, S. K., J. A. Grzybowski, Jr, S. I. Rothstein, L. J. Petit, M. C. Brittingham, and F. R.

Thompson, III (1993). Management implications of cowbird parasitism on neotropical migratory birds. In: *Status and management of neotropical migratory birds* (eds D. M. Finch and P. W. Stangel), pp. 93–102. USDA-USFS Gen. Tech. Rep. RM-229, Fort Collins, Colorado.

Robinson, S. K., S. I. Rothstein, M. C. Brittingham, L. J. Petit, and J. A. Grzybowski (1995*a*). Ecology and behavior of cowbirds and their impact on host populations. In: *Ecology and management of neotropical migratory birds* (eds T. E. Martin and D. M. Finch), pp. 428–460. Oxford University Press, New York.

Robinson, S. K., F. R. Thompson, III, T. M. Donovan, D. R. Whitehead, and J. Faaborg (1995*b*). Regional forest fragmentation and the nesting success of migratory birds. *Science*, **267**, 1987–1990.

Roth, R. R. and R. K. Johnson (1993). Long-term dynamics of a wood thrush population breeding in a forest fragment. *Auk*, **110**, 37–48.

Rothstein, S. I. (1975). Evolutionary rates and host defenses against avian brood parasitism. *Amer. Natur.*, **109**, 161–176.

Rothstein, S. I. (1976). Cowbird parasitism of the cedar waxwing and its evolutionary implications. *Auk*, **93**, 498–509.

Rothstein, S. I. (1982). Successes and failures in avian egg and nestling recognition with comments on the utility of optimality reasoning. *Amer. Zool.*, **22**, 547–560.

Rothstein, S. I. (1993). An experimental test of the Hamilton–Orians hypothesis for the origin of avian brood parasitism. *Condor*, **95**, 1000–1005.

Sherry, T. W. and R. T. Holmes (1995). Summer versus winter limitation of populations: what are the issues and what is the evidence? In: *Ecology and management of neotropical migratory birds* (eds T. E. Martin and D. M. Finch), pp. 85–120. Oxford University Press, Oxford.

Smith, J. N. M., T. Cook, S. K. Robinson, S. I. Rothstein, and S. G. Sealy (eds) *Ecology and management of cowbirds*. University of Texas Press, Austin, Texas (in press).

Soler, M., J. J. Soler, J. G. Martinez, and A. P. Møller (1995). Host manipulation by great spotted cuckoos: evidence for an avian mafia? *Evolution*, **49**, 770–775.

Takasu, F., K. Kawasaki, H. Nakamura, J. E. Cohen, and N. Shigesada (1993). Modeling the population dynamics of a cuckoo–host association and the evolution of host defenses. *Amer. Natur.*, **142**, 819–839.

Wiedenfeld, D. A. Cowbird population changes and their relationship to changes in some host species. In: *Ecology and management of cowbirds* (eds J. N. M. Smith, T. L. Cook, S. K. Robinson, S. I. Rothstein, and S. G. Sealy). University of Texas Press, Austin, Texas (in press).

Winker, K., J. H. Rappole, and M. A. Ramos (1990). Population dynamics of the wood thrush in southern Veracruz, Mexico. *Condor*, **92**, 444–460.

Yamauchi, A. (1993). Theory of intraspecific nest parasitism in birds. *Anim. Behav.*, **46**, 335–345.

Yamauchi, A. (1995). Theory of evolution of nest parasitism in birds. *Amer. Natur.*, **145**, 434–456.

Zahavi, A. (1979). Parasitism and nest predation in parasitic cuckoos. *Amer. Natur.*, **113**, 157–159.

Index

Birds and other animals are listed under their common names but cross referenced according to their Latin names (unless only a Latin name is given in the text). Major subjects such as coevolution, mating systems, nest desertion, etc., have entries of their own but additional material on these subjects may be found by checking the species entries for the parasites and hosts featured in this book. The featured **cuckoos** are: common; great spotted; Himalayan; Horsfield's bronze; Horsfield's hawk; little (lesser); and shining. The featured **cuckoo hosts** are: azure-winged magpie; dunnock; gray warble (gerygone); great reed warbler; magpie; meadow pipit; New Zealand tit; and reed warbler. The featured **cowbirds** are: brown-headed; screaming; and shiny. The featured **cowbird hosts** are: Acadian flycatcher; American robin; bay-winged cowbird; house wren; red-eyed vireo; red-winged blackbird; song sparrow; wood thrush; and yellow warbler. The featured species that engage in **conspecific parasitism** are: barn swallow; canvasback; European starling; northern masked weaver; and redhead.